AF518221

Advances in Plant Biopesticides

Editor
Dwijendra Singh
Microbial Technology & Entomology
Department, Crop Protection Division
CSIR-Central Institute of Medicinal
and Aromatic Plants
Lucknow, Uttar Pradesh, India

ISBN 978-81-322-2005-3 ISBN 978-81-322-2006-0 (eBook)
DOI 10.1007/978-81-322-2006-0
Springer New Delhi Heidelberg New York Dordrecht London

Library of Congress Control Number: 2014950834

Printed on acid-free paper

Springer is part of Springer Science+Business Media (www.springer.com)

Dwijendra Singh
Editor

Advances in Plant Biopesticides

Springer

Foreword

The term biopesticides has been mentioned and talked about for the past five to six decades or so. It was generally used with reference to bioactive principles of plant origin with potential use as pest control agents. These agents were considered relatively safe and environmentally compatible. During the past three decades, the term biopesticides has been evolving with broader inclusion of bioactive agents produced by viruses, bacteria, fungi, plants, and etc. Some experts even include antagonistic bacteria and fungi for plant disease control. It is a general trend to make the term biopesticides all inclusive covering all natural products from plants and living entities.

Notwithstanding this broad definition of the term "biopesticides", plant based bioactive agents provide large and varied sources of locally available plants, to be explored for the development of biopesticides for plant protection, public health and household insects and other pests. Bioactive agents from various parts of plants if properly extracted and formulated could provide effective and sustainable strategies for a variety of insect pests. These strategies will require low input to develop, necessitating simple pieces of equipment for application. These plant-based formulations with the exception of a few are relatively safe to use for pest and disease control on a variety of crops.

To date, plant-based biopesticides, their formulations, bioefficacy, selectivity and safety with the exception of a few agents have been studied mostly in the laboratory without significant fieldwork. There are literally thousands of plant species that have been extracted and evaluated against a variety of insects, nematodes, etc. Now we are at the stage to draw upon the knowledge available on the potential of plant based bioactive agents and mixtures discussed in the current book and literature cited in this book. This book will be extremely beneficial to experts and professionals in the realm of plant protection to be able to proceed and critically assess the comparative efficacy data and select a few agents to study rigorously to the stage where the products could be used in practical field applications. Stakeholders in key research institutions and plant protection field and other biological and chemical specialties should form "Task Forces" with responsibility to advance Biopesticides Research.

The Editor, Dwijendra Singh, of the book *Advances in Plant Biopesticides*, has sought the help of renowned research scientists around the world with expertise in basic research, plant protection and management of crop pests and diseases. The current book has 19 chapters discussing the stages in the development and use of biopesticides. Space does not allow to discuss the significance of each chapter, but I can in passing remarks mention chapters on evaluation, field testing, emergence of resistance, safety to non-target organisms, and developing sustainable technology. The last two chapters on production and use of biopesticides, formulation, registration and quality regulation of biopesticides, as well as other chapters are full of information and have been contributed by internationally reputed scientists.

I firmly believe that this book will make a lasting contribution in the field of crop protection and promotion of human health around the globe.

Mir S. Mulla
Department of Entomology
University of California, Riverside, CA, USA
April 23, 2014

Preface

The perils of pesticides developed on the basis of petro-chemical derivatives have drawn the attention of public domain extensively towards its environmental dangers and chemical pollution in view of survival and health of living ones and human on planet. World wide attention on chemical pollution in soil, air, water and other eco-systems through pesticides have led our focus to change the current mindset to investigate and innovate new approaches. In this context, a number of documentations have been made globally. However, the major focus has been Asian continent and particularly one of the thickly populated locality like India. Apart from residue in food and environment and resistance problem in target pests, detrimental effect of these petro-chemical pesticides on beneficial organisms and focussed cautions by various global programmes inspired me to utilize my three and half decades' working experience on plant biopesticides to publish a book along with potential subject experts. The book is intended for the benefit of students, researchers, teachers, enterpreneures and common persons including policy makers to have one hand knowledge and for the advancement in the area of plant biopesticides that could be used in future to manage the field and storage of arable crops, veterinary pests, disease carrying arthropods, storage pests, honeybee mite pests, utility of biologically active plants in transforming farming system, phytochemical pesticides, influence on non-target organisms, possible mode of action, limitations, registration and regulation of plant biopesticides along with the focus on potential bioresources as examples to develop useful and eco-friendly biopesticides etc.

With the above views and current worldwide society demand, I proposed to take the lead in identifying leading potential contributors globally whose contributions are noteworthy in the area of plant biopesticides to prepare quality write-ups on advances in this area based on research reviews and their experiences in the form of a book to get it published through an international publisher – Springer. I am kind enough to Springer (India) officials, New Delhi, India, who got my proposal internationally reviewed and accepted my proposal to publish the important document on burning issues of pesticides from biologically active plant species and associated major problems to overcome the losses and its side effects. The book contains 19 chapters on different issues of plant biopesticides. I am grateful to the staff of Springer (India) who cooperated me and gave replies to all my queries during preparation of the book manuscript. I am especially grateful to Springer (India) staff; Dr. Richa Sharma who in 1st instance gave green signal for the approval of publication of the book after completing their Springer's international review system; Dr. Mamta Kapila who supported, cooperated and inspired me to be in close contact with global contributors for enhancing the speed of the writeup and its technical editing, and Ms. Raman Shukla who put the manuscript to production. I am also grateful to Distinguished Professor Emeritus, Dr. Mir S. Mulla of University of California, Riverside, CA, USA, who accepted my request heartly to review and write up the foreword of the book. I am very thankful to all the contributors for submitting their chapters within the stipulated time limit. I am thankful to my beloved family members, support and regular inspiration especially from my wife – Mrs. Shail Singh, and several other colleagues, friends and well wishers who supported directly/indirectly asking about the status and progress of the book regularly till completion and manuscript submission to publisher.

I hope that this book will provide new knowledge to our present and future researchers who are interested to develop novel, environmentally safe plant biopesticides.

Lucknow, India

Dwijendra Singh
Editor

Contents

Contributors

Totan Adak Division of Crop Protection, Central Rice Research Institute, Cuttack, India

Sarah E.J. Arnold Natural Resources Institute, University of Greenwich, Kent, UK

Christos G. Athanassiou Laboratory of Entomology and Agricultural Zoology, Department of Agriculture, Crop Production and Rural Environment, University of Thessaly, Magnesia, Greece

Steven R. Belmain Natural Resources Institute, University of Greenwich, Kent, UK

J. Berliner Division of Crop Protection, Central Rice Research Institute, Cuttack, India

Someshwar Bhagat National Centre for Integrated Pest Management (ICAR), New Delhi, India

Ajanta Birah National Centre for Integrated Pest Management (ICAR), New Delhi, India

C. Chattopadhyay National Centre for Integrated Pest Management (ICAR), New Delhi, India

Nadia Z. Dimetry Department of Pests and Plant Protection, National Research Centre, Cairo, Egypt

Srikant Ghosh Entomology Laboratory, Parasitology Division, Indian Veterinary Research Institute, Izatnagar, Uttar Pradesh, India

Ayhan Gökçe Plant Production and Technologies Department, Faculty of Agricultural Sciences and Technologies, University of Nigde, Nigde, Turkey

S. Gopalakrishnan Grain Legumes, International Crops Research Institute for the Semi-Arid Tropics (ICRISAT), Patancheru, Telangana, India

Errol Hassan School of Agriculture and Food Sciences, The University of Queensland Gatton, Gatton, QLD, Australia

D.K. Hazra AINP on Pesticide Residues, Department of Agricultural Chemicals, Bidhan Chandra Krishi Viswavidyalaya, Nadia, West Bengal, India

Murray B. Isman Faculty of Land and Food Systems, University of British Columbia, Vancouver, BC, Canada

M. Kannan Department of Agricultural Entomology, Tamil Nadu Agricultural University, Coimbatore, Tamil Nadu, India

Nickolas G. Kavallieratos Laboratory of Agricultural Zoology and Entomology, Department of Crop Science, Agricultural University of Athens, Athens, Attica, Greece

Laboratory of Agricultural Entomology, Department of Entomology and Agricultural Zoology, Benaki Phytopathological Institute, Kifissia, Attica, Greece

Zeyaur R. Khan Plant Health Division, International Centre of Insect Physiology and Ecology (*icipe*), Nairobi, Kenya

Rakesh Kumar National Centre for Integrated Pest Management (ICAR), New Delhi, India

B. Ratna Kumari Grain Legumes, International Crops Research Institute for the Semi-Arid Tropics (ICRISAT), Patancheru, Telangana, India

Charles A.O. Midega Plant Health Division, International Centre of Insect Physiology and Ecology (*icipe*), Nairobi, Kenya

Mir S. Mulla Department of Entomology, University of California, Riverside, CA, USA

P.K. Patanjali Institute of Pesticide Formulation Technology (IPFT), Gurgaon, India

Roman Pavela Crop Research Institute, Praha, Czech Republic

John A. Pickett Biological Chemistry and Crop Protection Department, Rothamsted Research, Harpenden, Hertfordshire, UK

Jimmy O. Pittchar Plant Health Division, International Centre of Insect Physiology and Ecology (*icipe*), Nairobi, Kenya

Somnath S. Pokhare Division of Crop Protection, Central Rice Research Institute, Cuttack, India

Anand Prakash Division of Crop Protection, Central Rice Research Institute, Cuttack, India

S. Raguraman Department of Agricultural Entomology, Tamil Nadu Agricultural University, Coimbatore, Tamil Nadu, India

Yallappa Rajashekar Animal Bioresources Programme, Institute of Bioresources and Sustainable Development (IBSD), Department of Biotechnology, Government of India, Imphal, Manipur, India

Virendra S. Rana Division of Agricultural Chemicals, Indian Agricultural Research Institute, New Delhi, India

Pathipati Usha Rani Biology and Biotechnology Division, Indian Institute of Chemical Technology, Taranaka, Hyderabad, AP, India

Jagadiswari Rao Division of Crop Protection, Central Rice Research Institute, Cuttack, India

G.V. Ranga Rao Grain Legumes, International Crops Research Institute for the Semi-Arid Tropics (ICRISAT), Patancheru, Telangana, India

Reghu Ravindran Department of Parasitology, College of Veterinary and Animal Sciences, Wayanad, Kerala, India

S.K. Raza Institute of Pesticide Formulation Technology (IPFT), Gurgaon, India

Gadi V.P. Reddy Western Triangle Ag Research Center, Montana State University, Conrad, MT, USA

Supradip Saha Division of Agricultural Chemicals, Indian Agricultural Research Institute, New Delhi, India

K. Sahayaraj Crop Protection Research Centre, St. Xavier's College, Palayamkottai, Tamil Nadu, India

Kanchan Saikia Division of Crop Protection, Central Rice Research Institute, Cuttack, India

A. Sathya Grain Legumes, International Crops Research Institute for the Semi-Arid Tropics (ICRISAT), Patancheru, Telangana, India

Rita Seffrin Faculty of Land and Food Systems, University of British Columbia, Vancouver, BC, Canada

Thimmappa Shivanandappa Department of Zoology, University of Mysore, Mysore, Karnataka, India

Dwijendra Singh Microbial Technology & Entomology Department, Crop Protection Division, CSIR-Central Institute of Medicinal and Aromatic Plants, Lucknow, Uttar Pradesh, India

Padet Siriyasatien Department of Parasitology, Faculty of Medicine, Chulalongkorn University, Bangkok, Thailand

V. Srinivas Grain Legumes, International Crops Research Institute for the Semi-Arid Tropics (ICRISAT), Patancheru, Telangana, India

Philip C. Stevenson Natural Resources Institute, University of Greenwich, Kent, UK

Royal Botanic Gardens, Surrey, UK

Khanobporn Tangtrakulwanich Western Triangle Ag Research Center, Montana State University, Conrad, MT, USA

Apiwat Tawatsin National Institute of Health, Department of Medical Sciences, Ministry of Public Health, Nonthaburi, Thailand

Usavadee Thavara National Institute of Health, Department of Medical Sciences, Ministry of Public Health, Nonthaburi, Thailand

R. Vijayabharathi Grain Legumes, International Crops Research Institute for the Semi-Arid Tropics (ICRISAT), Patancheru, Telangana, India

Suresh Walia Division of Agricultural Chemicals, Indian Agricultural Research Institute, New Delhi, India

M.S. Yadav National Centre for Integrated Pest Management (ICAR), New Delhi, India

About the Editor

Dr. Dwijendra Singh has served the CSIR-CIMAP, Lucknow, in various capacities of research scientist heading the entomology division and reached the level of Chief Scientist and Head of the Crop Protection Division, Microbial Technology and Entomology Department, CSIR-CIMAP, Lucknow. Dr. Dwijendra Singh possesses 39 years of research experience in the field of applied and basic research in the area of entomology.

During his career as a scientist, Dr. Dwijendra Singh studied the major pests of medicinal and aromatic plants and phenology of various arthropods associated with mint, wormwood, opium poppy, henbanes, and senna. He developed suitable sampling methods for major arthropods of mint species and organic method of white butterfly larvae management by altering sowing time in *Cassia angustifolia*, identified fennel as an allelopathic plant species for mustard aphid management, screened whitefly- and *Begomovirus*-resistant mint genotypes, and investigated the effect of synthetic pesticides on biosynthesis of tropane alkaloids in henbane and menthol in mint as tools for developing integrated pest management. He invented insecticidal principles, himachalol and ß-himachalene from Himalayan cedarwood oil against adzuki bean beetles and houseflies, tylophorine from Indian ipecac leaves as antifeedants against Bihar hairy caterpillar, and α-amyrin acetate and oleanolic acid from *Catharanthus roseus* as insect growth regulators against *Helicoverpa armigera*; developed the cheapest and safest pulse grain-protecting phytotablet formulation to manage adzuki bean beetles, and demonstrated azadirachtin and its formulations in inducing alteration in the protein profile of *H. armigera* larvae which may be utilized to develop novel biopesticides. He contributed significantly in the development of high-yielding pest- and disease-resistant variety of menthol mint—'Himalaya'—that brought revolution in indigenous mint oil production and menthol export. He has explored the possibility of developing novel biopesticides from medicinal and aromatic plant species. He developed the technology on pulse grain-protecting tablets, organic method of pest management in Indian senna and shared in the development of agrotechnology for the new variety of mint—'Himalaya'.

Dr. Singh's research findings are evident by his credit to several research publications in reputed scientific journals of national and international importance, 2 granted Indian patents, and 1 US patent. He has been one of the authors of three books, apart from popular book and farm bulletin, and has also served as an editorial board member in reputed journals.

Different Plant Families as Bioresource for Pesticides

1

Nadia Z. Dimetry

Abstract

Consistent and injudicious applications of pesticides lead to the development of resistance in insects, destruction of beneficial organisms and increases in residual problems, thereby posing a threat to human health and its ecological partners in the living biome. The need of the future is to develop an eco-friendly approach to combat insect pests that should be able to regulate pest populations by exploring naturally occurring botanicals including extracts of plants, insecticidal plants and plant essential oils which may serve as useful repellents, antifeedants, insecticides, fungicides, weedicides, nematicides, molluscicides, etc. The most promising botanicals for use are species of the families Meliaceae, Rutaceae, Malvaceae, Asteraceae and Canellaceae. The results obtained in this review overwhelmingly confirmed the activity of a reasonable percentage of plants. Thus, these results will serve as useful guides in the collection of plants for laboratory and field research studies which may lead to the commercialization of plant biopesticides in the future.

Keywords

Botanicals • Plant extracts • Insecticidal plants • Fungicidal & nematicidal plants • Molluscicides • Polyester of sugars • Essential oils • Repellent and antifeedant plants

1 Introduction

The estimate of the world population for 2001 was 6.134 billion inhabitants (http://apps.fao.org) according to the Food and Agriculture Organization (FAO). The projection towards 2025 is nearly 8.5 billion inhabitants. Such an increase, which will occur mainly in developing countries, will inevitably require an additional

N.Z. Dimetry, Ph.D. (✉)
Department of Pests and Plant Protection, National Research Centre, Dokki, 1231 Cairo, Egypt
e-mail: nadia_dimetry@yahoo.com

D. Singh (ed.), *Advances in Plant Biopesticides*,
DOI 10.1007/978-81-322-2006-0_1, © Springer India 2014

agricultural production of 2.4×10^9 t/year. However, this additional production should not be based on an increase in an arable surface taken from temperate or rain forest, but on the improvement of crop productivity. This can be achieved in part by suitable control of losses due to biotic agents (pests, diseases and weeds) which on average are estimated to be 38–42 % of the potential production (Agrios 1997; Oerke et al. 1994).

Thus, the world food production is adversely affected by insects and pests during crop growth, post-harvest and storage (Kulkarni et al. 2009). Currently, the control of plant pests, diseases and weeds is achieved mainly by spraying crops with a vast amount of synthetic chemical pesticides (Agrios 1997; Cook 2000). However, an increase in the use of chemical pesticides for pest control to support the derived increase in agricultural activity needed to sustain the expected population growth which may severely deteriorate the planet's health because of non-target effects. Insect pests are considered one of the major problems which farmers face. The insect pests cause great losses to many crops and fruits besides the low grade of infested products. Apart from the farm environment, insects and pests constitute serious menace in the home, gardens and bodies of water and also transmit a number of diseases by acting as hosts to some disease-causing parasites. Thus, elimination of these insects and pests or mitigation of their activities will go a long way in reducing world food crisis to improve human and animal health.

On the other hand, the use of synthetic organic pesticides, particularly the chlorinated hydrocarbons such as DDT and derivatives, has led to serious environmental pollution (water, air and soil), affecting human health and causing death of non-target organisms (animals, plants and fish). This situation led to the Stockholm Convention in 2001 and the eventual ban of DDT in 2004 (UNEP (2005)). Before the ban efforts were already made by scientists for alternative sources of pesticides and due to other reasons, including nonselectivity, ineffectiveness, high costs of synthetic chemicals and development of resistance, and that many of the synthetic compounds have not been successfully marketed (Duke 1990), natural products from plants have attracted researchers in recent years as potential sources of new insecticides. Thus, increasing reports about the negative effects of synthetic pesticides, often resulting from indiscriminate applications, have renewed the interest in natural pesticides as an eco-chemical approach in pest control (Dubey et al. 2010). The folklore use of higher terrestrial plants by the natives of various parts of the world as pesticidal and antimicrobial materials has been well known (Dalziel. 1937; Ayensu 1978). A complimented method for pest control is the use of biopesticides besides low concentrations of synthetic pesticides when necessary. The use of plant materials as traditional protectants of stored products is an old practice used all over the world (Tripathi and Dubey 2004; Rajendran and Sriranjini 2008).

Overall in the present economic and political environment, there seems to be excellent opportunities for the development of phytochemical-based biopesticides. So far, more than 6,000 plant species have been screened for anti-insect properties, and of these, nearly 2,500 species belonging to 235 plant families exhibit measurable to considerable pest control activity (Saxena 1998). Various isolated compounds from plants indicate that approximately 350 compounds are insecticidal (Dev and Koul 1997). So many compounds have been isolated, characterized and evaluated as anti-insect compounds (Reda et al. 1989; Dimetry et al. 1990, 2003, 2007, 2010; Prakash et al. 1990; Schmutterer 1990, 1995; El-Gengaihi et al. 1997, 2011; Koul 1997; Singh et al. 1998; Mendiki et al. 1999; Singh and Mehta 2010; Dimetry 2012).

This chapter emphasizes the experience and progress made in the potential and promise of biopesticides in general and with a special reference to new crop protection agents of plant origin. There is also a need to develop strategies based on the biochemical and ecological behaviour of the pests. Thus, to complement and eventually substitute synthetic pesticides with botanical pesticides would represent economically beneficial and ecologically sound alternatives.

2 Biopesticide Definition

Biopesticide is a term that includes many aspects of pest control such as microbial (viral, bacterial and fungal) organisms, entomophagous nematodes, plant-derived pesticides (botanicals), secondary metabolites from microorganisms (antibiotics), insect pheromones applied for mating disruption, monitoring or lure-and-kill strategies and genes used to transform crops to express resistance to insect, fungal and viral attacks or to render them tolerant of herbicide application (Copping and Menn 2000; Montesinos and Bonaterra 2009). According to the FAO definition, biopesticides include those biocontrol agents that are passive agents in contrast to biocontrol agents that actively seek out the pest, such as parasitoids, predators and many species of entomopathogenic nematodes. Thus, biopesticides cover a wide spectrum of potential products that can be classified as follows.

2.1 Microbial Pesticides and Other Entomopathogens

These are pesticides that contain microorganisms, like bacteria, fungi, virus and protozoan which attack specific pest species, or entomopathogenic nematodes as active ingredients. Most of these agents "entomopathogens" attack insect species referred to as bioinsecticides. These pathogenic organisms are isolated from diseased insects during naturally occurring epidemics. Over 400 species of fungi and more than 90 species of bacteria which infect insects have been reported. These microbial control agents have a range of properties that make them desirable for integrated crop management (Hajek 2004).

2.2 Plant-Incorporated Protectants (PIPs)

PIPs are pesticidal substances that plants produce from genetic material that has been added to the plant. For example, scientists can take the gene for the Bt pesticidal protein and introduce the gene into the plant's own genetic material. Then the plant, instead of the Bt bacterium, manufactures the substance that destroys the pest. The protein and its genetic material, but not the plant itself, are regulated by EPA. These crystalline proteins are highly insecticidal at very low concentrations (Schnepf et al. 1998). As these proteins are non-toxic to mammals and other organisms, Bt strains and their insecticidal crystal proteins (ICPs) have acquired acceptability as eco-friendly biopesticides all over the world and have been under extensive use in agriculture, horticulture, forestry, animal health and mosquito control for the past four decades (Schnepf and Curr 1995). Bt strains and ICPs were first found to affect a range of lepidopteran insects, which are recognized worldwide as major agricultural pests on crops. Subsequently, the discovery of new strains expanded the host range. Strains are now available which are toxic to coleopterans, dipterans, lice, mites and nematodes (Kumar et al. 1996).

2.3 Biochemical Pesticides

These are pesticides based on naturally occurring substances that control pests by non-toxic mechanisms, in contrast to chemical pesticides that contain synthetic molecules that directly kill the pest. Biochemical pesticides fall into different biologically functional classes including pheromones and other semiochemicals, plant extracts and natural insect growth regulators (Montesinos and Bonaterra 2009). When used as a component of integrated pest management (IPM) programs, biopesticides can greatly decrease the use of conventional pesticides, while crop yields remain high. There is now a growing concern among consumers towards food safety and environmentally sound practices, giving more and more importance on the use of biofertilizers and biopesticides (bioagents) as an alternative to farm chemicals. The total world production of biopesticides is over 3,000 t/year, which is increasing at a rapid rate. The market share of biopesticides is

Table 1.1 Global biopesticide and synthetic pesticide market

	Status of pesticides in different years (Million USD)				Average annual
Category	2003	2004	2005	2010	growth (%)
Biopesticides	468	562	672	1,075	9.9
Synthetic	27,144	26,600	26,076	24,205	−1.5
Total	27,612	27,162	26,748	25,280	−1.1
Biopesticides as % of total	4.25	4.25	4.25	4.25	4.25

Source: BCC Research (2006)

only 2.5 % of the total pesticide market (Gupta and Dikshit 2010).

Global assessments of biocontrol markets show that the percentage of biopesticides has steadily grown since 1997, and the momentum is projected to continue at a rate of 10 % per year. Global sales of total biopesticides were estimated at US $460 million in 2000 and have continued to increase annually (Table 1.1) with projections to reach over $1 billion by 2010.

2.3.1 Botanical Pesticides

Historical Overview

Botanical pesticides are an important group of naturally occurring, often slow-acting crop protectants that are usually safer to humans and the environment than conventional pesticides and with minimal residual effects. Moreover, botanical pesticides contain mixtures of biologically active substances, so no resistance is developed in pests and pathogens (Saxena 1983). Therefore, the use of plant pesticides has been recommended ever more as a suitable alternative of plant protection with minimum negative risks (Isman 2006). Especially, botanical insecticides have long been a subject of research in an effort to develop alternatives to conventional insecticides. The use of plant insecticides has a long-term tradition in Europe. The first known written reference to the application of plant extracts against pests comes from Rome and dates back to about 400 B.C. (Dayan et al. 2009). Also, botanical insecticides were considered important products for pest management in Ancient China (Long et al. 2006) Egypt, Greece and India (Isman 2006). Even in the United States and some European countries, botanical insecticides were predominantly used before the discovery of chlorinated hydrocarbons and organophosphorous insecticides in the late 1930s and early 1940s. In the 1980s and 1990s, interest in the use of botanical insecticides started to increase significantly, though not dramatically. This interest was driven primarily by concerns over the longer-lasting synthetic insecticides, environmental impact and their residues on the different crops. At present, several dozens of plant insecticides are used worldwide, based on various extracts, especially of the families Rutaceae, Lamiaceae, Meliaceae, Asteraceae, Annonaceae, Malvaceae and Labiatae. The survey of (Mwine et al. 2011) which established that thirty four species belonging to eighteen families are used in traditional agricultural practices for pest control in Southern Uganda. Botanical insecticides in their simplest form may be crude preparations of plant parts ground to produce a dust or powder that may be used full strength or distilled in a carrier such as clay, talc or diatomaceous earth (which itself is also insecticidal): such preparations include dusts from pyrethrum daisy flowers, cube roots (rotenone), sabadilla seeds, ryania stems or neem leaves, seeds or bark. Only slightly more sophisticated are water extracts or organic solvent extracts of insecticidal components of plants. These extracts or resins are then prepared as liquid concentrations or applied to talc or clays to be used as insecticidal dusts. Pyrethrins, rotenone, neem, citronella and other essential oils commonly are formulated as extracts or liquid concentrates. The most processed forms of botanical insecticides are purified insecticidal compounds that are isolated from plant materials by a series of extractions and distillations. Nicotine and limonene are distilled from plant materials or their extracts (Orozco et al. 2006; Recheigl and Recheigl 1999).

The plants that present biological activities against insects owe this feature to the presence of secondary metabolites, some of which have been widely investigated (Chandra et al. 2008).

The process of knowing and obtaining secondary metabolites against insects is by means of plant extracts, which can have variations according to the solvents used. Most of the secondary metabolites such as terpenoids and alkaloids are reported as candidates for insecticidal compounds that could be an effective alternative for insect pest management (Joseph et al. 2012).

Essential oils can also be obtained or dehydrated plant parts to be used as powders (Bobadilla et al. 2005). Secondary metabolites feature several properties against insects, like insecticidal activity, considered as that substance or mixture of substances that exert biocide action due to the nature of their chemical structure (Celis et al. 2008). However, most of the plants used against insects have an insectistatic rather than an insecticidal effect. This refers to the inhibition of the insect's development and behaviour, and it is divided into: repellence, antifeeding and deterrent activity (Eriksson et al. 2008), growth regulation, feed deterrence and oviposition deterrence (Banchio et al. 2003; Dimetry et al. 2007; Dimetry 2012). Repellent activity is presented in plants that have compounds with foul odour or irritating effects, which cause insects to get away from them (Peterson and Coats 2001). Antifeeding activity is exerted by compounds that, once ingested by the insect, cause it to stop feeding and eventually die of starvation (Isman 2006). Growth-regulating compounds inhibit metamorphosis or provoke precocious moulting. They alter the growth-regulating hormones and cause malformations, sterility or death in insects (Celis et al. 2008; Dimetry 2012). On the basis of the results of pesticidal screenings, it has been established that a number of plants have broad pesticidal activity, and those commonly used in traditional agricultural applications in many parts of developing countries, particularly in the tropical areas, are shown in Table 1.2. More than 2,000 plant species are known to have insecticidal properties, where Euphorbiaceae, Asteraceae, Labiatae, Fabaceae, Meliaceae and Solanaceae families

Table 1.2 Species, families and parts used and evaluated

Species	Families	Parts
Abrus precatorius L.	Fabaceae	L, S
Allium sativum L.	Alliaceae	L
Anacardium occidentale L.	Anacardiaceae	L
Annona senegalensis Pers.	Asteraceae	S, B
Artemisia annua L.	Asteraceae	L, B
Azadirachta indica A. Juss.	Meliaceae	L, B, R, F
Balanites aegyptiaca Linn. Bel.	Zygophyllaceae	R
Bidens pilosa L.	Asteraceae	L
Cannabis sativa L.	Cannabaceae	L, S, F
Capsicum frutescens L.	Solanaceae	F
Carica papaya L.	Caricaceae	R, B
Chrysanthemum coccineum Wild	Asteraceae	L, F
Clausena anisata	Rutaceae	L, R
Dalbergia saxatilis	Fabaceae	L, B
Dannettia tripetala	Annonaceae	L.
Eucalyptus globules	Myrtaceae	L, B
Gmelina arborea Juss.	Verbenaceae	L
Hyptis sauvcolens Poit.	Labiatae	Shoot
Jatropha curcas L.	Euphorbiaceae	Sap, F, S, B
Khaya senegalensis A. Juss.	Meliaceae	S, B
Lannea acida	Anacardiaceae	B
Lawsonia inermis	Lythraceae	L
Melia azedarach L.	Meliaceae	L, R, B
Mitracarpus scaber Zucc.	Rubiaceae	Shoot
Nicotiana tabacum L.	Solanaceae	L
Ocimum gratissimum L.	Liminaceae	L
Parkia clappentoniana Keay.	Mimosaceae	S, B
Phytolacca dodecandra L'Herit	Phytolaceae	L, F
Piper guineense Schum &Thonn	Piperaceae	F
Piliostigma thonningii	Caesalpiniaceae	R, B
Prosopis africana Linn.	Mimosaceae	S, B
Sphenoclea zeylanica Gearth	Sphenocleceae	Shoot
Tagetes minuta L.	Asteraceae	L
Tephrosa vogelii Hook	Fabaceae	L
Vernovia amygdalina L.	Asteraceae	L

Source: Simon Koma Okwute (2012)
Key: *L* leaf, *B* bark, *S* seed, *R* root, *F* fruit

stand out (Garcia et al. 2004). Among the metabolites with biological activities against insects are flavonoids, terpenoids, alkaloids, steroids and phenols (Orozco et al. 2006).

Botanical insecticides though natural in origin are not non-toxic or nonchemical. Some of the plant-derived compounds, particularly rotenone and nicotine, are as toxic or more toxic to humans than many common synthetic insecticides. In general, they are toxic to pests and beneficial insects alike, and if they are used repeatedly, botanical insecticides can disrupt natural biotic control of insect pests by their natural enemies. Their limited persistence in the environment helps to minimize their adverse effects, but plant-derived toxins with the exception of some neem formulations, e.g. NeemAzal T/S (Dimetry et al. 2013), certainly should not be viewed as harmless.

Selection Criteria of Botanical Pesticides

1. It should be perennial.
2. It should have a wide distribution and be present in large numbers in nature; otherwise, it should be possible to be grown by agricultural procedures like tissue cultures and genetic engineering.
3. The plant parts to be used should be removable: leaves, flowers or fruits.
4. Harvesting does not mean destruction of the plant (avoid the use of roots or barks).
5. The plants should require small space, reduced management and little water and fertilization.
6. The plants could have additional uses (e.g. medicinal use).
7. The plants used should not otherwise have a high economic value.
8. The active ingredients preferably are effective at low concentrations.

3 Current Botanicals in Use

On the basis of the results of pesticidal screenings, it has been established that a number of plants have broad pesticidal activity, and those commonly used in traditional agricultural applications in many parts of the developing countries, particularly in the tropical areas, are shown in Table 1.2, which are only representative but not exhaustive of the thousands of plants so far screened (Rajapake and Ratnaseka 2008). From various investigations, it has been established that activity is usually distributed in most cases among the various parts of the same plant though the lethality and quantities of the active components may vary (Rajapake and Ratnaseka 2008).

At present, there are four major types of botanical products used for insect control (pyrethrum, rotenone, neem and essential oils) along with three others in limited use (ryania, nicotine and sabadilla). Additional plant extracts and oils (garlic oil, *Capsicum oleoresin* and many constituent compounds) have not been produced for commercial use are known to have insecticidal or repellent characteristics (Jacobson and Crosby 1971; Grainger and Ahmed 1988; Jacobson 1989; Berenbaum 1989; Hedin 1991, 1997).

3.1 Botanicals with Insecticidal Effects

Throughout history, plant products have been successfully exploited as insecticides, insect repellents and insect antifeedants. Probably the most successful use of a plant product as an insecticide is that of the pyrethroids. The insecticidal properties of the several *Chrysanthemum* species were known for centuries in Asia. Crude pyrethrum powders were first introduced to Europe around 1800, and they were in use worldwide by around 1850 (Casida 1973; Matsumura 1985; Casida and Quistad 1995). Even today, powders of the dried flowers of these plants are sold as insecticides. After elucidation of the chemical structures of the six terpenoid esters (pyrethrins) (Fig. 1.2) responsible for the insecticidal activity of these plants, many synthetic analogues have been patented and marketed. Synthetic pyrethroids have better photostability and are generally more active than their natural counterparts.

Pyrethrins poison insects and mammals in similar manners; they interfere with nerve transmission by slowing or preventing the shutting of sodium channels in nerve axons (Bloomquist 1996). The results in insects are hyperactivity and convulsions; whole body tremors occur in mammals. The mode of action and resulting symptoms of poisoning are generally similar for pyrethrins and synthetic organochlorine insecticide. Although plant-derived pyrethrins are very toxic and fast acting against insects, they are not very toxic to mammals by oral or dermal routes at least in comparison with other insecticides (reviewed by Hayes 1982). When ingested, they are not readily absorbed from the digestive tract, and they are readily hydrolyzed in the acidic conditions of the gut and the liver. As a result, pyrethrins are more toxic to mammals via inhalation than ingestion because inhalation provides a more direct route to the blood (Hayes 1982). Pyrethrum and, to a lesser extent, the extracts that contain pyrethrins can, however, cause irritation and allergic reactions in humans (Barthel 1973). Pyrethrins are moderately to highly toxic to fish, but their labile nature in the environment greatly reduces their potential hazard (Pilmore 1973).

In an attempt to counteract their instability in the environment, pyrethrins usually are combined with antioxidants to extend their persistence. Insecticidal formulations usually contain synergists to slow pyrethrin detoxification by oxidative enzymes in target pests. Common synergists include piperonyl butoxide (PBO) and N-octyl bicycloheptene dicarboximide (MGK 264). Synergists often are added to pyrethrin formulations at a ratio of 2:1 to 10:1 (synergist to insecticide).

Pyrethrum and pyrethrins are used most commonly in ectoparasiticides for humans and pets, aerosols for fly control in livestock and in closed spaces such as greenhouses and grain storages. Pyrethrins are also combined with slower-acting botanicals such as rotenone or ryania in gardens. In their uses, the instability of pyrethrins is desirable so that unwanted residues do not persist on treated products or individuals.

Camphene, a plant terpenoid (Fig. 1.1), was a very successful herbicide in its polyhalogenated form. Sold as toxaphene, this product was the leading insecticide in the United States before it was removed from the market. Although this product was a mixture of over two hundred chlorinated forms of camphene, certain specific compounds in the mixture were found to be much more active than the mixture on a unit weight basis. Many other terpenoids have been demonstrated to have insecticidal or other insect-inhibiting activities. For instance, azadirachtin and other terpenoids of the limonoid group from the families Meliaceae and Rutaceae are potent growth inhibitors of several insect species (Schmutterer 1990, 1995; Dimetry 2012).

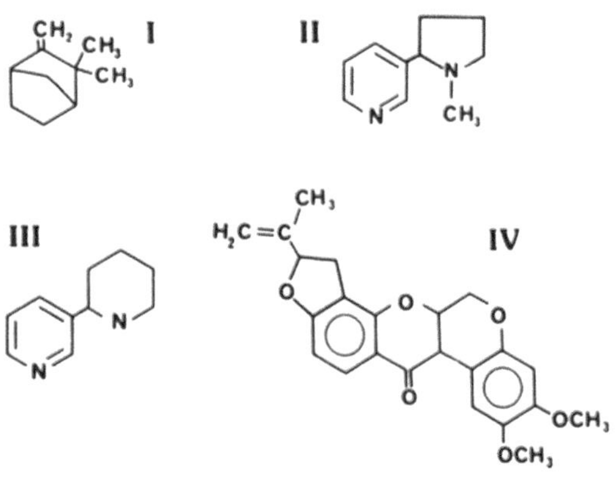

Fig. 1.1 Some plant-produced compounds with insecticidal activity. I-camphene, II-nicotine, III-anabasine, IV-rotenone

Nicotine, nornicotine and anabasine are related alkaloids derived from tobacco and other plant species in which they may comprise 2–8 % of the dry weight of leaves. Hayes (1982) reported that nicotine is usually derived from *Nicotiana tabacum*.

Nicotine (Fig. 1.1) and nornicotine, components of several members of the genus *Nicotiana*, have been used commercially as insecticides. *N. rustica* is the chief commercial source. Other natural analogues of nicotine have been shown to have significant insecticidal properties, and one, anabasine or nornicotine (Fig. 1.1), has been produced as an insecticide from the shrub, *Anabasis aphylla*, in the Soviet Union. Synthetic variations of nicotine such as 5-methylnornicotine have been demonstrated to be effective insecticides. Ware (1988) reported that steam distillation is the most common current method for the preparation of commercial insecticides. Free basic nicotine is very unstable and it is used more often for greenhouse fumigation.

Nicotine poisons insects and mammals by a similar mode of action. The overall symptoms of nicotine poisoning resemble those of poisoning by organophosphate or carbamate insecticides. Oral and dermal LD_{50} values have been estimated in the range of 59–60 mg/kg, though a great range in toxicity estimates have been published (Hayes 1982; Ware 1988; Coats 1994).

Ryanodine (Fig. 1.2), an alkaloid from the tropical shrub *Ryania speciosa*, has been used as a commercial insecticide against European corn borer. Physostigmine, an alkaloid from *Physostigma venenosum*, was the compound upon which carbamate insecticides were designed. Furoquinoline and beta-carboline alkaloids such as dictamine and harmaline, respectively, are potent photosensitizing compounds that are highly toxic to insect larvae in sunlight. The relative high toxicity to mammals and limited efficacy have limited the use of natural alkaloid insecticides.

Preparations of roots from the genera *Derris*, *Lonchocarpus* and *Tephrosia* containing rotenone (Fig. 1.1) were commercial insecticides in the 1930s. Rotenone is a flavonoid derivative that strongly inhibits mitochondrial respiration. No other phenolic compound has been used commercially as an insecticide, although the content of certain phenolic compounds in plant tissues has been correlated with host plant resistance to insects and many have been demonstrated to be strong insect growth inhibitors and antifeedants.

As in plants, delta amino levulinic acid (ALA), in combination with 2,2′-dipyridyl, can cause accumulation of toxic levels of photodynamic porphyrin compounds. Larvae of several insect species, when fed these compounds and exposed to light, were rapidly killed (Arnason et al. 1983). Protoporphyrin IX, the same compound caused to accumulate in plants by certain photobleaching herbicides, is the porphyrin responsible for the toxicity of these compounds to insects. Other photodynamic compounds from plants such as polyacetylenes are acutely toxic to insects; however, their general toxicity would probably preclude them from commercial use.

Sabadilla is derived from the seeds of sabadilla lily, *Schoenocaulon officinale* Gray (Fig. 1.2). Sabadilla insecticides have been used for hundreds of years. The chemistry and uses of sabadilla have been reviewed by Jacobson and Crosby (1971). The major insecticidal components of sabadilla are the alkaloids cevadine and veratridine (Fig. 1.2) which occur in the seeds (2–4 %). The extracted alkaloids are highly poisonous, e.g. LD_{50} for veratridine 1.35 mg/kg mouse intraperitonially. Ground seeds, on the other hand, appear to be quite safe (LD_{50}, 5,000 mg/kg rat, oral) and are used after mixing with sodium carbonate solution as a contact and stomach poison to insects on food crops. No residues are left after the application of sabadilla because it breaks down rapidly in sunlight. Sabadilla acts as a stomach poison against caterpillars, leafhoppers, thrips, sting bugs and squash bugs.

Citrullus colocynthis is related to the Cucrbitaceae family. Its fruits were collected from Hagol Valley. This is a desert road between Cairo and Suez Governorates. The crude extracts of this plant have a deterrent effect against different pests. Different formulations of *Citrullus colocynthis* seeds were manufactured either from chloroform or alcohol extract in the form of dusting powder or emulsifiable concentrate. Chloroform extract contains free cucurbitacins, aglycone, but the alcohol extract contains cucurbitacins and flavonoids in the form of glycosides. *Citrullus* powder was formulated with use of 10 % extract, 10 % fine silica (silicon dioxide) and 80 % talc powder, while the emulsifiable concentrate was prepared with 10 % extract, 10 % emulsifiable and 80 % solvent (mostly xylenes). The different formulations have insecticidal as well as anti-ovipositional responses against *Callosobruchus maculatus* (Dimetry et al. 2007). The fruits have been used for the protection of clothes from insect pests. One company has produced special hangers with citrullus seeds to repel insects.

Nicandra physaloides is related to the family Solanaceae. It is called Shoo plant. It has a potent

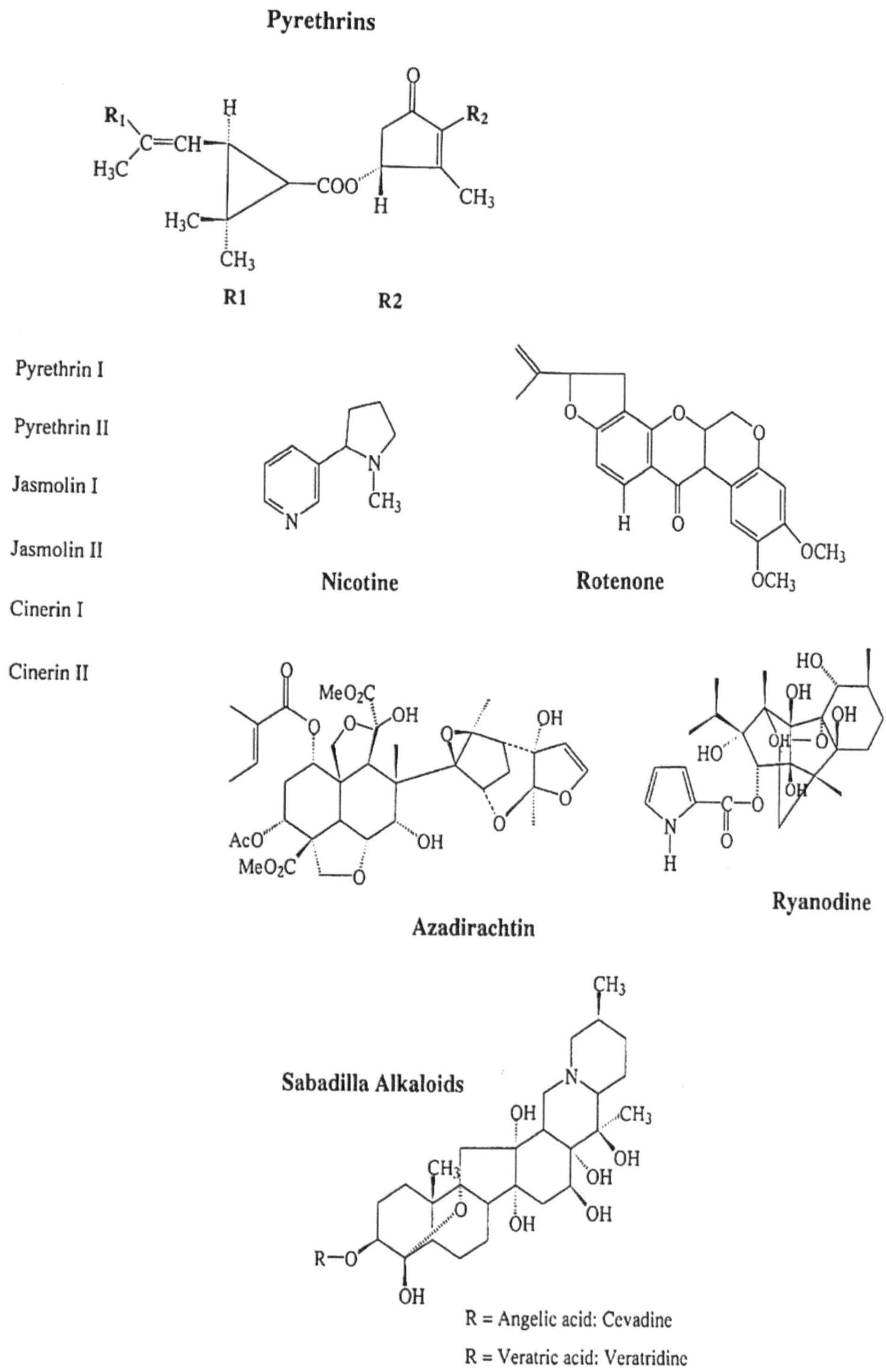

Pyrethrins

R1 R2

Pyrethrin I

Pyrethrin II

Jasmolin I

Jasmolin II

Cinerin I

Cinerin II

Nicotine

Rotenone

Azadirachtin

Ryanodine

Sabadilla Alkaloids

R = Angelic acid: Cevadine

R = Veratric acid: Veratridine

Fig. 1.2 Some plant-produced compounds with insecticidal activity discussed in the review

insecticidal effect against whiteflies. It is native in China. Its seeds *were obtained from the United States (New York). It is cultivated successfully in Egypt. It* has been reported to contain an antifeedant steroid (Gill et al. 1986) and shows insecticidal activity against aphids (Dimetry and El-Hawary 1995).

Himalayan cedar wood oil (*Cedrus deodara*) has a potent insecticidal effect against the pulse beetle (*Callosobruchus analis* F.) and the housefly (*Musca domestica* L). Almost all fractions isolated showed insecticidal activity against both test species (Singh and Agarwal 1988). Also, Singh et al. (1984) reported the insecticidal potential of Himalayan cedar wood oil (*C. deodara*) against Indian mosquitoes *Anopheles stephensi* under laboratory conditions.

One of the most fantastic and important botanical insecticides is neem. Neem insecticide is derived from the tropical and subtropical *Azadirachta indica*, a tree in the family Meliaceae (the mahogany family). This tree is native to southern and southeastern Asia and is now grown in different countries in Africa, America and Australia (Saxena 1989; Schmutterer 1990). Azadirachtin (Fig. 1.2) is the principal active ingredient in neem extracts. Neem's value as an insect deterrent and as an insecticide has been known in India for centuries. Insecticidal products include teas or dusts made from leaves and bark extracts prepared from fruits, seeds or seed kernels and oils pressed from the seeds. Although all plant parts show some characteristics of insect repellence or suppression, seeds or seed kernels provide the greatest amounts of insecticidal liminoids (Schmutterer 1990). Extraction in methanol is a common method of obtaining azadirachtin from seeds.

The numerous reported effects of neem on insects include repellence, feeding deterrence, oviposition deterrence, reduced growth and development and interference with reproduction (Schmutterer 1990; Dimetry 2012), but the biochemical or molecular mode of action for such effects remain untouched. Neem and its principal component "azadirachtin" are extremely low in mammalian toxicity (Schmutterer 1990), and most forms are nonirritating to skin and mucous membranes. Neem extracts have been used medicinally for centuries in Asia and India to lower blood pressure, reduce inflammation and reduce fevers.

Neem's broad activity against plant-eating insects and its virtual non-toxicity to mammals make it an extremely appealing insecticide, and various products registered and approved by the US Environmental Protection Agency (EPA) are currently marketed for use as sprays, dusts or soil-applied systemic insecticides. Schmutterer (1990) indexed research on target pests and reviewed neem's uses. It is generally considered to be most effective against the soft-bodied immature stages of plant pests including whiteflies (Dimetry et al. 1996), leaf miners, aphids and various caterpillars (Dimetry et al. 1995; Dimetry and Schmidt 1992; Dimetry and El-Hawary 1995), and it is low in toxicity to insect predators and parasitoids (Dimetry et al. 2013).

3.2 Polyester of Sugars

A new class of insecticides was recently discovered by Beltsville researchers led by Puterka in the United States that offers a safe and effective alternative to commercial insecticides. They are polyesters of sugars and include *sucrose* and *sorbitol octanoates*. They were isolated from the poisonous hairs on the tobacco leaves which hitherto were assumed to contain nicotine, a popular insecticide. When insects were contaminated by rubbing, they caused death of the insects by a dehydration process, and rapidly degraded to harmless sugars and fatty acids. These polyesters are known to be effective against a variety of farm and domestic insect pests and the deadly parasitic *varroa* mite which usually settles on the back of honeybees (Bliss 2005).

3.3 Repellent and Antifeedant Plants

Closely related to the insecticidal agents and sometimes used in combination with insecticides in pest management strategies are some classes of pesticidal agents with interesting and peculiar biological activities. They include insect *repellents, antifeedants* or *deterrents* and *attractants*. These classes are far less common in plant sources than the insecticides but will be given some attention. Sometimes, a given insecticide may act as an insecticide or as a *repellent* depending on the concentration. The major difference between the two is that a repellent does not kill insects but keeps them away by exuding pungent vapours or exhibiting slightly toxic effects (Rajapake and Ratnaseka 2008). By these activities, a *repellent* prevents insects from perching or landing on the surfaces of targets. Thus, *repellents* can be used to prevent and control the outbreak of insect-borne diseases such as malaria. The insects of interests in this regard include the mosquito, flea, fly and *arachnid tick* (Maia and Moore 2011). The use of plant materials as *insect repellents* is increasingly receiving attention, particularly in the developing countries. For example, Seyoun et al. (2002) reported that in western Kenya, the natives employ direct burning of the species *Ocimum americanum* L., *Lantana camara* L., *Tagetes minuta and Azadirachta indica*. Some recent studies on repellent plants have led to the isolation and characterization of some active components. Prominent among these *compounds are callicarpenal* and *intermedeol* from the *Cymbopogon nardus* species, which showed promising alternative in the control of infestations by *Amblyomma cajennense* (Soares et al. 2010); *nepetalactone*, a catnip *compoun*d for the control of the Asian adult male and female lady beetle as well as cockroaches, flies, termites and mosquitoes (Suszkiw 2009; McElvain et al. 1941); and *geraniol* and *p-menthane-3,8-diol* (PMD), monoterpenoid alcohols from the *citronella* and *lemon oils*, respectively (Bernad and Xue 2004). Some researchers have found that products containing 40 % lemon *eucalyptus* oil are as effective as products containing high concentrations of DEET and that neem oil can give up to 12 h of protection against mosquitoes in cage experiments (Carroll and Loye 2006; Mishra et al. 1995). Among few plants studied, some and their extracts (*Xylopia aethiopica*) have feeding deterrency or antifeedency (Soares et al. 2010). The hexane and methanol extracts of the fruits and seeds have been shown to possess strong termite *antifeedant* activity, and *entkauranes* and some *phenolic amides* have been implicated. Among the *entkauranes*, the activity was significantly dependent on the structures and that *(−)-kau-16-en-19-oic acid* had the strongest antifeedant activity (Lajide et al. 1995).

Another species with promise is different formulations from *Citrullus colocynthis* seeds: chloroform and alcohol extracts of *Citrullus colocynthis* either as dusting powder or emulsifiable concentrate are effective bioinsecticides against *Callosobruchus maculatus* (F.). The chloroform extract contains free aglycones of cucurbitacins, while an alcohol extract constitutes the cucurbitacins and flavonones in glycosidic nature. They gave considerable protection for the stored cowpea seeds for different periods, depending on the kinds of storage sacks and the rate of formulations used. Both damour and polyethylene sacks proved to be highly suitable for the storage of treated cowpea seeds. They protected the seeds completely from infestation for seven months, and no weight loss was detected in the stored seeds in the majority of treatments. This is due to the fact that *C. maculatus* was deterred completely from treated seeds and no infestation appeared (Dimetry et al. 2007; Dinan et al. 2001). Also, Juneja and Patel (2002) found that seeds of green gram treated with 1 % of either powdered custard apple or black pepper seeds were totally protected from the pulse beetle for up to five and four months, respectively. Moreover, Raja et al. (2001) and Dimetry et al. (2002) stated that the weight loss was reduced in stored pulses protected with a plant extract such as *Melia azadirachta*, oil extracted from sweet flag (*Acorus calamus* L.) and different formulations of neem (NeemAzal T/S, NeemAzal F and NeemAzal T). All parts of the neem tree possess anti-insect activities, but the seed kernel is the most active. Neem bark, leaf, fruit and oil as well as extracts with various solvents, especially ethanol, have been found to exhibit activity against different pests. About 413 insect species are reportedly susceptible to differ-

Table 1.3 List of insect pests susceptible to neem products

Insect order	Number of susceptible species
Orthoptera	24
Dictyoptera	6
Dermapters	1
Phasmida	1
Isoptera	6
Thysanoptera	13
Phthoraptera	4
Hemiptera	82
Hymenoptera	8
Coleoptera	79
Lepidoptera	136
Diptera	49
Siphonaptera	4
Total	413

Source: Schmutterer and Singh (1995)

ent concentrations of neem preparations (Table 1.3). The repellent and antifeedant effects of neem have been reported against a wide range of different pests (Ketkar 1976; El-Sayed 1982–1983; Saxena 1989; Koul et al. 1990; Dimetry and Schmidt 1992). Another species with promise is the *Jatropha podagrica* plant cultivated in West Africa. The organic extracts showed reasonable *antifeedant* activity against *Chilo partellus*, the maize stem borer, at concentrations of 100 %/leaf disc, the chloroform extract being the strongest. The most active compound isolated was *15-epi-4 Z-jatrogrossidentadion* (Aiyelaagbe et al. 2011).

Attractants are *semiochemicals* produced usually by some insects with effect on other insects as a communication tool and can be used to determine or control insect populations, particularly by disrupting their mating patterns. Rarely do plants produce chemicals that attract insects that are natural enemies of other insects that feed on the plants except the tea tree (Weinzierl et al. 2009). Thus, field application of this phenomenon is not common and therefore will not be discussed further.

3.4 Botanicals with Fungicidal Activity

Extracts from many plant species have been found to be active against many phytopathogenic fungi without imposing ill side effects (Lyon et al. 1995). In some cases, the active components have been identified and tested directly. Some plant extracts act as contact fungicides. Some disrupt cell membrane integrity at different stages of fungal development, while others inactivate key enzymes and interfere with metabolic processes. The results of various investigations have implicated essential oils of many species as possessing fungitoxic activity. They are therefore agents of protection in plants against diseases. Consequently, since the leaves, resins and lattices of plants contain essential oils more commonly than other parts of plants, they have been more commonly investigated for fungicidal activity (Rai et al. 1999; Rai and Acharya 2000).

Without an immune system to combat pathogenic microorganisms, plants rely primarily on chemical protection with secondary compounds. Compounds that inhibit the establishment of and growth of plant pathogens are termed phytoalexins. Many of these secondary compounds have been chemically characterized, and the proof that these compounds have such a role in plant disease prevention and control is being developed. In fact, there is some evidence that certain synthetic fungicides used in plant protection act by inducing the production of phytoalexins in plants.

Several plant-derived compounds have been demonstrated to be strong elicitors of phytoalexins. For instance, certain oligosaccharide components of cell walls from stressed or dying higher plant cells will act as *elicitors*. Further knowledge of plant-derived phytoalexin elicitors could lead to their use as fungicides. Several isoflavonoid compounds, such as glyceollin, phaseolin and pisatin (Fig. 1.3) in soybean, garden bean and pea, respectively, have been implicated in the protection of these crops from pathogens. Many others confirmed or suspected phytoalexins have been identified. Some of these compounds have demonstrated utility against fungi under field conditions. Foliar application of the phenolic lactone juglone (Fig. 1.3), a product of several walnut species, provides better protection of bean seedlings from rust than some commercial fungicides. Terpenoid phytoalexins and fungicides are known and some have been tested for commercial efficacy. Wyerone, an acetylenic acid derivative produced by legumes as

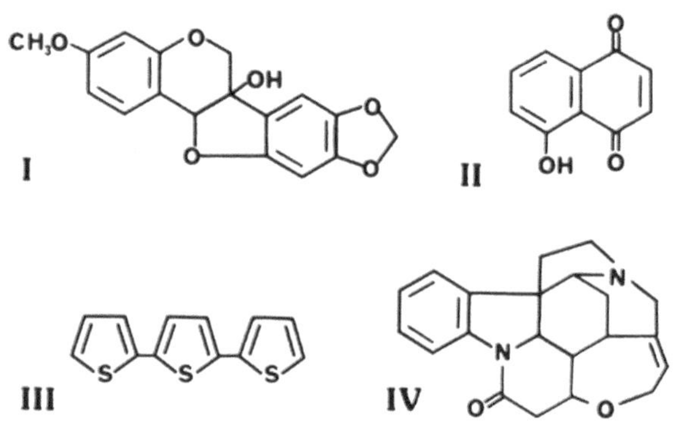

Fig. 1.3 Some plant-produced compounds with fungicidal, nematicidal and rodenticidal activity. I-pisatin II-juglone, III-alpha-terthienyl, IV-strychnine

a phytoalexin, has a wide fungicidal spectrum against plant pathogens and has been successfully tested against fungal infection of crop plants. Despite a repertoire of many *antifungal* and *antibacterial* compounds, plant products have not been used to any significant extent in the development of *antimicrobial* pesticides.

3.5 Botanicals with Nematicidal Activity

Many plant species are known to be highly resistant to nematodes. The most well documented of these include marigolds (*Tagetes* spp.), rattlebox (*Crotalaria spectabilis*), chrysanthemums (*Chrysanthemum* spp.), castor bean (*Ricinus communis*), margosa (*Azadirachta indica*) and many members of the family Asteraceae and family Compositae (Duke 1990). The active principle(s) for this nematicidal activity has not been discovered in all of these examples, and no plant-derived products are sold commercially for the control of nematodes. In the case of the Asteraceae, the photodynamic compound alpha-terthienyl (Fig. 1.3) has been shown to account for the strong nematicidal activity of the roots.

3.6 Botanicals with Molluscicidal Activity

The plant-derived saponins are generally highly toxic to snails. Cyanogenic glucosides are responsible for the resistance of certain legumes to snails and slugs. *Bilharzia* affects millions of people, particularly children who play or swim in infected freshwaters in the developing countries of Africa, Asia and Latin America. The disease was discovered in 1851 by Theodor Bilharz as the cause of urinary *schistosomiasis*. It is associated with certain species of aquatic snails of the genera *Biomphalaria*, *Bulinus* and *Oncomelania*. Therefore, one way of attacking the disease is to eliminate the host snails (Brown 1980; Dalton and Pole 1978). Chemicals that kill snails are called *molluscicidal agents*. Most of the *molluscicidal agents* in use today are synthetic and, like most synthetic pesticides, are harmful to man and the environment. *Molluscicidal agents* of natural origin are important in the widespread control of *schistosomiasis*. Mirazid, an Egyptian drug from *myrrh*, was being developed as an oral drug until 2005 when it was found to be only eight times as effective as *praziquantel*, a synthetic chemical, and has therefore not been recommended by the WHO. However, other plants have been studied, and some have demonstrated potential activity that may provide leads for future drugs, but more important are the searches for molluscicidal agents from plants to eliminate the host snails (Vet Parasitology 2009). This is the focus of the presentation in this chapter.

Adesina, Adewunmi, Kloos and McCullough have separately investigated the species *Clausena anisata* and found it to possess *molluscicidal activity* which is distributed among the root, leaves, bark and stem in a decreasing order of potency (Adesina and Adewunmi 1981; Kloos and McCullough 1987). Adedotun and Alexander (2008) evaluated the *molluscicidal activity* of the aqueous and ethanolic extracts of fruits and roots of *Dalbergia sissoo* against the egg mass and adults of *Biomphalaria pfeifferi* and found that only the ethanolic extracts showed significant

Table 1.4 Molluscicidal activity of *Clausena anisata* and *Tetrapleura tetraptera*

Plant parts	Concentration	Mortality	Solvent
Clausena anisata	Root 6–10 ppm	100	Methanol
Leaves	1,000 ppm	53.3	Water
Stem	1,000 ppm	7	Water
Bark	1,000 ppm	40	Water
Tetrapleura tetrptera fruit	100 %	100	Water
Fruit	10 %	100	Methanol

Source: Adesina and Adewunmi (1981)

activities. Thus, the active constituents of ethanol extracts are more potent than water (Adedotun and Alexander 2008; Dos Santos and Sant'Ana 1999). Similar observations have been recorded for *Clausena anisata* parts and *Tetrapleura tetraptera* fruits, particularly when the active components are *glycosides* (Table 1.4) (Adesina and Adewunmi 1981; Adesina et al. 1980).

3.7 Rodenticidal Plants

Plants produce a myriad of compounds that are poisonous to mammals. Some of these, such as strychnine (Fig. 1.3), are used in commercial rodenticides. The chronic poison warfarin and several analogues are coumarin derivatives. This chemistry led to the discovery of indanediones and 4-hydroxy-2H-1-benzopyran-2-ones as rodenticides (Duke 1990).

3.8 Herbicidal Plants

Apart from insects and diseases disturbing crop plants, weeds also need to be controlled because they retard plant growth and reduce crop yields. Herbicides, also commonly known as weed killers, are pesticides used to kill unwanted plants. Selected herbicides kill specific targets while leaving the desired crop relatively unharmed. Inhibition of plant growth and production of phytotoxic symptoms by certain plants and their residues are a well-established phenomenon. In searching for potential herbicides from plants, screening of compounds known to function in plant-plant interactions is a logical strategy (Duke and Lydon 1987). Not only would they offer novel sites of action but

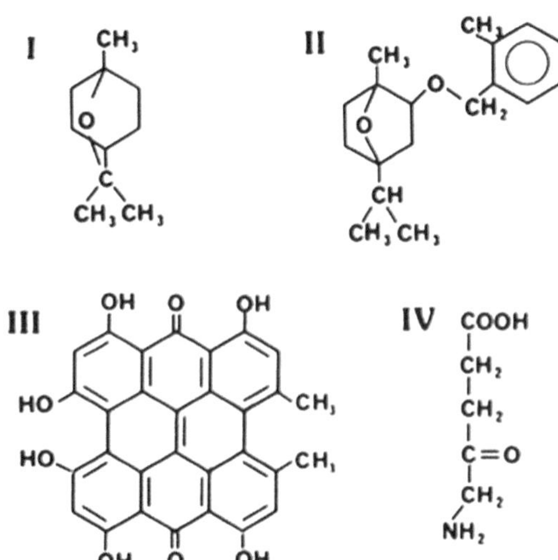

Fig. 1.4 Some plant-produced compounds and derivatives with herbicidal activity. I-1,8-cineole, II-cinmethylin, III-hypericin, IV-delta amino levulinic acid

these natural compounds would show strong selection for target species. All plants produce secondary compounds that are phytotoxic to some degree. However, in only a relatively few cases has it been established that particular compounds provide the producing species a competitive advantage over other species that are less tolerant to the compound. Only a few of these allelochemicals have been actively pursued as herbicides, and in these cases, the natural compound has been modified. A derivative of the terpenoid allelochemical 1,8-cineole (Fig. 1.4), with the common name of cinmethylin (Fig. 1.4), is being commercially developed. ToxapheneReg., a mixture of chlorinated camphene derivatives, was sold as an herbicide and an insecticide, but was removed from the market in 1982 by the EPA. Other very weakly phytotoxic compounds from plants such as benzoic acids can be made much more herbicidally active by halogen substitutions. Several benzoic acid derivatives such as dicamba (3,6-dichloro-2-methoxybenzoic acid) are widely used as herbicides.

A few highly phytotoxic plant-produced compounds have been discovered. However, none have been developed as herbicides. The sesquiterpenoid lactone, artemisinin from *Artemisia annua* L., was found to inhibit plant growth as well as the commercial herbicide cinmethylin (Duke et al. 1988). Other compounds, such as 2,4-dihydroxy-1,4-benzoxazin-3-one, are as active as plant growth inhibitors as many herbicides. Plants produce many photodynamic compounds, such as hypericin (Fig. 1.4), that are strongly phytotoxic, provided they can be introduced into the plant cell. These compounds are unlikely to be developed as pesticides because, in the presence of light, they are toxic to all living organisms. However, any plant can be caused to generate phytotoxic levels of photodynamic porphyrin compounds by treating the plant with *d*-aminolevulinic acid (Fig. 1.4), a natural porphyrin precursor, and 2,2′-dipyridyl, a synthetic compound. This relatively safe combination of compounds is being developed as the "laser" herbicide (Duke 1990). Several classes of commercial herbicides have been shown to act by causing target plant species to accumulate phytotoxic levels of protoporhyrin IX, a photodynamic chlorophyll and heme precursor (Lydon and Duke 1988). Thus, a natural product, not the synthetic herbicides, is the acutely toxic compound in these cases. Application of protoporhyrin IX alone to plant tissues, however, is not effective, apparently because it does not reach the proper cellular compartment in sufficient quantity.

A problem with plant-produced phytotoxins as potential herbicides is that in the native state, they are generally only weakly active compared to commercial herbicides. Most known allelochemicals would have to be applied at rates of more than 10 kg/ha to achieve significant weed control, whereas most recently marketed herbicides would achieve the same level of control at levels three orders of magnitude smaller. This is not unexpected, because the production of highly phytotoxic compounds would lead to strong autotoxicity unless the producing plant develops metabolic or physical mechanisms to cope with its own phytotoxins. Some of the more potent allelochemicals are toxic to the producing species, and this autotoxicity has been implicated in vegetation shifts. Microbial conversion of relatively non-phytotoxic compounds in the soil to highly phytotoxic derivatives has been documented (Duke 1986).

Plants have been much more successfully exploited as sources of pesticides for pests other than weeds. This is probably due to several factors. The selection pressure caused by pathogens and herbivores has probably been more acute and intense than that caused by plant competitors. A plant species can effectively compete with plant foes in many ways other than by poisoning them and having to cope with autotoxicity. Pathogens and herbivores have many potential physiological and biochemical sites of action for pesticides that the plant does not share. Biosynthesis of a compound to affect one of these sites reduces the chance of autotoxicity. Thus, the chemical option is generally a more attractive option in responding to an herbivore or pathogen that can rapidly devour or invade the plant than it is in responding to a plant competitor.

3.9 Essential Oils

Essential oils (EOs), or their constituents, provide viable alternatives to synthetic pesticides with regard to efficacy/dose, environmental impact and human health concerns. For example, based on the suggested seeding rate for corn of 70,000–81,000 seeds per hectare (Nielson 2005), 28–32 g thymol per hectare or 56–65 g of citronellal per hectare would be required to protect seeds from wire worm damage, while 9–113 g of thiamethoxam per hectare (Cruiser 5FS, Syngenta Crop Protection, Greensboro, NC, USA) would be needed. Environmentally, the tested EOs are biodegradable and photodegradable, thereby greatly reducing their persistence in the environment. In addition, their relative lipophilicity and high vapour pressure reduce leaching into groundwater or their persistence in soil or sediments (Isman 1999).

In recent years, several researchers have reported the acute toxic effect of essential oils (EOs) on plant or human pests (Isman 1999; Hummelbrunner and Isman 2001). Of the EOs evaluated, the oil from the citronella plant (*Cymbopogon nardus* L.) and its major constituent, citronellal, have shown the greatest promise for pest control (Isman 1999). Citronellal is often used as a non-toxic alternative to DEET (*N,N-diethyl-m*-toluamide) for protection against mosquitoes and other biting insects. In addition to citronellal, the acute toxicities (i.e. acaricidal and larvicidal) of other monoterpeniods have been reported. These include thujone from *Artemisia absinthium* L. on western corn root worm (*Diabrotica virgifera virgifera* LeConte), eugenol (from *Syzygium aromaticum* L.) on the two-spotted spider mite (*Tetranychus urticae* Koch) and thymol from *Thymus vulgaris* L. on the common housefly, *Musca domestica* L. (Lee et al. 1997). In addition, the oils from rosemary (*Rosmarinus officinalis* L.), thyme (*Thymus vulgaris* L.) pepper mint (*Mentha piperita* L.), lavender (*Lavandula angustifolia* Mill.) and spearmint (*Mentha spicata* L.) have all shown repellent activities against the two-spotted spider mite (*T. urticae*) (Hori 1998), while the EOs from thyme, peppermint and spearmint have also shown larvicidal activity against tobacco cut worm (*Spodoptera litura* L.) (Isman et al. 2001). Ngoh et al. (1998) demonstrated both contact and fumigant toxicities of eugenol against the American cockroach (*Periplaneta Americana* L.). In addition, Waliwitiya et al. (2005) demonstrated both contact and volatile toxicities from thymol, citronellal, eugenol and rosemary oil against late instar larvae of *Agriotes obscurus* (L.). Also, the general compatibility of EOs with organic farming practices make them attractive topics for researchers interested in developing sustainable agricultural practices, minimizing chemical use, regardless of the type or form of the compound is desirable and can be achieved by a more targeted use of these compounds, as seed/seedling protection agents, and not as broad-based soil or plant treatments. Despite these positive indications, relatively few EOs have been tested for their pesticidal efficacy under greenhouse or field conditions, and currently none have gained widespread use (Bernays 1983; Landis and Gould 1988).

3.9.1 Mode of Action

Symptoms of insects acutely poisoned by certain essential oils or their pure constituents have shown similarity to neurotoxins, irrespective of the route of administration (Isman 1999). Insects poisoned with EOs often display hyperactivity, convulsions and tremors followed by paralysis (Isman 1999; Enan 2001). Enan (2001) reported that eugenol, terpenoid and cinnamic alcohol block octopamine receptor binding sites in American cockroach, therefore negatively affecting the nervous system.

4 Future Prospective of Botanical Pesticides

Reports on negative effects of synthetic pesticides and environmental risks resulting from their indiscriminate application have renewed interest towards botanical pesticides as an eco-chemical approach in pest management.

Natural plant chemicals will play a significant role in the future for pest control in both industrialized and developing countries. Biodiversity-rich countries should quickly bioprospect their traditionally used flora to document pesticide plants in order to check future cases of biopiracy and establish their sovereign right on the botanical pesticides developed from such plants.

5 Expected Growth Rate of Botanicals

The main market for plant-origin insecticides is in parks and gardens. This is due to their low environmental persistence so that people will be less exposed to the toxic compounds. It is expected that within 10–15 years, these compounds will increase 25 % in insecticide market share, and they will not be limited to garden areas but there may be a massive growth towards urban and agricultural uses. However, even though plant-origin insecticides are advantageous options from an ecological point of view, it would be unreasonable to think they will completely replace synthetic insecticides. On the contrary, it will be quite logical to expect a complementary use and the coexistence of the two kinds of compounds as it happens today with pyrethrum and synthetic pyrethroids in some IPM programs. On the other hand, organic agriculture is a market with a high demand for plant-origin insecticides as organic growers cannot use conventional agrichemicals. At this time, this market is expanding, and as a rule, rates of return are high, so this is an important "niche" to take care of.

References

Adedotun AA, Alexander BO (2008) Laboratory assessment of molluscicidal activity of crude aqueous and ethanolic extracts of *Dalbergia sissoo* plant parts against *Biomphalaria pfeifferi*. Travel Med Infect Dis 4:219–227

Adesina SK, Adewunmi CO (1981) The isolation of molluscicidal agents from the root of *Clausena anisata* (Wild) Oliv. In: Abstracts 4th international symposium on medicinal plants, Ile-Ife. Elsevier Ireland Ltd. Nigeria, pp 44–45

Adesina SK, Adewunmi CO, Marquis VO (1980) Phytochemical investigations of the molluscicidal properties of *Tetrapleura tetraptera* (Taub). J Afr Med Plants 3:7–15

Agrios NG (1997) Plant pathology. Academic, San Diego

Aiyelaagbe OO et al (2011) Insect anti-feedant and growth regulatory activities of *Jatropha podagrica* Hook. Abstract 008 at 1st PACN/RSC Congress on Agricultural Productivity, Accra, Ghana, 21–23 Nov 2011

Arnason JT, Towers GHN, Philogene BJR, Lambert JDH (1983) The role of natural photosensitizers in plant resistant to insects. In: Hedin P (ed) Plant resistance to insects. Symposium, vol. 208. American Chemical Society, Washington, DC, pp 139–151

Ayensu S (1978) Medical plants of West Africa. Reference Publications, Algonac

Banchio E, Valladares G, Defago M, Palasios SY, Carpinella C (2003) Effects of *Melia azedarach* (Meliaceae) fruit extracts on the leaf miner *Liriomyza huidobrensis* (Diptera, Agromyzidae) Assessment in laboratory and field experiments. Ann Appl Biol 143:187–193

Barthel WF (1973) Toxicity of pyrethrum and its constituents to mammals. In: Casida JE (ed) Pyrethrum. The natural insecticides. Academic, New York, pp 123–142, 329 pp

BCC Research (2006) The new biopesticide market. Available from: http://www.Bccresearch.com/Chemicals

Berenbaum MR (1989) North American ethnobotanicals as sources of novel plant-based insecticides. In: Ransom JT, Philogene BJR, Morand P (eds) Insecticides of plant origin ACS symp. series 387. American Chemical Society, Washington, DC

Bernad DR, Xue R (2004) Laboratory evaluation of mosquito repellents against *Aedes albopictus, Culex nigripalpus and Ochlerotatus triseriatus* (Diptera: Culicidae). J Med Entomol 41(4):726–730

Bernays EA (1983) Antifeedants in crop pest management. In: Whitehead DL, Bowers WS (eds) Current themes in tropical science, vol 2, Natural products for innovative pest management. Pergamon Press, New York, pp 259–269

Bliss RM (2005) Death by desiccation: sugar esters dry out insect pests of flowers and ornaments. US Agricultural Research Services (ARS) Report. Agricultural Research Magazine

Bloomquist JR (1996) Ion channels as targets for insecticides. Annu Rev Entomol 41:163–190

Bobadilla M, Zavala F, Sisniegas M, Zavaleta G, Mostacero JY, Taramona L (2005) Evaluacion larvicida de suspensions acuosas de *Annona muricata* Linnaeus (guanabana) sobre *Aedes aegypti* Linnaeus (Diptera, Culicidae). Rev Peru Biol 12:15–52

Brown DS (1980) Fresh water snails of Africa and their medical importance. Taylor & Francis, London

Carroll SP, Loye J (2006) Mosquito repellent with deet-like efficacy. J Am Mosq Control Assoc 22(3): 507–514

Casida JE (1973) Pyrethrum: the natural insecticide. Academic, New York

Casida JE, Quistad GB (1995) Pyrethrum flowers: production, chemistry, toxicology, and uses. Oxford University Press, Oxford

Celis A, Mendoz C, Pachon M, Cardona J, Delgado WY, Cuca L (2008) Extractos vegetales utilizados como biocontroladores con enfasis en la familia Piperacea. Una Rev Agron Colomb 26:97–106

Chandra B, Kumar PY, Matiur M (2008) Aphicidal activity of some indigenous plant extracts against bean aphid "*Aphis craccivora* Koch" Homoptera : Aphididae). J Pestic Sci 81:153–159

Coats JR (1994) Risks from natural versus synthetic insecticides. Annu Rev Entomol 39:489–515

Cook RJ (2000) Advances in plant health management in the 20th century. Annu Rev Phytopathol 38:95–116

Copping LG, Menn JJ (2000) Biopesticides: a review of their action, supplications and efficacy. Pest Manag Sci 56(8):651–676

Dalton PR, Pole D (1978) Water contact patterns in relation to *Schistosoma haematobium* infection. Bull World Health Org 56:417–426

Dalziel JM (1937) The useful plants of West Tropical Africa. Crown Agents for Overseas Governments, London

Dayan FE, Cantrell CL, Duke SO (2009) Natural products in crop protection. Bioorg Med Chem 17:422–434, ISSN: 0968–0896

Dev S, Koul O (1997) Insecticides of natural origin. Harwood Academic, Amsterdam, p 365

Dimetry NZ (2012) Prospects of botanical pesticides for the future in integrated pest management programme

(IPM) with special reference to neem uses in Egypt. Arch Phytopathol Plant Prot 45(9–12):1138–1161, June–July 2012

Dimetry NZ, El-Hawary FM (1995) Neem Azal-F as an inhibitor of growth and reproduction in the cowpea aphid *Aphis craccivora* Koch. J Appl Entomol 119(1):67–71

Dimetry NZ, Schmidt GH (1992) Efficacy of Neem-Azal S and Margosan –O against the bean aphid "*Aphis fabae* Scop". Anz Schadlingskde Pflanzenschut Umweltschutz 65:75–79

Dimetry NZ, El-Gengaihi SE, Reda AS, Amer SAA (1990) Toxicity of some compounds isolated from *Abrus precatorious* L. seeds towards the two spotted spider mite *Tetranychus urticae* Koch. Acarologia 4:361–366

Dimetry NZ, Barakat AA, Abdalla EF, Metwally HE, Abd El-Salam AE (1995) Evaluation of two neem seed kernel extracts against *Liriomyza trifolii* (Burg.) (Diptera: Agromyzidae). Anz Schadlingskde Pflanzenshutz Umweltschutz 68(2):39–41

Dimetry NZ, Gomaa AA, Salem AA, Abd El-Moniem ASH (1996) Bioactivity of some formulations of neem seed extracts against the white fly "*Bemisia tabaci* (Genn.). Anz Schadlingskde Pflanzenschutz Umweltschutz 69(6):140–141

Dimetry NZ, Hafez M, Abbass MH (2002) Efficiency of some oils and neem extract formulations against the cowpea beetle *Callosobruchus maculatus* (F.) (Coleoptera: Bruchidae). In: Koul O, Dhaliwal GS, Marwaha SS, Arora JK (eds) Proceedings of biopesticide conference, Biopesticides Emerging Trends (BET 2001), Chandigarh, 7–9 Feb 2001. Campus Books Int., New Delhi, pp 1–10

Dimetry NZ, Amer SAA, El-Gengaihi SE (2003) Toxicological evaluation and biological potency of petroleum ether extract of two plants and their isolates towards the two spotted spider mite *Tetranychus urticae* Koch. Acarologia XLIII 1:67–73

Dimetry NZ, El-Gengaihi S, Abd El-Salam AEM (2007) Protection of stored cowpea from *Callosobruchus maculatus* (F.). Herba Pol 53(1):71–84

Dimetry NZ, Abd El-Salam AME, El-Hawary FMA (2010) Importance of plant extract formulations in managing different pests attacking beans in new reclaimed area and under storage conditions. Arch Phytopathol Plant Prot 43(7):700–711

Dimetry NZ, El-laithy AY, Abd El-Salam AME, El Saiedy AE (2013) Management of the major piercing sucking pests infesting cucumber under plastic house conditions. Arch Phytopathol Plant Prot 46(2): 158–171

Dinan L, Harmatha J, Lafont R (2001) Chromatographic procedures for the isolation of plant steroids. J Chromatogr A 935:105–123

Dos Santos AF, Sant'Ana AE (1999) Molluscicidal activity of the diterpenoids jatrophone and jatropholones A and B from *Jatropha elliptica* (Pohl) Muel, Arg. Phytother Res 13(8):660–664

Dubey NK, Shukla R, Kulmar A, Singh P, Prakash B (2010) Prospects of botanical pesticides in sustainable agriculture. Curr Sci 98(4):479–480

Duke SO (1986) Microbial phytotoxins as herbicides. A perspective. In: Putnam AR, Tang CS (eds) The science of allelopathy. Wiley, New York, pp 287–304

Duke SO (1990) Natural pesticides from plants. In: Janick J, Simon JE (eds) Advances in new crops. Timber Press, Portland, pp 511–517

Duke SO, Lydon J (1987) Herbicides from natural compounds. Weed Technol 1:122–128

Duke SO, Paul RN, Lee SM (1988) Terpenoids from the genus *Artemisia* as potential pesticides. Am Chem Soc Symp Ser 380:318–334

El-Gengaihi SE, Dimetry NZ, Mohamed SM (1997) Chemical and biological studies of *Peganum harmala* L. plant (alkaloid) J. Appl Entomol 121:165–167

El-Gengaihi SE, Dimetry NZ, Amer SAA (2011) Phytochemical, acaricidal, and biological activity of *Dodonaea viscosa* L. against the two spotted spider mite *Tetranychus urticae* Koch. Arch Phytopathol Pflanzenschutz 44(5):419–425, March 2011

El-Sayed EI (1982–1983) Neem *Azadirachta indica* (A. Juss) seeds as anti-feedant and ovipositional repellents for the Egyptian cotton leaf worm *Spodoptera littoralis* (Boisd.). Bull Entomol Soc Egypt Econ Ser 13:49–58

Enan E (2001) Insecticidal activity of essential oils: octopaminergic sites of action. Comp Biochem Physiol 130:325–337

Eriksson C, Mansson P, Sjodin KY, Schlyter F (2008) Antifeedants and feeding stimulants in bark extracts of ten woody non host species of the pine weevil, *Hylobius abietis*. J Chem Ecol 34:1290–1297

Garcia MR, Perez PR, Rodriguez HCY, Soto HM (2004) Toxicidad de alcaloides de Eythrina americana en larvas de mosquito *Culex quinquefasciatus*. Fitotec Mex 27:297–303

Gill HK, Smith RW, Whiting DA (1986) Biosynthesis of the insect antifeedant steroid Nic.-I. J Chem Soc Chem Commun 18:1457–1458

Grainger M, Ahmed S (1988) Handbook of plants with pest control properties. Wiley, New York, 470 pp

Gupta S, Dikshit AK (2010) Biopesticides: an ecofriendly approach for pest control. J Biopestic 3(1):186–188

Hajek A (2004) Natural enemies an introduction to biological control. Cambridge University Press, Cambridge

Hayes WJ (1982) Pesticides derived from plants and other organisms. In: Hayes WJ (ed) Pesticides studied in man. Williamson and Williamson, Baltimore, pp 75–111

Hedin PA (ed) (1991) Naturally occurring pest bio regulators. American Chemical Society, Washington, DC, 456 pp

Hedin PA (ed) (1997) Phytochemicals for pest control. American Chemical Society, Washington, DC

Hori M (1998) Repellency of rosemary oil against *Myzus persicae* in a laboratory and in a screen house. J Chem Ecol 24:1425–1432

Hummelbrunner LA, Isman MB (2001) Acute, sublethal, antifeedant and synergistic effects of monoterpenoid essential oil compounds on the tobacco cutworm, *Spodoptera litura* (Lepidoptera: Noctuidae). J Agric Food Chem 49:715–720

Isman MB (1999) Pesticides based on plant essential oils. Pestic Outlook 10:68–72

Isman MB (2006) Botanical insecticides, deterrents and repellents in modern agriculture and an increasingly regulated world. Annu Rev Entomol 51:45–66

Isman MB, Wan AJ, Passreiter CM (2001) Insecticidal activity of essential oils to the tobacco cutworm. Spodoptera Lit Fitoter 72:65–68

Jacobson M (1989) Botanical pesticides, past, present and future. In: Arnason JT, Philogene BJR, Morand P (eds) Insecticides of plant origin. American Chemical Society, Washington, DC, pp 1–10

Jacobson M, Crosby DG (1971) Naturally occurring insecticides. Marcel Dekker, New York

Joseph B, Sowmya, Sujatha S (2012) Insight of botanical based biopesticides against economically important pest. Int J Pharm Life Sci 3(1):2138–2148

Juneja RP, Patel JR (2002) Persistence of botanical materials as protectant of green gram *Vigna radiate* (L.) Wilczek against pulse beetle (*Callosobruchus analis* Fabricus). Seed Res 30(2):294–297

Ketkar CM (1976) Utilization of neem (*Azadirachta indica*) and its byproducts. Directorate of Non-edible Oils and Soap Industry, Khadi and Village Industries Commission, Bombay, 234 pp

Kloos H, McCullough FS (1987) Plants with recognized molluscicidal activity (chapter 3). In: Mott KE (eds) Plant molluscicides. UNDP World Bank/ WHO Special Programme for research and Training in Tropical Diseases. Belo Horizonte, pp 45–108

Koul O (1997) Molecular targets from feeding deterrents in phytophagous in natural and manmade environment. Int. Sci. Publication, New Delhi; Backhuys Publishers, Leden, pp 123–134

Koul O, Isman MB, Ketkar CM (1990) Properties and uses of neem *Azadirachta indica* A. Juss. Can J Bot 68:1–11

Kulkarni J, Kapse N, Kulkarni OK (2009) Plant based pesticides for control of *Helicoverpa armigera* on cucumis. Asian Agric Hist 13(4):327–332

Kumar PA, Sharma RP, Malik VS (1996) Adv Appl Microbiol 42:1–43

Lajide L et al (1995) Termite anti-feedant activity in *Xylopia aethiopica* (Annonaceae). Phytochemistry 40(4):1105–1112

Landis DA, Gould F (1988) Screening for phyto-protectants to guard corn seeds/seedlings from southern corn root worm feeding injury. J Entomol Sci 23:201–211

Lee S, Tsao R, Peterson C, Coats JR (1997) Insecticidal activity of monoterpenoids to western corn rootworm (Coleoptera: Chrysomelidae), two-spotted spider mite (Acari: Tetranichidae), and house fly (Diptera: Muscidae). J Econ Entomol 90:883–892

Long Z, Hock SY, Hung S (2006) Screening of Chinese medicinal herbs for bioactivity against *Sitophilus zeamais* Motschulsky and *Tribolium castaneum* (Herbst). J Stored Prod Res 43:290–296

Lydon J, Duke SO (1988) Porphyrin synthesis is required for photo-bleaching activity of the p-nitro substituted diphenyl ether herbicides. Pestic Biochem Physiol 31:74–93

Lyon GD, Beglinski T, Newton AC (1995) Novel disease control compounds: the potential to "immunize plants against infection. Plant Pathol 44:407–427

Maia MF, Moore SJ (2011) Plant-based insect repellents: a review of the efficacy, development and testing. Malaria J 10(1)

Matsumura F (1985) Toxicology of insecticides, 2nd edn. Plenum Press, New York

McElvain SM et al (1941) The constituents of the volatile oil of catnip. 1. Nepetalic acid, Nepetalactone and related compounds. J Am Chem Soc 63(6): 1558–1563

Mendiki M et al (1999) Pesticidal activity of certain plant extracts to control stored grain pest *Callosobruchus chinensis*. Pestology XXIII:64–67

Mishra AK, Singh N, Sharma VP (1995) Use of neem oil as a mosquito repellent in tribal villages of Mandela district, Madhya Pradesh. Indian J Malar 32(3):99–103

Montesinos E, Bonaterra A (2009) Microbial pesticide. In: Schaechter M (ed) Encyclopedia of microbiology, 3rd edn. Elsevier, New York, pp 110–120

Mwine J et al (2011) Ethno botanical survey of pesticidal plants used in South Uganda: case study of Masaka district. J Med Plant Res 5(7):1155–1163

Ngoh PH, Lionel EWC, Fung PY, Kini MR, Ho SH (1998) Insecticidal and repellent properties of nine volatile constituents of essential oils against the American cockroach, *Periplaneta americana* (L.). Pestic Sci 54:261–268

Nielson RL (2005) Site specific seeding rates for corn agronomy. Department, Purdue University. http://www.agry.purdue.Edu/ext/corn/news/articles.95/p&c9504.htm

Oerke EC, Dehne HW, Shoenbeck F, Weber A (1994) Estimated losses in major food and cash crops. Elsevier, London

Okwute SK (2012) Plants as potential sources of pesticidal agents: a review. http://dx.doi.org/10.5772/46225

Orozco J, Soto AY, Hipolito A (2006) Efecto de repelencia de Crotalaria juncea, Galactia striata y Cymbopgon nardus para el manejo de Cyrtomenus bergi (Hemiptera: Cydnidae). Rev Biol Cienc Tierra 6:179–185

Peterson CY, Coats J (2001) Insect repellents. Past, present and future. Pestic Outlook 12:154–158

Pilmore RE (1973) Toxicity of pyrethrum to fish and wild life. In: Casida JE (ed) Pyrethrum: the natural insecticide. Academic Press, New York, pp 143–165, 329 pp

Prakash A, Tiwari SN, Rao J (1990) Exploitation of natural plant products for management of pests and diseases in rice ecosystems. In: Proceedings of symposium on growth, development and resource conservation. Springer-Verlag Pergamon Press, Oxford, pp 23–36

Rai M, Acharya D (2000) Search for fungitoxic potential in essential oils of Asteraceous plants. Compos Lett 35:18–23

Rai MK et al (1999) In vitro susceptibility of opportunistic *Fusarium* spp. to essential oils. Mycoses 42(1, 2):97–101

Raja N, Babu A, Dorn S, Ignacimurhu S (2001) Potential of plants for protecting stored pulses from *Callosobruchus maculatus* (Coleoptera: Bruchidae) infestation. Biol Agric Hortic 19(1):19–27

Rajapake RHS, Ratnaseka D (2008) Pesticidal potential of some selected tropical plant extracts against *Callosobruchus maculatus* (F.) and *Callosobruchus chinensis* L. Trop Agric Res Ext 11:69–71

Rajendran S, Sriranjini V (2008) Plant products as fumigants for stored product insect control. J Stored Prod Res 44:126–135

Recheigl JE, Recheigl NA (1999) Biological and biotechnological control of insect pests. Lewis Publishers, Boca Raton

Reda AS, Dimetry NZ, Amer SAA, Motawe HM (1989) Activity of *Abrus precatorious* L. extracts and compounds isolated on orientation and oviposition behavior of the two spotted spider mite *Tetranychus urticae* Koch. Z Ang Entomol 107:395–400

Saxena RC (1983) Naturally, occurring pesticides and their potential. In: Shemilt LE (ed) Conference on chemistry applied world needs, Manila, Philippines. Pergamon, Oxford, pp 143–161

Saxena RC (1989) Insecticides from neem. In: Arnason JT, Philogene BJR, Morand P (eds) Insecticides of plant origin. American Chemical Society, Washington, DC, pp 110–135

Saxena RC (1998) Green revolutions without blues: botanicals for pest management. In: Dhaliwal GS, Randhawa NS, Arora R, Dhawan AK (eds) Ecological agriculture and sustainable development, vol 2. Indian Ecological Society, Punjab Agricultural University, Ludhiana and Centre for Research in Rural and Industrial Development, Chandigarh, pp 111–127

Schmutterer H (1990) Properties and potential of natural pesticides from the neem tree, *Azadirachta indica*. Annu Rev Entomol 35:271–297

Schmutterer H (1995) The neem tree *Azadirachta indica* A. Juss and other Meliaceous plants. Sources of unique natural products for integrated pest management, Medicine, Industry and other purposes. VCH Verlags gesellschaft, Weinheim, FRG, pp 375–384

Schmutterer H, Singh RP (1995) Lists of insect pests susceptible to neem products. In: Schmutterer H (ed) The neem tree. VCH, Weinheim, pp 598–604

Schnepf HE, Curr OP (1995) *Bacillus thuringiensis* toxins: regulation, activities and structural diversity. Curr Opin Biotechnol 6:305–312

Schnepf E, Crickmore N, Van Rie J, Lereclus D (1998) *Bacillus thuringiensis* and its pesticidal crystal proteins. Microbiol Mol Biol Rev 62:775–806

Seyoun A et al (2002) Traditional use of mosquito repellent plants in Western Kenya and their evaluation in semi-field experimental huts against *Anopheles gambiae:* ethnobotanical studies and application by thermal expulsion and direct burning. Trans R Soc Trop Med Hygiene 96(3):225–231

Singh D, Agarwal SK (1988) Himachalol and β – Himachalene: Insecticidal principles of Himalayan Cedarwood oil. J Chem Ecol 14:1145–1151

Singh D, Mehta SS (2010) Menthol containing formulation inhibits Adzuki Bean Beetle, *Callosobruchus chiensis* L. (Coleoptera: Bruchidae) population in pulse grain storage. J Biopestic 3(3):523–530

Singh D, Roa SM, Tripathi AK (1984) Cedarwood oil as a potential insecticidal agent against mosquitoes. Naturwissenschaften 71:265–266

Singh D, Mehta SS, Ram RD, Kumar S (1998) A novel formulation useful as pest repellent for stored grains. Granted Indian Patent No. 189730 dated 27-1-2004 Application No. 1974/Del/198 dated 10-07-98

Soares SF et al (2010) Repellent activity of plant derived compounds against *Amblyomma cajennense* (Ascari: Ixodidae) nymphs. Vet Parasitol 167(1):67–73

Suszkiw Jan (2009) Catnip compounds curb Asian Lady Beetles. US Department of Agriculture, Agricultural Research Services (ARS) Report. Belo Horizonte

Tripathi P, Dubey NK (2004) Exploitation of natural products as an alternative strategy to control post harvest fungal rotting of fruits and vegetables. Postharvest Biol Technol 32:235–245

UNEP (2005) Ridding the world of POPs. In: A guide for the Stockholm convention on persistent organic pollutants. UNEP

Vet Parasitology (2009) Molluscicidal activity of *Saraca soca* and *Thuja orientalis* against the freshwater snail *Lymnaea acuminate*. 164 ((2–4): 206–210

Waliwitiya R, Isman MB, Vernon RS, Riseman A (2005) Insecticidal Activity of Selected Monoterpenoids and Rosemary oil to *Agriotes obscures* (Coleoptera: Elateridae). J Econ Entomol 98:1560–1565

Ware GW (1988) Complete guide to pest control, 2nd edn. Thompson Publication, Fresno

Weinzierl R et al (2009) Insect attractants and traps, In: Alternatives to insect management. Office of the Agricultural Entomology, University of Illinois at Urbana-Champaign, Champaign

Natural Insecticides from the Annonaceae: A Unique Example for Developing Biopesticides

2

Murray B. Isman and Rita Seffrin

Abstract

The documented negative impacts of synthetic insecticides on the environment as well as on human health, the consumers' concerns over insecticide residues in foods, and the emergence of resistant insects call for new approaches to manage insect pests. In this context, research on the potential use of plant extracts and their constituents has grown dramatically in the past decade. Among terrestrial plant families, Annonaceae has drawn considerable attention since the 1980s, owing to the presence of acetogenins, a class of natural products with a broad range of insecticidal bioactivities. Crude extracts from seeds, leaves, bark, twigs, and fruits obtained from the plant species of Annonaceae have been extensively tested in recent years for bioactivity to pest insects and related arthropods worldwide. *Asimina triloba*, *Annona muricata*, and *Annona squamosa* are the species that have been most frequently examined for their insecticidal effects.

Keywords

Annonaceae • Acetogenins • *Annona squamosa* • Squamocin • Annonicin • Mitochondrial poison • Botanical insecticide

1 Introduction

Annonaceae is the largest plant family in the order Magnoliales (Westra and Maas 2012) and comprises around 2,500 species and 130 genera (Pirie et al. 2005). Except for two related North American genera (*Asimina* and *Deeringothamnus*), the family is entirely tropical (Thomas and Doyle 1996). Annonaceae enjoyed considerable attention from plant systematists in the twentieth century. The Swedish botanist Robert Fries spent a lifelong career studying herbarium specimens, mainly originating from the Neotropics. He contributed greatly to the flora of Central America, South America, and the West Indies, especially to the knowledge of the family of Annonaceae (Erkens 2007).

M.B. Isman, Ph.D. (✉) • R. Seffrin, Ph.D.
Faculty of Land and Food Systems, University of British Columbia, Vancouver V6T1Z4, BC, Canada
e-mail: murray.isman@ubc.ca

D. Singh (ed.), *Advances in Plant Biopesticides*,
DOI 10.1007/978-81-322-2006-0_2, © Springer India 2014

Given the problems that synthetic insecticides have caused to the environment as well as to human health, there has been a surge of research on plant extracts and plant natural products for insect control (Castillo-Sánchez et al. 2010). The practice of using plant extracts in agriculture for pest control is not new. They have been used for at least two millennia, when botanical insecticides were considered important products for pest management in ancient China (Long et al. 2006), Egypt, Greece, and India (Isman 2006). Even in the United States and some European countries, botanical insecticides were widely used before the discovery of organochlorine and organophosphate insecticides in the late 1930s and early 1940s (Isman 1997). The Annonaceae family has drawn a lot of attention since the 1980s, due to the presence of acetogenins, a class of natural products with a broad range of biological activities, among which insecticidal activity stands out (Ocampo and Ocampo 2006).

2 Brief History of the Acetogenins and Their Pesticidal Activity

In the 1970s, the National Cancer Institute started funding Dr. Jerry McLaughlin at Purdue University to find botanical substances with anticancer potential. He tested and screened over 3,500 species of plants and found that the acetogenin compounds of the Annonaceae family had the most potential activity. McLaughlin (2008) thus focused his studies on 14 annonaceous species that yielded these novel substances. These were *Asimina triloba* (pawpaw), *Goniothalamus giganteus*, *Annona squamosa* L. (sugar apple or sweetsop), *Annona muricata* L. (soursop, graviola, guanabana), *Annona bullata* Rich., *Asimina parviflora* (Michx.) Dunal (dwarf pawpaw), *Annona longifolia* Kral. (long-leafed dwarf pawpaw), *Annona reticulata* L. (custard apple), *Annona glabra* L. (pond apple), *Annona jahnii* Saff., *Annona cherimola* Mill. (cherimoya), *Xylopia aromatica* (Mart.) Lam., *Rollinia mucosa* (Jacq.) Baill. (biriba), and *Rollinia emarginata* Schlecht.

Arguably, his most important contributions have been to our knowledge of the annonaceous acetogenins found in the pawpaw tree. From the pawpaw and these additional species, his research team isolated and characterized over 200 new annonaceous acetogenins. Following his observations in 1982 that the North American pawpaw tree was potently bioactive, his group identified some 50 acetogenins in its seeds and bark with antitumor and pesticidal properties. The acetogenins are found in the leaves and branches and, predominantly, in the seeds of annonaceous plants. McLaughlin worked with Eli Lilly and Company (Greenfield, IN) and the USDA (Peoria, IL), demonstrating that the pawpaw acetogenins are potent inhibitors of a number of agricultural and other pests: mosquito larvae, two-spotted spider mites, Mexican bean beetles, striped cucumber beetles, European corn borers, melon or cotton aphids, blowfly larvae, and a nematode (*Caenorhabditis elegans*) (Alkofahi et al. 1989; Mikolajczak et al. 1989).

However, despite its relatively large size, this plant family had chemically been one of the least explored. The discovery of uvaricin, a *bis* (tetrahydrofuranoid) fatty acid lactone that was first isolated in 1982 from the roots of *Uvaria accuminata* – the first of the acetogenins – invigorated wide interest in this family (Jolad et al. 1982). The temperate American pawpaw (*A. triloba*; Fig. 2.1) and the tropical soursop (*A. muricata*; Fig. 2.2) and sweetsop (*A. squamosa* L.; Fig. 2.3) have been the species most intensively examined for their insecticidal effects. Each of these species contains complex mixtures of acetogenins comprising at least 30 compounds. Sesquiterpenes and monoterpenes are the main types of compounds present in essential oils of *Annona* species (Rios et al. 2003). From the wide variety of acetogenins, squamocin and annonacin have shown the greatest impact on insects (Álvarez et al. 2008) (Fig. 2.4). The annonaceous acetogenins are an important group of long-chain fatty acid derivatives found exclusively in the plant family Annonaceae. Nearly 400 compounds from this class have been published in the literature since the discovery of uvaricin. The potential application of acetogenin molecules is linked

Fig. 2.1 *Asimina triloba* (location unknown; Scott Bauer photo, courtesy of the United States Department of Agriculture)

also to their pesticidal properties (e.g., asimicin and annonin) (Gupta et al. 2011).

3 Mode of Action of Acetogenins

Acetogenins are mitochondrial poisons, inhibiting cellular energy production through a mode of action identical to that of the well-known botanical insecticide and fish poison, rotenone (Londershausen et al. 1991). More specifically, acetogenins block the respiratory chain at NADH-ubiquinone reductase (complex I) and cause a decrease in ATP levels, directly affecting electron transport in the mitochondria, causing apoptosis (Alali et al. 1999). Acetogenins also inhibit insect development and behavior (Table 2.1).

4 Research on Natural Insecticides from the Annonaceae in Canada

The majority of research on natural insecticides from Annonaceae conducted in Canada has taken place in our laboratory at the University of British

Fig. 2.2 *Annona muricata* growing in Saba, Netherlands Antilles (Mary Roduner photo)

Fig. 2.3 *Annona squamosa* growing at Cabocla Farm, Limoeiro do Norte County, State of Ceara, Brazil (Dr. Antonio Lindembergue Martins Mesquita photos)

Columbia. The first of our research in this area was published in 2004. Toxicity and antifeedant activities of crude seed extracts of *A. squamosa* from Maluku, Indonesia, against the diamondback moth, *Plutella xylostella* L. (Lepidoptera: Plutellidae), and the cabbage looper, *Trichoplusia ni* (Hübner) (Lepidoptera: Noctuidae), were determined using different bioassays. Aqueous seed extracts and an aqueous emulsion of ethanolic seed extracts were toxic to both species. A crude aqueous extract also deterred feeding of fourth-instar *P. xylostella* in a leaf disc choice bioassay. Toxicities of crude aqueous extracts to natural enemies, *Chrysoperla carnea* (Stephens) (Neuroptera: Chrysopidae) and *Orius insidiosus* (Say) (Hemiptera: Anthocoridae), were investigated using direct spray and residual contact tests. *C. carnea* larvae were less susceptible to

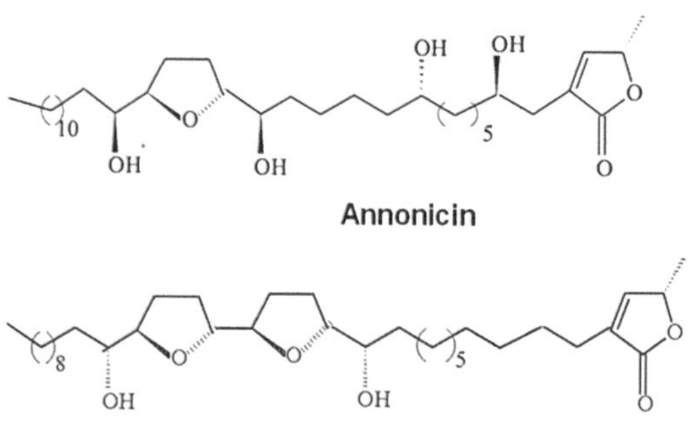

Annonicin

Squamocin-I

Fig. 2.4 Chemical structures of annonacin and squamocin, major insecticidal acetogenins from the seeds of commonly cultivated *Annona* species

the extracts than were *O. insidiosus* adults (Leatemia and Isman 2004a).

Leatemia and Isman (2004b) assessed the efficacy of crude seed extracts of *A. squamosa* against larvae of the diamondback moth, *P. xylostella* L., feeding on cabbage. Three greenhouse trials were carried out using aqueous seed extracts and an aqueous emulsion of ethanolic seed extracts. At a concentration of 0.5 % (w/v), an aqueous emulsion of an ethanolic seed extract was 2.5-fold more effective than 1 % rotenone, a commercial botanical insecticide. Crude aqueous seed extracts showed efficacy compared to pyrethrum, the most widely used botanical insecticide. Crude ethanolic seed extracts of *A. muricata*, *A. squamosa* (Annonaceae), *Lansium domesticum*, and *Sandoricum koetjape* (Meliaceae) collected from different locations and years in Maluku, Indonesia, were screened for inhibition of larval growth against the polyphagous lepidopteran *Spodoptera litura* (Noctuidae). Extracts of *A. squamosa* were significantly more active (20-fold) than those of *A. muricata* (Leatemia and Isman 2004c). A similar study of *A. squamosa* and *Annona atemoya* from Brazil indicated that crude methanolic seed extracts of *A. squamosa* were ~10 times more active as a feeding deterrent than *A. atemoya* against third-instar *T. ni* larvae in a leaf disc choice bioassay. *A. squamosa* was ~ three times more active as a growth inhibitor than *A. atemoya*. Methanolic seed extracts of *A. squamosa* and *A. atemoya* were toxic to third-instar *T. ni* larvae following either topical or oral application. *A. squamosa* was more toxic through feeding (LC_{50} = 167.5 ppm vs. 382.4 ppm), whereas *A. atemoya* exerted greater toxicity via topical application (LC_{50} = 301.3 mg/larva vs. 197.7 mg/larva). Both *A. squamosa* and *A. atemoya* extracts reduced leaf area consumption and larval growth in a greenhouse experiment (De Seffrin et al. 2010) (Fig. 2.5).

Regarding public health pests, a crude ethanolic extract obtained from seeds of *A. squamosa* was evaluated for its larvicidal effect against the mosquitoes *Aedes aegypti* and *Aedes atropalpus*. The extract produced >70 % mortality in both young (late 1st to 2nd instar) and older larvae (3rd to early 4th instar) of *A. aegypti* at concentrations of 250–500 ppm. In most cases, mortality was greater at 48 h compared to 24 h. At a concentration of 100 ppm, the extract produced complete mortality of young instar larvae of *A. atropalpus* at 24 h. *A. atropalpus* is significantly more susceptible than *A. aegypti* (Srikrishnaraj and Isman 2006).

5 Research on Natural Insecticide from the Annonaceae Worldwide

Crude extracts from seeds, leaves, bark, twigs, and fruits from Annonaceae have been extensively tested in recent years for bioactivity to pest insects and related arthropods.

Among agricultural pests, Dadang and Prijono (2009) assessed the effectiveness of two botanical insecticide formulations: mixtures of *Piper retrofractum* (Piperaceae) and *A. squamosa* extracts and *Aglaia odorata* (Meliaceae) and *A. squamosa* extracts at 0.05 and 0.1 %. These concentrates were compared to the synthetic pyrethroid insecticide deltamethrin at 0.04 % and the microbial insecticide *Bacillus thuringiensis* at 0.15 % in the field of two major cabbage insect pest populations. The application of both mixtures and conventional formulations decreased populations of *Crocidolomia pavonana* (F.) (Lepidoptera: Pyralidae) and *P. xylostella* (L.) (Lepidoptera: Yponomeutidae).

Table 2.1 Biological activity of different plant species from the Annonaceae family on arthropods

Plant species	Arthropod species	Biological activity	Plant part tested	Lethal dosage (LD_{50}/LC_{50}/EC_{50})	Concentration	References
Annona squamosa	*Plutella xylostella*	T, FD	Seeds		0.5 %	Leatemia and Isman (2004a, b, c)
	Trichoplusia ni	FD	Seeds	167.5 ppm		De Seffrin et al. (2010)
	Spodoptera litura	GI	Seeds	191 ppm		Leatemia and Isman (2004a, b, c)
	Aedes atropalpus	T	Seeds		100 ppm	Srikrishnaraj and Isman (2006)
	Macrosiphum rosaeformes	T	Leaves		20 %	Dhembare et al. (2011)
	Bemisia argentifolii	T	Seeds		0.25 %	Lin et al. (2009)
	Aphis gossypii	T	Seeds		0.25 %	Lin et al. (2009)
	Tetranychus kanzawai	T	Seeds		0.125 %	Lin et al. (2009)
	Tribolium castaneum	T	Leaves		20 %	Anita et al. (2012)
	Callosobruchus chinensis	GI, OD	(+)-O-Methylarmepavine		1–4 μg/μL/larva	Konkala et al. (2012)
	Anopheles subpictus	T	Bark	93.80 mg/L		Kamaraj et al. (2011)
	Culex tritaeniorhynchus	T	Bark	104.94 mg/L		Kamaraj et al. (2011)
	Culex quinquefasciatus	T	Leaves	11.01 μg/mL		Magadula et al. (2009)
	Pediculus humanus capitis	T	Fruits		0.1, 1, and 10 % w/w	Kosalge and Fursule (2009)
	Musca domestica	GI	Seeds	345 mg/L		Begum et al. (2010)
	Sitophilus oryzae	T	Leaves		1 %	Kumar et al. (2010)
Asimina triloba	*Acalymma vittatum*	FD	Fruits		1 %–5 %	Sedlacek et al. (2010)
Annona muricata	*Plutella xylostella*	T	Seeds, leaves		5 ppm	Trindade et al. (2011)
	Anastrepha ludens	T	Stem, leaves		2,000 μg/mL	González-Esquinca et al. (2012)
	Bactericera cockerelli	GI	Seeds		2,500–5,000 ppm	Flores-Davila et al. (2011)
	Sitophilus zeamais	T	Seeds		0.4 %	Asmanizar and Idris (2012)
	Aedes aegypti	T	Seeds	93.48 μg/mL		Grzybowski et al. (2013)
Annona montana	*Spodoptera frugiperda*	T	Acetogenins		100 ppm	Di Toto Blessing et al. (2012)

FD feeding deterrence, *T* toxicity, *GI* growth inhibition, *R* repellency, *OD* oviposition deterrence

a

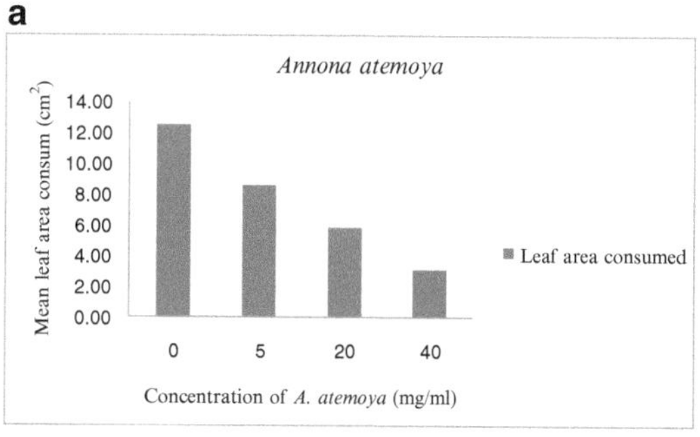

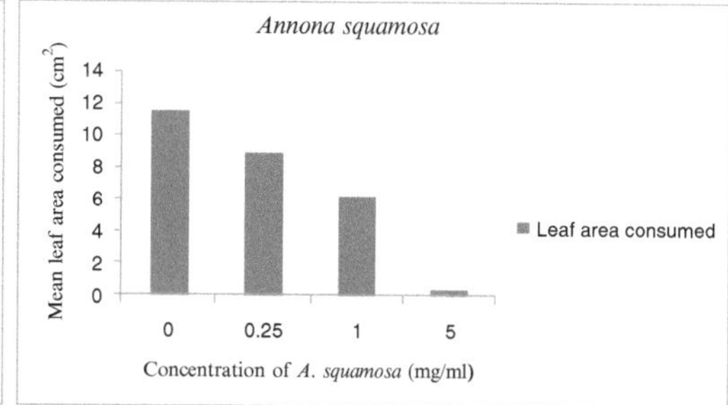

b

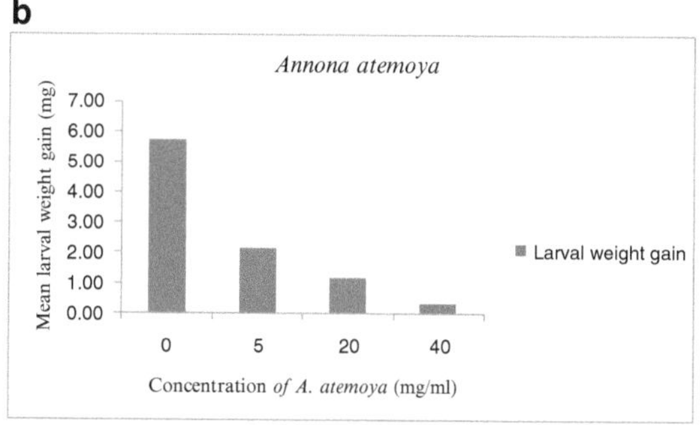

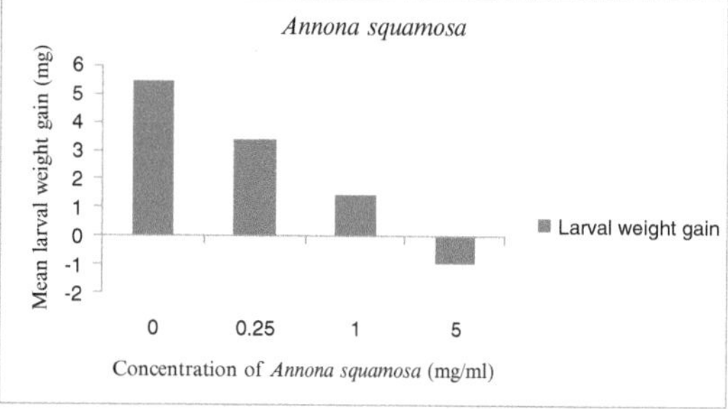

Fig. 2.5 Greenhouse trial measuring mean leaf area consumed (**a**) and mean larval weight gained (**b**) by third-instar *T. ni* when placed on cabbage plants sprayed with an increasing concentration of *A. atemoya* and *A. squamosa* crude extract

The mixtures of botanicals at 0.1 % were more effective than the synthetic insecticide, and they did not affect the performance of insect pest natural enemies. Effects of plant extracts were tested on the tomato leaf miner (*Tuta absoluta* [Meyrick]) (Lepidoptera: Gelechiidae), under laboratory and greenhouse conditions. Anosom EC (annonin), azadirachtin, and mixtures thereof were very effective in controlling this pest (Durmusoglu et al. 2011). Laboratory experiments examined the effects of pawpaw *A. triloba* (L.) Dunal, fruit extract on mortality, and feeding deterrence of the striped cucumber beetle, *Acalymma vittatum* (F). Pawpaw fruit extract reduced feeding by 89 % and 97 % at concentrations of 1 % and 5 %, respectively. The calculated LC_{50} value was 5.05 %, whereas the LCF_{10} (concentration at which only 10 % of the leaves were consumed) was 0.20 %. At 10 %, only 10 % of the beetles were killed; however, only 3 % of the leaf tissue was consumed. Thus, pawpaw fruit extract may be an effective insect feeding deterrent (Sedlacek et al. 2010).

The bioactivity of an ethanolic leaf extract of *A. muricata* on the development of the larvae and pupae of the diamondback moth *P. xylostella* was evaluated by Trindade et al. (2011). At the highest concentration tested (5 ppm), the most active extract caused 100 % larval mortality; at lower concentrations, the duration of the larval phase was increased by up to 2.6 days, and larval survival was significantly reduced. The pupal stage was far less affected by exposure to the extracts, although the duration was increased by up to 1 day in the presence of nonlethal concentrations. The major acetogenins from a Bolivian collection of *A. montana* (Fig. 2.6) – annonacin, cis-annonacin-10-one, densicomacin-1, gigantetronenin, murihexocin B, and tucupentol – were evaluated for their antifeedant and toxic effects on *Spodoptera frugiperda* Smith (Lepidoptera: Noctuidae), a serious pest affecting corn crops in Argentina and throughout the Americas. All the acetogenins produced 100 % mortality during the larval or pupal stages at 100 ppm in diet. In addition, the compounds annonacin, cis-annonacin-10-one,

Fig. 2.6 *Annona montana* growing at Gaspar, SC, Brazil (Anestor Mezzomo photo)

and densicomacin-1 deterred feeding by more than 80 % at the same concentration (Di Toto Blessing et al. 2012).

Among sucking insects, aqueous extracts of various annonaceous and other plants were tested for their insecticidal efficacy against the rose aphid, *Macrosiphum rosaeformes* (Davis), under laboratory conditions. Thirteen plant extracts were tested. Garlic, *Allium sativum* (Linn), was found to be the most effective, followed by custard apple, *A. squamosa* (Linn), and bullock heart, *A. reticulata* (Linn.), closely followed by neem, *Azadirachta indica* A. Juss. (Dhembare et al. 2011). Lin et al. (2009) tested the cold-pressed oil from the seeds of *A. squamosa*. The oil was effective in controlling the silverleaf whitefly, *Bemisia argentifolii* Bellows and Perring (Homoptera: Aleyrodidae), infesting the leaves of tomato plants under greenhouse conditions. Sugar apple seed oil was also very effective in controlling the cotton aphid, *Aphis gossypii* Glover (Homoptera: Aphididae), on melon leaves and the Kanzawa spider mite, *Tetranychus kanzawai* Kishida (Acari: Tetranychidae), on soybean leaves. Nymphs of the potato psyllid, *Bactericera cockerelli* (Sulc), were treated with extracts of *A. muricata*, *Carica papaya*, *Euphorbia dentata*, *Thuja occidentalis*, *Sapindus saponaria*, and A. *indica*. After 72 h, *A. muricata* seed extract, at concentrations of 2,500 and 5,000 ppm, produced 98 and 100 % mortality of potato psyllid nymphs, followed by *A. indica* oil, causing 91 and 100 % mortality at 2,000 and 2,500 ppm, respectively. *A. muricata* seed extract was the most effective insecticide in the study (Flores-Davila et al. 2011).

Increased concern by consumers over insecticide residues in food products, the occurrence of insecticide-resistant insects, and the precautions necessary to work with traditional chemical insecticides call for new approaches to control stored product insect pests (Konkala et al. 2012). Management of stored product pests using materials of plant origin is the subject that has received considerable research attention because of their minimal environmental hazards and low mammalian toxicity (Isman 1994). Chemical pesticides have often provided the first line of defense against insect pests of grain. Synthetic pesticide treadmills and inefficiencies have resulted in increased input costs for resource-poor farmers in developing countries. The need for cheaper but effective options for combating insect pests has resulted in the resurgent use of plant materials where the majority of the farmers in developing countries are resource constrained (Chikukura et al. 2011). Asmanizar and Idris (2012) tested the bioactivity of *Jatropha curcas* and *A. muricata* crude seed extracts against the maize weevil *Sitophilus zeamais* (Coleoptera: Curculionidae) by using dipping and surface protectant methods. Both *J. curcas* and *A. muricata* extracts had contact and stomach poison activities against *S. zeamais*. Using a dipping method, weevil mortality was 90 and 70 %, respectively, at a concentration of 20 % (v/v), whereas using a surface protectant method, weevil mortality was 100 % at 0.4 % (v/w) concentration for both crude extracts. *J. curcas* and *A. muricata* extracts applied to rice grain (surface protectant method) can reduce F_1 progeny production, weight loss, and rice grain damage.

The insecticidal and repellent activities of fruit extracts of *Xylopia aethiopica* (Dunal) A. Rich. and *Dennettia tripetala* (Baker f.) G.E. Schatz (both belonging to the family Annonaceae) were studied against the rice weevil *Sitophilus oryzae* (L.), an economic, primary postharvest pest of rice and other cereal products. The extracts of both plants caused significant adult weevil mortality and a reduction in F1 progeny emergence. Extracts were also significantly ($P<0.001$) toxic to *S. oryzae* after 24 h with the highest dose (2 mg cm^{-2}) producing 100 % weevil mortality. Similarly, both male and female weevils significantly avoided the test arm compared to the control arm in a Y-tube olfactometer repellence test. These results suggest that *X. aethiopica* and *D. tripetala* natural extracts have potential use as part of an integrated pest management system for stored product protection against *S. oryzae* (Ukeh et al. 2012).

Pulverized leaves of *A. squamosa* (L.), *Moringa oleifera* (Lam.), and *Eucalyptus globulus* (Labill.) were tested for their insecticidal and seed protective effect against the confused flour beetle, *Tribolium castaneum* (Herbst), a stored grain pest of wheat. When larvae were introduced to pulverized leaves of *A. squamosa*, *M. oleifera*, and *E. globulus* separately, mortality increased with increasing concentrations and resulted in 100 % mortality within a short period of 8–10 days at the highest concentration. Larvae that were introduced to these cultures did not grow well or molt to the next developmental stage. These plants are also very effective at preventing seed damage. The seed protective effect ranged from 39 to 82 % for *A. squamosa*, 34–78 % for *M. oleifera*, and 42–88 % for *E. globulus*. Considering the insecticidal and seed protective effect, these three plant powders could be employed as alternatives to chemical and synthetic pesticides for smallholder farmers (Anita et al. 2012).

Topical application of (+)-O-methylarmepavine, a juvenile hormone analogue isolated from the leaves of *A. squamosa* L., caused inhibition of growth and development of fifth-instar larvae of the bruchid beetle *Callosobruchus chinensis*. Production of larval-pupal intermediates and pupal-adult intermediates was observed as well as adults with various ovarian abnormalities (Konkala et al. 2012).

Kamanula et al. (2011) conducted a survey on farmer ethno-ecological knowledge of pests of stored maize and bean and their pest management practices including pesticidal plant use in eastern Zambia and northern Malawi. They concluded that the rational use of pesticidal plants for insect pest management needs a constructive collaboration between scientists and farmers. Scientists can develop guidelines to ensure efficacious use of the pesticidal plants widely used by farmers. Some plants are required in large quantities, so cultivation may be an important consideration to increase their availability. Researchers can also develop guidelines for the propagation and cultivation of those plant species. This would have a positive impact on the availability of such plants and encourage more farmers to use them.

In the field of public health, mosquitoes such as *A. aegypti* L. are important pantropical vectors of dengue and yellow fever (Komansilan et al. 2012). Essential oils of *Guatteria hispida*, *G. blepharophylla*, and *G. friesiana* were tested against *A. aegypti*. GC-MS and NMR analyses confirmed the presence of caryophyllene oxide as the main constituent of the leaves of *G. blepharophylla*; in *G. friesiana*, the α- and β-geudesmols prevail; and in *G. hispida* α- and β-pinene and (*E*)-caryophyllene are the predominant compounds. The lethal concentrations LC_{50}, LC_{95}, and LC_{99} were, respectively, 85.74, 199.35, and 282.76 ppm for *G. hispida*; 58.72, 107.6, and 138.37 ppm for *G. blepharophylla*; and 52.6, 94.37, and 120.22 ppm for *G. friesiana*. The oil extracted from *G. friesiana* presented the best insecticidal effect (Aciole et al. 2011).

Costa et al. (2012) described morphological changes that occur in the midgut of third-instar *A. aegypti* L. (Diptera: Culicidae) following treatment with a methanolic extract of *Annona coriacea*. Insects exposed to the extract displayed intense, destructive cytoplasmic vacuolization in columnar and regenerative midgut cells. The apical surfaces of columnar cells exhibited cytoplasmic protrusions oriented toward the lumen, suggesting that these cells could be involved in apocrine secretory processes and/or apoptosis. *A. coriacea* extracts induced morphological alterations in the midgut of *A. aegypti* midgut larvae, supporting the use of plant extracts for control of this disease vector.

Cytotoxicity and larvicidal properties of the leaf extracts of *A. muricata*, *A. senegalensis* Pers, and *A. squamosa* L. were tested against brine shrimp larva and late 3rd instar of *Culex quinquefasciatus* Say. With the larvicidal properties of *A. senegalensis* being described for the first time, its value together with that of *A. squamosa* may prove to be the best natural source of larvicidal agents. The LC_{50} values for crude extracts of *A. senegalensis* and *A. squamosa* were 0.67 and 0.64 μg/mL, respectively, for shrimp larvae and 23.42 and 11.01 μg/mL, respectively, for *C. quinquefasciatus* (Magadula et al. 2009). An extract obtained from the fruits of Indian neem, *A. indica*, and seeds of *A. squamosa* L. was tested against the head louse *Pediculus humanus capitis*. Petroleum ether extracts of these plants produced high levels of mortality in adult lice. *A. squamosa* L. extract showed more potent activity than *A. indica* extracts at all concentrations (0.1, 1, and 10 % w/w) (Kosalge and Fursule 2009).

6 Solvents Used for Extracting Acetogenins

An ethanolic extract of *A. squamosa* leaves showed potent activity against the rice weevil *S. oryzae*. The extract produced significant knockdown (KDT_{50}) at 1 % (23.1 min) and 5 % w/v (11.4 min). Complete mortality was achieved at 39.6 ± 1.4 and 14.5 ± 1.1 min for 1 % and 5 % w/v, respectively (Kumar et al. 2010). Kamaraj et al. (2011) assessed the larvicidal activities of hexane, chloroform, ethyl acetate, acetone, and methanol dried leaf and bark extracts of *A. squamosa*, *Chrysanthemum indicum*, and *Tridax procumbens* against fourth-instar larvae of the malaria vector, *Anopheles subpictus* Grassi, and the Japanese encephalitis vector, *Culex tritaeniorhynchus* Giles (Diptera: Culicidae). All plant extracts showed moderate effects after 24 h of exposure; however, the most toxic were the methanolic bark extract of *A. squamosa*, leaf ethyl acetate extract of *C. indicum*, and leaf acetone extract of *T. procumbens* against the larvae of *A. subpictus* (LC_{50} = 93.80, 39.98, and 51.57 mg/L) and methanolic bark extract of *A. squamosa*, leaf methanol extract of *C. indicum*, and leaf ethyl acetate extract of *T. procumbens* against the larvae *of Cx. tritaeniorhynchus* (LC50 = 104.94, 42.29, and 69.16 mg/L) respectively. The acetone, chloroform, hexane, petroleum ether, and ethanol extracts *of A. squamosa* foliage were studied against the early fourth-instar larvae of *A. aegypti*, *Anopheles stephensi*, and *Culex quinquefasciatus*. Larval mortality was observed after 24 h exposure. All extracts showed moderate larvicidal effects; however, the greatest larval mortality was obtained with a petroleum ether extract (Kumar et al. 2011).

González-Esquinca et al. (2012) used water and ethanolic extracts to determine the activity of stem and leaf extracts of *A. muricata* L., *A. diversifolia* Saff., and *A. lutescens* Saff. against larvae of *Anastrepha ludens* (Mexican fruit fly). Extracts of the three *Annona* species showed time-dependent larvicidal activity against *A. ludens*, with variable mortality rates at 72 h of exposure as follows: *A. lutescens* 87–94 %, *A. diversifolia* 70–90 %, *and A. muricata* 63–74 %. Grzybowski et al. (2013) tested crude ethanolic extracts of *A. muricata* L. seeds and *Piper nigrum* L. fruits against *A. aegypti* larvae. The LC_{50} value for *A. muricata* was 93.48 μg/mL and for *P. nigrum* 1.84 μg/mL. Begum et al. (2010) investigated the toxic effects of ethanol extracts of seeds of *A. squamosa* and *Calotropis procera* (Asclepiadaceae) against different developmental stages of the housefly *Musca domestica* L. (Diptera: Muscidae). LC_{50} values for the extracts of *C. procera* and *A. squamosa* seeds were 870 and 345 mg/L, respectively. The high concentration (10 %) of extract from the seeds of *A. squamosa* exhibited maximum inhibitory effects (56 %) on acetylcholinesterase activity from all three developmental stages of the fly. The extracts can be dissolved in other solvents such as dichloromethane, dimethyl sulfoxide, or emulsified in water with Tween 20® (Castillo-Sánchez et al. 2010). From this information, it can be inferred that acetogenins can range from very polar, such as those extracted by water and ethanol, to nonpolar, i.e., those extracted by hexane; however, environmental considerations would suggest the use of more polar solvents (Bobadilla et al. 2005).

7 Commercial Pesticides Based on the Annonaceae

Acetogenins for insect control will probably continue to be based on crude or partially refined extracts obtained from plant sources (Leatemia and Isman 2004c), at least in developing countries or for use in organic food production in industrialized countries. The seeds are powdered and mixed with water or alcohol for application in Indian tea plantations. They can be useful against stem borers, sucking pests, and scale insects (Mamun and Ahmed 2011). The North American pawpaw *A. triloba* is a tree fruit in the early stages of commercial production in the United States and Canada. This plant contains acetogenins with pesticidal properties in the twigs, unripe fruit, seeds, roots, and bark tissues. However, commercial development of these compounds, based on twig extracts, has been problematic due to limited availability of biomass for extraction (Pomper et al. 2009). Commercially formulated botanicals are often more expensive than synthetic insecticides and not as widely available (Rajashekar et al. 2012).

Globally, the rapid increase in the human population and limited availability of arable land are becoming key factors stimulating market growth for agrochemicals. Also since 1990, the world market for organic produce has grown. Pressure from consumers for organic produce and other foods is leading even conventional growers to reduce their use of synthetic pesticides and consider alternatives, creating greater market space for botanical insecticides.

8 Conclusion

Asimina triloba, *Annona muricata*, and *Annona squamosa L.* are the species that have been most frequently examined for their insecticidal effects. Extracts from Annonaceae have been tested for control of Lepidoptera, Hymenoptera, Coleoptera, and Diptera, especially against *Spodoptera frugiperda*, *Plutella xylostella*, *Aedes aegypti*, and stored grain insects. Crude extracts from seeds, leaves, and fruits have been frequently tested as biopesticides to control arthropods. At present, there are two commercial insecticides in India based on an *Annona*: Anosom™ (seed extracts of *A. squamosa* and *A. reticulata*, containing 1 % squamocin as the active ingredient) and Bio Rakshak™ (a seed extract of *A. squamosa*). The registration and large-scale production of standardized botanical pesticide products are important barriers to commercialization of botanical pesticides. Smallholder farmers in developing countries are using their empirical familiarity with plant properties to protect their crops from pests (Perez et al. 2008).

References

Aciole SDG, Picollil CF, Duque L, Jonny E, Costa EV, Navarro-Silva MA, Marques FA, Sales Maia BHLN, Pinheiro MLB, Rebelo MT (2011) Actividad insecticida de tres especies de *Guatteria* (Annonaceae) contra *Aedes aegypti* (Diptera: Culicidae). Rev Colomb Entomol 37:262–268

Alali FQ, Xiao-Xi L, McLaughlin J (1999) Annonaceous acetogenins: recent progress. J Nat Prod 62:504–540

Alkofahi A, Rupprecht JK, Anderson JE, Mclaughlin JL, Mikolajczak KL, Scott BA (1989) Search for new pesticides from higher plants. In: Arnason JT, Philogene PJR, Morand P (eds) Insecticides of plant origin. American Chemical Society, Washington, DC

Álvarez O, Barrachina I, Ayala I, Goncalvez M, Moya P, Neske A, Bardón A (2008) Toxic effects of annonaceous acetogenins on *Oncopeltus fasciatus*. J Pest Sci 81:85–89

Anita S, Sujatha P, Prabhudas P (2012) Efficacy of pulverised leaves of *Annona squamosa* (L.), *Moringa oleifera* (Lam.) and *Eucalyptus globulus* (Labill.) against the stored grain pest, *Tribolium castaneum* (Herbst.). Recent Res Sci Technol 4:19–23

Asmanizar AD, Idris AB (2012) Evaluation of *Jatropha curcas* and *Annona muricata* seed crude extracts against *Sitophilus zeamais* infesting stored rice. J Entomol 9:13–22

Begum N, Sharma B, Pandey RS (2010) Toxicity potential and anti AchE activity of some plant extracts in *Musca domestica*. J Biofert Biopestic 2:108–114

Bobadilla M, Zavala F, Sisniegas M, Zavaleta G, Mostacero J, Taramona L (2005) Evaluación larvicida de suspensiones acuosas de *Annona muricata* Linnaeus (guanábana) sobre *Aedes aegypti* Linnaeus (Diptera, Culicidae). Rev Peru Biol 12:15–152

Castillo-Sánchez LE, Jiménez- Osornio JJ, Delgado-Herrera MA (2010) Secondary metabolites of the Annonaceae, Solanaceae and Meliaceae families used as biological control of insects. Trop Subtrop Agroecosyst 12:445–462

Chikukura L, Myum BM, Chikonzo R, Chenzara C (2011) Evaluation of selected indigenous pesticidal plant powders against stored maize and cowpeas insect pests. Afr Crop Sci Conf Proc 10:189–192

Costa MS, Pinheiro DO, Serrão JE, Pereira MJB (2012) Morphological changes in the midgut of *Aedes aegypti* L. (Diptera: Culicidae) larvae following exposure to an *Annona coriacea* (Magnoliales: Annonaceae) extract. Neotrop Entomol 41:311–314

Dadang EDF, Prijono D (2009) Effectiveness of two botanical insecticides formulations to major cabbage insect pests on field application. J Int Soc Southeast Asian Agric Sci 15:42–51

De Seffrin RC, Shikano I, Akhtar Y, Isman MB (2010) Effects of crude seed extracts of *Annona atemoya* and *A. squamosa* L. against the cabbage looper, *Trichoplusia ni* in the laboratory and greenhouse. Crop Prot 29:20–24

Dhembare AJ, Deokar M, Kolte P (2011) Efficacy of some plant extracts against rose aphid *Macrosiphum rosaeformes* (Davis). J Exp Zool 14:325–327

Di Toto Blessing L, Álvarez Colom O, Popich S, Neske A, Bardón A (2012) Antifeedant and toxic effects of acetogenins from *Annona montana* on *Spodoptera frugiperda*. J Pest Sci 83:307–310

Durmusoglu E, Hatipoglu A, Hasan B (2011) Efficiency of some plant extracts against *Tuta absoluta* (Meyrick, 1917) (Lepidoptera: Gelechiidae) under laboratory conditions. Turk J Entomol 35:651–663

Erkens RAJ (2007) From morphological nightmare to molecular conundrum. Phylogenic, evolutionary and taxonomic studies on *Guatteria* (Annonaceae). Dissertation, Utrecht University

Flores-Davila M, Gonzalez-Villegas R, Guerrero-Rodriguez E, Mendoza-Villarreal R, Cardenas-Elizondo A, Cerna-Chavez E, Aguirre-Uribe L (2011) Insecticidal effect of plant extracts on *Bactericera cockerelli* (Hemiptera: Psyllidae) nymphs. Southwest Entomol 36:2137–2144

González-Esquinca AR, Luna-Cazáres LM, Schlie Guzmán MA, De la Cruz CI, Hernández GL, Breceda SF, Gerardo PM (2012) In Vitro larvicidal evaluation of Annona muricata L., A. diversifolia Saff. and A. lutescens Saff. extracts against Anastrepha ludens larvae (Diptera, Tephritidae). Interciencia 37:264–289

Grzybowski A, Tiboni M, Silva MA, Chitolina RF, Passos M, Fontana JD (2013) Synergistic larvicidal effect and morphological alterations induced by ethanolic extracts of *Annona muricata* and *Piper nigrum* against the dengue fever vector *Aedes aegypti*. Pest Manag Sci 69:589–601

Gupta A, Pandey S, Shah DR, Yadav JS, Seth NR (2011) Annonaceous acetogenins: the unrevealed area for cytotoxic and pesticidal activities. Syst Rev Pharm 2:104–109

Isman MB (1994) Botanical insecticides and antifeedants: new sources. Pest Res J 6:11–19

Isman MB (1997) Neem and other botanical insecticides: barriers to commercialization. Phytoparasitica 25:339–344

Isman MB (2006) Botanical insecticides, deterrents, and repellents in modern agriculture and an increasingly regulated world. Annu Rev Entomol 51:45–66

Jolad SD, Hoffmann JJ, Schram KH, Cole JR, Tempesta MS, Kriek GR, Bates RB (1982) Uvaricin, a new antitumor agent from *Uvaria accuminata* (Annonaceae). J Org Chem 47:3151–3153

Kamanula J, Sileshi GW, Belmain SR, Sola P, Mvumi BM, Nyirenda GKC, Nyirenda SPN, Stevenson PC (2011) Farmers' insect pest management practices and pesticidal plant use for protection of stored maize and beans in Southern Africa. Int J Pest Manag 57:41–49

Kamaraj C, Bagavan A, Elango G, Zahir AA, Rajakumar G, Marimuthu S, Santhoshkumar T, Rahuman AA (2011) Larvicidal activity of medicinal plant extracts against *Anopheles subpictus* and *Culex tritaeniorhynchus*. Indian J Med Res 134:101–106

Komansilan A, Abadi AL, Yanuwiadi B, Kaligis DA (2012) Isolation and identification of a biolarvicide from soursop (*Annona muricata* Linn) seeds to mosquito (*Aedes aegypti*) larvae. Int J Eng Technol 12:28–32

Konkala A, Sabita Raja S, Kaur A (2012) Insect growth regulating activity of the phytochemical (+)-O-methylarmepavine, against the stored pulse pest *Callosobruchus chinensis*. J Pharm Bio Sci 2:25–29

Kosalge SB, Fursule RA (2009) Investigation of licicidal activity of some plants from Satpura Hills. Int J Pharm Tech Res 1:564–567

Kumar JA, Rekha T, Shyamala DS, Kannan M, Jaswanth A, Gopal V (2010) Insecticidal activity of ethanolic extract of leaves of *Annona*. J Chem Pharm Res 2:177–180

Kumar SV, Mani P, John Bastin TMM, Arun Kumar R, Ravikumar G (2011) Larvicidal, oviposition deterrent and repellent activity of *Annona squamosa* extracts against hazardous mosquito vector. Int J Pharm Tech 3:3143–3155

Leatemia JA, Isman MB (2004a) Toxicity and antifeedant activity of crude seed extracts of *Annona squamosa* (Annonaceae) against lepidopteran pests and natural enemies. Int J Trop Insect Sci 24:150–158

Leatemia JA, Isman MB (2004b) Efficacy of crude seed extracts of *Annona squamosa* against diamondback moth, *Plutella xylostella* L. in the greenhouse. Int J Pest Manag 50:129–133

Leatemia JA, Isman MB (2004c) Insecticidal activity of crude seed extracts of *Annona* spp., *Lansium domesticum* and *Sandoricum koetjape* against lepidopteran larvae. Phytoparasitica 32:30–37

Lin CY, Wu DC, Ko WH (2009) Control of silverleaf whitefly, cotton aphid and kanzawa spider mite with oil and extracts from seeds of sugar apple. Neotrop Entomol 38:531–536

Londershausen M, Leicht W, Lieb F, Moesschler H (1991) Molecular mode of action of annonins. Pest Manag Sci 33:427–433

Long Z, Hock S, Hung S (2006) Screening of Chinese medicinal herbs for bioactivity against *Sitophilus zeamais* Motschulsky and *Tribolium castaneum* (Herbst). J Stored Prod Res 43:290–296

Magadula JJ, Innocent E, Otieno JN (2009) Mosquito larvicidal and cytotoxic activities of three *Annona* species and isolation of active principles. J Med Plant Res 3:674–680

Mamun MSA, Ahmed M (2011) Prospectus of indigenous plant extracts in tea pest management. J Agric Res Innov Technol 1:16–23

McLaughlin JL (2008) Paw paw and cancer: Annonaceous acetogenins from discovery to commercial products. J Nat Prod 71:1311–1321

Mikolajczak L, McLaughlin JL, Rupprecht JK (1989) US Patent 4,855,319

Ocampo D, Ocampo R (2006) Bioactividad de la familia Annonaceae. Rev Univ Caldas 26:135–155

Perez OP, Lazo FJ, Diaz OL, Khambay B, Brandorf-White C (2008) Cuban flora as a source of bioactive compounds. Int J Cuban Stud 1:67–75

Pirie MD, Chatrou LW, Erkens RHJ, Maas JW, Van der Niet T, Mols JB, Richardson JE (2005) Phylogeny reconstruction and molecular dating in four Neotropical genera of Annonaceae: the effect of taxon sampling in age estimations. Plant Species Level Syst New Perspect Pattern Process 143:149–174

Pomper KW, Lowe JD, Crabtree SB, Keller W (2009) Identification of annonaceous acetogenins in the ripe fruit of the North American pawpaw (*Asimina triloba*). J Agric Food Chem 57:8339–8343

Rajashekar Y, Bakthavatsalam N, Shivanandappa T (2012) Botanicals as grain protectants. Psyche. doi:10.1155/2012/646740

Rıos MY, Castrejun F, Robledo N, Léon I, Rojas G, Navarro V (2003) Chemical composition and antimicrobial activity of the essential oils from *Annona cherimola* (Annonaceae). J Mex Chem Soc 47:134–142

Sedlacek JD, Friley KL, Lowe JD, Pomper KW (2010) Potential of ripe pawpaw fruit extract as an insecticide and feeding deterrent for striped cucumber beetle (Coleoptera: Chrysomelidae) on squash. J Entomol Sci 45:378–384

Srikrishnaraj KA, Isman MB (2006) Toxicity of a crude seed extract of *Annona squamosa* (*Annonaceae*) to larvae of *Aedes aegypti* and *A. atropalpus* (Diptera: Culicidae). Biopestic Int 2:73–78

Thomas A, Doyle JA (1996) Geographic relationships of Malagasy Annonaceae. Biogéogr Madag: 85–94

Trindade RCP, Luna JS, Lima MRF, Silva PP, Sant"Ana AEG (2011) Actividad larvicida y variación estacional del extracto de Annona muricata en *Plutella xylostella* (Lepidoptera: Plutellidae). Rev Colomb Entomol 37:223–227

Ukeh DA, Oku EE, Udo IA, Nta AI, Ukeh JA (2012) Insecticidal effect of fruit extracts from *Xylopia aethiopica* and *Dennettia tripetala* (Annonaceae) against *Sitophilus oryzae* (Coleoptera: Curculionidae). Chil J Agric Res 72:195–200

Westra LYT, Maas PJM (2012) *Tetrameranthus* (Annonaceae) revisited including a new species. PhytoKeys 12:1–21

Exploiting Phytochemicals for Developing Sustainable Crop Protection Strategies to Withstand Climate Change: Example from Africa

3

Zeyaur R. Khan, Charles A.O. Midega, Jimmy O. Pittchar, and John A. Pickett

Abstract

Africa suffers chronic food insecurity resulting from ravaging effects of insect pests, weeds and poor soil fertility, with rising poverty and increasingly dry and hot weather conditions associated with climate change further aggravating this situation. Scientists at the International Centre of Insect Physiology and Ecology (*icipe*) together with national and international partners have developed a platform technology, 'push–pull', based on locally available companion plants for integrated management of these constraints by exploiting innate plant defence systems including secondary metabolism. This involves intercropping cereal crops, the main staple and cash crops for millions of smallholder farmers in the continent, with forage legumes in the genus *Desmodium* and planting Napier grass as a trap plant around this intercrop. Stemborer pests are attracted to Napier grass (pull) and are repelled from the main cereal crop by the repellent desmodium (push). Desmodium root exudates effectively control the parasitic striga weed by causing abortive germination and also improve soil fertility through nitrogen fixation, provide natural mulching and improve biomass. Both companion plants provide high-value animal fodder, facilitate milk production and fetch additional income for farmers. The technology is appropriate to smallholder mixed cropping systems in sub-Saharan Africa (SSA) as it effectively addresses major production constraints and significantly increases cereal yields. It is currently being practiced by about 90,000 smallholder farmers in eastern Africa and has also been adapted

Z.R. Khan, Ph.D. (✉) • C.A.O. Midega, Ph.D.
J.O. Pittchar, Ph.D.
Plant Health Division, International Centre of Insect Physiology and Ecology (*icipe*), PO Box 30772, Nairobi 00100, Kenya
e-mail: zkhan@icipe.org

J.A. Pickett, Ph.D.
Biological Chemistry and Crop Protection Department, Rothamsted Research, Harpenden, Hertfordshire AL5 2JQ, UK

D. Singh (ed.), *Advances in Plant Biopesticides*,
DOI 10.1007/978-81-322-2006-0_3, © Springer India 2014

to harsh conditions associated with climate change by incorporating drought-tolerant companion plants. This chapter highlights the developmental process of the technology and its benefits in SSA in the face of climate change.

Keywords
Food insecurity • Cereals • Push–pull • Semiochemicals • Stemborer • Striga

1 Introduction

Food insecurity is a major setback to realisation of economic growth in Africa and is complicated by continuous decline in per capita food production in the recent past, making the continent a net importer of agricultural commodities. The situation is graver in sub-Saharan Africa (SSA) where land degradation, pests and weeds are major constraints to efficient production of crops and to human and livestock health. These constraints in addition to being partly responsible for the food insecurity in the region also affect nutrition and income thus causing abject poverty, with over 30 % of the population (close to 200 million people) being undernourished. Cereals, principally maize, *Zea mays* (L.); sorghum, *Sorghum bicolor* (L.) Moench; finger millet, *Eleusine coracana* (L.) Gaertn.; and rice, *Oryza sativa* (L.), are the most important food and cash crops for millions of rural farm families in SSA. Cereal production in the region is however severely affected by a series of constraints, mainly biotic and abiotic. Among the biotic constraints is a complex of about 20 economically important lepidopteran stemborers (Maes 1998), the most injurious insect pests attacking cereal crops in the region, with the indigenous *Busseola fusca* (Noctuidae) and the invasive *Chilo partellus* (Crambidae) being the most important. Attack by stemborer pests causes significant yield losses, ranging from about 10–80 % depending on the crop cultivar, stage of the crop at infestation and infestation level (Kfir et al. 2002). Effective control of stemborers is difficult, largely due to the cryptic and nocturnal habits of the adult moths and the protection provided by the host stem for immature pest stages (Ampofo 1986). Moreover, the conventionally recommended chemical control strategies are often impractical and uneconomical for smallholder farmers (Van den Berg and Nur 1998), while effectiveness of some of the cultural control methods, considered cheaper for resource-constrained farmers, is not empirically demonstrated (Van den Berg et al. 1998).

In addition to stemborers, parasitic weeds in the genus *Striga* (Orobanchaceae) (Oswald et al. 2001), commonly known as striga, are another group of serious biotic factors constraining cereal production in SSA. There are over 20 species of striga out of which *S. hermonthica* is by far the most important, infesting over 40 % of the arable land in the region (Lagoke et al. 1991). Infestations of cereal crops by striga result in severe grain yield losses, estimated at US$ 7 billion annually (Berner et al. 1995), with the most affected being the resource-poor subsistence farmers (Gurney et al. 2006). Striga weakens the host, wounding its outer root tissues and absorbing its supply of moisture, photosynthates and minerals (Tenebe and Kamara 2002), and is so ingeniously adapted to its environment (Bebawi and Metwali 1991) and integrated with the host that it will only germinate in response to specific chemical cues present in root exudates of its hosts or certain non-host plants (Yoder 1999; Parker and Riches 1993). It also causes 'phytotoxic' effects within days of attachment to its hosts (Frost et al. 1997; Gurney et al. 1999), whose underlying mechanism has not yet been elucidated (Gurney et al. 2006). These effects result in a large reduction in host plant height, biomass and eventual grain yield (Gurney et al. 1999). Various control strategies have been tried, some with partial or local success, but all have limitations and none has provided a complete solution (Oswald 2005). It has also been complicated by the abundant seed production by striga plants, longevity of the seed bank (Bebawi et al. 1984) and a complicated mode of parasitism. The effects of striga are most severe in degraded environments, with low soil fertility and low rainfall, and in subsistence farming systems where there are few options for purchasing external inputs (Sauerborn et al. 2003; Gurney et al. 2006). Unfortunately striga infestation continues to extend to new areas in the region as farmers abandon heavily infested fields for new ones (Khan et al. 2002; Gressel et al. 2004), a practice that is untenable due to consistent reduction in landholdings due to increases in human population.

The soils are also severely degraded, and these conditions are further aggravated by the effects of climate change. This results into high food and nutrition insecurity and poverty as the average cereal yields are less than 1 t/ha. The smallholder farmers are resource-constrained and therefore unable to invest in pest and soil fertility management. They are expected to increase during the next few decades as agriculture intensifies to meet the extra food demand from a growing population and as a result of increasingly dry and hot weather conditions associated with climate change.

2 Opportunities for Agricultural Development in Sub-Saharan Africa

It is recognised that economic growth led by the agricultural sector has a disproportionately positive impact on the poor and has a documented effect on reducing poverty (World Development Report 2008). Increasing cereal production is therefore an important challenge in addressing economic growth, alleviating poverty and arresting environmental degradation over most of SSA. Most of these cereals are produced by millions of smallholder farmers in predominantly mixed crop–livestock farming systems in the region (Romney et al. 2003). There is thus a need to enhance technical efficiency in the production systems in order to address the widening gap between food supply and food demand in the region. Given that SSA is the only region in the world where hunger and poverty are projected to worsen over the next two decades, there is a need for drastic action to improve agriculture and economic development. Attention should strongly focus on environmental sustainability of soil, crop and water resources and sustainable ways of managing weeds and pests through ecologically sound agronomic innovations. According to the poverty reduction strategy reports, growth in the agricultural sector, achievable by reducing major constraints to productivity mentioned above, in target countries is essential to reducing poverty and ensuring food security. This will benefit mainly poor cereal–livestock smallholders (about 80 % of the producers). Although in some cases insecticides and herbicides can help to alleviate these problems, complete control is seldom achieved. Moreover, the resource-constrained subsistence farmers in SSA cannot afford expensive chemicals. A large number of farmers, therefore, do not attempt to manage stemborers or striga, resulting in high grain yield losses and food insecurity (Chitere and Omolo 1993; Oswald 2005).

3 Exploiting Phytochemicals for Sustainable Cereal Production

To provide an effective and compatible management approach to smallholder farmers in SSA, a series of studies were conducted to identify companion plants that would naturally manipulate the pest behaviour while delivering additional benefits to the farmers. These studies led to development of a cropping strategy, known as 'push–pull', which exploits phytochemicals released by carefully chosen companion plants grown in between and around the main cereal crops. These companion plants release semiochemicals that (1) repel insect pests from the main crop using an intercrop which is the 'push' component and (2) attract insect pests away from the main crop using a trap crop which is the 'pull' component (Cook et al. 2007). Such a system requires a good understanding of the chemical ecology of plant–insect and plant-plant interactions on the different crops. In the development process of the technology, candidate crops needed to be systematically evaluated in field trials. During the process described herein that was specifically for the control of cereal stemborers, we discovered that certain intercrops had further benefits in terms of suppression of striga. However, the mechanism underpinning this was established to be an allelopathic effect of intercrop root exudates (nonvolatile) and hence only required the intercrop component of the push–pull system.

Development of the push–pull technology therefore began with an extensive field survey that involved over 500 grass species belonging to Poaceae, Cyperaceae and Typhaceae, as well as some leguminous crops in different agro-ecological zones in Kenya. The survey identified appropriate species that could be used as intercrop (push) and trap crop (pull) components of the push–pull mixed cropping system. Plant species selected as potential intercrops had to be repellent to stemborers and reduce their populations on the main cereal crop, maize, while those selected as potential trap crops had to be preferred by stemborers to maize and other cereal crops for oviposition. The best trap crops were those which were attractive but did not support development of the immature stages of the stemborer pest. Napier grass, *Pennisetum purpureum* Schumach, a forage crop, attracted considerably more oviposition by stemborer moths than maize but did not support development of the stemborer pest populations (Khan et al. 2000, 2006a). Stemborer larvae did not survive on Napier grass because it produces a gummy substance that immobilises the young larvae as they try to bore into the stem.

Molasses grass, *Melinis minutiflora* P. Beauv, another indigenous forage plant, attracted no stemborer oviposition at all and was identified as an effective repellent (push) plant. Since farmers in SSA practice multiple cropping where cereal crops are interplanted with legumes, selected legumes were also evaluated in these studies although they are not attacked by cereal stemborers. Two plants in the *Desmodium* genus, silverleaf, *D. uncinatum* DC and greenleaf, *D. intortum* (Mill) Urb, were observed to repel gravid stemborer moths as had been observed with molasses grass (Khan et al. 2000). Planting 3 rows of Napier grass as a trap crop around a plot of maize resulted in significant reductions in the infestation of maize by stemborers (Khan et al. 1997, 2000). Additionally, planting molasses grass or desmodium between the rows of maize resulted in >80 % reduction in stemborer infestation in maize (Khan et al. 2000).

While the putative trap and repellent intercrops were being evaluated for stemborer control, it was noticed that maize intercropped with *D. uncinatum* or *D. intortum* suffered far less striga infestation than maize in monoculture. This effect was confirmed by further field testing and shown to be significantly greater than that observed with other legumes widely recommended as intercropping solutions to striga problems, for example, cowpea, *Vigna unguiculata* (L.) Walp. as were the concomitant yield increases (Khan et al. 2002, 2007b).

4 Semiochemistry of Companion Plants

4.1 Semiochemistry of Companion Plants for Stemborer Control

Insects are attracted to their host plants through sophisticated detection of specific attractive semiochemicals (natural signal chemicals mediating changes in behaviour or development) (Nordlund and Lewis 1976) or specific ratios of semiochemicals (Bruce et al. 2005) emitted by these plants and other host organisms. Detection of specific semiochemicals or mixtures of semiochemicals associated with non-host taxa (Hardie et al. 1994) also guides avoidance of emitters of those chemical compounds by insects. From a series of studies, Napier grass trap crop was found to produce significantly higher amounts of volatile organic compounds (VOCs) used by gravid stemborer females to locate host plants, than maize or sorghum (Birkett et al. 2006). Additionally, it was established that there was also an increase of approximately 100-fold in the total amounts of these compounds produced in the first hour of nightfall (scotophase) by Napier grass (Chamberlain et al. 2006), the period during which stemborer moths seek host plants for oviposition (Päts 1991), causing the differential oviposition preference. However, about 80 % of the stemborer larvae did not survive (Khan et al. 2006a, 2007a) as Napier grass tissues produce sticky sap in response to feeding by the larvae which traps them causing their mortality. The intercrops, molasses grass and desmodium on the other hand were found to produce repellent VOCs that push away the stemborer moths. These include (*E*)-β-ocimene and (*E*)-4,8-dimethyl-1,3,7-nonatriene, semiochemicals typically produced during damage to plants by herbivorous insects and are responsible for the repellence of desmodium to stemborers (Khan et al. 2000) (Fig. 3.1).

4.2 Allelopathic Mechanism for Striga Control

Desmodium was found to effectively control striga, resulting in significant yield increases in maize from 1 to 3.5 t/ha per cropping season (Khan et al. 2008a). Similar results have also been observed with sorghum (Khan et al. 2006b), finger millet (Midega et al. 2010) and upland rice (Khan et al. 2010). In the elucidation of the mechanisms of striga suppression by *D. uncinatum*, it was found that, in addition to benefits derived from increased availability of nitrogen and soil shading, an allelopathic effect of the root exudates of the legume, produced independently of the presence of striga, is responsible for the dramatic reduction of striga in an intercrop with maize. Presence of blends of secondary metabolites with striga seed germination stimulatory, 4″,5″,-dihydro-5,2′,4′-trihydroxy-5′′,-isopropenylfurano-(2″,3″;7,6)-isoflavanone, and post-germination inhibitory, 4″,5″-dihydro-2′-methoxy-5,4′-dihydroxy-5″-isopropenylfurano-(2″,3″;7,6)-isoflavanone, activities in the root exudates of *D. uncinatum* which directly interferes with parasitism was observed (Tsanuo et al. 2003). This combination thus provides a novel means of *in situ* reduction of the striga seed bank in the soil through efficient suicidal germination even in the presence of graminaceous host plants in the proximity. Other *Desmodium* spp. have also been evaluated and have similar effects on stemborers and striga (Khan et al. 2006b) and are currently being used as intercrops in maize, sorghum and millets. Recently another key post-germination inhibitor, di-*C*-glycosylflavone 6-*C*-α-L-arabinopyranosyl-8-C-β-D-glucopyranosylapigenin, also known as isoschaftoside, as well as other *C*-glycosylflavones have been characterised from a more polar fraction of *D. uncinatum* root exudates and solvent extracts (Pickett et al. 2007; Hooper et al. 2009), and full chemical elucidation of other allelopathic agents is ongoing. Detailed studies on understanding of structure of chemicals, elucidation and understanding the mechanisms by which desmodium suppresses striga will ensure sustainability of desmodium-based cropping systems and provide an opportunity for exploitation of the biochemical pathways in desmodium root system beyond the smallholder cereal cropping systems.

5 Field Implementation and Benefits of the Push–Pull Technology

5.1 Agronomic and Environmental Benefits of the Technology

Following extensive research and development efforts, it was found that not only were stemborers and striga effectively controlled by the technology under farmers' conditions but farmers also reported additional benefits such as increased soil fertility and improved availability of animal fodder resulting in increased milk production (Khan et al. 2008b) and up to threefold increases in grain yields (Khan et al. 2008a). Desmodium also fixes atmospheric nitrogen (110 kg N/ha) (Whitney 1966), adds organic matter to the soil,

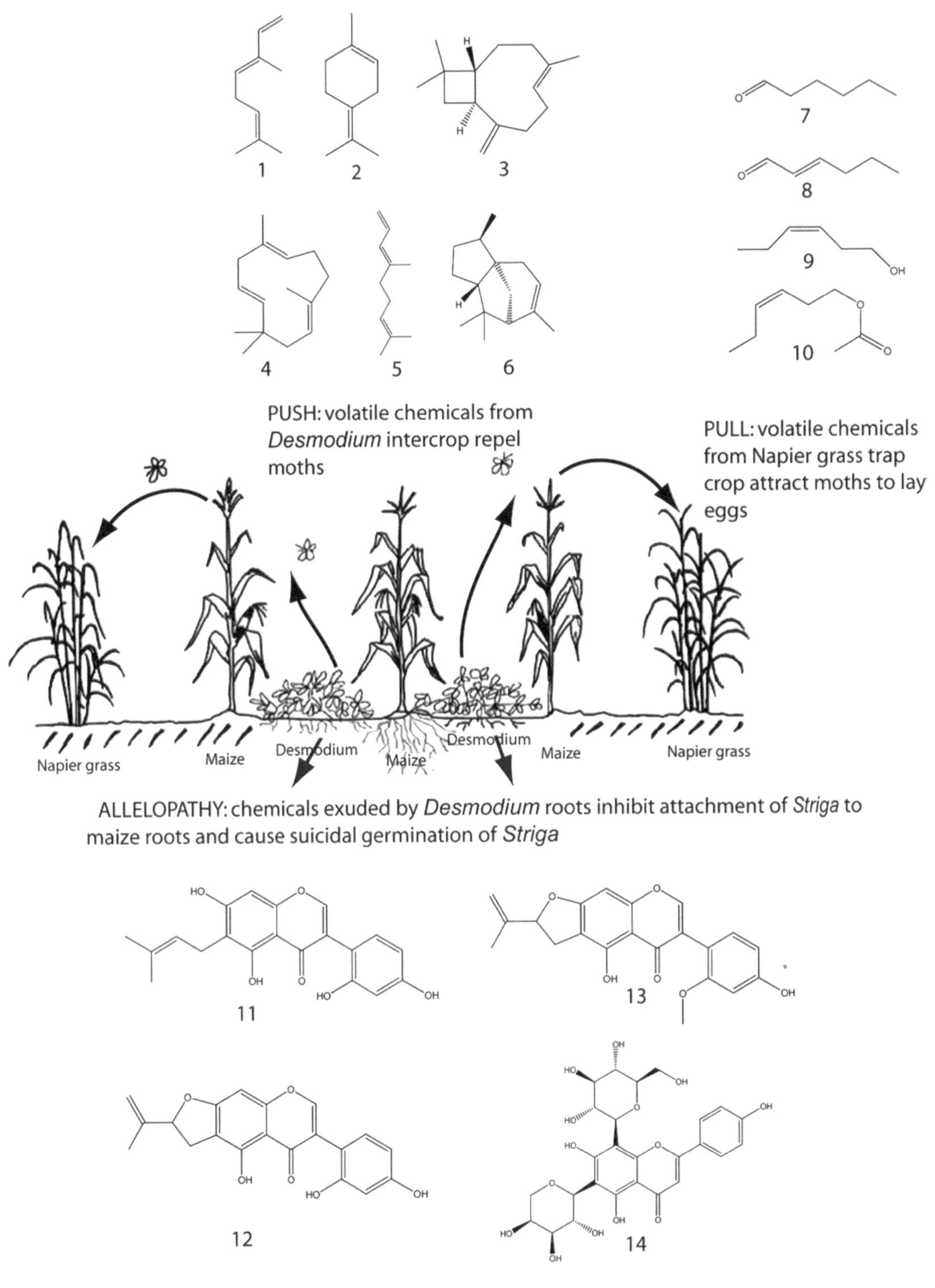

1 2 3 4 5 6 7 8 9 10

PUSH: volatile chemicals from *Desmodium* intercrop repel moths

PULL: volatile chemicals from Napier grass trap crop attract moths to lay eggs

Napier grass Maize Desmodium Maize Desmodium Maize Napier grass

ALLELOPATHY: chemicals exuded by *Desmodium* roots inhibit attachment of *Striga* to maize roots and cause suicidal germination of *Striga*

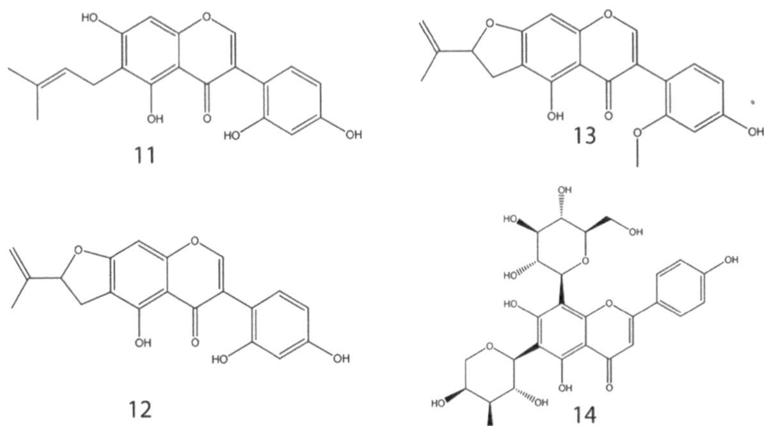

Fig. 3.1 How the push–pull system works: stemborer moths are repelled by intercrop volatiles while attracted to trap crop volatiles. Root exudates from the desmodium intercrop cause suicidal germination of striga and inhibit attachment to maize roots (*1* (*E*)-β-ocimene, *2* α-terpinolene, *3* β-caryophyllene, *4* humulene, *5* (*E*)-4,8-dimethyl-1,3,7-nonatriene, *6* α-cedrene, *7* hexanal, *8* (*E*)-2-hexenal, *9* (Z)-3-hexen-1-ol, *10* (Z)-3-hexen-1-yl acetate, *11* 5,7,2′,4′-tetrahydroxy-6-(3-methylbut-2-enyl) isoflavanone (uncinanone A), *12* 4″,5″-dihydro-5,2′,4′-trihydroxy-5″-isopropenylfurano-(2″,3″;7,6)-isoflavanone (uncinanone B), *13* 4″,5″-dihydro-2′-methoxy-5,4′-dihydroxy-5″-isopropenylfurano-(2″,3″;7,6)-isoflavanone (uncinanone C), and *14* di-*C*-glycosylflavone 6-*C*-α-L-arabinopyranosyl-8-*C*-β-D-glucopyranosylapigenin (Adapted with permission from Khan et al. (2010))

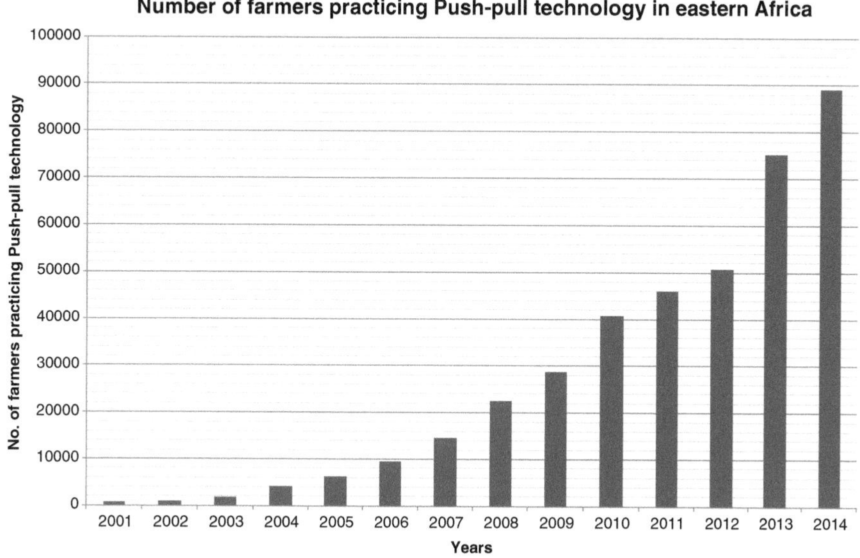

Fig. 3.2 Number of farmers practicing the push–pull technology in East Africa as of August 2014

conserves soil moisture and enhances soil biodiversity, thereby improving soil health and fertility (Khan et al. 2006b). Additionally, it provides ground cover and, together with surrounding Napier grass, protects the soil against erosion. The technology also enhances arthropod abundance and diversity, part of which is important in soil regeneration processes, pest regulation (Midega et al. 2008) and stabilisation of food webs. In deed there is also a clear demonstration of the value of biodiversity because of the important roles played by companion plants and beneficial insects in the system. It therefore improves agro-ecosystem sustainability and resilience, with great potential to mitigate the effects of climate change (Khan et al. 2014). Both desmodium and Napier grass provide valuable year-round quality animal forage while the sale of desmodium seeds generates additional income for the farmers. The push–pull technology thus opens up significant opportunities for smallholder growth and represents a platform technology around which new income generation and human nutritional components, such as livestock keeping, can be added. It therefore affords the smallholder farmers an opportunity to enter into the cash economy.

On-farm uptake of the technology by about 90,000 farmers in East Africa (Fig. 3.2) has confirmed the technology's effectiveness and significant impacts on food security, human and animal health, soil fertility, conservation of agro-biodiversity, agro-ecosystem services, empowerment of women and income generation for resource-poor farmers (Fig. 3.3). Upscaling of the technology to reach the current adopters and beyond has been achieved through deployment of a combination of dissemination pathways catering to different sociocultural contexts and literacy levels of farmers. This has been complimented by a multilevel collaboration with research institutions, national extension networks and non-governmental organisations (NGOs), and farmer groups, combined with extension efforts underpinned by a robust scientific base and continuous technical backstopping. Further involvement of a series of interventions (Khan et al. 2008b) including information bulletins (brochures, detailed practical manuals on how to plant push–pull) and mass media (radio programmes in local languages and newspaper articles) have boosted transfer of the technological information to a wider audience.

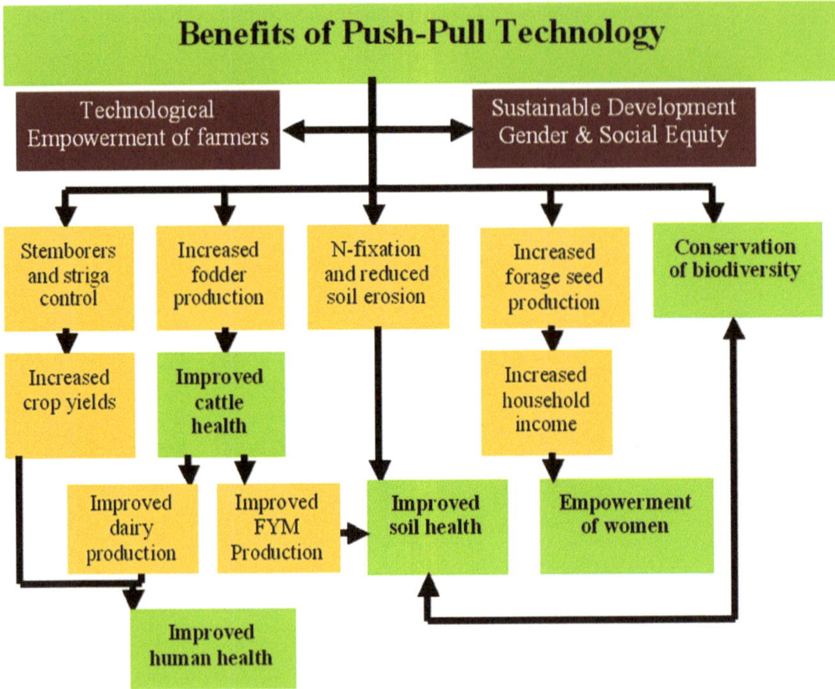

Fig. 3.3 Diagrammatic representation of benefits of the push–pull technology

5.2 Economic Benefits of the Technology

A number of studies have demonstrated that push–pull is more profitable than farmers' own practices, and some of the practices designed to improve soil fertility. Significantly higher benefit/cost ratio was realised with push–pull compared with maize monocrop and/or use of pesticides, posting a positive return on investment of over 2.2 compared with 0.8 obtained with the maize monocrop, and slightly less than 1.8 for pesticide use (Khan et al. 2001). Additionally, push–pull with no additional fertiliser had the best gross returns, while less profit was recorded with the use of fertiliser, implying it was economically propitious for poor smallholders who could not afford external inputs to invest in push–pull. Further economic analyses on returns to investment for the basic factors of production under push–pull showed these were significantly higher compared to those from maize–bean intercropping and maize monocrop systems (Khan et al. 2008c). Positive total revenues ranged from $351/ha in low potential areas to $957/ha in the high potential areas, with general increases in subsequent years. The returns to labour which were recovered within the first year of establishment of push–pull ranged from $ 0.5/man day in the low potential areas to $ 5.2/man day in the higher potential areas, whereas in the maize monocrop, this was negligible or even negative. Furthermore, the net present value (NPV) from push–pull was positive and consistent over several years. A more recent study (De Groote et al. 2010) that used discounted partial budget and marginal analysis corroborated these findings and concluded that push–pull earned the highest revenue compared to other soil fertility management technologies, including green manure rotation.

6 Adaptation of Push–Pull Technology

6.1 Adaptation of the Technology to Farmers' Needs

Smallholder farmers in SSA practice multiple cropping that closely integrates cereal crops and food legumes, with the latter forming an important part of diets of these farm families. Typically farmers plant edible legumes between rows of cereal crops or with the cereal crops in the same holes. To respond to farmers' need for food legumes, strategic research efforts were made that led to integrating beans into the technology thereby adapting it to farmers' need (Khan et al. 2009). Results indicated that integration of beans in the maize–desmodium intercrops and the planting arrangement did not compromise the striga and stemborer control efficacy of desmodium. While this integration significantly increased labour and total variable costs, total revenue, gross benefits and benefit/cost ratios did not significantly differ between the bean integration and maize–desmodium intercrops. Where labour is easily available, farmers are advised to plant cereal crops and beans in separate holes to avoid the risk of competition for moisture and nutrients where these might be limiting. This has increased the technology's appeal to the farmers as it guaranteed an additional protein source in the diet (Khan et al. 2009), resulting into higher technology adoption rates in eastern Africa ranging from 10,000 to about 90,000 farmers.

6.2 Adaptation of Push–Pull Technology to Climate Change

Climate change is anticipated to have far-reaching effects on the sustainable development of SSA, including the ability to attain the Millennium Development Goals (MDGs). The predictions indicate that atmospheric temperature will continue to increase, so will the incidences of flood and drought. These will result into progressively more serious land degradation and increased pest and weed pressure, increased incidences of crop failure and general increases in food and nutritional insecurity for resource poor farmers in many parts of SSA. To adapt to these adverse conditions, the resource-constrained smallholder farmers are moving to more drought-resilient cereal crops, such as sorghum and millet, and small ruminants for dairy production. The constraints will be addressed through wide-scale dissemination of climate-smart agricultural approaches like push–pull for smallholder cereal–livestock production in drier and hotter areas to withstand climate change.

With rising uncertainties in the region's rain-fed agriculture due to the continent's vulnerability to climate change, there was a demand and need to adapt conventional push–pull to withstand increasingly adverse and variable conditions. The trap and intercrops used in conventional push–pull were rainfall and temperature limited as the initial system was developed under average rainfall (800–1,200 mm) and moderate temperatures (15–30 °C). In order to ensure that push–pull continues to affect food security positively in Africa over the longer term, new drought-tolerant trap (*Brachiaria* cv. mulato) and intercrop (drought-tolerant species of desmodium, e.g. *D. intortum*) plants have been selected from research undertaken in collaboration with national and international partners. The new companion plants also have the appropriate chemistry in terms of natural enemy attractancy for the trap component, and stemborer repellence and striga suppression, and ability to improve soil fertility and soil moisture retention, for the intercrop component, and have been shown to significantly improve yields of maize and sorghum (Khan et al. 2014). Both trap and repellent plants provide high-quality livestock fodder over long periods of drought. In addition, they provide other ecosystem services such as biodiversity improvement and conservation, and organic matter improvement.

In identification of trap plants, *icipe* and partners screened about 400 grass species from which 21 drought-tolerant species were initially selected. Out of these, *Brachiaria* cv. mulato was chosen as the trap plant for the climate-adapted push–pull given also its ability to control stemborers, farmers' preference for it as livestock fodder and commercial availability of its seed that would allow faster dissemination and uptake. Additionally, it allowed minimal survival of stemborer larvae, a suitable characteristic of a trap plant that would support populations of natural enemies within season and when the cereal crop is not in season. Additionally, drought-tolerant species of desmodium that emit volatiles that repel stemborers, fix nitrogen to improve soil

Fig. 3.4 Sorghum planted in sole stand (**a**) and sorghum intercropped with greenleaf desmodium (**b**)

fertility, produce high biomass, cover the soil and improve soil health were identified. From a collection of 43 accessions collected from dry and hot areas in Africa and other arid environments, greenleaf (*D. intortum*) was observed to be more drought tolerant and was chosen as the intercrop species for immediate integration into a climate-adapted push–pull. Greenleaf desmodium was chosen given its known ability to control striga and stemborers (Khan et al. 2007b) (Fig. 3.4) coupled with commercial availability of its seed that would enable its wider testing by farmers within the project target areas. The work to isolate and purify all the active compounds in the desmodium root exudates and fully elucidate their effects on striga suppression is currently ongoing. Similarly, the full mechanism of stemborer control by the new companion plants is currently being elucidated with the aim of providing both sustainability and quality assurance as more companion plants are selected for new agro-ecologies.

Currently over 30,000 smallholder farmers in drier parts of Kenya, Tanzania and Ethiopia have taken up the climate-smart push–pull and have reported effective control of stemborers and striga weed resulting in significant increases, up to fivefold, in grain yields of both maize and sorghum (Khan et al. 2014). The new companion plants have also ensured availability of high quality fodder at the farms thereby increasing productivity of livestock. Validation of gross return from the adapted climate-smart push–pull showed $1075.8/ha and $1,289/ha gross benefits for sorghum and maize respectively and significantly higher marginal rates of return (MRR) implying that the net increase in benefits of climate-smart push–pull outweighs the net increase in costs compared to farmers' own practices.

7 Conclusions and Future Outlook

The push–pull system effectively addresses the constraints to production faced by the farmers and is an appropriate system because it uses locally available companion plants rather than expensive inputs. It has been adapted for drier areas vulnerable to climate change by identifying and incorporating drought-tolerant trap and repellent plants. This has made the technology more resilient in the face of climate change as rainfall becomes increasingly unpredictable. Moreover, the technology is being made 'smarter' through identification and incorporation of cereal crops with defence systems against stemborer pests that are inducible by egg deposition by the pests. Companion plants that are able to signal defence systems of the neighbouring smart cereals are also being identified. Accompanying these are efforts to elucidate full mechanisms of striga and stemborer control by conventional and the new companion plants. Science-based solutions to crop protection, which are environmentally sustainable and low cost, like push–pull, are urgently needed to address the real and increasing dangers of food insecurity without causing any ecological and social harm.

Acknowledgements The International Centre of Insect Physiology and Ecology (*icipe*) appreciates the core support from the Governments of Sweden, Germany, Switzerland, Denmark, Norway, Finland, France, Kenya and the UK. The work on push–pull technology has been primarily funded by the Gatsby Charitable Foundation, Kilimo Trust and the European Union, with additional support from the Rockefeller Foundation, Biovision Foundation, McKnight Foundation, Bill and Malinda Gates Foundation and DFID. Rothamsted Research receives grant-aided support from the Biotechnology and Biological Sciences Research Council (BBSRC), UK, with additional funding provided under the Biological Interactions in the Root Environment (BIRE) initiative.

References

Ampofo JKO (1986) Maize stalk borer (Lepidoptera: Pyralidae) damage and plant resistance. Environ Entomol 15:1124–1129

Bebawi FF, Metwali EM (1991) Witch-weed management by sorghum–Sudan grass seed size and stage of harvest. Agron J 83:781–785

Bebawi FF, Eplee RE, Harris CE, Norris RS (1984) Longevity of witchweed (*Striga asiatica*) seed. Weed Sci 32:494–507

Berner DK, Kling JG, Singh BB (1995) *Striga* research and control – a perspective from Africa. Plant Dis 79:652–660

Birkett MA, Chamberlain K, Khan ZR, Pickett JA, Toshova T, Wadhams LJ, Woodcock CM (2006) Electrophysiological responses of the lepidopterous stemborers *Chilo partellus* and *Busseola fusca* to volatiles from wild and cultivated host plants. J Chem Ecol 32:2475–2487

Bruce TJA, Wadhams LJ, Woodcock CM (2005) Insect host location: a volatile situation. Trends Plant Sci 10:269–274

Chamberlain K, Khan ZR, Pickett JA, Toshova T, Wadhams LJ (2006) Diel periodicity in the production of green leaf volatiles by wild and cultivated host plants of stemborer moths, *Chilo partellus* and *Busseola fusca*. J Chem Ecol 32:565–577

Chitere PO, Omolo BA (1993) Farmers' indigenous knowledge of crop pests and their damage in western Kenya. Int J Pest Manag 39:126–132

Cook SM, Khan ZR, Pickett JA (2007) The use of 'push–pull' strategies in integrated pest management. Annu Rev Entomol 52:375–400

De Groote H, Vanlauwe B, Rutto E, Odhiambo GD, Kanampiu F, Khan ZR (2010) Economic analysis of different options in integrated pest and soil fertility management in maize systems of Western Kenya. Agric Econ 41:471–482

Frost DL, Gurney AL, Press MC, Scholes JD (1997) *Striga hermonthica* reduces photosynthesis in sorghum: the importance of stomatal limitations and a potential role for ABA? Plant Cell Environ 20:483–492

Gressel J, Hanafi A, Head G, Marasas W, Obilana AB, Ochanda J, Souissi T, Tzotzos G (2004) Major heretofore intractable biotic constraints to African food security that may be amenable to novel biotechnological solutions. Crop Prot 23:661–689

Gurney AL, Press MC, Scholes JD (1999) Infection time and density influence the response of sorghum to the parasitic angiosperm *Striga hermonthica*. New Phytol 143:573–580

Gurney AL, Slate J, Press MC, Scholes JD (2006) A novel form of resistance in rice to the angiosperm parasite *Striga hermonthica*. New Phytol 169:199–208

Hardie J, Isaacs R, Pickett JA, Wadhams LJ, Woodcock CM (1994) Methyl salicylate and (−)-(1*R*,5S)-myrtenal are plant derived repellents for black bean aphid, *Aphis fabae* Scop. (Homoptera: Aphididae). J Chem Ecol 20:2847–2855

Hooper AM, Hassanali A, Chamberlain K et al (2009) New genetic opportunities from legume intercrops for controlling *Striga* spp. parasitic weeds. Pest Manag Sci 65:546–552

Kfir R, Overholt WA, Khan ZR, Polaszek A (2002) Biology and management of economically important lepidopteran cereal stemborers in Africa. Annu Rev Entomol 47:701–731

Khan ZR, Ampong-Nyarko K, Chilishwa P et al (1997) Intercropping increases parasitism of pests. Nature 388:631–632

Khan ZR, Pickett JA, Van den Berg J, Wadhams LJ, Woodcock CM (2000) Exploiting chemical ecology and species diversity: stemborer and striga control for maize and sorghum in Africa. Pest Manag Sci 56:957–962

Khan ZR, Pickett JA, Wadhams LJ et al (2001) Habitat management strategies for the control of cereal stemborers and Striga in maize in Kenya. Insect Sci Appl 21:375–380

Khan ZR, Hassanali A, Overholt W, Khamis TM, Hooper AM, Pickett AJ, Wadhams LJ, Woodcock CM (2002) Control of witchweed *Striga hermonthica* by intercropping with *Desmodium* spp., and the mechanism defined as allelopathic. J Chem Ecol 28:1871–1885

Khan ZR, Midega CAO, Hutter NJ, Wilkins RM, Wadhams LJ (2006a) Assessment of the potential of Napier grass (*Pennisetum purpureum*) varieties as trap plants for management of *Chilo partellus*. Entomol Exp Appl 119:15–22

Khan ZR, Hassanali A, Pickett JA (2006b) Managing polycropping to enhance soil system productivity: a case study from Africa. In: Uphoff N, Ball AS, Palm C, Fernandes E, Pretty J, Herren H, Sanchez P, Husson O, Sanginga N, Laing M, Thies J (eds) Biological approaches to sustainable soil systems.

CRC Press, Taylor and Francis, Boca Raton, pp 575–586
Khan ZR, Midega CAO, Hassanali A et al (2007a) Assessment of different legumes for the control of *Striga hermonthica* in maize and sorghum. Crop Sci 47:728–734
Khan ZR, Midega CAO, Wadhams LJ et al (2007b) Evaluation of Napier grass (*Pennisetum purpureum*) varieties for use as trap plants for the management of African stemborer (*Busseola fusca*) in a 'push–pull' strategy. Entomol Exp Appl 124:201–211
Khan ZR, Pickett JA, Hassanali A, Hooper AM, Midega CAO (2008a) *Desmodium* species and associated biochemical traits for controlling *Striga* species: present and future prospects. Weed Res 48:302–306
Khan ZR, James DG, Midega CAO, Pickett JA (2008b) Chemical ecology and conservation biological control. Biol Control 45:210–224
Khan ZR, Amudavi DM, Midega CAO et al (2008c) Farmers' perceptions of a 'push–pull' technology for control of cereal stemborers and striga weed in western Kenya. Crop Prot 27:976–987
Khan ZR, Midega CAO, Wanyama JM et al (2009) Integration of edible beans (*Phaseolus vulgaris* L.) into the push–pull technology developed for stemborer and striga control in maize-based cropping systems. Crop Prot 28:997–1006
Khan ZR, Midega CAO, Bruce TJA, Hooper AM, Pickett JA (2010) Exploiting phytochemicals for developing a push – pull crop protection strategy for cereal farmers in Africa. J Exp Bot 61:4185–4196
Khan ZR, Midega CAO, Pittchar JO, Murage AW, Birkett M, Bruce TJA, Pickett JA (2014) Achieving food security for one million sub-Saharan African poor through push-pull innovation by 2020. Philos Trans R Soc B 369(1639):20120284
Lagoke STO, Parkinson V, Agunbiade RM (1991) Parasitic weeds and control methods in Africa, in Combating striga in Africa. In: Kim SK (ed) Proceedings of an international workshop organized by IITA, ICRISAT and IDRC, 22–24 Aug 1988. IITA, Ibadan, pp 3–14
Maes K (1998) Pyraloidea: Crambidae, Pyralidae 87–98. In: Polaszek A (ed) African cereal stem borers: economic importance, taxonomy, natural enemies and control. CABI, Wallingford
Midega CAO, Khan ZR, Van den Berg J, Ogol CKPO, Dippenaar-Schoeman AS, Pickett JA, Wadhams LJ (2008) Response of ground-dwelling arthropods to a push–pull habitat management system: spiders as an indicator group. J Appl Entomol 132:248–254
Midega CAO, Khan ZR, Amudavi DA, Pittchar JO, Pickett JA (2010) Integrated management of *Striga hermonthica* and cereal stemborers in finger millet (*Eleusine coracana* (L.) Gaertn.), through intercropping with *Desmodium intortum*. Int J Pest Manag 56:145–151
Nordlund DA, Lewis WJ (1976) Terminology of chemical releasing stimuli in intraspecific and interspecific interactions. J Chem Ecol 2:211–220
Oswald A (2005) Striga control technologies and their dissemination. Crop Prot 24:333–342
Oswald A, Ransom JK, Kroschel J, Sauerborn J (2001) Transplanting maize and sorghum reduces *Striga hermonthica* damage. Weed Sci 49:346–353
Parker C, Riches CR (1993) Parasitic weeds of the world: biology and control. CAB International, Wallingford
Päts P (1991) Activity of Chilo partellus (Lepidoptera: Pyralidae): eclosion, mating and oviposition time. Bull Entomol Res 81:93–96
Pickett JA, Khan ZR, Hassanali A et al (2007) Chemicals involved in post-germination inhibition of Striga by Desmodium: opportunities for utilizing the associated allelopathic traits. In: Ejeta G, Gressel J (eds) Integrating new technologies for striga control: towards ending the witch-hunt. World Scientific, Singapore/Hackensack/London, pp 61–70
Romney DL, Thorne P, Lukuyu B, Thornton PK (2003) Maize as food and feed in intensive smallholder systems: management options for improved integration in mixed farming systems of east and southern Africa. Field Crop Res 84:159–168
Sauerborn J, Kranz B, Mercer-Quarshie H (2003) Organic amendments mitigate heterotrophic weed infestation in savannah agriculture. Appl Soil Ecol 23:181–186
Tenebe VA, Kamara HM (2002) Effect of *Striga hermonthica* on the growth characteristics of sorghum intercropped with groundnut varieties. J Agron Crop Sci 188:376–381
Tsanuo MK, Hassanali A, Hooper AM et al (2003) Isoflavanones from the allelopathic aqueous root exudates of *Desmodium uncinatum*. Phytochemistry 64:265–273
Van den Berg J, Nur AF (1998) Chemical control. In: Polaszek A (ed) African cereal stemborers: economic importance, taxonomy, natural enemies and control. CAB International, Wallingford, pp 319–332
Van den Berg J, Nur AF, Polaszek A (1998) Cultural control. In: Polaszek A (ed) African cereal stemborers: economic importance, taxonomy, natural enemies and control. CAB International, Wallingford, pp 333–347
Whitney AS (1966) Nitrogen fixation by three tropical forage legumes and the utilization of legume-fixed nitrogen by their associated grasses. Herb Abstr 38:143
World Bank (2008) World development report 2008: agriculture for development. The World Bank, Washington, DC
Yoder JI (1999) Parasitic plant responses to host plant signals: a model for subterranean plant–plant interactions. Curr Opin Plant Biol 2:65–70

Development of Insect Resistance to Plant Biopesticides: An Overview

4

Khanobporn Tangtrakulwanich and Gadi V.P. Reddy

Abstract

Plant-incorporated protectants (PIPs) and botanical biochemical pesticides have been widely used as alternative pesticides to synthetic chemical insecticides in the cropping industry. Usually, it takes longer for insects to develop a resistance to such alternative pesticides because these pesticides have broad or non-specific mode of actions. The fact that they are naturally derived pesticides and have multiple modes of action does not mean they are less susceptible to resistance. This review provides information about the resistance development in insect population against PIPs (*Bt* crop) and botanical pesticides. The mechanisms of resistance, evidence of resistance to these biopesticides, resistance management, and future trend for plant biopesticides are discussed.

Keywords

Insect resistance • Plant biopesticides • Microbial pesticides • Plant-incorporated protectants • Biochemical pesticides • Metabolic resistance • Target-site resistance • Penetration resistance • Behavioural resistance

1 Introduction

Pesticides used nowadays in cropping systems are classified in two groups: conventional and alternative (selective) pesticides. Conventional pesticides are synthetic chemical pesticides which usually belong to the chemical classes of organophosphates, carbamates, pyrethroids, and organochlorines (United States Environmental Protection Agency 1998). Biopesticides are types of alternative pesticides derived from natural products. The use of biopesticides in controlling insects or pests has been increased dramatically in recent years. According to the US Environmental Protection Agency (EPA), biopesticides have been categorized in three major classes: (1) microbial pesticides which consist of a microorganism (e.g. a bacterium, fungus, virus, or protozoan) as the active ingredient; (2) plant-incorporated protectants (PIPs)

K. Tangtrakulwanich, Ph.D.
G.V.P. Reddy, Ph.D. (✉)
Western Triangle Ag Research Center,
Montana State University, 9546 Old Shelby Rd,
Conrad, MT 59425, USA
e-mail: reddy@montana.edu

D. Singh (ed.), *Advances in Plant Biopesticides*,
DOI 10.1007/978-81-322-2006-0_4, © Springer India 2014

which are pesticidal substances that plants produce from genetic material that are added into plants to render them immune or tolerant of insects and pests; for example, the insertion of the gene for Bt insecticidal protein into the plant's genetic material, and (3) biochemical pesticides which are naturally occurring substances that control or regulate insect pest populations by nontoxic mechanisms, such as pheromones and (e.g. sex, aggregation, and alarm) scented plant extracts that attract insects for lure-and-kill strategies (US EPA 1998).

Biopesticides have been used for commercial, agricultural, and horticultural industries to reduce the risk of chemical pesticides for nontarget organisms short-residual activity, workers/applicators, and to natural enemies (e.g. parasitoids and predators). When incorporated into the integrated pest management programme, the biopesticides can delay the development of resistance in insects and pests (Khater 2012). However, the fact that some biopesticides are naturally derived compounds does not mean they solve all problems. Although the short-residual activity is considered a benefit, this means multiple applications are required (Cloyd 2012). This may increase selection pressure on pest populations, possibly leading to resistance. For this review, we only focused on insect resistance on plant biopesticides which are plant-incorporated protectants (PIPs), plant-derived pesticides, or plant-derived essential oils (botanicals).

2 Resistance Mechanisms

Resistance to pesticides or insecticides is characterized by rapid evolution under strong selection (Walker et al. 2012). Natural selection allows some very rare, naturally occurring insects with resistance genes to survive in the environment and pass this resistance trait on to the next generation. The long-term continued application of pesticides with the same mode of action (MOA) leads to selection for the resistant individuals. When continual selection pressure occurs, a recessive allele for resistance will increase to fixation in a single generation if the insecticide could eliminate all susceptible insects (Mallet 1989). However, this circumstance could hardly happen in the field because it is assumed that there is a group of insects being unexposed to the insecticides (Curtis 1987). If there is less than 100 % mortality of susceptible insects in treated plots in the fields, and the unexposed insects to insecticides exists, a rare mutation allele will an increase much more rapidly if dominant than if recessive to the wild type. Therefore, the effective dominance of a resistance allele is reduced with an increasing dose of insecticides (Mallet 1989). Anyhow, there are some factors that contribute to the existence of resistance genes. These factors are the facts that we can hardly spray the crop fields evenly with insecticides and, moreover, a group of insects with resistant genes that may exist outside the treated area but partially mix with the treated population by dispersal (Mallet 1989). It usually occurs that the resistant genes do not do very well among, the general populations under normal condition. Therefore, resistant individuals will hardly expand in the population above pretreatment level if the unexposed insect population is very large. Moreover, much less resistance will develop in the treated area than it would when migration does not occur. However, the migration can never be protected. There will be a migration rate that resistance increase quickly due to the treated population evolving independently (Commins 1977; May and Dobson 1986).

There are many explanations about the development of resistance in insects to pesticides. For example, extensive use of chemical insecticides for the control of whiteflies has caused whiteflies to develop a resistance to almost every major class of synthetic insecticides all over the world (Omer et al. 1992; Polumbo et al. 2001; Wardlow et al. 1976; Zou and Zheng 1988). For example, the green house whiteflies, *Trialeurodes vaporarium*, were found to be resistant to buprofezin, which could be related to high volatility of the active ingredient resulting in inadvertent selection pressure (Cloyd 2012). Because of that, there is an increased probability of resistance. Moreover, greenhouses and also nurseries usually restrict insect populations with minimal migration.

This results in interbreeding and passing on the resistance trait to the next generation. Consequently, the resistant genes will be increased in the gene pool. This leads to the increase of resistant individuals in the population. Furthermore, some insects do not move or disperse very far which means that concentrated populations of resistant insects are continually present and may be exposed to pesticide applications. Moreover, if the pesticides with similar modes of actions are applied frequently, this will enhance more resistant individuals in the population. According to Miller (1988), there are several factors responsible for insect resistance to insecticides in the crop industry, which are metabolic resistance, target-site resistance, penetration resistance, and behaviour resistance.

2.1 Metabolic Resistance

This resistance mechanism involves non-specific enzymes that normally detoxify foreign lipophilic chemicals such as monooxygenases, oxidase or cytochrome P-450-dependent oxidases, hydrolases, and transferases (Wood et al. 1981; Ahmad et al. 1986). Many of these enzymes are able to induce the temporary tolerances to pesticides. Inherited resistance often involves oxidases (Tsukamoto 1983; Brown 1985). Resistant insects may detoxify the toxin faster than susceptible insects or quickly rid their bodies of the toxic molecules. Metabolic resistance is the most common mechanism and often presents the greatest challenge. Insects use their internal enzyme systems to break down insecticides. Resistant strains may possess higher levels or more efficient forms of these enzymes. In addition to being more efficient, these enzyme systems also may have a broad spectrum of activity (i.e. they can degrade many different insecticides) (Maestre 2012). The example of this resistance was revealed by Krieger et al. (1971). From this study, they observed the polyphagous lepidopterans which had been exposed to a broad spectrum of plant toxins. The result showed that these lepidopterans had significantly higher levels of P450-dependent poly-substrate monooxygenase activity than in monophagous or oligophagous lepidopterans. Another example was that insecticide resistant aphids and mosquitoes were found to have a much amplification of esterase genes (Holloway and McCaffery 1988). For *Myzus persicae* aphids, this metabolic resistance is due to the increased esterase protein which acts as an insecticide reservoir as well as a catalytic enzyme, rather than increasing catalysis per molecule (Mallet 1989). Ahmad et al. (1986) also demonstrated that the deh gene of houseflies which regulate DDT-dehydrochlorinase was also an example of a glutathione-S-transferase which induced temporary tolerances to DDT.

2.2 Target-Site Resistance

The target site where the insecticide acts in the insect may be genetically modified to prevent the insecticide binding or interacting at its site of action, thereby reducing or eliminating the effect of the insecticide. Resistance caused by altered insecticide target molecules has been found to result from point mutations in their genes (Mutero et al. 1994; Newcomb et al. 1997). There are some examples of target-site resistance such as organophosphorus insecticides which bind to acetylcholinesterase (AChE), the enzyme responsible for breaking down the neurotransmitter acetylcholine (Ach) at synapses by competitively inhibiting Ach (Berenbaum 1986). While the Ach/AChE complex catalyzed very quickly (in a few seconds) to liberate the enzyme, the organophosphorus enzyme complex breaks down very slowly. Because the enzyme has already been preoccupied, Ach remains available in the synaptic space and thus binds to the receptor, continuing to depolarize the postsynaptic cell and therefore producing tremors, convulsions, and other physiological disturbances (Murray 1996). Seven mosquito species were reported to have AChEs that are insensitive to organophosphorus insecticides (Bonning 1990). Unrestricted overuse of these insecticides is thought to have caused the selection of an AChE with an altered catalytic subunit which leads to decreased activity and enhanced opportunities for detoxification enzymes to operate on the insecticides (Murray 1996).

Another example of target-site resistance was revealed in the milkweed bug, *Oncopeltus fasciatus*, which feed on plants containing cardiac glycoside (ouabain). Ouabain is a cardiac glycoside which inhibits the Na^+/K^+ ATPase that is very important for maintaining the correct distribution of Na^+ and K^+ across neuronal membranes (Murray 1996). Moore and Scudder (1986) demonstrated that the milkweed bug has a Na^+/K^+ ATPase that is resistant to ouabain, and as a result, milkweed bugs display 200-fold less sensitivity to inhibition by ouabain than other species not specialized to feed on ouabain-containing plants.

2.3 Penetration Resistance

This type of resistance is a morphological mechanism, which prevents toxic substances from entering the body. Resistant insects may absorb the toxin more slowly than susceptible insects. Penetration resistance occurs when the insect's outer cuticle develops barriers, which can slow down absorption of the chemicals into their bodies. This can protect insects from a wide range of insecticides. Penetration resistance is frequently present along with other forms of resistance, and reduced penetration intensifies the effects of other mechanisms (Mallet 1989). The examples of penetration resistance can be found in houseflies and lepidopterans (Plapp 1984; Little et al. 1989). Farnham (1973) showed that gene *pen* is responsible for reducing the rate of penetration of insecticides through the cuticle in *Musca domestica*.

2.4 Behavioural Resistance

Resistant insects may detect or recognize a danger and avoid the toxin. This mechanism of resistance has been reported for several classes of insecticides, including organochlorines, organophosphates, carbamates, and pyrethroids. Insects may simply stop feeding if they come across certain insecticides or leave the area where spraying has occurred (for instance, they may move to the underside of a sprayed leaf, move deeper in the crop canopy, or fly away from the target area) (www.irac-online). Several studies demonstrated that according to the DDT-synergist space-spray applications, resistant flies contact less spray than susceptible flies by reducing flights when this chemical presents (Kilpatrick and Quarternam 1952). Being exposed to residual DDT applications, these flies changed their behaviour of nocturnal resting places from indoor to outdoors during the warmer periods of the year (Kilpatrick and Quarternam 1952; Maier et al. 1952). Carroll and Hoffman (1980) have demonstrated that the beetles in genus *Epilachna* chew a circular trench which isolates the leaf area on which they are feeding. The beetles do that in order to prevent the rapid translocation of a bitter-tasting allelochemical, thus preserving the palatability and quality of the leaf (Murray 1996). Another example of behavioural resistance found in some species of Lepidoptera and Coleoptera is that the insects cut the leaf or petiole vein to prevent the flow of toxins into the leaf on which they are feeding (Compton 1987; Edwards and Wanjura 1989).

Since behavioural resistance is actually an avoidance of a particular chemical, no selection is involved. Therefore, it seems logical that the phenomenon of behavioural resistance may help maintain susceptibility to a particular chemical within a population. It had been proposed that by using insecticides with repellent properties, we can reduce the rate at which resistance develops in a particular population. However, it is unclear whether this type of approach would only serve to indicate susceptible individuals. Non-susceptible would continue to arise due to random mutations (Leatherman 1997).

Nevertheless, there have been reports that some species seem to have developed a resistance to any given compound in a variety of ways. For example, behavioural, penetrative, and metabolic resistance as well as target-site insensitivity to pyrethroids have been demonstrated in *Heliothis virescens* or the budworm moth (Curtis 1987; Dowd et al. 1987; Little et al. 1989; Mallet 1989). Thus, the multiple mechanisms of resistance can occur in the same time at once in insects.

3 Insect Resistance on Plant: Incorporated Protectants (PIPs)

PIPs are pesticidal substances produced by plants and are the genetic material necessary for the plant to produce the substance (U.S. Environmental Protection Agency 1998). For example, scientists can take the gene for a specific *Bt* pesticidal protein and introduce the gene into the plant's genetic material. Then the plant manufactures the pesticidal protein that controls the pest when it feeds on the plant. Both the protein and its genetic material are regulated by EPA; the plant itself is not regulated.

Bacillus thuringiensis or *Bt* is a gram-positive bacterium that exists naturally in soil all over the world. Over a few decades, scientists found that some strains of *Bt* kill particular insects and demonstrated that a specific protein was the toxic chemical which caused the death of these insects. When certain insects ingest either the bacterium or the protein produced by the bacterium (Cry protein or delta endotoxin), their digestive systems function abnormally and become disrupted. This leads to the death of insects. However, *B. thuringiensis* is not toxic to humans, mammals, most beneficial insects, and other nontarget organisms. Therefore, it does not cause serious environmental problems like conventional synthetic insecticides (Flexner et al. 1986; Wilcox et al. 1986). The families of insects that respond to *Bt* are Lepidoptera (caterpillars; e.g. European corn borer or cotton bollworm), Coleoptera (beetles; e.g. Colorado potato beetles), and Diptera (flies and mosquitoes).

As we know that pest resistance to conventional insecticides is widespread, this situation could also occur to *B. thuringiensis* which is quite threatening to the future of insect pest control (Gould 1988; Raffa 1989). However, *B. thuringiensis* has been used to control pests for over 20 years and there has been lack of reports of resistance to it. Scientists have made presumptions that resistance was unlikely, perhaps because of its unique mode of action (Bowman 1981; Briese 1981; Wilcox et al. 1986; Wilding 1986; De Barjac 1987). Several studies have been conducted and showed that laboratory selection enhanced resistance to *B. thuringiensis* in the Indian meal moth, *Plodia interpunctella* (Hübner) (McGaughey 1985; McGaughey and Beeman 1988) and the tobacco budworm, *Heliothis virescens* (F.) (Stone et al. 1989).

According to laboratory selection which shows that many pests maintain genetic variation in susceptibility to *Bt* toxins and therefore have the capability to develop resistance to *Bt* crops in the field, (Gould 1998; Tabashnik et al. 2003) *Bt* genes have been genetically engineered and inserted in many crops such as potato, tomato, corn, cotton, and sugar cane. The major pests targeted by *Bt* crops have been studied for the development of resistance which is a heritable decrease in a population's susceptibility to a toxin (Tabashnik et al. 2003).

There were studies in many countries, for example, China, Spain, Australia, and the United States, monitoring the resistance to *Bt* crops in field populations of six major insect pests which are *Helicoverpa armigera*, *H. zea*, *H. virescens*, *Ostrinia nubilalis*, *Pectinophora gossypiella* and *Sesamia monogrioides* (Tabashnik et al. 2008). Tabashnik et al. (2008) demonstrated that the resistant alleles have increased substantially in some field populations of *H. zea*, but not in the other five species. However, the absence of field resistance to *Bt* crops nowadays is presumed to depend on some factors such as large fitness costs or other disadvantages suffered by resistant individuals, an initial low frequency of resistant alleles, a dilution of resistant alleles with susceptible individuals from refuges (non-*Bt* plants) and a high dose of toxin released by plants (Bates et al. 2005).

4 Insect Resistance on Plant-Derived Pesticides or Plant-Derived Essential Oils (Botanicals)

Plant essential oils are highly volatile and usually obtained from nonwoody parts of the plant, particularly foliage (exceptionally sandalwood, agarwood, and cedarwood), through physical processes such as steam or hydrodistillation (Batish et al. 2008). Most plant essential oils are a complex mixture of mainly terpenoids, especially monoterpenes and sesquiterpenes, and a variety of aromatic phenols, oxides, ethers, alcohols, esters, aldehydes, and ketones that determine the characteristic aroma and odor of the plant (Batish et al. 2008). The plant-derived products might also be classified as repellents, feeding deterrents, toxicants, grain protectants, reproduction inhibitors, and insect growth inhibitors (Talukder 2006). Many extracts of plants have been evaluated for their activity against agriculturally important insects for over decades (Chiu 1989; Akhtar and Isaman 2004; Leatemia and Isman 2004). Moreover, the plant extracts are currently being evaluated further for use in integrated pest management because of their eco-friendly characteristics (Koul and Walia 2009). Many studies have shown strong evidences of the efficacy of chemical compounds extracted from plants in control insects (National Research Council 1986). For example, contact and residual toxicity of more than 30 plant extracts were experimented on larvae of the Colorado beetle (Gokce et al. 2006). The results were that certain plant extracts were toxic to the beetle larvae and may have potential for controlling this destructive pest under field condition (Gokce et al. 2007). Some plant extracts are able to inhibit oblique banded leaf roller insects from ovipositing (Gokce et al. 2005). Some plant extracts are found to be toxic to aphids (Das et al. 2008) and lepidopterans (Akhtar et al. 2008). Leaf and flower extracts of *Citrus sinensis*, *Ocimum canum*, *Ocimum sanctum*, and *Rhinacanthus nasutus* have been demonstrated to have antifeedant and larvicidal activities and were thought to have high potential as an ideal eco-friendly approach for the control of the agricultural pests (Kamaraj et al. 2008). There are also several essential oils which have been known for pest control properties, for example, lemongrass (*Cymbopogon winterianus*), eucalyptus (*Eucalyptus globules*), rosemary (*Rosmarinus officinalis*), vetiver (*Vetiveria zizanioides*), clove (*Eugenia caryophyllus*), and thyme (*Thymus vulgaris*) (Koul and Walia 2009).

Among the most promising of the plant extracts investigated to date are those derived from the Indian neem tree, *Azadirachta indica* (Meliaceae) (Feng and Isman 1995). The key active ingredient found in neem-based products is azadirachtin which is a naturally occurring substance that classifies to a group of an organic molecule class known as tetranortriterpenoids. The azadirachtin has an antifeedant, ecdysis disrupting, and reproductive inhibiting properties against more than 200 species of insects (Warthen 1989; Govindachari 1992; Champagne et al. 1992; Hassen et al. 1994; Schumutterer 1995). Moreover, the obvious advantages of azadirachtin, for example, extremely low mammalian toxicity, selective activity against pest insects, and systemic action in a variety of important crop plants, make the azadirachtin an attractive biopesticide (Salako 2002; Khalid and Shad 2002). As we know, many individual chemical compounds extracted from plants have high potential properties for the control of pests and insects. However, many studies demonstrated that the toxicity of these compounds can be enhanced in mixtures. Therefore, the efficacy of the mixture is higher than it would be expected by adding up the activities of its individual constituents (Koul and Walia 2009). When this enhanced effect of the mixture occurs, it is called synergism. The synergism has been demonstrated for mixtures of limonoids (Koul et al. 2004a, b) or essential oil constituents (Hummelbruner and Isman 2001; Singh et al. 2008, 2009).

Because the botanical products have broad or non-specific modes of action, meaning these compounds are active on multiple target sites of insect body, the active constituents should attack a variety of enzymatic or metabolic systems. According to this nature and diversity of the modes of actions, insects and mites would need to encounter extensive physical, biochemical, physiological, and genetic adaption to develop resistance to these plant-derived pesticides. However, it does not mean that resistance to these botanicals will never occur. For example, among plant extracts, feeding deterrents make the major category of extracts or allelochemicals that have been studied for the efficacy against many pests (Koul and Walia 2009) and thus, if used indiscriminately, may also result in the development of resistance (Koul and Walia 2009). This has been demonstrated in the study of selection of resistance to azadirachtin in the green peach aphid, *Myzus persicae* (Feng and Isman 1995). In this study, they experimented two lines of *Myzus persicae* of the same origin repeatedly treated with pure azadirachtin (aza) or a refined neem seed extract (NSE) at the same concentration of aza. After 40 generations, the aza-selected line had developed ninefold resistance to aza compared to a nonselected control line, whereas the NSE-selected line did not. The researchers concluded that a blend of active constituents in a botanical insecticide such as neem might diffuse the selection process, reducing the development of resistance compared to that expected with a single active ingredient. Another constraint of applying botanical insecticides is the potential for fast desensitization to a feeding deterrent. Individual insects primarily deterred by a feeding inhibitor later on become increasingly tolerant due to repeated or continuous exposure (Koul and Walia 2009), for example, the use of azadirachtin and toosendanin against tobacco cutworms (Bomford and Isman 1996).

5 Resistance Management

Insect resistance to insecticides either conventional synthetic chemical or bio-insecticide can be managed by slowing selection pressure by these insecticides on the insect population (Resistance Management for Sustainable Agriculture and Improved Public Health 2012). The situation when all susceptible individuals are totally eliminated by a given chemical but the most resistant ones still remain alive in the population leads to the increase in the selection for resistant genes. The way to prevent resistance can be achieved by avoiding unnecessary insecticide applications by using nonchemical control techniques and leaving untreated refuges where susceptible insects can survive (Boerboom 2001; Onstad 2008).

5.1 Resistance Management for Plant: Incorporated Protectants

Plant-incorporated protectants such as *Bt* crops have been found in laboratory studies that insects exposed to high doses of *Bt* over many generations have developed resistance (Tabashnik 1994; Gould 1998; Shelton et al. 2002; Tabashnik et al. 2003). Many of these induced resistances are recessive, few are additive or recessive and none so far has been dominant. Consequently, deployment management strategies have concentrated on ensuring that sufficient populations of susceptible insects are present to mate with possible resistant ones, making certain that the frequency of the resistant allele is not fixed in the population (Krattiger 1996). In order to make the prevention of resistance development to *Bt* crops applicable, there are several strategies which have been implemented those are refugia, high-dose and low-dose approaches, multiple genes deployment and targeted expression (Krattiger 1996).

5.1.1 Refugia

The approach for this strategy is to reduce the chances of an increase in the frequency of resistance genes by inhibiting resistant insects from mating with other resistant insects; thus, the creation of a resistant population could hardly occur.

This strategy has been established to reduce the long-term impact by preventing resistant insects from mating with other resistant population. This can be done by ensuring that there are always plenty of susceptible insects close by for the few resistant individuals to mate with. The methods of creating the refuge areas can be done by planting the *Bt* crops in only part of the field and for another part where is as close as possible should be maintained as an unimproved, conventionally treated area and totally untreated area. By doing this, it is hoped that this will prevent the mating of resistant insects among themselves which will consequently lead to the establishment of the resistance gene in the population (Mallet and Porter 1992; Krattiger 1996; Cullen et al. 2008).

5.1.2 High-Dose and Low-Dose Approaches

The high level of toxin is to deploy with the aim that it will take quite a long period of time for insects to overcome the toxin. As we know that most resistant individuals are recessive alleles, and most resistance carriers are heterozygous, the level of toxin will be made sure that it is high enough to kill all heterozygous resistance individuals (Roush 1997, 1998). In the field situation, the initial occurrence of individuals homozygous for resistance is very rare that it can be ignored; thus, the rate of resistance development is driven primarily by the frequency and survival of heterozygote (Roush 1997). High dose will ensure that crop damage should be maintained below an economic threshold. However, in order to make this approach applicable, many assumptions must be met (Gould 1998). These assumptions for the efficacy of high-dose strategy are as follows: (1) the initial allele frequency for resistance is low, (2) inheritance is recessive alleles, and (3) resistant and susceptible insects must mate randomly. However, there were reports that one or more of these assumptions may be violated with some insect species. For instance, the assortative mating occurred in pink bollworms according to the disparity in development time between resistant and susceptible individuals (Liu et al. 2001). There were incomplete or non-recessive inheritance to many *Bt* crops investigated in European corn borer (*Ostrinia nubilalis*), tobacco budworm (*H. virescens*), *H. armigera, H. zea, Plutella xylostella and Leptinotarsa decemlineata* (Frutos et al. 1999; Akhurst et al. 2003; Burd et al. 2003), even though we have learned that insect resistance to *Bt* in high-expressing transgenic plants has proven functionally recessive alleles (Tabashnik et al. 2003; Gould 1998).

For the strategy of low levels of toxin, this approach was implemented in order to make insects vulnerable to predators and parasites. However, this strategy has been discarded by companies because a considerable level of damage would still be inflicted on the crop which would not be acceptable from a commercial point of view (Krattiger 1996).

5.1.3 Gene Pyramiding

This method of control resistance development has been well known as a lasting *Bt* resistance management strategy (Shelton et al. 2002; Jackson et al. 2003). It requires more than one resistance gene with different modes of action for the range of insect species that were not being adequately controlled by a single toxin (Manyangarirwa et al. 2006). However, there are three assumptions required for this strategy in order to accomplish the effective resistance control (Manyangarirwa et al. 2006). The first assumption is that insects resistant to only one toxin can be effectively controlled by a second toxin produced in the same plant. The second assumption is that strains resistant to two toxins with independent actions cannot emerge through selection pressure with one toxin alone. Karim et al. (2000) demonstrated that the use of multiple toxins to delay development of resistance is based on the theory that if homozygous insects for one resistance gene are rare, the homozygous individuals for multiple resistance genes are even more extremely rare. By using multiple crystal proteins (e.g. Cry1 Ac, Cry 2Ac), not only insects homozygous for one or two resistance genes but also the heterozygous individuals for another resistance gene would still be overcome by crops with multiple *Bt* toxins (Schnepf et al. 1998; Sisterson et al. 2004). The third assumption is that a single gene will not confer resistance to two toxins that are immunologically distinct and have different binding target sites (Gaham et al. 2005).

5.1.4 Targeted Expression

This approach is done by ensuring that the toxin gene is expressed only specifically in a certain vulnerable part of the plant (e.g. stem) or is expressed both in a certain part of the plant and at a particularly critical time in the development of the plant (e.g. flowering) (Gould 1995). This method of resistance management will allow some susceptible insects to breed normally, therefore increasing their predators and parasitic populations, at the same time be prevented from causing damage in the critical plant parts or vulnerable life cycles (Krattiger 1996).

5.2 Resistance Management for Botanical Insecticides

As we know, most plant-derived insecticides (botanical origin) are extracts containing a group of active ingredients of many varieties of chemical nature. Plant extract insecticides usually have short-residual lives which hardly accumulate in the environment (Khater 2012). This property of plant insecticides could be considered an advantage in the way that there will be a very low probability that two extracts would always be exactly identical. Therefore, the selective pressure on the insects will not always become the same (Koul and Walia 2009). Even if all the same compounds are found in the extract, concentrations almost always will be different. In general, insects take longer time to develop resistance to a mixture or combination of natural active compounds than to any one individual allelochemical (Koul and Walia 2009). The common problem of using botanical insecticide is that the potential for rapid desensitization to these insecticides. For example, individual insects initially deterred by feeding inhibitor become increasingly tolerant upon repeated or continuous exposure (Koul and Walia 2009). Bomford and Isman (1996) demonstrated that tobacco cutworms became habituated and cross-habituated when the use of azadirachtin and toosendanin is being applied. The way to solve this operational problem can be mitigated by using mixtures of several compounds in a multicomponent strategy such as non-azadirachtin types of compounds or a combination of xanthotoxin and thymol to prevent individuals to develop tolerant (Koul et al. 2004a). Mixtures of several compounds will also ensure that the formulations have a variety of toxic, growth inhibitory and antifeedant effects. Such complexes are desirable in that the target spectrum is widened, because different species respond differently to individual compounds. These mixtures are also likely to reduce the potential for the development of genetic resistance or development of behavioural desensitization (Koul et al. 2004a).

6 Future Trend

The use of products from plant origins permits us to develop and exploit naturally occurring plant defence mechanisms, thus reducing the use of conventional synthetic pesticides. For the plant-incorporated protectants such as *Bt* crops, certain integrated pest management (IPM) has been exercised over decades and made the resistance management strategies effective thus far. However, several factors appear to limit the success of botanical pesticides such as regulatory barriers and the ability to compete synthetic pesticides. The efficacy of botanical pesticides is short when compared with synthetic pesticides even though there are specific pest contexts where the control is equivalent to that with conventional products has been observed (Koul and Walia 2009). Botanical pesticides currently do not play major role in crop protection. As a result, increasingly stringent regulatory requirements have prevented many botanical products from reaching the market place in North America and Europe in the last two decades (Isman 2006). The practice of using botanical pesticides need to be considered in the areas of organizing the natural sources, developing quality control, adopting standardization strategies, and modifying regulatory constraints (Koul and Walia 2009) (Figs. 4.1 and 4.2).

7 Conclusion

The use of plant-derived biopesticides has become a common practice in many crop protection programmes. These biopesticides are effective tools in integrated pest management (IPM) for delay resistance development to synthetic chemical pesticides and decrease environmental exposure to conventional synthetic pesticides. Moreover, the botanical and plant-induced protectant (PIP) biopesticides have been rarely found to develop resistance in the field compared to chemical synthetic pesticides. Since biopesticides deploy multiple modes of actions (MOAs)

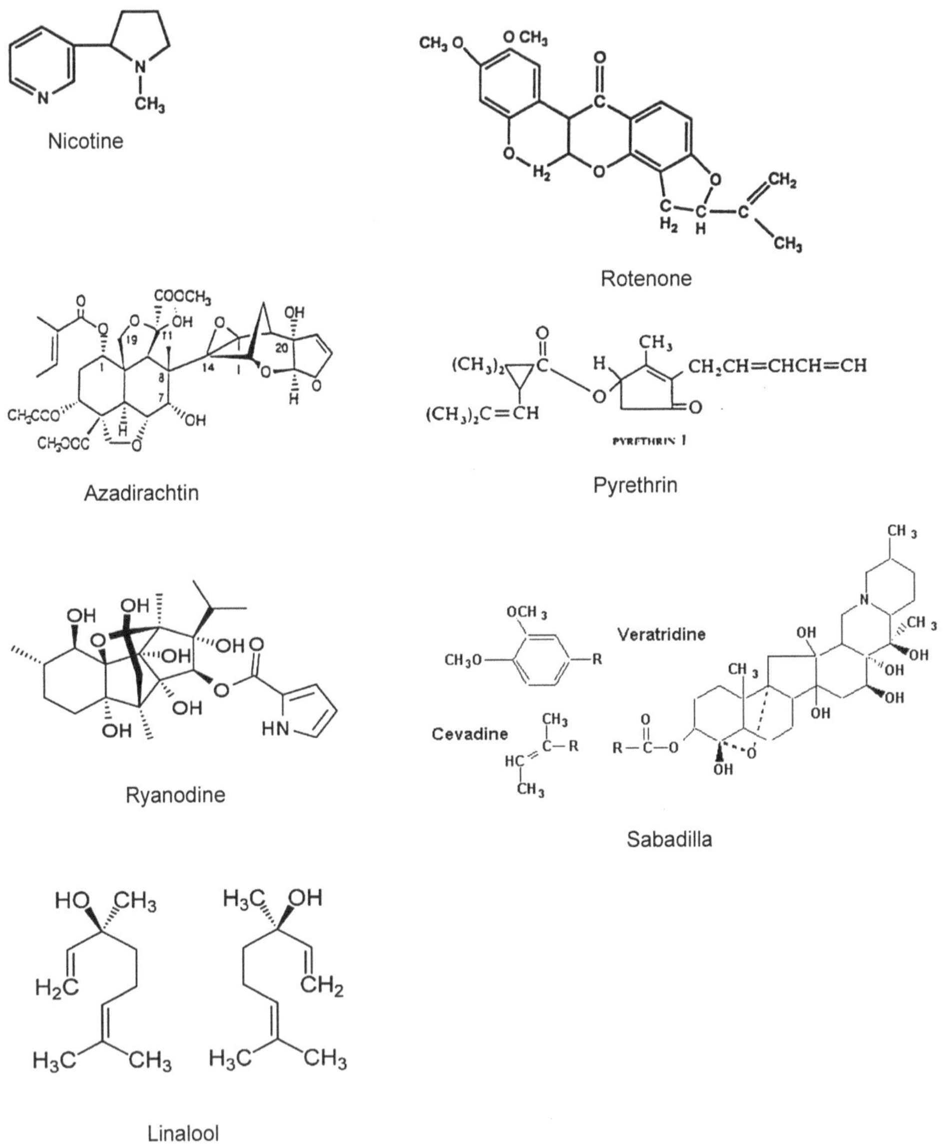

Nicotine

Rotenone

Azadirachtin

Pyrethrin

Ryanodine

Sabadilla

Linalool

Fig. 4.1 Chemical structure of few major botanical pesticides

to suppress pests, development of resistance to these multiple factors by target insects is extremely unlikely. However, plant-derived biopesticides should meet the same criteria as conventional synthetic chemical pesticides. This means they should be selective to the target insects and must have adequate residual action to protect the crop through its window of vulnerability to the key pests (Isman 2002). That these biopesticides are naturally derived does not mean they are safe to humans. In general, their essential oils and major constituents are relatively nontoxic to mammals, with acute oral LD50 values in rodents ranging from 800 to 3,000 mg kg^{-1} for pure compounds and >5,000 mg kg^{-1} for formulated products (Khater 2012). Despite their safety, some essential oils can be irritating to the skin (Barnard and Xue 2004). Although the short-residual activity is considered a benefit for the environment, this means multiple applications are required. Because of that, the selection pressure on insect populations will be increased and possibly lead

Fig. 4.2 Plants containing the pesticide chemical compounds

Table 4.1 A list of botanical chemicals exhibited resistance mechanism in insects

Chemical trade name	Chemical common name	Scientific name of plants containing the chemical compounds	Reference demonstrated evidence of resistance in insect
Align, Azatin, Turplex, Neemix, Neemolin	Azadirachtin	*Azadirachta indica*	Feng and Isman (1995) Bomford and Isman (1996)
Black Leaf 40, Tender leaf plant insect spray	Nicotine	*Nicotiana rustica*	Morris and Harrion (1984) Murray (1996)
Buhach, Chrysanthemum Cinerariaefolium, Ofirmotox, Insect Powder, Dalmatian Insect Flowers, Firmotox, Parexan and NA 9184	Pyrethrin	*Chrysanthemum cinerariaefolium*	Lloyd and Parkin (1963)
Chem-Fish, Cuberol, Fish Nox, Noxfire, Rotacide, Sinid Tox-R, Curex Flea Duster, Derrin, Cenol Garden Dust, Chem-Mite, Cibe Extract, Green Cross Warble Powder	Rotenone	*Derris*, *Lonchocarpus*, *Tephrosia* species	Mitsuhashi et al. (1970)
Ryanodine, Ryanicide	Ryania	*Ryania speciosa*	No report of resistance in insects
Sabadilla Dust®, Sabadilla Pest Control®	Sabadilla	*Schoenocaulon officinalis*	Humeres and Morse (2006)
Linalool 925	Linalool	Lamiaceae plant and herb family, which including mints and other scented herbs	Davoudi et al. (2011)

to the development of resistance. It is often said that when resistance occurs, this could lead to the failure to control insect population because this is generally directed toward synthetic chemical insecticides. As we know, growers have been facing problems of insect pests developing resistance to conventional pesticides and that biopesticides are the alternative methods believed to prevent insects from developing resistance. However, this is not always concordant with facts. Insects develop whatever means are necessary to survive and evolve in order to sustain existing populations no matter what the synthetic chemicals or biopesticides are. The continual use of the same types of any pesticides, either synthetic or alternative (plant-derived pesticides) with similar modes of action will probably lead to the development of resistant individuals in insect populations. Therefore, regardless of whether a pesticide is considered a conventional pesticide or a biopesticide, insect populations have the tendency to develop resistance due to selection pressure. This depends on how often the growers apply the pesticides and the amount or concentration of dosage used. In order to make botanical insecticides well commercialized and successfully competitive on a meaningful scale in the near future, the substantial effort is very much needed for organizing the natural sources, developing quality control, adopting standardization strategies, and modifying regulatory constraints (Table 4.1).

References

Ahmad S, Brattsten LB, Mullin CA, Yu SJ (1986) Enzymes involved in the metabolism of plant allelochemicals. In: Brattsten LB, Ah-mad S (eds) Molecular aspects of insect plant associations. Plenum Press, New York

Akhtar Y, Isaman MB (2004) Comparative growth inhibitory and antifeedant effects of plant extracts and pure allelochemicals on four phytophagous insect species. Appl Entomol 128:32–38

Akhtar Y, Yeoung YR, Isman MB (2008) Comparative bioactivity of selected extracts from Meliaceae and some commercial botanical insecticides against two noctuid caterpillars, *Trichoplusia ni* and *Pseudaletia unipluncta*. Phytochem Rev 9:77–88

Akhurst RJ, James JW, Bird L, Beard C (2003) Resistance to the Cry1Ac δ endotoxin of *Bacillus thuringiensis* in the cotton bollworm, *Helicoverpa armigera* (Lepidoptera: Noctuidae). J Econ Entomol 96:1290–1299

Barnard DR, Xue RD (2004) Laboratory evaluation of mosquito repel lents against *Aedes albopictus*, *Culex nigripalpus*, and *Ochlerotatus triseriatus* (Diptera: Culicidae). J Med Entomol 41(4):726–730

Bates SL, Zhao JZ, Roush RT, Shelton AM (2005) Insect resistance management in GM crops: past, present and future. Nat BioTechnol 23:57–62

Batish DR, Singh HP, Kohli RK, Kaur S (2008) Eucalyptus essential oil as a natural pesticide. For Ecol Manag 256:2166–2174

Berenbaum MR (1986) Target site insensitivity in insect-plant interaction. In: Brattsten LB, Ahmad S (eds) Molecular aspects of insect plant interaction, 1st edn. Plenum Press, New York

Boerboom C (2001) Glyphosate resistant weeds. http://128.104.239.6/uw_weeds/extension/articles/glyphres.htm. Accessed 5 Feb 2013

Bomford MK, Isman MB (1996) Desensitization of fifth instar Spodoptera litura to azadirachtin and neem. Entomol Exp Appl 81:301–313

Bonning BC (1990) Insensitive acetylcholinesterase in insecticide resistant mosquitoes. In: Cassida JE (ed) Pesticides and alternatives: innovative chemical and biological approach to pest control, 1st edn. Elsevier, New York

Bowman HG (1981) Insect responses to microbial infections. In: Burges HD (ed) Microbial control of pests and plant diseases 1970–1980, 1st edn. Academic, New York

Briese DT (1981) Resistance of insect species to microbial pathogens. In: Davidson W (ed) Pathogenesis of invertebrate microbial diseases. Allanheld Osmun, Totowa

Brown WM (1985) The mitochondrial genome of animals. In: MacIntyre RJ (ed) Molecular evolutionary genetics, 1st edn. Plenum, New York

Burd AD, Gould F, Bradley JR, Van Duyn JW, Moar WJ (2003) Estimated frequency of nonrecessive *Bt* resistance genes in bollworm, *Helicoverpa zea* (Boddie) (Lepidoptera: Noctuidae) in eastern North Carolina. J Econ Entomol 96:137–142

Caroll CR, Hoffman CA (1980) Chemical feeding deterrent mobilized in response to insect herbivory and counter adaptation by *Epilachna tredecimnotata*. Science 209:414–416

Champagne DE, Koul O, Isman MB, Scudder GGE, Towers GHN (1992) Biological activity of limonoids from the Rutales. Phytochemstry 31:377–394

Chiu SF (1989) Recent advances in research on botanical insecticides in China. In: Arnason JT, Philogene BJR, Morand P (eds) Insecticides of plant origin, 1st edn. ACS Symposium Series, No. 387, Washington, DC, USA

Cloyd RA (2012) Biopesticides: are they immune to resistance? http://www.greenhousegrower.com. Accessed 19 Jan 2013

Commins HN (1977) The development of insecticide resistance in the presence of migration. J Theor Biol 65:399–420

Compton SG (1987) *Aganais speciosa* and *Danaus chrysippus* (Lepidoptera) sabotage the latex defences of their host plants. Ecol Entomol 12:115–118

Cullen E, Prost R, Volenberg D (2008) Insect resistance management and refuge requirements for *Bt* corn. http://corn.agronomy.wisc.edu/Management/pdfs/A3857.pdf. Accessed 19 Jan 2013

Curtis CF (1987) Genetic aspects of selection for resistance. In: Ford MG, Khambay BPS, Sawicki RM (eds) Combating resistance to xenobiotics, 1st edn. Ellis Horwood, Chichester

Das B, Sarker P, Rahman MD (2008) Aphicidal activity of some in digenous plant extracts against bean aphid Aphis craccivora Koch (Homoptera: Aphididae). J Pest Sci 81:153–159

Davoudi A, Shayesteh N, Shirdel D, Hosseinzadeh A (2011) Effect of diethyl maleate on toxicity of linalool against two stored product in sects in laboratory condition. Afr J Biotechnol 10(48):9918–9921

De Barjac H (1987) Operational bacteria insecticides and their potential for future improvement. In: Aramorosch K (ed) Biotechnology in vertebrate pathology and cell culture, 1st edn. Academic, San Diego

Dowd PM, Gagne CC, Sparks TC (1987) Enhanced pyrethroid hydrolysis in pyrethroid resistant larvae of tobacco budworm, *Heliothis virescens* (F.). Pestic Biochem Physiol 28:9–16

Edwards PE, Wanjura WJ (1989) Eucalypt-feeding insects bite off more than they can chew: sabotage of induced defenses? Oikos 54:246–248

Farnham AW (1973) Genetic of resistance of pyrethroid-selected house flies, *Musca domestica* L. Pestic Sci 4(4):513–520

Feng R, Isman MB (1995) Selection for resistance to azadirachtin in the green peach aphid, *Myzus persicae*. Experientia 51:831–833

Flexner JL, Lighthard B, Croft BA (1986) The effects of microbial pesticides on non-target, beneficial arthropods. Agric Ecosyst Environ 3:515–518

Frutos R, Rang C, Royer M (1999) Managing resistance to plants producing *Bacillus thuringiensis* toxins. Crit Rev Biotechnol 19:227–276

Gaham LJ, Ma YT, Cobble MLM, Gould F, Moar WJ, Heckel DG (2005) Genetic basis of resistance to Cry 1Ac and Cry 2Aa in *Heliothis virescens* (Lepidoptera: Noctuidae). J Econ Entomol 98:1357–1368

Gokce A, Stelinski LL, Whalon ME (2005) Behavioral and electrophysiological responses of leafroller moths to selected plant extracts. Environ Entomol 34:1426–1432

Gokce A, Whalon ME, Cam HT, Yanar Y, Demirtas IM, Goren N (2006) Plant extract contact toxicities to various developmental stages of Colorado potato bee-

tles (Coleoptera: Chyrsomelidae). Ann Appl Biol 148:197–202
Gokce A, Whalon ME, Cam HT, Yanar Y, Demirtas IM, Goren N (2007) Contact and residual toxicities of 30 plant extracts to Colorado potato beetle larvae. Arch Phyopathol Plant Prot 40:441–450
Gould F (1988) Evolutionary biology and genetically engineered crops. BioScience 38(1):26–33
Gould F (1995) Strategies for developing and deploying pesticidal cotton, rice and corn/maize for use in poor and developing countries. Presented at the Bellagio meeting on *Bt* resistance, October 1995 (unpublished)
Gould F (1998) Sustainability of transgenic insecticidal cultivars: integrating pest genetics and ecology. Annu Rev Entomol 43:701–726
Govindachari TR (1992) Chemical and biological investigations on *Azadirachta indica* (neem tree). Curr Sci 63(3):117–122
Hassen DJ, Cuomo J, Khan M, Gallagher RT, Ellenberger WP (1994) Advances in neem and azadirachtin chemistry and bioactivity. In: Hedin J, Menn J, Hollingworth RM (eds) Natural and engineered pest management agents. ACS symposium series 551. American Chemical Society, Washington, DC
Holloway GJ, McCaffery AR (1988) Reactive and preventative strategies for the management of insecticide resistance. Brighton Crop Prot Conf 1988: 465–470
Humeres EC, Morse JG (2006) Resistance of avocado thrips (Thysanoptera: Thripidae) to sabadilla, a botanically derived bait. Pest Manag Sci 62(9):886–889
Hummelbruner LA, Isman MB (2001) Acute, sublethal, antifeedant and synergistic effects of monoterpenoid essential oil compounds on the tobacco cutworm, *Spodoptera litura* (Lepidoptera: Noctuidae). J Agric Food Chem 49:715–720
Isman MB (2002) Insect antifeedants. Pesticide Outlook, pp 152–157
Isman MB (2006) Botanical insecticides, deterrents, and repellents in modern agriculture and an increasingly regulated world. Annu Rev Entomol 51:45–66
Jackson RE, Bradley JR, Van Duyn JW (2003) Field performance of transgenic cottons expressing one or two Bacillus thuringiensis endotoxins against bollworm, *Helicoverpa zea* (Boddie). J Cotton Sci 7:57–64
Kamaraj C, Rahuman A, Bagavan A (2008) Antifeedant and larvicidal effects of plant extracts against *Spodoptera litura* (F.), *Aedes aegypti* L. and *Culex quinquefasciatus* Say. Parasitol Res 103:325–331
Karim S, Riazuddin S, Gould F, Dean DH (2000) Determination of receptor binding properties of *Bacillus thuringiensis* endotoxins to cotton bollworm (*Helicoverpa zea*) and pink bollworm (*Pectinophora gossypiella*) midgut brush border membrane vesicles. Pest Biochem Physiol 67:198–216
Khalid S, Shad RA (2002) Potential advantage of recent allelochemical discoveries and agro ecosystem. Prog Farm 11:30–35
Khater HF (2012) Prospects of botanical biopesticides in insect pest management. Pharmacology 3:641–655
Kilpatrick JW, Quarternam KD (1952) Field studies on the resting habits of flies in relation to chemical control part II in rural areas. Am J Trop Med Hyg 1:1026–1031
Koul O, Walia S (2009) Comparing impacts of plant extracts and pure allelochemicals and implications for pest control. http://www.cabi.org/cabreviews. Accessed 19 Jan 2013
Koul O, Singh G, Singh R, Singh J, Daniewski WM, Berlozecki S (2004a) Bioefficacy and mode of action of some limonoids of salannin group from *Azadirachta indica* A. Juss and their role in a multicomponent system against lepidopteran larvae. J Biosci 29:409–416
Koul O, Kaur H, Goomber S, Wahab S (2004b) Bioefficacy of rocaglamide from *Aglaia elaeagnoidea* (syn. *A. roxburghiana*) against gram pod borer, Helicoverpa armigera (Hubner). J Appl Entomol 128:177–181
Krattiger AF (1996) Insect resistance in crops: a case study of *Bacillus thuringiensis* (*Bt*) and its transfer to developing countries. http://www.isaaa.org/kc/Publications/pdfs/isaaabriefs/Briefs%202.pdf. Accessed 19 Jan 2013
Krieger RI, Feeny PP, Wilenson CF (1971) Detoxification enzymes in the guts of caterpillars: an evolutionary answer to plant defences. Science 172:579–581
Leatemia JA, Isman MB (2004) Insecticidal activity of crude seed extracts of *Annona* spp., *Lansium domesticum* and *Sandoricum koetjape* against Lepidopteran larvae. Phytoparasitica 32(1):30–37
Leatherman M (1997) Resistance. http://www.udel.edu/chem/C465/senior/fall97/insecticide/resist.html. Accessed 21 Jan 2013
Little EJ, McCaffery AR, Walker CH, Parker T (1989) Evidence for an enhanced metabolism of cypermethrin by a monooxygenase in a pyrethroid resistant strain of the tobacco budworm (*Heliothis virescens*). Pestic Biochem Physiol 34:58–68
Liu YB, Tabashnik BE, Dennehy TJ, Patin AL, Sims MA, Meyer SK, Carrière Y (2001) Effects of Bt cotton and Cry1Ac toxin on survival and development of pink bollworm (Lepidoptera: Gelechiidae). J Econ Entomol 94:1237–1242
Lloyd CJ, Parkin EA (1963) Further studies on a pyrethrum resistant strain of the granary weevil, *Sitophilus granarius* (L.). J Sci Food Agric 14(9):655–663
Maestre RH (2012) What causes insecticide resistance? http://www.magicexterminating.com/. Accessed 28 Dec 2012
Maier PP, Baker WC, Bogue MD, Kilparick JW, Quarterman KD (1952) Field studies on the resting habits of flies in relation to chemical control part I in urban areas. Am J Trop Med Hyg 1:1020–1025
Mallet J (1989) The evolution of insecticides resistance: have the insects won? Tree 4(11):336–340
Mallet J, Porter P (1992) Preventing insect adaptation to insect resistant crops: are seed mixtures or refugia the best strategy? Proc R Soc Lond B 250:165–169
Manyangarirwa W, Turnbull M, McCutcheon GS, Smith JP (2006) Gene pyramiding as a *Bt* resistance management strategy: how sustainable is this strategy? Afr J Biotechnol 5(10):781–785

May RM, Dobson AP (1986) Population dynamics and the rate of evolution of pesticide resistance. In: National Research Council (ed) Pesticide resistance: strategies and tactics for management, 1st edn. National Academy Press, Washington, DC

McGaughey WH (1985) Insect resistance to the biological insecticide *Bacillus thuringiensis*. Science 229:193–195

McGaughey WH, Beeman RW (1988) Resistance to *Bacillus thuringiensis* in colonies of Indian meal moth (Lepidoptera: Pyralidae). J Econ Entomol 81:28–33

Miller TA (1988) Mechanisms of resistance to pyrethroid insecticide. Parasitol Today 4(7):S8–S12

Mitsuhashi J, Grace TDC, Waterhouse DF (1970) Studies on the effect of rotenone on the growth of insect cells cultivated in vitro. Entomol Exp Appl 13(4):467–473

Moore LV, Scudder GGE (1986) Ouabain resistant Na, K-ATPases and cardenolide tolerance in the large milkweed bug, *Oncopeltus fasciatus*. J Insect Physiol 32:27–33

Morris CE, Harrion JB (1984) Central nervous system features of a nicotine resistant insect, the tobacco hornworm *Manduca sexta*. Tissue Cell 16(4):601–612

Murray CL (1996) A P-glycoprotein-like mechanism in the nicotine resistant insect, *Manduca sexta*. PhD dissertation, Ottawa-Carleton Institute of Biology, Ottawa

Mutero A, Pralavorio M, Bride JM, Fournier D (1994) Resistance-associated point mutations in insecticide insensitive acetylcholinesterase. Proc Natl Acad Sci U S A 91:5922–5926

National Research Council (1986) Pesticide resistance: strategies and tactics for management. National Academy of Sciences, Washington, DC

Newcomb RD, Campbell PM, Ollis DL, Cheah E, Russell RJ, Oakeshott JG (1997) A single amino acid substitution converts a carboxylesterase to an organophosphorus hydrolase and confers insecticide resistance on a blowfly. Proc Natl Acad Sci U S A 94:7464–7468

Omer AD, Leigh TF, Granett J (1992) Insecticide resistance in field populations of greenhouse whitefly (Homoptera: Aleyrodidae) in the San Joaquin Valley (California) cotton cropping system. J Econ Entomol 85:21–27

Onstad DW (2008) Insect resistance management. Elsevier, Amsterdam

Plapp FW (1984) The genetic basis of insecticide resistance in the housefly: evidence that a single locus plays a major role in metabolic resistance to insecticides. Pestic Biochem Physiol 22:194–201

Polumbo JC, Horowitz AR, Prabhaker N (2001) Insecticidal control and resistance management for *Bemisia tabaci*. Crop Prot 20:739–765

Raffa KF (1989) Genetic engineering of trees to enhance resistance to insects evaluating the risk of biotype evolution and secondary pest outbreak. Bioscience 39(8):524–534

Resistance Management for Sustainable Agriculture and Improved Public Health (2012) http://www.irac-online. Accessed 22 Dec 2012

Roush RT (1997) *Bt*-transgenic crops: just another pretty insecticide or a chance for a new start in resistance management? Pestic Sci 51:328–334

Roush RT (1998) Two-toxin strategies for management of insecticidal transgenic crops: can pyramiding succeed where pesticide mixtures have not? Philos Trans R Soc Lond B 353:1777–1786

Salako EA (2002) Plant protection for the resource-poor farmers. A key note address at Nigerian Society for Plant Protection. In: 30th annual conference, UNAAB, Abeokuta 1–4 Sept 2002

Schnepf E, Crickmore N, Van Rie J, Lereclus D, Baum J, Feitelson J, Zeigler DR, Dean DH (1998) *Bacillus thuringiensis* and its pesticidal crystal proteins. Microb Mol Biol Rev 62:775–806

Schumutterer H (1995) The neem tree. VCH, Weinheim

Shelton AM, Zhao JZ, Roush RT (2002) Economic, ecological, food safety, and social consequences of the deployment of *Bt* transgenic plants. Annu Rev Entomol 47:845–881

Singh R, Rup PJ, Koul O (2008) Bioefficacy of Eucalyptus camaldulensis var. *obtusa* and *Luvunga scandens* essential oils against *Spodoptera litura* (Lepidoptera: Nocutidae). Biopest Int 4:128–137

Singh R, Koul O, Rup PJ, Jindal J (2009) Toxicity of some essential oil constituents and their binary mixtures against *Chilo partellus* Swinhoe (Lepidoptera: Pyralidae). Int J Trop Insect Sci 29:93–101

Sisterson MS, Antilla L, Carriere Y, Ellers-Kirk C, Tabashnik BE (2004) Effects of insect population size on evolution of resistance to transgenic crops. J Econ Entomol 97:1413–1424

Stone TB, Sims SR, Marrone PG (1989) Selection of tobacco bud worm for resistance to a genetically engineered *Pseudomonas fluorescens* containing the delta-endotoxin of *Bacillus thuringiensis* subsp. *kurstaki*. J Invert Pathol 53:228–234

Tabashnik BE (1994) Evolution of resistance to *Bacillus thuringiensis*. Annu Rev Entomol 39:47–79

Tabashnik BE, Carriere Y, Dennehy TJ, Morin S, Sisterson MS, Roush RT, Shelton M, Zhao JZ (2003) Insect resistance to transgenic *Bt* crops: lessons from the laboratory and filed. J Econ Entomol 96:1031–1038

Tabashnik BE, Gassmann AJ, Crowder DW, Carriere Y (2008) Insect resistance to *Bt* crops: evidence versus theory. Nat Biotechnol 26:199–202

Talukder FA (2006) Plant products as potential stored-product insect management agents: a mini review. Emir J Agric Sci 18(1):17–32

Tsukamoto M (1983) Methods of genetic analysis of insecticide resistance. In: Georghiou GP, Saito T (eds) Pest resistance to pesticides, 1st edn. Plenum, New York

U.S. Environmental Protection Agency (1998) The environmental protection agency's white paper on *Bt* plant-pesticide resistance management. EPA, Washington, DC

Walker CH, Sibly RM, Hopkin SP, Peakall DB (2012) Principle of ecotoxicology. CRC Press, Boca Raton

Wardlow LR, Ludlam FAB, Bradley LF (1976) Pesticide resistance in glasshouse whitefly (*Trialeurodes vaporariorum* Westwood). Pestic Sci 7:320–324

Warthen JD (1989) Neem (*Azadirachta indica* A. Juss): organisms affected and reference list update. Proc Entomol Soc Wash 91:367–388

Wilcox DR, Shivakumar AG, Melin BE, Miller MF, Benson TA, Shopp CW, Casuto D, Gundling GJ, Bolling TJ, Spear BB, Fox JL (1986) Genetic engineering of bioinsecticides. In: Inouye M, Sarma R (eds) Protein engineering: applications in science, medicine, and industry, 1st edn. Academic, Orlando

Wilding N (1986) The pathogens of diamondback moth and their potential for its control a review. In: Talekar NS (ed) Diamondback moth management: proceedings of the first international workshop. Asian Vegetable Research and Development Center, Shanhua

Wood RJ, Bishop JA, Lees DR (1981) Genetic consequences of man-made change. Academic, New York

Zou YQ, Zheng BZ (1988) The toxicity of some insecticides to green house whitefly (*Trialeurodes vaporariorum* Westwood) and monitoring of resistance. Acta Phytophylacica Sin 15:277–281

Efficacy of Major Plant Extracts/ Molecules on Field Insect Pests

5

S. Gopalakrishnan, B. Ratna Kumari, R. Vijayabharathi, A. Sathya, V. Srinivas, and G.V. Ranga Rao

Abstract

Insect pests are considered the major hurdle in enhancing the production and productivity of any farming system. The use of conventional synthetic pesticides has led to the emergence of pesticide-resistant insects, environmental pollution, and negative effects on natural enemies, which have caused an ecological imbalance of the predator-prey ratio and human health hazards; therefore, eco-friendly alternative strategies are required. The plant kingdom, a rich repertoire of secondary metabolites, can be tapped as an alternative for insect pest management strategies. A number of plants have been documented to have insecticidal properties against various orders of insects in vitro by acting as antifeedants, repellents, sterilant and oviposition deterrents, etc. However, only a few plant compounds are applicable at the field level or presently commercialised. Here, we have provided an overview of the broad-spectrum insecticidal activity of plant compounds from neem, *Annona*, *Pongamia*, and *Jatropha*. Additionally, the impact of medicinal plants, herbs, spices, and essential oils has been reviewed briefly.

Keywords

Insect pests • Field crops • Botanicals • Plant extracts • Secondary metabolites • Insecticidal properties

S. Gopalakrishnan, Ph.D. (✉) • B.R. Kumari, Ph.D.
R. Vijayabharathi, Ph.D. • A. Sathya, Ph.D.
V. Srinivas, M.Sc. • G.V.R. Rao, Ph.D.
Grain Legumes, International Crops Research Institute for the Semi-Arid Tropics (ICRISAT), Patancheru 502 324, Telangana, India
e-mail: s.gopalakrishnan@cgiar.org

D. Singh (ed.), *Advances in Plant Biopesticides*,
DOI 10.1007/978-81-322-2006-0_5, © Springer India 2014

1 Introduction

The plant kingdom is recognised as the most efficient producer of chemical compounds that are used to defend plants against different insect pests (Isman and Akhtar 2007). The literature focusing on the effects of plant secondary compounds on insects is voluminous. As many as 2,121 plant species are reported to possess pest management properties, and 1,005 species of plants exhibit insecticidal properties, which includes 384 species with antifeedant properties, 297 species with repellent properties, 27 species with attractant properties, and 31 species with growth-inhibiting properties (Singh et al. 2008). The biological activity of plant extracts against bacteria, fungi, viruses, and insects has been discussed adequately (Bozsik 1996; Macedo et al. 1997; Unicini Manganelli et al. 2005; Gopalakrishnan et al. 2010, 2011). Botanical insecticides can be made from roots, flowers, seeds, stems, leaves, fruits, and bark in water or in organic solvents. Botanicals used as insecticides presently constitute 1 % of the world insecticide market (Rozman et al. 2007). In this chapter, the importance of botanicals in agriculture with an emphasis on field pests is reviewed.

2 History of Botanicals Used as Pesticides

Botanicals have been in nature for millions of years with no adverse effects on the ecosystem. The repellency of plant material has been exploited for thousands of years by mankind by hanging bruised plants in houses, a practice that is still in wide use throughout developing countries. The use of plant extracts and plant parts in the form of powder as insecticides dates back at least as far as the Roman Empire. For instance, during the reign of Persian king Xerxes (400 BC), children were deloused with a powder obtained from the dry flowers of a plant known as pyrethrum, *Tanacetum cinerariaefolium* (family – Compositae). In India, a poisonous plant is mentioned in the Rig Veda, the classic book of Hinduism, which was composed during the second millennium BC (Chopra et al. 1949). Today, in Mexico and several Central American countries, it is common practice to treat pests with plants known for their insecticidal properties. Crude botanical insecticides have been used for several centuries and have been known in tribal or traditional cultures around the world (Richard 2000). Plants have also been used for centuries in the form of crude fumigants, where plants were burned to drive away mosquitoes and later as oil formulations applied to the skin or clothes (Maia and Moore 2011). Mixing grain with plant oils is an ancient Indian and African approach of protecting grains against insect attack (Pereira 1983). The first botanical insecticide used as such, i.e., tobacco, dates back to the seventeenth century. A plant insecticide known as rotenone, which was obtained from the roots of the timbo plant, was introduced circa 1850. In 1965, Sláma and Williams made a surprising discovery that paper towels made from the wood of the balsam fir (*Abies balsamea*) released vapours that elicited a potent effect on hemipteran bugs of the Pyrrhocoridae family (Hodin 2009; Sláma and Williams 1965).

3 Insect Pest Management with Botanicals: A Depiction

3.1 Neem

Azadirachta indica A. Juss (syn: *Melia azadirachta, M. indica,* and *Antelaea azadirachta*), also known as the Indian neem tree, belongs to the family Meliaceae (mahogany) and was first described by the French botanist Adrian Henri Laurent de Jussieu in 1830. The botanical name *M. azadirachta* is sometimes confused with *M. azedarach*, a West Asian tree commonly known as chinaberry, Persian lilac, bakain, and dharak (National Research Council (NRC) 1992).

Every part of the neem tree has been used extensively as household remedies and also in ayurvedic, unani, and homeopathic medicine. Hence, it has been described as a 'cynosure of modern medicine' by Biswas et al. (2002). Broad-spectrum biological effects (antibacterial, antifungal, antiviral, antiplasmodial, antitrypanosomal, anthelminthic, molluscicidal, nematicidal, insecticidal, larvicidal, antifeedant, and insect repellent) and pharmacological activities (antioxidant, anticancer, antiulcer, spermicidal, antidiabetic, anti-implantation, immunomodulating and immunocontraceptive activities, etc.) attested by various parts and extracts of neem have been reviewed (Atawodi et al. 2009). Hence, the neem tree has received attention from the international scientific community, and authors from different countries have referred to this tree as a 'miracle tree/multipurpose crop/village dispensary/living pharmacy' (Biswas et al. 2002). The importance of the neem tree was been recognised years ago by the US National Academy of Sciences, which published a report entitled 'Neem – a tree for solving global problems' (NRC 1992).

3.1.1 Azadirachtin and Related Compounds

A. indica produces a plethora of triterpenoids. The first bitter compound isolated from neem oil is nimbin, and it is found to be one of the most abundant limonoids in seeds (Johnson et al. 1996). Subsequently, azadirachtin ($C_{35}H_{44}O_{16}$), a complex compound, was extracted from neem seeds by Butterworth and Morgan (1968, 1971) followed by its identification as a highly oxygenated tetranortriterpenoid by Kraus et al. (1985) and Broughton et al. (1986). In later years, Rembold (1989) isolated six related compounds (azadirachtins B–G), whereas Govindachari et al. (1991) reported seven compounds (azadirachtins A, B, D, F, H, I, and K) of closely related structures from neem kernels.

The ratio between these complex compounds has been reported differently by various authors. Klenk et al. (1986) and Rembold (1990) have stated that the ratio of azadirachtin B to azadirachtin A is 1:5, whereas the others compounds (azadirachtins C-G) occur at a ratio of 1 to 100 parts of azadirachtin A. Sidhu et al. (2003) studied the intra-provenance and inter-provenance variations of azadirachtin content in neem trees and found that azadirachtin A is 13–16-fold higher than azadirachtin B and that this difference varies among the natural populations. He also stated that climatic factors have no influence on azadirachtin content and that the observed differences might be due to the individual genetic compositional variations among the trees. Studies by Kumar and Parmar (1997) on neem ecotypes also affirm the variations in azadirachtin content and the non-impact of climatic factors on the same. In contrast, Ermel (1995) and Venkateswarlu et al. (1997) revealed the influence of humidity, rainfall, temperature, or season on azadirachtin content variations. It is also found that seasonal variations contribute to the synthesis of specific azadirachtins (Sidhu and Behl 1996). However, azadirachtins A and B are the major active metabolites of neem seeds/kernels. More than 135 compounds have been isolated from different parts of neem, and several reviews have been published on the chemistry and structural diversity of these compounds (Taylor 1984; Koul et al. 1990; Govindachari 1992; Chatterjee and Pakrashi 1994; Kraus 1995; Devakumar and Sukh Dev 1996).

Various methods have been reported for the isolation and purification of azadirachtins from neem seeds and/or kernels (Warthen et al. 1984; Schroeder and Nakanishi 1987; Govindachari et al. 1991; Sharma et al. 2003a, b). Yamasaki et al. (1986) and Schroeder and Nakanishi (1987) isolated azadirachtins by flash chromatography and reported different concentrations. Jarvis et al. (1999) isolated 11 triterpenoids including azadirachtins by supercritical fluid chromatography.

3.1.2 Neem Products as Pesticides/Insecticides

Voluminous reports on extracts, purified compounds/formulations, and traditional preparations of neem indicate their versatility and broad-spectrum insecticidal activity. The chemistry, environmental behaviour, and biological effects of neem products have been reported widely (Mordue and Blackwell 1993; Sundaram 1996; Williams and Mansingh 1996; Veitch et al. 2008). Because there is a vast amount of data, we have presented an overview of the potential of neem products. Laboratory testing has indicated that neem extract-based insecticides are effective against more than 400 pest species (Koul 1999), which has generated wide agricultural and environmental interests. Natural neem preparations have shown anti-insecticidal activity against the noctuid moths *Spodoptera littoralis* (Gelbic and Nemec 2001; Sharma et al. 2003a, b) and *S. litura* (Kumar and Parmar 1997; Govindachari et al. 2000), *Peridroma saucia*, the heteropteran bug *Oncopeltus fasciatus* (Isman et al. 1990), the leafhopper *Jacobiasca lybica*, and the whitefly *Bemisia tabaci* (El Shafie and Basedow 2003). Apart from the neem extract/formulations, published reports have also established that azadirachtin A, the major triterpenoid, and its related compounds possess insecticidal activity.

This broad-spectrum pesticidal/insecticidal activity is exerted by their phago and oviposition deterrent, repellent, antifeedant, growth retardant, moulting inhibitor, and sterilant properties. It was also reported that neem products prolong larval developmental times and prevent larval maturation. These effects might be influenced by neem products acting both as systemic and as contact poisons (Schmutterer 1990a, b; Mordue and Blackwell 1993), which was demonstrated by the treatment of *S. litura* with azadirachtin and other terpenoids (salannin, nimbinene, and nimbin) (Koul et al. 1996). Additionally, the broad-spectrum insecticidal attributes have been reported in *Plutella xylostella* (Schmutterer 1990a, b; Isman 1995; Liang et al. 2003), *Pieris brassicae* (Hasan and Ansari 2011), *S. littoralis* (Pineda et al. 2009), and other insects (Isman et al. 1990). Broad-spectrum insecticidal activity has also been demonstrated by neem seed kernel extracts under field conditions against Lepidoptera, Coleoptera, and Orthoptera insects (Schmutterer 1985).

However, insects from different orders differ markedly in their behavioural/physiological responses to azadirachtin/related compounds/neem extracts. This difference was demonstrated in a study conducted by Aerts and Mordue (1997), which showed strong toxicity and antifeedant activity of azadirachtin against *O. fasciatus* (Hemiptera), *S. littoralis* (Lepidoptera), and *Schistocerca gregaria* (Orthoptera). The study also stated that the lower structural forms of azadirachtin, such as azadirone, azadiradione, nimbin, and salannin, exhibit only the antifeedant activity, specifically on the lepidopteran pest *S. littoralis*. Therefore, the lepidopteran insect pest shows higher sensitivity than Orthopoda to neem compounds, and the combination of toxicity and antifeedant activity of azadirachtin renders it a strong insecticide. The malformation of *S. littoralis* at various developmental stages by azadirachtins has been substantiated by Martinez and Van Emden (2001), Gelbic and Nemec (2001), and Nathan and Kalaivani (2006). Inter-genus variation has also been demonstrated by neem compounds on lepidopteran members, where salannin was active against *S. littoralis* and nimbin was active against *Heliothis virescens* and *H. armigera* (Blaney et al. 1990). Interfamily variations on Noctuidae members, which includes the black army cutworm *Actebia fennica*, bertha armyworm *Mamestra configurata*, variegated cutworm *P. saucia*, zebra caterpillar *Melanchra picta*, Asian armyworm *S. litura*, and cabbage looper *Trichoplusia ni*, by azadirachtin have also been documented, where *A. fennica* and *S. litura* are less inhibited than the other species (Isman 1993). Nathan and colleagues recorded the antifeedant and growth inhibition activity of various neem limonoids (azadirachtin, salannin, deacetylgedunin, gedunin, 17-hydroxyazadiradione, and deacetylnimbin) against the rice leaf folder *Cnaphalocrocis medinalis* (Nathan et al. 2005) and legume pod borer *H. armigera* (Murugan et al. 1998). Apart from the major tetranortriterpenoids (nimbin and salannin) of neem seeds, the photo-oxidation products of tetranortriterpenoids, such as nimbinolide, isonimbinolide, salanninolide, and isosalanninolide, have also shown anti-insect properties, as demonstrated on *S. littoralis* (Jarvis et al. 1997).

Neem seed kernel extract (NSKE) has a profound effect on the rice leaf folder and sorghum shootfly *Atherigona soccata* (Shrinivas and Balikai 2009). The effect of NSKE was also reported on the potato tuber moth *Phthorimaea operculella* (Shelke et al. 1987), *H. armigera* (Sinha 1993), and *Lampides boeticus* (Irulandi and Balasubramanian 2000). Sap feeders, such as White Backed Plant

Hopper (WBPH) (Reddy et al. 2012), *Aphis craccivora*, *Empoasca kerri*, *Megalurothrips distalis* (Irulandi and Balasubramanian 2000; Dalwadi et al. 2008), pink mealy bug *Maconellicoccus hirsutus* (Sathyaseelan and Bhaskaran 2010), and the cotton stem weevil *Pempherulus affinis* (Ratnakumari and Chandrasekaran 2005), were subdued with NSKE. NSKE is effective not only against agriculturally important pests but also against parasites and pests that affect humans and animals, which further validates the broad-spectrum activity of NSKE (Schmahl et al. 2010).

Although azadirachtin is the major active constituent of neem extracts, the effects observed from such extracts could be due to the sum or synergy of azadirachtin and the other terpenoids present in the extract mixture (Boursier et al. 2011; Martinez and Van Emden 2001). Azadirachtin is one of the main active ingredients in neem extracts, but non-azadirachtin compounds (e.g., 6-β-hydroxygedunin or different volatiles from neem) isolated from *A. indica* seem to be involved in the toxicity and antifeedant activity of neem extracts (Reddy and Singh 1998; Koul et al. 2003).

3.1.3 Mechanism of Action

Schmutterer 1990a, b suggests that the well-established antifeedant and growth-inhibitory properties of azadirachtin is attributed to the disruption of endocrine events. Mordue and Blackwell (1993) stated that there are three modes of action of azadirachtin in insects: (i) feeding inhibition by the blockage of input receptors for phagostimulants or by the stimulation of deterrent receptor cells, or both; (ii) growth inhibition by the blockage of morphogenetic peptide hormone release, which affects ecdysteroid and juvenile hormone titers; and (iii) direct detrimental and histopathological effects on insect muscles, fat body, and gut cuticular epithelial cells. Azadirachtin has demonstrated its negative effects by reducing hemocyte count, degenerating organelles, and destroying plasma membranes (Sharma et al. 2003a, b). Recently, Nathan et al. (2006) revealed the impact of neem extracts on digestive enzymes, such as amylase, protease, and lipase.

3.1.4 Boundaries/Barriers for Commercialisation

In general, the principal barriers affecting the commercialisation of botanical pesticides include the scarcity of natural resources, standardisation, quality control, and registration (Murray and Isman 1997), which are each less of a concern for conventional insecticides. The limited stability of azadirachtins and related compounds under natural conditions, such as temperature, light, UV radiation, rainfall, pH, etc., has been addressed by studies conducted by Barnby et al. (1989), Schmutterer (1988 and 1990b), Stokes and Redfern (1982), Warthen et al. (1984), and Jarvis et al. (1998). Other azadirachtoids, such as deacetylnimbin, deacetylsalannin, nimbin, and salannin, disappeared rapidly when they were exposed to sunlight. Notably, azadirachtin A and B are less photostable in formulations than their corresponding pure forms (Cabone et al. 2009). This instability of azadirachtin has been attributed partly to the presence of unsaturation in the tiglate and enol ether moieties. Hence, the hydrogenation of the labile olefinic moieties is crucial to obtain more stable reduced products (Bilton et al. 1985; Yamasaki and Klocke 1987). Though structural change is an option to overcome this problem, structure-activity relationship studies have indicated that modifications of the basic molecule lead to altered insecticidal activity (Yamasaki and Klocke 1987; Rembold 1988). Alternatively, the use of stabilisers, such as antioxidants and UV/sunscreens, has been contemplated (Chowdhuri 1996; Sundaram and Curry 1996a, b).

Azadirachtin A is resistant to hydrogenation at ambient conditions of temperature and pressure. However, Bilton et al. (1985) and Ley et al. (1989) have reduced the azadirachtins using high pressure with selected catalysts to dihydroazadirachtin-A (reduction of the 22,23 double bond) and tetrahydroazadirachtin-A (reduction of the 2′,3′ double bond of tiglic acid of dihydroazadirachtin-A). Dihydroazadirachtin-based products have been registered by the Environment Protection Agency for use in the USA. Sharma et al. (2006) tested the structurally altered compound tetrahydroazadirachtin-A and the native compound azadirachtin against *H. armigera*. They found that tetrahydroazadirachtin has equal stability and effectiveness on comparison with azadirachtin, which validates its use as commercial neem biopesticides in the future.

Another barrier affecting neem compounds/formulations is the compatibility with biological control. Hoelmer et al. (1990) recorded no detrimental effects of the commercial neem product Margosan-O on *Encarsia formosa* and *E. transvena*, the parasitoids/natural enemies of *B. tabaci*. They observed no significant effect on the degree of parasitism on any of the natural enemies. Feldhege and Schmutterer (1993) recorded a considerable reduction in the degree of parasitism by *E. formosa* against azadirachtin-exposed *T. vaporariorum* pupae. A recent study by Scudeler and Dos Santos (2013) on *Ceraeochrysa claveri*, a natural enemy of several pests, such as whiteflies, thrips, lepidopteran pests, aphids, and mites (Pappas et al. 2011), demonstrated severe alterations in their midgut cells by the indirect ingestion of neem oil-treated prey. In contrast to these mild negative effects, neem products contribute to a favourable prey/predator ratio and help provide a healthy functioning ecosystem (El Shafie and Basedow 2003).

3.1.5 Commercialisation

Although purified neem compounds/formulations exhibit instability and produce negative effects on natural enemies, its role as a broad-spectrum insecticide at very low concentrations with desirable residual properties and reduced insect resistance has made these compounds valuable tools in insect pest management and commercialisation in countries such as the USA, Canada, Mexico, European Union, and New Zealand and in Asian countries, such as India and China. In fact, neem is the most commercially exploited plant for insect pest management (Schmutterer 2002). Some of the examples of neem products include NeemAzal-T/S®, Neemix, Neemexcel (Ahmad et al. 2012), Thai neem 111® (Yule and Srinivasan 2013), Margosan-O (Lindquist and Casey 1990), Neem-EC® (Sundaram et al. 1997), Neem gold (Sharma et al. 2003a, b), Tre-san®; MiteStop®; Wash Away Louse®, and Picksan LouseStop® (Schmahl et al. 2010), which have proved their efficacy against various pests and insects. Boursier et al. (2011) compared traditional neem preparations with commercial formulations of azadirachtin-A against the leafhopper *Macrosteles quadripunctulatus* and *S. littoralis* and found that there was equivalent activity between both of the preparations. However, in the same study, whitefly *B. tabaci* required a higher concentration of traditional neem extract, which indicated species specificity.

3.2 *Annona*

The largest species in the family of angiosperm is Annonaceae, which includes approximately 135 genera with 2,500 species (Chatrou et al. 2004). Annonaceae are tropical trees and shrubs found in Central America, Africa, and Southeast Asia and are one of the main sources of edible fruits and edible oils (Ngiefu et al. 1976; Heywood 1978) in those regions. Some of the fruits from Annona species include pawpaw, chirimoya, sweetsop, soursop, and custard apple (Isman 2006). Annona species also possess medicinal properties. The wood of the Annonaceous trees is used in alcohol production (Savard and Espil 1951). Studies conducted on Annona reveal diverse chemical compounds in various species (Leboeuf et al. 1982; Chang et al. 1998; Kotkar et al. 2001). A complex mixture of acetogenins with 30 compounds has been identified in each of the species. In particular, sesquiterpenes and monoterpenes are the main compounds of the oils of *Annona* spp. (Ríos et al. 2003). Many genera from the family viz. *Annona, Asimina, Goniothalamus, Rollinia*, and *Uvaria* have been studied, and, to date, approximately 400 compounds have been explored (Alali et al. 1999; Johnson et al. 2000). These compounds have been found to possess cytotoxic, antitumor, antimalarial, insecticidal, and parasiticidal properties (Rupprecht et al. 1990; Fang et al. 1993; Alali et al. 1999; Ocampo and Ocampo 2006). In small farms worldwide, some species of Annona have been used as botanical insecticides by traditional home made preparations (Secoy and Smith 1983; Okonkwo 2005; Castillo et al. 2010).

3.2.1 The Annonaceous Acetogenins (ACGs)

ACGs are the most rapidly growing class of natural products that possess a wide range of biological activities, such as anthelminthic, antimalarial, antimicrobial, antiprotozoal, antitumor, and cytotoxic actions (Fang et al. 1993; Gu et al. 1995; Zeng et al. 1996; Cavé et al. 1997). In 1982, uvaricin was the first ACG discovered, and it was found to possess antileukemic activity (Jolad et al. 1982). ACGs have a series of C-35/C-37 derived from C-32/C-34 fatty acids combined with a 2-propanol unit. 'A long aliphatic chain with a terminal methyl substituted R, an *â-unsaturated* ç-lactone ring with one, two, or three tetrahydrofuran rings along with hydrocarbon chain, and a number of oxygenated moieties (hydroxyls, acetoxyls, ketones, epoxides) and/or double bonds' is the characteristic feature of ACG (Shi et al. 1995, 1996; Alali et al. 1998; Chávez et al. 1998; Colman-Saizarbitoria et al. 1998; Paul et al. 2013).

Among the Annona genus, two species have outstanding insecticidal properties, *A. muricata* and *A. squamosa*. The following acetogenins are found in *A. muricata* and *A. squamosa*: annocatalin, annohexocin, annomonicin, annomontacin, annomuricatin, annomuricin, annonacin, coronin, corossolin, corossolone, gigantetrocin, gigantetronenin, montanancin, muracin, muricatalicin, muricin, robustosin, solamin, squamocin, and uvariamicin (Raintree Nutrition 2004). These compounds can range from polar to nonpolar and, hence, can be extracted using various solvents, such as water (Pérez-Pacheco et al. 2004), ethanol (Bobadilla et al. 2002), acetone (Khalequzzaman and Sultana 2006), chloroform (Parvin et al. 2003), petroleum ether (Álvarez et al. 2008), and hexane (Fontana et al. 1998).

3.2.2 Biology of ACGs

ACGs have been reported to be potent inhibitors of NADPH/ubiquinone oxidoreductase (complex I), an enzyme of the electron transport chain system of mitochondria. This inhibition deprives the cell's ATP and induces apoptosis (Tormo et al. 1999; Alali et al. 1999). These bioactive effects of ACGs have also been confirmed against other species of insects including sap-sucking Lepidoptera larvae such as *Myzus persicae*, spider mites, mosquito larvae, striped cucumber beetles, melon aphids, Colorado potato beetles, Mexican bean beetles, bean leaf beetles, European corn borers, blowfly larvae, and free-living nematodes (Ahammadsahib et al. 1993; He et al. 1997; McLaughlin et al. 1997; Guadaño et al. 2000; Leatemia and Isman 2004; Álvarez et al. 2007; Cólom et al. 2008). The incorporation of ACGs in the diet of *S. frugiperda* led to morphological changes, which reflected the interference of ACGs with hormonal activity (Di Toto Blessing et al. 2010). Annonaceous extracts have been evaluated in several groups of medically and agriculturally important insects. Thus, the biological activity of acetogenins is similar to that of limonoid azadirachtin isolated from seeds of *A. indica* (Mordue and Nisbet 2000). Similarly, there are various insecticidal ACGs (Table 5.1). Recent reports have also revealed that these ACGs are more effective against stored grain pests. In developing countries, postharvest grain loss is a major problem because of stored-grain pest infestations. Anita et al. (2012) and Ribeiro et al. (2013) showed that the extracts of *A. squamosa* and *A. mucosa* are toxic to *Tribolium castaneum* and *Sitophilus zeamais*, respectively. Isoquinoline alkaloids are another class of compounds associated with plant defence against herbivorous insects (Da Rocha et al. 1981; Cordell et al. 2001; Bermejo et al. 2005; Cólom et al. 2009). Isoquinoline alkaloids interact with neural signal transduction networks and interfere with neuroreceptors via enzymes involved in neurotransmission metabolism and ion channels (Wink et al. 1997; Wink 2000); however, no studies are available regarding the biological activity of isoquinoline alkaloids. To date, Annona products are not available commercially, and studies have not addressed the commercial use of these products. Hence, this is one area where more work needs to be performed.

Table 5.1 Active compounds with biological activity from the Annonaceae family (Modified from Castillo-Sanchez et al. 2010)

Botanical name	Plant part	Active compound	Insect	Biological activity	Reference
A. muricata	Seeds	Bullatalicin	*Anticarsia gemmatalis* *Pseudaletia sequax*	Feeding deterrence and growth inhibition	Fontana et al. (1998)
		Squamocin	*M. persicae*	Insecticide	Guadano et al. (2000)
			Leptinotarsa decemlineata		
		Annonacin	*Rhodnius pallescens*	Insecticide	Parra-Henao et al. (2007), Robledo-Reyes et al. (2008), Leatemia and Isman (2004)
			R. prolixus	Insecticide and repellency	
			Periplaneta americana	Insecticide	
			S. litura		
A. squamosa	Seeds	Squamocin	*T. castaneum*	Insecticide	Khalequzzaman and Sultana (2006), Seffrin et al. (2010), Khalequzzaman and Sultana (2006), Leatemia and Isman (2004)
		Methanolic extract	*T. ni*	Insecticide and feed deterrent	
		Crude extract	*S. litura* and *H. armigera*	Insecticide	
			T. castaneum	Insecticide	
		Annonins	*S. litura*	Insecticide	
		Annotemoyin-1	*T. castaneum*	Insecticide	Parvin et al. (2003), Laetamia and Isman (2004)
		Neoannonin	*P. xylostella*		
A. cherimolia	Seeds	Squamocin	*S. frugiperda*	Insecticide and antifeedant activity	Álvarez et al. (2007)
		Itrabin			
		Cherimolin-1			
		Neoannonin			
		Asimicin			
		Squamocin Almunequin	*O. fasciatus*	Insecticide	Álvarez et al. (2008)
		Itrabin			
		Molvizarin			
A. montana	Leaves and branch	Annonacin	*O. fasciatus*	Insecticide	Álvarez et al. (2008)
		Cis-annonacin 10-one			
		Densicomacin-1			
		Annonacin-a			
		Gigantetronenin	*S. frugiperda*	Antifeedant and toxic	Di Toto Blessing et al. (2010)
		Murihexocin			
		Tucupentol			
Oxandra cf xylopioides	Leaves	Berenjenol	*S. frugiperda*	Insecticide	Rojano et al. (2007)
A. atemoya	Seeds	Methanolic extract	*T. ni*	Insecticide and feed deterrent	Seffrin et al. (2010)

3.3 *Pongamia*

Pongamia pinnata (Linn.) Pierre (syn: *P. glabra* Vent; *Derris indica* (Lam) Bennett; *Millettia novo-guineensis* Kane & Hat) belongs to the family Fabaceae, which is an indigenous plant of the Indian subcontinent and Southeast Asia (Krishnamurthi 1969; CSIR 1969). It is known as the poonga oil tree, pongam tree, karanja, karanj, karum, kanji, Indian beech, etc. and has been used traditionally in ayurvedic, siddha, and folk medicines. The impact of chemical, biological, and pharmacological aspects of this tree has been reviewed (Meera et al. 2003). Pongam seed oil contains 5–6 % flavonoids. Karanjin, a furanoflavonoid compound, is the main constituent of the flavonoids (Brinji 1987) and known for its insecticidal properties (Kumar and Singh 2002). The isolation of this compound from oil and de-fatted oil cakes has been detailed in the work of Vismaya et al. (2010) and Katekhaye et al. (2012) and the work of Susarla et al. (2012), respectively. The insecticidal property of Karanjin was enhanced through structural modifications that converted Karanjin into karanj ketone, karanj ketone oxime esters, and karanj ketone oxime N-O-nonanoate. This enhanced insecticidal property was demonstrated on the aphid *Lipaphis erysimi* (Mondal et al. 2010).

For approximately 70 years, natural product chemists have studied the complex organic compounds of karanj, especially the flavone-like molecules kanjone, pongamol, pongapinnols A–D, pongagalabrone, pongapin, pinnatin, pongone, pongacoumestan, glabrachalcone, isopongachromene, isopongaflavone, galbone, pongalabol, pongagallone A and B, and 6-methoxyfuroflavone (Rao and Rao 1941; Pavanaram and Ramachandra Row 1955; Aneja et al. 1963; Mahey et al. 1972; Malik et al. 1976; Roy et al. 1977; Garg 1979; Talopatra et al. 1985; Shameel et al. 1996; Chauhan and Chauhan 2002; Simin et al. 2002; Carcache-Blanco et al. 2003; Ahmad et al. 2004; Alam et al. 2004; Yadav et al. 2004; Li et al. 2006; Yin et al. 2006).

The compounds, either as oil or as an extract from methanol/aqueous/chloroform/acetone, are biologically active against insect pests. They act as insecticides, repellents, oviposition deterrents, antifeedants, and larvicides (Parmar and Gulati 1969; Kumar and Singh 2002; Pavela and Herda 2007a, b). The extract of *P. pinnata* is also toxic against *S. litura* and *H. armigera* as well as the stored grain pests *Trogoderma granarium* and *T. castaneum* (Kumar et al. 2006; Pawar et al. 2011; Reena and Sinha 2012). Karanj oil and Karanj leaf extract have been used against mustard aphid *L. erysimi* in field conditions (Bunker et al. 2006; Singh 2007).

Karanj extract is a constituent of commercially available insecticidal formulations, such as Plexin, Karrich, Salotrap, RD Repelin, and RD9 Repelin, for the control of various insect pests. PONEEM, which is discussed further in the latter part of this chapter, is a patented insecticidal formulation of pongam oil and neem oil. However, compared with neem products, the reduced efficiency of Karanj extract in aqueous solutions is a limiting factor for their wider applications; therefore, this area requires further research attention.

3.4 *Jatropha*

Jatropha curcas, also known as physic nut, is a tropical plant that belongs to Euphorbiaceae and is native to North America but now thrives in Africa and Asia (Willis 1967). The seeds of this plant consist of 47 % fat and various antinutritional factors such as saponin, phytate, trypsin inhibitor, and cyanogenic glycosides (Makkar et al. 1997; Kumar and Sharma 2008; Rakshit et al. 2008). *Jatropha* have an immense potential to generate a large amount of feedstock oil for biodiesel production. The global jatropha oil production is proposed to be approximately 28 MT/annum and is the major oil produced in Asia (GEXSI 2008).

There are various terpenoids present as secondary metabolites of *Jatropha*. To date, 65 types of terpenes have been explored (Devappa et al. 2011). Phorbol esters (PEs), a group of tigilane diterpenes, are a major toxic constituent of *Jatropha* (Adolf et al. 1984; Makkar et al. 1997). Approximately six PEs have been identified so far, and 70–75 % of the PEs are in an extractable form in *J. curcas* kernels, whereas 25–30 % are non-extractable (Haas et al. 2002). The non-extractable PEs remain tightly bound to the matrix of the kernel (Makkar et al. 2009; Makkar and Becker 2009). Goel et al. (2007) have investigated the structure along with the biological activity and toxicity of PEs in animals, while the medicinal property, phytochemistry, and pharmacological properties of *Jatropha* spp. have been studied by Sabandar et al. (2013).

The toxic constituents of PEs present in the seed extract and leaf extract of *J. curcas* possess insecticidal, fungicidal, and molluscicidal properties (Nwosu and Okafor 1995; Liu et al. 1997; Solsoloy and Solsoloy 1997). The PEs contained in jatropha oil are effective against many insects and pests of both field crops and stored grains, including *Callosobruchus maculatus*, *C. chinensis*, *C. maculatus*, *Clavigralla tomentosicollis*, *Sitophilus zeamais*, *Rhyzopertha dominica*, *T. castaneum*, *Oryzaephilus surimanensis*, *L. erysimi*, *P. rapae*, *Phthorimaea operculella*, *Tetranychus urticae*, *O. fasciatus*, *Coptotermes vastator*, *Amrasca biguttula*, *Aphis gossypii*, and *Aphis fabae* (Shelke et al. 1985, 1987; Solsoloy 1995; Solsoloy and Solsoloy 1997; Wink et al. 1997; Solsoloy and Solsoloy 2000; Jing et al. 2005; Adabie–Gomez et al. 2006; Adebowale and Adedire 2006; Acda 2009; Devappa et al. 2010; Habou et al. 2011; Katoune et al. 2011; Ravindra and Kshirsagar 2010; Silva et al. 2012), by antifeedant, oviposition deterrent, ovicidal, and antibirth properties.

The PE-enriched fraction has been studied broadly against various insect pests, such as viz. *S. frugiperda* (Devappa et al. 2012), *C. maculatus* (Adebowale and Adedire 2006; Jadhau and Jadhua 1984), *Corcyra cephalonica* (Khani et al. 2012), *Busseola fusca*, *Sesamia calamistis*, *H. armigera*, and *Manduca sexta* (Sauerwein et al. 1993; Mengual 1997; Makkar et al. 2007; Ratnadass et al. 2009). The PE fractions are susceptible to oxidation and, hence, the addition of antioxidants/storage at cold temperatures is necessary for an increased shelf life (Devappa et al. 2013). The PE fraction has been found to be stable even after 2 years at 4 °C (Devappa et al. 2009). The extract from another species, *J. gossypiifolia*, has been shown to have toxic effects on three lepidopteran pests, *B. fusca*, *Ostrinia nubilalis*, and *S. nonagrioides* (Valencia et al. 2006). Additionally, the PE fraction has been shown to possess antifeedant properties and insecticidal activity against *S. exigua* (Khumrungsee et al. 2009) and *S. frugiperda* (Bullangpoti et al. 2012). Only a limited number of insects have been evaluated under controlled conditions, so the evaluation of broad-spectrum insecticidal activity is crucial. In addition, the fate of PEs under field conditions with the impact of water, soil, plants, and environmental risks has to be investigated. PE toxicity testing on natural enemies, mammalian systems, and modes of action of the insecticidal activity is another prerequisite. Apart from PEs, other secondary metabolites of *Jatropha* sp., such as saponins, lectins, and cyanogenic glycosides, should also be evaluated for their insecticidal activity.

3.5 Weeds as Plant Protection Tools

A plant considered undesirable, troublesome, and growing where it is not wanted is a weed. Here, it is worth considering the importance of weeds as botanical insecticides. *Lantana* is an invasive species in the tropical and subtropical regions of the world. However, the plant exhibits insecticidal properties on insects such as aphids, mites (Suliman et al. 2003), potato tuber moths (Lal 1987), and *S. obliqua* (Sharma et al. 1982). The most common weeds, such as *Parthenium* and *Cyperus*, were found to be successful in minimising the *Epilachna* beetle, diamond back moth, and cabbage head caterpillar in vegetables (Dhandapani et al. 1985; Venkataramireddy et al. 1990; Thebtaranonth et al. 1995; Prijono et al. 1997). *Calotropis* is reported to be active on rice plant hoppers (Prakash et al. 2008).

3.6 Medicinal Herbs as Plant Protectants

Medicinally valued herbs can also control agriculturally important insects. Herbs such as *Gynandropsis gynandra*, *Catharanthus roseus*, *Vitex*, *Ocimum*, and *Euphorbia royleana* were found to be effective against the *Epilachna* beetle, *H. armigera*, *M. hirsutus*, mustard aphid, and mesta hairy caterpillars (Sharma et al. 1982; Chandel et al. 1987; Rajasekaran et al. 1987;

Roy and Pande 1991; Prakash et al. 2008; Sathyaseelan and Bhaskaran 2010). Furthermore, the *Epilachna* beetle can be regulated by medicinal herbs such as *Solanum xanthocarpum* and *Strychnos nux-vomica* (Dhandapani et al. 1985; Chitra et al. 1991). Other plants, such as the bitter gourd and *Vernonia amygdalina*, known for their bitterness, have demonstrated efficacy against the flea beetle on okra and the coffee leaf miner *Leucoptera coffeella* (Alves et al. 2011; Onunkun 2012). *Passiflora mollissima*, popularly known as banana passion fruit, is not only used in the food industry for ice cream production but also for the control of insect pests such as the Mesta hairy caterpillar (Tripathi et al. 1987).

3.7 Spices and Condiments with Insecticidal Action

Homemade botanicals from spices and condiments, such as peppers and garlic, have shown particular promise as a source of botanical pesticides. For instance, powdered chilli pepper deters the onion fly, *Delia antiqua*, *Earias insulana*, *T. ni*, and *T. urticae* (Antonious et al. 2007). Garlic shows its effect on soft-bodied insects, such as aphids, and cumin (*Nigella sativa*) has been shown to be effective against the *Epilachna* beetle (Chandel et al. 1987).

3.8 Essential Oils

Essential oils (EOs) are complex mixtures of volatile organic compounds produced as secondary metabolites in plants to defend themselves against herbivores and pathogens. The major plant families from which EOs are extracted include Myrtaceae, Lauraceae, Lamiaceae, and Asteraceae. EOs have repellent, antifeedant, reproduction retardant, fumigant toxicity, and growth-reducing effects on a variety of insects (Singh et al. 1989; Singh and Singh 1991). EOs are neurotoxic, and there is evidence for interference with the neuromodulator octopamine (Enan 2005) or GABA-gated chloride channels (Priestley et al. 2003). Zoubiri and Baaliouamer (2011) performed a detailed literature survey on 230 plants and listed the insecticidal activity of their essential oils along with the major active compounds.

3.9 Miscellaneous

Plantago (Alves et al. 2011), *Zea mays* male flowers (not important after pollination; Al-Khafaji et al. 2003), and velvet bean (*Mucuna cochinensis*; Premchand 1989) are also considered to contain insecticidal activity. Other plants such as mahua, *Psoralea corylifolia,* and *Lindenbergia grandiflora* were indicated to possess insecticidal principles (Tripathi et al. 1987; Mohanty et al. 1988; Narasimhan and Mariappan 1988). Plant extracts of cassava, papaya, sweet potato, tea, *Solanum nigrum*, *Solanum incanum*, Mexican tea, Mexican marigold, blackjack (*Bidens pilosa*), thorn apple, and *Aloe* are known to have insecticidal properties.

3.10 Botanical Pesticides from Herbal Compost

In addition to insecticides derived from leaf and seed extracts of plants, the compost made out of plants also contributes to insect pest control. Biowashes of crude extracts of *Annona*, *Jatropha*, and *Pongamia* vermicompost were reported to kill *H. armigera* and *S. litura* (Gopalakrishnan et al. 2011). Maize stover compost demonstrated very good control of whiteflies, *Podagrica* species, *Zonocerus variegatus*, and *B. tabaci* with its efficacy ranging from 60 to 80 % control. Organic composts, especially maize stover compost, can be used as an insecticide in organic farming systems to raise okra plants (Alao et al. 2011). Akanbi et al. (2007) reported that the foliar application of organic composts effectively controlled the level of insect infestation of *Telfairia occidentalis.* In the field, organic compost extracts did not kill the observed insects but had a repellent and/or barrier effect.

4 Synergism

Biological activity can be enhanced by combinations of biomolecules. An extensive study has been carried by Packiam et al. (2012) with formulations of pongam oil and neem oil. This formulation, PONEEM, has been patented in India. The phytochemicals present in PONEEM karanjin and azadirachtin have controlled *Scirtothrips dorsalis* in chilli thrips efficiently by acting as a feeding deterrent (Packiam and Ignacimuthu 2013). The same phytopesticide formulation has been evaluated against *S. litura* and *H. armigera*. The oviposition deterrent activities at different concentrations with different formulations were studied extensively by Packiam et al. (2012). A formulation with pongam oil and *Thymus vulgaris*/*Foeniculum vulgare* has showed lower LC_{50} values against *P. xylostella* than pongam oil alone indicates the syndergism between the botanicals (Pavela 2012). A combination of *Bacillus thuringiensis* subsp. *kurstaki* (Btk) with extracts of *Acacia arabica*, *A. squamosa*, *Datura stramonium*, *Eucalyptus globulus*, *Ipomoea carnea*, *Lantana camara*, *Nicotiana tabacum* and *P. pinnata* was prepared by Rajguru et al. (2011). They investigated the efficacy and synergistic activity against *S. litura* larvae and revealed that high mortality using the fortified extracts was due to synergistic action. In particular, the leaf extracts of *N. tabacum* and the seed extracts of *A. arabica*, *A. squamosa*, and *D. stramonium* showed a promising result and were compatible with Btk. Hence, it might be important to explore microbial combinations with insecticides from plants in field conditions.

5 Research at ICRISAT on Botanical Pesticides

During the process of evolution, plants have developed some defence systems to compete against biotic (herbivores, insects, microorganisms, etc.) and also abiotic stresses. One such system is the production of secondary metabolites, such as protease inhibitors, lectins, terpenoids, nonprotein amino acids, alkaloids, cyanogenic glycosides, saponins, tannins, etc. (Mazid et al. 2011). In the past few years, ICRISAT research has been focused on the molecules responsible for host-plant resistance and their corresponding effect on pests to develop efficient biopesticides. In ICRISAT, several studies have been conducted on various botanicals to identify potential and cost-effective strategies for the control of insects. Sharma and colleagues have worked on various plant materials/metabolites for the control of the major devastating crop pest *H. armigera* and also against other insect pests. Earlier, they worked on various solvent extracts of unripened seeds, ripened seeds, kernels, and leaves of the neem tree against *Mythimna separata* and reported an antifeedant compound, which has differed from earlier reports on *A. indica* and *M. azedarach* (Sharma et al. 1983). In long-term collaboration with IICT (Indian Institute of Chemical Technology, Hyderabad, India), active fractions of neem and custard apple extracts were tested under laboratory and field conditions against various insect pests (Sharma et al. 1999). In the past few years, they have analysed the effect of lectins from leguminous and nonleguminous plants, such as field bean, pigeon pea, chickpea, soybean, peanut, lentil, canavalia (concanavalin A), garlic, snowdrop, and jackfruit (jacalin), and trypsin inhibitors from soybean against *H. armigera*. All of the tested plant metabolites affected survival and developmental parameters (Shukla et al. 2005). A phenolic compound (stilbene) from pigeon pea with antifeedant activity against *H. armigera* was also documented by their research group (Green et al. 2003).

Eleven indigenous plant materials (*Cleistanthus collinus*, *C. gigantea*, *P. glabra*, *Artemisia dubia*, *Sphaeranthus indicus*, *Cassia occidentalis*, *Chloroxylon swietenia*, *Vitex negundo*, *Madhuca indica*, *Strychnos nux-vomica*, and *S. potatorum*) known for insecticidal properties collected from Andhra Pradesh and Chhattisgarh states of India were evaluated against *S. litura* larvae. The water extract of these products against second/fourth instar larvae clearly indicated the superiority of *C. collinus*, *C. gigantea* (leaf extract), and *P. glabra* (seed extract) in suppressing the larval growth and development of *S. litura* (Fig. 5.1). The above list of plant materials, excluding *A. dubia*, against second instar larvae of *H. armigera* clearly indicated the superiority of *C. collinus* and *S. indicus* with 57 % larval mortality one week after exposure and with others recording 10–48 % mortality (Rao and Gopalakrishnan 2009). Further observations 2 weeks after exposure revealed a similar trend with a range of 17–63 % larval mortality (Gopalakrishnan et al. 2009).

Furthermore, experiments were conducted to evaluate 18 different botanical extracts against *S. litura* and *H. armigera*. The larvicidal activity of the botanical extracts produced a ranged of mortality between 52 % and 86 %, with the maximum found in neem fruit powder. Studies on larval mortality and oviposition deterrence of various botanicals against *H. armigera* produced the highest larval mortality in neem extracts, followed by *Datura*, rain tree pod, and *Chrysanthemum*. For *S. litura*, the maximum mortality was recorded with neem fruit extract followed by *Pongamia*, rain tree pod, *Datura*, and *Annona*. The oviposition of the two species was severely affected by the plant extract sprays, which reflected the potential of the botanicals in the suppression of key pests (Table 5.2) (Gopalakrishnan et al. 2009).

Bioactive compounds from 17 different botanicals, particularly *Annona, Datura, Pongamia, Parthenium, Gliricidia,* neem, and *Jatropha*, which are capable of managing

Fig. 5.1 Effect of botanical extracts on *Spodoptera litura*

Table 5.2 Evaluation of one percent botanical extract against neonates of *H. armigera* and *S. litura* (Gopalakrishnan et al. 2009)

	Mortality (%)		Repellency/ovipositional deterrence (%)	
Scientific name	*H. armigera*	*S. litura*	*H. armigera*	*S. litura*
A. squamosa	20	40	47	96
A. squamosa	28	46	79	90
A. squamosa	25	34	45	93
C. gigantea	18	12	33	66
C. domestica	38	32	65	55
D. metel	40	46	32	46
J. curcas	24	26	8	95
Tagetes erecta	35	34	66	96
M. azedarach	22	30	93	87
A. indica	42	48	9	73

(continued)

Table 5.2 (continued)

	Mortality (%)		Repellency/ovipositional deterrence (%)	
Scientific name	*H. armigera*	*S. litura*	*H. armigera*	*S. litura*
A. indica	21	74	84	100
P. hysterophorus	10	26	13	75
P. pinnata	30	64	92	56
P. pinnata	29	32	4	77
Prosopis juliflora	16	44	34	37
Samanea saman	31	44	43	ND
S. saman	39	52	21	73
Tridax procumbens	32	34	32	85
V. negundo	10	44	1	93

ND not determined

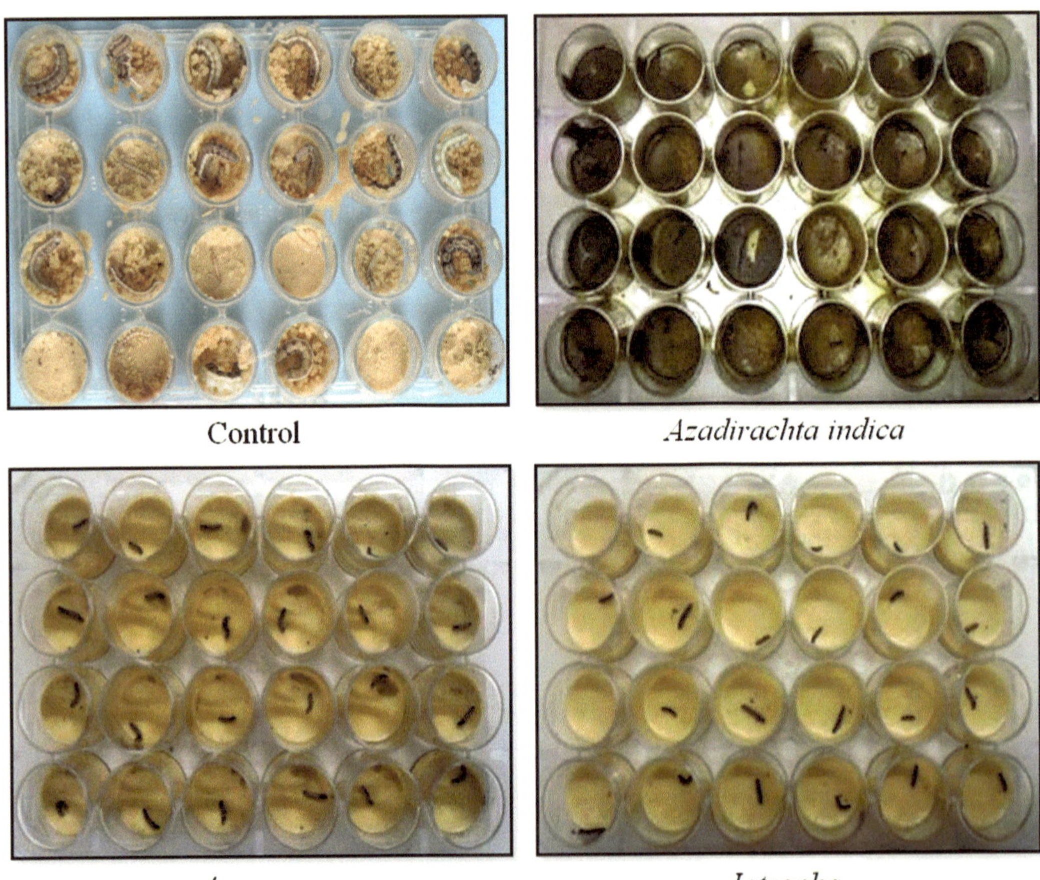

Fig. 5.2 Effect of botanical extracts on *Helicoverpa armigera* larvae

H. armigera and *S. litura*, were identified. When the feed was treated with crude biowash (Figs. 5.2 and 5.3) for healthy larvae (4-day old), 42 and 86 % mortality and 32 and 71 % weight reduction compared with the control were reported for *H. armigera*, while *S. litura* exhibited 46 and 74 % larval mortality and 47 and 77 % weight reduction compared with the untreated control (Fig. 5.4).

Fig. 5.3 Preparation of herbal biowash

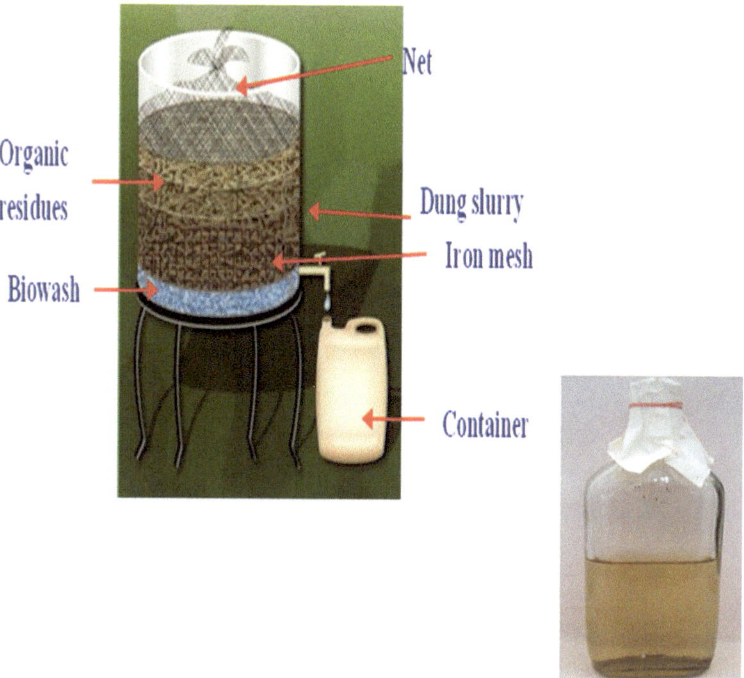

Fig. 5.4 The protocol for biowash preparation and the final product (*inset*)

Table 5.3 Influence of various botanical extracts on *H. armigera* and *S. litura* (Gopalakrishnan et al. 2011)

	H. armigera		*S. litura*	
Treatments	Mortality (%)	Weight reduction (%)	Mortality (%)	Weight reduction (%)
Annona	58	60	57	70
Chrysanthemum	43	43	49	47
Datura	42	56	58	77
Jatropha	86	53	62	53
Neem	71	60	60	52
Parthenium	69	32	57	59
Pongamia	76	58	74	56
Tridax	45	54	46	54
Vitex	45	71	52	58
SE±	13.1**	9.6**	5.5***	8.3***
CV%	42	34	24	36

SE standard error, *CV* coefficient of variance
** statistically significant at 0.01 (*p* values); *** statistically significant at 0.001 (*p* values)

Table 5.4 The effect of adsorbed and nonadsorbed fractions of potential crude biowash samples on *H. armigera* larvae (Gopalakrishnan et al. 2011)

	Adsorbed fraction		Nonadsorbed fraction	
Treatment	Mortality (%)	Weight reduction (%)	Mortality (%)	Weight reduction (%)
Annona	91	89	65	80
Datura	88	76	64	89
Jatropha	87	84	72	97
Neem	81	79	69	89
Parthenium	93	73	65	91
Pongamia	93	91	73	91
SE±	1.6*	6.0 (NS)	1.7*	2.9*
CV%	3	13	4	6

NS nonsignificant, *SE* standard error, *CV* coefficient of variance
*statistically significant at 0.001

The adsorbed and nonadsorbed fractions of crude biowash from open column chromatography of the promising botanicals (*Annona, Datura, Jatropha,* neem, *Parthenium,* and *Pongamia*) showed significant mortality on *H. armigera* (Tables 5.3 and 5.4) (Gopalakrishnan et al. 2011). The compatibility of botanical extracts (such as *Annona*, *Datura*, neem fruit, and *Parthenium*) and some selected entomopathogenic microorganisms (such as *Bacillus subtilis* [BCB-19] and *Metarhizium anisopliae*) were assessed. Neem fruit and *Datura* were found to be compatible with *B. subtilis* (BCB-19). None of the four botanical powder extracts suppressed *M. anisopliae* up to 8 days. In another study, three botanicals (*Annona, Datura*, and neem fruit powder) and three entomopathogens (viz. *Bacillus megaterium* (SB-9), *B. pumilus* (SB-21), and *Serratia marcescens* (HIB28)) were evaluated for their compatibility. There were no definite signs of suppression by any of the botanicals on the bacteria. However, there were some signs of improved growth in the case of SB-9 + neem fruit, SB-9 + *Annona*, and HIB-28 + *Datura* (Gopalakrishnan et al. 2011).

6 Conclusion

The inherent toxicity and unforeseen environmental problem intensified by the extended use of synthetic pesticides have been addressed by regulatory agencies from time to time via banning/restricting the use of toxic pesticides in agriculture. As a result, much attention is being paid to the exploration of plant biodiversity because plants produce a rich repertoire of phytochemicals, which evolved partly as defence molecules against attacking organisms. However, they are non-phytotoxic, are easily biodegradable, and have a minor impact on environmental and human health. Therefore, natural products are generally regarded as safe. Safe insecticides are presently in demand because insects are becoming resistant to existing products at a greater rate than new insecticides can be developed. However, the appropriate protection of species and ecological communities has to be considered to protect biodiverse resources from threats.

References

Acda MN (2009) Toxicity, tunneling and feeding behavior of the termite, Coptotermes vastator, in sand treated with oil of the Physic nut, *Jatropha curcas*. J Insect Sci 6:1–8

Adabie–Gomez DA, Monford KG, Agyir-Yawson A, Owusu-Biney A, Osae M (2006) Evaluation of four local plant species for insecticidal activity against *Sitophilus zeamais* Mots (Coleoptera:Curculionidae) and *Callosobruchus maculatus* (F) (Coleoptera: Bruchidae-). Ghana J Agric Sci 39:147–154

Adebowale KO, Adedire CO (2006) Chemical composition and insecticidal properties of the underutilized *Jatropha curcas* seed oil. African J Biotech 5(10):901–906

Adolf W, Opferkuch HJ, Hecker E (1984) Irritant phorbol derivatives from four *Jatropha* species. Phytochemistry 23:129–132

Aerts RJ, Mordue (Luntz) AJ (1997) Feeding deterrence and toxicity of neem triterpenoids. J Chem Ecol 23(9):2117–2132

Ahammadsahib KI, Hollingworth RM, McGovern JP, Hui YH, McLaughlin JL (1993) Mode of action of bullatacin: A potent antitumor and pesticidal annonaceous acetogenins. Life Sci 53:1113–1120

Ahmad G, Yadav PP, Maurya R (2004) Furanoflavonoid glycosides from *Pongamia pinnata* fruits. Phytochemistry 65:921–924

Ahmad N, Ansari MS, Hasan F (2012) Effects of neem based insecticides on *Plutella xylostella* (Linn.). Crop Prot 34:18–24

Akanbi WB, Adebayo TA, Togun OA, Adeyeye AS, Olaniran OA (2007) The use of compost extract as foliar spray nutrient source and botanical insecticide in *Telfairia occidentalis*. WJAS 3(5):642–652

Alali FQ, Rogers LL, Zhang Y, McLaughlin JL (1998) Unusual bioactive annonaceous acetogenins from *Goniothalamus giganteus*. Tetrahedron 54:5833–5844

Alali FQ, Liu X, McLaughlin JL (1999) Annonaceous acetogenins: recent progress. J Nat Prod 62:504–540

Alam S, Sarkar Z, Islam A (2004) Synthesis and studies of antibacterial activity of pongaglabol. J Chem Sci 116:29–32

Alao FO, Adebayo TA, Olaniran OA, Akanbi WB (2011) Preliminary evaluation of the insecticidal potential of organic compost extracts against insect pests of Okra (*Abelmoschus* esculentus (L.) Moench). Asian J Plant Sci Res 1(3):123–130

Al-Khafaji AS, Omar AD, Eskander T (2003) Effect of plant extracts on the control of some important agricultural pests. In: Eighth Arab congress of plant protection, 12–16 October, El-Beida, Libya, pp 87–91

Álvarez O, Neske A, Popich S, Bardón A (2007) Toxic effects of annonaceous acetogenins from *Annona cherimolia* (Magnoliales: Annonaceae) on *Spodoptera frugiperda* (Lepidoptera: Noctuidae). J Pestic Sci 80:63–67

Álvarez O, Barrachina I, Ayala I, Goncalvez M, Moya P, Neske A, Bardón A (2008) Toxic effects annonaceous acetogenins on *Oncopeltus fasciatus*. J Pestic Sci 81:85–89

Alves DS, Oliviera DF, Carvalho GA, Dos Santos JRHM, Carvalho DA, Santos MAI, De Carvalho HWP (2011) Plant extracts as an alternative to control *Leucoptera coffeella* (Guérin-Mèneville) (Lepidoptera: Lyonetiidae). Neotrop Entomol 40(1):123–128

Aneja R, Khanna RN, Seshadri TR (1963) 6-Methoxyfuroflavone, a new component of the seeds of *Pongamia glabra*. J Chem Soc: 163–168

Anita S, Sujatha P, Prabhudas P (2012) Efficacy of pulverised leaves of *Annona squamosa* (L.), *Moringa oleifera* (Lam.) and *Eucalyptus globulus* (Labill.) against the stored grain pest, *Tribolium castaneum* (Herbst.). Recent Res Sci Technol 4(2):19–23

Antonious GF, Meyer JE, Rogers JA, Hu YH (2007) Growing hot pepper for cabbage looper, *Trichoplusia ni* (Hübner) and spider mite, *Tetranychus urticae* (Koch) control. J Environ Sci Health B 42(5): 559–567

Atawodi ES, Joy C, Atawodi JC (2009) *Azadirachta indica* (neem): a plant of multiple biological and pharmacological activities. Phytochem Rev 8:601–620

Barnby MA, Yamasaki R, Klocke JA (1989) Biological activity of azadirachtin, three derivatives, and their ultraviolet radiation degradation products against tobacco budworm (Lepidoptera, Noctuidae) larvae. J Econ Entomol 82:58–63

Bermejo A, Figadére B, Zafra-Polo MC, Barrachina I, Estornell E, Cortes D (2005) Acetogenins from Annonaceae: recent progress in isolation, synthesis and mechanisms of action. Nat Prod Rep 22:269–303

Bilton JN, Broughton HB, Ley SV, Lidert A, Morgan ED, Rzepa HS, Sehppard RN (1985) Structural reappraisal of the limonoid insect antifeedant azadirachtin. J Chem Soc Chem Commun 14:968–971

Biswas K, Chattopadhyay I, Banerjee RK, Bandyopadhyay U (2002) Biological activities and medicinal properties of neem (*Azadirachta indica*). Curr Sci 82(11): 1336–1345

Blaney WM, Simmonds MSJ, Ley SV, Anderson JC, Toogood PL (1990) Antifeedant effects of azadirachtin and structurally related compounds on lepidopterous larvae. Entomol Exp Appl 55:149–160

Bobadilla M, Zavaleta G, Gil F, Pollack L, Sisniegas M (2002) Efecto bioinsecticida del extracto etanólico de las semillas de *Annona cherimolia* Miller (chirimoya) y *A. muricata* Linnaeus (guanábana) sobre larvas del IV estadio de *Anopheles sp.*. Rev Peru Biol 9:64–73

Boursier CM, Bosco D, Coulibaly A, Negre M (2011) Are traditional neem extract preparations as efficient as a commercial formulation of azadirachtin A? Crop Prot 30(3):318–322

Bozsik A (1996) Studies on aphicidal efficiency of different stinging nettle extracts. Anz Schädl kd Pflanzenschutz Umweltschutz Appita 69:21–22

Brinji NV (1987) Non-traditional oil seeds and oils in India. Oxford/IBH Publishing, New Delhi, pp 143–166

Broughton HB, Ley SV, Slawin AMZ, Williams DJ, Morgan ED (1986) X-ray crystallographic structure determination of detigloyldihydroazadirachtin and reassignment of the structure of the limonoid insect antifeedant azadirachtin. J Chem Soc 1:46–47

Bullangpoti V, Wajnberg E, Feyereisen PAR (2012) Antifeedant activity of *Jatropha gossypiifolia* and *Melia azedarach* senescent leaf extracts on *Spodoptera frugiperda* (Lepidoptera: Noctuidae) and their potential use as synergists. Pest Manag Sci 68:1255–1264

Bunker GK, Rana BS, Ameta OP (2006) Efficacy of some plant products against mustard aphid, *Lipaphis erysimi* (Kaltenbach). Pestology 30:28–32

Butterworth JH, Morgan ED (1968) Isolation of a substance that suppresses feeding in locusts. J Chem Soc Chem Commun 1:23–24

Butterworth JH, Morgan ED (1971) Investigation of the locust feeding inhibition of the seeds of the neem tree, *Azadirachta indica*. J Insect Physiol 17:969–977

Cabone P, Sarais G, Angioni A, Lai F, Dedola F, Paolo Cabras P (2009) Fate of azadirachtin A and related azadirachtoids on tomatoes after greenhouse treatment. J Environ Sci Health B 44:598–605

Carcache-Blanco EJ, Kang YH, Park EJ, Su BN, Kardono LB, Riswan S, Fong HH, Pezzuto JM, Kinghorn AD (2003) Constituents of the stem bark of *Pongamia pinnata* with the potential to induce quinone reductase. J Nat Prod 66:1197–1202

Castillo LE, Jimenez JJ, Delgado MA (2010) Secondary metabolites of the Annonaceae, Solanaceae and Meliaceae families used as biological control of insects. Trop Subtrop Agroecosyst 12:445–462

Cavé A, Figadefe B, Laurens A, Cortes D (1997) Acetogenins from annonaceae. In: Herz W, Kirby GW, Moore RE, Steglish W, Tamm C (eds) Progress in the chemistry of organic natural products. Springer, New York, pp 81–287

Chandel BS, Pandey UK, Kumar A (1987) Insecticidal evaluation of some plant extracts against *Epilachna vigintioctopunctata* Fabr. (Coleoptera: Coccinellidae). Indian J Entomol 49(2):294–296

Chang FR, Yang PY, Lin JY, Lee KH, Wu YC (1998) Bioactive kaurane diterpenoids from *Annona glabra*. J Nat Prod 61:437–439

Chatrou LW, Rainer H, Maas PJM (2004) Annonaceae (Soursop family). In: Smith N et al (eds) Flowering plants of the neotropics. New York Botanical Garden, New York, pp 18–20

Chatterjee A, Pakrashi S (eds) (1994) The treatise on Indian medicinal plants, vol 3, Publications and information directorate. CSIR, New Delhi, p 76

Chauhan D, Chauhan JS (2002) Flavonoid glycosides from *Pongamia pinnata*. Pharm Biol 40:171–174

Chávez D, Acevedo LA, Jimenezin MR (1998) A novel annonaceous acetogenin from the seeds of *Rollinia mucosa* containing adjacent tetrahydrofuran-tetrahydropyran ring systems. J Nat Prod 61:419–421

Chitra KC, Reddy PVR, Rao KP (1991) Effect of certain plant extracts in the control of brinjal spotted leaf beetle, *Henosepilachna vigintioctopunctata* Fabr. J Appl Zool Res 2(1):37–38

Chopra RN, Bhadwan RL, Gosh S (1949) Poisonous plants in India. Scientific monograph no. 17, ICAR, New Delhi, pp 10–12

Chowdhuri H (1996) Effect of curcumin and turmeric oil on photostability and efficacy of azadirachtin. Ph.D. thesis, PG School, Indian Agricultural Research Institute, New Delhi, India

Colman-Saizarbitoria T, Johnson HA, Alali FQ, Hopp DC, Rogers LL, McLaughlin JL (1998) Annojahnin from *Annona jahnii*: a possible precursor of mono-tetrahydrofuran acetogenins. Phytochemistry 49:1609–1616

Cólom OA, Barrachina I, Mingol IA, Mass CG, Sanz PM, Neske A, Bardon A (2008) Toxic effects of annonaceous acetogenins on *Oncopeltus fasciatus*. J Pest Sci 81:81–85

Cólom OA, Neske A, Chahboune N, Zafra-Polo MC, Bardón A (2009) Tucupentol, a novel mono-tetrahydrofuranic acetogenin from *Annona montana*, as potent inhibitor of mitochondrial complex I. Chem Biodivers 6:335–340

Cordell GA, Quinn-Beattie ML, Farnsworth NR (2001) The potential of alkaloids in drugs discovery. Phytother Res 15:183–205

CSIR (1969) The wealth of India – a dictionary of Indian raw materials, vol 8. Publications and Information Directorate, Council of Scientific and Industrial Research, New Delhi, pp 206–211

Da Rocha AI, Luz AIR, Rodrigues WA (1981) A presença de alcaloides em espécies botânicas da Amazônia. III – Annonaceae. Acta Amazon 11:537–546

Dalwadi MM, Korat DM, Tank BD (2008) Bio-efficacy of some botanical insecticides against major insect pests of Indian bean, *Lablab purpureus* L. Karnataka J Agric Sci 21(2):295–296

Devakumar C, Sukh Dev (1996) In: Randhawa NS, Parmar BS (eds) Neem, 2nd edn. New Age, pp. 77–110. ISBN 10: 812242046X/ISBN 13:9788122420463

Devappa RK, Maes J, Makkar HPS, Greyt WD, Becker K (2009) Isolation of phorbol esters from *Jatropha curcas* oil and quality of produced biodiesel. In: 2nd international congress on biodiesel: the science and the technologies, Munich, Germany.

Devappa RK, Makkar HPS, Becker K (2010) Jatropha toxicity – a review. J Toxicol Environ Health B Crit Rev 13:476–507

Devappa RK, Makkar HPS, Becker K (2011) Jatropha diterpenes: a review. J Am Oil Chem Soc 88:301–322

Devappa RK, Makkar HPS, Becker K (2012) Localisation of antinutrients and qualitative identification of toxic components in *Jatropha curcas* seed. J Sci Food Agric 92:1519–1525. doi:10.1002/jsfa.4736

Devappa RK, Bingham JP, Khana SK (2013) High performance liquid chromatography method for rapid quantification of phorbol esters in *Jatropha curcas* seed. Ind Crops Prod 49:211–219

Dhandapani N, Rajendran NS, Abul KA (1985) Plant products as anti-feedants to control insects. Pesticides 19(11):53–60

Di Toto Blessing L, Álvarez Colom O, Popich S, Neske A, Bardón A (2010) Antifeedant and toxic effects of acetogenins from *Annona montana* on *Spodoptera frugiperda* L. J Pest Sci 83:307–310

El Shafie HAF, Basedow T (2003) The efficacy of different neem preparation for control of insects damaging potatoes and eggplants in the Sudan. Crop Prot 22:1015–1021

Enan EE (2005) Molecular and pharmacological analysis of an octopamine receptor from American cockroach and fruit fly in response to plant essential oils. Arch Insect Biochem Physiol 59:161–171

Ermel K (1995) Azadirachtin content of neem seed kernels from different regions of the world. In: Schmutterer H (ed) The neem tree, source of unique natural products for integrated pest management, medicine, industry and other purposes. Weinheim, VCH, pp 89–92

Fang XP, Rieser MJ, Gu GX, McLaughlin JL (1993) Annonaceous acetogenins: an updated review. Phytochem Anal 4:27–49

Feldhege M, Schmutterer H (1993) Investigations on side effects of Margosan-O on *Encarsia Formosa* Gah. (Hym., Aphelenidae), parasitoid of the greenhouse whitefly, *Trialeurodes vaporariorum* Westw. (Hom., Aleyrodidae). J Appl Entomol 115:37–42

Fontana J, Lancas F, Pasos M, Cappelaro E, Villegas J, Baron M, Noseda M, Pomiiio M, Vitale A, Webber A, Maul A, Foerster LPW (1998) Selective polarity- and adsorption-guided extraction/purification of *Annona* sp. polar acetogenins and biological assay against agricultural pests. Appl Biochem Biotechnol 70:67–76

Garg GP (1979) New component from leaves of *Pongamia glabra*. Planta Med 37:73–74

Gelbic I, Nemec V (2001) Developmental changes caused by metyrapone and azadirachtin in *Spodoptera littoralis* (Boisd.) (Lep., Noctuidae) and *Galleria melonella* (L.) (Lep., Pyralidae). J Appl Entomol 125:417–422

GEXSI (2008) http://www.Jatropha-platform.org/documents/GEXSI

Goel G, Makkar HPS, Francis G, Becker K (2007) Phorbol esters: structure, biological activity, and toxicity in animals. Int J Toxicol 26:279–288

Gopalakrishnan S, Rupela OP, Humanyun P, Kiran BK, Sailasree J, Alekhya G, Sandeep D (2009) Bio-active metabolites from PGPR and botanicals. In: Reddy MS, Sayyed RZ, Sarma YR, Reddy KRK, Desai S, Rao VK, Reddy BC, Podile AR, Kloepper JW (eds) Proceedings of first Asian PGPR congress on plant growth promotion by rhizobacteria for sustainable agriculture, Andhra Pradesh, India. Scientific Publishers, Jodhpur, pp 44–51

Gopalakrishnan S, Kannan IGK, Alekhya G, Humayun P, Sree Vidya M, Deepthi K (2010) Efficacy of *Jatropha*, *Annona* and *Parthenium* biowash on *Sclerotium rolfsii*, *Fusarium oxysporum* f. sp. *ciceri* and *Macrophomina phaseolina*, pathogens of chickpea and sorghum. Afr J Biotechnol 9(47):8048–8057

Gopalakrishnan S, Ranga Rao GV, Humayun P, Rameshwar Rao V, Alekhya G, Simi J, Deepthi K, Sree Vidya M, Srinivas V, Mamatha L, Rupela O (2011) Efficacy of botanical extracts and entomopathogens on control of *Helicoverpa armigera* and *Spodoptera litura*. Afr J Biotechnol 10(73):16667–16673

Govindachari TR (1992) Chemical and biological investigations on *Azadirachta indica* (the neem tree). Curr Sci 63:117–122

Govindachari TR, Sandhya G, Ganeshraj SP (1991) Isolation of novel azadirachtin H and I by high-performance liquid chromatography. Chromatographia 31:303–305

Govindachari TR, Suresh G, Gopalakrishnan G, Wesley SD (2000) Insect antifeedant and growth regulating activities of neem seed oil the role of major tetrenotriterpenoids. J Appl Entomol 124:287–291

Green PWC, Stevenson PC, Simmonds MSJ, Sharma HC (2003) Phenolic compounds on the pod-surface of pigeonpea, *Cajanus cajan,* mediate feeding behavior of *Helicoverpa armigera* larvae. J Chem Ecol 29(4):811–821

Gu ZM, Zhao GX, Oberlies NH, Zeng L, McLaughlin JL (1995) Annonaceous acetogenins: potent mitochondrial inhibitors with diverse applications. In: Arnason JT, Mata R, Romeo JT (eds) Recent advances in phytochemistry, vol 29. Plenum Press, New York, pp 249–310

Guadaño A, Gutiérrez C, De la Peña E, Cortés D, González A (2000) Insecticidal and mutagenic evaluation of two annonaceous acetogenins. J Nat Prod 63:773–776

Haas W, Strerk H, Mittelbach M (2002) Novel 12 deoxy-16-hydroxyphorbol diesters isolates from the seed oil of *Jatropha curcas*. J Nat Prod 65:1434–1440

Habou ZA, Haougui A, Mergeai G, Haubruge E, Toudou A, Verheggen FJ (2011) Insecticidal effect of *Jatropha curcas* oil on the aphid *Aphis fabae* (Hemiptera: Aphididae) and on the main insect pests. associated with cowpeas (*Vigna unguiculata*) in Niger. Tropicultura 29(4):225–229

Hasan F, Ansari MS (2011) Toxic effects of neem-based insecticides on *Pieris brassicae* (Linn.). Crop Prot 30:502–507

He K, Zeng L, Ye Q, Shi G, Oberlies NH, Zhao GX, Njoku CJ, McLaughlin JL (1997) Comparative SAR evaluations of annonaceous acetogenins for pesticidal activity. Pestic Sci 49:372–378

Heywood VH (1978) Flowering plants of the world. University Press, Oxford (Cortes D 1991)

Hodin J (2009) On the origins of insect hormone signaling, phenotypic plasticity of insects. Science Publishers, Enfield, New Hampshire, USA, pp 817–839

Hoelmer KA, Osborne LS, Yokomu RK (1990) Effect of neem extracts on beneficial insects in greenhouse culture. In: USDA neem workshop on neem's potential in pest management programs. USDA, Beltsville, pp 100–105

Irulandi S, Balasubramanian G (2000) Report on the effect of botanicals against *Megalurothrips distalis* (Karny) (Thripidae: Thysanoptera) and *Lampides boeticus* Linn. (Lycaenidae: Lepidoptera) on greengram. Insect Environ 5(4):175–176

Isman MB (1993) Growth inhibitory and antifeedant effects of azadirachtin on six noctuids of regional economic importance. Pestic Sci 38:57–63

Isman MB (1995) Lepidoptera: butterflies and moths. In: Schmutterer H (ed) The neem tree, source of unique natural products for integrated pest management, medicine, industry and other purposes. VCH Verlagsgesellschaft, Weinheim, pp 229–318

Isman MB (2006) Botanical insecticides, deterrents, and repellents in modern agriculture and an increasingly regulated world. Annu Rev Entomol 51:45–66

Isman MB, Akhtar Y (2007) Plant natural products as a source for developing environmentally acceptable insecticides. In: Shaaya I, Nauen R, Horowitz AR (eds) Insecticides design using advanced technologies. Springer, Berlin/Heidelberg, pp 235–248

Isman MB, Koul O, Luczynski A, Kaminski J (1990) Insecticidal and antifeedant bioactivities of neem oils and their relationship to azadirachtin content. J Agric Food Chem 38:1408–1411

Jadhau KB, Jadhua LD (1984) Use of vegetable oils, plant extracts and synthetic products as protectants from pulse beetle. *Callosobruchus maculatus* in stored grain. J Food Sci Technol 21:110–113

Jarvis AP, Johnson S, Morgan ED, Simmonds MSJ, Blaney WM (1997) Photoxidation of nimbin and salannin, tertranortriterpenoids from the neem tree (*Azadirachta indica*). J Chem Ecol 23(12):2841–2860

Jarvis AP, Johnson S, Morgan ED (1998) Stability of the natural insecticide azadirachtin in aqueous and organic solvents. Pestic Sci 53:217–222

Jarvis AP, Morgan ED, Edwards C (1999) Rapid separation of triterpenoids from neem seed extracts. Phytochem Anal 10:39–43

Jing L, Fang Y, Ying X, Wenxing H, Meng X, Syed MN, Fang C (2005) Toxic impact of ingested jatropherol-I on selected enzymatic activities and the ultrastructure of midgut cells in silkworm, *Bomboxy mori* L. J Appl Entomol 129:98–104

Johnson S, Morgan ED, Peiris CN (1996) Development of the major triterpenoids and oil in the oil and seeds of neem (*Azadirachta indica*). Ann Bot 78:383–388

Johnson HA, Oberlies NH, Alali FQ, McLaughlin JE (2000) Thwarting resistance: annonaceous acetogenins as new pesticidal and antitumor agents. In: Cutler SJ, Cutler JG (eds) Biological active natural products: pharmaceuticals. CRC Press, Boca Raton, pp 173–183

Jolad SD, Hoffman JJ, Schram KH, Cole JR, Tempesta MS, Kriek GR, Bates RB (1982) Uvaricin a new antitumor agent from *Uvaria accuminata*, (Annonaceae). J Org Chem 47:3151–3153

Katekhaye SD, Kale MS, Laddha KS (2012) A simple and improved method for isolation of karanjin from *Pongamia pinnata* Linn. seed oil. Indian J Natl Prod Resour 3(1):131–143

Katoune HI, Malam Lafia D, Salha H, Doumma A, Drame AY, Pasternak D, Ratnadass A (2011) Physic nut (*Jatropha curcas*) oil as a protectant against field insect pests of cowpea in Sudano-Sahelian cropping systems. J SAT Agric Res 9:1–6

Khalequzzaman M, Nahar J, (2003) Toxicity of azadirachtin to larvae and adults of *Tribolium castaneum* (Herbst). J Biol Sci 11:19–24

Khalequzzaman M, Sultana S (2006) Insecticidal activity of *Annona squamosa* L. seed extracts against the red flour beetle, *Tribolium castaneum* (Herbst). J Bio-Sci 14:107–112

Khani M, Awang RM, Omar D, Rahmani M (2012) Bioactivity effect of *Piper nigrum* L. and *Jatropha curcas* L. extracts against *Corcyra cephalonica* [Stainton]. Agrotechnol 2:1

Khumrungsee N, Bullangpoti V, Pluempanupat W (2009) Efficiency of *Jatropha gossypiifolia* L. (Euphorbiaceae) against *Spodoptera exigua* HÜbner (Lepidoptera: Noctuidae): toxicity and its detoxifying enzyme activities. KKU Sci J 37:50–55

Klenk A, Bokel M, Kraus W (1986) 3-Tigloylazadiractl, an insect growth regulating constituent of *Azadirachta indica*. J Chem Soc Chem Commun 7:523–524

Kotkar HM, Prashant SM, Sangeetha VGSS, Shipra RJ, Shripad MU, Maheshwari VL (2001) Antimicrobial and pesticidal activity of partially purified flavonoids of *Annona squamosa*. Pest Manag Sci 58:33–37

Koul O (1999) Insect growth regulating and antifeedant effects of neem extracts and azadirachtin on two aphid species of ornamental plants. J Biosci 1:85–90

Koul O, Isman MB, Ketkar CM (1990) Properties and uses of neem, *Azadirachta indica*. Can J Bot 68:1–11

Koul O, Shankar JS, Kapil RS (1996) The effect of neem allelochemicals on nutritional physiology of larval *Spodoptera litura*. Entomol Exp Appl 79:43–50

Koul O, Multani JS, Singh G, Daniewski WM, Berlozecki S (2003) 6 beta-hydroxygedunin from *Azadirachta indica*. Its potentiation effects with some non-azadirachtin limonoids in neem against lepidopteran larvae. J Agric Food Chem 51:2937–2942

Kraus W (1995) In: Schmutterer H, Ascher KRS (eds) The neem tree: source of unique natural products for integrated pest management, medicine, industry and other purposes. VCH Publishers Inc., New York, pp 35–88. ISBN 3-527-30054-6

Kraus W, Bokel M, Cramer R, Klenk A, Poehnl H (1985) Constituents of neem and related species. A revised structure of azadirachtin. Abstr 3rd Int Conf Chem Biotechnol Biol Act Nat Prod 4:446–450

Krishnamurthi A (1969) The wealth of India, vol VIII. Publication and Information Directorate CSIR, New Delhi

Kumar J, Parmar BS (1997) Neem oil content and its key chemical constituents in relation to the agro-ecological factors and regions of India. Pestic Res J 9:216–225

Kumar A, Sharma S (2008) An evaluation of multipurpose oil seed crop for industrial uses (*J. curcas* L.): a review. Ind Crops Prod 28:1–10

Kumar M, Singh R (2002) Potential of *Pongamia glabra* Vent as an insecticide of plant origin. Biol Agric Hortic 20:29–50

Kumar V, Chandrashekar K, Sidhu OP (2006) Efficacy of karanjin and different extracts of *Pongamia pinnata* against selected insect pests. J Entomol Res 30(2):103–108

Lal L (1987) Studies on natural repellents against potato tuber moth. *Phthorimaea operculella* Zell in country stores. Potato Res 30(2):329–334

Leatemia J, Isman M (2004) Efficacy of crude seed extracts of *Annona squamosa* against diamondback moth, *Plutella xylostella* L. in the greenhouse. Int J Pest Manag 50:129–133

Leboeuf M, Cavé A, Bhaumik PK, Mukherjee B, Bukherjee R (1982) The phytochemistry of Annonaceae. Phytochemistry 21:2783–2813

Ley SV, Anderson JC, Blaney WM, Jones PS, Lidert Z, Morgan ED, Robinson NG, Santafianos D, Simmonds MSJ, Toogood PL (1989) Insect antifeedants from *Azadirachta indica*: chemical modification and structure–activity relationships of azadirachtin and some related limonoids. Tetrahedron 45:5175–5192

Li L, Li X, Shi C, Deng Z, Fu H, Proksch P, Lin W (2006) Pongamone A–E, five flavonoids from the stems of a mangrove plant, *Pongamia pinnata*. Phytochemistry 67:1347–1352

Liang GM, Chen W, Liu TX (2003) Effects of three neem-based insecticides on diamondback moth (Lepidoptera: Plutellidae). Crop Prot 22:333–340

Lindquist RK, Casey ML (1990) Evaluation of oils, soaps and natural product derivatives for miner, foxglove aphid, western flower thrips and greenhouse whitefly control. Ohio Flor Assoc Bull 727:3–5

Liu SY, Sporer F, Wink M, Jouurdane J, Henning R, Li YL, Ruppel A (1997) Anthraquinones in *Rheum palmatum* and *Rumex dentatus* (Polygonaceae) and phorbol esters in *J. curcas* (Euphorbiaceae) with molluscicidal activity against the schistosome vector snails *Oncomelania*, *Biomphalaria* and *Bulinus*. Trop Med Int Health 2:179–188

Macedo ME, Consoli RA, Grandi TS, Dos Anjos AM, Oliveira AB, Mendes NM, Queiroz RO, Zani CL (1997) Screening of Asteraceae (Compositae) plant extracts for larvicidal activity against *Aedes fluviatilis* (Diptera: Culicidae). Mem Inst Oswaldo Cruz 92:565–570

Mahey S, Sharma P, Seshadri TR (1972) Structure and synthesis of globrachromene, a new constituent of *Pongamia glabra*. Indian J Chem 10:585–588

Maia MF, Moore SJ (2011) Plant-based insect repellents: a review of their efficacy, development and testing. Malaria J 10(1):S11

Makkar HPS, Becker K (2009) *Jatropha curcas*, a promising crop for the generation of biodiesel and value-added coproducts. Eur J Lipid Sci Technol 111:773–787

Makkar HPS, Becker K, Sporer F, Wink M (1997) Studies on nutritive potential and toxic constituents of different provenances of *J. curcas*. J Agric Food Chem 45:3152–3157

Makkar HPS, Francis G, Becker K (2007) Bioactivity of phytochemicals in some lesser known plants and their effects and potential applications in livestock and aquaculture production systems. Animal 1:1371–1391

Makkar HPS, Maes J, Becker K (2009) Removal and degradation of phorbol esters during pre-treatment and transesterification of *Jatropha curcas* oil. J Am Oil Chem Soc 86:173–181

Malik SB, Seshadri TR, Sharma P (1976) Minor component of the leaves of *Pongamia glabra*. Indian J Chem 14B:229–230

Martinez SS, Van Emden HF (2001) Growth disruption, abnormalities and mortality of *Spodoptera littoralis* (Boisduval) (Lepidoptera: Noctuidae) caused by azadirachtin. Neotrop Entomol 30:113–125

Mazid M, Khan TA, Mohammad F (2011) Role of secondary metabolites in defense mechanisms of plants. Biol Med 3(2):232–249

McLaughlin JL, Zeng L, Oberlies NH, Alfonso D, Johnson HA, Cummings BA (1997) In: Hedin PA, Hollingworth RM, Miyamoto J, Thompson DG (eds) Phytochemicals for pest control, ACS symposium series 658. American Chemical Society, Washington, DC, pp 117–133

Meera B, Kumar S, Kalidhar SB (2003) A review of the chemistry and biological activity of *Pongamia pinnata*. J Med Aromat Plant Sci 25:441–465

Mengual L (1997) Extraction of bioactive substances from *J. curcas* L. and bioassays on *Zonocerus variegatus*, *Sesamia calamistis* and *Busseola fusca* for characterisation of insecticidal properties. In: Gübitz GM, Mittelbach M, Trabi M (eds) Biofuel and industrial products from *Jatropha curcas*. Dbv-Verlag University, Graz, pp 211–215

Mohanty KK, Chakraborty DP, Roy S (1988) Antifeedant activity of oil fractions of some leguminous plants against *Diacrisia obliqua*. Indian J Agric Sci 58(7):579–580

Mondal A, Walia S, Shrivastava C, Kumar B, Kumar J (2010) Synthesis and insecticidal activity of karanj ketone oxime and its ester derivatives against the mustard aphid (*Lipaphis erysimi*). Pestic Res J 22(1):39–43

Mordue AJ, Blackwell A (1993) Azadirachtin, an update. J Insect Physiol 39:903–924

Mordue AJ, Nisbet AJ (2000) Azadirachtin form the neem tree *Azadirachta indica*: its action against insects. An Soc Entomol Bras 29:615–632

Murray B, Isman (1997) Neem and other botanical insecticides: barriers to commercialization. Phytoparasitica 25(4):339–344

Murugan K, Jeyabalan D, Kumar SN, Babu R, Sivaramakrishnan S, Nathan SS (1998) Antifeedant and growth inhibitory potency of neem limonoids against *Helicoverpa armigera* Hubner (Lepidoptera: Nocutuidae). Insect Sci Appl 1:157–162

Narasimhan V, Mariappan V (1988) Effect of plant derivatives on green leaf hopper (GLH) and rice tungro (RTV) transmissions. Int Rice Res Newsl 13(1):28–29

Nathan SS, Kalaivani K (2006) Combined effects of azadirachtin and nucleopolyhedrovirus (SpltNPV) on *Spodoptera littoralis* Fabricius (Lepidoptera: Noctuidae) larvae. Biol Control 39:96–104

Nathan SS, Kalaivani K, Murugan K (2005) Paul Gene Chung. Efficacy of neem limonoids on *Cnaphalocrocis medinalis* (Guenée) (Lepidoptera: Pyralidae) the rice leaffolder. Crop Prot 24:760–763

Nathan SS, Chunga PG, Murugan K (2006) Combined effect of biopesticides on the digestive enzymatic profiles of *Cnaphalocrocis medinalis* (Guenée) (the rice leaffolder) (Insecta: Lepidoptera: Pyralidae). Ecotoxicol Environ Saf 64:382–389

Ngiefu CK, Paquot C, Vieux A (1976) Oil-bearing plants of Zaire. II. Botanical families providing oils of medium unsaturation. Oleagineux 31(12):545–547

National Research Council (NRC) (1992) Neem: tree for solving global problems. National Academy Press, Washington, DC

Nwosu MO, Okafor JI (1995) Preliminary studies of the antifungal activities of some medicinal plants against Basidiobolus and some other pathogenic fungi. Mycoses 38:191–195

Ocampo D, Ocampo R (2006) Bioactividad de la família Annonaceae. Rev Univ Caldas 5:135–155

Okonkwo EO (2005) Plant materials used for controlling insect pests of stored products in Nigeria, Families Annonaceae, Piperaceae, and Rutaceae. J Herbs Spices Med Plants 11:47–69

Onunkun O (2012) Evaluation of aqueous extracts of five plants in the control of flea beetles on okra (*Abelmoschus esculentus* (L.) Moench). J Biopest 5(Suppl):62–67

Packiam SM, Ignacimuthu S (2013) Effect of botanical pesticide formulations against the chilli trips (*Scirtothrips dorsalis* Hood) on peanut ecosystem. Int J Nat Appl Sci 2(1):1–5

Packiam SM, Anbalagan V, Ignacimuthu S, Vendan SE (2012) Formulation of a novel phytopesticide PONNEEM and its potentiality to control generalist herbivorous Lepidopteran insect pests, *Spodoptera litura* (Fabricius) and *Helicoverpa armigera* (Hübner) (Lepidoptera: Noctuidae). Asian Pac J Trop Dis 2(2):S720–S723

Pappas ML, Broufas GD, Koveos DS (2011) Chrysopid predators and their role in biological control. J Entomol 8(3):301–326

Parmar BS, Gulati KC (1969) Synergists for pyrethrins (II)-karanjin. Indian J Entomol 31:239–243

Parra-Henao G, García C, Cotes J (2007) Actividad insecticida de extractos vegetales sobre *Rhodnius prolixus* y *Rhodnius pallescens* (Hemiptera: Reduviidae). Bol Malariol Salud Ambient 47:125–137

Parvin S, Islam E, Rahman M, Haque E (2003) Pesticidal activity of pure compound Annotemoyin-1 isolated from chloroform extract of the plant *Annona squamosa* Linn. against *Tribolium castaneum* (Herbst) (Coleoptera: Tenebrionidae). Pak J Biol Sci 6:1088–1091

Paul J, Gnanam R, Jayadeepa RM, Arul L (2013) Anti cancer activity on graviola, an exciting medicinal plant extract vs various cancer cell lines and a detailed computational study on its potent anti-cancerous leads. Curr Top Med Chem 13(14):1666–1673

Pavanaram SK, Ramachandra Row L (1955) New flavones from *Pongamia pinnata* (L.) Merr.: identification of compound D. Nature 176:1177

Pavela R (2012) Efficacy of three newly developed botanical insecticides based on pongam oil against *Plutella xylostella* L. larvae. J Biopest 5(1):62–70

Pavela R, Herda G (2007a) Effect of pongam oil on adults of the greenhouse whitefly *Trialeurodes vaporariorum* (Homoptera: Trialeurodidae). Entomol Gen 30:193–201

Pavela R, Herda G (2007b) Repellent effects of pongam oil on settlement and oviposition of the common greenhouse whitefly *Trialeurodes vaporariorum* on *Chrysanthemum*. Insect Sci 14:219–224

Pawar PV, Joseph M, Sen A, Joshi SP (2011) Growth regulatory and toxic effects of non-edible oil seed extracts and purified extracts against *Helicoverpa armigera* (Hŭbner). J Agric Technol 7(5):1275–1282

Pereira J (1983) The effectiveness of six vegetable oils as protectants of cowpeas and banbara groundnuts against infestation by *Callosobruchus maculatus* (F.). Indian J Agric Sci 51:910–912

Pérez-Pacheco R, Rodríguez C, Lara J, Montes R, Valverde G (2004) Toxicidad de aceites, esencias y extractos vegetales en larvas de mosquito *Culex quin-*

quefasciatus Say (Diptera: Culicidae). Acta Zool Mex 20:141–152

Pineda S, Martínez AM, Figueroa JI, Schneider MI, Del Estal P, Viñuela E, Gómez B, Smagghe G, Budia F (2009) Influence of azadirachtin and methoxyfenozide on life parameters of *Spodoptera littoralis* (Lepidoptera: Noctuidae). J Econ Entomol 102(4):1490–1496

Prakash A, Rao J, Nandagopal VN (2008) Future of botanicals in rice, wheat, pulses and vegetables pest management. J Biopest 1(2):154–169

Premchand (1989) Presence of feeding deterrent in the velvet bean, *Mucuna cochinensis* Roxb. Indian J Entomol 51(2):217

Priestley CM, Williamson EM, Wafford KA, Sattelle DB (2003) Thymol, a constituent of thyme essential oil, is a positive allosteric modulator of human GABA receptors and a homo-oligomeric GABA receptor from *Drosophila melanogaster*. Braz J Pharmacol 140:1363–1372

Prijono D, Gani MS, Syahputra E (1997) Insecticidal activity of annonaceous seed extracts against *Crocidolomia binotalis* Zeller (Lepidoptera: Pyralidae). Bull Plant Pest Dis 9:1–6

Raintree Nutrition (2004) Graviola monograph. www.rain-tree.com/Graviola-Monograph.pdf. Visitado 26 Feb 2009

Rajasekaran B, Jayraj S, Raghuramman S, Narayanswamy T (1987) Use of neem products for the management of certain rice pests and diseases. In: Mid-term appraisal works on botanical pest control of rice based cropping system, p 13

Rajguru M, Sharma AN, Banerjee S (2011) Assessment of plant extracts fortified with *Bacillus thuringiensis* (Bacillales: Bacillaceae) for management of *Spodoptera litura* (Lepidoptera: Noctuidae). Int J Trop Insect Sci 31(1–2):92–97

Rakshit KD, Darukeshwara J, Rathina RK, Narasimhamurthy K, Saibaba P, Bhagya S (2008) Toxicity studies of detoxified *J. curcas* meal (J. curcas) in rats. Food Chem Toxicol 46:3621–3625

Rao GVR, Gopalakrishnan S (2009) Biopesticides research at ICRISAT: a consortium model. In: "Expert consultation on biopesticides and biofertilizers for sustainable agriculture" APAARI and council of research, Taiwan, 22–29 Oct

Rao NVS, Rao JV (1941) A note on glabrin: a new component of the seeds of *Pongamia glabra*. Proc Indian Acad Sci 14:123–125

Ratnadass A, Togola M, Cissé B, Vassal JM (2009) Potential of sorghum and physic nut (*Jatropha curcas*) for management of plant bugs (Hemiptera: Miridae) and cotton bollworm (*Helicoverpa armigera*) on cotton in an assisted trap-cropping strategy. http://ejournal.icrisat.org/Volume7/Sorghum_Millets/SG703.pdf. Accessed 13 Dec 2011

Ratnakumari B, Chandrasekaran S (2005) Efficacy of neem products against cotton stem weevil (*Pempherulus affinis* Faust) in Coimbatore. Andhra Agric J 52(1,2):295–297

Ravindra V, Kshirsagar (2010) Insecticidal activity of *Jatropha* seed oil against *Callosobruchus maculatus* (Fabricius) infesting *Phaseolus aconitifolius* Jacq. Bioscan 5(3):415–418

Reddy AV, Singh RP (1998) Fumigant toxicity of neem (*Azadirachta indica* A. Juss.) seed oil volatiles against pulse beetle, *Callosobruchus maculatus* Fab. (Col., Bruchidiae). J Appl Entomol 122:607–611

Reddy AV, Devi RS, Reddy DVV (2012) Evaluation of botanical and other extracts against plant hoppers in rice. J Biopest 5(1):57–61

Reena SR, Sinha BK (2012) Evaluation of *Pongamia pinnata* seed extracts as an insecticide against American bollworm *Helicoverpa armigera* (Hübner). Int J Agric Sci 4(6):257–261

Rembold H (1988) Azadirachtins: their structures and mode of action. In: Jacobson M (ed) Focus on phytochemical pesticides: the neem tree. CRC Press, Boca Raton, pp 47–67

Rembold H (1989) Azadirachtins, their structure and mode of action. In: Arnason JT, Philogene BJR, Morand P (eds) Insecticides of plant origin, ACS symposium series 387. American Chemical Society, Washington, DC, pp 150–163

Rembold H (1990) Isomeric azadirachtins and their mode of action. In: Jacobson M (ed) Focus on phytochemical pesticides 1: the neem tree. CRC Press, Boca Raton, pp 47–67

Ribeiro LP, Vendramim JD, Bicalho KU, Andrade MS, Fernandes JB, Moral RA, Demétrio CGB (2013) *Annona mucosa* Jacq. (Annonaceae): A promising source of bioactive compounds against *Sitophilus zeamais* Mots. (Coleoptera: Curculionidae). J Stored Prod Res 55:6–14

Richard AW (2000) Botanical insecticides, soaps and oils. In: Recheigl JE, Recheigl NA (eds) Biological and biotechnological control of insect pests. Lewis Publishers, London, pp 101–121

Ríos MY, Castrejun F, Robledo N, Léon I, Rojas G, Navarro V (2003) Chemical composition and antimicrobial activity of the essential oils from *Annona cheromola* (Annonnaceae). J Mex Chem Soc 47:134–142

Robledo-Reyes P, González R, Jaramillo G, Restrepo J (2008) Evaluación de la toxicidad de acetogeninas annonáceas sobre ninfas de *Periplaneta americana* L. (Dyctioptera: Blattidae). Bol Museo Entomol Univ Valle 9:54–61

Rojano B, Gaviria C, Sáez J, Yepes F, Muñoz F, Ossa F (2007) Berenjenol aislado de *Oxandra cf xylopioides* (Annonaceae) como insecticida. Vitae 14:95–100

Roy DC, Pande YD (1991) Effect of *hyptis suaveolens* (L.) Poit. leaf extract on the population of *Lipaphis erysimi* Kalt. In: IV National symposium on growth, development and control techniques of insect pests. Zoological Society, Muzaffarnagar, p 37

Roy D, Sharma NN, Khanna RN (1977) Structure and synthesis of iso-pongaflavone, a new component of the seeds of *Pongamia glabra*. Indian J Chem 15:1138–1139

Rozman V, Kalinovic I, Korunic Z (2007) Toxicity of naturally occurring compounds of Lamiaceae and Lauraceae to three stored product insects. J Stored Prod Res 43:349–355

Rupprecht JK, Hui YH, McLaughlin JL (1990) Annonaceous acetogenins: a review. J Nat Prod 53(2):237–278

Sabandar CW, Ahmat N, Jaafar FM, Sahidin I (2013) Medicinal property, phytochemistry and pharmacology of several *Jatropha* species (Euphorbiaceae): a review. Phytochemistry 85:7–29

Sathyaseelan V, Bhaskaran V (2010) Efficacy of some native botanical extracts on the repellency property against the pink mealy bug, *Maconellicoccus hirsutus* (Green). in Mulberry crop. Recent Res Sci Technol 2(10):35–38

Sauerwein M, Sporer F, Wink M (1993) Insect-toxicity of phorbol esters from *Jatropha curcas* seed oil. Planta Med 59:A686

Savard J, Espil L (1951) Centre Tech For Trop Nogent Marne Publ 3:7

Schmahl G, Al-Rasheid KAS, Abdel-Ghaffar F, Klimpel S, Mehlhorn H (2010) The efficacy of neem seed extracts (Tre-san®, MiteStop®) on a broad spectrum of pests and parasites. Parasitol Res 107:261–269

Schmutterer H (1985) Which insect pests can be controlled by application of neem seed kernel extract under field conditions. Z Angew Entomol 100:468–475

Schmutterer H (1988) Potential of azadirachtin-containing pesticides for integrated pest control in developing and industrialized countries. J Insect Physiol 34:713–719

Schmutterer H (1990a) Future tasks of neem research in relation to agricultural needs worldwide. In: Locke JC, Lawson RH (eds) Proceedings of workshop on neem's potential in pest management programs. USDA–ARS, Beltsville, ARS-86, pp 15–22

Schmutterer H (1990b) Properties and potential of natural pesticides from the neem tree, *Azadirachta indica*. Annu Rev Entomol 35:271–297

Schmutterer H (2002) The neem tree. Neem Foundation, Mumbai

Schroeder DR, Nakanishi KA (1987) Simplified isolation procedure for azadirachtin. J Nat Prod 50:241–244

Scudeler EL, Dos Santos DC (2013) Effects of neem oil (*Azadirachta indica* A. Juss) on midgut cells of predatory larvae *Ceraeochrysa claveri* (Navás, 1911) (Neuroptera: Chrysopidae). Micron 44:125–132

Secoy DM, Smith AE (1983) Use of plants in control of agricultural and domestic pests. Econ Bot 37:28–57

Seffrin RC, Shikano I, Akhtar Y, Isman MB (2010) Effects of crude seed extracts of *Annona atemoya* and *Annona squamosa* L. against the cabbage looper, *Trichoplusia ni* in the laboratory and greenhouse. Crop Prot 29:20–24

Shameel S, Usmanghani K, Ali MS (1996) Chemical constituents from seeds of *Pongamia pinnata* (L.) Pierre. Pak J Pharm Sci 9:11–20

Sharma INS, Singh AK, Singh SP (1982) Allelopathic potential of some plant substances as anti-feedants against insect pests of jute. In: Proc. first national symposium on allelopathy in agro-ecosystems. Indian Society of Allelopathy, HAU, Hisar, pp 157–176

Sharma HC, Leuschner K, Sankaram AVB, Gunasekhar D, Marthandamurthi M, Bhaskaraiah K, Subrahmanyam M, Sultana N (1983) Insect antifeedants and growth inhibitors from *Azadirachta indica* and *Plumbago zeylanica*. In: Schmutterer H, Ascher KRS (eds) Natural pesticides from neem tree and other tropical plants: proceedings of the 2nd international neem conference, 1986, GTZ. Eschborn. German Society for Technical Cooperation, Eschborn

Sharma HC, Sankaram AVB, Nwanze KF (1999) Utilization of natural pesticides derived from neem and custard apple in integrated pest management. In: Juss A, Singh RP, Saxena RC (eds) *Azadirachta Indica A. Juss*. Oxford/IBH, New Delhi, pp 199–213

Sharma PR, Sharma OP, Saxena BP (2003a) Effect of Neem gold on haemocytes of tobacco armyworm, *Spodoptera littoralis* (Fabricius) (Lepidoptera; Noctuidae). Curr Sci 84:690–695

Sharma V, Walia S, Kumar J, Nair MG, Parmar BS (2003b) An efficient method for the purification and characterization of nematicidal azadirachtins A, B and H using MPLC and ESIMS. J Agric Food Chem 51:3966–3972

Sharma V, Walia S, Dhingra S, Kumar J, Balraj SP (2006) Azadirachtin-A and tetrahydroazadirachtin-A concentrates: preparation, LC-MS characterization and insect antifeedant/IGR activity against *Helicoverpa armigera* (Hűbner). Pest Manag Sci 62:965–975

Shelke SS, Jadhav LD, Salunkhe GN (1985) Ovipositional and adult repellent action of some vegetable oils/extracts against potato tuber moth. Maharashtra Agric Univ 10:284–286

Shelke SS, Jadhav LD, Salunkhe GN (1987) Ovicidal action of some vegetable oils and extracts in the storage pest of potato, *Phthorimaea operculella*. Zell Biovigyanam 13:40–41

Shi GE, Alfonso D, Fatope MO, Zeng L, Gu ZM, Zhao GX, He K, MacDougal JM, McLaughlin JL (1995) Mucocin: a new annonaceous acetogenin bearing a tetrahydropyran ring. J Am Chem Soc 117: 10409–10410

Shi G, Kozlowski JF, Schwedler JT, Wood KV, MacDougal JM, McLaughlin JL (1996) Muconin and mucoxin: additional non-classical bioactive acetogenins from *Rollinia mucosa*. J Org Chem 61:7988–7989

Shrinivas M, Balikai S (2009) Evaluation of plant products in combination with cow urine and panchagavya against sorghum shootfly, *Atherigona soccata* Rondani. Karnataka J Agric Sci 22:618–620

Shukla S, Arora R, Sharma HC (2005) Biological activity of soybean trypsin inhibitor and plant lectins against cotton bollworm/legume pod borer, *Helicoverpa armigera*. Plant Biotechnol 22(1):1–6

Sidhu OP, Behl HM (1996) Seasonal variations in azadirachtins in seeds of *Azadirachta indica*. Curr Sci 70(12):1084–1086

Sidhu OP, Kumar V, Behl HM (2003) Variability in neem (*Azadirachta indica*) with respect to azadirachtin content. J Agric Food Chem 51:910–915

Silva GN, Faroni LRA, Sousa AH, Freitas RS (2012) Bioactivity of *Jatropha curcas* L. to insect pests of stored products. J Stored Prod Res 48:111–113

Simin K, Ali Z, Khaliq-Uz-Zaman SM, Ahmad VU (2002) Structure and biological activity of a new rotenoid from *Pongamia pinnata*. Nat Prod Res 16:351–357

Singh YP (2007) Efficacy of plant extracts against mustard aphid, *Lipaphis erysimi* on mustard. Indian J Plant Prot 35(1):116–117

Singh D, Singh AK (1991) Repellent and insecticidal properties of essential oils against housefly *Musca domestica* L. Insect Sci Appl 12(4):487–491

Singh D, Siddiqui MS, Sharma S (1989) Reproduction retardant and fumigant properties in essential oils against rice weevil (Coleoptera: Curculionidae) in stored wheat. J Econ Entomol 83(3):727–733

Singh R, Rup PJ, Koul O (2008) Bioefficacy of *Eucalyptus camaldulensis* var. *obtusa* and *Luvanga scandens* essential oils against *Spodoptera litura* (Lepidoptera: Noctuidae). Biopestic Int 4:128–137

Sinha SH (1993) Neem in the integrated management of *Helicoverpa armigera* Hubn. in chickpea. In: World neem conference. Indian Society of Tobacco Science, Bangalore, p 6

Sláma K, Williams CM (1965) Juvenile hormone activity for the bug *Pyrrhocoris apterus*. Proc Natl Acad Sci U S A 54(2):411–414

Solsoloy AD (1995) Pesticidal efficacy of the formulated physic nut, *Jatropha curcas* L. oil on pests of selected field crops. Philipp J Sci 124:59–74

Solsoloy AD, Solsoloy TS (1997) Pesticidal efficacy of formulated *J. curcas* oil on pests of selected field crops. In: Gubitz GM, Mittelbach M, Trabi M (eds) Biofuels and industrial products from *J. curcas*. DBV Graz, Graz, pp 216–226

Solsoloy AD, Solsoloy TS (2000) Insecticide resistance management in cotton in the Philippines. Philipp J Crop Sci 25:26

Stokes JB, Redfern RE (1982) Effect of sunlight on azadirachtin antifeeding potency. J Environ Sci Health A 17:57–65

Suliman R, Saker I, Namora D (2003) Importance of plant extracts in managing acarids damaging to crops. In: Eighth Arab congress of plant protection, 12–16 October, El-Beida, Libya, pp 87–91

Sundaram KMS (1996) Azadirachtin biopesticide: a review of studies conducted on its analytical chemistry, environmental behaviour and biological effects. J Environ Sci Health Part B 31:913–948

Sundaram KMS, Curry J (1996a) Effect of some UV-light absorbers on the photostability of azadirachtin of a neem based pesticide. Chemosphere 32:649–659

Sundaram KMS, Curry J (1996b) Photostabilization of the botanical insecticide azadirachtin in the presence of lecithin as UV protectants. J Environ Sci Health B 31:1041–1060

Sundaram KMS, Sundaram A, Curry J, Sloane L (1997) Formulation selection, and investigation of azadirachtin – a persistence in some terrestrial and aquatic components of a forest environment. Pestic Sci 51:74–90

Susarla RS, Murthy MM, Rao BVSK, Chakrabarti PP, Prasad RBN, Kanjilal S (2012) A method for isolation of karanjin from the expelled cake of *Pongamia glabra*. Eur J Lipid Sci Technol 114:1097–1101

Talopatra B, Malik AK, Talapatra SK (1985) Triterpenoids and flavonoids from the leaves of *Pongamia glabra* Vent.Demethylationstudieson5-methoxyfuranoflavone. J Indian Chem Soc 62:408–409

Taylor DAH (1984) The chemistry of the Limonoids from Meliaceae Prog Chem Org Nat Prod 45:1–101

Thebtaranonth C, Thebtaranonth Y, Wanauppathamkul S, Yuthavong Y (1995) Antimalarial sesquiterpenes from tubers of *Cyperus rotundus*: structure of 10, 12-peroxycalamenene, a sesquiterpene endoperoxide. Phytochemistry 540:125–128

Tormo JR, Gallardo T, González MC, Bermejo A, Cabedo N, Andreu I, Estornell E (1999) Annonaceous acetogenins as inhibitors of mitochondrial complex I. Curr Top Phytochem 2:69–90

Tripathi AK, Rao SM, Singh D, Chakraborty RB, Bhakuni DS (1987) Antifeedant activity of plant extracts against *Diacrisia oblique* walk. Curr Sci 56(12):607–608

Unicini Manganelli RE, Zaccaro L, Tomei PE (2005) Antiviral activity in vitro of *Urtica dioica* L., *Parietaria diffusa* M. and K., and *Sambucus nigra* L. J Ethnopharmacol 98:323–327

Valencia A, Frérot B, Guénego H, Múnera DF, Grossi De Sá MF, Calatayud PA (2006) Effect of *Jatropha gossypiifolia* leaf extracts on three lepidoptera species. Rev Colomb Entomol 32(1): 45–48

Veitch GE, Boyer A, Ley SV (2008) The azadirachtin story. Angew Chem Int Ed 47:9402–9429

Venkataramireddy P, Chitra KC, Rao PK (1990) Efficacy of plant extracts in the control of brinjal spotted leaf beetle, *Henosepilachna vigintioctopunctata* F. In: Proc of symposium of botanical pesticides in IPM, Rajahmundry, pp 225–227

Venkateswarlu B, Katyal JC, Choudhari J, Mukhopadhyay K (1997) Azadirachtin content in the neem seed samples collected from different dry land regions. Neem Newsl (IARI) 14(1):7–11

Vismayaa SEW, Manjunatha JR, Srinivas P, Kanyaa S TC (2010) Extraction and recovery of karanjin: a value addition to karanja (*Pongamia pinnata*) seed oil. Ind Crops Prod : 118–122 (32pp)

Warthen JD, Stokes JR, Jacobson M, Kozempel MF (1984) Estimation of azadirachtin content in neem extract and formulations. J Liq Chromatogr 7:591–598

Williams LAD, Mansingh A (1996) The insecticidal and acaricidal actions of compounds from *Azadirachta indica* (A. Juss.) and their use in tropical pest management. Integr Pest Manag Rev 1(3):133–145

Willis TE (1967) Textbook of pharmacognosy. J. and A Churchill Ltd, pp 513–524

Wink M (2000) Interference of alkaloids with neuroreceptors and ion channels. In: Rahman XA (ed) Bioactive natural products. Elsevier, Amsterdam, Netherlands, pp 3–129

Wink M, Koschmieder C, Sauerweien M, Sporer F (1997) Phorbol esters of *J. curcas* – biological activities and potential applications. In: Gubitz GM, Mittelbach M, Trabi M (eds) Biofuel and industrial products from *Jatropha curcas*. DBV, Graz, pp 160–166

Yadav PP, Ahmad G, Maurya R (2004) Furanoflavonoids from *Pongamia pinnata* fruits. Phytochemistry 65:439–443

Yamasaki RB, Klocke JA (1987) Structure bioactivity relationship of azadirachtin, a potent insect control agent. J Agric Food Chem 35:467–471

Yamasaki RB, Klocke JA, Lee SM, Stones GA (1986) Darlington, M. V. Isolation and purification of azadirachtin from neem (*Azadirachta indica*) seeds using flash chromatography and high performance liquid chromatography. J Chromatogr 18:467

Yin H, Zhang S, Wu J, Nan H, Long L, Yang J, Li Q (2006) *Ponga flavonol*: a prenylated flavonoid from *Pongamia pinnata* with a modified ring A. Molecules 11:786–791

Yule S, Srinivasan R (2013) Evaluation of bio-pesticides against legume pod borer, *Maruca vitrata* Fabricius (Lepidoptera: Pyralidae), in laboratory and field conditions in Thailand. J Asia Pac Entomol 16:357–360

Zeng L, Ye Q, Oberlies NH, Shi G, Gu ZM, He K, McLaughlin JL (1996) Recent advances in Annonaceous acetogenins. Nat Prod Rep 13:275–306

Zoubiri S, Baaliouamer A (2011) Potentiality of plants as source of insecticide principles. J Saudi Chem Soc. doi:10.1016/j.jscs.2011.11.015

Botanical Pesticides for the Management of Plant Nematode and Mite Pests

6

Anand Prakash, Jagadiswari Rao, J. Berliner, Somnath S. Pokhare, Totan Adak, and Kanchan Saikia

Abstract

Plant-infesting nematodes and mites are some of the biotic stresses known to cause considerable yield losses in crops of agricultural importance every year. These bio-agents during feeding plant parts synergize infection of many fungal diseases and also transmit a number of viral diseases. Therefore, management of both these pests is essential to avoid such losses. Plant products/botanical origin pesticides, considered first-generation pesticides, are still important to manage plant nematodes in many vegetable and food grain crops where use of synthetic pesticides is more hazardous to human health. This chapter deals with plant products found effective against the plant nematodes and mites, especially the higher plants, which have been arranged in alphabetical order in tabular form. The active components/active principles isolated and identified from the effective botanical formulations have also been scanned and mentioned against both the plant nematodes and phytophagous mites. Currently, many potential plant products are being used under organic agriculture globally. Of course, for pest management in the large areas, availability of large quantities of these products is a limitation. But soil amendments of the oil cakes/green leaves, interculture of these plants with main crops and crop rotations having intelligently grown the biopesticidal plants are useful strategies for the management of the plant nematodes. For the management of plant mites, no doubt, botanical pesticides have been reported to show toxicity to a number of phytophagous mites, but their applications have not been popularized yet. Thus, there is a need to explore more effective botanical miticides, their formulations and application methods.

A. Prakash (✉) • J. Rao • J. Berliner • S.S. Pokhare
T. Adak • K. Saikia
Division of Crop Protection, Central Rice Research Institute, Cuttack 753 006, India
e-mail: prakas52@yahoo.com

D. Singh (ed.), *Advances in Plant Biopesticides*,
DOI 10.1007/978-81-322-2006-0_6, © Springer India 2014

Keywords
Meloidogyne incognita • *Tetranychus* spp. • Azadirachtin • Nematicide • *Tegetes erecta* • Miticide • Botanical pesticides • Neem • Groundnut • *Trichilia cuneata*

1 Introduction

Plant parasitic nematodes and mites are considered as one of the important pests of agricultural crops in spite of their microscopic nature. They have the potential to devastate the crop, if not controlled in time. Nematodes are threadlike organisms, characterized by triploblastic, bilaterally symmetrical, colourless, unsegmented, pseudocoelomate and vermiform invertebrate animals with well-developed excretory, reproductive and nervous systems and ill-developed circulatory and respiratory system. These are the most abundant group of soil biota next to arthropods. Plant parasitic nematodes are considered as one of the limiting factors for agricultural production. According to world estimates derived in the USA on the basis of international opinion and data, the 20 major life-sustaining crops suffer 10.7 % yield losses due to nematodes, and another 20 economically important crops suffer 14 % yield loss globally; under Indian condition, the average yield losses amount to 12.3 % (Gaur and Pankaj 2009). Further, AICRP (Nematodes) reported annual loss due to nematodes in monetary terms to the tune of Rs. 21,000 million (Jain et al. 2007).

Similarly, a mite is a microscopic arthropod (0.2–1 mm body length) having no eyes, no antennae, no wings, unsegmented body divided in two regions proterosoma (cephalothorax) and hysterosoma (abdomen), short life cycle (4–8 days) and high reproductive potential. Mites generally escape attention of those involved in the production process, even though they play a significant role in reducing crop yields (Rao et al. 1999). Phytophagous mites cause qualitative and quantitative losses directly and also in association with several pathogenic fungi and bacteria in the forms of grain sterility, ill-filled grains and discolouration of the grains/seeds and necrosis on/in the plant leaf and leaf-sheath tissues and are also known as vectors for several diseases.

Phytophagous mites cause injuries/wounds to different plant tissues wherever they feed and thus act as wounding agents. In addition, these mites are also known to carry several spores and mycelia of pathogenic fungi and bacteria on their body parts and contaminate the plants and thus act as carrier, through which these mites enhance the spread of the fungal/bacterial diseases as these mites are fast moving arthropods on the plant surfaces. These two simultaneous acts of the phytophagous mites, first causing injuries to the plant tissues and then contaminating the tissues with the pathogenic microorganisms, help the pathogen to penetrate into the plant tissues, establish the pathogen and easily cause disease more quicker as compared to infection by the pathogen alone. Thus, these mites are observed to break the plant's resistance to pathogens (Rao et al. 1999). Plant mites also act as vectors for a number of viral diseases. Several eriophyid and tetranychid mites are reported as vectors of many viruses causing viral diseases, resulting into potential crop/yield losses. Wheat Streak Mosaic Virus (WSMV) transmitted by *Aceria tulipae* (Jeppson et al. 1975), Sugarcane Streak Mosaic Virus (MMMV) by *Aceria sacchari* (Sithanatham et al. 1979), Pigeonpea Sterility Mosaic Virus (PPSMV) by *A. cajani* (Seth 1962), Dolichos Enation Mosaic Virus (DEMV) by *Tetranychus ludeni* (Rajagopan 1974), Rye Grass Mosaic Virus (RGMV) by *Abacarus hystrix* and Potato Virus Y by *T. cinnabarinus* are common examples of the mites transmitting viruses.

2 Botanical Pesticides

Botanical pesticides are first-generation pesticides being used in traditional agriculture for more than a century. These are higher plant origin pesticides which can directly or indirectly kill or reduce the target pest population. These are the important alternatives to minimize the use of synthetic pesticides as they possess an array of properties including toxicity to the pest, repellency, antifeedancy and insect growth regulatory activities against pests of agricultural importance (Parmar and Dev Kumar 1993; Prakash and Rao 1997). Advantages of botanicals over synthetic pesticides are as follows:

- In general, possess low mammalian toxicity and thus constitute least/no health hazards and environmental pollution
- Practically, no risk of developing pest resistance to these products, when used in natural forms
- Less hazards to nontarget organisms and pest resurgence
- No adverse effect on plant growth, seed viability and cooking quality
- Less expensive and easily available

Thousand seventy-five species of higher plants have been found to possess pesticidal property against insects, mites, nematodes, molluscs, birds and rodent pests of agricultural importance. Some of the botanicals like neem, bel, ocimum, senwar, pyrethrum, tobacco, karanj, mahua, cymbopogon and sweet flag have already attained the status of potential pesticides of plant origin against field pests including phytonematodes and plant mites and also against insect pests in storage ecosystems (Prakash and Rao 1997).

3 Botanicals in Plant Nematode Management

Botanicals/plant products have been extensively used earlier for the management of plant nematodes especially in the form of soil application of the oil cakes of different plant seeds like mustard, mahua, neem, safflower, sunflower, etc. Leaves, seed and whole plant extracts have also been reported to possess nematicidal activity (Table 6.1).

4 *Azadirachta indica* A. Juss (Meliaceae): Indian Neem

Leaf, kernel extracts, kernel powder and neem oil having nematicidal activity are briefed as follows:

4.1 Neem Leaf as Nematicide

The application of 5–10 % leaves in the infested soil reduced the incidences of root-knot nematodes in tomato and okra (Singh and Sitaramaiah 1967). Similarly, chopped leaves incorporated with the soil also reduced populations of *M. incognita* in okra and eggplant (Lall and Hameed 1969; Hussain and Masood 1975). Egumjobi and Afolami (1976) found neem leaf extract to reduce the populations of *Pratylenchus brachyurus* in maize roots. Alam (1976) reported neem seedlings planted in the periphery of tomato, eggplant and chillies in naturally infested soil in a mixed cropping pattern to reduce populations of the nematodes. Agbenin and co-workers in 2005 found that *M. incognita* egg masses exposed to 20 and 30 % concentration of fresh neem leaf extract did not hatch and showed 50 and 90 % mortality within 15 min of exposure, respectively. By 30 min, mortality had reached almost 100 %, while at 10 % concentration of the neem leaf extract, an egg mass ruptured, but the released larvae were all found dead or moribund. All unhatched egg masses teased up contained dead or moribund larvae. The cumulative effect of the extracts of dry and fresh neem leaves on mortality of larvae and hatchability of egg masses was similar after 24 h of exposure regardless of the concentrations of the extracts. Neem leaf powder incorporated in the soil at 40 g/pot was found to reduce the incidences of the root-knot nematode, *M. incognita*, in okra seedlings (Bhatnagar et al. 1980). Kaliram and Gupta (1980) reported neem leaf extract in water to reduce root galls caused by *M. javanica* in infested chickpea.

Table 6.1 Nematicidal activity of the botanicals against plant nematode

Generic name/common name	Plant part/formulation	Nematicidal activity	Reference
Acacia auriculiformis A. Cunn. (Mimosaceae): Australian wattle	Alcoholic extract and air-dried powdered funicles	Larval mortality to root-knot nematode, *Meloidogyne incognita*	Sinhababu et al. (1992)
Adhatoda vasica Nees. (Acanthaceae): Malabar nut tree	Powdered leaves at 100 kg/kg soil	Reduced multiplication of *M. incognita*	Pandey (1995)
Agastache rugosa (Lamiaceae): Korean mint	Essential oils	Exhibited strong nematicidal activity against *M. incognita*	He Qin Li et al. (2013)
Ageratum conyzoides Linn. (Asteraceae): Goat weed	Whole plant aqueous extract	Toxicity to *Meloidogyne javanica*	Mukherjee (1983)
Allium cepa Linn. (Amaryllidaceae): Onion	Bulb aqueous extract	Toxicity to *M. incognita* and *Radopholus similis*	Sarra-Guzman, (1984)
	Root exudates	Toxicity to 2nd and 5th juveniles of *M. incognita*	Hasan (1992)
Allium sativum Linn. (Amaryllidaceae): Garlic	Crude aqueous extracts of leaves and bulbs	Toxicity to sugarcane nematode, *Aphelenchoides sacchari*, and citrus nematode, *Tylenchorhynchus semipenetrans*	Nath et al. (1982)
	Aqueous bulb extract	Toxicity to *M. incognita* and *M. javanica*	Sukul and Ghosh (1974), Mukherjee (1983)
	Aqueous bulb extract	Toxicity to *Radopholus similis*	Sarra-Guzman, (1984)
	Bulb extract	Inhibited the hatching of egg masses in *M. incognita*	Agbenin et al. (2005)
	Aqueous bulb extract applied to soil	Toxicity to *M. incognita* in vegetables	Gupta et al. (1985)
	Garlic oil	Reduced fertility and larval toxicity to *M. incognita*	Gupta and Sharma (1985, 1990b)
	Aqueous bulb extract	Reduced fertility and larval toxicity to *M. incognita*	Gupta and Sharma (1990a, b)
	Aqueous bulb extract	Larvicidal to *M. javanica* in tobacco nursery	Krishnamurthy and Murthy (1990)
Aloe barbadensis Mill. (Liliaceae): Barbados aloe	Whole plant extract	Ovicidal activity to *M. incognita* in vegetable crops	Hussain and Masood (1975)
	Leaf extract in water	Toxicity to *M. incognita* in potted vegetable plants	Mahmood et al. (1979)
	Aqueous root and shoot extracts	Ovicidal activity to *M. incognita*	Pandey and Haseeb (1988)
Amaranthus gracilis Linn. (Amaranthaceae): Slender amaranth	Whole plant aqueous extract	Toxicity to *M. javanica*	Mandal and Bhatti (1983)
Anagallis arvensis Linn. (Primulaceae): Poor man's weather grass	Aqueous leaf extract	Toxicity to root nematode, *Rotylenchulus reniformis*	Mahmood et al. (1982)
Andrographis paniculata (Burm.f.) Wall. Ex. Nees (Acanthaceae): King of bitters	Leaf and whole plant extract	Toxicity to *M. incognita*	Mukherjee and Sukul (1978)
	Dried plant amended in soil	Reduced incidences of *M. incognita* in vegetables	Goswami and Vijayalakshmi (1985)
	Plant extracted applied in root zone of brinjal	Reduced populations of *M. incognita, R. reniformis* and *Pratylenchus delattrei*	Poornima and Vodivelu (1993)

(continued)

Table 6.1 (continued)

Generic name/common name	Plant part/formulation	Nematicidal activity	Reference
Angelica pubescens Linn. (Umbelliferae): Shishiudo	Whole plant aqueous extract	Toxicity to rice white-tip nematode, *Aphelenchoides besseyi*	Katsura (1981)
Annona squamosa Linn. (Annonaceae): Sweetsop	Aqueous and ethanolic leaf extracts	Ovicidal and toxicity to *M. incognita* in vegetable roots	Hussain and Masood (1975), Mahmood et al. (1979)
Anthocephalus chinensis (Lamb.) Rich. ex. Walp. (Naucleaceae): Kadam	Aqueous root extracts and its purified fractions	Toxicity to *M. incognita* and *Radopholus similis*	Sarra-Guzman (1984)
	Leaf extract	Toxicity to *M. incognita*	Chatterjee and Sukul (1979)
Araucaria cookii Don. (Araucariaceae): Cook pine	Aqueous leaf extract	Toxicity to *M. incognita* in vegetable crops	Jain et al. (1986)
Argemone mexicana Linn. (Papaveraceae): Mexican prickly poppy	Aqueous root and leaf extracts	Toxicity to *M. incognita* and *M. javanica*	Ramanath et al. (1982), Mukherjee (1983)
	Chopped leaves mixed with soil	Reduced gall incidences of *M. incognita* in papaya	Reddy and Khan (1990)
	Root exudates	Reduced adult female population and gall incidences and enhanced life cycle span of *M. incognita*	Hasan (1992)
	Aqueous shoot extract	Reduced penetration of *M. incognita* in brinjal and soybean crops	Sharma and Prasad (1995)
Artabotrys odoratissimus R. Br. (Annonaceae): Climbing ylang-ylang	Aqueous leaf extract	Reduced larval hatching of *M. incognita* tested under controlled conditions	Chattopadhaya and Mukhopadhaya (1989b)
Artemisia vulgaris Auct. Non. Linn. (Asteraceae): Indian wormwood fleabane	Aqueous root extract	Toxicity to *M. incognita* and *Radopholus similis*	Sarra-Guzman (1984), Sherif et al. (1987)
Artemisia herba alba (Compositae): Wormwood	Leaf extract in methanol	Increase mortality of *M. incognita*	Al-Banna et al. (2003)
Asparagus officinalis Linn. (Liliaceae): Garden asparagus	Aqueous root extract	Toxicity to *Rotylenchulus reniformis* and *M. incognita*	Das (1983)
Azadirachta indica A. Juss (Meliaceae): Indian neem	*Its leaves, kernel extracts, oil, oil cake, etc. are the most exploited among the botanicals for nematicidal activities and have been described elsewhere in the same chapter*		
Azolla pinnata R. Br. (Azollaceae): Azolla	Aqueous leaf extract	Ovicidal to *M. incognita*	Thaker (1988)
	Azolla biofertilizer	Reduced incidences of *M. incognita* in okra	Thaker et al. (1987)
	Dried leaf powder applied in soil	Reduced incidences of *M. incognita* and *M. javanica* in okra	Patel et al. (1990a, b)
Bocconia cordata Linn. (Papaveraceae): Plume poppy	Aqueous root extract	Toxicity to *Rhabditis* sp.	Rohde (1972)
Bomarea nivea (Linn.) Gauel (Urticaceae)	Oil cake and leaves amended in soil	Reduced population of *Meloidogyne arenaria*	Mian and Rodriguez (1982)
Borrelia sp.	Leaf extract	hatching inhibition in *M. incognita*	Agbenin et al. (2005)

(continued)

Table 6.1 (continued)

Generic name/common name	Plant part/formulation	Nematicidal activity	Reference
Brassica spp.	As a cover crop	Nematicidal activity against *M. incognita*	Monforta et al. (2007)
Brassica campestris Linn. Var. sarson Prain. (Brassicaceae): Yellow mustard/Indian colza	Oil cake amended in soil at 2.5 t/ha	Reduced root-gall intensity due to *M. javanica* in tomato and okra	Hameed (1968), Singh and Sitaramaiah (1971)
	Oil cake water extract	Toxicity to *M. incognita*	Mishra and Prasad (1973)
	Oil cake emended in soil	Suppressed populations of *H. indicus, M. incognita, A. avenae* and *T. brassicae*	Khan et al. (1974)
	Oil cake at 100 kgN/ha amended in soil	Reduced development of root-knot nematodes in potato and promoted plant growth	Azam and Khan (1976)
	Oil cake amended in soil	Reduced populations of *H. indicus, M. incognita, Tylenchus filiformis* and *T. brassicae*	Alam et al. (1977)
	Oil cake amended in soil	Reduced cyst populations and hatching in *Heterodera avenae*	Sharma et al. (1981)
	Oil cake water-soluble fractions applied in soil	Reduced incidences of *M. incognita* and *M. javanica*	Patel (1985)
Brassica hirta Moench (Brassicaceae): White mustard	Oil cake aqueous extract	Toxicity to golden nematode, *Heterodera rostochiensis*	Ellenby (1945a, b)
Brassica integrifolia Linn. (Brassicaceae): Mustard	Oil cake application in soil	Reduced the incidences of root-knot nematode, *M. incognita*	Varma (1976)
Brassica nigra (Linn.) Koeh. (Brassicaceae): Black mustard/ black rye	Water extract of oil cake	Inhibited the larval emergence of *M. incognita*	Khan et al. (1967)
	Oil cake amended in soil 3 weeks before tomato transplanting	Reduced the root galls by *M. javanica*	Singh and Sitaramaiah 1971; Siddiqui (1976), Alam et al. (1976)
	Oil cake at 100 kg N/ha amended in soil	Reduced populations of stylet-bearing nematodes in soil	Mukherjee (1983)
	Oil cake water extract	Toxicity to oat nematode, *Aphelenchus avenae*; stem and bulb nematode, *Ditylenchus cypei*; *H. rostochinesis*, *T. brassicae* and *M. incognita*	Krishnamurthy and Murthy (1990)
Caesalpinia crista Linn. (Caesalpiniaceae): Molucca bean	Leaf extract (1:10) dilution applied in soil of tobacco nursery	Reduced incidences of root galls by *M. incognita*	Krishnamurthy and Murthy (1990)
Calendula officinalis Linn. (Asteraceae): Pot marigold	Dried plant amended in soil	Reduced incidences of root galls by *M. incognita*	Goswami and Vijayalakshmi (1985)
Calophyllum inophyllum Linn. (Clusiaceae): Indian laurel	Oil cake amended in soil	Significantly controlled *M. javanica* in tobacco	Pillai and Desai (1976)
	Oil cake water extract	Toxicity to *M. incognita* and *M. javanica*	Mukherjee (1983), Goswami and Vijayalakshmi (1987)
	Oil cake amended in soil	Reduced incidences of root-knot nematode, *M. graminicola*	Prakash et al. (1990)

(continued)

Table 6.1 (continued)

Generic name/common name	Plant part/formulation	Nematicidal activity	Reference
Calotropis gigantea Ait. (Asclepiadaceae): Crown plant/Madar	Leaves amended in soil	Reduced incidences of *M. incognita*	Kumar and Nair (1976)
	Leaf powder incorporated in soil	Reduced incidences of *M. incognita*	Bhatnagar et al. (1980)
	Flower/plant 1 00 g/Kg soil	Reduced larval penetration of *M. incognita* in tomato	Vijayalakshmi and Goswami (1985)
	Dried leaves amended in soil	Reduced population of *M. incognita, H. indicus* and *R. reniformis* in betelvine	Sivakumar and Marimuthu (1986)
Calotropis procera Willd. (Asclepiadaceae): Swallow-wort	Aqueous leaf extract applied in potted vegetables	Toxicity to *M. incognita*	Jain et al. (1986)
	Chopped leaves admixed with infested soil	Reduced gall formation by *M. incognita* in papaya	Reddy and Khan (1990)
	Chopped leaves admixed with soil for potted tomato	Reduced gall formation by *M. incognita* and also its population	Sundarbabu et al. (1993)
	Aqueous leaf extract	Ovicidal activity to cyst nematode *Heterodera avenae* on barley	
	Chopped leaves admixed with soil at 1.5 kg/sq.m in pointed gourd fields	Reduced the populations of *M. incognita, H. indicus, T. vulgaris* and *H. dihystera*	Verma and Anwar (1995)
Camellia sinensis (Linn.) O. Kuntz. (Theaceae): Tea	Decaffeinated tea waste	Inhibited oviposition of *M. graminicola* and also its larval penetration in rice root	Roy (1976)
Cannabis sativa Linn. (Moraceae): Hemp	Root and shoot aqueous extract	Toxicity to *T. brassicae, H. indicus* and *R. reniformis*	Haseeb et al. (1982)
	Soil amendment of whole plant	Reduced incidences of *M. incognita*	Goswami and Vijayalakshmi (1986a, b)
Capsicum annum Linn. (Solanaceae): Red pepper	Whole plant extract (1:10 dilution) applied to soil in tobacco nursery	Reduced gall formation by *M. incognita*	Krishnamurthy and Murthy (1990)
Carica papaya Linn. (Caricaceae): Papaya	Unripe fruit extract in water	Toxicity to *M. incognita*	Mukherjee (1983)
Carthamus oxyacantha M. Bieb (Asteraceae): Wild safflower	Whole plant and leaf extracts in water	Toxicity to nematodes of agricultural importance in general	Mukherjee (1983)
Carthamus tinctorius Linn. (Asteraceae): Safflower/Kusum	Flower and root aqueous extract	Toxicity to white-tip nematode, *Aphelenchoides besseyi,* and also to other nematodes	Katsura (1981)
	Oil admixed with soil	Toxicity to *Pratylenchus penetrans*	Mukherjee (1983)
	Oil cake amended in soil in tomato	Reduced incidences of *M. incognita* and *M. javanica*	Miller (1979), Goswami and Vijayalakshmi (1986a, b)
Carum carvi (Apiaceae or Umbelliferae): Caraway	Umbels	Highly effective at J_2 immobilization and hatching inhibition of *Meloidogyne javanica*	Oka et al. (2000)

(continued)

Table 6.1 (continued)

Generic name/common name	Plant part/formulation	Nematicidal activity	Reference
Carissa carandas Linn. (Caesalpiniaceae): Karonda	Aqueous leaf extract	Toxicity to larvae of *M. incognita* and adults of *R. reniformis*	Haseeb et al. (1982)
Cassia fistula Linn. (Caesalpiniaceae): Amaltas	Leaves as soil amendment at 5 % (w/w)	Reduced incidences of *M. javanica*	Singh and Sitaramaiah (1967)
	Aqueous leaf extract	Toxicity to *R. reniformis* under lab. test	Haseeb et al. (1982)
Cassia occidentalis Linn. (Caesalpiniaceae): Coffee senna	Leaves as soil amendment at 5–10 % (w/w)	Reduced incidences of *M. javanica*	Singh and Sitaramaiah (1967)
Catharanthus roseus (Linn.) G. Don. (Apocynaceae): Madagascar periwinkle	Whole plant and leaves applied in rows,1–2 days before seedling transplanting	Inhibited development of *R. reniformis* and *T. vulgaris* in okra fields	Patel et al. (1990a, b)
	Plant extract (1:10 dilution)	Reduced root galling due to *M. incognita*	Krishnamurthy and Murthy (1990)
Centrosema pubescens Benth (Fabaceae): Centro	Aqueous leaf extract	Toxicity to *M. incognita*	Hoan and Davide (1979)
Chenopodium album Linn. (Chenopodiaceae): Pigweed	Whole plant extract in water	Toxicity to *M. incognita*	Nandal and Bhatti (1983)
Chenopodium ambrosioides Linn. (Chenopodiaceae): Wormwood seed	Aqueous leaf extract	Toxicity to larvae of *M. incognita*	Hussain and Masood (1975)
	Root and shoot extract in water	Toxicity to *T. brassicae, H. indica* and *R. reniformis*	
	Whole plant extract	Potential nematicide to control plant nematodes	Garciaspinosa (1980)
Chromolaena odorata D.C. (Asteraceae): Hagonoy	Aqueous leaf extract	Toxicity to root-knot nematode, *Meloidogyne incognita*	Hoan and Davide (1979)
Chrysanthemum indicum Linn. (Asteraceae): Japanese chrysanthemum	Whole plant extract (1:10 dilution) applied in soil of tobacco nurseries	Toxicity to *M. incognita*	Krishnamurthy and Murthy (1990)
Cirsium arvense Mill. (Asteraceae): Canadian thistle	Whole plant extract	Showed nematicidal activity to nematodes of agricultural importance	Mukherjee (1983)
Citrus reticulata Blanco (Rutaceae): Mandarin orange	Fruit peel and leaf extract	Toxicity to *M. incognita* and *M. javanica*	Mukherjee (1983)
Clerodendrum inerme (Linn.) Garetin (Verbenaceae): Glory Bower, Indian privet Vilayati mehndi	Whole plant extract	Reduced the incidences of *M. incognita* and *M. javanica* in okra roots	Patel et al. (1985)
	Leaf powder applied in soil before seeding okra	Reduced the incidences of *M. incognita* and *M. javanica* in okra roots	Patel et al. (1990a)
	Leaf extract applied in soil	Reduced root galling of *M. incognita* in tobacco nurseries	Krishnamurthy and Murthy (1990)
	Leaves amended in soil at 5 kg/plant	Reduced nematode population in plantain and increased bunch weight	Jacob et al. (1990)
Colocasia antiquorum Remy. (Asteraceae): Cocoyam	Leaf water extract	Toxicity to *R. reniformis* under laboratory test	Haseeb et al. (1982)

(continued)

Table 6.1 (continued)

Generic name/common name	Plant part/formulation	Nematicidal activity	Reference
Coffea robusta Linden (Rubiaceae): Congo coffee	Coffee powder	Toxicity to root-knot nematode, *Meloidogyne arenaria*	Mian and Rodriguez (1982)
Cosmos bipinnatus Cav. (Asteraceae): garden cosmos or Mexican aster	Aqueous flower extract	Toxicity to *M. incognita* and inhibited oviposition	Bano et al. (1986)
Cordia myxa Linn. (Boraginaceae): Glue berry	Aqueous leaf extract	Toxicity to *Rotylenchus reniformis*	Haseeb et al. (1982)
Crotalaria juncea Linn. (Fabaceae): Sunn hemp	Leaves amended in soil at 5–10 % w/w	Reduced the populations of *M. javanica* in potted tomato and okra	Singh and Sitaramaiah (1967)
	Whole plant aqueous extract	Toxicity to *M. arenaria* and *M. incognita*	Mian and Rodriguez (1982), Das (1983)
	Chopped leaves applied in soil	Reduced population of *Heterodera avenae*	Dahlya et al. (1986)
Crotalaria spectabilis Medik. (Fabaceae): Rattlebox	Aqueous leaf and seed extract	Toxicity to *M. incognita*	
Croton sparsiflorus Marong. (Euphorbiaceae): Croton	Aqueous leaf extract	Toxicity to *M. incognita*	
Cymbopogon citratus (D.C.) Stapf. (Poaceae): Lemon grass	Aqueous leaf extract	Toxicity to *M. incognita*	Kumar and Nair (1976), Hoan and Davide (1979)
Cynodon dactylon (Linn.) Pers. (Poaceae): Bermuda grass	Whole plant aqueous extract	Toxicity to *M. incognita*	Hoan and davide (1979)
Cyperus rotundus Linn. (Cyperaceae): Nut grass	Whole plant aqueous extract	Toxicity to *M. incognita*	Hoan and Davide (1979)
Dalbergia sissoo Roxb. (Fabaceae): Sissoo	Saw dust amended in soil alone or with neem kernel powder	Reduced populations of *M. javanica*	Sitaramaiah and Singh (1978)
	Aqueous leaf extract	Toxicity to *R. reniformis*	Haseeb et al. (1982)
Daphne odora Linn. (Thymelaeaceae): Sweet daphne	Aqueous root extract	Toxicity to *A. besseyi*	Munakata (1978), Katsura (1981)
Datura metel Linn. (Solanaceae): Angel's trumpet	Leaves amended in soil	Reduced gall incidences by *M. javanica* in chickpea	Kaliram and Gupta (1980, 1982), Gupta and Kaliran (1981)
	Aqueous leaf extract	Toxicity to *M. incognita* and *M. javanica*	Mukherjee (1983)
Datura stramonium Linn. (Solanaceae): Jimson weed	Seed and leaf aqueous extract	Toxicity to *M. incognita* and *M. javanica*	Mukherjee (1983)
	Leaf, bulb and stem water and methanolic extracts	Toxicity to J2 stage of *Tylenchulus semipenetrans* and *Anguina tritici*	Kumari et al. (1986)
Derris elliptica (Wall.) Benth (Fabaceae): Tuba root	Root and seed extracts in water	Toxicity to *M. incognita*	Hoan and Davide (1979)
Desmodium zangelicum (Linn.) D.C. (Leguminosae): Jimson weed or thorn apple	Root extract in water	Toxicity to *M. incognita* and *Radopholus similis*	Sarra-Guzman, (1984)
Digitaria documbens (Linn.) Nees. (Poaceae): Pangola grass	Whole plant aqueous extract	Ovicidal to *M. incognita*	Haroon et al. (1982)
	Soil planted with this grass	Reduced populations of *M. incognita* in tomato and toxicity to its larvae	Haroon and Smart (1983a, b)

(continued)

Table 6.1 (continued)

Generic name/common name	Plant part/formulation	Nematicidal activity	Reference
Eclipta alba Hassk. (Asteraceae): Morchand	Root, shoot and whole plant aqueous extract	Toxicity to *M. graminicola* in rice and other spp. of *Meloidogyne* and *Rotylenchulus*	Prasad and Rao (1979)
	Whole plant extract	Ovicidal and toxic to *M. incognita*	Bano et al. (1986)
	Leaves amended in soil	Reduced incidences of root knots in tomato	Goswami and Vijaylakshmi (1986a, b)
Eichornia crassipes (Mart.) Salms (Pontedariaceae): Water hyacinth	Compost amendment in soil	Reduced hatching of eggs of *M. graminicola*	Roy (1976)
	Root extract and its pure fractions	Toxic to *M. incognita* and *R. similis*	Sarra-Guzman, (1984)
	Aqueous leaf extract	Toxicity to *M. incognita* in vegetable crops	Jain et al. (1986)
Emblica officinalis Gaertn. (Euphorbiaceae): Indian gooseberry	Aqueous leaf extract	Toxicity to *M. incognita*	Haseeb et al. (1982)
Eragrostis amabilis Jacq. (Poaceae): Bug's egg grass	Aqueous leaf extract	Toxicity to *M. incognita*	Hoan and Davide (1979)
Eruca vesicaria (L.) Cav. Subsp. *sativa* (Mill.) Thell. (Cruciferae): Roquette/taramira	Water-soluble fractions of its oil cake extract	Toxicity to root-knot nematode, *M. incognita*	Bhatnagar et al. (1978)
Erythrina indica Lamk. var. *parcellii* Hort. (Papilionaceae): Indian coral tree	Aqueous leaf extract	Toxicity to *M. incognita* in the potted vegetable plants	Mohanty and Das (1987, 1988)
Eucalyptus melliodora	Essential oils	High nematicidal activity against *M. javanica*	Ntalli and Caboni (2012)
Eupatorium odoratum Linn. (Asteraceae): Bitterbush	Leaves amended in soil	Reduced gall incidences by *M. incognita* in okra	Kumari et al. (1986), Subramaniyan (1985)
Euphorbia cyparissias Lindl. (Euphorbiaceae): Cypress spurge	Whole plant extract in water	Toxicity to nematodes in general	McIndoo (1982)
Ficus elastica Roxb. (Moraceae): Indian rubber tree	Chopped leaves of seedlings and early plants amended in soil	Toxicity to *M. incognita, R. reniformis* and *T. brassicae* in tomato and eggplants	Siddiqui et al. (1987, 1992)
Fleurya interrupta Gandich. (Urticaceae): Indian birthwort	Aqueous leaf extract	Toxicity to *M. incognita* and *M. javanica*	Mukherjee and Sukul (1978), Mukherjee (1983)
Foeniculum vulgare (Umbelliferae): Fennel	*Umbels*	Highly effective at J2 immobilization and hatching inhibition of *M. javanica*	Oka et al. (2000)
Gaillardia picta Foug. (Asteraceae): Gaillardia	Flower, stem and leaf extracts in water	Toxicity to *M. incognita*	Tiyagi et al. (1985)
Gliricidia sepium (Jacq.) Walp. (Papilionaceae): Madre tree	Leaves amended in soil	Reduced incidences of *M. incognita* in okra	
	Chopped leaves amended in soil at1 0g/kg per pot	Reduced incidences of nematodes in black pepper	Jasy and Koshy (1992)
Gloriosa superba Linn. (Liliaceae): Tiger's claws	Aqueous root and shoot extracts	Toxicity to *M. incognita*	Pandey and Haseeb (1988)

(continued)

Table 6.1 (continued)

Generic name/common name	Plant part/formulation	Nematicidal activity	Reference
Gossypium hirsutum Linn. (Malvaceae): Cotton	Seed oil mixed with soil	Toxic to *Heterodera tabacum* and also to *Pratylenchus penetrans*	Miller (1979)
	Oil cake amended in soil	Reduced incidences of *M. incognita*	Lanjeswar and Shukla (1979)
Guizotia abyssinica Cass. (Astreaceae): Niger seed	Seed oil cake amended in soil	Reduced incidences of *M. incognita* in tomato	Gowda (1972)
Hannoa klaineana Planch. (Simaroubaceae)	Aqueous seed extract	Inhibited penetration of *M. javanica*	Proct and Comprobst (1983)
Hannoa undulata Planch. (Simaroubaceae)	Seed extract mixed with soil	Inhibited penetration of *M. javanica*	Proct and Comprobst (1983)
Helenium (hybrid) Linn. (Asteraceae): Moerheim beauty	Leaf extract in water	Toxicity to *Pratylenchus penetrans*	Gommeres (1971)
Helianthus annuus Linn. (Asteraceae): Sunflower	Root, flower and seed extracts	Toxicity to *M. incognita*	Das (1983)
	Oil cake amended in soil	Reduced root-knot index in eggplants	Singh and Singh (1988)
Holarrhena antidysenterica R. Br. (Apocynaceae): Tellicherry bark	Leaf, root and bark extracts in water	Toxicity to *M. incognita* and *M. javanica*	Mukherjee (1983)
Hydnocarpus laurifolia (Donnstr.) Sleumer (Flacourtiaceae): Marotti	Oil cake amended in soil	Toxicity to *M. incognita*	Desai et al. (1973)
	Flower, fruit and seed extracts	Toxicity to *M. incognita* and *M. javanica*	Mukherjee (1983)
Hypericum androsaemum (Hypericaceae): Tutsan	Whole plant extract in methanol	Increase mortality *Meloidogyne javanica*	Al-Banna et al. (2003)
Imperata cylindrica (L.) P. Beauv. var. *major* (Nees.) C.E. Hubb. ex. Hubb. and Vaughan (Poaceae)	Leaf extract in water	Toxicity to *M. incognita*	Hoan and Davide (1979)
Ipomoea palmata Forsk. (Convolvulaceae): Railway creeper	Leaf and stem extracts in water and alcohol	Toxicity to 2nd stage juveniles of *T. semipenetrans* and *Anguina tritici*	Kumari et al. (1986)
Iris japonica Linn. (Iridaceae): Japanese iris	Aqueous leaf extract	Toxicity to white-tip nematode, *A. besseyi*	Katsura (1981)
Jasminum arborescens Roxb. (Oleaceae): Tree jasmine	Leaf extract in water	Toxicity to *M. incognita* in vegetable crops	Hussain and Masood (1975)
Jatropha gossypifolia Linn. (Euphorbiaceae)	Leaf extract in water	Toxicity to eggs of *Heterodera avenae*	
Lantana camara Linn. (Verbenaceae): Lantana	Chopped leaves mixed with infested soil	Reduced root galling in papaya by *M. incognita*	Reddy and Khan (1990)
Lawsonia inermis Linn. (Lathyraceae): Egyptian privet/henna	Leaf extracts in water and methanol	Toxicity to 2nd stage juveniles of *T. semipenetrans* and *Anguina tritici*	Kumari et al. (1986)
Leucaena leucocephala (Lamk.) de Wit. (Mimosaceae): White popinac/horse tamarind	Root extract in water	Toxicity to *M. incognita* and also to *R. similis*	Hoan and Davide (1979), Sarra-Guzman (1984)
	Chopped leaves amended in soil	Reduced population of cyst nematode, *Heterodera avenae*, in wheat	Dahlya et al. (1986)
	Aqueous leaf extract	Toxicity to *Aphelenchoides composticola*	Grewal and Sohi (1988)
	Chopped leaves admixed with soil at 5 g/kg	Reduced gall index of *M. incognita* in potted tomato	Sundarbabu et al. (1993)

(continued)

Table 6.1 (continued)

Generic name/common name	Plant part/formulation	Nematicidal activity	Reference
Linum usitatissimum Linn (Linaceae): Linseed/flax	Oil cake applied in soil grown with tomato and okra seedling	Reduced root gall incidences of *M. incognita* and *M. javanica*	Hameed (1968), Singh and Sitaramaiah (1969, 1971, 1976), Srivastava (1971), Langeswar and Shukla (1986)
	Aqueous leaf extract	Toxic to *R. reniformis*	Mahmood et al. (1982)
	Oil admixed with soil	Toxicity to *P. penetrans* and *M. incognita*	Miller (1979), Sen and Das-Gupta (1985)
	Oil cake amended in soil	Reduced root-knot index and populations of *M. incognita*	Singh and Singh (1988)
Litsea cubeba litsea	Essential oils	Nematicidal activity against pinewood nematode *Bursaphelenchus xylophilus*	Park et al. (2007)
Madhuca indica J.F. Gmel. (Sapotaceae): Mahua	Oil cake water extract	Inhibited larval emergence of *M. javanica* and *M. incognita*	Khan et al. (1967), Khan (1974)
	Oil cake applied in soil grown with tomato and okra seedling	Reduced root gall incidences of *M. incognita*	Singh and Sitaramaiah (1969)
	Oil cake amended in soil	Reduced populations of stylet-bearing nematodes in/around the roots of eggplant, guava and citrus	Mobin and Khan (1969)
	Oil cake amended in soil at 100 Kg N/ha	Reduced population of *M. incognita* in tomato and other vegetable crops	Khan et al. (1973), Azam and Khan (1976), Langeswar and Shukla (1986)
	Oil cake water extract	Toxic to *A. avenae* and *Ditylenchus cypei*	Mukherjee (1983)
	Oil cake amended in soil at 1 g/kg	Reduced root galling by *M. incognita* in *Ocimum basilicum*	Haseeb et al. (1988)
	Oil cake amended in soil at 100 Kg N/ha	Reduced population of *M. incognita*, *R. reniformis* and *P. delattrei* in brinjal	Poornima and Vodivelu (1993)
Mangifera indica Linn. (Anacardiaceae): Mango	Leaf extract (1:10 dilution) amended in soil	Reduced root galling by *M. incognita* in tobacco nurseries	Krishnamurthy and Murthy (1990))
	Green leaves admixed with soil at 5 t/ha	Controlled *M. incognita* in okra	Kumar and Nair (1976)
	Leaf extract in water	Toxicity to 2nd stage juveniles of *M. incognita*	Vijayalakshmi et al. (1979)
	Solution of dried leaves soaked in water	Showed 88 % mortality to *Criconemoides annulatum*	Jain and Saxena (1993)
Melia azedarach Linn. (Meliaceae): Chinaberry/dharek	Leaves amended in soil	Reduced root-knot incidences by *M. incognita* in tomato and okra	Singh and Sitaramaiah (1969)
	Root and shoot aqueous extracts	Toxicity to *T. brassicae*, *H. indicus* and *R. reniformis*	
	Its interculture with tomato and okra	Reduced incidences of *M. incognita* and *R. reniformis*	Siddiqui and Alam (1987)
	Oil cake amended in soil at 5 t/ha	Reduced root-knot index of *M. incognita*	Verma (1993)

(continued)

Table 6.1 (continued)

Generic name/common name	Plant part/formulation	Nematicidal activity	Reference
Mentha spicata (Linn.) Huds. (Lamiaceae): Spearmint	Aqueous leaf extract	Toxicity to *M. incognita*	Haseeb et al. (1982)
		Hatching inhibition of *M. javanica*	Oka et al. (2000)
Mentha rotundifolia: Apple mint	Foliage	Highly effective in immobilizing juveniles and hatching inhibition of *Meloidogyne javanica*	Oka et al. (2000)
Mimosa pudica Linn. (Mimosaceae): Touch me not	Aqueous leaf extract	Toxicity to *M. incognita*	Hoan and Davide (1979)
Momordica charantia Linn. (Cucurbitaceae): Bitter gourd	Aqueous and alcoholic leaf extracts	Toxicity to *M. incognita* and *M. javanica*	Mahmood et al. (1979), Mukherjee (1983)
Moringa pterygosperma Gaertn. (Rubiaceae): Horse tree	Aqueous leaf extract	Toxicity to *M. incognita* and *R. similis*	Hoan and Davide (1979), Sarra-Guzman (1984)
Moringa oleifera Lamk. (Moringaceae): Drumstick tree	Aqueous leaf extract (1:10 dilution) in soil	Reduced root galling by *M. incognita* in tobacco nursery	Krishnamurthy and Murthy (1990)
Nasturtium officinale R. Br. (Brassicaceae): Watercress	Aqueous leaf extract	Toxicity to *Heterodera rostochiensis*	Ellenby (1945a)
Nicotiana tabacum Linn. (Solanaceae): Tobacco	Root and shoot aqueous extracts	Toxicity to *T. brassicae*, *H. Indicus* and *R. reniformis*	
	Leaf extract	Highly toxic to *M. incognita*	Wiratno et al. (2009)
Ocimum sanctum Linn. (Lamiaceae): Holy basil	Crude leaf aqueous extract	Toxicity to *M. incognita*	Chatterjee et al. (1982)
	Leaf extracts in ethanol and methanol	Toxicity to *M. incognita* and *M. javanica*	Mukherjee (1983)
	Root and shoot aqueous extracts	Ovicidal to *M. incognita*	Haseeb and Butool (1990)
	Aqueous leaf extract	Ovicidal to *H. avenae*	
Origanum syriacum	Leaf extract in methanol	Causes mortality *M. javanica*	AL-Banna et al. (2003)
Syrian oregano	Foliage	Effective at reducing root galling on cucumber by *Meloidogyne javanica*	Oka et al. (2000)
Origanum vulgare (C-type) Oregano (Lamiaceae): Wild marjoram	Foliage	Effective at reducing root galling on cucumber by *Meloidogyne javanica*	Oka et al. (2000)
Oryza sativa Linn (Poaceae): Rice	Aqueous leaf extract	Toxicity to *M. incognita*	Hoan and Davide (1979)
Papaver rhoeas Linn. (Papaveraceae): Corn poppy	Aqueous root and shoot extract	Toxicity to *T. brassicae*, *H. indicus* and *R. reniformis*	
Parthenium hysterophorus Linn. (Asteraceae): Carrot grass	Whole plant extract	Toxicity to *M. incognita* and *H. dihystera* in vegetable crops	Hasan and Jain (1984)
	Chopped leaves mixed with infested soil	Reduced root galling by *M. incognita* in papaya	Reddy and Khan (1990)
Peristrophe bicalyculata (Retz.) Nees. (Acanthaceae)	Leaves amended in soil	Reduced incidences of *M. incognita* in okra	Chatterjee and Sukul (1979)
Phaseolus lunatus Linn. (Papilionaceae): Lima bean/ Sieva bean	Aqueous seed extract	Inhibited larval hatching and found toxic to *M. incognita*	Hussain and Masood (1975), Mahmood et al. (1979)
Pimenta dioica allspice (Myrtaceae): Jamaica pepper, pepper	Essential oils	Nematicidal activities against the pinewood nematode, *Bursaphelenchus xylophilus*	Park et al. (2007)

(continued)

Table 6.1 (continued)

Generic name/common name	Plant part/formulation	Nematicidal activity	Reference
Piper nigrum Linn. (Piperaceae): Black pepper	Whole plant extract (1:10 dilution) in soil	Reduced root galling by *M. incognita* in tobacco nursery	Krishnamurthy and Murthy (1990)
Piper betle (Piperaceae): Betelvine, betel pepper	Leaf extract	Highly toxic to *M. incognita*	Wiratno et al. (2009)
Polyalthia longifolia (Sonn.) Thw. (Annonaceae)	Aqueous leaf extract	Toxic to eggs of *Heterodera avenae*	
Pongamia glabra Vent. (Fabaceae): Puna oil tree/ Karanja	Water extract of oil cake	Toxicity to *M. incognita*	Mishra and Prasad (1973)
	Oil cake amended is soil	Reduced population of *M. javanica* in tomato and *M. incognita* in tobacco and in betelvine	Singh and Sitaramaiah (1976): Desai et al. (1979): Jagadale et al. (1985a, b)
	Aqueous leaf extract	Reduced populations of *Aphelenchoides composticola*	Rao and Pandey (1982)
	Mulching of green leaves with soil	Reduced incidences of *M. incognita* in mulberry	Govindaiah et al. (1989)
Populus deltoides Marsh. (Salicaceae): Caroline poplar	Aqueous leaf extract	Toxicity to *Aphelenchoides composticola*	Grewal and Sohi (1988)
Portulaca oleracea Linn. (Portulacaceae): Common purslane	Aqueous leaf and stem extracts	Toxicity to *M. incognita*	Hoan and Davide (1979)
Prosopis juliflora (Sw.) D.C. (Mimosaceae): Mesquite/ Algaroba	Chopped leaves admixed with soil at 5 g/kg	Reduced gall index of *M. incognita* in potted tomato	Sundarbabu et al. (1993)
Pueraria lobata (Willd.) Ohwi. (Fabaceae): Thumberry/ Kudzu vine	Aqueous leaf extract	Toxicity to *M. incognita*	Mian and Rodriguez (1982)
Raphanus sativus Linn. (Brassicaceae): Radish	Aqueous root extract	Toxicity to *M. incognita*	Das (1983)
Ricinus communis Linn. (Euphorbiaceae): Castor	Oil cake amended in soil	Reduced root galls in okra and tomato by *M. incognita*	Singh and Sitaramaiah (1976)
	Water extract from oil cake	Toxicity to larvae of *M. incognita*	Sharma et al. (1985)
	Water extract from oil cake	Toxicity to adults of *H. indicus, R. reniformis* and *T. brassicae* Reduced	Khan et al. (1973)
	Oil cake amended in soil for citrus, brinjal and guava and oil cake extract	Populations of *H. indicus, R. reniformis, M. incognita, A. avenae* and *T. brassicae* and showed toxicity to these nematodes	Alam et al. (1976, 1977)
	Oil cake amended in soil for okra and brinjal	Reduced incidences of *M. incognita* and *M. javanica*	Sikdar et al. (1986), Reddy and Khan (1990)
	Leaf powder amended in soil at 40 g/pot	Reduced incidences of nematodes in okra and tomato	Bhatnagar et al. (1980)
	Chopped leaves mixed with soil	Reduced root galls by *M. incognita* and *M. javanica* in tomato and papaya	Ravidutta and Bhatti (1986a, b), Zaki and Bhatti (1989), Reddy and Khan (1990)
	Aqueous leaf extract at44 g/kg soil	Reduced larval hatching and penetration of *M. incognita* in tomato	Ravidutta and Bhatti (1986a, b)

(continued)

Table 6.1 (continued)

Generic name/common name	Plant part/formulation	Nematicidal activity	Reference
	Castor cropping in between vegetables	Reduced root galling by *M. incognita*	Prasad et al. (1987)
	Root exudates	Reduced gall formation by *M. incognita*	Hasan (1992)
	Chopped leaves mixed with soil at 1.5 kg/m^2 to grow pointed gourd	Reduced populations of *M. incognita, H. indicus, H. dihystera* and *T. vulgaris*	Verma and Anwar (1995)
Scilla indica Roxb. (Liliaceae): Indian squill	Aqueous root and shoot extracts	Toxicity to *M. incognita*	Pandey and Haseeb (1988)
Sesamum indicum Linn. (Pedaliaceae): Sesame/gingilly	Oil cake amended in soil	Toxicity to nematodes in general and also *M. incognita*	Midha and Chabra (1974), Sikdar et al. (1986)
	Water-soluble fractions of oil	Reduced incidences of *M. incognita* in okra	Bhatnagar et al. (1978)
	Chopped leaves mixed with infested soil	Reduced incidences of *M. incognita* in vegetable crops	Reddy and Khan (1990)
Sesbania aculeata (Willd.) Poir. (Papilionaceae): Dhaincha	Leaves amended in soil at 5–10%w/w in pot experiments	Reduced incidences of *M. javanica* in tomato	Singh and Sitaramaiah (1967)
Shorea robusta Gaertn. f. (Dipterocarpaceae): Sal	Saw dust amended in soil along with neem kernel powder	Reduced incidences of *M. javanica* in tomato	Singh and Sitaramaiah (1971)
	Oil cake amended in soil for wheat crop	Reduced populations of *M. incognita*	Sharma et al. (1985)
Sida cordifolia Linn. (Malvaceae): Country mallow	Aqueous leaf extract	Toxicity to *M. incognita* and *R. reniformis*	Mahmood et al. (1982)
Solanum hispidum Linn. (Solanaceae)	Aqueous leaf and root extracts	Toxicity to *T. brassicae* and *H. indicus*	
Solanum pampasense Linn. (Solanaceae)	Aqueous root extracts	Toxicity to *H. rostochiensis*	Jacobson (1958)
Solanum sucrense (Linn.) Mill. (Solanaceae)	Aqueous root extracts	Toxicity to *H. rostochiensis*	Jacobson (1958)
Solanum xanthocarpum Schrad. and Wendl. (Solanaceae)	Aqueous root extracts	Toxicity to *M. incognita*	Hasan (1992)
Sonchus oleraceus Linn. (Asteraceae)	Aqueous leaf and root extracts	Toxicity to *M. incognita*	Bano et al. (1986)
Sophora flavescens Linn. (Fabaceae)	Aqueous root extracts	Toxicity to pinewood nematode, *Bursaphelenchus xylophilus*	Masuda et al. (1991)
Sorghum vulgaris Pers. (Gramineae): Sorghum	Whole plant extract (1:10 dilution) in soil	Reduced root galling by *M. incognita* in tobacco nursery	Krishnamurthy and Murthy (1990)
Sphenoclea zeylanica Gaertn. (Sphenocleaceae): Gooseweed	Green manuring in rice fields and aqueous leaf extract	Reduced populations of *Hirschmanniella oryzae* and showed toxicity to this nematode	Mohandas et al. (1981)
Syzygium aromaticum (Myrtaceae): Clove	Bud extract	Toxicity to *M. incognita*	Wiratno et al. (2009)
Tagetes erecta Linn. (Asteraceae): Aztec marigold/ big marigold	Intercropping with tomato, brinjal and wheat and also with mulberry	Reduced incidences of *M. incognita, H. rostochiensis, P. penetrans*	Ruelo (1976), Ruelo and Davide (1979): Govindaiah et al. (1990)
	Aqueous leaf and root extracts	Inhibited larval hatching of *M. incognita*	Hussain and Masood (1975), Hoan and Davide (1979)

(continued)

Table 6.1 (continued)

Generic name/common name	Plant part/formulation	Nematicidal activity	Reference
	Seedlings of tomato, brinjal and chillies surrounded by marigold	Reduced populations of *H. indicus, R. reniformis, T. brassicae* and *M. incognita* and inhibited larval hatching	Alam et al. (1975, 1976)
	Methanolic leaf and stem extracts	Up to 100 % mortality to 2nd juveniles of *T. penetrans* and *A. tritici*	Kumari et al. (1986)
	Chopped leaves mixed with infested soil	Reduced root galling by *M. incognita* in papaya and by *H. indicus, T. vulgaris* and *H. dihystera* in pointed gourd and by *M. incognita* in tomato	Reddy and Khan (1990), Verma and Anwar (1995), Siyanand et al. (1995)
Tagetes lucida Cav. (Asteraceae): sweet scented marigold	Root extract in water	Toxicity to *M. javanica* and other nematodes infesting vegetables	Mukherjee (1983)
	Aqueous flower extract	Toxicity and inhibiting hatching of juvenile of *M. incognita, R. reniformis, T. brassicae, H. indicus* and *T. filiformis*	Siddiqui and Alam (1988a, b)
Tagetes minuta Linn. (Asteraceae): Stinking Roger	Oil extracted from its leaves	Toxicity to nematodes of agric importance	Mukherjee (1983)
	Root exudates	Toxicity to *H. indicus*	Siddiqui and Alam (1988b)
Tagetes patula Linn. (Asteraceae): French marigold	Alcoholic root extract	Toxicity to *M. incognita*	Morallo-Rajessus and Eroles (1978)
	Root exudates	Toxicity to *H. indicus*	Siddiqui and Alam (1988b)
	Aqueous leaf extract	Toxicity to *Radopholus similis*	Subramaniyan and Selvaraj (1988)
Tagetes tenuifolia Cav. (Asteraceae): striped marigold	Aqueous root extract	Toxicity to nematodes of agric importance	Mukherjee (1983)
	Root exudates	Toxicity to *H. indicus*	Siddiqui and Alam (1988b)
Tamarindus indica Linn. (Caesalpiniaceae): Tamarind	Leaf extract in water admixed with soil in pots	Inhibited larval hatching of *M. incognita* in vegetables	Hussain and Masood (1975)
	Leaf extract (1:10 dilution) in soil	Reduced root galling by *M. incognita* in tobacco	Krishnamurthy and Murthy (1990)
Thuja orientalis Linn. (Cupressaceae): Oriental arborvitae	Aqueous leaf extract	Toxicity to *M. incognita* in vegetable crops	Jain et al. (1986)
Tithonia diversifolia A. Grar (Asteraceae): Wild sunflower	Flower extract in water	Toxicity to root-knot nematode, *M. incognita*	Tiyagi et al. (1985)
Trichosanthes anguina Linn. (Cucurbitaceae): Snake gourd	Seed extract in water admixed with soil	Toxicity to *M. incognita* in potted vegetable crops	Mahmood et al. (1979)
Trachyspermum ammi (Apiaceae): Ajwain	Essential oils	Nematicidal activities against the pinewood nematode, *Bursaphelenchus xylophilus*	Park et al. (2007)
Typhonium trilobatum (Linn.) Schott. (Araceae)	Green plant extract in water	Reduced incidences of *M. incognita* in eggplant	Mukhopadhyaya and Chattopadhayaya (1979)
	Corn powder admixed with soil	Reduced populations of *M. incognita*	Chattopadhaya and Mukhopadhaya (1989a, b)
Vernonia anthelmintica Willd. (Compositae): Purple fleabane	Leaf and root extracts in water	Toxicity to *M. incognita* and *M. javanica*	Mukherjee (1983)

(continued)

Table 6.1 (continued)

Generic name/common name	Plant part/formulation	Nematicidal activity	Reference
Vinca rosea Linn. (Apocynaceae): Old maid	Whole plant extract in water	Toxicity to *M. incognita* and *M. javanica*	Mukherjee (1983)
	Chopped leaves admixed with infested soil	Reduced root galling by *M. incognita* in papaya	Reddy and Khan (1990)
Withania somnifera (Linn.) Dunal. (Solanaceae): Ashwagandha, Indian ginseng	Aqueous root and shoot extracts	Toxicity to *T. brassicae, H. indicus* and *R. reniformis*	
Xanthium strumarium Linn. (Asteraceae): Clotbur	Aqueous leaf extract	Toxicity to *M. incognita*	Nandal and Bhatti (1983), Bano et al. (1986)
	Essential oil extracted from seed	Toxicity to 3rd instar juvenile of *M. incognita*	Ghosh and Sukul (1992)
Zinnia elegans Jacq. (Asteraceae): Youth and old age	Whole plant extract in water	Toxicity to juveniles and adult of *M. incognita* and also inhibited hatching of its larvae	Bano et al. (1986)

Neem leaves amended in the soil also reduced the incidences of *M. javanica* in chickpea (Singh and Sitaramaiah 1967; Kaliram and Gupta 1980, 1982; Gupta and Kaliran 1981) and *M. incognita* in papaya (Reddy and Khan 1990) and also reduced the nematode populations and enhanced average bunch weight in plant (Jacob et al. 1990). Saxena et al. (1993) found neem leaf aqueous extract to show mortality to *Criconemoides annulatum* and *Xiphinema basiri* when tested under laboratory conditions. Neem leaves mixed at 20 g/kg in soil showed 85 % increase in the yield of wheat over control when tested against cyst nematode, *Heterodera avenae* (Singh et al. 1995). Similarly chopped neem leaves admixed with the soil in the plant pit at 1,500 g/m^2 area of the pointed gourd under field conditions against *M. incognita, H. indicus, T. vulgaris* and *H. dihystera* showed significant reduction in the populations of these nematodes in the roots (Verma and Anwar 1995).

4.2 Neem Root and Kernel as Nematicide

Root exudates of 1-month-old neem seedlings were toxic to adults of *Helicotylenchus indicus, R. reniformis, T. brassicae* and *Tylenchus filiformis* and larvae of *M. incognita* (Alam et al. 1975). Hasan (1992) found neem root exudates to reduce root galling by *M. incognita.* Proct and Kornprobst (1983) found that the pretreatment of tomato seedlings with the neem seed extract inhibits the penetration of nematode, *M. javanica,* juveniles to its roots. Neem oil fraction in hexane was also found to be toxic to *M. incognita* (Devakumar et al. 1985). Dried neem seed kernel powder applied to wheat seeds with gum at 800 mg/100 g seeds successfully inhibited the penetration and development of wheat gall nematode, *Anguina tritici,* in the seedlings (Gokte and Swarup 1985). Mishra and Mojumder (1993) found decomposed neem kernel to reduce populations of *M. incognita* when amended in the soil at 10 %w/w. Mojumder and Mishra (1995) reported 20 % w/w neem kernel powder coated on chickpea seed to effectively reduce the population of *M. incognita* and *R. reniformis* and enhance grain yield.

4.3 Neem Oil Cake as Nematicide

Aqueous extract of neem oil cake inhibited larval emergence of *M. incognita* (Khan et al. 1967). Singh and Sitaramaiah (1969) reported neem oil cake incorporated in infested soil in field plots before planting of tomato and okra to reduce incidences and populations of the root-knot nematodes. Mobin and Khan (1969) found suppressed populations of *Tylenchorynchus* sp., *Hoplolaimus* sp., *Helicotylenchus* sp., *R. reniformis* and *M. incognita* around roots of eggplant, guava and citrus plants when oil cake was applied in the soil. Amendment of oil cake in the soil 1 month before sowing of wheat also reduced parasitic nematode populations (Gour and Prasad 1970). Further, oil cake amended in the soil reduced the populations of *M. incognita* and *M. javanica* and also root galls in okra, tomato and other vegetables (Singh and Sitaramaiah 1971; Sharma et al. 1971; Gowda 1972; Desai et al. 1973; Khan et al. 1973; Mishra and Prasad 1973). However, neem oil cake influenced and modified the relationship between plant and nematode and the aqueous extract of oil cake showed toxicity to *M. incognita* and *M. javanica* (Mishra and Prasad 1973).

Water-soluble fractions of the oil cake were toxic to *H. indicus, R. reniformis, T. brassicae* and *M. incognita* (Khan et al. 1973; Singh 1980). Alam and Khan (1974) found its oil cake reducing populations of *H. indicus, T. brassicae, M. incognita, Ditylenchus cypei* and *Aphelenchus avenae*, when incorporated in soil at 110 kg N/ha. Neem oil cake applied at 1 g/49 g soil reported reduction in the nematodes fauna and suggested librated ammonia to be detrimental to these nematodes (Khan et al. 1974). Azam and Khan (1976) found that oil cake applied at 100 Kg N/ha soil promotes plant growth and reduces root-knot populations alone or in combination with nemagon. Khan et al. (1976) and Siddiqui (1976) reported suppressed populations of the nematodes, namely, *T. brassicae, H. indicus, H. erythrinae* and *M. incognita,* in rhizosphere of vegetable plants amended with its oil cake in soil. Singh and Sitaramaiah (1976) found its oil cake to reduce populations of *M. incognita* and *M. javanica* in tomato, when applied as organic manure.

Water extract of its oil cake amended in soil inhibited oviposition and hatching of *M. incognita, H. indicus* and *A. avenae* (Sitaramaiah and Singh 1977). Khan (1979) and Vijayalakshmi and Goswami (1983) reported inhibition in larval hatching of *M. incognita* when water extract of its oil cake was tested in controlled conditions. Neem oil cake amended in the soil reduced incidences of *M. incognita* in mung bean (Vijayalakshmi and Prasad 1979), in tomato (Desai et al. 1979) and in betelvine (Acharya and Padhi 1988; Jagadale et al. 1985a, b) and in mulberry (Sikdar et al. 1986) and populations of *M. incognita* and *M. javanica* in tomato and okra (Srivastava 1971; Routray and Sahoo 1985; Shukla and Banker 1985; Goswami and Vijayalakshmi 1987; Gupta and Bhattacharya 1995) and also decreased the population of *T. elegans* when applied at 1 % w/w (Sitaramaiah and Singh 1977). Neem oil cake extract in hot water and methyl alcohol also reduced the populations of *M. incognita* in a laboratory test (Sharma et al. 1985).

Cowpea seeds treated with cake extract reduced incidences of *M. incognita* and *R. reniformis* after sowing (Khan and Hussain 1988a, b). Neem oil cake amended in infested soil at 1 g/kg reduced root galling by *M. incognita* in potted tulasi (Haseeb et al. 1988). Similarly, neem cake at 1.7 t/ha controlled population of *Helicotylenchus* sp., *Hoplolaimus indicus, R. reniformis* and *Hemicriconemoides* sp. in soil (Gaur and Mishra 1989). Reddy and Khan (1990) found that oil cake amended in soil at 1 t/ha reduces incidences of root galls in okra by *M. incognita*. Prasad and Khan (1990) reported that oil cake amended in soil alone or in combination with aldicarb eliminates the populations of *H. indicus, H. elegans, Pratylenchus zeae* and *T. vulgaris* in soybean. Post-pruning top dressing of neem cake at 1–2 Kg/vine to *Vitis* sp. crop was effective in suppressing the populations of *M. incognita* (Jayaraj 1991). Haseeb and Butool (1991) also reported that neem oil cake amendment in soil at 1 g/kg soil suppresses *M. incognita* in roots of Davana, *Artemisia pollens.* Hasan (1992) found neem cake reduces root galling by *M. incognita.*

Amendment of powdered neem cake in soil at 400 g/m^2 prevented root-knot nematode incidences in tobacco nurseries (Hussaini et al. 1993) and oil cake incorporated in soil showed toxicity to *M. incognita* and *R. reniformis* in mung bean (Tiyagi and Alam 1993). Neem cake amended at 1 g/kg soil reduced root-knot index of *M. incognita* on *Trachyspermum ammi.*

Application of Neem cake at 100 kg N/ha in pots significantly reduced the populations of *M. incognita, R. reniformis* and *P. delattrei* in brinjal (Poornima and Vodivelu 1993). In Japanese mint cultivation, population of *M. incognita* was effectively reduced when neem cake was admixed with soil at 20 % w/w alone (Pandey 1995) and in combination with carbofuran (Singh and Singh 1995).

5 Active Components of the Botanicals Having Nematicidal Activity

Two triterpenoid saponins, acaciaside A and B isolated from funicles of *Acacia auriculiformis* killed about 63 % juveniles of *M. incognita in cowpea*, when soil drenched (10 mg/ml) as well as foliar sprayed at 4 mg/ml concentration. These also reduced root galling and nematode populations in roots of cowpea (Roy et al. 1993). Nath et al. (1982) isolated and identified active component as diallyl disulphide from the bulb of garlic, *Allium sativum*, which showed toxicity to nematodes. Asparagusic acid, an active component isolated from *Asparagus officinalis* root extract, found to be toxic to *H. rostochiensis, M. halpa, P. penetrans* and *P. curvitatus* (Das 1983).

Isolated from neem extracts, two active components, namely, nimbidin and thionimone, showed toxicity to adults of *H. indicus, R. reniformis, T. brassicae* and *M. incognita* and also inhibited the growth of their larvae (Khan et al. 1973). Khan et al. (1974) found ammonia librated through decomposition of neem oil cake in the soil as toxic to the nematodes. Alam et al. (1976) reported formaldehyde in the oil cake manure as a chemical responsible for killing the nematodes. However, considerable amount of phenol was detected from neem oil cake in the soil as toxicant to nematodes (Alam et al. 1979). Limonoids extracted from neem kernels were also reported to inhibit the hatching and showed high larval mortality to *M. incognita* (Devakumar et al. 1985). Rao and Reddy (1993) reported repelin and welgro at 2.5 and 5 % to significantly reduce populations of *M. incognita* in tomato nursery, when drenched with their solutions. Neem allelochemicals like azadirachtin, nimbidic acid and quercetin were also found to show nematicidal activity when tested against *M. incognita, R. reniformis, H. indicus* and *T. brassicae* under controlled conditions (Siddiqui and Alam 1993).

Khan et al. (1974) reported ammonia liberated from oil cake of *Brassica campestris* as active component to show toxicity to *T. brassicae, M. incognita, H. indicus* and *A. avenae*. Phenolic contents in mustard oil cake also showed toxicity to soil inhabiting nematodes (Alam et al. 1977, 1979). Allyl isothiocyanates as main components isolated from black mustard seed oil showed toxicity to *Heterodera rostochiensis* in potato (Ellenby 1945a, b). Two isomeric polyacetylene compounds isolated from the root and flower extracts of *Carthamus tinctorius* were found to be) toxic to *Aphelenchoides besseyi* (Shigefumi et al. 1976). From the root extracts of *Daphne odora*, two nematicidal components as diterpenes having an ortho-ester group named odoracin and odoratrin have been isolated and tested to be toxic to the nematodes (Munakata 1978).

The alkaloids, namely, atropine, nicotine and scopolamine, isolated from leaf, bulb and root extracts of *Datura stramonium* showed toxicity to *T. semipenetrans* and *A. tritici* (Winoto 1969; Kumari et al. 1986). The whole plant extract of *Eclipta alba* was found to show nematicidal activity (Prasad and Rao 1979). From the methanolic extract of *Helenium* (hybrid) sp., two nematicidal active components, i.e. 1-tridecanene-3,5,7,9,11-pentayne and 2-3-dihydro-2-hydroxy-3-methyl -one-6-methyl benzofuran, were isolated and found to show toxicity to *Pratylenchus penetrans* (Gommeres 1971). From the leaves of *Lawsonia inermis*, active components, i.e. palmitic, linoleic and oleic acids, were isolated and found to show toxicity to larvae of *T. penetrans* (Badami and Patil 1975). Alam et al. (1976, 1979) reported two nematicidal components, formaldehyde and phenol from *Ricinus communis*. From the root extract of *Tagetes erecta*, two active components, namely, alpha-terthienyl and 5-(e-butens-1-ynyl)-2,2'-bithienyl, were isolated and reported to be toxic to *H. rostochiensis* and *P. penetrans* (Uhlenbrock and Bijoo 1959a). Similarly, chemicals like thiophene and alpha-terthienyl derivatives isolated from leaf and stem extracts of *Tagetes patula* are found to show toxicity to *T. semipenetrans* and *A. tritici* (Uhlenbrock and Bijoo 1959b; Winoto 1969; Kumari et al. 1986).

6 Acaricidal and Mite Growth Regulating Activities in Botanicals

A number of higher plants possessing miticidal and other biological activities against plant mites have been given in Table 6.2. Results reviewed on botanicals and their active formulations revealed that biological activities of the test products varied with the dose and also with host plants which is briefed in Table 6.2. Patel et al. (1993) evaluated neem-based formulations against spider mites, *Tetranychus macfarlanei* and *Tetranychus cinnabarinus*, and found that repelin 1 %, margocide CK 1 %, margocide OK 0.8 % and neemark 0.5 % were effective against *T. cinnabarinus* infesting brinjal crops and Indian bean but failed to kill *T. macfarlanei* infestation in okra. However, their effectiveness varied with different plant hosts (Table 6.3).

7 Conclusion

Botanical origin pesticides have potential to control plant nematodes and mites. Botanical pesticides have been extensively used before onset of the synthetic chemical pesticides in

Table 6.2 Biological activities of botanicals against plant mites

Generic name of plant	Plant part and its formulation	Miticidal activity	Reference
Ajuga remota Linn. (Lamiaceae)	Juice and rude extract of the leaves	Toxicity to spider mite, *T. urticae*	Schauer and Schmutterer (1981)
Ailanthus altissima L. (leaves) (Simaroubaceae)	Leaf extract	Caused mortality of two-spotted spider mite (*T. urticae* Koch) and also reduced reproductive potential and deterred from feeding on suitable host plants	Chermenskaya et al. (2010)
Aloysia triphylla Ort. and Palau (Verbenaceae)	Oils of its seed and leaf	Toxicity to red spider mite, *T. telarius*	Jacobson (1975)
Allium sativum Linn. (Liliaceae)	Acetone extract of the bulb with deltamethrin	Increased miticidal activity to *T. urticae*	Barkat et al. (1986b)
Aloe vera L. (Liliaceae)	Leaf extract	Acaricidal activity against female adults of carmine spider mite (*T. cinnabarinus*) (Boisduval)	Wei et al. (2011)
Artemisia sp. (Asteraceae)	Worm wood powder	Reduced the attack of *Acarapis woodi* to bee colonies	Abu-Zaid and Salem (1988)
Artemisia annua (Asteraceae)	Leaf acetone extract	Contact action activity against carmine spider mite (*T. cinnabarinus* Boisduval)	Zhang et al. (2008)
Artemisia saissanica (Asteraceae)	Chloroform extracts of leaves, buds and twigs	Repellency to *T. urticae*	Adekenov et al. (1990)
Anabasis aphylla Linn. (Chenopodiaceae)	Aqueous leaf extract	Toxicity of mites in general	
Azadirachta indica A. Juss (Meliaceae)	Kernel extract	Repellency to carmine spider mite, *T. cinnabarinus*	Mansour and Ascher (1983)
Azadirachta indica A. Juss (Meliaceae)	Kernel extract in acetone, methanol and ethanol	Repellency to carmine spider mite, *T. cinnabarinus*	Mansour et al. (1987)

(continued)

Table 6.2 (continued)

Generic name of plant	Plant part and its formulation	Miticidal activity	Reference
Azadirachta indica A. Juss (Meliaceae)	Leaves (1 %) and kernel (0.5 %) extracts	Controlled *T. urticae* on brinjal	Vinothkumar et al. (2009)
Azadirachta indica A. Juss (Meliaceae)	Neem seed extract (NeemAzal T/S, 1 % of azadirachtin)	Controlled broad mite (*Polyphagotarsonemus latus*) on chilli pepper	Venzon et al. (2008)
Boswellia dalzielli Roxb. (Burseraceae)	Aqueous bark extract	Toxicity to mites in general	Jacobson (1975)
Brassica oleracea var. *rapifera* Metz. (Brassicaceae)	Aqueous root extract	Toxicity to spider mite, *T. atlanticus*	Jacobson (1975)
Cardaria draba Desv. (Brassicaceae)	Aqueous leaf and root extract	Toxicity to mites in general	McIndoo (1982)
Chamaecyparis pisifera (Sieb. and Zucc.) (Cupressaceae)	Aqueous leaf and root extract	Regulated the growth of *T. pisifera*	Ahn et al. (1984)
Chrysanthemum spp. (Asteraceae)	Flower head aqueous extract	Toxicity to red spider mite, *T. telarius*	Markkula et al. (1969)
Citrullus colocynthis (Linn.) Kuntze (Cucurbitaceae)	Aqueous leaf and fruit extract	Toxicity to red spider mite, *T. telarius*	McIndoo (1982)
Consolida regalis Gilib. (Ranunculaceae)	Aqueous root, leaf and stem extracts	Toxicity to red spider mite, *T. telarius*	McIndoo (1982)
Convolvulus krauseanus Regel and Schmalh. (Convolvulaceae)	Root extract	Caused mortality of two-spotted spider mite (*T. urticae* Koch) and also reduced reproductive potential and deterred from feeding on suitable host plants	Chermenskaya et al. (2010)
Coriandrum sativum Linn. (Umbelliferae)	Oil emulsion (2 %) spray	Toxicity to spider mites	Feinstein (1952)
Croton bonplandianum Roxb. (Euphorbiaceae)	Aqueous leaf extract	Toxicity to red spider mite, *T. telarius*	McIndoo (1982)
Curcuma longa Linn. (Zingiberaceae)	Turmeric powder extract in water	Toxicity to red spider mite, *T. telarius*	McIndoo (1982)
Cymbopogon citratus (D.C.) Stapf. (Poaceae)	Aqueous leaf extract	Toxicity to red spider mite, *T. telarius*	McIndoo (1982)
Cynodon dactylon (Linn.) Pers. (Poaceae)	Aqueous leaf extract	Toxicity to red spider mite, *T. telarius*	McIndoo (1982)
Daphne odora B. Don. (Thymelaeaceae)	Root and bark extracts	Toxicity to *T. urticae*	Inamori et al. (1987)
Delphinium staphisagria Linn. (Ranunculaceae)	Aqueous, leaf and seed extracts	Toxicity to red spider mite, *T. telarius*	McIndoo (1982)
Hyptis suaveolens (Linn.) Poit. (Labiaceae)	Aqueous leaf extract	Toxicity to *T. neocaledonicus*	Roy and Pande (1991)
Lepidium ruderale Linn. (Brassicaceae)	Aqueous leaf extract	Toxicity to plant mites in general	McIndoo (1982)
Leucas cephalotes (Roth.) Spreng. (Lamiaceae)	Aqueous leaf extract	Toxicity to spider mites	McIndoo (1982)
Lippia sidoides Cham. (Verbenaceae)	Essential oil extract from leaves	Exhibited potent acaricidal activity against two-spotted spider mite (*T. urticae* Koch)	Cavalcanti et al. (2010)

(continued)

Table 6.2 (continued)

Generic name of plant	Plant part and its formulation	Miticidal activity	Reference
Melia azedarach Linn. (Meliaceae)	Oil and aqueous seed extracts	Toxicity to citrus red spider mite, *Panonychus citri*	Chiu (1982)
Melia toosendan Linn. (Meliaceae)	Oil and aqueous seed extracts	Toxicity to mites in general	Chiu (1982, 1989)
Micromeria fruticosa L (Lamiaceae)	Essential oil vapour	Toxic effect on *T. urticae* Koch	Calmasur et al. (2006)
Nepeta racemosa L. (Lamiaceae)	Essential oil vapour	Toxic effect on *T. urticae* Koch	Calmasur et al. (2006)
Nicotiana tabacum Linn. (Solanaceae)	Leaf and whole plant extracts	Toxicity to red spider mite, *T. telarius*	McIndoo (1982)
Ocimum basilicum Linn. (Lamiaceae)	Aqueous leaf extract	Toxicity to red spider mite, *T. telarius*	McIndoo (1982)
Oregano (*Origanum onites* L.) and thyme (*Thymbra spicata* L. subsp. spicata)	Essential oils	Acaricidal activity against carmine mite (*T. cinnabarinus* Boisd.) adults	Sertkaya et al. (2010)
Origanum vulgare L. (Lamiaceae)	Essential oil vapour	Toxic effect on *T. urticae* Koch	Calmasur et al. (2006)
Piper nigrum Linn. (Piperaceae)	Seed extracts in acetone and diethyl ether	Toxicity to adults of *T. urticae*	Barakat et al. (1986a)
Prangos pabularia Hybrid (Celasteraceae)	Aqueous whole plant extract	Toxicity to mites in general	Jacobson (1958)
Syzygium cumini (Pomposia)	Fruit extract on ethanol	Acaricidal activity against *T. urticae*	Abd et al. 2011
Trichilia cuneata P.Br. (Meliaceae)	Aqueous leaf and fruit extracts	Toxicity to mites in general	McIndoo (1982)
Trichilia trifolia P.Br. (Meliaceae)	Ethanolic seed extract and its fractions	Toxicity to rice mite, *Caloglyphus berlesei*	Rao and Prakash (1995)
Trigonella foenum-graecum (Papilionaceae)	Diethyl ether extract of its seed with deltamethrin	Toxicity to adults of *T. urticae*	Barakat et al. (1986b)

agriculture and are still important to manage plant nematodes in many vegetable and food grain crops especially for organic farming where use of synthetic pesticides is being avoided as these are hazardous to human health. A number of plant products were found effective against the plant nematodes and mites, especially the higher plants, which have been arranged in alphabetical order in tabular form in this book chapter. A number of the active principles have also been isolated and identified from the effective botanical formulations have been scanned and their toxicity is described against both the plant nematodes and mites. Botanical nematicides and miticides have potential to manage the nematode and mite pests, respectively, and most effectively in small farming systems and kitchen gardens, etc., if applied correctly using their right formulations and doses. Of course, for pest management in the large areas availability of large quantities of these products is a limitation. Soil amendments with oil cakes or green leaves, intercropping with biopesticidal plants are some of the useful strategies for the management of the plant nematodes. For management of plant mites, no doubt, botanical pesticides have been reported to show toxicity to a number of phytophagous mites, but yet use of the botanicals for the management of plant mites is not popularized. Therefore, there is a need to explore more effective botanical miticides, their formulations and application methods.

Table 6.3 Active components isolated from the plants and its formulations showing toxicity to plant mites

Active component	Source (plant/plant parts)	Biological activity	Reference
6-Methoxy analogue of pisiferic acid	*Chamaecyparis pisifera/* root and leaf extracts	Inhibited the feeding of adults and hatching of eggs of *T. urticae*	Ahn et al. (1984)
Daphnodorous A, B, C: three flavones	*Daphne odora*: root and bark	Toxicity to *T. urticae*	Inamori et al. (1987)
Asimicin	Ethanol extract of *Daphne odora*	Toxicity to *T. urticae*	Alkofahi et al. (1989)
Clerodane terpenes, isomers of ajugarins	*Ajuga remota* leaf extract	Toxicity to *T. urticae*	Jacobson (1989)
Alkaloids	*Abrus precatorius*, leaf extract	Repellency to adults of *T. urticae* and also reduced its life span and oviposition	
Repelin, having azadirachtin: a neem-based formulation	*Azadirachta indica,* kernel powder	Toxicity to phytophagous mite, *Tetranychus cinnabarinus*	Mansour et al. (1993)
Neem-based formulation: neemark 0.5 % and margocide OK 0.8 % both having azadirachtin	*Azadirachta indica,* kernel powder	Toxicity to *Tetranychus cinnabarinus* infesting brinjal and Indian bean	Patel et al. (1993)
Plumbagin (5-hydroxy-2-methyl-1,4-naphthoquinone)	*Diospyros kaki* roots	Toxicity to *Tetranychus urticae*	Akhtara et al. (2012)
Monoterpenes (thymol and carvacrol)	Essential oil of *Lippia sidoides* Cham.	Exhibited potent acaricidal activity against two-spotted spider mite (*T. urticae* Koch)	Cavalcanti et al. (2010)

References

Abd EMA, Hossam SEB, Sayed AF, Emad AS (2011) Acaricidal activity of different extracts from *Syzygium cumini* L. Skeels (Pomposia) against *Tetranychus urticae* Koch. Asian Pacific J Trop Biomed 1(5):359–364. doi:10.1016/S2221-1691(11)60080-4

Abu Zaid MI, Salem MM (1988) Evaluation of certain dosages of wormwood as bioactive agent against the acarine mite, Acarapis woodi Rennie. Bull Entomol Soc Egypt Econ Ser 1989, 1991(17):121–125

Acharya A, Padhi NN (1988) Effect of neem oil cake and sawdust against root-knot nematode, *Meloidogyne incognita* on betelvine. Indian J Nematol 18(1):105–107

Adekenov SM, Kupriyanov AN, Gafurov NM, Kurmanova RS (1990) Sesquiterpene lactones of *Artemisia saissanica*. Chem Nat Compd 26:716–717

Agbenin NO, Emechebe AM, Marley PS, Akpa AD (2005) Evaluation of nematicidal action of some botanicals on *Meloidogyne incognita* in vivo and in vitro. J Agric Rural Dev Trop Subtrop 106(1):29–39

Ahn JW, Wada K, Marumo S (1984) Pisiferdiol, a novel phenolic diterpenoid from Chamaecyparis pisifera Endl. Agric Biol Chem 48:3177–3179

Akhtara Y, Ismana MB, Lee CH, Lee SG, Lee HS (2012) Toxicity of quinones against two-spotted spider mite and three species of aphids in laboratory and greenhouse conditions. Ind Crop Prod 37:536–541

Alam MM (1976) Effect of mix cropping of marigold and margosa with some vegetables on plant growth and parasitic nematodes. In: All India symposium on modern concept in plant protection at Udaipur, 6–9 Mar 1976, p 97

Alam MM, Khan AM (1974) Control of phytonematodes with oil cakes amendment in spinach fields. Indian J Nematol 4(3):239–240

Alam MM, Masood A, Hussain SI (1975) Effect of margosa and marigold root exudates on mortality and larval hatch of certain nematodes. Indian J Expt Biol 13(4):412–414

Alam MM, Khan M, and Saxena SK (1976) Mechanism of control of plant parasitic nematodes as a result of application of oil cakes in soil. Role of aldehyde and ketones. All India symposium on modern concept in plant protection at Udaipur, 6–9 Mar 1976, pp 104–105

Alam MM, Khan M, Saxena SK (1977) Persistent action of oil cake on the population of nematode in the field. Botyu-Kagitu 42:119–124

Alam MM, Khan M, Saxena SK (1979) Mechanism of control of plant parasitic nematodes as a result of application of oil cakes in soil. Role of phenolic compounds. Indian J Nematol 9(2):126–142

Al-Banna L, Darwish RM, Aburjai T (2003) Effect of plant extracts and essential oils on root-knot nematode. Phytopathol Mediterr 42:123–128

Alkofahi A, Rupprecht JK, Anderson JE, McLaughlin JL, Mikolajczak KL, Scott BA (1989) Search for new pesticides from higher plants. In: Arnason JT, Philogene BJR, Morand P (eds) Insecticides of plant origin. American Chemical Society, Washington, DC, pp 25–43

Azam MF, Khan AM (1976) Combined effect of organic additives and nematicides in controlling root-knot nematodes in tomato. All India symposium on modern concept in plant protection at Udaipur, 6–9 Mar 1976, p 96

Badami RC, Patil K (1975) Minor seed oils X. Physico-chemical characteristic composition of seven oils. J Oil Technol Assoc India 7:82–84

Bano M, Anver S, Tyagi SA, Alam MM (1986) Evaluation of nematicidal properties of some members of compositae. Int Nematol Netw Newl 3(1):10

Barkat A, Sharref GM, Abdallah SS, Amer SA (1986a) Toxic action of some plant extracts against *Tetranychus urticae* Koch. Bull Ent Soc Egypt Econ Ser 14:233–242

Barakat AA, Shereef GM, Abdallah SA, Amer SAA (1986b) Joint action of some pesticides and plant extracts against Tetranychus urticae Koch. Bull Entomol Soc Egypt Econ Ser 14:243–249

Bhatnagar A, Mukherjee TK, Tiagi B (1978) Control of root-knot nematodes on okra by application of water soluble fraction of oil cakes in soil. Acta Bot Ind 6:188–189

Bhatnagar A, Mukherjee TK, Tiagi B (1980) Effect of leaves of some plants on *Meloidogyne incognita*. Andhra Agric J 27(5&6):312–314

Calmasur O, Aslan I, Fikrettin S (2006) Insecticidal and acaricidal effect of three Lamiaceae plant essential oils against *Tetranychus urticae* Koch and *Bemisia tabaci* Genn. Ind Crop Prod 23:140–146

Cavalcanti SCH, Niculau ES, Blank AF, Câmara CAG, Araújo IN, Alves PB (2010) Composition and acaricidal activity of *Lippia sidoides* essential oil against two-spotted spider mite (*Tetranychus urticae* Koch). Bioresour Technol 101:829–832

Chatterjee A, Sukul NC (1979) Total protein content of the galled roots: an index of root-knot nematode infestation of lady's finger plants. Phytopathology 71(4):372–374

Chatterjee A, Sukul NC, Laskar S, Ghoshmajumdar S (1982) Nematicidal principles from two species of Lamiaceae. J Nematol 14(1):118–120

Chattopadhaya PR, Mukhopadhaya MC (1989a) Comparative studies of the nematicidal properties of *Typhonium trilobatum* and *Melia azedarach*. Indian J Nematol 19(1):5–9

Chattopadhaya PR, Mukhopadhaya MC (1989b) Effect of leaf extract of *Artabotrys odoratissimus* R. Br. (Annonaceae) on hatching of eggs of *Meloidogyne incognita*. Indian J Nematol 19(1):28–31

Chermenskaya TD, Stepanycheva EA, Shchenikova AV, Chakaeva AS (2010) Insectoacaricidal and deterrent activities of extracts of Kyrgyzstan plants against three agricultural pests. Ind Crop Prod 32:157–163

Chiu S-F (1982) Experiments on insecticidal plants as a source of insect feeding inhibitors and growth regulation with special reference to Meliaceae. Plant Protection. South China Agriculture College, Cuangzhon, 42 p

Chiu SF (1989) Recent advances in research on botanical insecticides in China. In: Arnason JT, Phlogene BJR, Morand P (eds) Insecticides of plant origin. American Chemical Society symposium series 387, Washington, DC, pp 69–77

Dahlya RS, Dutta R, Gupta DC (1986) Effect of subabul and sunnhemp leaves on cereal cyst nematode, *Heterodera avenae* (Wolt, 192) in wheat. Haryana Agric Univ J Res 16:119–122

Das SN (1983) Nematodes management using plant products: Department of Nematology, OUAT, Bhubaneswar, Pers. Commn. In Michael et al. (1985) Plant species reportedly possessing pest control properties. An EWCUH DATABASE, University of Hawaii, Hawaii, pp 249

Desai MV, Shah HM, Pillai SN, Patel AS (1973) Nematicidal properties of some plant species. Indian J Nematol 3:77–78

Desai MV, Shah HM, Pillai SN, Patel AS (1979) Oil cakes in control of root-knot nematodes. Tropic Res 5(1):105–108

Devakumar C, Goswami BK, Mukherjee SK (1985) Nematicidal properties from neem (*Azadirachta indica* A. Juss.) Screening of neem kernel fractions against *Meloidogyne incognita*. Indian J Nematol 15:121–124

Egumjobi OA, Afolami SO (1976) Effect of neem (*Azadirachta indica* A. Juss.) leaf extract on population of *Pratylenchus brachyurus* on the growth of maize. Nematologica 22:125–132

Ellenby C (1945a) The influence of crucifers and mustard oil on the emergence of larvae of the potato root eelworm, *Heterodera rostochiensis* Wollenw. Ann Appl Biol 32:67–70

Ellenby C (1945b) Control of the potato root eelworm, *Heterodera rostochiensis* Wollenw. by allyl-isothiocyanate from mustard oil, *Brassica nigra*. Ann Appl Biol 32:237–239

Feinstein L (1952) Insecticides from plants. In: Insects: the year book of agriculture. USDA, Washington, DC, pp 222–229

Garciaspinosa R (1980) *Chenopodium ambrosioides* Linn.: a potential use combating phyto-parasitic nematodes. Agric Trop 2:92–97

Gaur HS, Mishra SD (1989) Integrated control of nematodes in lentil with aldicarb, neem cake and treatment with thimet and its residual effect on the subsequent mung bean. Indian J Entomol 51(3):283–287

Gaur HS, Pankaj (2009) 'Nematodes' chapter: handbook of agriculture. ICAR Publications, New Delhi

Ghosh SK, Sukul (1992) Nematicidal principles in the leaves of *Xanthium strumarium* Linn. In: Tauro P, Narwal SS (eds) Proceedings of national symposium. Allelopathy in Agro-ecosystems. Indian Society of Allelopathy, H.A.U., Hisar, India, pp 178–151

Gokte RT, Swarup G (1985) Effect of neem (*Azadirachta indica* A. Juss.) seed kernel powder treatment on penetration and development of *Anguina tritici* on wheat. Indian J Nematol 15(1):149–152

Gommeres FJ (1971) Nematicidal active principles isolated from the root of Helenium hybrid plants. Phytochemistry 10:1945

Goswami BK, Vijayalakshmi K (1985) In vitro and pot culture studies on the effect of dry products of *Andrographis paniculata, Calendula officinalis, Enhydra fluctuans* and *Solanum khasianum* on *M. incognita*. Indian J Nematol 15(2):264–266

Goswami BK, Vijayalakshmi K (1986a) Effect of some indigenous plant materials and oil cakes amended soil on the growth of tomato and root-knot nematode populations. Ann Agric Res 7(2):263–266

Goswami BK, Vijayalakshmi K (1986b) Efficacy of some indigenous plant materials and non-edible oil seed cakes against *Meloidogyne incognita* in tomato. Indian J Nematol 16(2):280–281

Goswami BK, Vijayalakshmi K (1987) Effect of period of decomposition of oil seed cakes in soil on *M. incognita*. Indian J Nematol 17(1):84–86

Gour AC, Prasad SK (1970) Effect of organic matter and inorganic fertilizers on soil and plant nematodes. Indian J Entomol 32:186–188

Govindaiah NS, Sharma DD, Gargi (1989) Effect of mulching of green leaves for the control of root-knot nematode in mulberry. Indian J Nematol 19(1):25–28

Govindaiah SB, Dandin TP, Datia RK (1990) Effect of marigold (*Tagetes patula*) intercropping against *Meloidogyne incognita* infecting mulberry. Indian J Nematol 20(1):96–99

Gowda DN (1972) Studies on comparative efficacy of oil cakes on control of root-knot nematodes, *Meloidogyne incognita* on tomato. Mysore J Agric Sci 6(4):524–525

Grewal PS, Sohi HS (1988) Toxicity of some plant extracts to Aphelenchoides compsticola. Indian J Nematol 18:354–355

Gupta P, Bhattacharya S (1995) Efficacy and economics of using neem products in management of root-knot nematodes in tomato. National symposium on nematode management with eco-friendly approaches and bio-components, 24–26 Mar 1995, IARI, New Delhi, (abstract) pp 90

Gupta DC, Kaliran (1981) Studies on the control of *Meloidogyne javanica* infesting chickpea in different types of soil. Indian J Nematol 11(1):77–80

Gupta R, Sharma NK (1985) Efficacy of garlic as nematicide against *Meloidogyne incognita*. Presented in IV nematology symposium at Udaipur, India, 15 May 1985

Gupta R, Sharma NK (1990a) Nematicidal property of garlic (*Allium sativum* Linn.). Presented in national symposium on problems and prospects on botanical pesticides in integrated pest management, at CTRI, Rajahmundry, India, 21–22 Jan 1990 (abstract), p 29

Gupta R, Sharma NK (1990b) Ovicidal activity of garlic extract to *Meloidogyne incognita* (Kofoid and White) Chitwood. J Appl Zoolog Res 1(2):90–92

Gupta R, Sharma NK, Kaur DJ (1985) Efficacy of garlic as nematicide against *Meloidogyne incognita*. Indian J Nematol 15:266–267

Hameed SF (1968) Relative efficacy of different oil cakes on the incidences of root-knot nematodes. Madras J Agric Sci 55:168–171

Haroon SA, Smart GC Jr (1983a) Development of *Meloidogyne incognita* inhibited by *Digitaria decumbens* c.v. *pangola*. J Nematol 15(1):102–105

Haroon SA, Smart GC Jr (1983b) Effect of root extracts of pangola digital grass on egg hatching and larval survival of *Meloidogyne incognita* inhibited by *Digitaria decumbens* c.v. *pangola*. J Nematol 15(4):646–649

Haroon SA, Nation JN, Smart GC Jr (1982) Isolation of natural nematicides from pangola digital grass, *Digitaria decumbens c.v. pangola* that effect hatching and larval survival of *M. incognita*. J Nematol 14:444

Hasan A (1992) Effect of certain plant exudates and byproducts on the development of root-knot nematode. In: Tauro P, Narwal SS (eds) Proceedings of national symposium on allelopathy in agro-ecosystems. Indian Society of Allelopathy, HAU, Hisar, India, pp 188–186

Hasan A, Jain RK (1984) Bio-toxicity of *Parthenium hysterophorus* extract against Meloidogyne incognita and *Helicotylenchus dihystera*. Nematological Mediterranea 12:239–242

Haseeb A, Butool F (1990) Evaluation of nematicidal properties in some members of family Lamiaceae. Int Nematol Netw Newl 7(2):24–26

Haseeb A, Butool F (1991) Evaluation of different pesticides and oil cakes for the control of root-knot nematode, *Meloidogyne incognita* infesting on davana, *Artemisia pallens*. Curr Entomol 2(2):201–204

Haseeb A, Khan KM, Saxena SK (1982) Toxicity of leaf extracts of plants to root-knot and reniform nematodes. Indian J Parasitol 6(1):119–121

Haseeb A, Pandey R, Hussain A (1988) A comparison of nematicides and oil cakes for control of *Meloidogyne incognita* on *Ocimum basilicum*. Nematropica 18(1):65–69

He Qin Li, Qi Zhi Liu, Zhi Long Liu, Du Shu Shan, Zhi Wei Deng (2013) Chemical composition and nematicidal activity of essential oil of *Agastache rugosa* against *Meloidogyne incognita*. Molecules 18:4170–4180

Hoan LT, Davide RG (1979) Nematicidal properties of root extracts of seventeen plant species on *Meloidogyne incognita*. Philipp Agric 62:285–295

Hussain SI, Masood A (1975) Effect of some plant extracts on larval hatching of *Meloidogyne incognita*. Acta Bot Ind 3:142–146

Hussaini SS, Bharata KS, Prasad Rao PVV (1993) Combined efficacy of soil solarization and neem cake application against root-knot nematode, *Meloidogyne incognita* and weeds in tobacco nurseries. World neem conference, Bangalore, 24–28 Feb 1993 (abstract), pp 43

Inamori Y, Takeuchi K, Baba K, Kozawa M (1987) Chem Pharm Bull 35:3931

Jacob A, Bai HJ, Kuriyan KJ (1990) Plant products as an alternate to nematicide for nematode management in banana. Presented in national symposium on problems and prospects on botanical pesticides in integrated pest management, at CTRI, Rajahmundry, India, 21–22 Jan 1990 (abstract), pp 45–46

Jacobson M (1958) Insecticides from plants: a review of the literature, 1941–53. USDA handbook 154. USDA, Washington, DC

Jacobson M (1975) Insecticides from plants: a review of the literature, 1952–72. USDA handbook 461. USDA, Washington, DC

Jacobson M (1989) Botanical pesticides: past, present and future. In: Insecticides of plant origin, ACS Symposium series 387 of III Chemical Congress of North America, 195th National meeting of American Chemical Association, held at Toronto, Canada, 5–11 June 1988, pp 1–10

Jagadale GB, Panwar AB, Darekar KS (1985a) Effect of organic amendments and antagonistic plants on control of the root-knot nematode infesting betelvine, *Piper betel* Linn. Presented in IV nematology symposium at Udaipur, India, 15 May 1985 (abstract), p 35

Jagadale GB, Panwar AB, Darekar KS (1985b) Effect of organic amendments and antagonistic plants on control of the root-knot nematode infesting betelvine, *Piper betel* Linn. Indian J Nematol 15:264

Jain SK, Saxena R (1993) Evaluation of nematicidal potential of *Mangifera indica*. Indian J Nematol 23(1):131–132

Jain U, Dutta S, Trivedi PC, Tiagi B (1986) Allelochemic effects of some plants on hatching of *M. incognita*. Indian J Nematol 16(2):275–276

Jain RK, Mathur KN, Singh RV (2007) Estimation of losses due to plant parasitic nematodes on different crops in India. Indian J Nematol 39:219–221

Jasy T, Koshy PK (1992) Effect of certain leaf extracts and leaves of *Gliricidia maculata* (H.B & K.) Steud. As green manure on Radopholus similis. Indian J Nematol 22:117–121

Jayaraj S (1991) An overview of botanical pest control research in TNGDNAU in Tamil Nadu, India. In: Proceedings of the botanical pest control project – phase II. Midterm Project Review Meeting, 28–31 July 1991, Dhaka, Bangladesh. International Rice Research Institute, Los Baños, pp 36–38

Jeppson LR, Keifer HH, Baker EW (1975) Mites injurious to economic plants. University of California Press, Berkley, p 614

Kaliram D, Gupta C (1980) A note on efficacy of fresh neem leaf extract in the control of Meloidogyne javanica infesting chickpea Cicer arietinum. Indian J Nematol 10:96–98

Kaliram, Gupta DC (1982) Efficacy of neem leaves, nematicides and fertilizers alone or in combination against *Meloidogynae javanica* infesting chickpea. Indian J Nematol 12(1):221–225

Katsura M (1981) Nematicidal natural products. ICIPE conference, May 1980, Department of Agricultural Chemistry, Nagoya University, Japan, (abstract), p 23

Khan MW (1974) Effect of water soluble fractions of oil cakes and bitter principles of neem on some fungi and nematodes. Acta Bot Ind 2(2):120–128

Khan MW (1979) Suppression of phytophagous nematodes and certain fungi in rhizosphere of okra due to oil cakes amendments. Acta Bot Ind 7(1):51–56

Khan TA, Hussain SI (1988a) Studies on the efficacy of the seed treatments with pesticides, oil cakes, neem leaves and culture filtrate of *P. lilacinus* for control of disease caused by *R. reniformis, M. incognita* and *Rhizoctonia solani* occurring individually or concomitantly on cowpea. Indian J Nematol 18(2):182–198

Khan TA, Hussain SI (1988b) Studies on the efficacy of *P. lilacinus* against a disease complex caused by *R. reniformis, M. incognita* and *Rhizoctonia solani* on cowpea. Nematol Mediterr 16(2):229–231

Khan AM, Alam MM, Siddiqi ZA, Saxena SJ (1967) Effect of different oil cakes on hatching of larva and development of root-knot caused by *Meloidogyne incognita*. International Symposium Plant Pathology, IARI, New Delhi (abstract), p 37

Khan WM, Khan AM, Saxena SK (1973) Influence of certain oil cakes amendments on nematodes and fungi in tomato fields. Acta Bot Ind 1:49–54

Khan AM, Alam MM, Rais A (1974) Mechanism of control of plant parasitic nematodes as a result of the application of the oil cakes to the soil. Indian J Nematol 4:93–96

Khan WM, Huq SAM, Saxena SK (1976) Suppression of phytophagous nematodes and certain fungi in rhizosphere in okra due to oil cakes amendments. All India symposium on modern concept in plant protection, 23–26 Mar 1976 (abstract), pp 88–89

Krishnamurthy GVG, Murthy PSN (1990) Further studies with plant extracts on larvae of root-knot nematode, *Meloidogyne incognita*. Presented in national symposium on problems and prospects on botanical pesticides in integrated pest management, at CTRI, Rajahmundry, India, 21–22 Jan 1990 (abstract), pp 43–44

Kumar TP, Nair MRG (1976) Effect of some green leaves and organic wastes on root-knot infection on bhindi. Agric Res J Kerala 14(1):64–67

Kumari R, Verma KK, Dhindsa KS, Bhatti DS (1986) Datura, Ipomea, Tagetes and Lawsonia as a control of Tylenchulus semipenetrans and Anguina tritici. Indian J Nematol 16(2):236–240

Lall SB, Hameed SF (1969) Studies on the biology of the root-knot nematodes, *Meloidogyne* spp. with special reference to host resistance and manuring. First all India nematological symposium. IARI, New Delhi, p 6

Lanjeswar RD, Shukla VN (1979) Vulnerability of larvae and eggs of Meloidogyne incognita to some oil cakes and fungicides. Indian J Nematol 16:69–73

Langeswar RD, Shukla VN (1986) Vulnerability of larvae and eggs of *M. incognita* to some oil cakes and fungicides. Indian J Nematol 16(1):69–73

Mahmood I, Saxena SK, Zakirddin R (1979) Effect of some plant extract on the mortality of *Meloidogyne incognita* and *Rotylenchus reniformis*. Acta Bot Ind 7(2):121–132

Mahmood I, Saxena SK, Zakirddin R (1982) Effect of some plant extract on the mortality of *Rotylenchus reniformis* and *Meloidogyne incognita*. Bangl J Bot 11(2):154–157

Mandal SB, Bhatti DS (1983) Preliminary screening of some weed shrubs for their nematicidal activity against *M. incognita*. Indian J Nematol 13(1):123–127

Mansour FA, Ascher KRS (1983) Effect of neem (*Azadirachta indica*) seed kernel extracts from different solvents on the carmine spider mite (*Tetranychus cinnabarinus*). Phytoparasitica 11(3–4):177–185

Mansour F, Ascher KRS, Omari N (1987) Effects of Neem (Azadirachta indica) seed kernel extracts from different solvents on the predacious mite Phytoseiulus persimilis and the phytophagous mite Tetranychus cinnabarinus. Phytoparasitica 15(2):125–130

Mansour FA, Ascher KRS, Abo-Moch F (1993) Effects of Margosan-OTM, AzatinTM, and RD9-Repelin ® on spiders, and on predacious and phytophagous mites. Phytoparasitica 21:205–211

Markkula M, Roukka K, Tiittanen K (1969) Reproduction of *Myzus persicae* (Sulz.) and *Tetranychus telarius* (L.) on different chrysanthenum cultivars. Ann Agric Fenn 8:175–183

Masuda K, Yamada K, Kimura M, Hamada M (1991) Nematicidal activity of matrine and its derivatives against pine wood nematodes. J Agric Food Chem 39:189–191

McIndoo I (1982) USDA Bur Entomol. & Pl. Quar. Pers. Commn. In Michael et al. 1985. Plant species reportedly possessing pest control properties. An EWC/UH DATABASE, University of Hawaii, pp 249

Mian IM, Rodriguez R (1982) Survey of nematicidal properties of some organic materials available in Alabama as amendment to soil for control of *Meloidogyne arenaria*. Nematropica 12:235–246

Midha SK, Chabra HK (1974) Effect of *Tagetes erecta* and *Sesamum indicum* on population of plant parasitic nematodes. Indian J Horticul 31(2):196–197

Miller PM (1979) Vegetable oils as protectants against nematode infestation. J Nematol 11(4):402–403

Mishra SD, Mojunder V (1993) Nematicidal efficacy of decomposed products on Meloidogyne incognita infesting mungbean and their residual effects on chickpea. World neem conference, Bangalore, India, 24–28 Feb 1993, (abstract), p 37

Mishra SD, Prasad SK (1973) Effect of water extract of oil seed cakes on the 2nd stage larvae of *M. incognita* at different concentrations and exposure times. Indian J Entomol 35(2):101–106

Mobin MP, Khan AM (1969) Effect of organic amendment on the population of rhizosphere fungi and nematodes around roots of some fruit trees. All India nematology symposium. IARI, New Delhi, pp 11–13

Mohandas C, Rao YS, Sahu SC (1981) Cultural control of rice root nematodes (Hirschmanniella spp.) with Splzenoclea zeylanica. Proc Indian Acad Sci (Anim Sci) 90:373–376

Mohanty KC, Das SR (1987) In vitro toxicity of Erythrina indica extract on Meloidogyne incognita. Indian J Plant Prot 15(2):172–173

Mohanty KC, Das SN (1988) Nematicidal properties of *Erythrina indica* against *Meloidogyne incognita* and *Tylenchorhynchus mashoodi*. Indian J Nematol 18(1):138

Mojumder V, Mishra SD (1995) Use of neem products as seed treatments for the management of root-knot, reniform and other plant parasitic nematodes. Presented in national symposium on nematode problems in India-an appraisal of the nematode management with eco-friendly approaches and bio-components. IARI, New Delhi, 24–26 Mar 1995 (abstract), p 57

Monforta WS, Csinosa AS, Desaegerb J, Seeboldc K, Websterd TM, Diaz-Perez JC (2007) Evaluating Brassica species as an alternative control measure for root knot nematode (*M. incognita*) in Georgia vegetable plasticulture. Crop Prot 26:1359–1368

Morallo-Rajessus B, Eroles LC (1978) Two insecticidal principles from marigold (*Tagetes* spp) roots. Philipp Entomol 4(1/2):87–97

Mukherjee SK (1983) ICAR, New Delhi, Pers. Commun. In Michael et al. 1985. Plant species reportedly possessing pest control properties. An EWC/UH DATABASE, University of Hawaii, pp 249

Mukherjee SN, Sukul NS (1978) Nematicidal action of three species of wild herbs. J Res India 2(2):12

Mukhopadhyaya MC, Chattopadhayaya PR (1979) Control of root-knot nematode, *Meloidogyne incognita* (Kofoid and White) Chitwood with *Typhonium trilobatum,* an annual herb. Indian J Nematol 9(1):53–58

Munakata K (1978) Nematicidal substances from plants. Advances in pesticide sciences Part-2. Pergamon Press, Oxford, pp 295–302

Nandal SN, Bhatti DS (1983) Preliminary screenings of some weeds and shrubs for their nematicidal activity against *Meloidogyne javanica*. Indian J Nematol 13(1):123–127

Nath A, Sharma NK, Bhardwaj S, Thappa CN (1982) Nematicidal property of garlic. Nematologia 28(2):253–255

Ntalli NG, Caboni P (2012) Botanical nematicides: a review. J Agric Food Chem 60(40):9929–9940

Oka Y, Nacar S, Putievsky E, Ravid U, Yaniv Z, Spiegel Y (2000) Nematicidal activity of essential oils and their components against the root-knot nematode. Phytopathology 90:710–715

Pandey R (1995) Performance of oil seed cake, pesticides and dry leaf matter on the reproduction potential of Meloidogyne incognita and yield for Japanese mint HY-77. National symposiyum on nematode management with eco-friendly approaches and bio-components, 24–26 Mar 1995. IARI, New Delhi (abstract), p 37

Pandey R, Haseeb A (1988) Studies on the toxicity of certain medicinal plants to root-knot nematode, *Meloidogyne incognita*. Indian J Pl Path 6(2):184–186

Park Il-Kwon, Kim Junheon, Lee Sang-Gil, Shin SangChul (2007) Nematicidal activity of plant essential oils and components from Ajowan (*Trachyspermum ammi*), Allspice (*Pimenta dioica*) and Litsea (*Litsea cubeba*) essential oils against pine wood nematode (*Bursaphelenchus Xylophilus*). J Nematol 39(3):275–279

Parmar BS, Dev Kumar C (1993) Botanicals and biopesticides. Westville Publishing House, New Delhi, pp 199

Patel HR, Thakar NA, Patel CC (1985) Azolla and mustard cake against nematicides for root-knot management in okra. Sour Madras Agric J 72(10):593–594

Patel HR, Patel DJ, Patel CC, Thaker NA (1990a) Effectivity of *Clerodendrum inerme* Linn. and periwinkle, *Catharanthus roseus* (Linn.) G. Don. for management of root-knot nematode on okra. Presented in national symposium on problems and prospects on botanical pesticides in integrated pest management, at CTRI, Rajahmundry, India, 21–22 Jan 1990 (abstract), pp 40–41

Patel HR, Makwana MG, Patel BN (1990b) Effect of periwinkle, *Catharanthus roseus* (Linn.) G. Don. on reniform nematode. Presented in national symposium on problems and prospects on botanical pesticides in integrated pest management, at CTRI, Rajahmundry, India, 21–22 Jan 1990 (abstract), p 42

Patel CB, Rai AB, Patel MB, Patel AJ, Shah AA (1993) Acaricidal tests of botanical pesticides in comparison to conventional acaricides/pesticides against red spider mites (Acari: Tetranychidae) on okra, brinjal and Indian bean. Indian J Entomol 55(2):184–190

Pillai SN, Desai MV (1976) Punnakkai' cake on the control of root-knot nematode. Indian J Mycol Plant Pathol 6(1):14–17

Poornima K, Vodivelu S (1993) Comparative efficacy of nematicide, oil cakes and plant extracts in the management of *Meloidogyne incognita, Pratylenchus delattrei* and *Rotylenchus reniformis* on brinjal. Indian J Nematol 23(2):170–173

Prakash A, Rao J (1997) Botanical pesticides in agriculture. CRC Lewis Publs, Boca Raton, p 481

Prakash A, Tiwari SN, Rao J (1990) Exploitation of natural plant products for management of pests and diseases in rice ecosystems. In: Proceedings, symposium of growth, development, resource conservation, 23–36 pp

Prasad D, Khan E (1990) Studies on integrated control of phytonematodes infesting soybean. Presented in national symposium on problems and prospects on botanical pesticides in integrated pest management, at CTRI, Rajahmundry, India, 21–22 Jan 1990 (abstract), pp 44–45

Prasad JS, Rao YS (1979) Nematicidal property of the weed, *Eclipta alba* Hassk. (Compositae). Revista di Parasitologia XL: 87–90

Prasad D, Agrawal AK, Buchae YM, Sethi CL (1987) Biotoxicity of groundnut testas extract against *Meloidogyne incognita*. Indian J Nematol 17(2):120–124

Proct JC, Kornprobst JM (1983) Effects of *Azadiracta indica, Hannoa undulata* and *Hannoa klaineana* seed extracts on the ability of *Meloidogyne javanica* juveniles to penetrate tomato roots. Revue de Nematologie 6(2):330–332

Rajagopan K (1974) First record of spider mite, *Tetranychus ludeni* Zacher transmitting Dolichos Enation Mosai Virus. Curr Sci 43(15):488–489

Ramanath MN, Khan NH, Kamalwanshi RS, Trivedi RP (1982) Effect of *Argemone mexicana* on M. incognita in okra. Indian J Nematol 12:205–208

Rao MS, Pandey M (1982) Comparative efficacy of karanja leaf and carbofuran on the management of *Aphelenchoides composticola*. Indian J Nematol 12:258–159

Rao MS, Reddy PP (1993) Effective utilization of welgro and repelin-9 for the management of root-knot nematode in tomato. World neem conference, Bangalore, India, 24–28 Feb 1993 (abstract), p 38

Rao J, Prakash A (1995) Insects, mites and nematode to deteriorate seed quality in the paddy fields in Orissa. J Appl Zoo Res 6(2):89–92

Rao J, Prakash A, Ghosh SK (1999) Mites in rice ecosystems. AZRA, Cuttack, India, pp 109

Ravidutta B, Bhatti DS (1986a) Effect of chemicals and phytotherapeutic substances on biological phenomenon of *M. incognita* infesting tomato. Indian J Nematol 16(1):19–22

Ravidutta B, Bhatti DS (1986b) Effective doses and time of application of insecticides and castor leaves for controlling *M. incognita* infesting tomato. Indian J Nematol 16(1):8–11

Reddy RP, Khan RN (1990) Management of root-knot nematode infesting papaya by soil incorporation of some plant leaves. National symposium on problems and prospects of botanical pesticides in IPM, 42 pp

Rohde RA (1972) Expression of resistance in plants to the nematodes. Annu Rev Plant Physiol Plant Mol Biol 10:23–33

Routray BN, Sahoo H (1985) Integrated control of root-knot nematode, M. incognita with neem cake and granular insecticides on tomato. Indian J Nematol 15(2):261

Roy AK (1976) Effect of de-caffeinated and water hyacinth compost on the control of *Meloidogyne graminicola*. Indian J Nematol 6:73–77

Roy DC, Pande YD (1991) Effect of *Hyptis suaveolens* (L.) Poit. leaf extract on the population of *Lipaphis erysimi* Kalt. IV National symposium on growth, development and control technology of insect pests, p 37

Roy D, Sinhababu SP, Sukul NC, Mahato SB (1993) Nematicide principles from the funicles of *Acacia auriculiformis*. Indian J Nematol 23(2):152–153

Ruelo JS (1976) Chemical biological and integrated biochemical control of *Meloidogyne incognita*. PhD thesis. University of Philliphines Los Banos, Laguna, pp 188

Ruelo JS, Davide (1979) Studies on the control of *Meloidogyne incognita* II integration of biological and chemical control. Phillip Agric 62:159–165

Sarra-Guzman R (1984) Toxicity screening of various plant extracts against *Meloidogyne incognita* Chitwood and *Radopholus similis* Cobb. and characterization of their nematicidal components. PhD thesis, University of Philippines, Los Banos College, Laguna, Philippines, pp 139

Saxena R, Jain SK, Sharma (1993) Nematicidal effect of neem leaf extract. World neem conference, Bangalore, India, 24–28 Feb 1993 (abstract), p 39

Schauer M, Schmutterer H (1981) The effect of squeezed fresh juices and crude extracts of *Ajuga remota* on the two-spotted spider mite *Tetranychus urticae* Koch. Z Angew Entomol 91(5):425–433

Sen K, Das-Gupta MK (1985) Effect of different substances on larval hatching of *M. incognita*. Indian J Nematol 15(1):100–103

Sertkaya E, Kaya K, Soylu S (2010) Acaricidal activities of the essential oils from several medicinal plants against the carmine spider mite (*Tetranychus cinnabarinus* Boisd.) (Acarina: Tetranychidae). Ind Crop Prod 31:107–112

Seth ML (1962) Transmission of pigeon pea sterility by an eriophyid mite. Indian Phytopathol 15:317–319

Sharma HK, Prasad D (1995) Potentialities of a few botanicals in management of root-knot, *Meloidogyne incognita* on soybean. Ann Agric Res 16(4): 473–476

Sharma AN, Sharma UK, Singh RS (1971) Oil cakes and wood sawdust control root-knot. Indian Farming, pp 17–20

Sharma SK, Sakhuja PK, Singh I (1981) Effect of mustard cake on the hatching and cyst population of *Heterodera avenae* on wheat. Indian J Nematol 11:11–14

Sharma HK, Vinal OP, Prasad D (1985) Nematicidal and manorial values of some oil cakes. Int J Crop Pl Disease 3:51–55

Sherif A, Hall RG, El Amangi M (1987) Drugs, insecticides and other agents from *Artemisia*. Med Hypoth 23:187–193

Shigefumi K, Kojiro W, Katsura M (1976) Isolation of nematicidal polyacetylenes from Carthamus tinctorius L. Agric Biol Chem 40(10):2085–2089

Shukla VN, Banker (1985) Studies on the nematicidal properties of some plant species against *Meloidogyne incognita*. P K V Res J India 9(2):46–52

Siddiqui ZA (1976) Nematode populations and yields of certain vegetable crops influenced by oil cake amendments. Indian J Nematol 6(2):179–182

Siddiqui MA, Alam MM (1987) Efficacy of seed dressing with extracts of neem and Persian lilac against *Meloidogyne incognita* and *Rotylenchus reniformis*. Nematol Mediterr 15(2):399–403

Siddiqui MA, Alam MM (1988a) Toxicity of different plant parts of *Tagetes lucida* to plant parasitic nematodes. Indian J Nematol 18(2):181–185

Siddiqui MA, Alam MM (1988b) Studies on the nematotoxicity of root exudates of certain species of *Tagetes*. Indian J Nematol 18(2):335–337

Siddiqui MA, Alam MM (1993) Evaluation of nematicidal potential in neem allele-chemicals. World Neem Conference, Bangalore, India, 24–28 Feb 1993 (abstract), p 39

Siddiqui MA, Haseeb A, Alam MM (1987) Evaluation of nematicidal properties of some latex bearing plants. Indian J Nematol 17(1):99–102

Siddiqui MA, Haseeb A, Alam MM (1992) Control of plant parasitic nematodes by soil amendments with latex bearing plants. Indian J Nematol 22(1):25–28

Sikdar AK, Govindaiah Madhava Rao YR, Jolly MS, Giridhar K (1986) Control of root-knot nematode disease of mulberry by Temik 10G and oil-cakes. Indian J Seric 25(1):33–35

Singh I (1980) Effect of oil cake extracts on hatching of root-knot nematode, *Meloidogyne incognita*. Indian J Mycol 12:115–116

Singh SP, Singh V (1988) Evaluation of oil cakes and nematicides for the control of *Meloidogyne incognita* infesting eggplants. Indian J Nematol 18(2):338–342

Singh RS, Sitaramaiah K (1967) Effect of decomposition of green leaves, sawdust, and urea on the incidence of root-knot nematodes in okra. Indian Phytopathol 20:149–155

Singh RS, Sitaramaiah K (1969) Control of root-knot nematode through organic and inorganic amendments in the soil I Effect of oil cakes and saw dusts. World neem conference, Bangalore, India, 24–28 Feb 1993 (abstract), p 10

Singh RS, Sitaramaiah K (1971) Control of root-knot nematode through organic and inorganic amendments in the soil. II. Effect of oil cakes and saw dusts. Indian J Mycol Pl Path 1(1):20–29

Singh RS, Sitaramaiah K (1976) Incidences of root-knot of okra and tomato in oil cake amended soil. Plant Disease Report 59(9):668–672

Singh D, Dahiya RS, Dalal MM (1995) Nematicidal efficacy of some plant leaves to *Heterodera avenae* in wheat. National symp on nematode management with eco-friendly approaches and bio-components, 24–26 Mar 1995. IARI, New Delhi, (abstract), p 5

Sinhababu SP, Ray D, Sukul NC, Mahato SR (1992) Nematicidal principles from Acacia auriculiformis. In: Tauro P, Narwal SS (eds) Proceedings of first national symposium on allelopathy in agro-ecosystems. Indian Society of Allelopathy, HAU, Hisar, India, pp 181–183

Sitaramaiah K, Singh RS (1977) Response of plant parasitic and soil nematodes to plant extracts in amended soil. Pantanagar J Res 2(2):153–157

Sitaramaiah K, Singh RS (1978) Effect of organic amendment on phenolic content in soil and plant in response to *Meloidogyne javanica* and its host to related compounds. Plant Soil 50(3):671–679

Sithanatham S, Muthuswamy S, Bhaskaran TL (1979) A note on the role of the eriophyid mite, *Aceria sacchari* in the transmission of streak virus disease of sugarcane. Sci Cutl 38:248–250

Sivakumar M, Marimuthu T (1986) Efficacy of different organic amendments against phytonematodes associated with betel vine (Piper betle L.). Indian J Nematol 16:278

Siyanand, Meher HC, Kaushal KK, Singh G (1995) Induction of *Tagetes patula* Linn. As mixed rotation crop-A bubble technology protocol for managing root-knot nematode, *Meloidogyne incognita* in vegetables. National symposium on nematode management with eco-friendly approaches and bio-components, 24–26 Mar 1995, IARI, New Delhi (abstract), p 19

Srivastava AS (1971) Application of organic amendments for control of root-knot nematode, *Meloidogyne javanica* Treuh. Labdev J Sci Tech 9B(3/4):203–204

Subramaniyan S (1985) Effect of *Eupatorium odoratum* extracts on *Meloidogyne incognita*. Indian J Nematol 15(2):247

Subramaniyan S, Selvaraj (1988) Effect of *Tagetes patula* Linn. Leaf extract on *Radopholus similis* (Cobb, 1933) Thome, 1949. Indian J Nematol 18(2):337–338

Sukul NC, Ghosh T (1974) Nematicidal action of some edible crops. Nematologia 20(2):187–191

Sundarbabu R, Sankaranarayan C, Vadivelu S (1993) Nematode management with plant products. Indian J Nematol 23(2):177–178

Thaker NA (1988) Effect of extracts of Azolla pinnata on eggs hatching of root-knot nematodes, *Meloidogyne incognita* and *M. javanica*. Madras Agric J 75:297–299

Thaker NA, Patel HR, Patel CC (1987) Azolla in management of root-knot disease in okra. Indian J Nematol 17(1):136–137

Tiyagi SA, Alam MM (1993) Efficacy of neem cake against soil nematodes and fungi. World neem conference, Bangalore, India, 24–28 Feb 1993(abstract): 39–40

Tiyagi SA, Muktar J, Alam MM (1985) Preliminary studies on the nematicidal nature of two plants of family Compositae. Int Nematolog Netw Newl 2(3):19–0021

Uhlenbrock JH, Bijoo JD (1959a) Nematicidal substances in marigold, *Tagetes erecta* Linn. Recl Trav Chem Pays Bas Belgium 78:289

Uhlenbrock JH, Bijoo JD (1959b) Isolation and structure of nematicidal principle occurring in *Tagetes* roots. In: Proceedings of IV International Congress Crop protection, Hamburg 1957, pp 579–581

Varma MK (1976) Organic amendments in soil and incidences of root-knot nematode, *Meloidogyne incognita* in tomato, *Lycopersicon esculentum* Mill. Udyanika 2:39–40

Venzon M, Consolac-ao RM, Molina-Rugama AJ, Duarte VS, Dias R, Angelo P (2008) Acaricidal efficacy of neem against *Polyphagotarsonemus latus* (Banks) (Acari: Tarsonemidae). Crop Prot 27:869–872

Verma RR (1993) Effect of different materials on the populations of root-knot nematode, *Meloidogyne incognita* infesting tomato crops. Indian J Nematol 23(1):135–136

Verma AC, Anwar A (1995) Integrated nematode management by using leaves of medicinal plants in pointed gourd, *Trichosanthes dioica* field. National symposiyum on nematode management with eco-friendly approaches and bio-components, 24–26 Mar 1995, IARI, New Delhi, (abstract): 18

Vijayalakshmi K, Goswami BK (1983) Effect of aqueous extract of various plants and non-edible oil seed cakes on the hatching of *Meloidogyne incognita* eggs. III. Nematological symposium, Himachal Pradesh Krishi Vishwa Vidyalaya, Solan, 24–29 May 1983 (abstract): 32

Vijayalakshmi K, Goswami BK (1985) Effect of aqueous extract of madar, *Calotropis gigantea* and amarbel, *Cuscuta reflexa* on larval mortality, hatching and penetration of *Meloidogyne incognita*. Indian J Nematol 15(2):265

Vijayalakshmi K, Prasad SK (1979) Nematicidal property of some indigenous plant materials against 2nd stage larvae of *Meloidogyne incognita*. Indian J Nematol 9(1):81–82

Vijayalakshmi K, Mishra DS, Prasad SK (1979) Nematicidal property of some indigenous plant materials against 2nd stage larvae of *Meloidogyne incognita* (Kofoid and White) Chitwood. Indian J Entomol 41(4):326–331

Vinothkumar S, Chinniah C, Muthiah C, Sadasakthi A (2009) Biorationals in the management of two spotted spider mite, *Tetranychus urticae* Koch in brinjal. Karnataka J Agric Sci 22(3):682–684

Wei J, Ding W, Zhao YG, Vanichpakorn P (2011) Acaricidal activity of Aloe vera L. leaf extracts against *Tetranychus cinnabarinus* (Boisduval) (Acarina: Tetranychidae). J Asia Pac Entomol 14:353–356

Winoto SR (1969) Studies on the effects of *Tagete*s species on plant parasitic nematodes. Dissertation, Agricultural University, Wageningen, 133 p

Wiratno D, Taniwiryono H, Van den Berg J, Riksen AG, Rietjens IMCM, Djiwanti SR, Kammenga JE, Murk AJ (2009) Nematicidal activity of plant extracts against the root-knot nematode, *Meloidogyne incognita*. Open Nat Product J 2:77–85

Zaki FA, Bhatti DS (1989) Effect of castor (*Ricinus communis*) leaves in combination with different fertilizer doses on *Meloidogyne incognita* infesting tomato. Indian J Nematol 19(2):171–176

Zhang Y, Ding W, Zhao Z, Wu J, Fan Y (2008) Studies on acaricidal bioactivities of Artemisia annua L. Extracts against Tetranychus cinnabarinus Bois. (Acari: Tetranychidae). Agric Sci China 7(5):577–584

Plant Disease Management: Prospects of Pesticides of Plant Origin

7

Someshwar Bhagat, Ajanta Birah, Rakesh Kumar, M.S. Yadav, and C. Chattopadhyay

Abstract

The indiscriminate use of chemical fungicides led to pesticide residues in food products, risk of development of new pathotypes and pollution of soil and water ecosystem. This resulted in several ill effects on human beings, flora and fauna. To overcome the ill effects of chemical pesticides, attention had been paid to explore into products of higher plants for developing novel biopesticides in plant disease management. Our ancestors had been using these botanicals for the management of plant diseases, before the era of conventional fungicides. But the popularity of pesticides of plant origin has again been increasing due to its potential fungicidal action against several plant pathogens without any deleterious effect to the crop plants as well as environment. Several plants have been identified for antimicrobial properties which can suppress the growth and multiplication of plant pathogens, reduction in storage decay and spoilage of food products. The potential plant origin pesticides, viz. neem (*Azadirachta indica*), garlic bulb (*Allium sativum*), eucalyptus (*Eucalyptus globulus*), turmeric (*Curcuma longa*), tobacco (*Nicotiana tabacum*), ginger (*Zingiber officinale*), etc., have been successfully used for the management of several plant diseases. Moreover, seed treatment+foliar spray of freshly prepared garlic bulb extract has resulted into the reduction of *Alternaria* blight (35.6 %), white rust (50.4 %), powdery mildew (67.7 %) and *Sclerotinia* rot (80.3 %) in mustard with 27.3 % increase in yield over untreated control. These pesticides can suitably fit in any integrated pest management framework as well as in organic farming system which is a necessity in the current situation. Keeping in view the ever-increasing demand for safe food, pesticides of plant origin have a pivotal role to play in the management of plant diseases in comparison to the

S. Bhagat, Ph.D. • A. Birah, Ph.D. • R. Kumar, Ph.D.
M.S. Yadav, Ph.D. • C. Chattopadhyay, Ph.D. (✉)
National Centre for Integrated Pest Management (ICAR), LBS Building, Pusa Campus, New Delhi 110012, India
e-mail: chirantan_cha@hotmail.com

D. Singh (ed.), *Advances in Plant Biopesticides*,
DOI 10.1007/978-81-322-2006-0_7, © Springer India 2014

conventional chemical pesticides. These pesticides are not only useful to the developing countries due to their easy availability, being relatively cheap, easy sustenance in any crop protection programme and having direct relevance to the developed countries for healthy and quality produce of foodstuffs.

Keywords
Plant products • Pesticides • Plant disease management • Chemotherapeutants

1 Introduction

Indiscriminate use of chemical pesticides leads to pesticide residues in food as well as in soil and water ecosystem that has encouraged researchers to look for appropriate alternatives. This has also led to the development of resistant strains of plant pathogens as well as different environmental and human health hazards. Recently, plant origin pesticides have got more impetus towards the development of novel chemotherapeutants in plant protection. Because of non-phytotoxicity, systemic in nature to some extent, easy biodegradability and stimulatory nature of host metabolism, plant products possess the potential in pest management. These botanical products were used widely before the 1950s. These products were displaced by modern synthetic pesticides in view of cheaper, easier and long-lasting properties during that period. The popularity of botanical pesticides is gaining importance, and some plant products are being used globally as green pesticides. However, only a handful of botanicals are currently used in agriculture. Pyrethroids and neem products are well established as commercial botanical pesticides, and recently some essential oils of higher plants have also been used as antimicrobials against storage diseases because of their relatively safe and wide acceptance by the consumers. The secondary metabolites of plants like phenols, phenolic acids, quinones, flavones, flavonoids, flavonols, tannins, coumarins, etc., can be used for the management of several plant pathogens. The allicin, a volatile antimicrobial substance, obtained from freshly prepared garlic bulb extract, effectively controlled *Sclerotinia* stem rot, *Alternaria* leaf blight, club root and white rust diseases of mustard (Meena et al. 2013), and *Alternaria* blight of sunflower (Chattopadhyay 1999) and safflower (Chattopadhyay 2001). Allicin was also found effective against seed-borne *Alternaria* sp. in carrot, *Phytophthora* leaf blight in tomato, tuber blight in potato, rice blast pathogen, downy mildew of *Arabidopsis thaliana* (Slusarenko et al. 2008) and *Sclerotinia* rot of mustard (Chattopadhyay et al. 2004). Similarly, the oil extracted from lemongrass was found effective against post-harvest anthracnose of mango fruit (Salomone et al. 2008); the extracts from *Bougainvillea spectabilis* and *Prosopis chilensis* seem to be antiviral protein which reduced the sunflower necrosis virus infection in cowpea and sunflower (Lavanya et al. 2009).

The use of chemical fungicides is considered as the most effective method of plant disease management and often practised worldwide. However, repeated use of certain chemical fungicides has led to the appearance of fungicide-resistant pathotypes of several pathogens. In recent years, there has been considerable pressure by consumers to reduce or eliminate chemical fungicides in food products. Further, the use of chemical fungicides for the management of plant diseases has its limitations due to their carcinogenic, teratogenic properties, high and acute residual toxicity, hormonal imbalance, slow and long degradation period, environmental pollution and deterioration of food quality and adverse effects on human health (Brent and Hollomon 1998; Dubey et al. 2007; Kumar et al. 2007). The use of synthetic chemicals as antimicrobials for the management of plant pathogens has undoubtedly increased but with some deterioration of environmental quality and human health (Cutler and Cutler 1999). Their uninterrupted and indiscriminate use has not only led to the development of resistant strains, but the accumulation of toxic residues on food grains used for human consumption has also led to health problems (Sharma and Meshram 2006).

1.1 Plant Origin Pesticides as Alternatives

Concerns related to chemical pesticides have encouraged researchers to look for alternative solutions to synthetic pesticides. Food safety is receiving increased attention worldwide as the important link between food and health is being increasingly recognized. Improving food safety is an essential element of improving food security which exists when populations have access to sufficient and healthy food. Numerous studies have documented the antifungal (Suhr and Nielsen 2003; Mishra and Dubey 1994; Elgayyar et al. 2001) and antibacterial (Canillac and Mourey 2001) effect of essential oils from plant. The examination of indigenous local herbs and plant materials has also been reported from different parts of the world, viz. India (Ahmad and Beg 2001), Australia (Cox et al. 1998), Argentina (Penna et al. 2001) and Finland (Rauha et al. 2000). Higher plants contain essential oils and a wide spectrum of secondary metabolites such as phenols, flavonoids, quinones, tannins, alkaloids, saponins and sterols. Such plant-derived chemicals may be exploited for their different biological properties. Because of their natural/plant origin, they are biodegradable and do not usually leave toxic residues or by-products.

2 Commonly Used Plants as Pesticides of Botanical Origin

2.1 Plant Extracts

Neem (*Azadirachta indica* A. Juss), garlic (*Allium sativum* Linn.), eucalyptus (*Eucalyptus globulus* Labill.), turmeric (*Curcuma longa* Linn.), tobacco (*Nicotiana tabacum* Linn.) and ginger (*Zingiber officinale* Rosc.).

Plants and/or their parts are collected randomly from their habitat areas. Fresh or dried plant materials can be used as a source for the extraction of secondary plant metabolites. Several researchers had reported about plant extract preparation from the fresh plant tissues (Chattopadhyay et al. 2004; Yadav et al. 2012; Meena et al. 2013). The ethnomedicinal use of fresh plant materials among the traditional and tribal people had encouraged the researchers to use the pesticides of plant origin for the management of plant diseases. Some plants are used in the dry form or as an aqueous extract. Plants/plant parts are usually air-dried to a constant weight before extraction. Normally researchers had dried the plants in the oven at about 40 °C for 72 h (Salie et al. 1996). Evidently, most of the essential oils/flavonoids are found evaporated during drying at higher temperature above 40 °C. The underground parts (root, tuber, rhizome, bulb, etc.) of a plant are more commonly used than other above-ground parts for bioactive compounds possessing antimicrobial properties. The basic principle is to grind the plant material (dry or wet) finer, which increases the surface area for extraction, thereby increasing the rate of extraction.

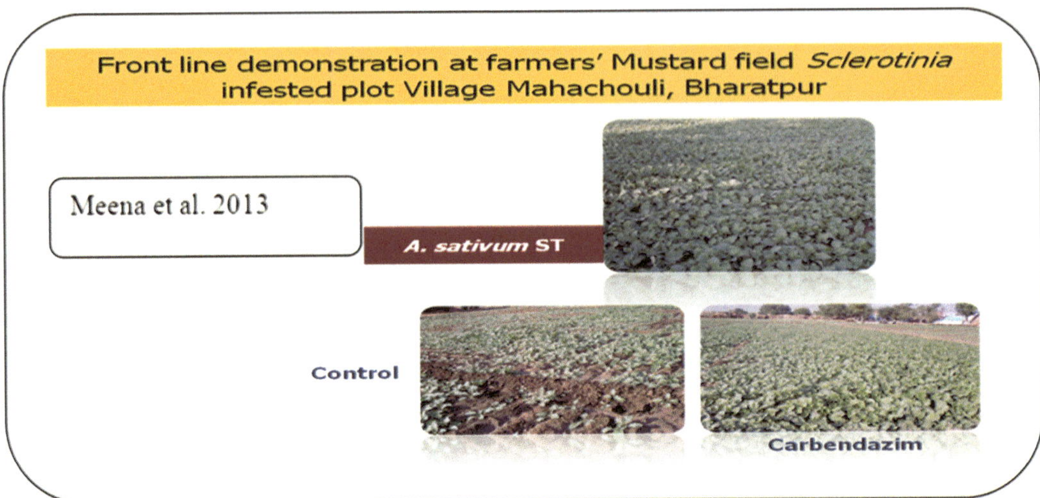

2.2 Essential Oils

Nettle oil (*Urtica* spp.), thyme oil (*Thymus vulgaris* Linn.), eucalyptus oil (*Eucalyptus globulus* Labill.), rue oil (*Ruta graveolens* Linn.), lemongrass oil [*Cymbopogon flexuosus* (Steud.) Wats] and tea tree oil (*Melaleuca alternifolia*).

The essential oils obtained from plant origin are used against postharvest pathogens. These oils are terpenoids and aromatic compounds acting as fungistatic substances. The major components of essential oils are carvacrol, thymol, cymene, terpene, phenylpropene derivatives, eucalyptol and anisole.

2.3 Gel and Latex

2.3.1 *Aloe vera* (Tourn. Ex Linn.)

Plant products are important sources of new agrochemicals for the management of plant diseases (Cardellina 1988). Biocides of plant origin are non-phytotoxic, systemic and easily biodegradable (Mason and Mathew 1996). There are a number of reports on resistance-inducing activities of plant extracts as well as plant extract components. Leaf extracts of spinach and rhubarb (Doubrava et al. 1988), barley (Yokoyama et al. 1991), giant knotweed (Daayf et al. 1997), *Bougainvillea xbuttiana* (Narwal et al. 2000), *Azadirachta indica* (Paul and Sharma 2002) and seaweed extract (Jayaraj et al. 2008) induced resistance against bacterial and fungal pathogens. In most of the reports, the induction of resistance was either through foliar sprays or soil drench. However, the seed treatment is preferred economic option, as ideally a compound or inducer applied to seed may provide protection during germination, emergence and the early establishment phase of the crop plants (Jensen et al. 1998).

3 Plant Origin Pesticides: Photoactivated Pesticides

Some plants can produce different types of chemicals which can be stimulated by sunlight, and these photosensitive metabolites become toxic to plant pathogenic organisms in the presence of sunlight (320–400 nm). For example, UVA-mediated alpha-terthienyl has both bactericidal (*Agrobacterium*, *Bacillus*, *E. coli*, *Pseudomonas*, *Staphylococcus*) and fungicidal (*Alternaria*, *Aspergillus*, *Candida*, *Pythium*, *Cladosporium*, *Colletotrichum*, *Fusarium*, *Rhizoctonia*, *Rhizopus*, *Saccharomyces*, *Saprolegnia*) activity (Downum 1986).

4 Plant Origin Pesticides: Antimicrobial Secondary Metabolites

Several plants have ability to synthesize aromatic metabolites. Some metabolites with their mechanism of action are as follows:

1. Flavonoids – inactivate enzymes, form complex with cell wall
2. Phenolic acids – bind to adhesions, form complex with cell wall, inactivate enzymes
3. Phenolics – membrane disruption, substrate deprivation
4. Terpenoids, essential oils – membrane disruption
5. Alkaloids – intercalate with cell wall
6. Tannins – bind to proteins, inhibit enzymes, non-availability of substrate
7. Coumarins – interaction with eukaryotic DNA
8. Lectins and polypeptides – form disulphide bridges

Some of the good examples of plant diseases managed successfully by pesticides of plant origin are presented in Table 7.1.

5 Advantages of Plant Origin Pesticides

- Promote sustainable agriculture: It does not cause ill effect on the crop plants, soil health and environment.
- Reduce crop losses: Several plant diseases/ plant pathogens can be effectively managed by reducing disease incidence and related losses in the crop plants.
- Eco-friendly: It does not cause ecological imbalance and suitably fit in any agroecosystem.
- Biodegradable: It rapidly degrades under the exposure of sunlight.
- Organic farming: It suitably fits in the organic farming system as it is eco-friendly in nature.
- Cheaper: It is relatively cheaper than conventional chemical fungicides.
- Integrated disease management: It can be suitably incorporated in the framework of integrated disease management.

6 Limitations of Plant Origin Pesticides

- Extraction methods for plant origin pesticides are still not standardized.
- Rapid degradation, hence, the fungicidal effect persists for a very short period.
- Most studies are on in vitro efficacy, and field efficacy of these pesticides is often very low or ineffective due to rapid degradation under the exposure of natural factors.
- Need development of novel formulations with enhanced bioefficacy and longer shelf life.
- Some chemical compounds are harmful to humans and plants, which could be identified with proper expertise. The precursors of possible harmful compounds need to be identified to check their conversion to toxic molecules.
- Mostly pesticides of plant origin, efficacy is slow and lower compared to chemical pesticides.
- Less availability of formulations due to instability in pure compound under normal storage condition.

7 Plant Disease Management Strategies

Strategies, tactics and techniques used in plant disease management can be grouped under one or more very broad principles of action. Differences between these principles often are not clear. The simplest system consists of two principles, prevention and therapy (treatment or cure). The first principle (prevention) includes disease management tactics applied before infection (i.e. the plant is protected from disease), and the second principle (therapy or curative action) functions with any measure applied after the plant is infected (i.e. the plant is treated for the disease). An example of the first principle is the enforcement of quarantines to prevent the introduction of a disease agent (pathogen) into a region where it does not occur.

The second principle is illustrated by heat or chemical treatment of vegetative material such as bulbs, corms and woody cuttings to eliminate fungi, bacteria, nematodes or viruses that are established within the plant material. Chemotherapy is the application of chemicals to an infected or diseased plant that stops (i.e. eradicates) the infection. Although many attempts have been made to utilize chemotherapy, few have been successful. In a few diseases of ornamental or other high value trees, chemotherapy has served required purpose with the need of repetition at intervals of one to several years. For example, antibiotics

Table 7.1 Utilization of botanicals in plant disease management

Plant disease(s)/pathogens	Plant material(s) used	Biosource/plant species	Author(s) (year)
Alternaria blight, white rust and *Sclerotinia* rot of mustard	Aqueous extract of garlic and onion bulb	Garlic (*Allium sativum*) and onion (*Allium cepa*)	Chattopadhyay et al. (2004), Yadav (2009), Yadav et al. (2012), and Meena et al. (2004, 2013)
Alternaria blight of sunflower	Aqueous extract of garlic bulb	Garlic (*Allium sativum*)	Chattopadhyay et al. (2005)
Curvularia lunata	Ethanol extract of garlic bulb	Garlic (*Allium sativum*)	Upadhyaya and Gupta (1990)
Downy mildew and powdery mildew	Aqueous extract of garlic bulb	Garlic (*Allium sativum*)	Meena et al. (2013)
Gram-negative and Gram-positive bacteria	Volatile oil from the leaf of black pepper, clove, geranium, nutmeg	Black pepper (*Piper nigrum*), clove (*Syzygium aromaticum*), geranium (*Geranium* sp.), nutmeg (*Myristica fragrans*)	Dorman and Deans (2000)
Powdery mildew of pea	Neem seed kernel extract	Neem (*Azadirachta indica*)	Surwase et al. (2009)
Anthracnose of pepper	Crude extract of neem leaf	Neem (*Azadirachta indica*)	Nduagu et al. (2008)
A. alternata	Neem seed kernel oil	Neem (*Azadirachta indica*)	Dharam and Sharma (1985)
Early blight of tomato	Crude extract of neem seed/leaf	Neem (*Azadirachta indica*)	Patil et al. (2001)
Sheath blight of rice	Achook formulation of neem leaf	Neem (*Azadirachta indica*)	Kandhari (2007)
Rice tungro virus	Neem oil from seed kernel	Neem (*Azadirachta indica*)	Muthamilan and Revathy (2007)
Fusarium oxysporum, *Aspergillus flavus*	Essential oil from the seeds of neem, black cumin and asafoetida	Neem (*Azadirachta indica*), black cumin (*Nigella sativa*) and asafoetida (*Ferula assafoetida*)	Sitara et al. (2008)
Curvularia lunata	Extract of root, stem, leaf and flowers of *Datura* (*D. stramonium*) (*Calotropis procera/Ocimum* spp.)	*Datura* (*Datura stramonium/Calotropis procera/Ocimum* spp.)	Manoharachary and Gourinath (1988)
Phytophthora infestans, *Fusarium solani*, *Pyricularia oryzae*	Crude extract of ginger rhizome	Ginger (*Zingiber officinale*)	Bandara et al. (1989)
Helminthosporium maydis	Crude leaf extract of purslane (*Portulaca oleracea*)	Purslane (*Portulaca oleracea*)	Noriel and Robles (1990)
Drechslera oryzae	Crude leaf extract of henna (*Lawsonia inermis*)	Henna (*Lawsonia inermis*)	Natarajan and Lalithakumari (1987)
Aspergillus flavus	Essential oil from the leaf of *Ocimum/Chenopodium ambrosioides*	*Ocimum/Chenopodium ambrosioides*	Mishra et al. (1989)
Rhizoctonia solani/Sclerotium rolfsii	Essential oils from the leaf of spearmint (*Mentha spictata*)/Salvia (*Salvia fruticosa*)/Thyme (*Thymbra* spp.)	Spearmint (*Mentha spictata/Salvia fruticosa/Thymbra* spp.)	Yegen et al. (1992)
Botrytis ricini	Crude leaf extract of *Lantana camara*	*Lantana camara*	Bhattiprolu and Bhattiprolu (2006)

Tikka disease of groundnut	Crude leaf extract of *Calotropis procera*	*Calotropis procera*	Srinivas et al. (1997)
Leaf blight of onion	Leaf extract of *Lantana camara/Pongamia pinnata*	*Lantana camara/Pongamia pinnata*	Bhosale et al. (2008)
Botrytis cinerea in grapes	Essential oil from the leaf of *Ocimum sanctum/Prunus persica*	*Ocimum sanctum/Prunus persica*	Tripathi et al. (2008)
Bacterial blight of rice	Achook/neemazal from the leaf/seed of neem	Neem (*Azadirachta indica*)	Sunder et al. (2005)
Fusarium oxysporum/Phoma tracheiphila	Essential oil from leaf *Origanum hercleoticum*	*Origanum hercleoticum*	Salomone et al. (2008)
Anthracnose of strawberry	Volatile compounds from fruit of strawberry (*Fragaria* spp.)	Strawberry (*Fragaria* spp.)	Arroyo et al. (2007)
A. alternata/F. oxysporum	Phenolic compound from leaf of garden cotton	Garden cotton (*Gossypium* sp.)	Naidu (1988)
Brown spot of rice	Crude leaf extract of *Nerium oleander*	*Nerium oleander*	Harish et al. (2008)
F. oxysporum/F. solani/P. capsici/ C. capsici/Sclerotinia sclerotiorum/ B. cinerea/R. solani	Essential oil from the leaf of *Metasequoia glyptostroboides*	*Metasequoia glyptostroboides*	Bajpai and Kang (2010)
Aspergillus flavus	Leaf/seed/fruit extract of clove/turmeric/garlic/holy basil	Clove (*Syzygium aromaticum*)/turmeric (*Curcuma longa*)/garlic (*Allium sativum*)/ holy basil (*Ocimum tenuiflorum*)	Reddy et al. (2009)
Cercospora leaf spot of sesame	Aqueous leaf extract of *Aspilia africana/Chromolaena odorata/Musa paradisiaca/Tithonia diversifolia*	*Aspilia africana/Chromolaena odorata/Musa paradisiaca/Tithonia diversifolia*	Enikuomehin (2005)
Black mould disease of onion bulbs	Volatile compounds from the leaf/stem of *Brassica napus/Lycopersicon esculentum*	*Brassica napus/Lycopersicon esculentum*	Abd-Alla et al. (2006)
Root rot disease of cowpea	Crude leaf/fruit/seed extract of ginger/aloe/bitter kola/neem	Ginger (*Zingiber officinale*)/aloe (*Aloe* spp.)/bitter kola (*Garcinia kola*)/neem (*Azadirachta indica*)	Suleiman and Emua (2009)
Rice blast/sheath blight and wheat leaf rust	Crude root/stem extract of *Chloranthus japonicus/Paulownia coreana*	*Chloranthus japonicus/Paulownia coreana*	Choi et al. (2004)
Stem rot of vanilla	Crude leaf extract of *Eugenia aromatica/Alpinia galanga/Sphaeranthus indicus*	*Eugenia aromatica/Alpinia galanga/Sphaeranthus indicus*	Suprapta and Khalimi (2009)
Soilborne pathogens of tomato and black pepper	Aqueous extract of Burma dhania (*Eryngium foetidum*)	Burma dhania (*Eryngium foetidum*)	Bhagat (2010)

have been infused into plants to reduce severity of phytoplasma diseases of palms (lethal yellowing) and pears (pear decline), and fungicides have been injected into elms to reduce severity of Dutch elm disease (caused by *Ophiostoma ulmi*). But in all cases, the chemotherapeutant must be reapplied periodically. There are some "systemic" fungicides such as the sterol biosynthesis inhibiting (SBI) and demethylation inhibiting (DMI) fungicides that diffuse into the plant tissues to some extent and eliminate recently established infections. The pesticides of plant origin may be used for preventive as well as curative purposes against plant diseases (Table 7.2).

8 The Way Forward

The future of plant origin pesticides as alternatives to synthetic pesticides is bright as more people are using them due to the availability and accessibility of the raw materials and easy use of the botanicals, with minimum demand for specialized application and equipment characteristic of synthetic pesticides. Most of the research on disease management in both developed and developing countries is now centred on the use of botanicals as alternatives to synthetic pesticides. The World Health Organization has also encouraged the use of plant origin pesticides because there is a prevalence of diseases

Table 7.2 Economic analysis of different treatments in Indian mustard

Treatment	Yield[a] (q/ha)	Additional cost (INR)	Total cost (INR)	Total returns (INR)	Benefit to cost ratio	Percent change in benefit to cost ratio over control
Garlic bulb extract 1 % w/v (ST)	13.5	250	18,400	23,531	1.28	10.97
Apron 35 SD 6 g/kg (ST)	13.1	105	18,255	22,833	1.25	8.54
Carbendazim 1 g a.i. (ST)	13.1	7	18,157	22,833	1.26	9.12
Apron 35 SD 6 g/kg + carbendazim 1 g a.i. (ST)	13.6	112	18,262	23,705	1.30	12.64
Trichoderma harzianum 10 g/kg (ST)	13.2	75	18,225	23,008	1.26	9.55
T. harzianum (ST) + *P. fluorescens* 10 ml/l (FS)	14.0	430	18,580	24,402	1.31	13.97
T. harzianum (ST) + *T. harzianum* 10 ml/l (FS)	14.4	430	18,580	25,099	1.35	17.22
Garlic bulb extract (ST) + garlic bulb extract (FS)	15.2	1,470	19,620	26,494	1.35	17.18
Apron 35 SD (ST) + Ridomil MZ 72 WP 2 g/l (FS)	14.8	3,285	21,435	25,796	1.20	4.43
Carbendazim (ST) + Ridomil MZ 72 WP 2 g/l (FS)	13.9	3,187	21,337	24,228	1.14	−1.47
Control	12.0	0	18,150	20,916	1.15	NA

[a]The aggregated mean of the yield over all locations; *ST* seed treatment, *FS* foliar spray, *INR* Indian Rupees
Source: Meena et al. (2013)

traceable to chemicals used for the management of plant diseases because the consumed plant materials possess high degree of residues that cause carcinogenic diseases and other health issues. Recently, several food commodities have been rejected by importing countries at the point of entry due to the presence of traceable chemical residues, which has sensitized the larger part of population worldwide. It is also an issue with environmentalists concerned about the threat of global warming because most of the pollution accredited to cause global warming is traced to the pollutants produced and discharged in the developing countries (Olufolaji 2006).

Only a handful of plant origin pesticides are currently used in agriculture in the industrialized world, and there are prospects for commercial development of new plant origin products. Accordingly, the use of green pesticides is being recommended globally, and the use of essential oils seems to be the best choice. Efforts should be made to scientifically document the pesticidal plants and to investigate the biocontrol efficacy of plant diseases of the plant products. Field trials are required to assess the practical applicability of the botanical pesticides. Biosafety studies need to be conducted to ascertain their toxicity to humans, animals and crop plants.

References

Abd-Alla MA, El Mohamedi RSR, Badeaa RI (2006) Effect on some volatile compounds on black mould disease on onions bulbs during storage. Res J Agric Biol Sci 2:384–390

Ahmad I, Beg AZ (2001) Antimicrobial and phytochemical studies on 45 Indian medicinal plants against multi-drug resistant human pathogens. J Ethnopharmacol 74:113–123

Arroyo FT, Moreno J, Daza P, Boianova L, Ro-mero F (2007) Antifungal activity of strawberry fruit volatile compounds against *Colletotrichum acutatum*. J Agric Food Chem 14:5701–5707. doi:10.1021/jf0703957

Bajpai VK, Kang SC (2010) Antifungal activity of leaf essential oil and extract of *Metasequoia glyptostroboides* Miki ex Hu. J Am Oil Chem Soc 87:327–336. doi:10.1007/s11746-009-1500-6

Bandara BMR, Kumar NS, Samaranayake KMS (1989) An antifungal constituent from the stem bark of *Butea monosperma*. J Ethnopharmacol 25:73–75. doi:10.1016/0378-8741(89)90046-9

Bhagat S (2010) *In vitro* efficacy of Burma dhania (*Eryngium foetidum*) against soil borne and foliar plant pathogens of tomato. CARI annual report (2010–11), Port Blair

Bhattiprolu SL, Bhattiprolu GR (2006) Management of castor grey rot disease using botanical and biological agents. Indian J Plant Prot 34:101–104

Bhosale MV, Perne RR, Shete MH, Pawar NB (2008) Efficacy of botanicals, bioagents and fungicides on management of leaf blight (*Alternaria porri*) of onion. J Plant Dis Sci 3:88–89

Brent KJ, Hollomon DW (1998) Fungicide resistance: the assessment of risk. FRAC, Global Crop Protection Federation, Brussels 2:1–48

Canillac N, Mourey A (2001) Antibacterial activity of the essential oil of *Picea excelsa* on *Listeria, Staphylococcus aureus* and coliform bacteria. Food Microbiol 18:261–268

Cardellina JH (1988) Biologically natural products in the search for new agrochemicals. In: Gulter HG (ed) Biologically active natural products. Potential use in agriculture. American Chemical Society, Washington, DC

Chattopadhyay C (1999) Yield loss attributable to Alternaria blight of sunflower (*Helianthus annuus* L.) in India and some potentially effective control measures. Int J Pest Manag 45:15–21. doi:10.1080/096708799227996

Chattopadhyay C (2001) Yield loss attributable to Alternaria blight of safflower (*Carthamus tinctorius* L.) and some effective control measures. J Mycol Plant Pathol 31:298–302

Chattopadhyay C, Meena PD, Meena RL (2004) Integrated management of Sclerotinia rot of Mustard. Indian J Plant Prot 32:88–92

Chattopadhyay C, Meena PD, Godika S, Yadav MS, Meena RL, Bhunia CK (2005) Garlic bulb extract-A better choice than chemical fungicides in managing oilseed crop diseases. J Mycol Plant Pathol 35:574 (Abstract)

Choi GJ, Jang KS, Kim JS, Lee SW, Cho JY, Cho KY, Kim JC (2004) *In vivo* fungal activity of 57 plant extracts against six plant pathogenic fungi. Plant Pathol J 20:184–191

Cox SD, Gustafson JE, Mann CM, Markham L, Liew YC, Hartland RP, Bell HC, Warmington JR, Wyllie SG (1998) Tea tree oil causes K+ leakage and inhibits respiration in *Escherichia coli*. Lett Appl Microbiol 26:355–358

Cutler HG, Cutler SJ (1999) Biological active natural products: agrochemicals. CRS Press, Boca Raton

Daayf F, Schmitt A, Belanger RR (1997) Evidence of phytoalexins in cucumber leaves infected with powdery mildew following treatment with leaf extracts of *Reynoutria sachalinensis*. Plant Physiol 113:719–727

Dharam V, Sharma RK (1985) Efficacy of fungicides XXIX studies on the fungicidal properties of neem oil. Indian J Plant Pathol 3:241–242

Dorman HJD, Deans SG (2000) Antimicrobial agents from plants: antibacterial activity of plant volatile oils. J Appl Microbiol 88:308–316. doi:10.1046/j.1365-2672.2000.00969.x

Doubrava NS, Dean RA, Kuc J (1988) Induction of systemic resistance to anthracnose caused by *Colletotrichum lagenarium* in cucumber by oxalate and extracts from spinach and rhubarb leaves. Physiol Mol Plant Pathol 33:69–79

Downum KR (1986) Photoactivated biocides from higher plants. In: Green MB, Hedin PA (eds) Natural resistance of plants: roles of allelochemicals, vol 296, ACS symposium services. American Chemical Society, Washington, DC

Dubey SC, Suresh M, Singh B (2007) Evaluation of *Trichoderma* species against *Fusarium oxysporum* f. sp. *ciceris* for integrated management of chickpea. Biol Control 40:118–127

Elgayyar M, Draughon FA, Golden DA, Mount JR (2001) Antimicrobial activity of essential oils from plants against selected pathogenic and saprophytic microorganisms. J Food Prot 64:1019–1024

Enikuomehin OA (2005) Cercospora leaf spot disease management in Sesame (*Sesamum indicum* L.) with plant extracts. J Trop Agric 43:19–23

Harish S, Saravankumar D, Ebenezar EG, Seetharaman K (2008) Use plant extracts and biocontrol agents for the management of brown spot disease in rice. BioControl 53:555–567. doi:10.1007/s10526-007-9098-9

Jayaraj J, Wan A, Rahman M, Punja ZK (2008) Seaweed extract reduces foliar diseases on carrot. Crop Prot 27:1360–1366

Jensen BD, Latunde-Dada AO, Hudson D, Lucas JA (1998) Protection of *Brassica* seedlings against downy mildew and damping-off by seed treatment with CGA 245704, an activator of systemic acquired resistance. Pestic Sci 52:63–69

Kandhari J (2007) Management of sheath blight of rice through fungicides and botanicals. Indian Phytopathol 60:214–217

Kumar R, Mishra AK, Dubey NK, Tripathi YB (2007) Evaluation of *Chenopodium ambrosioides* oil as a potential source of antifungal, anti-aflatoxigenic and antioxidant activity. Int J Food Microbiol 115:159–164

Lavanya N, Saravanakumar D, Rajendran L, Ramiah M, Raguchander T, Samiyappan R (2009) Management of sunflower necrosis virus through anti-viral substances. Arch Phytopathol Plant Prot 42:265–276. doi:10.1080/03235400601037149

Manoharachary C, Gourinath A (1988) Effects of plant extracts on four pathogenic fungi. In: 5th international congress of plant pathology, Kyoto

Mason JR, Mathew DN (1996) Evaluation of neem as a bird repellent chemical. Int J Pest Manag 42:47–49

Meena PD, Meena RL, Chattopadhyay C, Kumar A (2004) Identification of critical stage for disease development and biocontrol of Alternaria blight of Indian mustard (*Brassica juncea*). J Phytopathol 152:204–209

Meena PD, Gour RB, Gupta JC, Singh HK, Awasthi RP, Netam RS, Godika S, Sandhu PS, Prasad R, Rathi AS, Rai D, Thomas L, Patel GA, Chattopadhyay C (2013) Non-chemical agents provide tenable, eco-friendly alternatives for the management of the major diseases devastating Indian mustard (*Brassica juncea*) in India. Crop Prot 53:169–174. doi.org/10.1016/j.cropro.2013.07.006

Mishra AK, Dubey NK (1994) Evaluation of some essential oils for their toxicity against fungi causing deterioration of stored food commodities. Appl Environ Microbiol 60:1101–1105

Mishra AK, Dwivedi SK, Kishore N (1989) Antifungal activity of some essential oils. Natl Acad Sci Lett 12:335–336

Muthamilan M, Revathy N (2007) Management of tungro virus disease of rice with antagonists and botanicals. J Ecobiol 19:81–88

Naidu GP (1988) Antifungal activity in *Codiaeum variegatum* leaf extract. Curr Sci 57:502–503

Narwal S, Balasubrahmanyam A, Sadhna P, Kapoor H, Lodha ML (2000) A systemic resistance inducing antiviral protein with N-glycosidase activity from *Bougainvillea xbuttiana* leaves. Indian J Exp Biol 39:600–603

Natarajan MR, Lalithakumari D (1987) Antifungal activity of the leaf extract of *Lawsonia inermis* on *Dreschslera oryzae*. Indian Phytopathol 40:390–395

Nduagu C, Ekefan EJ, Nwankiti AO (2008) Effect of some crude plant extracts on growth of *Colletotrichum capsici* (Synd) Butler & Bisby, causal agent of pepper anthracnose. J Appl Biosci 6:184–190

Noriel LM, Robles RP (1990) Fungicidal activity of *Portulaca oleracea* extract against *Helminthosporium maydis* Wisik and Miyake in corn (*Zea mays* L.). Philipp J Weed Sci 17:26–32

Olufolaji DB (2006) Effects of crude extracts of *Eichhornia crassipes* and *Chromolaena odorata* on the control of red-rot disease of sugarcane in Nigeria. In: Yang-Rui L, Solomon S (eds) Proceedings of the international conference of professionals in sugar and integrated technologies. International Association of Professionals in Sugar and Integrated Technologies, Nanning

Patil MJ, Ukey SP, Raut BT (2001) Evaluation of fungicides and botanicals for the management of early blight (*Alternaria solani*) of tomato. PKV Res J 2:49–51

Paul PK, Sharma PD (2002) *Azadirachta indica* leaf extract induces resistance in barley against leaf stripe disease. Physiol Mol Plant Pathol 61:3–13

Penna C, Marino S, Vivot E, Cruanes MC, Munoz JD, Cruanes J, Ferraro G, Gutkind G, Martino V (2001) Antimicrobial activity of Argentine plants used in the treatment of infectious diseases. Isolation of active compounds from *Sebastiania brasiliensis*. J Ethnopharmacol 77:37–40

Rauha JP, Remes S, Heinonen M, Hopia A, Kahkonen M, Kujala T, Pihlaja K, Vuorela H, Vuorela P (2000) Antimicrobial effects of Finnish plant extracts containing flavonoids and other phenolic compounds. Int J Food Microbiol 56:3–12

Reddy KRN, Reddy CS, Muralidharan K (2009) Potential of botanicals and biocontrol agents on growth and aflatoxin production by *Aspergillus flavus* infecting rice grains. Food Control 20:173–178. doi:10.1016/j.foodcont.2008.03.009

Salie F, Eagles PFK, Lens HMJ (1996) Preliminary antimicrobial screening of four South African *Asteraceae* species. J Ethnopharmacol 52:27–33. doi:10.1016/0378-741(96)01381-5

Salomone A, Scaritto G, Sacco A, Cabras G, Angioni A (2008) Inhibitory effects of the main compounds of oregano essential oil against some pathogenic fungi. Modern fungicide and Antifungal compounds V. Deut Phytomediz Gesellsch 44:345–360

Sharma K, Meshram NM (2006) Bioactivity of essential oils from *Acorus calamus* Linn. and *Syzygium aromaticum* Linn. against *Sitophilus oryzae* Linn. in stored wheat. Biopestic Int 2:144–152

Sitara U, Naseem J, Sultana N (2008) Antifungal effect of essential oils on *in vitro* growth of pathogenic fungi. Pak J Bot 40:409–414

Slusarenko AJ, Patel A, Portz D (2008) Control of plant diseases by natural products: allicin from garlic as a case study. Eur J Plant Pathol 121:313–322. doi:10.1007/s10658-007-9232-7

Srinivas T, Rao MS, Reddy PS, Reddy PN (1997) Integrated management of leaf spot of groundnut (*Arachis hypogaea* L.) with botanicals and chemicals. Z Pflanzenk Pflanzen 104:528–530

Suhr KI, Nielsen PV (2003) Antifungal activity of essential oils evaluated by two different application techniques against rye bread spoilage fungi. J Appl Microbiol 94:665–674

Suleiman MN, Emua SA (2009) Efficacy of four plant extract in the control of root rot disease of cow pea (*Vigna unguiculata* Linn. Walp). Afr J Biotechnol 8:3806–3808

Sunder S, Singh R, Dodan DS (2005) Management of bacterial blight of rice with botanical extracts and non-conventional chemicals. Plant Dis Res 20:12–17

Suprapta DN, Khalimi K (2009) Efficacy of plant extract formulation to suppress stem rot disease on vanilla seedlings. ISSA J 15:34–41

Surwase AG, Badgire DR, Suryawanshi AP (2009) Management of pea powdery mildew by fungicides, botanicals and bio-agents. Ann Plant Prot Sci 17:384–388

Tripathi P, Dubey NK, Shukla AK (2008) Use of some essential oils as post harvest botanical fungicides in the management of grey mould of grapes caused by *Botrytis cinerea*. World J Microbiol Biotechnol 24:39–46. doi:10.1007/s11274-007-9435-2

Upadhyaya MR, Gupta RC (1990) Effect of extracts of some medicinal plants on the growth of *Curvularia lunata*. Indian J Mycol Plant Pathol 20:144–145

Yadav MS (2009) Biopesticidal effect of botanicals on the management of mustard diseases. Indian Phytopathol 62:488–492

Yadav MS, Ahmad N, Singh S, Yadava DK, Godika S, Gaur RB (2012) Multilocational validation of integrated management practices for *Sclerotinia* rot of Indian mustard (*Brassica juncea*). Indian J Agric Sci 82:972–977

Yegen O, Begger B, Heitefuss R (1992) Studies on the fungitoxic effect for extracts of six selected plants from Turkey on phytopathogenic fungi. Z Pflanzenk Pflanzen 99:349–359

Yokoyama K, Aist JR, Baylse CJ (1991) A papilla-regulating extract that induces resistance to barley powdery mildew. Physiol Mol Plant Pathol 39:71–78

The Use of Plant Extracts for Stored Product Protection

8

Christos G. Athanassiou, Pathipati Usha Rani, and Nickolas G. Kavallieratos

Abstract

A wide range of plant extracts have been used alternatively to chemical insecticides against stored product insect pest species. These substances could be used alone or in combination with other alternatives or synthetic insecticides in order to enhance their action. Botanicals can be applied with the same techniques that are used for the application of traditional contact insecticides or fumigants. Given that stored product insects exhibit different levels of sensitivity or tolerance to plant extracts, a wide screening of these substances has been carried out to widen the available tools against these harmful organisms with interesting results.

Keywords

Stored product protection • Post-harvest control • Botanicals • Phytochemicals • Post-harvest applications • Pest management • Stored product pests

C.G. Athanassiou, Ph.D. (✉)
Laboratory of Entomology and Agricultural Zoology, Department of Agriculture, Crop Production and Rural Environment, University of Thessaly, Phytokou str., N. Ionia, 38446 Magnesia, Greece
e-mail: athanassiou@agr.uth.gr

P.U. Rani, Ph.D.
Biology and Biotechnology Division, Indian Institute of Chemical Technology, Taranaka, Hyderabad 500 007, AP, India

N.G. Kavallieratos, Ph.D.
Laboratory of Agricultural Zoology and Entomology, Department of Crop Science, Agricultural University of Athens, 75 Iera Odos str., 11855 Athens, Attica, Greece

Laboratory of Agricultural Entomology, Department of Entomology and Agricultural Zoology, Benaki Phytopathological Institute, 8 Stefanou Delta str., 14561 Kifissia, Attica, Greece

1 Background and Current Status

As in the case of field crops, the use of botanicals in stored product protection has been extensively evaluated for the control of several insect species in different commodities. Almost 15 years ago, Weaver and Subramanyam (2000) reviewed the use of plant extracts and related derivatives at the post-harvest stages of agricultural products, especially in the case of durable commodities. Basically, botanicals are considered as pesticides, which can be applied with the same techniques that are used for the application of traditional contact insecticides or fumigants. Shaaya et al. (1997) summarize the mode of action of botanicals

D. Singh (ed.), *Advances in Plant Biopesticides*,
DOI 10.1007/978-81-322-2006-0_8, © Springer India 2014

in six mechanisms: (a) act through respiration like a fumigant, (b) act through contact or digestion as a contact or oral insecticide, (c) prevent reproduction (also causing sterilization), (d) have an antifeedant effect, (e) have a repulsive effect or alter insect behavior, and (f) have a combination of the modes of action that are mentioned above. It is true that for most of the botanicals that have been tested for stored product protection, the only data available concern strictly efficacy, while the mode of action for many of them remains unknown. Also, even in the case of a specific mode of action, botanicals vary remarkably in their ways that affect insects. For instance, several botanicals that act as contact insecticides have neurotoxic action, while others act as insect growth regulators (IGRs) (Weaver and Subramanyam 2000). Prakash and Rao (1997) have reported a list of approx. 160 plant species which have been tested, with various results, against major insects of durable stored products. The most studied plant for this purpose is the neem tree, *Azadirachta indica* A. Juss. (Sapindales: Meliaceae), or its major insecticidal component, azadirachtin (Athanassiou et al. 2005; Kavallieratos et al. 2007). Other well-studied species are *Acorus calamus* L. (Acorales: Acoraceae), *Allium sativum* L. (Asparagales: Amaryllidaceae), *Arachis hypogaea* L. (Fabales: Fabaceae), *Cocos nucifera* L. (Arecales: Arecaceae), and *Vitex negundo* L. (Lamiales: Lamiaceae) (Weaver and Subramanyam 2000). In many of these plant species, the insecticidal value is located only in some plant parts, while for others (i.e., neem) merely all plant parts have an insecticidal activity. All these compounds fall in the category of "alternatives to pesticides" in stored product protection; nevertheless, very few ingredients are able to directly compete with the traditional contact protectants and fumigants. Currently, even in the case where some compounds have some certain insecticidal value, and some of them are registered for the control of field pests, their use in stored products is mainly restricted in the developing world. Apparently, their wider adoption is directly related with changes in registration and legislation policies, in conjunction with their thorough evaluation.

Since 1997 numerous studies available focused upon the use of plant extracts as contact insecticides alternatively to chemical insecticides against stored product insect pest species. These substances could be used alone or in combination with other alternatives or synthetic insecticides in order to enhance their action. Given that stored product insects exhibit different levels of sensitivity or tolerance to plant extracts, a wide screening of these substances has been carried out to widen the available tools against these harmful organisms with interesting results.

Huang et al. (1997) found that *Tribolium castaneum* (Herbst) (Coleoptera: Tenebrionidae) adults were approximately ten times more tolerant than *Sitophilus zeamais* Motschulsky (Coleoptera: Curculionidae) adults to the contact toxicity of the extraction (essential oil) of *Myristica fragrans* Houtt. (Magnoliales: Myristicaceae) (18 vs. 1.7 mg/cm^2 LC_{50} values, respectively) on filter papers. Younger larvae (10–16 days) of *T. castaneum* were more susceptible to *M. fragrans* essential oil in comparison to adults, whereas the oldest ones (18 days) were even more tolerant. *M. fragrans* essential oil was also effective against the F1 progeny production of both *T. castaneum* and *S. zeamais*. Obeng-Ofori et al. (1997) reported complete adult mortality of *Sitophilus granarius* (L.) (Coleoptera: Curculionidae), *S. zeamais*, *T. castaneum*, and *Prostephanus truncatus* (Horn) (Coleoptera: Bostrichidae) exposed to 1,8 cineole, the major component of the essential oil of *Ocimum kenyense* Ayob. ex A. J. Paton (Lamiales: Lamiaceae), at 10 μl/insect on filter papers or as grain protectant at 0.5 μl/kg of grain. According to Paneru et al. (1997) the powder of *A. calamus* rhizomes collected from low altitude in eastern Nepal provided 100 % mortality of *Sitophilus oryzae* (L.) (Coleoptera: Curculionidae) and *S. granarius* adults after 7 days of exposure in wheat treated with 2 % w/w of the powder. Furthermore, the formulation provided adequate control of wheat for a period of 7–8 weeks. Another plant extract from *Thujopsis dolabrata* (Thunb. ex L. f.) Siebold and Zucc. var. *hondai* Makino (Pinales: Cupressaceae), the terpenoid carvacrol, exhibited significant contact toxicity against *Lasioderma serricorne* (F.) (Coleoptera: Anobiidae), *Callosobruchus chinensis* (L.) (Coleoptera: Bruchidae), and *S. oryzae* adults on filter papers after 48 h of exposure (Ahn et al. 1998). Obeng-Ofori et al. (1998) reported high contact toxicity of camphor, the major component of the essential oil of *Ocimum kilimandscharicum* Baker ex Gurke (Lamiales: Lamiaceae) on filter paper against *S. granarius, S. zeamais*, and *P. truncatus*, i.e., 96, 94, and 98 %, respectively, after 24 h of exposure at the concentration of 100 μg/insect. With topical application, camphor provided complete mortality of *S. granarius, S. zeamais, T. castaneum*, and *P. truncatus* after 48 h of exposure at the concentration of 60 μg/insect. Similar results were obtained at the dose of 1 mg camphor/kg of grain against the same species after 24 h of exposure, whereas it suppressed the progeny production of *S. granarius* and *S. zeamais*. Kim et al. (2003) screened the contact toxicity of extracts of 30 plants belonging to 23 families against *L. serricorne* adults on filter papers. They reported 100 % mortality of the individuals exposed to *Cinnamomum cassia* (Nees and T. Nees) J. Presl (Laurales: Lauraceae) bark and oil, *Cocholeria aroracia* L. (Brassicales: Brassicaceae) oil, *Brassica juncea* (L.) Vassiliĭ Matveievitch Czernajew (Brassicales: Brassicaceae) oil, *Illicium verum* Hook. f. (Austrobaileyales: Schisandraceae) fruit, and *Foeniculum vulgare* Mill. (Apiales: Apiaceae) fruit 24 h after treatment at the rate of 3.5 mg/cm^2. Similarly, all adults were dead at the same rate when exposed to extracts of *Agastache rugosa* (Fisch. and C. A. Mey.) Kuntze (Lamiales: Lamiaceae) whole plant and *Acorus calamus* L. var. *angustatus* Besser (Acorales: Acoraceae) rhizome after 2 and 4 days post treatment. When the rate was reduced to 0.7 mg/cm^2 *C. cassia* oil, *C. aroracia* oil and *B. juncea* oil caused complete mortality of *L. serricorne* adults after 24 h of exposure. According to Akob and Ewete (2009), ethanolic extracts of *Vetiveria zizanioides* (L.) Nash (Poales: Poaceae) roots and *Cupressus arizonica* Greene (Pinales: Cupressaceae), *Ocimum gratissimum* (L.) (Lamiales: Lamiaceae), and *Eucalyptus grandis* W. Hill ex Maiden (Myrtales: Myrtaceae) leaves provided significant contact toxicity against the F1 progeny of *S. zeamais* in comparison to the controls in maize treated with a range (12,500–100,000 ppm) of concentrations 37 days after infection. Epidi and Udo (2009) tested the contact toxicity of extracts of leaves of *Dracaena arborea* (Willd.) Link (Asparagales: Asparagaceae) against *S. zeamais* and *Callosobruchus maculatus* (F.) (Coleoptera: Chrysomelidae) adults and recorded complete mortality for both species when they were exposed for 2 days at 20 μl/ml of aqueous fraction of the plant extract by topical application. When *S. zeamais* adults were exposed on filter papers which previously had been treated with 200 μl/ml of ethyl acetate fraction of the plant extract, mortality reached 80 % after 96 days of exposure. Among powders of 20 plant species which were tested as grain protectants against *S. oryzae* at 2 % w/w, *V. negundo* had the best performance by providing 99.1 % mortality followed by *Alpinia officinarum* Hance (Zingiberales: Zingiberaceae) (96.6 %) and *Nelumbium speciosum* Willd. (Proteales: Nelumbonaceae) (94.4 %) after 7 days of exposure. Owolabi et al. (2009) investigated the toxic contact effect of the essential oils of *Cymbopogon citratus* (DC.) Stapf. (Poales: Poaceae) and *Monodora myristica* Dunal (Magnoliales: Annonaceae) as maize and cowpea protectants against *S. zeamais* and *C. maculatus*. Both essential oils were strongly toxic performing LD_{50} values of 0.560 and 0.346, respectively. The contact application on filter paper of acetonic extracts of *Cucurbita maxima* Duchesne (Cucurbitales: Cucurbitaceae), *Citrus sinensis* (L.) Osbeck (Sapindales: Rutaceae), and *Citrus aurantium* L. (Sapindales: Rutaceae) provided 100 % mortality of *S. oryzae* and *Rhyzopertha dominica* (F.) (Coleoptera: Bostrichidae) at 8.5 mg/cm^2 after 3 days of exposure (Rajasekharreddy and Usha Rani 2010). The crude extracts of *C. sinensis* and *C. aurantium* caused 89 and 76 % mortality of *S. oryzae* and *R. dominica*, respectively, under the same method of application after 3 days of exposure. According to Usha Rani and Rajasekharreddy (2010), the seed extract of *Sterculia foetida* L. (Malvales: Malvaceae), cyclopropene fatty acid, provided complete mortality to *S. oryzae*, *C. sinensis*, and *T. castaneum* adults under topical application on filter papers 2 days after treatment at 0.2 mg/cm^2. Yang et al. (2010) combined the essential oil of *A. sativum* with diatomaceous earth against *S. oryzae* and *T. confusum* and reported that their combination was significantly more effective than the use of either essential oil or diatomaceous earth alone. Similarly, Saiful Islam et al. (2010) showed that when two diatomaceous earth formulations were combined with the monoterpenoids eugenol and cinnamaldehyde, toxicity against *S. oryzae* was higher than with the diatomaceous earths alone. The contact toxicity of the essential oil from leaves, stems, and flowers of *Dracocephalum moldavica* L. (Lamiales: Lamiaceae) exhibited strong contact toxicity through topical application against *S. zeamais* and *T. castaneum* adults with LD_{50} 22.10 and 18.28 μg/insect (Chu et al. 2011). Khani et al. (2011) reported that when rice kernels were treated with 6–10 μl/g of petroleum ether seed extract of *Piper nigrum* L. (Piperales: Piperaceae), the mortality of the exposed *S. oryzae* adults was 100 % after 3 days of exposure. The leaf crude extract of *Capsicum annuum* L. (Solanales: Solanaceae) applied as grain protectant at the concentration of 100 mg/20 g of extract resulted in complete mortality of *S. oryzae* and *T. castaneum* adults after 3 days of exposure. At the same dose and exposure, the leaf crude extract *Momordica charantia* Descourt (Cucurbitales: Cucurbitaceae) provided 97.6 and 94.0 of *S. oryzae* and *T. castaneum* adults (Usha Rani and Devanand 2011). Usha Rani et al. (2011) reported suppression of the F1 progeny production of *R. dominica*, *T. castaneum*, and *C. sinensis* in grain treated with 15 and 30 mg/100 ml of leaf crude extracts of *C. nucifera* and *Terminalia catalpa* L. (Myrtales: Combretaceae). *Datura alba* Nees (Solanales: Solanaceae) leaf extract provided also long-term protection of rice kernels from *Trogoderma granarium* Everts (Coleoptera: Dermestidae) and *S. oryzae* based on the significant reduction of F1 and F2 progeny production in comparison to the controls (Ali et al. 2012). Usha Rani et al. (2013) examined several compounds which were chloroform extracted by the powder of roots of *Derris scandens* (Roxb.) Benth. (Fabales: Fabaceae) against larvae of *T. castaneum* and *Corcyra cephalonica* (Stainton) (Lepidoptera: Pyralidae). Osajin, lupalbigenin, scandinone, and genistein exhibit 100 % mortality of *Tribolium confusum* Jacquelin du Val (Coleoptera: Tenebrionidae) larvae 10 days after exposure in wheat flour at 400 μg/g of diet. Complete mortality was also recorded when larvae of *C. cephalonica* exposed in the same diet were treated with the same substances at 600 μg/g of flour after 14 days of exposure. Athanassiou et al. (2013) indicated that 48 h of exposure of *S. oryzae* adults in wheat treated with 500 ppm of silica gel enhanced with the essential oil of *Juniperus oxycedrus* L. ssp. *oxycedrus* (Pinales: Cupressaceae) caused 82 % mortality. For 7 days of exposure, 100 and 98 % of *S. oryzae* adults died when treated with 500 and 250 ppm of enhanced silica gel, respectively. At 14 days of exposure, all adults died both at 250 and 500 ppm of enhanced silica gel. De Souza Tavares et al. (2013) showed that extracts of *Psychotria capitata* Ruiz and Pavon, *Psychotria goyazensis* Müll. Arg., *Psychotria hoffmannseggiana* (Willd. ex Roem. and Schult.) Müll. Arg., and *Psychotria prunifolia* (Kunth) Steyerm (Gentianales: Rubiaceae) were toxic as contact insecticides against *S. zeamais*. Rajashekar et al. (2014) found that leaf methanol extract of *Lantana camara* L. (Lamiales: Verbenaceae) caused > 89 % mortality of *S. oryzae*, *C. sinensis*, and *T. castaneum* as a grain protectant 7 days after treatment with a concentration of 500 mg/l.

Other important series of extracts have been derived from the neem tree, *A. indica*. These extracts contain several compounds which have insecticidal properties against several stored product species (Dunkel et al. 1991; Jilani et al. 1988; Jilani and Saxena 1990; Makanjuola 1989; Pereira and Wohlgemuth 1982; Prakash and Rao 1997; Rahim 1998; Weaver and Subramanyam 2000; Xie et al. 1995). The main active constituent is the triterpenoid azadirachtin (Butterworth and Morgan 1968). Athanassiou et al. (2005) found that azadirachtin A at rates of 100 and

200 ppm was equally effective against *S. oryzae* on rye and oats, where mortality was complete after 7 and 14 days of exposure, respectively. Furthermore, for *R. dominica* and *S. oryzae*, significantly less progeny were recorded in rye and oats compared to untreated seeds. Kavallieratos et al. (2007) showed that >91 % *S. oryzae* and *T. confusum* adults were found dead after 14 days of exposure in wheat or maize treated with azadirachtin A at 200 mg/kg. Adarkwah et al. (2010) reported 90 % mortality of *T. castaneum* exposed to wheat treated with Calneem® oil, an extract from the seeds of *A. indica* that contains 0.3 % azadirachtin as a major active ingredient, after 3 days of exposure at 3 % v/v. When *T. castaneum* adults were exposed to filter paper treated with the same dose of biopesticide the mortality resulted was 88 % after 3 days of exposure.

2 Prospects for Stored Product Protection

Solid and liquid preparations are the types that are used in bioassays, usually as admixtures with stored products (Weaver et al. 1995). Essential oils are the most common approach for testing botanicals in stored product protection, and this is why a major part of the studies available concern the fumigant toxicity of the tested ingredients. Also, aqueous plant extracts, in many cases, can be considered as another way for the application of essential oils (Ranjendran and Sriranjini 2008; Shaaya et al. 1991, 1997; Weaver and Subramanyam 2000). One recent example is allicin, which component is obtained from garlic, *Allium sativum* L. (Asparagales: Amaryllidaceae). Lu et al. (2013) noted that allicin has a considerable fumigant toxicity against three major stored product insect species. One other example is the use of plant derivatives from Lamiaceae and Lauraceae, which have been thoroughly examined for their insecticidal activity (mainly *Lavandula*, *Rosmarinus*, *Laurus*, and *Thymus*), through their essential oils that are associated with the effect of terpenoids, mainly monoterpenes (Weaver and Subramanyam 2000; Lee et al. 2003; Rozman et al. 2007). The basic stages of experimentation in these cases are (a) testing the way that a certain plant extract of plant derivative acts against insects (i.e., as a fumigant, as a contact insecticide, as an antifeedant, etc.), (b) indicating which plant part has this type of activity, (c) indicating the insects that are susceptible after exposure to the specific extracts/derivatives, (d) identifying the major components (usually by HPLC, GCMS, etc.), (e) isolating the major component (or components), and (f) repeating the testing with these isolated compounds. Newer studies, however, add new dimensions on the evaluation of certain botanicals against stored product insect pests, such as (a) the formulation and the application of these botanicals (i.e., though silica or other inert materials), (b) the clarification of the mode of action (i.e., amylase inhibitors, etc.), and (c) the evaluation of parallel activity against pathogens, such as fungi. In this context, Magro et al. (2011) showed that the essential oils of clove *Syzygium aromaticum* (L.) Merrill and Perry (Myrtales: Myrtaceae) and laurel *Laurus nobilis* L. (Laurales: Lauraceae) were able species of the genera *Fusarium* and *Aspergillus*. The exploration of naturally occurring antimicrobials at the post-harvest stages of agricultural products receives increasing attention, and it is estimated that it will be evaluated more thoroughly in the near future, given that many researchers recognized the antibacterial, antifungal, antiviral, and antioxidant properties of essential oils, along with their insecticidal activities (Kordali et al. 2005; Magro et al. 2011; Pezo et al. 2006; Schuenzel and Harrison 2002; Skandamis et al. 2001).

Regarding the overall context for the use of botanicals in stored product protection, there are an enormous number of papers that have been published on the subject throughout the last years. For example, at the *Journal of Stored Products Research*, which is one of the main journals in this area, currently (2013) there are many more papers on botanicals published 20 years earlier (1993). Similar figures can be also drawn from additional journals on the area. However, despite this extensive research, the use of botanicals and plant derivatives is still limited, while there are no data so far that document their potentials for a wider adoption for use at the post-harvest stages. In the following, we will refer to the reasons that negatively affect their wider use as insecticides in stored products and, to some extent, as ingredients their wider utilization as antimicrobials and fungicides.

One of the basic reasons is that most of these derivatives, in order to obtain a satisfactory level of insecticidal effect, should be applied at high doses, which are notably higher than the application rates of the currently used insecticides. Apart from the plant extracts per se, the same holds for specific ingredients that have been isolated and identified. For instance, Athanassiou et al. (2005) for azadirachtin noted that the dose rates for control *S. oryzae* and *R. dominica* were as high as 500 ppm or even higher.

One additional reason is that, for most of the botanicals, there are no clear indications on their mammalian toxicity, i.e., most of the papers available focus on their toxicity to insects and do not proceed further for mammalian toxicity, or their toxicity to nontarget organisms. Usually, the main argument in evaluating botanicals as insecticides at the post-harvest stages of durable commodities is that some of these compounds are food additives, and thus, it is postulated that mammalian toxicity will not occur. For example, the oral LD_{50} rat oral values the some essential oils are extremely low and can be generally recognized as safe (Ranjendran and Sriranjini 2008). However, there are many aspects for human health that are expanded far beyond acute toxicity, but they are directly related with human health, such as the evaluation of botanicals as allergens, skin reactions, reproductive parameters, etc. Moreover, there are certain molecules that have been classified as carcinogens. For instance, estragole and (+)-fenchone are carcinogenic (Kim and Ahn 2001; Ranjendran and Sriranjini 2008). The same evaluation should be performed in the case of the effect of certain botanicals on nontarget organisms (i.e., soil fauna, etc.). Consequently, when evaluating plant extracts for stored product protection against pests (and diseases), it is essential to assess additional aspects, apart from the insecticidal effect, repulsive effect, etc., through a widely adopted standardized testing protocol. Besides, the use of botanicals as insecticides in stored products should be regarded as purely and qualitatively different than the application of botanicals in field pests, given that stored products are practically close to the final stage in food processing. Hence, any ingredient that is added to food, even in low concentrations, should be carefully examined.

One other issue of major importance is the standardization of plant extracts. It is true that, even within the same species, extracts may vary remarkably according to the geographical area, season of plant harvest, etc. (Barton et al. 1989; Singh and Upadhyay 1993; Shaaya et al. 1997; Ranjendran and Sriranjini 2008). Also, the condition of the harvest and the preharvest period are critical parameters, since "stressed" plants often give quantitatively different compounds. Therefore, it is difficult to standardize the insecticidal properties.

The commercial utilization of plant extracts meets with several drawbacks, such as patent and intellectual rights, differences in extraction methods, and formulation of the active ingredients. Although these difficulties can be anticipated to a major degree, the standardization of specific compounds remains a high priority, especially in the case of essential oils. Stability during the formulation preparation also remains an issue of major importance.

Paradoxically, the information available on the effect of botanicals on the organoleptic and other properties of the treated commodities is scarce. Nevertheless, it is expected that some of the tested compounds may leave some residues on the product. From a critical review of the publications on the subjects, it becomes evident that, at least in most of the cases studied thoroughly, there is no effect on the nutritional properties of the treated commodities. For instance, Singh et al. (1995) found no effect on the nutritional value of sorghum exposed to *Mentha arvensis* L. (Lamiales: Lamiaceae), after 3 months of storage. Rozman et al. (2006) found that there was no effect on the major properties of flour that was treated with monoterpenoids (camphor, 1,8-cineole, and carvacrol). However, in this study, it was recorded that odor was evident in flour prepared from treated wheat. Moreover, Liu and Ho (1999) indicated that essential oils may have a serious impact on treated commodities, by altering some of the organoleptic characteristics. Seed germination could be also seriously affected in some cases (Vokou et al. 2003). Generally, oxygenated compounds are considered to have more serious effects on seed germination in comparison with non-oxygenated ones (Ranjendran and Sriranjini 2008).

Fumigants from plant sources are considered less penetrative to the commodities than traditional fumigants, such as phosphine. This may affect the insecticidal efficacy of some compounds, especially in the case of difficult to control life stages, such as eggs and pupae; this is directly related with diffusion and permeation properties (Dunkel and Sears 1998; Prates et al. 1999; Ranjendran and Sriranjini 2008; Shaaya et al. 1997). At the same time, there are no indications.

However, this has not been established in large-scale trials, which consist one additional disadvantage on the wider adoption, registration, and use of plant extracts. In fact, the lack of large-scale data is probably the most important limitation on the extensive use of botanicals in the "real world" of stored products.

3 Pesticides Made with Essential Oils

Essential oils are the plant volatile secondary metabolites, produced as by-products of plant metabolism, which consist of strong aromatic components that are spread in almost all parts of the plant. These secondary metabolite chemicals occur in cavities of plant, cell walls, or glandular hairs or in the other parts of plant such as leaves, flowers, bark, fruits, or flowers or even roots as droplets of fluids. Although they are commonly present in liquid form (Fig. 8.1), a slight increase in the temperature transforms them into gaseous form without any breakdown of the active components. These constituents are primarily lipophilic compounds that function as efficient insecticides, repellents, antifeedants, and oviposition inhibitors to a wide variety of insect pests. Himachalol and ß-himachalene obtained under biodirected isolation from a cheap essential oil, *Cedrus deodara* (Roxb.) G. Don (Pinales: Pinaceae), have resulted in insecticidal principles against *C. chinensis* (Singh and Agrawal 1988). The oils are composite mixtures of natural organic compounds which are primarily composed of terpenes (hydrocarbons) such as myrcene, pinene, terpinene, limonene, p-cymene, α- and β-phellandrene, etc., and terpenoids (oxygen-containing hydrocarbons) such as acyclic monoterpene alcohols (geraniol, linalool), monocyclic alcohols (menthol, 4-carvomenthenol, terpineol, carveol, borneol), aliphatic aldehydes (citral, citronellal, perillaldehyde), aromatic phenols (carvacrol, thymol, safrole, eugenol), bicyclic alcohol (verbenol), monocyclic ketones (menthone, pulegone, carvone), bicyclic monoterpenic ketones (thujone, verbenone, fenchone), acids (citronellic acid, cinnamic acid), and esters (linalyl acetate) (Koul et al. 2008). Some essential oils may also contain oxides (1, 8-cineole), sulfur-containing constituents, methyl anthranilate, coumarins, etc. Zingiberene, curcumene, farnesol, sesquiphellandrene, turmerone, nerolidol, etc., are examples of sesquiterpenes (C15) isolated from essential oils. Mono- and sesquiterpenoidal essential oil constituents are formed by the condensation of isopentenyl pyrophosphate units. Diterpenes usually do not occur in essential oils but are sometimes encountered as by-products. Asteraceae, Lamiaceae, Lauraceae, and Myrtaceae (Fig. 8.2) are some of the important plant families from which essential oils are extracted.

All essential oils break down fast and degrade readily in sunlight, air, and moisture and by detoxification enzymes, hence less persistence and reduced risks to nontarget organisms; therefore, more frequent applications and precise timings are needed. Recent investigations indicate that some chemical constituents of these oils interfere with the octopaminergic nervous system in insects. As this target site is not shared with mammals, most essential oil chemicals are relatively nontoxic to mammals and fish in toxicological tests (Koul et al. 2008).

It is evident that aromatic plant oils are superior in killing the stored pests, and most of these oils have at least a couple of common chemicals, and probably these chemicals represent the toxic action of the oils. The work reported by

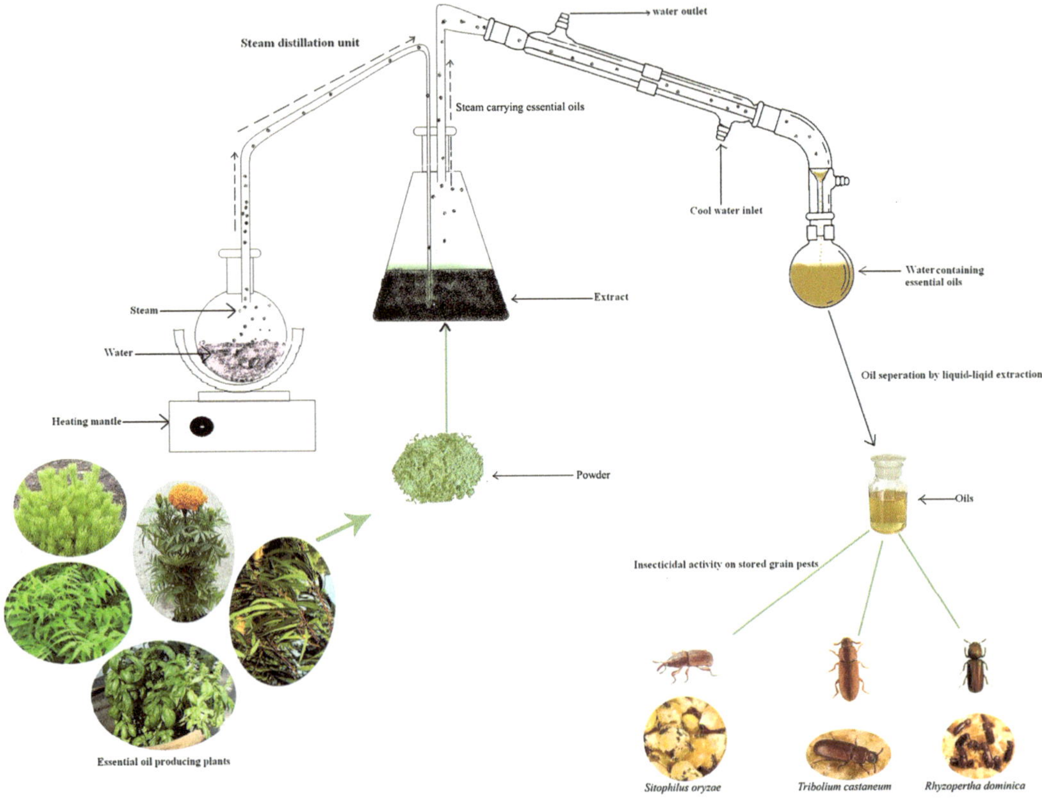

Fig. 8.1 Theoretical flow of extraction and assessment of botanical pesticides

Fig. 8.2 Some of the plant species and their derivatives that have insecticidal value for stored product pest control

Karabörklü et al. (2011) on the fumigant toxicity of the following aromatic oils, savory, *Satureja thymbra* L. (Lamiales: Lamiaceae); Turkish oregano, *Origanum onites* L. (Lamiales: Lamiaceae); myrtle, *Myrtus communis* L. (Myrtales: Myrtaceae); marjoram, *Origanum majorana* L. (Lamiales: Lamiaceae); laurel, *L. nobilis*; lemon, *Citrus limon* (L.) Burm. fil.

(Sapindales: Rutaceae); sticky goosefoot, *Chenopodium botrys* L. (Caryophyllales: Chenopodiaceae); and tansy, *Tanacetum armenum* (DC.) Sch. Bip., indicated that marjoram, laurel, lemon, goosefoot, and tansy oils which have a few common chemical components such as linalool, 1.8-cineole, citral, 2-(4a.8-dimethyl-1.2.3.4.4a.5.6.7-octahydro-naphthalen-2-yl)-prop-2-en-l-ol, and p-cymene also were commonly toxic to the adults of *Ephestia kuehniella* Zeller (Lepidoptera: Pyralidae). The oil of mountain savory, a member of the mint family having medicinal value, and oregano oil from the plant *Origanum vulgare* L. (Lamiales: Lamiaceae) can kill several flour moths including Indian meal moth *Plodia interpunctella* (Hübner) (Lepidoptera: Pyralidae) and the bean weevil *Acanthoscelides obtectus* (Say) (Coleoptera: Bruchidae) very effectively (Ayvaz et al. 2010). Apart from having a potentially high bioactivity against a range of insect pests, certain plant essential oils and their active constituents, mainly terpenoids, are highly selective to insects, since they are probably targeted to the insect-selective octopaminergic receptor, a nonmammalian target (Kostyukovsky et al. 2002). Apart from having a potentially high bioactivity against a range of insect pests, certain plant essential oils and their active constituents, mainly terpenoids, are highly selective to insects, since they are probably targeted to the insect-selective octopaminergic receptor, a nonmammalian target (Kostyukovsky et al. 2002). Many essential oils are used in food and beverages and as flavoring agents which is a good indication of their nontoxicity or least effects on mammals and other warm-blooded animals, and this relative safety makes them good candidates as alternatives to methyl bromide and phosphine for the preservation of stored grain.

Many insects are averse to the smell of essential oils, and thus they can repel or even kill the insects depending on their susceptibility to the chemical. Orange tree produces different composition of oils in their blossoms, citrus fruits, and/or leaves. In certain plants, one main essential oil constituent may predominate, while others are cocktails of various terpenes.

Some oils are significantly more active than others based on their constituents. Limonene found in the essential oil of various citrus leaves and fruit peels has exhibited significant insect control properties (Karr and Coats 1988). Few essential oils such as lemongrass, *Cymbopogon winteriana* Jowitt (Poales: Poaceae); eucalyptus, *Eucalyptus globulus* Labill (Myrtales: Myrtaceae); rosemary, *Rosmarinus officinalis* L. (Lamiales: Lamiaceae); vetiver, *V. zizanioides*; clove, *Eugenia caryophyllus* (Spreng.) Bullock and S. G. Harrison (Myrtales: Myrtaceae); and thyme, *Thymus vulgaris* L. (Lamiales: Lamiaceae), are known for their pest control properties. Among thirty-one essential oils obtained from different plant species evaluated for reproduction retardant and fumigant properties at 1,000 ppm in acetone against *S. oryzae* in laboratory, oil of *Pinus longifolia* Roxb. showed reproduction retardant, and oil of *Mentha citrata* Ehrh. (Lamiales: Lamiaceae) showed significant fumigant toxicity, while *Amomum subulatum* Roxb. (Zingiberales: Zingiberaceae), *Artemisia maritima* L. (Asterales: Asteraceae), and *C. deodara*, or *C. winteriana* oils protected the wheat grains from damage (Singh et al. 1989). When it comes to the control of stored product pests, the oils from edible plants are more preferred as most of them consist of acceptable flavors to human and their safety. During the investigations on the toxic properties of *Elettaria cardamomum* L. (Maton) (Zingiberales: Zingiberaceae), the essential oil extracted from cardamom was found lethal to *C. maculatus*, *T. castaneum*, and *E. kuehniella* (Abbasipour et al. 2011). During these studies, it was also found that the lepidopteran pest *E. kuehniella* was more susceptible than the coleopteran insects and the GC Mass analysis showed the 1,8-cineol, α-terpinyl acetate, terpinene, and fenchyl alcohol as major ingredients. Since cardamom is one of the most favorite spices, it makes a good choice for control of pests on stored grains and stored food products and is a low-risk insecticide.

Pesticides made with essential oils are derived from plants that are known to have insecticidal properties, which may not always be safe for humans and other mammals or that it cannot kill a wide variety of other life. Neem oil, garlic oil, and sabadilla are some least toxic botanical pesticides, whereas citrus oils, mint oil, pine oil, pepper extracts, and tree oils are moderately toxic. Many pesticides made with essential oils are formulated with synergists. These have no insecticidal effect of their own but serve to enhance the insecticidal effect of the botanicals. As many of these essential oils are nontoxic to humans and are widely available in some countries, the process of commercialization of essential oil-based pesticides is faster. All these properties make them effective in agricultural utilization, particularly for organic food production. Neem oil, derived from the neem tree scientifically known as *A. indica*, contains a complex mixture of biologically active compounds that can act as repellents, antifeedants, toxicants, oviposition deterrents, and growth retardant properties against various insect pest populations. Its active ingredients can act as direct insecticides, and it has both contact and systemic toxicity toward crop as well as domestic pests, but a very low toxicity to mammals. The active ingredients biodegrade rapidly in the presence of sunlight and within a few weeks in the soil. It is used as a grain protectant for centuries without apparent harm to humans. Oil extracted from garlic, *A. sativum*, also possesses insecticidal properties besides antifungal and antibacterial nature. Eugenol from cloves, *E. caryophyllus*; 1,8- cineole from *E. globulus*; citronellal from lemongrass, *Cymbopogon nardus* (L.) Rendle (Poales: Poaceae); pulegone from *Mentha pulegium* L. (Lamiaceae: Lamiales); and thymol and carvacrol from *T. vulgaris* are among the most active constituents against insects. Limonene found in the essential oil of various citrus leaves and fruit peels has exhibited significant insect control properties (Karr and Coats 1988). Another important plant species is *O. gratissimum*. The essential oil extracted from this plant caused 74 % mortality in *S. zeamais* when allowed to feed on essential oil-treated diet (Ngamo et al. 2001). The oil from *Xylopia aethiopica* (Dunal) A. Rich. (Magnoliales: Annonaceae) and *P. nigrum* also showed potent insecticidal activity to the maize weevil in food treatment method showing almost 100 % toxicity when the weevils were left to be in contact with the treated grain for 48 h (Ngamo et al. 2001). The acetone diluted pure essential oils from plant species currently found in Cameroon; it exhibited good fumigant toxicity to *S. zeamais*. Among these *Hyptis spicigera* Lam. (Lamiales: Lamiaceae) was the most toxic (fumigant) after 48 h of fumigation followed by *Annona senegalensis* Pers. (Magnoliales: Annonaceae) and *X. aethiopica* 96 and 95 % of mortality, respectively. Kouninki et al. (2005) and Ngamo Tinkeu et al. (2004) also reported the insecticidal activity of the essential oils *Eucalyptus citriodora* Hook. and *Eucalyptus saligna* Sm. (Myrtales: Myrtaceae), *Lippia rugosa* A. Chev. (Lamiales: Verbenaceae), and *O. gratissimum* to *S. zeamais*.

The turmeric *Curcuma longa* L. (Zingiberales: Zingiberaceae) plant is extremely important and most versatile and is widely used for medicinal and culinary purpose. It is well known for its vibrant color and abundant healing powers. Particularly in India, it has a lot of significance not only in medicinal use, but since thousands of years, it is also used in various religious ceremonies. The plant leaves and certain other parts of turmeric plant, on hydrodistillation, yield oil rich in α-phellandrene (70 %). It has moderate knockdown effect to *T. castaneum*. Essential oils found in the *Eucalyptus* plant foliage possess a wide spectrum of biological activity including antimicrobial, fungicidal, insecticidal/insect repellent, herbicidal, acaricidal, and nematicidal. The use of eucalyptus oil as a natural pesticide is of immense significance in view of the environmental and toxicological implications of the indiscriminate use of synthetic pesticides and overcoming/reducing the problem of increasing pest resistance (Batish et al. 2008).

4 Fumigants

Generally essential oils are toxic to stored product pests in the vapor form, acting as fumigants, which are an added advantage, particularly in storage conditions. Constituents of many essential oils, particularly monoterpenes being volatile in nature, are more useful as fumigants, and in recent years several monoterpenoids have been considered as potential alternates to synthetic pesticides. However, the degree of toxicity of different compounds to one species differs considerably. The active molecules can penetrate between the grains, thus enhancing the pest's exposure to the treatment. Several of the chemical components in essential oils are good toxicants in the vapor form also. Pulegone, linalool, and limonene the common chemical constituents of certain essential oils are effective fumigants against *S. oryzae*, while *M. citrata* oil containing linalool and linalyl acetate exhibits significant fumigant toxicity against this species (Singh et al. 1989). The *l*-carvone has been reported to cause 24 times more fumigant toxicity than its contact toxicity to *R. dominica* (Tripathi et al. 2003). Carvon can be obtained from the essential oil of dill plant, *Anethum sowa* Roxb. ex Fleming (Apiales: Apiaceae). The other major constituent of *A. sowa*, namely, dillapiole, is well known for its insecticide synergistic properties. This compound is present in the oil of *Anethum graveolens* L. (Apiales: Apiaceae) seeds (40–60 %) as well as from spearmint oil, *Mentha spicata* L. (Lamiales: Lamiaceae) (about 50 %). The volatile oil constituents of *Mentha* species are shown to be toxic to the common stored grain pests, *C. maculatus* and *T. castaneum* (Tripathi et al. 2000). Essential oils isolated from pine, *P. longifolia*; eucalyptus, *Eucalyptus obliqua* L. (Myrtales: Myrtaceae); and coriander, *Coriandrum sativum* L. (Apiales: Apiaceae), were reported to have toxicity in both contact and fumigation modes against three major stored grain pests, i.e., *S. oryzae*, *C. chinensis*, and *C. cephalonica* (Usha Rani 2012). However, the direct toxicity results varied with test material, dosage applied, insect species tested, and exposure time. Essential oils were much more potential in fumigation bioassays than in contact mode. The common monoterpenoids (1,8-cineole, α-pinene, carvone, linalool, etc.), phenolic acids, quercetin, caffeic acid, and protocatechuic acid present in these essential oils are responsible for their insecticidal activity against stored grain pests which have potential for applications in IPM program (Usha Rani 2012). The remarkable efficiency of the essential oil of African marigold, *Tagetes erecta* L. (Asterales: Asteraceae), against *S. oryzae*, *T. castaneum*, and *C. chinensis* was reported by Usha Rani and Udaya Lakshmi (2007). Unlike eucalyptus, pine, and coriander oils, the *Tagetes* oil showed similar toxicity in both contact and fumigation studies to stored product pests. The essential oils from *A. graveolens*, *Cuminum cyminum* L. (Apiales: Apiaceae), *I. verum*, *M. fragrans*, *Nigella sativa* L. (Ranunculales: Ranunculaceae), *P. nigrum*, and *Trachyspermum ammi* Sprague (Apiales: Apiaceae) isolated by hydrodistillation method produced mortality in adults and larvae of *C. chinensis* when fumigated (Chaubey 2008).

5 Repellents

Since essential oils contain aromatic volatile compounds, they evaporate quickly and many of them act as repellents to insects. These compounds are more useful for the use of domestic pests than the crop pests. Many essential oils and their monoterpenic constituents are known for their repellent activity against stored product pests. Citronella (*C. nardus*) essential oil has been used for over 50 years both as an insect repellent and an animal repellent. The steam distillation of aromatic roots of vetiver (*V. zizanioides*) yields an essential oil containing a large number of oxygenated sesquiterpenes which can protect clothes and other valuable materials from insect attack. Jaenson et al. (2006) demonstrated the repellency of oils of lemon, eucalyptus, geranium, and lavender against *Ixodes ricinus* (L.) (Acari: Ixodidae) in the laboratory and field. The essential oils extracted from the flowers of *T. erecta* have been effective repellent against insects (Ray et al. 2000). The chemical components isolated from turmeric rhizome powder oil, turmerone and *ar*-turmerone (dehydroturmerone), exhibited strong repellent activity toward certain stored grain pests, and the turmeric oil has been reported to provide protection to wheat grains against *T. castaneum* (Chahal et al. 2005). The volatile components from the essential oils of *Artemisia* species showed repellent activity to several coleopteran beetles, *Sitophilus* sp., *T. castaneum*, *C. maculatus*, and *R. dominica* (Negahban et al. 2007).

6 Insect Growth Regulators

Another important role played by essential oils in the life of insects is altering their growth parameters. Several essential oil constituents can alter the growth and affect the reproduction of the treated insects. In this case, Lepidoptera are the most susceptible insects. The oils of *A. graveolens*, *C. cyminum*, *I. verum*, *M. fragrans*, *N. sativa*, *P. nigrum*, and *T. ammi* reduced the oviposition potential, egg-hatching rate, pupal formation, and emergence of adults of F1 progeny of *C. chinensis* when fumigated with sublethal concentrations (Chaubey 2008). The growth inhibitory activities of these essential oils, which also showed good fumigant toxicity, could be due to the inhibition of various biosynthetic processes of the insects at different developmental stages. Essential oils extracted from peels of *Citrus reticulata* Blanco (Sapindales: Rutaceae) caused growth inhibition in *R. dominica* (Abbas et al. 2012). Some essential oils are good inhibitors of pest's oviposition, thereby affecting overall growth of the populations. Essential oil of *E. cardamomum* had a good efficacy on oviposition deterrence of *C. maculatus* females, and treatments reduced pest numbers in the treated grain (Abbasipour et al. 2011).

7 Technology for Application in Stored Product Protection

Considerable success has been achieved in terms of the development of botanical pesticides, products, and formulations and cost-effective viable technologies. The major properties of the plant-based pesticides, namely, effective, biodegradable, and considerably low mammalian toxicity, made them highly desirable (sought-after) environmental friendly alternatives. Till now methyl bromide and phosphine are used as broad spectrum fumigants against stored product pests, and it is very difficult to achieve the introduction of such effective and broad spectrum chemicals, but essential oils have got good prospects to be developed as alternative fumigants, if not against broad range but to particular species of insects, or to be used for a specific food product commodity. Insecticidal activities of *Annona squamosa* L. (Magnoliales: Annonaceae), *A. calamus*, *Melia azedarach* L. (Sapindales: Meliaceae), *A. indica*, or *V. negundo* have already been determined and commercialized. Photostable neem formulations will also be available soon.

Several essential oils and neem seed limonoids are identified as pesticides and have been applied for crop pest as well as stored product pest control. Though there are ample chances of essential oils/botanical compounds becoming alternates to chemical or synthetic pesticides, these compounds need to be available in a range of formulations that can be used in a variety of pests and in different situations, i.e., wettable powders, liquid concentrates, dusts, and aerosols. Since these botanical compounds are generally regarded as environmentally safe, they can be directly used even for the stored pest control. Particularly because most of the botanicals are often effective in small doses and decompose faster, the exposure will be lower, hence safer to use in food grain protection. Recent research has shown that certain combinations of essential oils such as peppermint oil, clove oil, citrus oils, lavender oil, thyme oil, and rosemary oil sometimes also combined with natural pyrethrins are effective. Until recently the availability of plant-based insecticides was limited to a few products like neem oil and pyrethrum. While these products are effective and exhibit very low environmental impact, they have a limited range of uses. Interest in organic farming is increasing rapidly all over the world, and efforts to produce and commercialize plant-based pesticides are also increasing enormously. Apart from being economical and eco-friendly, they also provide an opportunity for income generation in lower-income groups in some countries. The production of plant-based biopesticides on commercial scales can be an economically viable option for employment generation in the rural areas. The technology of insecticides application in storages has not changed much in the last 3–4 years. The application in empty storage and grain products is conducted by spraying, dusting, and fumigation and with different combinations of physical and chemical measures. Limiting insect infestation in grain storage must be a primary consideration as these pests not only cause considerable damage but also make the food unacceptable by leaving their frass, cadavers, etc. Generally fumigation continues to play a valuable role in stored pest control and is still widely used for the control of insects because these pesticides can reach even the interstitial places in the grain and kill insects. Several modern technological developments, including instrumentation for gas detection and analysis, improved formulations as well as increased demand for effective and economical pest control measures and have done much to improve fumigation procedures.

Another method proved to be efficient in botanical application is admixing where suitable formulations mostly botanicals made from edible plant materials are mixed with the stored grain to prevent insect infestation. Since usually botanicals are nonpersistent, hence, it may not stay longer on the treated grain. In rural India, farmers still use jute bags for storing their produce. Treatment to these bags is one method in which often the pests are killed or repelled, resulting in the grain protection. A new method was reported for the application of chemicals for protecting cereals and pulses by Usha Rani (1997) in which the required chemical is mixed uniformly with rice bran or rice husk in various dilutions which in turn was admixed with the grain. In this way the chemical is not in direct contact with the grain, but still the insects are in close contact with the chemical. A single application of deltamethrin applied in this method prevented insect infestation for a period of up to 20 months. Registration has been the main bottleneck in putting new products on the market, but more essential oils have been approved for use in the United States than elsewhere owing to reduced-risk processes for these materials (Regnault-Roger et al. 2012).

8 The Future

The future of botanicals in stored product protection remains uncertain. As noted above, for many ingredients, the toxicological/ecotoxicological aspects are poorly understood and definitely require additional assessment. Hence, despite the fact that some ingredients are now commercially available as plant protection products and biocides, the registration for direct application on a given stored commodity requires attention and further considerations in conjunction with a food safety-orientated strategy. Moreover, the economics involved in the identification of some compounds are often in contradiction with the production of large quantities regardless of the fact that these quantities can be also synthesized. At the same time, paradoxically, the majority of these compounds, according to the current legislation, in most counties need a "regular" registration procedure, just like in the case of conventional plant protection products. This is a major implication as an investment toward that direction, in comparison with the registration of the conventional active ingredients. A drastic and realistic change in legislation is needed globally in order to promote further registration and use of certain botanicals. Still, above all, the consumers' demand for residue-free food is not always compatible with the use of botanicals in stored product protection, unless it is proven that these compounds are safe and do not affect food properties.

From the techniques mentioned above, it seems that the use of botanicals as essential oils has a certain merit for further commercial development, taking into account that some of these compounds are also registered as biocides (against urban pests). One other promising category is the use of botanicals with the use of certain types of nanoparticles. For example, the use of botanicals with silica or diatomaceous earths has been evaluated with success and may appear in the future at a commercial scale, especially in durable stored products, such as grains and related amylaceous commodities. Other inert materials such as zeolites or electrostatically charged dusts can play a role toward this direction.

References

Abbas SK, Ahmad F, Sagheer M, Hasan M, Yasir M, Saeed A, Wali M (2012) Insecticidal and growth inhibition activities of *Citrus paradisi* and *Citrus reticulata* essential oils against lesser grain borer, *Rhyzopertha dominica* (F.) (Coleoptera: Bostrichidae). World J Zool 7:289–294

Abbasipour H, Mahmoud M, Fahimeh R, Hosseinpour MH (2011) Fumigant toxicity and oviposition deterrency of the essential oil from cardamom, *Elettaria cardamomum*, against three stored–product insects. J Insect Sci 11:165

Adarkwah C, Obeng-Ofori D, Büttner C, Reichmuth C, Schöller M (2010) Bio-rational control of red flour beetle *Tribolium castaneum* (Herbst) (Coleoptera: Tenebrionidae) in stored wheat with Calneem® oil derived from neem seeds. J Pest Sci 83:471–479

Ahn YJ, Lee SB, Lee HS, Kim GH (1998) Insecticidal and acaricidal activity of carvacrol and β-thujaplicine derived from *Thujopsis dolabrata* var. *hondai* sawdust. J Chem Ecol 24:81–90

Akob CA, Ewete FK (2009) Laboratory evaluation of bioactivity of ethanolic extracts of plants used for protection of stored maize against *Sitophilus zeamais* Motschulsky in Cameroon. Afr Entomol 17:90–94

Ali A, Ahmad F, Biondi A, Wang Y, Desneu N (2012) Potential for using *Datura alba* leaf extracts against two major stored grain pests, the khapra beetle *Trogoderma granarium* and the rice weevil *Sitophilus oryzae*. J Pest Sci 85:359–366

Athanassiou CG, Kontodimas DC, Kavallieratos NG, Anagnou-Veroniki M (2005) Insecticidal effect of NeemAzal against three stored-product beetle species on rye and oats. J Econ Entomol 98:1733–1738

Athanassiou CG, Kavallieratos NG, Evergetis E, Katsoula AM, Haroutounian SA (2013) Insecticidal efficacy of silica gel with *Juniperus oxycedrus* ssp. *oxycedrus* (Pinales: Cupressaceae) essential oil against *Sitophilus oryzae* (Coleoptera: Curculionidae) and *Tribolium confusum* (Coleoptera: Tenebrionidae). J Econ Entomol 106:1902–1910

Ayvaz A, Sagdic O, Karaborklu S, Ozturk I (2010) Insecticidal activity of the essential oils from different plants against three stored product insects. J Insect Sci 10:21

Barton AFM, Tjandra J, Nicholas PG (1989) Chemical evaluation of volatile oils in *Eucalyptus* species. J Agric Food Chem 37:1253–1257

Batish DR, Singh HP, Kohli RK, Kaur S (2008) *Eucalyptus* essential oil as a natural pesticide. For Ecol Manag 256:2166–2174

Butterworth JH, Morgan ED (1968) Isolation of a substance that suppress feeding in locusts. Chem Commun 1:23–24

Chahal KK, Arora M, Joia BS, Chhabra BR (2005) Bioefficacy of turmeric oil against *Tribolium castaneum* (Herbst) under laboratory conditions. In: Dilawari VK, Deol GS, Joia BS, Chuneja PK (eds) Proceedings of the 1st Congress on insect science: contributed papers. PAU, Ludhiana, pp 147–148

Chaubey MK (2008) Fumigant toxicity of essential oils from some common spices against pulse beetle, *Callosobruchus chinensis* (Coleoptera: Bruchidae). J Oleo Sci 57:171–179

Chu SS, Liu SL, Liu QZ, Liu ZL, Du SS (2011) Composition and toxicity of Chinese *Dracocephalum moldavica* (Labiatae) essential oil against two grain storage insects. J Med Plants Res 5:4621–4626

De Souza TW, Grazziotti GH, Jr De Souza AA, De Souza FS, Consolaro HN, De Aquino Ribeiro PE, Zanuncio JC (2013) Screening of extracts of leaves and stems of *Psychotria* spp. (Rubiaceae) and *Spodoptera frugiperda* (Lepidoptera: Noctuidae) for maize protection. J Food Prot 76:1892–1901

Dunkel FV, Sears LJ (1998) Fumigant properties of physical preparations from mountain big sagebrush, *Artemisia tridentata* Nutt. ssp. *vaseyana* (Rydb.) beetle for stored grain insects. J Stored Prod Res 34:307–321

Dunkel FV, Sriharan S, Niziyimana E, Serugendo A (1991) Evaluation of neem kernel extract (Margosan-O) against major stored insect pests of beans and sorghum in Rwanda. In: Fleurat-Lessard F, Ducom P (eds) Proceedings of the 5th international working conference on stored-product protection, Bordeaux, France. Imprimerie du Médoc, Bordeaux, pp 527–536

Epidi TT, Udo IO (2009) Biological activity of ethanolic extract fractions of *Dracaena arborea* against infestation of stored grains by two insect pests. Pak J Biol Sci 12:976–980

Huang Y, Tan JMWL, Kini RM, Ho SH (1997) Toxic and antifeedant action of nutmeg oil against *Tribolium castaneum* (Herbst) and *Sitophilus zeamais* Motschulsky. J Stored Prod Res 4:289–298

Jaenson TGT, Garboui S, Pålsson K (2006) Repellency of oils of lemon, eucalyptus, geranium, and lavender and the mosquito repellent MyggA natural to *Ixodes ricinus* (Acari: Ixodidae) in the laboratory and field. J Med Entomol 43:731–736

Jilani G, Saxena RC (1990) Repellent and feeding deterrent effects of turmeric oil, sweetflag oil, neem oil, and a neem-based insecticide against lesser grain borer (Coleoptera: Bostrychidae). J Econ Entomol 83:629–634

Jilani C, Saxena RC, Rueda BP (1988) Repellent and growth-inhibiting effects or turmeric oil, sweetflag oil, neem oil and "Margosan-O" on red flour beetle (Coleoptera: Tenebrionidae). J Econ Entomol 81:1226–1230

Karabörklü S, Ayvaz A, Yilmaz S, Akbulut M (2011) Chemical composition and fumigant toxicity of some essential oils against *Ephestia kuehniella*. J Econ Entomol 104:1212–1219

Karr LL, Coats JR (1988) Insecticidal properties of *d*-Limonene. J Pestic Sci 13:287–290

Kavallieratos NG, Athanassiou CG, Saitanis CJ, Kondodimas DC, Roussos AN, Tsoutsa MS, Anastassopoulou UA (2007) Effect of two azadirachtin formulations against adults of *Sitophilus oryzae* and *Tribolium confusum* on different grain commodities. J Food Prot 70:1627–1632

Khani M, Awang RM, Omar D, Rahmani M, Rezazadeh S (2011) Tropical medicinal plant extracts against rice weevil, *Sitophilus oryzae* L. J Med Plants Res 5:259–265

Kim DH, Ahn YJ (2001) Contact and fumigant activities of constituents of *Foeniculum vulgare* fruit against three coleopteran stored-product insects. Pest Manag Sci 57:301–306

Kim SI, Park C, Ohh MH, Cho HC, Ahn YJ (2003) Contact and fumigant activities of aromatic plant extracts and essential oils against *Lasioderma serricorne* (Coleoptera: Anobiidae). J Stored Prod Res 39:11–19

Kordali S, Kotan R, Mavi A, Cakir A, Ala A, Yildirim A (2005) Determination of the chemical composition and antioxidant activity of the essential oil of *Artemisia dracunculus* and of the antifungal and antibacterial activities of Turkish *Artemisia absinthium*, *Artemisia dracunculus*, *Artemisia santonicum* and *Artemisia spicigera* essential oils. J Agric Food Chem 53:9452–9458

Kostyukovsky M, Rafaeli A, Gileadi C, Demchenko N, Shaaya E (2002) Activation of octopaminergic receptors by essential oils constituents isolated from aromatic plants: possible mode of action against insect pest. Pest Manag Sci 58:1101–1106

Koul O, Walia S, Dhaliwal GS (2008) Essential oils as green pesticides: potential and constraints. Biopestic Int 4:63–84

Kouninki H, Haubruge E, Noudjou FE, Lognay G, Malaisse F, Ngassoum MB, Goudoum A, Mapongmetsem PM, Ngamo LS, Hance T (2005) Potential use of essential oils from Cameroon applied as fumigant or contact insecticides against *Sitophilus zeamais* Motsch. (Coleoptera: Curculionidae). Commun Agric Appl Biol Sci 70:787–792

Lee SE, Peterson CJ, Coats JR (2003) Fumigation toxicity of monoterpenoids to several stored product insects. J Stored Prod Res 39:77–85

Liu ZL, Ho SH (1999) Bioactivity of the essential oil extracted from *Evodia rutaecarpa* Hook f. et Thomas against the grain storage insects, *Sitophilus zeamais* Motsch. and *Tribolium castaneum* (Herbst). J Stored Prod Res 35:317–328

Lu Y, Zhong J, Wang Z, Liu F, Wan Z (2013) Fumigation toxicity of allicin against three stored product pests. J Stored Prod Res 55:48–54

Magro A, Matos O, Bastos M, Carolino M, Mexia A (2011) The fungicide activity of clove and laurel essentials oils against storage fungi. IOBC/WPRS Bull 69:230–235

Makanjuola WA (1989) Evaluation of extracts of neem (*Azadirachta indica* Adr. Juss.) for the control of some stored-product pests. J Stored Prod Res 25:231–238

Negahban M, Moharramipour S, Sefidkon F (2007) Fumigant toxicity of essential oil from *Artemisia sieberi* Besser against three stored-product insects. J Stored Prod Res 43:123–128

Ngamo LS, Ngassoum MB, Jirovetz L, Ousman A, Nukenine EC, Mukala OE (2001) Protection of stored maize against *Sitophilus zeamais* (Motsch.). Mededelingen van de Rijksuniversiteit Landbouwkundige te Gent 66:473–478

Ngamo Tinkeu LS, Goudoum A, Ngassoum MB, Mapongmetsem PM, Kouninki H, Hance T (2004) Persistence of the insecticidal activity of five essential oils on the maize weevil *Sitophilus zeamais* (Motsch.) (Coleoptera: Curculionidae). Commun Agric Appl Biol Sci 69:145–147

Obeng-Ofori D, Reichmuth C, Bekele J, Hassanali A (1997) Biological activity of 1,8 cineole, a major component of essential oil of *Ocimum kenyense* (Ayobangira) against stored product beetles. J Appl Entomol 121:237–243

Obeng-Ofori D, Reichmuth C, Bekele J, Hassanali A (1998) Toxicity and protectant potential of camphor, a major component of essential oil of *Ocimum kilimandscharicum*, against four stored product beetles. Int J Pest Manag 44:203–209

Owolabi MS, Oladimeji MO, Lajide L, Singh G, Marimuthu P, Isidorov VA (2009) Bioactivity of three plant derived essential oils against the maize weevils *Sitophilus zeamais* (Motschulsky) and cowpea weevils *Callosobruchus maculatus* (Fabricius). Electron J Environ Agric Food Chem 8:828–835

Paneru RB, le Patourel GNJ, Kennedy SH (1997) Toxicity of *Acorus calamus* rhizome powder from eastern Nepal to *Sitophilus granarius* (L.) and *Sitophilus oryzae* (L.) (Coleoptera: Curculionidae). Crop Prot 16:759–763

Pereira J, Wohlgemuth R (1982) Neem (*Azadirachta indica* A. Juss.) of West African origin as a protectant of stored maize. Z Angew Entomol 94:208–214

Pezo D, Salafranca J, Nerin C (2006) Design of a method for generation of gas-phase hydroxyl radicals, and use of HPLC with fluorescence detection to assess the antioxidant capacity of natural essential oils. Anal Bioanal Chem 385:1241–1246

Prakash A, Rao J (1997) Botanical pesticides in agriculture. CRC Press, Boca Raton

Prates HT, Santos JP, Waquil JM, Oliveira AB (1999) The potential use of plant substances extracted from Brazilian flora to control stored grain pests. In: Zuxun J, Quan L, Yongsheng L, Xianchang T, Lianghua G (eds) Proceedings of the 7th international working conference on stored-product protection, Beijing, China. Sichuan Publishing House of Science and Technology, Chengdu, pp 820–825

Rahim M (1998) Biological activity of azadirachtin-enriched neem kernel extracts against *Rhyzopertha dominica* (F.) (Coleoptera: Bostrychidae) in stored-wheat. J Stored Prod Res 34:123–128

Rajasekharreddy P, Usha Rani P (2010) Toxic properties of certain botanical extracts against three major stored product pests. J Biopestic 3:586–589

Rajashekar Y, Ravindra KV, Bakthavatsalam N (2014) Leaves of *Lantana camara* Linn. (Verbenaceae) as a potential insecticide for the management of three species of stored grain insect pests. J Food Sci Technol 10/2012 doi: 10.1007/s13197-012-0884-8

Ranjendran S, Sriranjini V (2008) Plant extracts as fumigants for stored product insect control. J Stored Prod Res 44:126–135

Ray DP, Walia S, Dureja P, Singh RP (2000) Composition and repellent activity of the essential oil of marigold (*Tagetes erecta* L.) flower. Indian Perfum 44:267–270

Regnault-Roger C, Vincent C, Arnason JT (2012) Essential oils in insect control: low-risk products in a high-stakes world. Annu Rev Entomol 57:405–424

Rozman V, Kalinovic I, Liska A (2006) Bioactivity of 1,8-cineole, camphor and carvacrol against rusty grain beetle (*Cryptolestes ferrugineus* Steph.) on stored wheat. In: Lorini I, Bacaltchuk B, Beckel H, Deckers D, Sundfeld E, Dos Santos P, Biagi JD, Celaro JC, Faroni LRD'A, Bartolini LOF, Sartori MR, Elias MC, Guedes RNC, Da Fonseca RG, Scussel VM (eds) Proceedings of the 9th international working conference on stored product protection, Campinas, São Paulo, Brazil. Brazilian Post-Harvest Association, ABRAPOS, Passo Fundo, pp 687–694

Rozman V, Kalinovic I, Korunic Z (2007) Toxicity of naturally occurring compounds of Lamiaceae and Lauraceae to three stored product insects. J Stored Prod Res 43:349–355

Saiful Islam MS, Hasan MM, Lei C, Mucha Pelzer T, Mewis I, Ulrichs C (2010) Direct and admixture toxicity of diatomaceous earth and monoterpenoids against the storage pests *Callosobruchus maculatus* (F.) and *Sitophilus oryzae* (L.). J Pest Sci 83:105–112

Schuenzel KM, Harrison MA (2002) Microbial antagonists of foodborne pathogens on fresh minimally processed vegetables. J Food Prot 65:1909–1915

Shaaya E, Ravid U, Paster N, Juven B, Zisman U, Pissarev V (1991) Fumigant toxicity of essential oils against four major stored product insects. J Chem Ecol 17:499–501

Shaaya E, Kostjukovski M, Eilberg J, Sukprakarn C (1997) Plant oils as fumigants and contact insecticides for the control of stored-product insects. J Stored Prod Res 33:7–15

Singh D, Agrawal SK (1988) Himachalol and β-Himachalene: insecticidal principles of Himalayan cedarwood oil. J Chem Ecol 14:1145–1151

Singh G, Upadhyay RK (1993) Essential oils: a potent source of natural pesticides. J Sci Ind Res 52:676–683

Singh D, Siddiqui MS, Sharma S (1989) Reproduction retardant and fumigant properties in essential oils against rice weevil (Coleoptera: Curculionidae) in stored wheat. J Econ Entomol 82:727–733

Singh M, Srivastava S, Srivastava RP, Chauhan SS (1995) Effect of Japanese mint (*Mentha arvensis*) oil as fumigant on nutritional quality of stored sorghum. Plant Foods Hum Nutr 47:109–114

Skandamis P, Koutsoumanis K, Fasseas K, Nychas GJE (2001) Inhibition of oregano essential oil and EDTA on *E. coli* O157:H7. Italian J Food Sci 13:55–65

Tripathi AK, Prajapati V, Aggarwal KK, Kukreja AK, Dwivedi S, Singh AK (2000) Effects of volatile oil constituents of *Mentha* species against stored grain pests, *Callosobruchus maculatus* and *Tribolium castaneum*. J Med Aromat Plant Sci 22:549–556

Tripathi AK, Prajapati V, Kumar S (2003) Bioactivity of l-carvone, d-carvone and dihydrocarvone toward three stored product beetles. J Econ Entomol 96:159–1601

Usha Rani P (1997) Bioactivity and residual toxicity of deltamethrin incorporated into rice bran and rice husk, against the pests of stored products. Int Pest Control 39:186–188

Usha Rani P (2012) Fumigant and contact toxic potential of essential oils from plant extracts against stored product pests. J Biopestic 5:120–128

Usha Rani P, Devanand P (2011) Efficiency of different plant foliar extracts on grain protection and seed germination in maize. Res J Seed Sci 4:1–14

Usha Rani P, Rajasekharreddy P (2010) Insecticidal activity of (2*n*-octyl*cyclo*prop-1-enyl)-octanoic acid (I) against three coleopteran stored product insects from *Sterculia foetida* (L.). J Pest Sci 83:273–279

Usha Rani P, Udaya Lakshmi S (2007) Evaluation of *Tagetes erecta* essential oil, floral and leaf extracts as grain protectants: efficacy against four species of stored product pests. Uttar Pradesh J Zool 27:159–169

Usha Rani P, Venkateshwaramma T, Devanand P (2011) Bioactivities of *Cocos nucifera* L. (Arecales: Arecaceae) and *Terminalia catappa* L. (Myrtales: Combretaceae) leaf extracts as post-harvest grain protectants against four major stored product pests. J Pest Sci 84:235–247

Usha Rani P, Hymavathi A, Suresh Babu K, Sreedhar Rao A (2013) Bioactivity evaluation of prenylated isoflavones derived from *Derris scandens* against two stored pest larvae. J Biopestic 6:14–21

Vokou D, Douvli P, Blionis GJ, Halley JM (2003) Effects of monoterpenoids, acting alone or in pairs, on seed germination and subsequent seedling growth. J Chem Ecol 29:2281–2301

Weaver DK, Subramanyam B (2000) Botanicals. In: Subramanyam B, Hagstrum DW (eds) Alternatives to pesticides in stored-product IPM. Kluwer Academic Publishers, Dordrecht, pp 303–320

Weaver DK, Phillips TW, Dunkel FV, Weaver T, Grubb RT, Nance EL (1995) Dried leaves from rocky mountain plants decrease infestation by stored-product beetles. J Chem Ecol 21:127–142

Xie YS, Fields PG, Isman MB (1995) Repellency and toxicity of azadirachtin and neem concentrates to three stored-product beetles. J Econ Entomol 88: 1024–1031

Yang FL, Liang GW, Xu YJ, Lu YY, Zeng L (2010) Diatomaceous earth enhances the toxicity of garlic, *Allium sativum*, essential oil against stored product pests. J Stored Prod Res 46:118–123

Pesticidal Plants for Stored Product Pests on Small-holder Farms in Africa

9

Philip C. Stevenson, Sarah E.J. Arnold, and Steven R. Belmain

Abstract

Despite the near elimination of pests from food stores in industrialised nations, insects are still the most important challenge to food security for small-holder farmers in less developed nations. Losses are frequently as high as 20 %. Synthetic products provide effective control when used correctly but are not sustainable or universally appropriate and present many challenges for farmers, not least of all their cost. Pesticidal plants offer an economic, effective and often the only alternative. Much published research, however, overlooks critical knowledge gaps providing outputs that are unlikely to improve pesticidal plant use or improve food security. This chapter identifies opportunities for better targeted research and improvements for uptake and use of pesticidal plants. We also highlight how a deeper understanding of different morphs, gender and age of insect can influence experimental results and should be considered more carefully.

To be effective plant materials need to show low animal and environmental toxicity at typical application levels but at the same time be effective against a wide range of target species, at low doses and with longevity. They must also be low cost, safe, compatible with other pest management technologies and stable and have no consequences for the stored products such as impairing flavour. Research should be targeted at optimising the efficacy of the pesticidal plants already known to have potential, and this should be supported by chemistry to fully understand spatial, temporal and phenotypic variability and nontarget impacts. Availability of plants is a

P.C. Stevenson, Ph.D. (✉)
Natural Resources Institute, University of Greenwich, Chatham Maritime, Kent, UK

Royal Botanic Gardens, Richmond, Surrey, UK
e-mail: p.stevenson@kew.org

S.E.J. Arnold, Ph.D. • S.R. Belmain, Ph.D.
Natural Resources Institute, University of Greenwich, Chatham Maritime, Kent, UK

D. Singh (ed.), *Advances in Plant Biopesticides*,
DOI 10.1007/978-81-322-2006-0_9, © Springer India 2014

limiting factor to uptake so propagation and cultivation of elite provenances would alleviate pressure on natural ecosystems and improve reliability of efficacy and supply when supported by improved harvesting techniques. The large-scale commercialisation of plants may not compete with synthetic products globally but local production may foster a mechanism to support and encourage uptake through local markets and value chains.

Keywords
Pesticidal plants • Botanical insecticides • *Tephrosia* • *Securidaca* • Bruchids • *Sitophilus* • Maize • Beans

1 Introduction

In recent decades, the postharvest sector for durable crops (grains and legumes) has undergone a massive divergence between developed and developing economies. Relatively advanced large-scale technologies such as refrigerated silos, fumigation and controlled atmosphere storage have virtually eliminated the problems of insect attack after harvest across Europe, North America and Australia. However, the situation in developing countries could not be more different. Particularly in the Tropics and for the relatively smaller-scale crop production systems found in sub-Saharan Africa and South Asia, postharvest losses routinely result in 10–20 % of food produced being lost along the value chain after harvest. In countries where subsistence and small-scale farmers dominate, postharvest losses from insects mean that farmers are often obliged to sell their crops soon after harvest because they do not have access to affordable technology that will protect their stored crops at the small-scale level. For subsistence farmers, the rapid deterioration caused by storage insects can mean seasonally extended food insecurity, lower quality and nutritive content of their produce and the need to account for loss expectations by expanding the amount of land cultivated. Some of the best estimates of postharvest loss that include small-holder production are now being collected in southeastern Africa with data made available by the African Postharvest Losses Information System (APHLIS).[1] This interactive map and database show postharvest losses can vary across seasons, regions and commodity type and gives some indication of how severe postharvest losses can be for subsistence farmers, e.g. during 2007 losses were up to 35 % for maize in most of Zimbabwe, while up to 25 % for most of Mozambique, Tanzania and eastern Kenya (see Fig. 9.1).

Although many initiatives are trying to develop and distribute appropriate technology for on-farm storage, e.g. hermetically sealed bags, insect-proof containers, etc., most subsistence farmers continue to store produce in woven polythene sacks or indigenously designed granaries. As subsistence farmers are, by definition, poor in resources and constitute some of the most marginalised people globally, providing them with new technology is fraught with difficulty to overcome vested interests in maintaining the status quo (e.g. market traders, pesticide manufacturers) as well as convincing and enabling farmers to invest in storage technology. Typically, most subsistence farmers do nothing to prevent postharvest losses. Their experience tells them to sell their produce early to avoid losses because they cannot maintain grain quality later in the season when market prices are high for good-quality grain. If farmers do plan on storing longer than 3 months, they usually treat their grain with a commercially available synthetic pesticide, particularly if they hope to sell the grain.

[1] http://www.aphlis.net

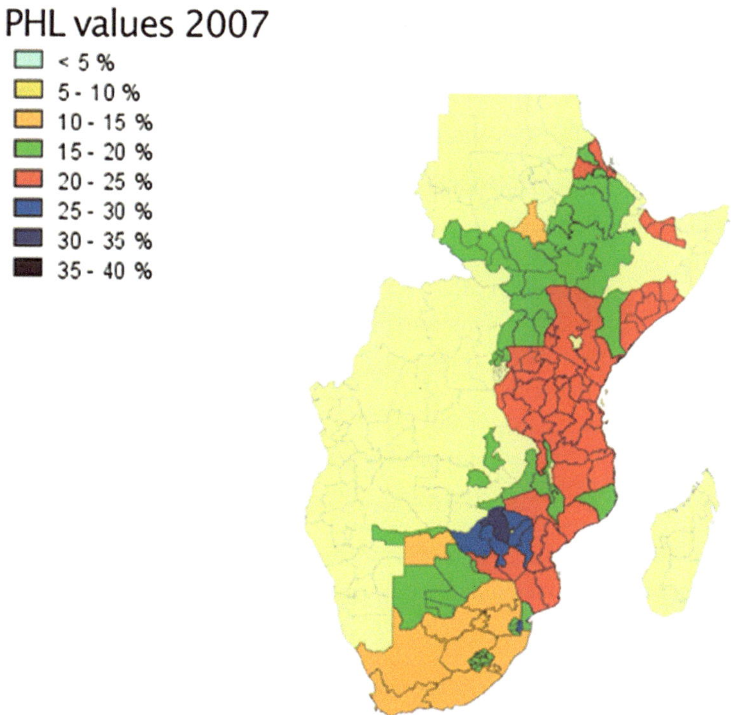

Fig. 9.1 Postharvest losses (PHL) for maize collected at the provincial level across southeastern African states during 2007 (Data derived from http://www.aphlis.net)

The most abundant pesticide used in sub-Saharan Africa to control insects in grain storage is the fumigant phosphine, which can really only be effectively employed at large scale as it requires gas-impermeable sheeting and monitoring equipment, the costs of which are beyond the reach of small-holder farmers. In some countries, farmers can legally (or illegally) obtain phosphine and attempt to fumigate their grain themselves at home but through poor knowledge do not maintain a gas-tight environment, often leading to human poisoning. For small-holders Actellic (pirimiphos) dust or its sister product Actellic Super (pirimiphos-methyl and permethrin) dust is used most commonly (Obeng-Ofori 2010). These are typically highly effective when used correctly but distribution may be restricted in some rural areas, and perhaps the most difficult problem with Actellic, as with most other pesticides, is that they are frequently adulterated by unscrupulous traders or poorly stored/expired product is used. Inappropriate application, the development of resistance among insect species and the limited number of insecticides available in remote areas are just some of the problems facing small-scale use of insecticides for stored production protection. Insecticides are typically applied without adhering to manufacturer advice about safe handling, and this can place farm workers, often women and children, at risk. In addition, the safety of food for consumers is not a high priority, and there is no mechanism for assessing chronic/acute outcomes from exposure. In 2013, 23 children died in an Indian school after eating a meal contaminated by pesticide.[2] Pesticide dumping and the impact on wildlife and beneficial organisms are also serious, while the prohibitive cost of pesticides for poor farmers can be a strong disincentive to use them. Recent surveys highlight these issues and report that the

[2] http://www.bbc.co.uk/news/world-asia-india-23436348

problems are well understood by farmers (Kamanula et al. 2011; Nyirenda et al. 2011) and have led to farmers simply avoiding synthetic pesticides altogether but at the cost of stored food losses.

Storage losses can be avoided using improved management and local materials. Using simple techniques, even the poorest resourced farmers can have some direct control over postharvest pests. The use of biorational approaches to control stored product pests has been reviewed recently (Phillips and Throne 2010), so this chapter will focus more specifically on the potential for plants to be effective alternatives to pesticides. Plant species have been studied for decades as potential alternatives to commercial synthetic chemicals owing to their abundance of defence chemicals, and it has been estimated that around 10 % of all plant species (~28,000) have some pesticidal chemical qualities but fewer than 500 have robust evidence supporting the reported activity (Simmonds 2003). Plants have not fulfilled their commercial promise and are perhaps best suited to providing important, and for some of the poorest resourced farmers the only, source of pest control material in developing nations (Isman 2006, 2008). Plants offer an economically viable alternative for small-holder farmers, but the research to develop them is lacking robust chemical backstopping, and their effective promotion, using applications that are improved based on scientific research, could improve their impact and sustainability for storage pest management for insect pests.

In most cases, the idea of using plant material to control pests is a familiar concept to farmers (Kamanula et al. 2011; Nyirenda et al. 2011), and many recognise their wider benefits as environmentally benign, less toxic and cost-effective compared to synthetic pesticides. Plant materials are also more difficult to adulterate especially if they are harvested by the farmers themselves, and while they may not require any significant financial outlay, their use does consume one important resource to farmers – time.

Farmer surveys carried out in Ghana have highlighted that many farmers do not use commercial synthetics (Belmain and Stevenson 2001) and, instead, use plant-based products. While there is a general acceptance that plants offer a useful alternative to commercial pesticides for small-holder farmers, some recent surveys reveal that elsewhere in Africa surprisingly few actually use them (Kamanula et al. 2011). This could be explained by knowledge gaps or failing policies to drive uptake or commercialisation. Availability of plant material and variability of that material are major limiting factors, and the propagation and commercialised cultivation of key verified or validated provenances could drive the change needed to enable more farmers to use this technology and establish plant materials as a genuine alternative to pesticides.

This chapter draws attention to knowledge gaps in aspects of pesticidal plant research and promotion and provides guidance on how scientists can better target their research to ensure that outcomes offer useful knowledge that can improve uptake and use of pesticidal plants. Much of this work has been carried out under the African Dryland Alliance for Pesticidal Plant Technologies (ADAPPT Project) and continues through propagation and outreach activities on a follow-up strategy Optimising Pesticidal Plants: Technology Innovation, Outreach and Networks Project (OPTIONs).

An analysis of published science based in Africa on pesticidal plants indicates that much research overlooks the critical knowledge gaps and is repeated work or delivers outputs that are unlikely to have any impact on agricultural development and poverty alleviation. Most research on pesticidal plants does not address the problems that limit the development, uptake and promotion of pesticidal plants. Furthermore, >80 % of published work in this research area in Africa reports laboratory bioassays that simply evaluate pesticidal plants against insects with little consideration of the information needed to optimise use, facilitate upscaling, enhance field efficacy or address health and safety issues (vertebrate toxicity), propagation, cultivation or conservation. This is particularly well illustrated by the example of biological activity reported in nonpolar extracts of a plant against a pest species that farmers will attempt to control in the field using crude water extracts that will unlikely contain any of the active components reported in the scientific study.

Ideally to be effective and useful in stored product protection, plant materials need to show low toxicity to mammals and the environment, show toxicity or repellency to a wide range of target species, have efficacy at low doses and with a long period of activity, have low cost in terms of time to collect process and apply, be compatible with other pest management technologies, be stable chemically and have no consequences for the stored product such as impairing flavour.

2 Pesticidal Plants as Alternatives to Synthetic Products in Africa

Generations of African farmers have used pesticidal plants to control stored product pests (Thacker 2002), and prior to the synthetic chemistry revolution, plants were the only technology available to farmers for insect pest control (Isman 2008). There are already many literature reviews about pesticidal plants and here we do not attempt to repeat this. Prakash and Rao (1997) have reviewed botanical insecticides describing biological activities and applications for >100 species, while details of plant chemistries and their role in integrated approaches to pest control pest are also reviewed (Koul and Dhaliwal 2004).

The main objective of this chapter is to highlight the importance of pesticidal plants as alternative pest control technologies for stored products and draw attention to knowledge gaps for future research based on what is absent from recent published work. Despite thousands of research articles in the literature, there are only a handful of successful commercial botanical products used for insect control including pyrethrum, rotenoid-based products, neem and essential oils. Thus, plants as commercial products have fallen short of their potential (Isman 2006). Indeed they comprise only a small fraction of the total pesticide used in high volume agriculture in Europe, Australasia and the Americas. However, in Africa they still have a major contribution to make for farmers, and it has been argued convincingly that it is for the rural poor that pesticidal plants have the greatest value in agriculture (Isman 2008). Many of the species available and used in Africa are described in detail by Stoll (2000) along with low-tech practical application procedures for a variety of pests and crops that require the minimum preparation. This is distinct from the aspiration of many scientists whose aim is a commercial product in a bottle that replaces a synthetic product. Here, we focus on pesticidal plants that comprise crude plant materials including leaves, stem bark, roots, fruits and seeds that are effective but need only rudimentary preparation and are thus more suitable for the resource-poor small-holder farmers across Africa. This processing includes drying, crushing and mixing with stored products (Belmain and Stevenson 2001) or producing crude extracts in water for application with a sprayer or applied using a brush (Stoll 2000). Indeed the lower the level of preparation that is required, the more universally appropriate the plant material. This is not to say that the materials do not require effective and robust backstopping with scientific knowledge and research. Indeed we argue below, using examples from our own work and recent work of others, that scientific support is essential to provide farmers with reliable products and materials that have predictable and measurable effects.

Kamanula et al. (2011) reported that farmers are well aware of the potential pesticidal value of plants but, depending upon the country, only between 20 and 50 % actually used them. This may be because efficacy is perceived to be lower than with the other available pesticides, or the inconsistency in efficacy could explain it. We expand below on how this has impacted on the use of pesticidal plants and how it might be overcome. However, for small-holder farmers who have no alternatives, some efficacy is better than none even if only a moderate reduction in damage can be achieved. Optimising efficacy through adaptations of the application based on understanding the chemical mechanisms and developing processes for their application that exploit this knowledge could improve uptake. There are key research areas that need attention in the optimization of the use of plants in pest control and are discussed later below.

The last 10 years has seen an explosion in the productivity of entomologists and related scientists identifying new plant materials with new biological activities against old storage insects pests, e.g. Ali et al. (2013), Padin et al. (2013). While this is useful and adds to our ever increasing resources of plant materials, the majority of these research outcomes do not provide any chemical basis for the biological activities reported and as such lose much potential impact. In some cases this is a serious oversight but perhaps more often simply indicates the limitations of laboratories working in this research domain.

A chemical comprehension of biological activity in pesticidal plants is important because chemistry can show extreme variation in plant material from the same species (Stevenson et al. 2012). Consequently it can be a serious limiting factor in its effective and reliable use as a pesticide whether in commercialised products or simply for crude use by poor farmers (Belmain et al. 2012). We will discuss how important this can be in more detail in subsequent sections of this chapter, but it is sufficient to say that the natural variation within populations of plants manifests itself in chemical as well as morphological diversity. One way to control for this is to carry out bioassays using plant materials from several different locations within the target region in field trials compared to a commercial synthetic pesticide which can be carried out in different locations and indicate the level of variation in efficacy that farmers might expect in plant materials being tested (Amoabeng et al. 2013).

A recent review of plant feeding deterrents for stored product pests provides a summary of 200 compounds primarily sesquiterpenes identified to date and illustrates the depth of options available to researchers of plant pesticides for stored products pests (Nawrot and Harmatha 2013). However, most research published in the last 10 years does not determine the chemical basis of the biological activity, and those that do report the identification of chemicals are investigating plants for their essential oils (Zhao et al. 2013; Liu et al. 2012). This is likely in many cases because target molecules in essential oils research can be identified using routine equipment (Gas Chromatography-Mass Spectrometry), and the now universal National Institute of Standards and Technology (NIST) mass spectrometry library can be used to assert a high probability of correct identification of many small molecules based on their fragmentation pattern from easily sampled material. Essential oils are also studied with a view to commercialisation since they are Generally Regarded as Safe, and so circumvent regulatory procedures in North America (Isman et al. 2011) may perhaps prove to be more commercially viable botanical insecticides than many plant pesticides that rely on more intrinsically toxic chemicals for their activities. One major issue working with essential oils is that where the biological activity of plant material is reported to be associated with the essential oil content, the pure compounds are tested at much higher concentrations than occur in the plant and so can be misleading about their potential effect in unprocessed plant material. Because the yields of essential oils are often very poor from plants, the costs of using essential oils at some of the concentrations proposed are high for large-scale agriculture (Jiang et al. 2012b).

Determining the structures of compounds from other main groups of natural products in plants that might account for the biological activity of plants is otherwise overlooked. This may be because the specialised spectroscopic instrumentation including LC-MS, Prep HPLC and NMR required to isolate and determine structures of large and often structurally complex molecules is not widely available nor the expertise to use the equipment, where available. This is particularly problematic if compounds are new or there are no literature data to compare. The consequence is that some of the most interesting new plant-based pesticidal discoveries do not advance the subject as far as they might have done as we do not know what the active components are. Another oversight that reduces the value of much recent research is that the plant material is not always botanically verified by a qualified botanist, and researchers do not deposit specimens in registered herbaria so that the work can be later revisited, verified and further studied by future researchers who might have access to the material tested in the published reports. The importance of herbaria is well presented by Funk (2004). Herbarium specimens necessarily require associated information including locality including the country, habitat, altitude, field identification and location (including GPS coordinates), date of collection and plant description all of which are essential information for anyone who might wish to use a specimen at later date or inform about the research published about the specimen in the first place (Bridson and Fornan 2000). A specimen collected at one time of the year or a specific location might be chemically and biologically different to one collected at another time or location. The New York Botanical Garden Index Herbariorum provides a list of accredited herbaria around the world, and this can be helpful. The value of herbarium material cannot be overstated and can even provide chemically important and reliable material for many decades. For example, a study analysing fresh leaf material of *Tetradium daniellii* compared extracts with a herbarium specimen collected in Yunnan province, China, in 1917 to verify the material and the chemistry of furanocoumarins was almost unchanged after 90 years (Stevenson et al. 2003).

Correct identification is critical but not always robust. The authors of this chapter have encountered many examples of incorrectly identified plant material being used in internationally funded research. For example, work published by Kestenholz et al. (2007) provided important work understanding how *Cassia sophera* (Leguminosae) could be used in storage protection against *Sitophilus oryzae,* but the plant material when originally sourced was described at *Cissus populnea* (Vitaceae), a quite different plant. Similarly, Stevenson et al. (2012) report perhaps the most widely used pesticidal plant material in Africa, *Tephrosia vogelii*, being incorrectly identified and promoted for use by farmers as *Tephrosia candida* and reported as such in numerous scientific papers over the past 20 years that is potentially highly confusing (Jama et al. 2008; Sileshi et al. 2005).

Increasingly, the application of organic solvent extracts is considered to adequately provide a step towards a better understanding of biological activities (Ortiz et al. 2012) but in reality does not necessarily inform any more than simply using powdered plant material and may even confuse our knowledge. More frustratingly some recent research indicates 'phytochemical screening' in the title but then actually simply assays extracts (Adeniyi et al. 2010; Udo 2012).

There are few reports of plant materials that do not work, but this is also important since the impression from the literature is that otherwise it seems all plants tested have activity (Baoua et al. 2012), while the majority are focused on laboratory bioassays and not on field trials.

3 Research Priorities for Improving Uptake and Use of Pesticidal Plants for Stored Product Pests

3.1 Understanding the Chemistry of Activity

As mentioned earlier, the use of pesticidal plants requires a strong understanding of the chemistry underlying the biological activities. At the very least this will help to understand variability in efficacy as determined by the testing of plants from different locations, but for most published research, this has been overlooked. This is required to enable intelligent enhancement of activities by improved application, a stronger understanding of the potential toxicities associated with plant material, improved harvesting strategies and potentially identifying new sources of chemicals where plants currently used are threatened through overharvesting or ecological limitations. Abundant plants with similar chemistry to a scarce but over harvested species could be promoted as an environmentally benign alternative.

The importance of understanding the chemistry is well illustrated by some recent research on *Tephrosia vogelii. Tephrosia* Pers. (Leguminosae) is a large genus of more than 350 species, many of which have important uses (Schrire 2005). *Tephrosia vogelii* Hook. f. is one of the best known of these species and is widely used in Africa (Kamanula et al. 2011; Nyirenda et al. 2011). The species is promoted for its ability to enrich soil both through biological nitrogen fixation but also as a green manure (Mafongoya and Kuntashula 2005; Sileshi et al. 2005; Sirrine et al. 2010) as well as for its pesticidal use and fish poisoning properties (Burkill 1995; Neuwinger 2004), although the latter application is now prohibited. As a consequence, it is cultivated widely by farmers on fallow land.

Much literature, particularly in unrefereed articles and on-line pamphlets, ascribe the biological activity of the species against storage pests to rotenoids and specifically to rotenone. Much of this is, however, assumed and unsubstantiated based on historic research identifying components in the plant that show the major rotenoids in the leaves are deguelin and rotenone (Irvine and Freyre 1959). While much 'grey' literature cites rotenoids in the leaves of *Tephrosia* to be insecticidal against stored product pests, surprisingly there is no published work to corroborate this. Indeed the bioactivity of *T. vogelii* against bruchids and weevils was even reportedly *not* associated with rotenoids according to Koona and Dorn (2005) although this was based solely on their finding that the biological activity of *T. vogelii* was associated with hexane extracts, and the authors assumed that hexane would not extract rotenoids.

Belmain et al. (2012) have, however, now shown categorically that rotenoids are indeed the biologically active compounds in the species to bruchids but report that rotenone itself plays only a relatively minor part in this effect compared to the much more abundant compound deguelin (Fig. 9.2). Interestingly, the structurally related rotenoid tephrosin was much less active than deguelin despite differing only in one additional hydroxylation indicating that not all rotenoids have similar activities, so assumptions about potential pesticidal efficacy must be avoided without evidence. Two other rotenoids identified in *T. vogelii* (sarcolobine and toxicarol) (Fig. 9.2) had similar biological activities to deguelin (Stevenson et al. 2012). The highest concentration of the active compounds is found in the leaf which makes it ideal for use since the foliage is the most abundant and sustainably harvested plant part. Besides this, it is relatively easily cultivated so is a species least likely to have any environmental impact if wild harvesting is not required. It is clear that the potential livelihood impact of this multiple use plant is compelling, and while it is well suited to resource-poor farming and is cultivated widely in southern and eastern Africa for soil improving qualities and as a pesticide, not all farmers reported that *Tephrosia* was a reliable pest control agent (Nyirenda et al. 2011). This could be because chemical content of leaves varied and so influences the pesticidal efficacy of *T. vogelii* since the activity of rotenoids is concentration dependent (Fang and Casida 1999). The occurrence of these compounds did vary dramatically among plant material sampled from 13 different locations in Malawi while deguelin and rotenone were in fact absent from approximately 25 % of sampled material (Stevenson et al. 2012). Indeed two distinct chemotypes have been proposed – one contains rotenoids which was the pesticidal one, while the second did not contain rotenoids but instead contained flavanones and flavones such as obovatin 5-*O*-methyl ether (Fig. 9.2), which was not pesticidal. This distinct chemical variation within a species likely explains the experience of some farmers who reported no pesticidal effect and highlights how important it is to understand chemistry since this enables us to understand why efficacy varies or is lost. Also – as has been the experience of farmers in Malawi – promotion of material for which an ascribed use is associated or simply presumed can seriously backfire if the material is not verified first.

This is as true for bioassays as assumptions made about chemistry. Earlier work asserting that the biological activity of *Tephrosia vogelii* was not associated with rotenoids based on an assumption that hexane extracts, which were active, could not contain rotenoids is misleading (Koona and Dorn 2005). Indeed, unpublished work in our laboratory indicates that hexane does extract rotenoids from *T. vogelii* leaves, and so rotenoids in this work almost certainly did account for the biological activities reported.

T. vogelii - chemotype 1 compounds

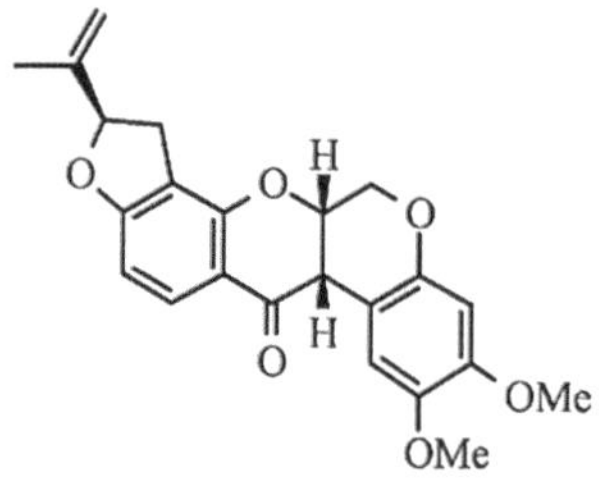

1 Rotenone

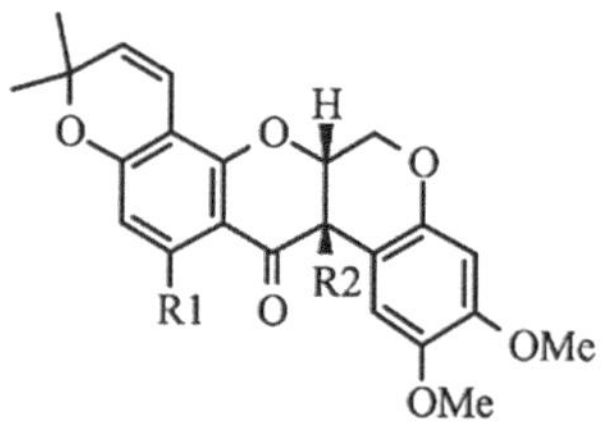

2 $R_1 = H, R_2 = OH$; Tephrosin
3 $R_1 = R_2 = H$; Deguelin
4 $R_1 = OH, R_2 = H$; α-Toxicarol

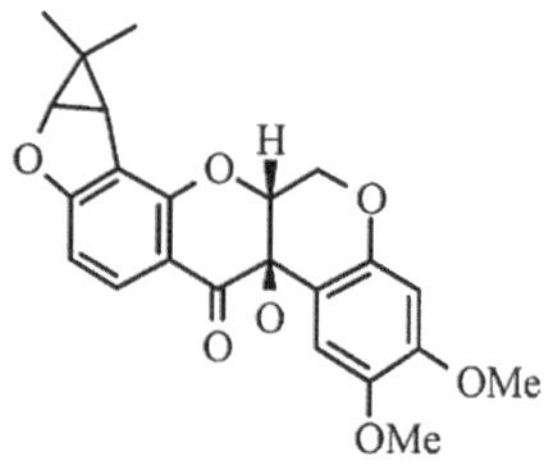

5 Sarcolobine

T. vogelii - chemotype 2 compounds

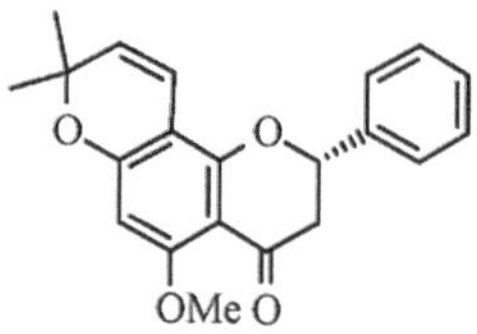

6 Obovatin-5-*O*-methylether

Fig. 9.2 The principal rotenoids and flavanones identified in chemotypes of *Tephrosia vogelii*

The solubility of the compounds from *Tephrosia vogelii* raises another important point. One of the benefits of understanding chemistry of activity is that current uses can be better understood and adapted with this new information. For example, some farmers report using *Tephrosia* extracts as a pesticidal spray but only have access to water as an extraction solvent. Rotenoids are only sparingly soluble in water and so this process is highly inefficient. Of course organic solvents are out of the question as they are unavailable and expensive and could potentially damage plants. Comparative analysis of the extraction efficiency of water carried out by Belmain et al. (2012) compared the extraction efficiencies of methanol with water and showed that it extracted 10.0 times as much deguelin as water. However, the extraction with Tween 20 increased the efficiency of water for rotenoids by almost three times using 1 % Tween 20 and by near 5 times using 5 % Tween 20. Thus, if farmers use *T. vogelii* extracts to treat grain or grain sacks, efficacy could be improved, and the amount of plant material required could be reduced by simply incorporating a detergent such as liquid soap which might additionally act as a surfactant and spreader and further optimise application and improve efficiency.

The compounds identified as biologically active in some plants have been reported to be water soluble such as triterpeneglycosides (saponins). These might be more sustainably exploited and more efficiently applied as water extracts than the typical application process in stored products – powdered plant material. In their search for new bioactive plant compounds against *Sitophilus oryzae*, Taylor et al. (2004) reported saponins from field pea extracts that were at least partly responsible for the anti-feedant effects of pea flour extracts. Elsewhere, saponins present in *Securidaca longepedunculata* (Polygalaceae) (Stevenson et al. 2009) have also been identified as contributing to the long-term biological activity of the roots of this species. *S. longepedunculata* (African violet tree) is a small tree occurring throughout sub-Saharan Africa. Farmers reported using the roots of this species for stored products in Ghana, and evidence indicates that *S. longepedunculata* was effective against *Rhyzopertha dominica*, *Callosobruchus*

Fig. 9.3 Securidacasides from roots of *Securidaca longepedunculata*

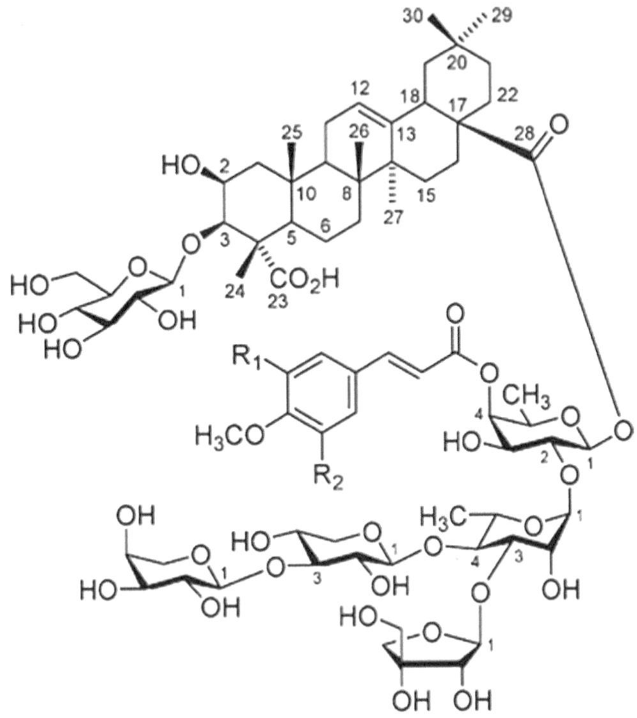

1 $R_1 = R_2 = H$; Securidacaside A
2 $R_1 = R_2 = OCH_3$; Securidacaside B

maculatus, *Sitophilus zeamais* and *Prostephanus truncatus* when compared to *Cassia sophera*, *Chamaecrista nigricans*, *Mitragyna inermis*, *Ocimum americanum* and *Synedrella nodiflora* which are other species reported in the Northern Region of Ghana to be used traditionally for pest control in stores (Belmain and Stevenson, 2001), and this is now supported by laboratory data (Jayasekera et al. 2003). When evaluating 33 West African species for toxicity to *C. maculatus*, Boeke et al. (2004a, b) also showed *S. longepedunculata* along with *Nicotiana tabacum* and *T. vogelii* to reduce F1 progeny of the beetles, while *Clausena anisata*, *Dracaena arborea*, *T. vogelii*, *Momordica charantia* and *Blumea aurita* were shown to be repellent. Surprisingly this study reported *Azadirachta indica* was attractive along with *Chamaecrista nigricans* and *Hyptis*. The roots of *S. longepedunculata* also have a characteristic odour of wintergreen, and the principal volatile compound in the root was shown to be methyl salicylate (Jayasekera et al. 2002), and this compound was shown to have toxic effects against several beetle species (Jayasekera et al. 2005) and could enhance the effects of the saponins reported above. Water-soluble compounds could be used to reduce the amount of plant material required by extracting the chemicals and so using plants more efficiently. We conducted a trial on a farm in Zambia to determine whether applying *S. longepedunculata* root bark was more efficient. Farmers reported they needed around half as much plant material but also insisted that they would never use the method as it was too time consuming compared to simply grinding up the roots and adding to their grain stores (Zulu and Stevenson, unpublished). This highlights another important condition of optimising the use of plant materials. Any improvement must be trialled and accepted by farmers to have any chance of being adopted (Fig. 9.3).

The chemistry of some plants varies according to the season, the plant age and location, so chemical analysis is essential to determine the best time to harvest or for identifying elite material for propagation (Sarasan et al. 2011). Of course the need to use fresh plant material over dry might determine harvesting priorities but chemical knowledge of the active components would certainly optimise harvesting times. The ability of many African laboratories to carry out this analytical work is limited to a handful of institutes. Cooperation through research networks is essential to move this area of work forward, and expertise and equipment in natural product chemistry need to be expanded.

Chemical analysis using tandem-linked detection techniques such as mass spectrometry can maximise information acquisition of compounds in each plant and can identify many components quickly, particularly seasonal variations and genotypic differences between specimens of a single species to better understand the potential variation in efficacy. These techniques can also be used to authenticate specimens before promoting widely. While these facilities are beyond the reach of many research institutes, many materials could be authenticated by less costly techniques such as thin layer chromatography.

3.2 Understanding the Target Pest

As with ensuring correct identification of plant material, it is important to ensure that the insects on which a biopesticide or potentially efficacious plant preparation is being tested really are the species they are supposed to be. Although in many cases, similar insects (e.g. *Sitophilus zeamais* and *S. oryzae*) will mostly respond similarly to biopesticides, subtle differences in ecological niches between species can result in differential behavioural or physiological responses to control measures. For example, Champ and Cribb (1965) found that *S. oryzae* was more susceptible to the insecticide diazinon and ronnel than *S. zeamais*, and *S. zeamais* is also more resistant to low temperatures than *S. oryzae* (Nakakita and Ikenaga 1997). It follows that testing of plant-derived pesticidal products may show similar susceptibility differences between closely related species.

Understanding the biology of the pest is key to understanding how to control it. In the case of storage pests, the insect species predate human agriculture and grain or legume storage in man-made structures, so the insect's ecology before that will provide keys to management strategies. For example, some cereal pests such as *Prostephanus truncatus* and *Rhyzopertha dominica* were originally wood-boring beetles which adapted to exploit grain stores, so their responses to odours have more in common with other wood-borers (e.g. no innate response to cereal odours (Fadamiro et al. 1998; Nguyen et al. 2008)).

It is easy to look at a sack of infested grain, legumes or other stored product and see only a mass of insects to be tackled in bulk, and indeed this is the approach taken in many studies. However, when devising control strategies it is more helpful to visualise the population as a collection of *individuals*, some male, some female, all at different life stages, and thus not all responding identically. They interact with one another but also behave independently – and diversely – in response to some cues. This is especially true in terms of behavioural syndromes associated with responses to odours, and odour-based repellency can be a key component of effectiveness for some pesticidal plants. It is well known that males and females of many insect species exhibit differential responses to sex pheromones; this has been observed in many groups of insects, including moths (Matsumoto and Hildebrand 1981) (though see Palanaswamy and Seabrook (1978)), beetles (Ukeh et al. 2008), Hemiptera (Ondarza et al. 1986; Manrique and Lazzari 1995) and braconid wasps (Kimani and Overholt 1995). However, equally, males and females do not always respond the same way to aggregation pheromones (Byers 1983; Walgenbach et al. 1983; Ondarza et al. 1986) or odours associated with host material (e.g. fresh plant matter, cereals, mammal odours). Sex differences in responses to host material have been recorded across the Insecta, in moths (Hansson et al. 1989), parasitoid wasps (Bouchard and Cloutier 1985), Hemiptera (Chinta et al. 1994; Wenninger et al. 2009), beetles (Zhu et al. 1999; Ukeh et al. 2008), tsetse (Otter et al. 1991), etc.

Sometimes this is related to differences in lifestyle habits between the sexes, for example, if the females seek host material for oviposition rather than personal consumption or in species where the female requires a blood meal, but it can also occur in species where both sexes feed on the same material. Consequently, bioassays testing unsexed individuals run the risk of assuming that males and females are controlled equally well without actually proving this to be the case.

Recent research has found differences in male and female responses of the cowpea weevil, *Callosobruchus maculatus*, to both host odours and the plant-derived repellent, methyl salicylate (Arnold et al. 2012). Females showed a significantly stronger preference than males for infested cowpea material (Fig. 9.4). These findings make evolutionary sense as females must seek safe host material for their offspring to consume during development and therefore are under pressure to use olfactory cues to aid this process. Males, conversely, seek virgin females and so have no particular reason to seek out uninfested cowpea as adults; however, they may target heavily infested material as virgin females may emerge from it.

Differential responses to the plant-derived repellent methyl salicylate are also apparent in the findings, with older, inactive-morph females repelled by methyl salicylate but weaker responses from some other subgroups of *C. maculatus*. As methyl salicylate is an active component of pesticidal plants such as *Securidaca longepedunculata*, etc., the finding that it is not equally reliable as a repellent of male and female *C. maculatus* is a notable one and should be taken into account when devising application strategies.

However, it is not just the sex of an insect that can determine its odour-mediated responses. Several insect species are dimorphic, with morphs better adapted for static breeding or for dispersal – this is particularly common in aphids (alate and apterous morphs) and also bruchid beetles (in which the morphs are termed 'active' and 'inactive' or 'flight' and 'flightless', etc.). Jaba et al. (2010) found responses of *Aphis craccivora* alates were higher to odours of fresh cowpea but lower to odours of *Lablab dolichos* compared with the responses of apterous individuals (Jaba et al. 2010). Equally, different sensory physiology has been observed in solitary and gregarious locusts, with a physical change in the numbers of sensilla on antennae when they switch morphs (Greenwood and Chapman 1984).

C. maculatus has both an 'inactive' morph, which is virtually flightless and short-lived, but the females have high fecundity, and an 'active' morph, which can fly and has a longer lifespan, but has limited fecundity (Caswell 1960; Utida 1972). Production typically switches to the active morph in response to extremes of temperature, photoperiod or humidity, degradation of the larval host material or overcrowding (Messina and Renwick 1985; Utida 1972). While both forms can potentially be the source of an infestation, it is typically expected that the active form will infest beans still in the field by flying in, whereas the inactive form is more likely to be introduced on infested material or emerge later from beans infested in the field. In spite of this, many laboratory tests on this species do not differentiate the morphs (and some long-term lab cultures are believed to produce only active morph adults). The differences in their typical lifestyles suggest that it might be expected that the morphs would respond differentially to host odours. Arnold et al. (2012), confirmed this, discovering that morph interacted with age as a factor to affect preferences for host odours and that inactive individuals were more attracted to dried cowpea than were the active morph individuals (Fig. 9.5). What is less easily anticipated but also extremely relevant for control based in part on repellency is that we found a similar differential response between morphs to the repellent odour, methyl salicylate. In their four-arm olfactometer experiment, the inactive morph was strongly repelled by the odour, whereas the active morph was indifferent (Fig. 9.6). This means that control strategies based upon the repellent properties of methyl salicylate or plant material containing it are best deployed to target the inactive morph, for example, within the stores and at ground level, and other methods may be required to ensure control of the active morph in fields and above-ground.

Further work also showed that sensitivity to odours can change depending on the age of the

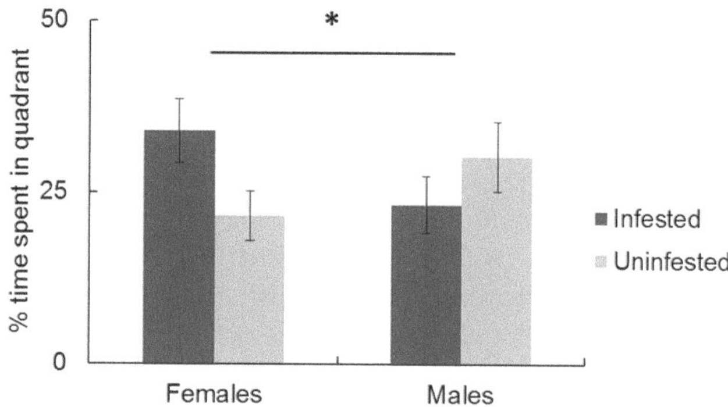

Fig. 9.4 Preferences of female and male *Callosobruchus maculatus* for insect-infested and uninfested cowpea material, tested in a four-arm olfactometer. 25 % indicates no preference (Data from Arnold et al. (2012))

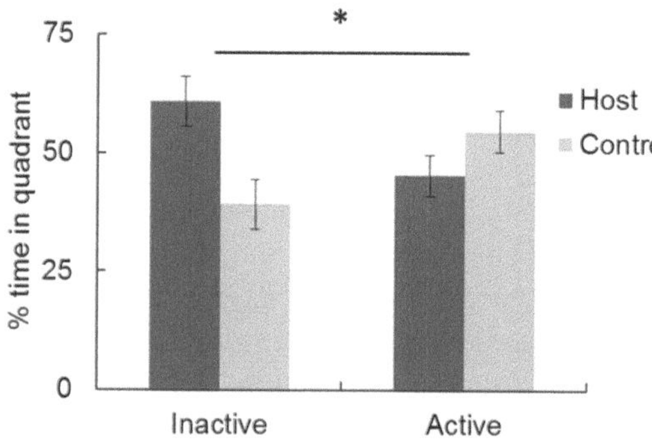

Fig. 9.5 Time spent in the presence of host (cowpea) and control (clean air) odours by inactive and active morph individuals of *C. maculatus* in a four-arm olfactometer with two arms containing host odours and two arms control odours. 50 % indicates no preference

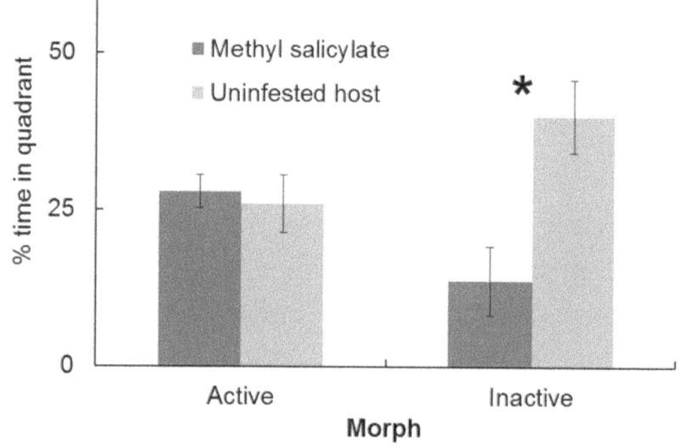

Fig. 9.6 Responses of active and inactive morphs of *C. maculatus* to the odour of 1 mg/ml methyl salicylate versus host odours in an olfactometer experiment. 25 % indicates no preference

insect, and therefore older as well as newly emerged individuals should always be tested when evaluating the effectiveness of a repellent or bait in order to ensure that acclimation to an odour does not occur. Furthermore, it is worthwhile comparing a variety of strains of the pest insect to evaluate whether behavioural and physiological susceptibility to a biopesticide is consistent for all subpopulations within the species or whether it is variable. Given the capacity of various stored product pests to evolve insecticide resistance (Champ and Cribb 1965; Champ and Dyte 1976; Collins et al. 1993), the possibility of some strains eventually evolving either physiological resistance or behavioural tolerance to biopesticides or plant-based repellents is not unthinkable. Characterising interstrain sensitivity to biopesticides will permit geographical optimisation of application of the plant material – as a hypothetical example, there would be little point promoting neem for control of *P. truncatus* in India if only the African strain of the insect were proven susceptible. This can equally apply to plant-derived repellents and deterrents: different strains of mosquitoes have been found to show diverse levels of repellency when exposed to diethyl toluamide (DEET) (Rutledge et al. 1978), so we may see strains of insect for which some ordinarily effective plant volatiles or essential oils are simply not deterrent, and it may be short-sighted to assume geographical homogeneity in odour responses.

Finally, behavioural responses can also be determined by aspects of physiological status. Malaria-infected mosquitoes have been shown to be more responsive to host odours (Smallegange et al. 2013), and likewise, feeding and parity status have been shown to affect both their attraction to host odours (Klowden et al. 1996) and their repellency responses to pyrethroid insecticides

(Chareonviriyaphap et al. 2006). Mated and unmated individuals of several insect species (Mechaber et al. 2002; Wenninger et al. 2009) likewise have been shown to exhibit differential responses to odours. This includes the stored product pest *Tribolium castaneum* – in one study (Fedina and Lewis 2007), unmated individuals consistently responded more strongly to a lure based on synthetic aggregation pheromone than did mated individuals. Feeding status also influenced their olfactory responses. Various bruchids (Leroy et al. 1999) are found to have changes in olfactory sensitivity according to mating status, and furthermore, there were found to be annual cycles in the beetles' responses to host odours. It is therefore advisable to take into account possible effects of mating status, for example, by testing both mated and virgin individuals, and of feeding status by rearing all insects on consistent medium and controlling for the period of starvation before testing.

Mass-testing protocols have their value: they allow a large number of insects to be tested in a short space of time and, under some circumstances, including the conspecific interactions may be a more accurate reflection of insect behaviour in the field. However, it is not always so simple. If, within a species, thresholds for responding to an attractive odour vary, are those 100 insects all sitting in the area of odour because they are all attracted to the test odour, or did only one of them truly respond to the test odour and the rest simply flocked to be close to their conspecific, as a result of aggregation pheromone or otherwise? This would have implications for trapping, as a lure to which only 1 % of insects respond, would be of limited use. One could argue that if the other insects follow the first into the trap, it does not matter what they are following, but what if the first insect dies and then lacks the attractiveness, so the remaining 99 % do not respond at all? Consistent individual responses may be a more reliable clue to the effectiveness of a strategy in many cases. Although time consuming, when testing individual insects the quality of the data will be higher, and it is possible to control for multiple factors (age, sex, strain, morph, mating status).

Care must be taken when performing laboratory evaluations of the pests' responses to odours using some items of equipment. It is easy to predict that when an insect is presented with an airstream containing an attractive odour, the insect will walk or fly upstream towards the odour source. However, how will the same insect react if the odour is aversive, such as repellent pesticidal plant material? Will it simply stop walking? Will it move perpendicular to the direction of airflow? Will it turn around and fly downstream? The unpredictability of these responses makes it hard to use some pieces of equipment to investigate repellent odours. An example is the locomotion compensator. While *S. zeamais* shows clear aversion to methyl salicylate in the four-arm olfactometer (Jayasekera et al. 2005) and clear attraction to maize odours in both the olfactometer (Fig. 9.7) (Ukeh et al. 2010, 2012) and on a locomotion compensator (Fig. 9.8a), it does not respond at all to methyl salicylate on the same piece of equipment (Fig. 9.8b). Similarly, in a Y-tube olfactometer, while if one arm is attractive and the other a blank control, it can be assumed that most insects will fairly reliably walk up towards the fork in the tube and then make a decision, ordinarily turning towards the attractive odour; what if one of the arms has a repellent odour? In that case, the repellent odour is drawn through the 'error' arm, but then onwards through the approach arm, so even at the start of the experiment when the insect is in the approach arm, unless it is established that airflow is perfectly laminar, it is receiving a stream of air tainted with a repellent odour. Why, then, should we expect it to meekly walk into this repellent odour stream all the way to the decision point at which the olfactometer branches before deciding to turn away from the scent?

3.3 Safety of Use of Pesticidal Plants

Natural does not necessarily equal safe, despite frequent implication of this in the modern comprehension of 'healthy'. Some plants are, in fact,

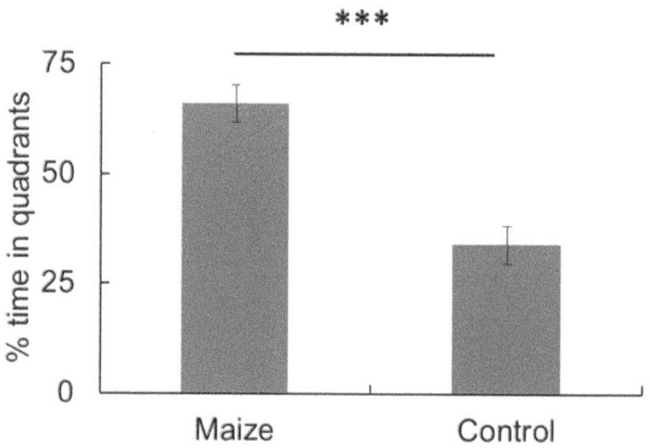

Fig. 9.7 Responses of *Sitophilus zeamais* to maize odours in a 4-arm olfactometer (2 arms with odour, 2 arms with clean air only; 50 % indicates no preference)

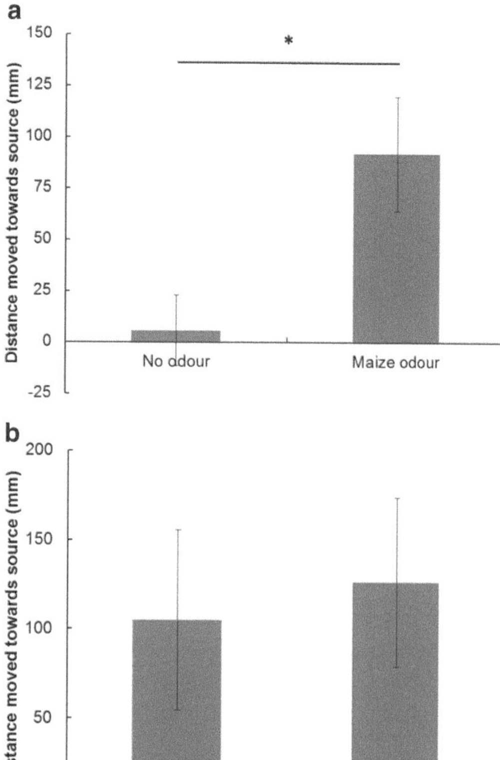

Fig. 9.8 Responses by two groups of *S. zeamais* to (**a**) maize odours and (**b**) methyl salicylate on a locomotion compensator, showing the distance walked by individuals towards the odour source either in the absence or presence of the odour

extremely toxic, such as *Taxus* spp., *Aconitum* spp. and, perhaps most notorious of all, *Ricinus communis* (Bonnici et al. 2010; Hernandez et al. 2010; Kuca and Pohanka 2010; Kolev et al. 1996). Plant materials used as pesticides are toxic, at least to insects, so safe use is particularly important in this respect as they are used to treat stored foods when used to control storage pest insects. Yet inadequate work has been published internationally on the vertebrate toxicity of African pesticidal plants. This may be because the costs of commercialising pesticidal plants are prohibitive, so there are no requirements to test them officially. Some commonly found plants used by farmers in Ghana have been shown to affect mammalian growth and development (Belmain et al. 2001).

Some plants consumed as an ingredient in food and drinks indicate a degree of safety and are the basis for the promotion of essential oil products in North America (Isman 2006). However, even where plants are used as food or drink, they may still show some toxicity. *Lippia javanica,* for example, is a popular treatment for fever in Southern Africa where it is drunk as a green tea. This species also showed acute oral toxicity in mice at very high concentrations (Madzimure et al. 2011). Nyahangare et al. (2012) also evaluated toxicity associated with *Strychnos spinosa* and *Bobgunnia madagascariensis* fruits and the foliage of *Vernonia amygdalina* and *Cissus quadrangularis* and found the first two to be toxic when administered at high doses but the latter two much less so.

Plants are typically applied as pesticides using amounts that present toxic components at low levels and so rarely pose any significant acute toxicity to users (Isman 2008) even when this is intended (Chesneau et al. 2009). Belmain et al. (2012) report that to suffer a lethal dose from consuming *Tephrosia vogelii,* a fully grown man would need to consume between 2 and 20 kg of dried leaf at one sitting.

Postharvest insect pest control in small-holder farming in Africa relies primarily on pirimiphos-methyl, an organophosphate that targets anti-cholinesterase. Where pest complexes include *Prostephanus truncatus*, the larger grain borer, pirimiphos-methyl is combined with permethrin (Sekyembe et al. 1993). Safety sheets for Actellic EC50 (pirimiphos-methyl) indicate moderate toxicity to mammals and extreme toxicity to aquatic invertebrates and are highly toxic to algae and fish (Syngenta 2006). Thus, fears over the toxicity of plant species like *Tephrosia vogelii* that are very popular alternatives to pesticides in southeastern Africa (Kamanula et al. 2011; Nyirenda et al. 2011), but have known toxicity to fish, are perhaps over cautious. A no cost but effective plant-based alternative should be considered a viable option for very poor farmers, but without commercialisation – which is unlikely in the current climate (Sola et al. 2013) – the wide-scale uptake seems unlikely. The best approach could be through the development of better propagation protocols and widespread cultivation.

A possible solution may be to develop ways to apply pesticidal plants that minimise their contact with the actual foodstuff, either by layering the plant and the food alternately or by applying the plant to the container or cover, rather than mixing it into the food itself. Novel strategies for application or methods of synergising more and less toxic plants to ensure continued efficacy with low risk are promising directions for future research.

3.4 Sustainable Harvesting and Local Production of Pesticidal Plants

Demand for some plants – particularly medicinal species – is outstripping supply, driven largely by the fact that, even today, over 80 % of the world's population are dependent upon medicinal plants for at least part of their healthcare (Harnischfeger 2000; Shackleton et al. 2005). Overgrazing, conversion to farmland and bush fires also reduce potentially productive land for harvesting useful plants from the wild. The use of pesticidal plants, which typically requires very much larger quantities of material than medicinal uses, when collected from the wild is only sustainable if small numbers of people use the plant or if the plant is abundant and ubiquitous or propagated easily. Access to sufficient material is perhaps the greatest constraint to wider uptake of pesticidal plant technologies among farmers; thus, ways to propagate and cultivate plant material will address this gap. If knowledge about efficacy can be determined and provenance selected based on chemical analysis, then propagated material can also be provided with a greater expectation of consistency in efficacy – so overcoming one the other major hurdles to wider uptake (Sarasan et al. 2011). Surprisingly little research is invested in aspects of propagation for pesticidal plants, with the exception of Pyrethrum, and this is largely for processing the active ingredient. However, medicinal plants and ornamental species are well studied to improve propagation (Gupta et al. 2012; Pijut et al. 2012), provide a greater pool of knowledge and be sourced for useful information where a species may provide pesticidal as well as ornamental or medicinal properties.

Propagation of medicinal plants is not always straightforward since conditions for successful germination vary considerably among species, and seeds may need some ecological hardship to induce germination such as long periods of dry weather or fire, in which cases artificial mechanisms may need to be used to overcome this – such as cutting the seed testa. Recent work has shown that some species can be propagated from seeds which could be important to establish large-scale cultivation of importantly pesticidal plants, particularly the tree species *Securidaca longepedunculata* and *Bobgunnia madagascariensis* (Thokozani et al. 2011; Zulu et al. 2011). Informed timing of seed collections is required and knowledge about seed drying, storage and germination are also important considerations.

In many cases, the part of the plant collected makes the technology highly unsuited to sustainable ecosystem management – for example, roots, bark, seeds or entire plants (Belmain and Stevenson 2001; Stoll 2000). Harvesting root bark, which may be as a consequence of following a scientifically unvalidated tradition, will likely kill the plant. Some research may identify active compounds in other plant parts such as the stem bark, which could be harvested more sustainably. Modified root harvesting such as only taking lateral roots could reduce the impacts on abundance. Harvesting times can be important too. The chemistry of plants varies according to location and season (Stevenson et al. 2012; Belmain et al. 2012), and, since their efficacy as pest control agents depends on these variable chemistries, harvesting times can be crucial in optimising their efficacy.

Some plants produce the secondary metabolites that account for their pesticidal activity only when subjected to biotic or abiotic stressors, and this has not been investigated for many indigenous pesticidal plants. As an example from wider research, *Nicotiana attenuata* produces trypsin protease inhibitors to protect against insect attack, but only at high levels in its vegetative tissues when attacked by herbivores (Dinh et al. 2013; Skibbe et al. 2008). Knocking out the genes involved in inducing this defence leads to extreme vulnerability to pest attack. Such inducible defences allow the plant to save energy, as production of secondary metabolites is costly, so breeding a pesticidal plant for constitutive expression of the compounds in such cases may result in less vigorous plants. In other cases, the induction of production of active ingredient may be initiated by an abiotic stressor such as drought or photoperiod change (Ramakrishna and Ravishankar 2011).

If propagation can be established for species, then marketing plants as an income-generating opportunity for small-scale farmers may be a realistic ambition (Sarasan et al. 2011) while developing a way to upscale and promote a technology (Moyo et al. 2011). Technology outreach is limited by funding typically of short projects. If sustainable promotion can be driven by the incentive of income generation and a formalisation of pesticidal plant use for agricultural pest management in Africa, then the chance of wider uptake is increased (Grzywacz et al. 2014), particularly if government and non-governmental organisations continue to promote the use of wild plants. Demand for pesticidal plants will continue to grow, which can only realistically be met through their cultivation and marketing (Sola et al. 2013). The limitations of regulatory procedures could be the undoing of any potential pesticidal plants as commercial products other than for those already established such as Pyrethrum. The regulatory processes are prohibitively expensive for small- and even medium-scale facilities and require any pesticidal product to satisfy the same level of interrogation regarding efficacy, toxicity and environmental hazard as synthetic products. This is, of course, right but the costs of producing most of the information are prohibitive and will continue to hold back wide-scale uptake. Conversely, the sale of herbal remedies in South Africa which are in many cases materials prescribed for oral consumption is unregulated, whereas a plant sold as a pest control product needs to be registered.

4 Pesticidal Plants for Stored Products in Southern Africa: Recent Developments for Selected Pesticidal Plants

Our recent work has combined surveys (Kamanula et al. 2011; Nyirenda et al. 2011) and databases (e.g. http://epic.kew.org) and has identified pesticidal plants of value as indicated by farmers themselves with interest particularly in Caesalpinioid (Miombo) woodlands. This list is not intended to be comprehensive but updates the research on some of the many species of interest in stored products protection while illustrating issues of importance to all pesticidal plant species. Some of the species reported by farmers are not supported by literature and so present new research opportunities. Having said that, there is an increasing argument towards better use of what we already have rather than investing time in evaluating more and more new material.

For example, farmers have reported to the authors (Stevenson, P. unpublished) that *Euphorbia tirucalli*, is effective against stored product pests and this activity is widely known among farmers in southern Africa. However, the scientific literature only reports activity in this plant against two species of mosquitoes (Rahuman et al. 2008; Yadav et al. 2002) suggesting that its potential to be use against storage pest insects has been overlooked by the research scientists. The latex of this species has potent skin irritant properties (Kinghorn 1978) which may be associated with the plant's activity and in practice may also dissuade use. Elsewhere, *Aloe ferox,* is reportedly burned and the leaf ash used to treat stored grain in Zimbabwe and Zambia (Phosiso Sola, Pers Comm.), but there is no evidence to support the efficacy of this use. Another species reportedly used for stored product pest control is *Solanum incanum* but only recently has any evidence to support activity against arthropods published. Nyahangare et al. (2013) reports that the plant has biological activity against cattle ticks in field trials which show a potentially promising opportunity, but the species is also identified as potentially toxic to mammals (Nyahangare et al. 2012).

Many species reportedly used among farmers include non-native species such as *Cymbopogon nardus* and *C. citratus* and neem (*Azadirachta indica*). These species are some of the most well-studied species in the literature for their anti-insect activities and still provide research outcomes today (Chebet et al. 2013; Jiang et al. 2012a). *A. indica,* however, provides an important lesson in promotion of plant materials for pesticidal use. Most farmers we have encountered who report using neem use leaves presumably because the leaves are easily available and around all the time. But they are very low in their bioactive components (Koul and Dhaliwal 2004). Furthermore, in some highland regions of Southern Africa, the climate is too cold to allow flowering in this species *Azadirachta indica*; thus, the trees do not produce seeds (kernels) where the greatest quantities and diversity of pesticidal and deterrent compounds can be found. *Lantana camara* is another exotic species which is popular in some parts of Africa, e.g. Kenya and Algeria (Chebet et al. 2013; Zoubiri and Baaliouamer 2012). However, this plant does not appear to be used in Miombo regions of Southern Africa where it is considered highly invasive and receives high priority for local eradication programmes. Perhaps its use more widely could help control its spread by giving farmers a reason to harvest the plants.

Powders and essential oils of *Cinnamomum camphora*, *Ocimum basilicum* and *Chenopodium ambrosioides* and seeds of *Pimpinella anisum* were shown to be insecticidal against *Tribolium granarium* and *T. castaneum* (Nenaah and Ibrahim 2011), while Chu et al. (2011) have reported fumigant activity of *Chenopodium ambrosioides* L. against *Sitophilus zeamais* and further report five compounds ((Z)-ascaridole, 2-carene,rho-cymene, isoascaridole and alpha-terpinene) of which (Z)-ascaridole recorded a LD50 against *S. zeamais* adults of 0.84 mg L^{-1} air with contact toxicity of 0.86 μg g^{-1} body. This species is easily propagated and as the active compounds are volatile (ascaridole) despite their being toxic should be easily evaporated from commodities and therefore pose little risk to consumers. However, applicators may need to take care.

Neorautanenia mitis (Leguminosae) is related chemically to *Derris*, *Lonchocarpus* and *Tephrosia* spp. The large underground tuber, which can weigh tens of kilos, has high concentrations of rotenoids, isoflavones and pterocarpans (Sakurai et al. 2006) and is relatively easy to propagate, thus like *Tephrosia* is conducive to cultivation by farmers. The principal active component is rotenone (Stevenson PC unpublished) although earlier studies have identified a variety of pterocarpans with potential antifungal activity (Sakurai et al. 2006) and earlier still reporting neotenone as the principal isoflavonoid component (Vanpuyvelde et al. 1987). Rotenone is well studied and there is much safety data on this compound, so in theory *N. mitis* is a good candidate for commercialisation. *N. mitis* roots are effective against a wide range of insects including the important mosquito species *Anopheles gambiae* and *C. quinquefasciatus* mosquitoes, with activities comparable to deltamethrin and cypermethrin (Joseph et al. 2004). It is therefore surprisingly underutilised in pest control. Chimbe and Galley (1996) reported the petroleum extract of the plant material against *Sitophilus oryzae* and *Prostephanus truncatus,* but it was less effective than *Dicoma sessiliflora*. Rotenoids are soluble in this solvent so probably account for the activity.

Bobgunnia madagascariensis (Leguminosae) is reportedly used for protection of stored products from beetles in Zambia and other parts of Southern Africa. There is little scientific evidence to support this, however, despite convincing evidence for its effects against molluscs (Borel and

Hostettmann 1987; Kone et al. 2004; Marston et al. 1993). Elsewhere ethyl acetate extracts of the pods were effective against whiteflies and mosquitoes (Georges et al. 2008; Minjas and Sarda 1986), while its anti-feedant and toxic effects are reported against *Heliothine* moth larvae showed and repellent to termites (Crombie et al. 1971). The activity is most likely caused by the presence of saponins which occur in the pods and bark (Marston et al. 1993; Stevenson et al. 2010) since the only other components found in the pods are highly glycosylated flavonoids which are not biologically active to insects (Stevenson et al. 2010). The presence of these saponins, however, does vary between locations or provenance; thus, elite materials are required for propagation (Sarasan et al. 2011).

The genus *Lippia* (Lamiaceae) is used as a medicinal tea against the symptoms of fever, flu and cold (Viljoen et al. 2005). Because this oral use suggests a low acute toxicity to mammals this species is likely to be a good material for upscaling. Madzimure et al. (2011) reported its activity against cattle ticks, while repellency to mosquitoes has been reported (Lukwa et al. 2009; Omolo et al. 2004). The repellency was reported to be due to essential oils including perillyl alcohol, cis-verbenol, cis-carveol, geraniol, citronellal, perillaldehyde and caryophyllene oxide. Several related species are also known to be effective. For example, *Lippia alba* has been reported to be a potent deterrent to *Tribolium castaneum* with benzyl benzoate, β-myrcene and carvone reported to be responsible for the effects (Caballero-Gallardo et al. 2011).

5 Conclusion

Pesticidal plants offer a traditional and economically viable and effective alternative to pesticides for the control of insect pests in stored products. Much research in the last decade has increased our knowledge about new materials and some details about mechanisms of activity, and this work should focus on fewer effective species to develop improved use. However, to enable the uptake and facilitate wide adoption, several aspects of the sector need to be developed. An evaluation of science and technology policies towards pesticidal plants needs to be conducted through multi-stakeholder networks on pesticidal plants. This should provide clear policy guidelines outlining opportunities and hurdles to upscaling the use of optimised pesticidal plant technologies in stored products.

Sustainable production of plant pesticides through commercialised propagation and cultivation needs promotion and scientific support particularly with the selection of elite propagating material. Scientists and nursery growers need to be trained in propagation and innovative application protocols for indigenous pesticidal trees and shrubs while optimising chemical consistency for established species such as Pyrethrum. Harvesting protocols and optimised preparations are required with clear guidelines on which insects certain plants are effective against. Without strong guidance, assumptions about efficacies can lead to ineffective use.

Science and technology innovations for safer and more effective application of pesticidal plants need to be developed and promoted to farmers. Elite materials need to be identified through analysis and biological evaluations, while application procedures need to be developed and promoted to farmers along with safe handling guidance. Finally scientific networks need to work together to ensure that as broad a skill base is in place to drive the research forward.

Pesticidal plants have been an important part of traditional pest management practice by farmers in Africa. We believe they should remain so and be made available as widely as possible. To ensure a future for pesticidal plants, the many bottlenecks need to be addressed by the scientific community, policy makers and institutions involved in research and implementation. Better information is needed to explain how plants work, which pests are appropriate targets and how variability may be overcome with respect to season, locality or variety and best practice for harvesting and application. Furthermore, scientists need to engage with policy makers to tackle conservation issues. If African Governments wish to see the widespread use of pesticidal plants, they need to be encouraged through policy. A strong future for indigenous knowledge and use of pesticidal plants in Africa will need policy changes to be made to current regulatory frameworks. Scientists and policy makers must work together for safety and to develop simple reasonably priced regulation, particularly for plants already being widely used. Improving our knowledge on variability of efficacy, conservation and regulation remains the big challenge, but we are sure this will increase farmer use and ultimately improve livelihoods through improved food security.

References

Adeniyi SA, Orjiekwe CL, Ehiagbonare JE, Arimah BD (2010) Preliminary phytochemical analysis and insecticidal activity of ethanolic extracts of four tropical plants (*Vernonia amygdalina, Sida acuta, Ocimum gratissimum* and *Telfaria occidentalis*) against beans weevil (*Acanthoscelides obtectus*). Int J Phys Sci 5:753–762

Ali A, Ahmad F, Biondi A, Wang YS, Desneux N (2013) Potential for using *Datura alba* leaf extracts against two major stored grain pests, the khapra beetle *Trogoderma granarium* and the rice weevil *Sitophilus oryzae*. J Pest Sci 85:359–366

Amoabeng BW, Gurr GM, Gitau CW, Munyakazi L, Stevenson PC (2013) Tri-trophic insecticidal effects of African plants against cabbage pests. PLoS One 8(10):e78651

Arnold SEJ, Stevenson PC, Belmain SR (2012) Odour-mediated orientation of beetles is influenced by age, sex and morph. PLoS One 7:e49071

Baoua IB, Amadou L, Margam V, Murdock LL (2012) Comparative evaluation of six storage methods for postharvest preservation of cowpea grain. J Stored Prod Res 49:71–175

Belmain SR, Stevenson PC (2001) Ethnobotanicals in Ghana: reviving and modernising an age-old practise. Pestic Outlook 6:233–238

Belmain SR, Neal GE, Ray DE, Golob P (2001) Insecticidal and vertebrate toxicity associated with ethnobotanicals used as post-harvest protectants in Ghana. Food Chem Toxicol 39:287–291

Belmain SR, Amoah BA, Nyirenda SP, Kamanula JF, Stevenson PC (2012) Highly variable insect control efficacy of *Tephrosia vogelii* chemotypes. J Agric Food Chem 60:10055–10063

Boeke SJ, Barnaud C, Van Loon JJA, Kossou DK, Van Huis A, Dicke M (2004a) Efficacy of plant extracts against the cowpea beetle, *Callosobruchus maculatus*. Int J Pest Manag 50:251–258

Boeke SJ, Baumgart IR, Van Loon JJA, Van Huis A, Dicke M, Kossou DK (2004b) Toxicity and repellence of African plants traditionally used for the protection of stored cowpea against *Callosobruchus maculatus*. J Stored Prod Res 40:423–438

Bonnici K, Stanworth D, Simmonds MSJ, Mukherjee E, Ferner RE (2010) Flowers of evil. Lancet 376:1616

Borel C, Hostettmann K (1987) Molluscicidal saponins from *Swartzia madagascariensis* Desvaux. Helv Chim Acta 70:570–576

Bouchard Y, Cloutier C (1985) Role of olfaction in host finding by aphid parasitoid *Aphidius nigripes* (Hymenoptera: Aphidiidae). J Chem Ecol 11: 801–808

Bridson D, Fornan L (2000) The herbarium handbook. Royal Botanic Gardens, Kew

Burkill HM (1995) The useful plants of West Tropical Africa, vol 3, 2nd edn. Royal Botanic Gardens, Kew

Byers J (1983) Sex-specific responses to aggregation pheromone: regulation of colonization density in the bark beetle *Ips paraconfusus*. J Chem Ecol 9:129–142

Caballero-Gallardo K, Olivero-Verbel J, Stashenko EE (2011) Repellent activity of essential oils and some of their individual constituents against *Tribolium castaneum* Herbst. J Agric Food Chem 59:1690–1696

Caswell GH (1960) Observations on an abnormal form of *Callosobruchus maculatus* (F.). Bull Entomol Res 50:671–680

Champ BR, Cribb JN (1965) Lindane resistance in *Sitophilus oryzae* (L.) and *Sitophilus zeamais* Motsch. (Coleoptera, Curculionidae) in Queensland. J Stored Prod Res 1:9–24

Champ BR, Dyte CE (1976) Food and Agriculture Organization of the United Nations, Report of the FAO global survey of pesticide susceptibility of stored grain pests. Food and Agriculture Organization of the United Nations, Rome. http://trove.nla.gov.au/work/9217354?selectedversion=NBD1565626

Chareonviriyaphap T, Kongmee M, Bangs MJ, Sathantriphop S, Meunworn V, Parbaripai A, Suwonkerd W, Akratanakul P (2006) Influence of nutritional and physiological status on behavioral responses of *Aedes aegypti* (Diptera: Culicidae) to deltamethrin and cypermethrin. J Vector Ecol 31:89–101

Chebet F, Deng AL, Ogendo JO, Al K, Bett PK (2013) Bioactivity of selected plant powders against *Prostephanus truncatus* (Coleoptera: Bostrichidae) in stored maize grains. Plant Protect Sci 49:34–43

Chesneau P, Knibiehly M, Tichadou L, Calvez M, Joubert M, Hayek-Lanthois M, De Haro L (2009) Suicide attempt by ingestion of rotenone-containing plant extracts: one case report in French Guiana. Clin Toxicol 47:830–833

Chimbe CM, Galley DJ (1996) Evaluation of material from plants of medicinal importance in Malawi as protectants of stored grain against insects. Crop Prot 15:289–294

Chinta S, Dickens J, Aldrich J (1994) Olfactory reception of potential pheromones and plant odors by tarnished plant bug, *Lygus lineolaris* (Hemiptera: Miridae). J Chem Ecol 20:3251–3267

Chu SS, Hu JF, Liu ZL (2011) Composition of essential oil of Chinese *Chenopodium ambrosioides* and insecticidal activity against maize weevil, *Sitophilus zeamais*. Pest Manag Sci 67:714–718

Collins PJ, Lambkin TM, Bridgeman BW, Pulvirenti C (1993) Resistance to grain-protectant insecticides in coleopterous pests of stored cereals in Queensland, Australia. J Econ Entomol 86:239–245

Crombie L, Ham PJ, Whiting DA (1971) Sapogenin of termite-repellent fruit of *Swartzia madagascariensis*. Chem Ind 6:176–177

Dinh ST, Baldwin IT, Galis I (2013) The Herbivore Elicitor-Regulated1 gene enhances abscisic acid levels and defenses against herbivores in *Nicotiana attenuata* plants. Plant Physiol 162:2106–2124

Fadamiro HY, Gudrups I, Hodges RJ (1998) Upwind flight of *Prostephanus truncatus* is mediated by aggregation pheromone but not food volatiles. J Stored Prod Res 34:151–158

Fang NB, Casida JE (1999) Cubé resin insecticide: identification and biological activity of 29 rotenoid constituents. J Agric Food Chem 47:2130–2136

Fedina T, Lewis S (2007) Effect of *Tribolium castaneum* (Coleoptera: Tenebrionidae) nutritional environment, sex, and mating status on response to commercial pheromone traps. J Econ Entomol 100:1924–1927

Funk V (2004) 100 uses for an Herbarium. Peabody Museum of Natural History. http://peabody.yale.edu/sites/default/files/documents/botany/100_uses.pdf. Accessed Aug 2013

Georges K, Jayaprakasam B, Dalavoy SS, Nair MG (2008) Pest-managing activities of plant extracts and anthraquinones from *Cassia nigricans* from Burkina Faso. Biores Technol 99:2037–2045

Greenwood M, Chapman RF (1984) Differences in numbers of sensilla on the antennae of solitarius and gregarious *Locusta migratoria* L. (Orthoptera: Acrididae). Int J Insect Morph Embryol 13:295–301

Grzywacz D, Stevenson PC, Belmain SR, Wilson K (2014) Improving food security in Africa: a new approach using indigenous ecological resources for pest control. Food Secur 6(1):71–86

Gupta SK, Kuo C-L, Chang HC, Chan HS, Chen ECF, Chueh FS, Tsay HS (2012) In vitro propagation and approaches for metabolites production in medicinal plants. In: Shyur LF, Lau ASY (eds) Recent trends in medicinal plants research. *Adv Bot Res* 62:35–55

Hansson BS, Pers J, Löfqvist JAN (1989) Comparison of male and female olfactory cell response to pheromone compounds and plant volatiles in the turnip moth, *Agrotis segetum*. Physiol Entomol 14:147–155

Harnischfeger G (2000) Proposed guidelines for commercial collection of medicinal plant material. J Herb Spice Med Plant 7:43–50

Hernandez JLH, Teran FQ, Macias JG (2010) *Taxus* poisoning. Med Clin 135:575–576

Isman MB (2006) Botanical insecticides, deterrents, and repellents in modern agriculture and an increasingly regulated world. Annu Rev Entomol 51:45–66

Isman MB (2008) Botanical insecticides: for richer, for poorer. Pest Manag Sci 64:8–11

Isman MB, Miresmailli S, Machial C (2011) Commercial opportunities for pesticides based on plant essential oils in agriculture, industry and consumer products. Phytochem Rev 10:197–204

Irvine JE, Freyre RH (1959) Source materials for rotenone – occurrence of rotenoids in some species of the genus *Tephrosia*. J Agric Food Chem 7:106–107

Jaba J, Haseena B, Tripathy S, Hosamani AC, Amaresh YS (2010) Olfactory response of cowpea aphid, *Aphis craccivora* Koch, to host odours and population of conspecifics. J Biopestic 3:405–407

Jama BA, Mutegi JK, Njui AN (2008) Potential of improved fallows to increase household and regional fuelwood supply: evidence from western Kenya. Agrofor Syst 73:155–166

Jayasekera TK, Stevenson PC, Belmain SR, Hall DR (2002) Methylsalicylate isomers in the roots of *Securidaca longepedunculata*. J Mass Spectrom 37:577–580

Jayasekera TK, Belmain SR, Stevenson PC, Hall DR (2003) *Securidaca longepedunculata* as a control for stored product insects pests. In: Credland PF, Armitage DM, Bell CH, Cogan PM, Highley E (eds) Advances on stored product protection. CAB International, Wallingford, pp 596–599

Jayasekera TK, Stevenson PC, Hall DR, Belmain SR (2005) Effect of volatile constituents from *Securidaca longepedunculata* on stored grain insect pests. J Chem Ecol 31:303–313

Jiang ZL, Akhtar Y, Zhang X, Bradbury R, Isman MB (2012a) Insecticidal and feeding deterrent activities of essential oils in the cabbage looper, *Trichoplusia ni* (Lepidoptera: Noctuidae). J Appl Entomol 136:191–202

Jiang GH, Liu QR, Chu SS, Liu ZL (2012b) Chemical composition and insecticidal activity of the essential oil of *Artemisia eriopoda* against maize weevil, *Sitophilus zeamais*. Nat Prod Commun 7:267–268

Joseph CC, Ndoile MM, Malima RC, Nkunya MHH (2004) Larvicidal and mosquitocidal extracts, a coumarin, isoflavonoids and pterocarpans from *Neorautanenia mitis*. Trans R Soc Trop Med Hyg 98:451–455

Kamanula J, Sileshi GW, Belmain SR, Sola P, Mvumi BM, Nyirenda GKC, Nyirenda SP, Stevenson PC (2011) Farmers' insect pest management practices and pesticidal plant use in the protection of stored maize and beans in Southern Africa. Int J Pest Manag 57:41–49

Kestenholz C, Stevenson PC, Belmain SR (2007) Comparative effects of field and laboratory evaluations of the ethnobotanical *Cassia sophera* L. (Leguminosae) for bioactivity against the storage pests *Callosobruchus maculatus* F. (Coleoptera: Bruchidae) and *Sitophilus oryzae* L. (Coleoptera: Curculionidae). J Stored Prod Res 43:78–86

Kimani SW, Overholt WA (1995) Biosystematics of the *Cotesia flavipes* complex (Hymenoptera: Braconidae): interspecific hybridization, sex pheromone and mating behaviour studies. B Entomol Res 85:379–386

Kinghorn AD (1978) Isolation of a novel irritant from *Euphorbia tirucalli*. J Nat Prod 41:648–649

Klowden MJ, Bock G, Cardew G (1996) Endogenous factors regulating mosquito host-seeking behaviour. CIBA Foundation symposium: olfaction in mosquito-host interactions 200:212

Kolev ST, Leman P, Kite GC, Stevenson PC, Shaw D, Murray VSG (1996) Toxicity following accidental ingestion of Aconitum containing Chinese remedy. Human Exp Toxicol 15:839–842

Kone WM, Atindehou KK, Terreaux C, Hostettmann K, Traore D, Dosso M (2004) Traditional medicine in North Cote-d'Ivoire: screening of 50 medicinal plants for antibacterial activity. J Ethnopharmacol 93:43–49

Koona P, Dorn S (2005) Extracts from *Tephrosia vogelii* for the protection of stored legume seeds against damage by three bruchid species. Ann Appl Biol 147:43–48

Koul O, Dhaliwal P (2004) Phytochemical biopesticides. CRC Press, Boca Raton

Kuca K, Pohanka M (2010) Chemical warfare agents. In: Luch A (ed) Medical, clinical and environmental toxicology, vol 2, Clinical toxicology. Birkhäuser, Basel, pp 543–558

Leroy T, Pouzat J, Biemont JC, Pierre D (1999) Variation in relation with physiological status of olfactory peripheral sensitivity in four Coleoptera, *Callosobruchus maculatus* Fab., *Bruchidius atrolineatus* Pic, *Acanthoscelides obtectus* Say (Bruchidae) and *Sitona lineatus* L. (Curculionidae). Ann Soc Entomol Fr 35:159–164

Liu P, Liu X-C, Dong HW, Liu Z-L, Du S-S, Deng Z-W (2012) Chemical composition and insecticidal activity of the essential oil of Illicium pachyphyllum fruits against two grain storage insects. Molecules 17:14870–14881

Lukwa N, Molgaard P, Furu P, Bogh C (2009) *Lippia javanica* (Burm F) Spreng: its general constituents and bioactivity on mosquitoes. Trop Biomed 26:85–91

Madzimure J, Nyahangare ET, Hamudikuwanda H, Hove T, Stevenson PC, Belmain SR, Mvumi BM (2011) Acaricidal efficacy against cattle ticks and acute oral toxicity of *Lippia javanica* (Burm F.). Trop Anim Health Prod 43:481–489

Mafongoya PL, Kuntashula E (2005) Participatory evaluation of *Tephrosia* species and provenances for soil fertility improvement and other uses using farmer criteria in eastern Zambia. Exp Agric 41:69–80

Manrique G, Lazzari CR (1995) Existence of a sex pheromone in *Triatoma infestans* (Hemiptera: Reduviidae): I. Behavioural evidence. Mem Inst Oswaldo Cruz 90:645–648

Marston A, Maillard M, Hostettmann K (1993) Search for antifungal, molluscicidal and larvicidal compounds from African medicinal-plants. J Ethnopharmacol 38:215–223

Matsumoto SG, Hildebrand JG (1981) Olfactory mechanisms in the moth *Manduca sexta*: response characteristics and morphology of central neurons in the antennal lobes. Proc R Soc Lond B Biol Sci 213:249–277

Mechaber W, Capaldo C, Hildebrand J (2002) Behavioral responses of adult female tobacco hornworms, *Manduca sexta*, to host plant volatiles change with age and mating status. J Insect Sci 2:1–8

Messina FJ, Renwick JAA (1985) Dispersal polymorphism of *Callosobruchus maculatus* (Coleoptera: Bruchidae): variation among populations in response to crowding. Ann Entomol Soc Am 78:201–206

Minjas JN, Sarda RK (1986) Laboratory observations on the toxicity of *Swartzia madagascariensis* (Leguminosae) extract to mosquito larvae. Trans R Soc Trop Med Hyg 80:460–461

Moyo M, Bairu MW, Amoo SO, Van Staden J (2011) Plant biotechnology in South Africa: micropropagation research endeavours, prospects and challenges. S Afr J Bot 77:996–1011

Nakakita H, Ikenaga H (1997) Action of low temperature on physiology of *Sitophilus zeamais* Motschulsky and *Sitophilus oryzae* L. (Coleoptera: Curculionidae) in rice storage. J Stored Prod Res 33:31–38

Nawrot J, Harmatha J (2013) Phytochemical feeding deterrents for stored product insect pests. Phytochem Rev 11:543–566

Nenaah GE, Ibrahim SIA (2011) Chemical composition and the insecticidal activity of certain plants applied as powders and essential oils against two stored-products coleopteran beetles. J Pest Sci 8:393–402

Neuwinger HD (2004) Plants used for poison fishing in tropical Africa. Toxicon 44:417–430

Nguyen DT, Hodges RJ, Belmain SR (2008) Do walking *Rhyzopertha dominica* (F.) locate cereal hosts by chance? J Stored Prod Res 44:90–99

Nyahangare ET, Hove T, Mvumi BM, Hamudikuwanda H, Belmain SR, Madzimure J, Stevenson PC (2012) Acute mammalian toxicity of four pesticidal plants. J Med Plant Res 6:2674–2680

Nyahangare ET, Hove T, Hamudikuwanda H, Belmain SR, Stevenson PC, Mvumi BM (2013) Acaricidal efficacy of *Strychnos spinosa* (Lam.) and *Solanum incanum* fruit-extracts against cattle ticks. Trop Anim Health Prod 45:1341–1347

Nyirenda SPN, Sileshi G, Belmain SR, Kamanula JF, Mvumi BM, Sola P, Nyirenda GKC, Stevenson PC (2011) Farmers' ethno-ecological knowledge of vegetable pests and their management using pesticidal plants in northern Malawi and eastern Zambia. Afr J Agric Res 6:1525–1537

Obeng-Ofori D (2010) Residual insecticides, inert dusts and botanicals for the protection of durable stored products against pest infestation in developing countries. In: Carvalho MO, Fields PG, Adler CS et al (eds) Proceedings of the 10th international working conference on stored product protection, Estoril, Portugal

Omolo MO, Okinyo D, Ndiege IO, Lwande W, Hassanali A (2004) Repellency of essential oils of some Kenyan plants against *Anopheles gambiae*. Phytochemistry 65:2797–2802

Ondarza RN, Gutiérrez-Martínez A, Malo EA (1986) Evidence for the presence of sex and aggregation pheromones from *Triatoma mazzottii* (Hemiptera: Reduviidae). J Econ Entomol 79:688–692

Ortiz UM, Silva AG, Tapia VM, Rodriguez MJC, Lagunes TA, Santillán-Ortega C, Robles-Bermúdez A, Aguilar-Medel S (2012) Toxicity of Boldo *Peumus Boldus* Molina for *Sitophilus zeamais* Motschulsky and *Tribolium castaneum* Herbst. Chil J Agric Res 72:345–349

Otter CJD, Tchicaya T, Schutte AM (1991) Effects of age, sex and hunger on the antennal olfactory sensitivity of tsetse flies. Physiol Entomol 16:173–182

Padin SB, Fuse C, Urrutia MI, Dal Bello GM (2013) Toxicity and repellency of nine medicinal plants against *Tribolium castaneum* in stored wheat. Bull Insectol 66:45–49

Palanaswamy P, Seabrook WD (1978) Behavioral responses of the female eastern spruce budworm *Choristoneura fumiferana* (Lepidoptera, Tortricidae) to the sex pheromone of her own species. J Chem Ecol 4:649–655

Phillips TW, Throne JE (2010) Biorational approaches to managing stored-product insects. Annu Rev Entomol 55:375–397

Pijut PM, Beasley RR, Lawson SS, Palla KJ, Stevens ME, Wang Y (2012) In vitro propagation of tropical hardwood tree species – a review (2001–2011). Propag Ornam Plant 12:25–51

Prakash A, Rao J (1997) Botanical pesticides in agriculture. CRC Press, Boca Raton

Rahuman AA, Gopalakrishnan G, Venkatesan P, Geetha K (2008) Larvicidal activity of some Euphorbiaceae plant extracts against *Aedes aegypti* and *Culex quinquefasciatus* (Diptera: Culicidae). Parasitol Res 102:867–873

Ramakrishna A, Ravishankar GA (2011) Influence of abiotic stress signals on secondary metabolites in plants. Plant Signal Behav 6:1720–1731

Rutledge LC, Moussa MA, Lowe CA, Sofield RK (1978) Comparative sensitivity of mosquito species and strains to the repellent diethyl toluamide. J Med Entomol 14:536–541

Sakurai Y, Sakurai N, Taniguchi M, Nakanishi Y, Bastow KF, Wang XH, Cragg GM, Lee KH (2006) Rautandiols A and B, pterocarpans and cytotoxic constituents from *Neorautanenia mitis*. J Nat Prod 69:397–399

Sarasan V, Kite GC, Sileshi GW, Stevenson PC (2011) The application of phytochemistry and in vitro tools to the sustainable utilisation of medicinal and pesticidal plants for income generation and poverty alleviation. Plant Cell Rep 30:1163–1172

Schrire BD (2005) Tribe Millettieae. In: Lewis G, Schrire B, Mackinder B, Lock M (eds) Legumes of the world. Royal Botanic Gardens, Kew

Sekyembe S, Maurer G, Gatimu J, Nkanya J, Injairu S (1993) Training manual for the control of the greater (larger) grain borer (*Prostephanus truncatus*, Horn). FAO, Rome. http://www.fao.org/wairdocs/X5421E/x5421e00.htm#Contents

Shackleton CM, Guthrie G, Main R (2005) Estimating the potential role of commercial over-harvesting in resource viability: a case study of five useful tree species in South Africa. Land Degrad Dev 16:273–286

Sileshi G, Mafongoya PL, Kwesiga F, Nkunika P (2005) Termite damage to maize grown in agroforestry systems, traditional fallows and monoculture on nitrogen-limited soils in eastern Zambia. Agric For Entomol 7:61–69

Simmonds MSJ (2003) Novel drugs from botanical sources. Drug Discov Today 8:721–722

Sirrine D, Shennan C, Snapp S, Kanyama-Phiri G, Kamanga B, Sirrine JR (2010) Improving recommendations resulting from on-farm research: agroforestry, risk, profitability and vulnerability in southern Malawi. Int J Agric Sustain 8:290–304

Skibbe M, Qu N, Galis I, Baldwin IT (2008) Induced plant defenses in the natural environment: *Nicotiana attenuata* WRKY3 and WRKY6 coordinate responses to herbivory. Plant Physiol 20:1984–2000

Smallegange RC, Van Gemert G-J, Van de Vegte-Bolmer M, Gezan S, Takken W, Sauerwein RW, Logan JG (2013) Malaria infected mosquitoes express enhanced attraction to human odor. PLoS One 8:63602

Sola P, Mvumi BM, Nyirenda SP, Ogendo JO, Mponda O, Andan FPH, Kamanula JF, Belmain SR, Stevenson PC (2013) Bio pesticides production, trade and regulatory mechanisms in African drylands: making a case for plant pesticide products. Food Sec 6:369–384

Stevenson PC, Simmonds MSJ, Yule M, Veitch NC, Kite GC, Irwin D, Legg M (2003) Insect antifeedant furanocoumarins from *Tetradium daniellii*. Phytochemistry 63:41–46

Stevenson PC, Jayasekera TK, Belmain SR, Veitch NC (2009) Bisdesmosidic saponins from *Securidaca longepedunculata* (Polygalaceae) with deterrent and toxic properties to Coleopteran storage pests. J Agric Food Chem 57:8860–8867

Stevenson PC, Nyirenda SP, Veitch NC (2010) Highly glycosylated flavonoid glycosides from *Bobgunnia madagascariensis*. Tetrahedron Lett 51:4727–4730

Stevenson PC, Kite GC, Lewis GP, Nyirenda SP, Forest F, Belmain SR, Sileshi G, Veitch NC (2012) Distinct chemotypes of *Tephrosia vogelii* and implications for their use in pest control and soil enrichment. Phytochemistry 78:135–146

Stoll G (2000) Natural crop protection in the tropics, 2nd edn. Margraf Publishers, Weikersheim

Syngenta (2006) Actellic 50 EC safety data sheet. Version 7. http://www.syngenta.com/country/ma/SiteCollectionDocuments/Products%20CP/Actellic/SDS.pdf. Accessed Aug 2013

Taylor WG, Fields PG, Sutherland DH (2004) Insecticidal components from field pea extracts: Soyasaponins and lysolecithins. J Agric Food Chem 52:7484–7490

Thacker JRM (2002) An introduction to arthropod pest control. Cambridge University Press, Cambridge, p 343

Thokozani BLK, Zulu D, Sileshi GW, Teklehaimanot Z, Gondwe DSB, Sarsan V, Stevenson PC (2011) Seed germination and in vitro regeneration of the pan African medicinal and pesticidal plant *Bobgunnia madagascariensis*. Afr J Biotechnol 10: 5959–5966

Udo IO (2012) Phytochemical screening of *Dracaena arborea* (Asparagaceae) for insecticidal activity in the control of *Sitophilus zeamais* (Coleoptera: Curculionidae) and *Callosobruchus maculatus* (Coleoptera: Chrysomelidae). Int J Trop Insect Sci 33:136–143

Ukeh DA, Udo IA, Ogban EI (2008) Trapping of stored-product insects using flight traps outside traditional African storage granaries. J Food Agric Environ 6:399–401

Ukeh DA, Birkett MA, Bruce TJA, Allan EJ, Pickett JA, Mordue AJ (2010) Behavioural responses of the maize weevil, *Sitophilus zeamais*, to host (stored-grain) and non-host plant volatiles. Pest Manag Sci 66:44–50

Ukeh DA, Woodcock CM, Pickett JA, Birkett MA (2012) Identification of host kairomones from maize, *Zea mays*, for the maize weevil, *Sitophilus zeamais*. J Chem Ecol 38:1402–1409

Utida S (1972) Density dependent polymorphism in the adult of *Callosobruchus maculatus* (Coleoptera, Bruchidae). J Stored Prod Res 8:111–125

Vanpuyvelde L, Dekimpe N, Mudaheranwa JP, Gasiga A, Schamp N, Declercq JP, Vanmeerssche M (1987) Isolation and structural elucidation of potentially insecticidal and acaricidal isoflavone-type compounds from *Neorautanenia mitis*. J Nat Prod 50:349–356

Viljoen AM, Subramoney S, Van Vuuren SF, Baser KHC, Demirci B (2005) The composition, geographical variation and antimicrobial activity of *Lippia javanica* (Verbenaceae) leaf essential oils. J Ethnopharmacol 96:271–277

Walgenbach CA, Phillips JK, Faustini DL, Burkholder WE (1983) Male-produced aggregation pheromone of the maize weevil, *Sitophilus zeamais*, and interspecific attraction between three *Sitophilus* species. J Chem Ecol 9:831–841

Wenninger EJ, Stelinski LL, Hall DG (2009) Roles of olfactory cues, visual cues, and mating status in orientation of *Diaphorina citri* Kuwayama (Hemiptera: Psyllidae) to four different host plants. Environ Entomol 38:225–234

Yadav R, Srivastava VK, Chandra R, Singh A (2002) Larvicidal activity of latex and stem bark of *Euphorbia tirucalli* plant on the mosquito *Culex quinquefasciatus*. J Commun Dis 34:264–269

Zhao NN, Zhang H, Zhang XC, Luan XB, Zhou C, Liu QZ, Shi WP, Liu ZL (2013) Evaluation of acute toxicity of essential oil of garlic (*Allium sativum*) and its selected major constituent compounds against overwintering *Cacopsylla chinensis* (Hemiptera: Psyllidae). J Econ Entomol 106:1349–1354

Zhu J, Cossé AA, Obrycki JJ, Boo KS, Baker TC (1999) Olfactory reactions of the twelve-spotted lady beetle, *Coleomegilla maculata* and the green lacewing, *Chrysoperla carnea* to semiochemicals released from their prey and host plant: electroantennogram and behavioral responses. J Chem Ecol 25:1163–1177

Zoubiri S, Baaliouamer A (2012) Chemical composition and insecticidal properties of *Lantana camara* L. leaf essential oils from Algeria. J Essent Oil Res 24:377–383

Zulu D, Thokozani BLK, Sileshi GW, Teklehaimanot Z, Gondwe DSB, Sarsan V, Stevenson PC (2011) Towards successful propagation of the Africa medicinal and pesticidal plant *Securidaca longepedunculata*. Afr J Biotechnol 10:5988–5992

Non-target Effects of Botanicals on Beneficial Arthropods with Special Reference to *Azadirachta indica*

10

S. Raguraman and M. Kannan

Abstract

The 'green' agriculture through better ecological approaches in pest management is the stimulus for further research in alternate pesticides molecules barring synthetic ones, which led to increased search in plant sources. Among several options, neem, *Azadirachta indica*, has emerged as pinnacle of plant origin pesticides. Undoubtedly, the use of neem and its products in agriculture are increasing day by day, but its safety to the environment is always questioned. This review focuses on the biological activity of the plant origin insecticides on entomophages (parasitoids and predators), honeybees and the environment and provides ways and means to use them in integrated pest management. In general, the plant-derived pesticides, although effective against insect pests of agricultural importance, spare the beneficial fauna comparative to synthetic pesticide because of their capacities of biodegradable nature and innate low mammalian toxicity. Although some plant-derived products either in crude or in formulation showed slight to moderate ill effects to beneficial fauna including parasitoids, predators and honeybees during the application may greatly reduce the risk. No residual/ persistent toxicity of neem or other botanical pesticides in the environment has so far been reported. In the context of organic agriculture, plant origin pesticides especially the neem-derived extracts/formulations and other biocontrol agents are the befitting components to reduce the input costs and environmental risks posed by the synthetic chemical pesticides.

Keywords

Botanical insecticides • Non-target effects • Predator • Parasitoids • Honeybees • Environment

S. Raguraman, Ph.D. (✉) • M. Kannan, Ph.D.
Department of Agricultural Entomology,
Tamil Nadu Agricultural University,
Coimbatore 641 003, Tamil Nadu, India
e-mail: raguraman99@gmail.com

D. Singh (ed.), *Advances in Plant Biopesticides*,
DOI 10.1007/978-81-322-2006-0_10, © Springer India 2014

1 Introduction

Entomophages, namely, parasitoids and predators, are natural enemies of insect pests in agroecosystems. Most insecticides used to protect the cultivated crops are relatively broad spectrum, killing the target insect pests and the non-target beneficial insects as well. But substances derived from plant resources have been generally considered safe compared to synthetic chemical insecticides. All the same, not all substances derived from plant resources are always safe. Sometimes adverse effects are also produced. Safety aspect of plant-derived substances is very important since there are many beneficial insects which contribute immensely for crop productivity and agricultural sustainability (Schmutterer 1992; Koul et al. 2004; Raguraman 2009; Isman 2013). Different kinds of parasitoids and predators play a very important role in natural control of insect pests. Many kinds of pollinators contribute to cross-fertilization of plants. Hence, it is important to protect all these from any harm. Conservation of beneficial insects is achieved by modification in insect control practices that allow beneficial insects to survive and promote their potential to suppress insect pests. This review focuses the impact and safety of plant origin insecticides to parasitoids and predators both in laboratory and field studies and envisages the strategies and application methods to achieve better pest management. For easy reading and documenting by the researchers, the plants are discussed alphabetically by scientific names under effects on parasitoids, predators, honeybees and environment.

2 Effect on Egg Parasitoids

Aqueous extract of *Acorus calamus* was reported to be safe to *Trichogramma japonicum* (Burman et al. 2003). With reference to neem, *Azadirachta indica*, as early as 1982, the side effects on egg parasitoids were studied in India by Joshi et al. (1982). They reported that 2 % of neem seed kernel extract (NSKE) applied on the egg masses of *Spodoptera litura did* not repel the egg parasitoid, *Telenomus remus*. When the treatment was carried out of pre-oviposition of the parasitoid, the emergence of adult parasitoids was normal, but their duration of life was shorter than that of controls. On the other hand, spraying NSKE after oviposition, *T. remus* increased the fecundity of the wasps developed in treated eggs and prolonged their life as compared to that of untreated controls. Li et al. (1986) tested in the laboratory 29 insecticides including *Bacillus thuringiensis* and neem oil in order to study their side effects on *Trichogramma japonicum* and concluded that neem oil was the safest pesticide for the parasitoid.

Studies were made on the effect of host species on the sex ratio and parasitism rate of *Anastatus ramakrishnanae* in the laboratory and field in Tamil Nadu, India, with the pentatomid *Halys dentata* (on *Cassia marginata*, *Azadirachta indica* and *Casuarina equisetifolia*) and the coreid *Homoeocerus prominulus* (on *Cassia marginata*, *Prosopis spicigera* and *Acacia leucophloea*) as hosts. The order of parasitism was *C. marginata* > *A. indica*> *C. equisetifolia* with *H. dentata* and *C. marginata* > *P. spicigera* > *A. leucophloea* with *H. prominulus*. Higher rates of parasitism were recorded with *H. dentata* than *H. prominulus* throughout the year (Velayuddan et al. 1988).

Fernandez et al. (1992) conducted experiments with the eggs of yellow stem borer of rice, *Tryporyza incertulas*, by dipping for 30 s in Bordeaux, neem products and water. The eggs were exposed for 40 h of parasitization by the parasitoid, *Telenomus rowani*. The data revealed the highest mean number of parasitoid emergence of 65.8 % in water treatment, 59.90 % in 5 % aqueous NSKE, 31.4 % in 3 % neem oil and 38.2 % in Bordeaux treatment. Klemm and Schmutterer (1993) applied NSKE (2.5 and 3 %) against *Trichogramma* spp., egg parasitoids of *Plutella xylostella*. *T. principium* accepted neem-treated eggs in the laboratory and *T. pretiosum* in the field, but two treatments prevented the eclosion of adult parasitoids from treated *P. xylostella* eggs completely. Spraying of eggs with 0.2 % neem oil reduced the number of eggs parasitized per female wasp by 13.3 %. Neem oil also reduced the emergence of *T. principium* from treated eggs by 45.1 %. However, neem seed kernel suspension (5 %) and neem oil 50 EC (3 %) were safe to the parasitoid *T. japonicum* in cotton ecosystem (Jayaraj et al. 1993).

Lyons et al. (1996) used the neem-treated eggs of *Ephestia kuehniella* in shell vials and offered to single females of *T. minutum* for parasitization by fixing the eggs with adhesive strips and held until all parasitoids had emerged from them; variable results were obtained. Azatin, neem EC (4.6 % AZA) and pure AZA were tested at concentrations of 50 g and 500 g/ha. At 50 g/ha, no significant effect was observed, but at 500 g/ha, Azatin and neem EC reduced the female survival by 64 and 40 %, respectively, whereas pure AZA showed neem oil effect. Likewise, at 500 g/ha the number of parasitized eggs was reduced by 89 % by Azatin, 29 % by neem EC and no reduction by AZA. The parasitoid development was reduced by all treatments. Cano and Gladstone (1994) studied the influence of the NSK-based extract NIM-20 on parasitization of eggs of *Helicoverpa zea* in a melon field in Nicaragua. Mass-reared *T. pretiosum* were released at six weekly intervals 1, 2, 6 and 24 h after application of NIM-20 at 2.5 g/l. No negative effect was observed as up to 84 % of the eggs of the pest were parasitized. Oswald (1989) treated the eggs of the coconut bug, *Pseudotheraptus wayi*, with aqueous NSKE (5 % w/v). The eggs were offered to the parasitoid *Oencyrtus albicrus* (Encyrtidae). There was a significant reduction in the number of wasps emerging from the treated eggs in comparison with controls. Srinivasa Babu et al. (1996) studied the effects of neem-based commercial insecticides such as Repelin and Neemguard on *T. australicum* in laboratory and field conditions. They reported that both the insecticides were relatively safe at lower concentrations, but higher concentrations adversely affected the parasitoids both in laboratory and in field. *Trichogramma chilonis* was affected by *A. indica* extracts (Stansly and Liu 1997). The egg parasitoid, *Trichogramma chilonis*, parasitized 45 % of eggs when host eggs were treated with neem formulation. However, emergence of *T. chilonis* from the parasitized eggs treated anytime after parasitization was not affected (Markandeya and Divakar 1999).

Similar results were obtained against *T. japonicum* using Econeem and NeemAzal-T/S (0.1–1.0 %) (Lakshmi et al. 1998a), other neem-based pesticides in an IPM for rice pest management (Garg and Baranwal 1998) and egg parasitoid, *Tetrastichus pyrillae* (Deepak and Choudhary 1998). On the whole, it has been assessed that neem products were fairly safe to *Trichogramma* spp. (Sreenivasa and Patil 1998; Sarode and Sonalkar 1999b). However, some neem formulations such as Nimbecidine (0.25–4.0 %), Neemgold (2.0–4.0 %) and Rakshak (1.0 %) are reported to possess adverse effects on parasitism (Lakshmi et al. 1998a).

Raguraman and Singh (1999) tested in detail the neem seed oil at concentrations of 5.0, 2.5, 1.2, 0.6 and 0.3 % for oviposition deterrence, feeding deterrence, toxicity, sterility and insect growth regulator effects against *Trichogramma chilonis*. Neem seed oil at 0.3 % deterred oviposition (parasitization) by the parasitoid, but the sensitivity varied considerably both under choice and no-choice conditions. Neem seed oil also deterred feeding at or above 1.2 % concentration both in choice and no-choice tests. In feeding toxicity tests, neem seed oil at 5 % concentration caused <50 % mortality to both males and females, but in contact toxicity tests, females were affected sparing males. No sterility effect was observed when the parasitoid was fed with neem seed oil-treated honey. Both pre- and post-treatment of host eggs revealed neem oil adverse effects on the development of the parasitoid.

Thakur and Pawar (2000) tested two neem-based insecticides (3 g Achook/litre and 2 ml Neemactin/litre), two biopesticides [1 g Halt (cypermethrin)/litre and 1 ml Dipel (*Btk*)/litre] and endosulfan (1.5 ml/l) in the laboratory for their relative toxicity to newly emerged adults of *T. chilonis*. Results revealed that neem-based pesticides and biopesticides were harmless, while endosulfan was slightly toxic to egg parasitoid. These observations also get support from the studies on different groups of chemicals, viz., insecticides, moult inhibitors and biopesticides against rice leaffolder C. *medinalis* and its parasitoid *T. chilonis*. Sprays of monocrotophos 36 WSC applied at 2.0 ml/l caused 100 % larval mortality to C. *medinalis* 7 days after treatment followed by buprofezin 25 WP applied at 3.2 g/l with 66.66 % larval mortality and neem seed kernel extract (NSKE) 5 % (50 g/l) with 63.33 % larval mortality. Application of *Bacillus thuringiensis* subsp. *galleriae* (*Btg*) at 5.0 ml/l and NeemAzal-F 5 % at 1 g/l recorded 56.66 and 53.33 % larval mortality, respectively. More than 90 % emergence of *T. chilonis* was recorded from eggs treated with *Btg* and NeemAzal-F and NeemAzal-T/S followed by NSKE (89.80 %) and buprofezin (82.60 %). Only 73.80 % of adult *T. chilonis* emerged from monocrotophos treated host eggs after parasitization (Saikia and Parameswaran 2001). Similarly, in Thailand, Asian corn stem borer, *Ostrinia furnacalis*, was controlled by neem preparations, and it was observed that such treatments had no side effects on the parasitoid, *T. plasseyensis* wasps (Breithaupt 1995).

Rao et al. (2002) reported that Neemgold^{R} at 0.015 % had 50.33 % parasitism by *T. chilonis* on *C. cephalonica*. Jyothi and Sannaveerappanavar (2003) tested the aqueous NSKE and neem oil against *T. chilonis* in laboratory. NSKE at 2 and 4 % and neem seed oil at 4 % were moderately toxic to *T. chilonis* adults, causing 33.33, 31.24 and 29.58 % mortality, respectively. Neem oil at 2 and 5 % and NSKE at 4 % significantly reduced the parasitization of C. *cephalonica* to 49.22, 57.41 and 59.29 %, respectively, compared with 88.91 % in the control. NSKE at 2 % did not affect the rate of parasitization. NSKE at 2 and 4 % and neem seed oil at 2 % did not affect the emergence of the parasitoids from treated eggs. Solis et al. (2004) tested four concentrations of the neem seed oil (0.1, 0.5, 1.0 and 5.0 %) *Gryon gallardoi*, an egg parasitoid of *Leptoglossus zonatus* under laboratory conditions. The longevity of adults fed on sugar solution containing the neem oil concentrations was significantly reduced, the only exception being observed in females treated with 0.1 % of oil. Most likely, the vegetal extract caused antifeedant behaviour in adults, mainly when offered in high concentrations. However, the parasitoid emergence from the host eggs treated with the neem product before parasitism did not suffer any influence, evidencing host acceptance and absence of repellence. The duration time of the immature stages and sex ratio of the parasitoid inside the host eggs topically treated with the product remained unchanged as well.

Moosa-Saber et al. (2004) evaluated the effects of NeemAzal on adult emergence and life table parameters of *Trichogramma cacoeciae*. The effect of NeemAzal on three developmental stages of the parasitoid was tested by dipping parasitized *Sitotroga cerealella* and *Cydia pomonella* eggs at the field recommended concentration 3, 6 and 9 days after parasitization corresponding to larval, prepupal and pupal stages. The emergence of adult parasitoids was adversely affected in both hosts, but the adverse effect was more in *S. cerealella* eggs compared with *C. pomonella*. The adult emergence was reduced by 73.30 and 33.76 % in *Sitotroga* and *Cydia* eggs compared with controls, respectively. Mitchell et al. (2004) evaluated extracts of neem seed kernel on scelionid egg parasitoid, *Gryon fulviventre* of a coreid *Clavigralla scutellaris*. Feeding by newly emerged wasps was dramatically reduced when honey was mixed with aqueous neem suspension, but 6-day survivorship of adults did not differ significantly from that of the control. Wasp oviposition behaviour was altered slightly when coreid eggs were treated with neem: the period of antennation was significantly extended, but time for drilling, oviposition and marking was unaffected. Neem-dipped eggs were accepted for oviposition and progeny emerged successfully from these treated eggs. Exposure of already parasitized eggs to neem did not interfere with progeny emergence, longevity or sex ratio. Thus, neem extract and egg parasitoids seem to be compatible and promising control strategies for *C. scutellaris*.

A. indica extracts were detrimental to parasitoid *Trichogramma pintoi* (Iannacone and Lamas 2003a). Boomathi et al. (2005) reported that the ethonolic extracts of neem + sweetflag and neem + sweetflag + pongamia at 0.12 and 0.18 % exhibited 34 % mortality of adult parasitoids and 79–81 % adult emergence of *T. chilonis* from

C. cephalonica. From a field trial, Thirumurugan and Koodalingam (2005) reported that cane crop sprayed with 5 % (NSKE) + *T. chilonis* at weekly interval had substantially reduced the internode borer and also registered higher yields of sugar. Masood and Mamoon-ur-Rashid (2006) reported that water extract and oil of neem seeds significantly affected the egg parasitization of *Trichogramma* spp. but did not show any negative effect on the adult emergence.

Hohmann et al. (2010) demonstrated that treating the hosts eggs of *Anagasta kuehniella* with aqueous neem seed extract (ANSE) at 15, 3 and 1.5 % and of an emulsifiable concentrate neem oil (ECNO) at 2.5, 0.5 and 0.25 % before parasitism was less deleterious to wasp emergence, especially for *Trichogramma annulata*. Pretreatments (24 h) of the host eggs with ECNO at concentrations varying from 0.5 to 0.25 % did not affect *T. pretiosum* longevity, but 2.5 % reduced *T. annulata* survival. Feeding wasps with honey mixed with 0.25 % ECNO negatively affected *T. annulata* survival.

Eyyüp and Başpınar (2012) sprayed NeemAzal-T/S for the control of *Liriomyza trifolii* larvae in tomato and showed that it had less impact on the parasitization by egg parasitoid in laboratory and greenhouse conditions. Usman et al. (2012) investigated the efficiency of *T. chilonis* alone, *T. chilonis* in combination with *Chrysoperla carnea* and neem extract against tomato fruitworm, *Helicoverpa armigera*, in field experiments. They found that treatment with Trichocard^R having 300 parasitized eggs in combination with *Chrysoperla* and neem extract was the most promising for effective management of *H. armigera* on tomato. Tunca et al. (2012) reported that azadirachtin, pyrethrum, capsaicin and d-Limonene were repellent to parasitoid *V. canescens* adults and therefore not compatible with parasitoid.

Mamoon-ur–Rashid et al. (2013) reported that neem oil at 1.5 and 2 % and neem seed water extract at 3 % significantly reduced the number of spotted bollworm larvae on treated leaves. Percentage parasitism of bollworm eggs by *T. chilonis* was significantly reduced when they were placed on leaves treated with 2 % neem oil, and 3 % neem seed water extract, but adult emergence of *T. chilonis* was not affected by any of the neem treatments.

Rotenone from *Derris elliptica* was toxic to adults of *Edovum puttleri*, an egg parasitoid of *Leptinotarsa decemlineata* (Obrycki et al. 1986; Hamilton et al. 1996). Rotenone caused mortality to *Trichogramma koehlri* (Iannacone and Lamas 2003a, b). Leaf extract of *Lantana camara* was safe to egg parasitoid *Trichogramma japonicum* (Burman et al. 2003). Nicotine sulphate from *N. tabacum* did not affect much the adults of *Encarsia formosa* (Heyler et al. 1992) but toxic to *Telenomus remus* (Chari et al. 1996). Water extract of *Pongamia pinnata* did not affect the emergence of egg parasitoid, *Trichogramma japanicum* (Burman et al. 2003). Ryanodine isolated from *Ryania speciosa* was safe to *Ooencyrtus kuwanai* and *Telenomus terebrans* (Tadic 1979). Water extract of *Vitex negundo* showed normal emergence of parasitoid, *Trichogramma japonicum* (Burman et al. 2003).

3 Effect on Larval Parasitoids

Aqueous extract of *Annona squamosa* was reported safe to parasitoid *Cotesia flavipes* (Reddy and Srikanth 1996). Neem, *Azadirachta indica*, oil was sprayed at 50 % as low volume application against the rice folder *C. medinalis*. The pest larvae were parasitized by ichneumonids, braconids and encyrtid groups of parasitoids in the field. Surprisingly the parasitization of the leaf folder larvae in neem oil-treated plots was double than control. This was due to the fact that most of the larvae could not spin the leaves together due to high toxicity of neem oil, thereby giving enough opportunity for parasitization. However, the neem oil had no side effects on parasitoids (Saxena et al. 1981). Such an increase in parasitoid population was also observed for the parasitoid *Diadegma semiclausum* than in control after the treatment of neem-based product 'Biosol' in cabbage plots (Chandra Mohan and Nanjan 1990). Similarly, various endoparasitic hymenoptera pupated and emerged normally from parasitized 4th and 5th instar C. *medinalis* larvae that were reared on rice leaves treated with neem fractions or extracts (Schmutterer et al. 1983).

Schauer (1985) found that aphid mummies containing larvae or pupae of braconid parasitoids, *Diaeretiella rapae* and *Aphidius cerasicola*, were unaffected by 5 % neem seed kernel suspension. Neem seed oil was also quite safe for the natural enemies like *Lycosa pseudoannulata* and *Apanteles cypris* (Wu 1986); ichneumonid parasitoid, *Campoletis chlorideae* of *H. armigera* (Prasad et al. 1987); and external larval parasitoid, *Bracon hebetor* of pod borer, *Maruca testulalis* (Jhansi and Sundara Babu 1987). Other studies with *B. hebetor* also support the fact that neem is safer for this parasitoid as aqueous suspension, and an ethanolic extract of neem seed kernel (NSK) at 0.3, 0.6, 1.2, 2.5 and 5.0 % administered via food or by contact had no influence on the *B. hebetor* oviposition (parasitization) on *C. cephalonica.* Parasitoid eggs and pupae were also unaffected by the extracts tested. The parasitoid larvae, however, were killed by feeding on contaminated host larvae and also through contact with neem extracts. Thus, use of a minimum safety period is suggested for inundative release of *B. hebetor* in integrated pest management (Raguraman and Singh 1998). In order to determine the toxicity of oil extracts to *Chelonus blackburni* to explore the possibility of using parasitoid along with oils/extracts in integrated control programme of potato tuber moth, it was observed that all the vegetable oils including neem oil were safe to *C. blackburni*, an egg-larval parasitoid of *P. operculella* (Shilke et al. 1990).

Schneider and Madel (1992) reported that there was no adverse effect on adults of the braconid *Diadegma semiclausum* after exposure for 3 days or during their lifetime in cages to residues of an aqueous NSKE (0.1–5 %). The longevity of the wasps exposed to neem residues was even prolonged, but the difference between treated and untreated individuals was statistically not significant. Females of the braconid, derived from larvae developed in neem-treated larvae of *P. xylostella*, showed neem oil reduced fecundity or activity as compared to controls. Fresh extracts showed neem oil repellent effect. The influence of AZA on *Diadegma terebrans*, parasitoid of the European corn borer, *Ostrinia nubilalis*, was investigated in the laboratory by McCloskey et al. (1993). These authors added sublethal doses (0.1 and 0.3 ppm) of AZA or ethanol (carrier solvent) to diets of second instar larvae of the pyralid. Both AZA concentrations showed neem oil significant difference of the parasitization percentage; host acceptance by the parasitoids was also not influenced. However, significantly higher mortality of parasitoids was observed in AZA-treated groups compared to untreated groups, especially after emergence from the hosts. The durations of the larval instars in the hosts were prolonged and the weight of pupae and adults from treated groups was reduced.

Lowery and Isman (1996) tested the effects of extracts from neem on aphids and their natural enemies. In field trials, populations of aphid natural enemies (predators and parasitoids) were not affected by application of neem insecticides, suggesting the compatibility of neem with biological control agents. Safety of natural enemies after neem application is also shown by the studies of Mani and Krishnamoorthy (1996) where encyrtid *Tetracnemoidea indica*, a dominant parasitoid of the pseudococcid *Planococcus lilacinus* on acid lime were exposed to acid lime leaves treated with 34 pesticides at field recommended doses. Fenvalerate (0.01 %) and NSKE (2 %) were non-toxic to the adult parasitoids. Similarly, there was no adverse effect of neem seed kernel water extract (NSKWE) (25 g/l) on the adult of *A. plutellae*, a parasitoid of leaf-eating caterpillar complex of cabbage (Bandara and Kudagamage 1996). It was also observed that after the spraying of the NSKWE and the two insecticides on the cocoons, there was no significant reduction in the adult emergence. Thus, NSKWE (25 g/l) had no adverse effect on *A. plutellae.* Similar results were obtained for *A. africanus* and *Telenomus remus* (Chari et al. 1997). Dobelin (1997) studied the side effects of NeemAzal-T/S (1.0 % azadirachtin) against two parasitoids of aphids, viz., *Aphidius colemani* and *Aphidoletes aphidimyza.* It was observed that neem products had no effects on these natural enemies.

Michelakis and Vacante (1997) advocated Neemark as a safe device for the control of

Phyllocnistis citrella that would not affect the parasitoid *Pnigalia* sp. They also implemented a biological control programme using *Ageniaspis citricola*, *Citrostichus phyllocnistoides* and *Semielacher petiolatus*, along with Neemark. Stansly and Liu (1997) found that neem extract, insecticidal soap and sugar esters had little or no effect on *Encarsia pergandiella*, the most abundant parasitoid of *Bemisia argentifolii* in south Florida vegetable fields, and can contribute significantly to natural biological control of this and other whitefly species. Olivella and Vogt (1997) collected 10 species of leaf-mining Lepidoptera in apple orchards in southwestern Germany in 1996, the most abundant being *Phyllonorycter blancardella*, *Lyonetia clerkella* and *Stigmella malella* and a mining curculionid, *Rhamphus oxyacanthae*. Of these total parasitism by chalcidoidea and ichneumonoidea ranged from 10 to 29 %. Use of a neem preparation for pest control had no effect on the rate of parasitism.

Sharma et al. (1999) reported that the extracts from neem and custard apple kernels were effective against the spotted stem borer, *Chilo partellus*; Oriental armyworm, *Mythimna separata*; head bug, *Calocoris angustatus*; and the yellow sugarcane aphid, *Melanaphis sacchari* in sorghum, but neem extract was non-toxic to the parasitoids and predators of the sorghum midge with slight reduction in parasitism. But Sharma et al. (1984) reported that an active neem fraction of NSK had an adverse effect on larval parasitoid *Apanteles ruficrus* of Oriental armyworm *M. separata*. Injection of 2.5–10 μg of azadirachtin to newly ecdysed fourth and fifth instar larvae of host either partially inhibited or totally suppressed the first larval ecdysis of braconid *Cotesia congregata*, an internal larval parasitoid of tobacco hornworm *Manduca sexta* (Beckage et al. 1988). They also reported that the parasitoid growth was arrested, while the host larvae survived for 2 weeks or longer, following injection of azadirachtin, but their parasitoids never recovered and died encased within exuvial cuticle. Lamb and Saxena (1988) gave topical treatment to the females of ectoparasite *Goniozus triangulifer* at doses from 5 to 50 μg/l solution of neem seed bitters. The results indicated decreased fecundity at 50 μg per female. When the rice plants were sprayed with 1,000 ppm neem seed bitters, very few larvae of leaf folder *Marasmia patnalis* sustained the development of *G. triangulifer* up to pupation stage. However, when 1,000 ppm neem seed bitters were sprayed three times, there was negative influence on parasitization by *G. triangulifer.* The studies further reported that when *Tetrastichus howardi* parasitized pupae of *M. patnalis* were dipped in 1,000 ppm of neem seed bitters, the adult emergence decreased significantly. However, topical application of 1,000–10,000 ppm of the neem seed bitters had no effect on *T. howardi*.

Loke et al. (1992) gave topical treatment of cocoons of *C. plutellae* with neem oil in the laboratory and found that neem inhibited adult eclosion significantly at 2.5 % concentration and no adult emergence was observed at 10 % level. Treated cocoons produced adults with reduced longevity but no morphological deformities. However, Osman and Bradley (1993) reported high mortality of larvae and morphogenetic defects of adult parasitoid *C. glomerata* developed from hosts treated with NSKE. Neem products did not affect adult parasitoids even after spraying with higher concentration, i.e. AZT-VR-K 2,000 ppm. Srivastava et al. (1997) reported that alcohol and hexane extracts of 17 neem ecotypes in India were found to be toxic to the egg, larval and pupal stages of the *B. brevicornis*. In general, the hexane extracts showed higher toxicity against the egg and pupal stages, whereas the alcohol extracts were more toxic against the larvae. Azadirachtin content of the neem ecotypes revealed no apparent correlation with the observed toxicity against different stages of the parasitoid.

Varied responses of parasitoids to various neem preparations have also been reported by several workers. For instance, Hoelmer et al. (1990) did experiments with parasitoids of *B. tabaci* and *Aphis gossypii* with the neem product Margosan-O. It was found that the aphid parasitoids, namely, *Lysiphlebus testaceipes* and *Aphelinus asychis*, were more sensitive to neem-treated surface, whereas the survival of the aphid parasitoid, *Eretmocerus californicus*, was the same on treated and untreated *Hibiscus* foliage. The *E. californicus* pairs in sealed Petri dishes with treated and untreated foliage survived for 5 days. It was also observed that dipping of aphid mummies parasitized by *L. testaceipes* and also dipping the parasitized puparia of *B. tabaci* by *Encarsia formosa* and *E. transversa* did not affect the emergence of the parasitoids. However, when *E. californicus* parasitized white fly puparia was dipped, the emergence of parasitoids was reduced by more than 5 %. Similarly, Stark et al. (1990) on the other hand found that a highly purified and concentrated neem extract prevented adult of fruit flies emergence from puparia. However, the parasitoid *Opius* sp. emerged freely. Neem seed kernel extracts reduced the population of *Encarsia* sp. and *Aleurodiphilus* sp. (Price and Schuster 1991). *A. indica* leaf extract was safe *to Diaeretiella rapae* (Men et al. 2002). Loke et al. (1992) reported reduction in the rate of emergence of braconid, *Cotesia plutellae*, when cocoons were sprayed with *A. indica* oil.

Schmutterer (1992) found in laboratory experiments that concentrations of 10 and 20 ppm of azadirachtin or an azadirachtin-free fraction and of an enriched and formulated seed kernel extract of *A. indica* were only slightly harmful to *C. glomerata*, provided they were applied against the 5th instar of *Pieris brassicae.* Under these circumstances, numerous larvae of *C. glomerata* emerged from their hosts, pupated and hatched as normal adults. However, higher concentration (40 ppm) of azadirachtin and of the azadirachtin-free fraction as well as 50 and 100 ppm of the enriched product reduced the number of parasitoids considerably. The parasitoids were mainly killed by lack of food and died within their hosts. Larvae of *P. brassicae* under the influence of metamorphosis disturbance by neem products did not die immediately after uptake of active principles, but there was reduced food uptake, leading to increased intraspecific competition among the gregarious grubs of *C. glomerata.* Direct growth regulation effects of neem products against *C. glomerata* were not observed. Application of neem products against young (1st–3rd) larval instars of *P. brassicae* led to the death of the caterpillars together with the grubs of the parasitoid.

It is obvious from the studies available so far that neem and its various products/formulations do have some side effects especially against the larval parasitoids. We have some specific reports of such side effects available. Beitzen and Hofmann (1992) studied the side effects of neem product AZT-VR-NR on the endoparasitic tachinid fly *Drino inconspicua* when the fly was exposed for 7 days with residues of neem product (45 g a.i./ha); it did not harm the adult flies, but fecundity was reduced by 18.5 % in comparison with control. Similarly, Serra (1992) observed no or only side effects of neem products on the parasitoids of the genera *Ganaspidium*, *Desorygma* and *Opius*, which emerged from the tomato leaf miner *Liriomyza sativae.* He also obtained similar results on the genera *Pseudapanteles* and *Glyptapanteles* that emerged from tomato pinworm, *Keiferia lycopersicella.* Moser (1994) observed no side effects of aqueous NSE (2.5 and 5.0 %) among natural enemies (coccinellids, syrphids, chrysopids and braconids) of *Aphis gossypii* on okra in Dominican Republic fields, but at the same time by slight harmful effects, viz., morphogenetic defects, delay of larval and pupal development was recorded in the laboratory experiments. Mineo et al. (2000) tested the side effects of azadirachtin mixed with mineral paraffin oil and a surfactant against the natural parasitoids of *P. citrella.* The observations revealed 16.67 % parasitoid larvae showing teratological symptoms.

Stark et al. (1992) studied the effect of azadirachtin on survival, longevity and reproduction of the three braconid parasitoids, namely, *Psystallia incisi* and *Diachasmimorpha longicaudata* from *Bactrocera dorsalis* and *Diachasmimorpha tryoni* from *Ceratitis capitata.* The results revealed that all host larvae that were exposed to sand treated with azadirachtin pupated and adult eclosion was concentration dependent in both fly species, with little or no fly eclosion at 10 ppm. However, *P. incisi* and *D. longicaudata* successfully eclosed from pupae treated with 10 ppm azadirachtin. In all the cases after the exposure of azadirachtin, the adult eclosion was inhibited. Even life spans of parasitoids that emerged from treated flies were not significantly different from controls. The azadirachtin had no effect on the longevity of parasitoid species tested in this study, indicating that the parasitoids were less sensitive to this chemical than were their hosts. The reproduction of *P. incisi* that developed in flies exposed to azadirachtin concentration of >20 ppm was reduced by 63.88 %. The reproduction of *D. longicaudatus* and *D. tryoni* was unaffected. This implies that neem-based products are safer at lower concentrations but induce adverse affects at higher levels of treatment, also obvious from the studies on neem products like Repelin and Neemguard that were tested on *Bracon hebetor* in laboratory and field conditions by Srinivasa Babu et al. (1996) to reveal their safety at lower concentrations against larval parasitoids. But at higher concentrations, both preparations adversely affected the development.

It is, however, also possible that parasitoids may be adversely affected due to lack of appropriate food. This is clear from the results of Jakob and Dickler (1996) where adults of the ectoparasitic, gregarious eulophid *Colpoclypeus florus*, an important parasitoid of the tortricid *Adoxophyes orana*, were not adversely affected by application of NeemAzal-S (25 and 100 ppm) in the laboratory and in the field, but 100 % of the larvae died, apparently due to lack of appropriate food on the neem-treated decaying larvae of the host. Schmutterer (1996) also described the varying sensitivity of bioagents to neem products, like eggs of predators such as coccinellids and chrysopids are not sensitive, but ectoparasitic gregarious larvae of *Bracon* sp. and *Colpoclypeus* sp. showed high mortality after contact with neem. Endoparasitic solitary or gregarious hymenopteran larvae were less endangered as their hosts protected them. Schmutterer suggested that often, lack of food in neem-treated hosts resulted in the death of parasitoids due to starvation.

Another aspect of interest is the IPM compatibility of neem with other products vis-a-vis the safety of natural enemies. The relative toxicity of pesticides to *Phyllocnistis citrella* and its parasitoid *Ageniaspis citricola* was compared by several bioassay methods. Azadirachtin (Neemix) + oil, diflubenzuron (Micromite) + oil, fenoxycarb (Eclipse) + oil and oil alone (FC 435–66) were classified as IPM-compatible insecticides. Sprays of azadirachtin (Align) + oil, neem oil (Neemguard) and drenched imidacloprid (Admire) were ranked as semi-compatible insecticides (Villanueva and Hoy 1998).

Teggelli et al. (1998) studied the effects of Nimbecidine (5 ml/l) and Achook (5 ml/l), nuclear polyhedrosis virus (NPV) (1.5 ml/l) and some recommended insecticides on the emergence of *Campoletis chlorideae* from host larvae 3, 5, 8 and 11 days after parasitization (DAP). Among insecticides, Achook resulted in the highest adult emergence (42.33 %) at eight DAP, while fenvalerate, methomyl, malathion, chlorpyrifos and monocrotophos completely inhibited emergence. At 11 DAP, the biopesticides, namely, Nimbecidine, Achook and NPV, recorded the highest percentage of emergence (58.66, 56.33 and 53.33 %, respectively), while monocrotophos was most toxic (8.66 % adult emergence). The toxicity of all insecticides was lower on cocoons. Nimbecidine and NPV did not cause mortality 24 h after treatment. Similarly, *Hypomecis* sp. caused severe damage to *Azadirachta indica* in Akola, India, during October 1988. *Apanteles fabiae* and *Aleides [Aleiodes]* sp. were observed parasitizing *Hypomecis* sp. (Men 1999).

An IPM strategy for the control of *P. xylostella* was formulated by Facknath (1999) using neem with *C. plutellae.* Reddy and Guerro (2000) evaluated biorational and regular insecticide applications for the management of the diamondback moth *P. xylostella* in cabbage. The IPM programme, based on the pheromone trap catch threshold of 8 moths per trap per night, included the utilization of *C. plutellae* (250,000 adults/ha), *Chrysoperla carnea* (2,500 eggs/ha), Nimbecidine (625 ml/ha), *Bt* (500 ml/ha) and phosalone (2.8 l/ha). The IPM programme induced a reduction of trap catches, egg and larval populations and, therefore, a low level of damage to the crop. All neem concentrations gave poor to very slight control of *Myzus persicae* when applied as contact action foliar sprays, with Pirimor^{R} providing the greatest contact kill. Neem at 180 ppm, when applied as a soil drench, gave total aphid control within 24 h, apparently through systemic action. Aphid parasitoids and other beneficial insects were not affected by neem treatments, whereas Pirimor[R] treatments reduced beneficial insect numbers. Although Pirimor[R] would be the preferred choice for immediate aphid control through contact action in commercial crop production, neem still has a place in the control of aphids in situations such as organic crop production or in crops where resistance to other chemicals by aphids has resulted. Other uses may be in indoor and outdoor landscape situations where human health is of major concern and a long-lasting systemic method of aphid control desirable. In these cases, neem could be applied as a soil drench at concentrations of 180 ppm, possibly through existing irrigation systems (Holmes et al. 1999). Perera et al. (2000) studied the effect of three feeding deterrents: denatonium benzoate (5, 50 and 250 mg/l), azatin [azadirachtin] EC (0.01, 0.1 and 1 ml/l) and Pestistat[R] (0.1, 1 and 2 ml/l) on the fourth instar larvae of important cabbage pests, *Chrysodeixis eriosoma* and *P. xylostella*, and on the parasitoid, *C. plutellae*. Results suggested that the three antifeedants were effective in managing cabbage pests, *C. eriosoma* and *P. xylostella*, and could be used in integrated pest management programmes.

Babu and Babu (2003) investigated the effects of neem-based formulations, viz., Neemguard and RD9 Repelin, on the developmental stages of *B. hebetor* and found that substantial reduction in hatching of treated eggs and growth disturbances but parasitoid pupae were spared. Dinesh et al. (2003) evaluated extracts of some locally available plant materials and propriety botanical formulations in comparison with a recommended insecticide (quinalphos: Ekalux[R]) for their usefulness against the coffee mealy bug (*P. citri*) and their effect on natural enemies (parasitoid, *Leptomastix dactylopii*, and attendant ant, *Anoplolepis longipes* [*A. gracilipes*]). The treatments included extracts of Tulsi (*Ocimum sanctum* [*O. tenuiflorum*]), Bilva (*Aegle marmelos*), calotropis (*Calotropis gigantea*), marigold (*Tagetes erecta*), 'Universal biopesticide' formulation (containing *Aloe vera* [*A. barbadensis*], *Lantana camara*, *Calotropis gigantea*, neem (*Azadirachta indica*) and *Vitex negundo*) and garlic. The proprietary products used were GB+ (100 % garlic preparation) and GB AG (77 % garlic and 22 % neem preparation). The treatments were effective against the mealy bug, and the parasitoid was relatively safe from its effect. Basappa and Lingappa (2004) tested various neem preparations, i.e. NSKE (5 %), neem leaf extract (NLE, 5 %), Margocide CK 20 EC (0.1 %), Achook (0.3 % water soluble neem powder) and Jawan (0.15 % neem extract) along with *Bougainvillea glabra* cold alcohol extract (CAE 30 %) against castor semilooper *Achaea janata* and its larval parasitoid (*Microplitis maculipennis* [*Snellenius maculipennis*]) under field conditions in Dharwad, Karnataka, India, during 1993 and 1994. NSKE, Margocide and Jawan were superior in reducing the larval population of castor semilooper at 3 days after spraying. Neem-based preparations were superior over B. glabra in reducing castor semilooper larval population at 3, 7 and 12 days after spraying. These botanicals also gave significantly higher seed yield than the untreated control. The highest cost/benefit ratio was recorded in NLE (1:2.80), followed by Margocide (1:2.20) and NSKE (1:1.94). A steady increase in *M. maculipennis* population was observed in all treatments after 3 days of spraying. Ahmad et al. (2003) did not recommend the use of parasitoids in combination with *A. indica* preparations since the application to the soil had a long-lasting effect on the rate of parasitization and on survival (as demonstrated with the parasitoid, *Diaeretiella rapae).*

Haseeb et al. (2004) evaluated the field doses of neem insecticides against *C. plutellae* under laboratory conditions. Agroneem (4.8 mg a.i./l), Neemix (20 mg a.i./l) and Ecozin (20 mg a.i./l) caused only 11.1, 16.7 and 5.6 % adult mortality, respectively. Vijay Bhardwaj et al. (2005) studied the adults of *D. fenestralis* [*D. fenestrale*] and *C. plutellae* by transferring into vials containing filter paper soaked in solutions of malathion (0.05 %), fenvalerate (0.01 %), fipronil (0.007 %), cypermethrin (0.015 %), Achook [*Azadirachta indica* extract] (0.3 %) and cypermethrin (0.0075 %) + *Bt* [*Bacillus thuringiensis*] (0.15 %).

Parasitoid mortality was evaluated after a 6-h exposure period and after 24 and 48 h posttreatment periods. *C. plutellae* mortality was highest with malathion (74.44 %), followed by fipronil (57.77 %), cypermethrin (27.77 %), fenvalerate (15.55 %), cypermethrin + *Bt* (14.44 %) and Achook (4.44 %). Malathion was also the most toxic to *D. fenestralis*, followed by fipronil, cypermethrin, cypermethrin + *Bt*, fenvalerate and Achook (94.44, 68.88, 67.77, 25.55, 24.44 and 3.33 % mortality, respectively), proving the safety of neem-based insecticide (Achook). Charleston et al. (2005) tested the effect of two botanical pesticides, viz., aqueous leaf extracts from the syringa tree *Melia azedarach* and commercial formulation from the neem tree *Azadirachta indica*, Neemix 4.5, in the laboratory and in a glasshouse on two species of parasitoids, *C. plutellae* and *Diadromus collaris*. No direct negative effect was recorded on the longevity of the parasitoid species. However, hind tibia length was found to be significantly shorter in male *C. plutellae* that emerged from *P. xylostella* that had been exposed to syringa extracts. In the glasshouse, a significantly higher proportion of *P. xylostella* was parasitized by *C. plutellae* on plants treated with neem than on the control plants. Rowell et al. (2005) reared six parasitoid species on diamondback moth larvae and pupae collected in northern Thailand. These included the larval parasitoid *C. plutellae*, a larval-pupal parasitoid *Macromalon orientale* Kerrich and pupal parasitoids *D. collaris* and *Brachymeria excarinata*. These parasitoids were effectively integrated in the pest management protocol of *P. xylostella* using simple presence-absence sampling for lepidopterous larvae, and the exclusive use of *Bacillus thuringiensis* or neem resulted in the highest yields of undamaged cabbage compared with a control or weekly sprays of cypermethrin (local farmer practice).

Adarkwah et al. (2011) reported that the commercial preparation of neem, Calneem[R], was safe to the larval parasitoids *Habrobracon hebetor* and *Venturia canescens*, while it controlled their hosts, rice moth *Corcyra cephalonica* (Stainton) (Lepidoptera: Pyralidae) and tropical warehouse moth *Cadra cautella* Walker (Lepidoptera: Pyralidae) in stored rice and wheat, respectively, in the laboratory. Kumar et al. (2010) tested two commercial neem products, viz., NeemAzal-T/S (1 % azadirachtin) as foliar application and NeemAzal-U (17 % azadirachtin) as soil application, in laboratory bioassays against about-to-emerge adults and adults of an aphelinid, *Eretmocerus warrae*, an efficient parasitoid of the whitefly *Bemisia tabaci*. They found that the longevity of the adult parasitoids was only affected after 36 h contact with high-dose residues of NeemAzal-T/S (10–15 ml l^{-1}) in a dry-residue bioassay test. But the experiments indicate that ecto-/endoparasitoids are principally highly vulnerable to neem but show in addition that soil application could reduce negative side effects compared to plant spraying and hence improve selectivity.

Alvarenga et al. (2012) treated the neem seed cake (NSC) extract to control the Mediterranean fruit fly along with its larval parasitoid *Diachasmimorpha longicaudata*. They found that NSC affected parasitoid emergence negatively. The effect of parasitism coupled to NSC did not provide greater reduction in the medfly emergence than when parasitism was used alone. However, each of these 2 methods affect a different life stage of medfly larvae and pupae, respectively, and their joint use may increase the probability of controlling medfly populations in field.

Leaf extract of *Datura metel* was safe to *Cotesia flavipes* (Reddy and Srikanth 1996). Aqueous leaf extract of *M. azedarach* did not show direct negative effect on the longevity of the parasitoid species of *Cotesia plutellae* and *Diadromus collaris* (Charleston et al. 2005). Nicotine sulphate from *N. tabacum* was toxic to *Apanteles congregata* (Barbosa et al. 1991).

4 Effect on Predatory Insects

Acetone extracts of *Aframomum melegueta* exhibited adverse effect on eggs, reduced the efficiency of predation in larvae and showed poor rate of pupation of Coccinellids, *Cheilomenes lunata* and *Cheilomenes vicina*

(Ofuya and Okuku 1994). Leaf extract of *Ailanthus excelsa* did not affect the predators such as coccinellids, chrysopids and syrphids (Patel et al. 2003). Leaf extract of *Aloe vera was safe* to grubs, pupae and adults of predators such as coccinellids and syrphid fly (Balikai and Lingappa 2004). Water extract of *Annona squamosa* was safe to *Chrysoperla carnea* and *Onus insidiosus* (Isman and Leatemia 2004). Methanolic extracts of leaves of *Atlantia monophylla* were found to be safe to aquatic mosquito predator, *Diplonychus indicus* (Sivagnaname and Kalyanasundaram 2004).

Bioactivity of neem, *A. indica*, against predatory insects and spiders is discussed under subheadings: Earwigs, Crickets, True Bugs, Ants, Beetles, Syrphids, Cecidomyiids, Lacewings, and Predatory Spiders and Mites.

4.1 Earwigs

NeemAzal-F was tested against the European earwig *Forficula auricularia* by Sauphanor et al. (1995). This earwig is a polyphagous crop pest and a predator at the same time. In peach and apricot orchards, for instance, it causes serious damage by feeding on ripening fruit, whereas in apple orchards it can be used against harmful populations of various aphid species. Adults of *F. auricularia*, exposed to 50 ppm AZA on glass plates (standardized method of IOBC/WPRS) in the laboratory, did not show increased mortality or reduced ingestion of food; fecundity was also not adversely influenced. On the other hand, second instar nymphs treated with 25, 50 or 250 ppm of NeemAzal-F could not complete their metamorphosis and died. They also exhibited reduced food intake and extended stadia. Neither repellent nor phagodeterrent effects were observed. Under field conditions in a peach orchard, the nymphal population of the earwig was reduced by 70 % when sprayed with NeemAzal-F at a concentration of 50 ppm. Hence, NeemAzal-F could be applied in peach or apricot orchards when a reduction of the nymphal population of the earwig is required but avoided if high numbers of the predator are desirable, for instance, in apple orchards to obtain a significant reduction of aphids. Earlier, Schauer (1985) and Eisenlohr et al. (1992) observed that *F. auricularia* had no side effects by neem products.

4.2 Crickets

The cricket *Metioche vittaticollis*, preying upon eggs of rice leaffolders in Asia, was not affected by spraying neem seed bitters (containing AZA and other active ingredients) at 10,000 ppm (8 l/ha: ultra low volume spray) in field trials in the Philippines (Lamb and Saxena 1988).

4.3 True Bugs

Chelliah and Rajendran (1984) tested the toxicity of seven insecticides against *Cyrtorhinus lividipennis*. The least toxic of the sprays was 0.07 % endosulfan, which was effective against rice hoppers, followed by 5 % neem oil; the corrected mortality percentages were 34.72–36.90 on 1 day after spraying, 7.40–38.11 on the second day and 20.01–20.03 on the third day. The other sprays tested (0.075 % quinalphos, 0.04 % chlorpyrifos, 0.07 % phosalone, 0.08 % monocrotophos and 1.0 % carbaryl) were highly toxic to the bugs exhibiting 36.07–53.96 % mortality on the third day. Sharma et al. (1984) observed that *Orius* sp., a predator of sorghum midge, *Contarinia sorghicola*, was unaffected by an active neem fraction.

A slight harmful effect of neem oil was reported to the mirid bug, *C. lividipennis* by Saxena et al. (1984). But Fernandez et al. (1992) conducted a trial in greenhouse against *C. lividipennis* using four treatments, viz., neem oil 3 %, aqueous NSKE 5 %, endosulfan and water, and observed no mortality in the case of neem oil and aqueous NSKE, while endosulfan induced cent per cent mortality. Serra (1992) reported malformed nymphs of the predator, *Nesidiocoris* sp., after spraying of neem seed water extract 4 % and neem oil 2 % in the laboratory. It was observed that these neem products had no significant effect on the field population of this bug. The toxicity of all the sprays diminished 5 days after spraying. Similarly, a delayed moulting and morphogenetic defects after spraying of Margosan-O on third instar nymphs of the pentatomid predator *Perillus bioculatus* of the Colorado potato beetle in the USA have been recorded (Hough and Keil 1991). Krishniah and Kalode (1992) reported that the LC50 for neem oil was 50 % for the black mirid bug, *Tytthus parviceps* was 2.88 %, whereas for its prey, the green rice leafhopper, *Nephotettix virescens*, was only 1.39 %.

Drescher and Madel (1995) reported increased mortality in test populations of the anthocorid *Orius majusculus* after NeemAzal-T/S treatment (concentrations 1:50 to 1:200), and after oral intake, the rate of emergence of first instar nymphs was reduced by 3 %. Neem oil's repellent or phagodeterrent effect was observed when treated eggs of *Sitotroga cerealella* served as food for the bugs. According to the guidelines of IOBC/WPRS for standardized tests on the side effects of pesticides, NeemAzal-T/S at 1:50 was 'slightly harmful' under laboratory conditions. They also reported neem oil's negative effect on fecundity, sex ratio, rate of emergence or behaviour.

Kareem et al. (1988) reported the effect of neem seed kernel extract on population of predatory mirid and spiders in rice and compared with those of monocrotophos (0.75 kg a.i./ha). It was revealed that the populations of mirids and spiders were also lower in plots treated with monocrotophos than in plots treated with neem, 48 days after treatment. Safety of neem formulations and insecticides to *Microvelia douglasi atrolineata*, studied for a predator of planthopper in rice ecosystem, revealed that Neemix (2 and 4 %) and Rakshak (0.2 and 0.5 %) were the safest neem formulations, whereas phorate and carbofuran (1 kg a.i./ha) granular application and quinalphos spray at 0.5 % were the least toxic to the predator (Lakshmi et al. 1998b). Also neem formulations vis-à-vis insecticides were safe to *C. lividipennis* after the application of Neemgold at 0.5 % and Neemix at 2.0 % even after 72 h exposure (Lakshmi et al. 1998c), though chlorpyrifos and monocrotophos recommended for rice caused 100 % mortality within 24 h of exposure. Ghelani et al. (2000) tested various synthetic and botanical pesticides for their contact toxicity to the eggs and nymphs of *R. fuscipes*. The data on mortality of eggs and nymphs revealed that all the synthetic insecticides were more toxic than botanical insecticides. Among synthetic insecticides, quinalphos was highly toxic, while endosulfan was least toxic. However, among botanical insecticides, nicotine sulphate was least toxic to the eggs and nymphs of *R. fuscipes.*

Sahayaraj and Paulraj (1999a, b) observed toxic effects of leaf extracts of *Azadirachta indica*, *Vitex negundo*, *Pongamia glabra* and *Calotropis gigantea* on different life stages of reduviid predator *Rhynocoris marginatus* by contact and stomach toxicity studies. Tedeschi et al. (2001) studied the side effects of three neem formulations (Neem-Amin EC, Stardoor and B.P. 20/S) on the mirid predator, *M. caliginosus*, in the laboratory. Direct toxicity tests on first instar nymphs exposed to fresh dry residues on glass plates at different doses demonstrated that all the products were harmful to the insects with LD50 values much lower than the maximum recommended rate (1.217, 0.264 and 1.083 mg a.i./l instead of 15, 31.5 and 80 mg a.i./l for Neem-Amin EC, Stardoor and B.P. 20/S, respectively). Moreover, a reduction of fecundity of the surviving females was assessed with Neem-Amin EC and B.P. 20/S. High mortality was recorded when the insects were introduced onto the plants just after the treatment, but no significant differences compared with the controls were observed 5 days after the treatment. The experiments showed that azadirachtin being biodegradable, thus having short persistence, makes this active ingredient a promising component in integrated pest management programmes, if time gap is guaranteed between the treatment and the introduction of the predator.

Sahayaraj and Karthikraja (2003) studied the effect of azadirachtin on egg hatchability, nymphal mortality and biological control potential of *Rhynocoris marginatus* on the cotton pest *Aphis gossypii.* The biopesticide at specific concentration did not affect the biological control potential of *R. marginatus.* Jaastad et al. (2009) stated that azadirachtin did not significantly affect the most important predatory hemipterans, *Anthocoris nemorum* and *Pachydiplax longipennis*, while the omnivorous *Psallus ambiguus* and *Aphelinus mali*, mainly regarded as beneficials, were negatively affected by azadirachtin treatment.

4.4 Ants

Hellpap (1985) tested the neem product (AZT-VR-K) fed larvae of *Spodoptera frugiperda* to the colonies of the ant, *Ectatomma vuridum.* The ants accepted the neem-treated larvae. The predatory earwigs, *Doru taeniatum*, were also exposed to the armyworm larvae. It was observed that after 7 days of exposure, there was significant difference in mortality among earwigs fed with treated larvae of armyworm. Schmidt and Pesel (1987) reported that worker ants were resistant when sprayed with neem products. On the other hand, feeding of AZT-VR-K and MTB/H20-K-NR to the red forest ant *Formica polyctena* led to a stimulation of egg production when low concentrations were used. In contrast, higher concentrations reduced the number of eggs drastically, sometimes down to zero after a few weeks. This effect could be reversed if feeding of neem products was stopped and untreated food supplied instead. Use of neem-based products with predatory ants, *Oecophylla smaragdina*, gave excellent control of fruit flies, *Bactrocera cucurbitae*, in organic agriculture system, but it was not sufficiently active to manage *Aulacophora* spp. (Rohan 2000).

4.5 Beetles

In laboratory experiments, adult *C. septempunctata*, kept on neem oil-treated glass plates according to IOBC/WPRS guidelines, did not show increased mortality or reduction of fecundity compared to untreated control, but the metamorphosis of the larvae was interrupted (Schmutterer 1981). In the same insect species when treated in laboratory and semi-field trials with AZT-VR-K (1,000 ppm) and a combination of it with neem oil (250–30,000 ppm), there was no effect on the emergence of first instar larvae from treated eggs (Kaethner 1990). Spraying on adults had no adverse effects on fecundity and activity (fitness), whereas the same treatment on fourth instar larvae under laboratory conditions induced mortality, especially of pupae that developed from treated larvae. Numerous adults that emerged from surviving pupae exhibited morphogenetic defects of their wings. In contrast, spraying on two coccinellid species including *C. septempunctata* in field cages did not result in any side effects. In laboratory studies of Lowery and Isman (1996), topical treatment of early second instar larvae of *C. undecimpunctata*, using 1 % neem oil, did not result in reduced pupation or emergence of adults as compared to controls.

Margosan-O had no harmful effect on *Delphastus pusillus* predating on *Bemisia tabaci* and *Scymnus* sp. predating on *Aphis gossypii* and *Myzus persicae* (Hoelmer et al. 1990). Margosan-O also did not show any adverse affect against predatory carabid beetle *Platynus dorsalis* when their soil habitat was treated with the neem product (Forster 1991). Saleem and Matter (1991) observed that the neem oil acted as temporary repellent against the predatory staphylinid beetle *Paederus alfieri*, the coccinellid *C. undecimpunctata*, and the lacewing *Chrysoperla carnea* in cotton, but otherwise neem oil had no adverse effect on the predators of *Spodoptera littoralis.* That neem oil had no adverse effect on predators is also obvious from the studies of Kaethner (1991), as it was found harmless to the eggs, larvae or adults of *Chrysoperla carnea* and *C. septempunctata.*

Mohapatra et al. (1991) observed even 24 % concentration of neem oil had no significant adverse effect on the coleopteran predators in rice, and Matter et al. (1993) demonstrated that although neem oil had residual activity for up to 6 days, yet it had no effect on survival or behaviour of larvae of *C. undecimpunctata* except for a prolongation of the fourth instar larva. Consumption of the aphids by this predator was unaffected. Eisenlohr et al. (1992) reported that NeemAzal-F had neem oil effect on oviposition of coccinellids in peach orchards, though residual toxicity of some insecticides and neem seed kernel extracts against the predatory beetle *Brumoides suturalis* has been recorded (Chandrababu et al. 1997). It was found that NSKE extract and endosulfan exhibited low toxicity to *B. suturalis* larvae and adults.

An interesting study of Patel and Yadav (1993) on the toxicity of some botanical and chemical insecticides to *Cheilomenes sexmaculata* and its hyperparasite *Tetrastichus coccinellae* shows that among the botanicals, nicotine sulphate (0.05, 0.04 and 0.03 %), Repelin (0.5, 0.75 and 1.0 %) and Neemark (0.05, 0.2 and 0.4 %) were highly toxic to adults of *T. coccinellae*, whereas they were absolutely safe to *C. sexmaculata*. In a detailed study of mortality and predation efficiency of *Coleomegilla maculata* following applications of neem extracts, it was observed that the toxicity of the neem extracts to *C. sexmaculata* was almost 100 % when both neem formulations were used at 10 % concentrations. The azadirachtin contents in neem oil (v/v) and neem seed kernels (w/v) were 13.7 and 91.0 ppm, respectively. Malathion was also tested at the field rate of 2.85 g a.i./ha (Roger et al. 1995). Adult mortality rate of the coccinellids after 72 h was 100 % following malathion treatments. Neem oil toxicity was observed after the treatments with the aqueous suspension of ground neem seeds. The predation efficiency of *C. sexmaculata* was also evaluated after topical application of these three insecticides at sublethal doses. Fifteen minutes after treatment, adult coccinellids were provided with 30 aphids for 24 h. The aqueous suspension of ground neem seeds caused 50 % reduction in the number of aphids consumed.

Stark and Wennergren (1995) opined that toxicity of pesticides to bioagents might not be straightforward, but the susceptibility of various life stages should be estimated for noteworthy concussions. Banken and Stark (1997) studied the stage and age influence on the susceptibility of *C. septempunctata* after direct exposure to neem product 'Neemix', where first instars were treated by direct application of 0, 40, 100, 200, 400, 600 and 1,000 ppm and fourth instars were treated with 400, 600, 800 and 1,000 ppm azadirachtin, the active ingredient in Neemix. The LC50 for first and fourth instars were estimated as 1,120 ppm and 520 ppm azadirachtin, respectively. These values were much higher than the recommended rates for control of aphids (3 weekly applications of 20 ppm), suggesting that Neemix might be used in IPM programmes because application rates that control aphids should result in appreciable mortality of predators. Fourth instar larvae of *C. septempunctata* were innately more sensitive to the growth-disrupting effects of acute exposure to Neemix than 151 instars. It is possible for early instars to sustain the effects of Neemix as long as the pesticide is detoxified before the onset of pupation. These results suggest that it is extremely important to examine more than one life stage of a species to estimate the total effect of pesticides. Banken and Stark (1998) also studied the exposure and the risk of neem products against *C. septempunctata* using direct sprays, residues on leaves and pesticide-contaminated prey. The pesticide alone and the predator caused significant decrease in aphid population. However, no significant ($P<0.05$) interaction between the predator and the pesticide was detected, indicating that the chemical and biological control agents were not working synergistically. Furthermore, exposure to the pesticide in microcosms significantly reduced or completely eliminated oviposition in adult *C. septempunctata*, and all of the larvae exposed to 100 or 600 ppm died within 10 days of treatment. Although survivorship of adult ladybird beetles was unaffected, exposure to Neemix resulted in a severe reduction in fecundity or complete sterility depending on the concentration.

Mani et al. (1997) studied the effect of 5 % neem seed kernel extracts on the predator *Cryptolaemus montrouzieri* and observed no detrimental effect on the progeny production. Dhaliwal et al. (1998) tested Achook and Nimbecidine for the control of insect pests on cabbage. The neem formulations were evaluated at 1, 2 and 4 kg/ha and compared to 0.5 kg a.i./ha of endosulfan used as treated control. Among these, endosulfan was the most effective against all the insect pests, followed by Achook and Nimbecidine. The feeding efficiency of the *C. septempunctata* on *L. erysimi* treated with neem-based insecticides was higher than for aphids treated with endosulfan. Studies on

L. erysimi control by Neemol and nicotine sulphate applied alone or in combination with chemical insecticides dimethoate and methyl-O-demeton in mustard (Vekaria and Patel 2000) have also revealed that both plant products were less toxic to the predators, *Diaeretiella rapae* and *C. septempunctata*, than the chemical insecticides. Chakraborti and Chatterjee (1999) also found that all formulations of neem were safe to the ladybird predators even at the highest concentrations (9 ml a.i./l).

Prasad and Logiswaran (1998) compared the toxicity of different insecticides to the adult of *C. sexmaculata* and reported that the neem oil is the safest insecticide based on LT 50 values. It was concluded that the less toxic phosalone, monocrotophos or neem oil could be integrated with the release of C. *sexmaculata* in the field. Azadirachtin and dichlorvos also induced lowest toxicity to the predator *C. montrouzieri* (Sundari 1998), and Neemix and Multineem had least effect against predatory coccinellids (Mishra and Mishra 1998).

Singh and Singh (1996) tested different neem-based formulations and synthetic insecticides on aphidophagous coccinellids on *Brassica juncea.* Achook (WSP), RD-9 Repelin, NeemAzal-T/S, Neemgold, Neemta 2100 and Nimbecidine at 0.03 % were quite safe to coccinellids than the synthetic insecticides such as endosulfan 35EC, fenvalerate 20EC, dimethoate 30EC and Chess 25EC. The order of safety was maximum in Achook followed by RD-9 Repelin, NeemAzal--T/S, Neemgold, Neemta 2100, Nimbecidine, endosulfan, Chess, fenvalerate and dimethoate during the first experimental trial (1994–1995) and Neemgold followed by Achook, Annona 20 EC, Neemta 2100, Achook EC, NeemAzal-T/S, endosulfan, Nimbecidine, Chess 25, fenvalerate and dimethoate during second experimental year (1995–1996). However, Imtiaz et al. (1998) found two neem extracts (RB-a and RB-b at 6,7,8,9 and 10 %) and an extract of bakayan *(Melia* sp.) berries (1, 2, 3, 4 and 5 %) toxic to the coccinellid *Coccinella* sp. and reported that 10 % RB-a and RB-b induced the highest mortality (85.7 and 82.5, respectively). Neem oil was, however, quite safe for natural enemies *Aphytis melinus* and *Chilocorus nigrita* predating *Aonidiella aurantii* (Krishnamoorthy and Rajagopal 1998).

Simmonds et al. (2000) investigated the effect of crude neem seed extract, a formulation of azadirachtin (Azatin), a pyrethrum extract and one of the two naphthoquinones isolated from *Calceolaria andina* Benth on the foraging behaviour of the *C. montrouzieri* larvae and adults. All the botanicals influenced the foraging behaviour of *C. montrouzieri*, at one or more concentrations. Larval and adult foraging behaviour was influenced most by neem that also affected larval behaviour; the predators contacted fewer treated leaves and spent less time on treated than on untreated leaves. Larvae also consumed fewer mealy bugs treated with naphthoquinones.

Ma et al. (2000) assessed the toxicity of several biorational pesticides and chemicals to *H. armigera* and *H. punctigera* and also on the major predators in cotton ecosystem. Moderate dose-dependent control was obtained in plots treated with neem seed extract, azadirachtin (AZA) at rates of 30, 60 and 90 g/ha. Plots treated with Talstar EC (bifenthrin) applications achieved the best results, followed by treatment with alternation of chemicals (methomyl, bifenthrin, thiodicarb and endosulfan) and biorational insecticides (neem oil, azadirachtin and *Btk*). Predators, including coccinellids, chrysopids, Araneae and hemipterans, were insensitive to AZA, toosendanin (Tsdn) and *Bt* applications. In contrast, chemicals were very toxic to predators. The toxicity of azadirachtin to predaceous insects attacking bollworm, *H. armigera*, by exposing *Menochilus signatus* and lacewings, *Harmonia conformis*, to neem oil (50 and 200 ppm) and endosulfan (50 and 200 ppm) through prey, which had consumed one or the other of these compounds, showed that endosulfan decreased predation rates by *H. conformis* at 50 ppm. However, azadirachtin, when ingested with prey, did not affect predation rates between 50 and 200 ppm concentrations (Oi et al. 2001). Neither of these pesticides caused direct mortality to adult beetles or lacewing larvae at the tested concentrations. Azadirachtin at both concentrations delayed pupation of *M. signatus* and extended duration of the larval stage, which increased the number of prey consumed by the predator causing serious morality of the pupae. However, pupal lacewings were all killed by 200 ppm azadirachtin treatment and 50 % at 50 ppm azadirachtin treatment, distinctly reducing the population of the next generation.

Two ladybird beetles, *Cycloneda sanguinea* and *Harmonia axyridis*, were tested in the laboratory to eight fungicide formulations commonly used in citrus production in Florida, USA. Both benomyl and the combination of copper and petroleum oil proved toxic to larvae of *C. sanguinea* that were exposed to concentrations corresponding to recommended field rates, either as leaf residues or in topical spray applications. Larvae of *C. sanguinea* also suffered lethal effects when exposed to neem oil as a leaf residue, but not after topical application. No compound appeared repellent to adult beetles of either species (Michaud 2001). Jalali and Singh (2001) reported that at field recommended rates, endosulfan and neem-based product (Replin) were safe to adults of *C. sexmaculata* immediately after spraying, giving 20 and 0 % mortality in a semi-field test up to 25 days after spraying on cotton plants. Residues of Repelin^R were also safe to grubs.

Lok Nath and Singh (2003) evaluated the safety of four plant extracts, one neem formulation and a synthetic insecticide (dimethoate) to the ladybird beetle (*C. septempunctata*) and syrphid flies (Syrphidae) predating on *H. coriandri* infesting coriander (*Coriandrum sativum*) in Kumarganj, Faizabad, Uttar Pradesh, India, during 2001/2002. One day after spraying, Pride of India (*Lagerstroemia indica*) seed kernel extract (PSKE 1 %) was the safest to both predators, followed by karanj (*Pongamia glabra* seed kernel extract (1 %), neem (*A. indica*) seed kernel extract (1 %), buken (*Melia azedarach*) seed kernel extract (1 %) and Neemarin (1 %). Dimethoate (0.03 %) was highly toxic to the ladybird beetle and syrphid larvae. A similar trend was evident at 3, 5 and 7 days after spraying. Thus, the botanical extracts, especially PSKE, were safer and more environment-friendly insecticides when used on coriander crop for aphid control. Chakraborti (2004) reported that detopping of affected shoots with pests at 16 days after transplanting (DAT) followed by application of neem cake at 3 and 1 kg/m^2 at 20-day intervals or foliar application of neem oil at 10 ml/l + azadirachtin at 4 ml/l at 7-day intervals beginning at 17 DAT were found to give better control of *A. gossypii*, *Scirtothrips dorsalis* and *Polyphagotarsonemus latus* on chilli crop, and at the same time, they spared coccinellids, syrphids and spiders compared to chemical control.

Balikai and Lingappa (2004) tested the aqueous extracts of some selected plant products during post rainy seasons in Bijapur, Karnataka, India, against the potential predators of aphids (*C. sexmaculata*, *C. septempunctata*, *C. carnea*, *Syrphus* sp. and *Ischiodon scutellaris*). The plant products were 2–3 times less toxic than malathion. Endosulfan was as toxic as 5 % *Datura metel* whole plant extract, 0.05 % Neem soap and 2 % *Pongamia pinnata* kernels. The botanicals such as 5 % *Ricinus communis* leaves, 5 % *Argemone mexicana* whole plant and 2 % *Prosopis juliflora* leaves were less effective to the predators. The plant products 5 % *Catharanthus roseus* leaves, 5 % *Pongamia pinnata* leaves, 5 % *Azadirachta indica* kernels, 5 % *Vitex negundo* leaves and 5 % *Adhatoda vasica* leaves were safe to natural enemies, and hence they can be effectively utilized in sorghum ecosystem for aphid management. Abudulai et al. (2004) studied the effects of Neemix 4.5 EC (at 210.4 g azadirachtin/ha) on the predators of *N. viridula* eggs on cowpea plants in the fields. Egg predation was not significantly different between the neem-treated and water-treated eggs in 2000 and 2001. Similarly, the percentage of predation on eggs in treated and untreated plots was not significantly different in both years. During the study, red imported fire ants (*Solenopsis invicta*) were regularly seen preying on egg masses of *N. viridula* in the field. Also, *Coccinella septempunctata*, *Coleomegilla maculata lengi* and other coccinellid larvae (Coccinellidae) were observed preying on eggs. Other predators of eggs were *Geocoris punctipes*, *Conoderus falli*, *Oecanthus celerinictus* and *Gryllus* sp. Silva and Martinez (2004) studied the effects of the neem seed oil aqueous solution on survival and performance of egg, larva and adult stage of the coccinellid predator

C. sanguinea under laboratory conditions. In a first trial, eggs and 2nd instar larvae were sprayed with the neem solutions at 0, 0.5 and 2.25 ml/l. Spraying the eggs did not affect egg hatch or larvae survival and development. When the larvae were sprayed, significant mortality was observed only at the higher concentration, and larval development and predatory capacity were not affected. Also, adults that emerged from treated larvae showed no alterations on sex rate, fecundity, fertility and longevity, thus indicating that at the tested concentrations, the neem oil does not reduce the reproductive potential of the species.

Hamd et al. (2005) conducted greenhouse experiments at Khartoum (Sudan) with the aphid predator *Hippodamia variegata* and NSKE (25 g/1), NeemAzal-T/S (1 % azadirachtin A) and fenvalerate (Sumicidin^{R} 20EC). An equivalent of 400 l/ha was applied with a plastic hand sprayer, containing 0.2 l Sumicidin (40 g/ha fenvalerate), 1.6 l NeemAzal (16 g/ha azadirachtin A) or 10 kg of neem seed powder (ca. 30 g/ha azadirachtin A was applied as neem seed water extract). The beetles were fed on *Aphis gossypii* reared on cucumber leaves. Four different stages of the predator were sprayed topically with the test preparations. The preparations varied in their effects on the different stages of the predator. NSKE was less harmful to the predator than NeemAzal-T/S. In topical treatments with NeemAzal-T/S, the corrected mortality (%) was eggs 37.7, larvae 40.0, pupae 38.2 and adults 16.7; with neem seed water extract, eggs 15.1, larvae 26.7, pupae 29.4 and adults 10.0; and with fenvalerate, eggs 86.8, larvae 100, pupae 73.5 and adults 100. Feeding L2 larvae and adults on contaminated aphids resulted in the following corrected mortalities (%): Sumicidin 100, NeemAzal ca. 40 and neem seed water extract 20/17 (larvae/adults). Feeding on aphids treated via the soil with neem seed water extract resulted at maximum in 26.4/16.7 % corrected mortality (larvae/adults), while no effects on the longevity of adults could be observed. So, in contrast to Sumicidin, neem preparations (being effective against pests) proved to be harmless for the beneficial beetle.

Larvae of the seven spotted ladybird beetle *Coccinella septempunctata* were treated with azadirachtin, and the impact on haemogram was investigated for 1, 30 and 60 min after treatment. Total haemocyte count increased 1 min after treatment with azadirachtin. Overall, azadirachtin was relatively safe for *C. septempunctata* larvae (Anjum et al. 2007).

Swaminathan and Hussain (2010) tested the side effects of botanicals, viz., neem seed kernel extract, eucalyptus oil and neem oil, against aphidophagous coccinellids, *Adonia variegata*, and found that neem seed kernel botanicals NSKE 10 % caused the highest mortality followed by neem oil (5.0 %), and the posttreatment effect (1 day after) evinced maximum reduction in feeding for NSKE (10 %) followed by neem oil (5 %).

4.6 Syrphids

Field trial conducted using neem emulsifiable concentrate for the control of sorghum aphid *Melanaphis sacchari* did not show any adverse effect on syrphid larvae and adults of coccinellids (Srivastava and Parmar 1985). Third instar larvae of the hover fly *Episyrphus balteatus* were mostly killed when treated with 100 ppm of an enriched seed kernel extract MTB/H20-VR-K synergized with sesame oil combined in a ratio of 1:4 (Schauer 1985). The larvae/pupae of syrphid flies seem to be more sensitive to neem products than those of other predators. Eisenlohr et al. (1992) reported that the number of syrphid larvae was not reduced in the field after spraying of NeemAzal-F on peach trees infested by *Myzus persicae*, but the survival of adults derived from larvae collected in the field on treated trees and held afterwards in the laboratory was quite low. Lowery and Isman (1996) observed that adult emergence of *Eupeodes fumipennis* was reduced by neem oil (0.5, 1, 2 %) to 35, 24 and 0 %, respectively, in comparison with controls.

4.7 Cecidomyiids

Lowery and Isman (1996) reported that the number of larvae of predaceous cecidomyiids was reduced in the field after application of neem seed extract and neem oil (1 %) as compared to controls.

4.8 Lacewings

Neem seed kernel suspension 2 % sprayed on tobacco plants conserved *Chrysopa scelestes*, an egg and larval predator of *S. litura* (Joshi et al. 1982). The adults of the lacewing *Brinckochrysa scelestes* (*Chrysopa scelestes*) were repelled from egg laying on cotton plants after they were sprayed with various commercial neem products of Indian origin and aqueous NSKE (Yadav and Patel 1992). First instar larvae of the predator emerged normally from treated eggs. Polyphagous predator, *Chrysoperla carnea*, treated in laboratory and semi-field trials with AZT-VR-K (1,000 ppm) and with a mixture of this product with neem oil (250–30,000 ppm) induced neem oil toxicity on eggs or adults; the fecundity of the latter was also not significantly affected (Kaethner 1990, 1991). The number of eggs (fecundity) laid by adult females developed from treated larvae was normal. The mortality of larvae fed with neem-treated aphids did not differ from that of controls. On the other hand, 79 % mortality of larvae occurred after topical treatment in the laboratory. In contrast, spraying of potato plants together with larvae of *C. carnea* in screenhouses did not result in any toxic or morphological effects.

Vogt (1993) did not find any significant influence of NeemAzal-F on the larvae of the lacewing in field trials. In laboratory experiments of Hermann et al. (1997), high mortality of larvae and pupae of *C. carnea* occurred if larvae were kept on NeemAzal-T/S (0.3 and 0.6 %) contaminated glass plates (IOBC/WPRS standardized tests), but practically no mortality was found in semi-field trials. Vogt et al. (1997) also studied the effectiveness of NeemAzal-T/S at 0.3 % against *Dysaphis plantaginea* on apple and on its side effects on *C. carnea*. A single application of NeemAzal-T/S in April gave very good control of *D. plantaginea* for about 5–6 weeks. After this period, *D. plantaginea* built up new colonies and *Aphis pomi*, too, increased in abundance. Yield losses caused by *D. plantaginea* were significantly lower in the neem-treated plot than in the untreated control plot. The side-effect test revealed that in the field NeemAzal-T/S was harmless to larvae of *C. carnea*. Neem seed extract was also found safe to *C. carnea* in comparison to nine insecticidal products (Sarode and Sonalkar 1999a) where chlorpyrifos, deltamethrin and cypermethrin were found highly toxic to *Chrysoperla*. There was no mortality of *C. camea* due to neem-based pesticides like NSKE at 5 %; Neemark, Achook and Nimbecidine each at 0.003 % and neem oil at 1 % (Deole et al. 2000). On the contrary, Srinivasan and Babu (2000) evaluated NSKE and commercial neem products, viz., NeemAzal-T/S, NeemAzal-F, Nimecidine, Neemgold, TNAU neem product 0.03 % EC, TNAU neem product neem oil 60 EC and Indeem against eggs, grubs and adults of *C. carnea*. The products caused 14.66–25.33 % egg mortality compared to 8.00 % in untreated controls and 6.66–16.66 % grub mortality compared to 3.33 % in controls. The longevity of treated adults ranged from 18.66 to 20.66 days in treatments, while it was 23.66 days in control. Fecundity was also affected slightly by all neem products (599.66 to 741.66) as against 874.66 eggs in controls.

Ingawale et al. (2005) determined the effective and safer insecticides and plant products for the control of lucerne aphids (including *Acyrthosiphon pisum* and *A. kondoi*) and their effect on the pest predators (ladybirds and *Chrysoperla carnea*) in an experiment which was conducted during November 2002 in Rahuri, Maharashtra, India. The treatments comprised sprays of malathion 0.05 %, DDVP [dichlorvos] 0.05 %, deltamethrin 0.0075 %, dimethoate 0.03 %, neem [*Azadirachta indica*] seed extract 5 %, neem leaf extract 10 %, nirgudi leaf extract 10 %, Econeem 2 %, Nirma 2 % and untreated control. Deltamethrin, dimethoate, dichlorvos, malathion and Econeem gave consistently superior results over other treatments for the control of aphids at 2 and 7 days after spraying (DAS). At 15 DAS, only deltamethrin and dimethoate gave the best results. The untreated control recorded the maximum number of ladybirds followed by the botanical insecticides. As regards *C. carnea* larvae, Econeem at 2 DAS and nirgudi leaf extract at 7 and 15 DAS showed the highest population.

Khan et al. (2013) tested neem oil and *Chrysoperla carnea* in different combinations against aphids in canola. Among the treatments, they found that module consisting of neem oil 2 % + *C. carnea* proved very effective in reducing the aphid population and neem oil concentrations relatively safe to predators and suitable for use in integrated pest management of aphids in canola.

5 Effect on Predatory Spiders and Mites

Saxena et al. (1984) reported that the wolf spider *Lycosa (Pardosa) pseudoannulata*, an important predator of leafhoppers in rice fields in Asia, was not harmed by neem oil and alcoholic or aqueous NSKE. In fact, neem oil (3 %) and aqueous NSKE (5 %) were quite safe for the spiders, though endosulfan induced 100 % mortality of the predators (Fernandez et al. 1992). NSKE, neem oil or neem cake extract (10 %) treated rice plots had better recolonization of spider *L. pseudoannulata* than in monocrotophos (0.07 %) treated plots after 7 days of treatment (Raguraman 1987; Raguraman and Rajasekaran 1996). The same neem products also spared the predatory mirid bug, *C. lividipennis* (Mohan 1989). The population of *L. pseudoannulata* and *C. lividipennis* were reported to be unaffected by different neem seed kernel extracts in paddy crop (Saxena 1987, 1989; Shukla et al. 1988; Jayaraj et al. 1993; Mariappan et al. 1993). Similar observation on rice crop was made by Nirmala and Balasubramaniam (1999) who studied the effects of insecticides and neem-based formulations on the predatory spiders of rice ecosystem. It was observed that feeding efficiency of *L. pseudoannulata* was higher than *T. javana* in all the treatments except in NSKE against green leafhopper *Nephotettix virescens* as prey, whereas rise in body weight was obtained in both predator species when they were treated with neem products, indicating the safety of neem to spiders. Babu et al. (1998) also reported that a combination of seedling root dip in 1 % neem oil emulsion for 12 h + soil application of neem cake at 500 kg/ha + 1 % neem oil spray emulsion at weekly intervals gave an effective level of control of green leafhopper (*Nephotettix virescens*) infesting rice (var. Swarna). A combination of neem oil + urea at a ratio of 1:10 when applied three times at the basal, tillering and panicle initiation stages gave a superior level of control of brown planthopper (*Nilaparvata lugens*). The treatments, urea + nimin [neem seed extract] and a seedling root dip with 1 % neem oil emulsion + neem cake at 500 kg/ha + 1 % neem oil spray emulsion at weekly intervals was equally effective against *N. lugens*. All neem products had little effect on predators, *C. lividipennis* and *L. pseudoannulata* (Sontakke 1993; Babu et al. 1998). NSKE sprays at 5, 10 and 20 % were also substantially safe for spiders and ants in cowpea ecosystems (Sithanantham et al. 1997).

Mansour et al. (1986) studied the toxicity of NSKE from different solvents on the spider *Cheiracanthium mildei* and found that NSKE 2 % did not affect the spiders. But at 4 % concentration, the sequence of toxicity of the extracts was pentane > acetone > ethanol > methanol and water; the latter two solvent extracts were nontoxic. Mansour et al. (1993) reported that the commercial products, namely, Margosan-O, Azatin and RD9 Repelin, showed no toxicity to the spider. Wu (1986) and Serra (1992) observed that the neem products were not at all toxic to predatory spiders. Nanda et al. (1996) observed the activity of natural enemies in cucurbit fields, where neem-based pesticides were applied for the control of *Henosepilachna vigintioctopunctata*. Natural enemies observed in considerable numbers were *Tetrastichus* sp., *Chrysocoris johnsoni*, *Tetragnatha* sp., *Oxyopes* sp. and orb-web spiders, and neem product did not inflict any harm to them. Lynx spider, *Oxyopes javanus*, was less sensitive to neem oil (50 % EC) than *L. pseudoannulata* (LC50 values = 9.73 and 1.18 %, respectively) (Karim et al. 1992), thereby confirming that neem oil was the safest pesticide for spiders (Wu 1986). In corn (Breithaupt 1995) and cabbage fields (Saucke 1995) in Papua New Guinea, no significant effect was observed against *Oxyopes papuanus* from aqueous NSKE (2 %) or NeemAzal-S treatments. Serra (1992) did observe adverse effects from NSKE 4 % applied on unidentified spiders in tomato fields in the Caribbean.

The bioefficacy tests of neem derivatives against the predatory wolf spiders (*L. pseudoannulata*), jumping spider (*Phidippus* sp.), lynx spider (*Oxyopes* sp.), dwarf spider (*Callitrichia formosana*), orb spider (*Argiope* sp.), damselflies (*Agriocnemis* sp.) and mirid bug (*C. lividipennis*) showed that neem seed kernel extract and neem oil were relatively safer than the insecticides to *L. pseudoannulata*, *Phidippus* sp. and *C. lividipennis* in field conditions (Nanda et al. 1996). Markandeya and Divakar (1999) evaluated the effect of a commercial neem formulation (Margosan 1500 ppm) in the laboratory against two parasitoids and two predators. The formulation was tested at the field recommended dose of 10 ml/l. The neem formulation Margosan 1,500 ppm was safe to all the four bioagents studied, viz., *T. chilonis*, *B. brevicornis*, *L. pseudoannulata* and *C. sexmaculata.* Spider population in rice ecosystem was the lowest in carbofuran treatment and highest in neem cake treatments. The mean predator population of *Ophionea indica*, *Paederus fuscipes*, *Lycosa* sp. and coccinellid beetles was significantly higher in plots with *Azolla* at 5 t/ha, with or without neem cake at 1.51/ha, in field trials conducted in southern Tamil Nadu, India, under lowland rice irrigated conditions (Baitha et al. 2000).

Predatory mites and spiders also showed appreciable tolerance to neem-based insecticide sprays both in laboratory and field. The effect of Neemguard obtained from *A. indica* seed kernels on the predacious mite *Phytoseiulus persimilis* and the predatory spider *Cheiracanthium mildei* was investigated in laboratory experiments. Neemguard had no toxic effect on *C. mildei* or *P. persimilis* (Mansour et al. 1997). Azadirachtin (Nimbokill 60 EC)-based products were relatively safe to the predatory mite *Amblyseius fallacies* (Kain and Agnello 2002). Rao et al. (2003) reported that neem oil, Biobit[R] + flufenoxuron, profenofos + Biobit and lambda-cyhalothrin + Biobit showed the best control of the rice leaffolder and also conserved the spiders in the field. Biobit treatment alone resulted in a leaf folder damage of 17.55 % folded leaves with predator population 4.33–5.33 predator/five hills.

Punzo (2005) conducted an experiment to evaluate the effects of azadirachtin on the mortality, growth and immunological function of the whipscorpion, *Mastigoproctus giganteus*, a large arachnid predator commonly found in arid areas and agroecosystems in southwestern USA and Florida. Ingesting prey injected with 1.0 and 10.0 mg/l of neem seed extract containing azadirachtin resulted in significant mortality over the 30-day test period. Ingestion of prey injected with 10 ppm azadirachtin gave significant mortality in both protonymphs and adult females. The ingestion of prey treated with 1.0 and 10 mg azadirachtin significantly decreased the size of *M. giganteus* nymphs. The pupae of azadirachtin-treated insects often exhibit deformities to the head and thoracic appendages.

Water extract of *Calotropis gigantea* was evaluated to *Chrysoperla carnea* and reported to exhibit normal egg hatchability and less larval and adult mortality (Patil et al. 1997). Aqueous suspension of *C. gigantea* did not affect hatchability and incubation period of *Rhynocoris marginatus* (Sahayaraj and Paulraj 1999a). Leaf extract of *Catharanthus roseus* was safe to grubs, pupae and adults of predators such as coccinellids, chrysopids and syrphid fly (Balikai and Lingappa 2004). Patil et al. (1997) also reported the safety of *C. roseus* water extract to *C. carnea.* Methanol and ethanol crude extracts of *Chenopodium ficifolium* were harmful to beneficial insects (Quang et al. 2010). Matter et al. (1993) tested the oils from *Citrus aurantium*, *Melia azedarach* and *Melia volkinsis* on *Coccinella septempunctata* in the laboratory; none of them affected the adult survival and behaviour and also the consumption of aphids. *Chrysanthemum coronarium* whole plant extract was safe to beneficial natural enemies (Mourad et al. 2008). Acetone extracts of *Cymbopogon citrates* exhibited adverse effects on eggs, reduced efficiency of predation in larvae and induced poor rate of pupation on the Coccinellids,

Cheilomenes lunata and *Cheilomenes vicina* (Ofuya and Okuku 1994).

Whole plant extract of *Datura metel* was safe to grubs, pupae and adults of coccinellids, chrysopids and syrphid fly (Balikai and Lingappa 2004). Rotenone from *Derris elliptica* caused mortality of larvae and adults of *Coleomegilla maculata* and *Chrysoperla carnea* (Hamilton and Lashomb 1997), *Perillus bioculatus* (Goldstein and Keil 1991), *Dolichogenoidea* and *Chrysoperla externa* (Iannacone and Lamas 2003a, b). But rotenone was safe to all the three instars of generalist predator, *Metacanthus tenellus* (Oliver and Bringas 2000).

Leaf extract of *Ipomoea carnea* was safe to predators such as coccinellids and chrysopids (Patel et al. 2003). Patil et al. (1997) evaluated the effect of *Lantana camara* to *C. carnea* and reported normal egg hatchability and low mortality of larvae and adults. Leaf extracts of *L. camara* and *Jatropha curcas* were safe to predators such as coccinellids, chrysopids and syrphids (Patel et al. 2003). Leaf extract of *Justicia adhatoda* (syn. *Adhatoda vasica)* was safe to grubs, pupae and adult stages of predators such as coccinellids, chrysopids and syrphid fly (Balikai and Lingappa 2004).

Leaf extract of *Melia azedarach* showed low mortality of mirid predator (Jazzar etal. 1999). Leaf extract of *Mentha spicata and Thevetia neriifolia* was safe to coccinellids, chrysopids and syrphids fly (Patel et al. 2003). Acetone extract of *Momordica charantia* exhibited adverse effects on eggs, reduced efficiency of predation by larvae and induced poor rate of pupation on coccinellids, *Cheilomenes lunata* and *C. vicina* (Ofuya and Okuku 1994).

Nicotine sulphate from *Nicotiana tabacum* did not affect adults of *Leptomastidea abnormis*, *Scymnus* sp. (Viggiani 1973) and *C. septempunctata* (Singh et al. 1985). But nicotine sulphate was found to be toxic to *Phytoseiulus persimilis* (Stenseth 1990), to *Amblyseius eharai* (Kashio 1983), to *Hyposoter annulipes* (Heneidy et al. 1988), to *Campoletis chlorideae* (Gunasena et al. 1990), to *Dacnusa sibirica* and *Diglyphus isaea* (Heyler et al. 1992) and to *Apanteles africanus* (Chari et al. 1996). On the contrary the nicotine from *N. tabacum* was safe to *Curinus coeruleus* (Diraviam and Viraktamath 1993).

Patil et al. (1997) evaluated the effect of leaf and whole plant extracts of *Parthenium hysterophorus* and found that the extracts did not affect eggs, larvae and adults *C. carnea.* Water extract of *Pongamia pinnata* did not influence hatching and incubation periods of eggs of *Rhynocoris marginatus* (Sahayaraj and Paulraj 1999). Derivatives of *P. pinnata* were least harmful to beneficial insects (Jayaraj 1991; Saminathan and Jayaraj 2002). Leaf or seed extracts of *Parthenium hysterophorus, Pongamia pinnata, Prosopis juliflora and Ricinus communis and Vitex negundo* were safe to grubs, pupae and adults of predators such as coccinellids, chrysopids and syrphid fly (Balikai and Lingappa 2004).

Ryanodine isolated from *Ryania speciosa* was safe to predators such as *Tetranychus urticae* and *Amblyseius longispinosus* (Schicha 1975), *Stethorus loxtoni*, *Stethorus nigripes* and *Stethorus vagans* (Walters 1976) and *Typhlodromus helenae* (Schicha 1977). Koodalingam et al. (2009) tested the sublethal concentrations of *Sapindus emarginatus* and observed that the kernel extract was found to be safe for two non-target aquatic larvae of *Chironomus costatus* and the nymphs of *Diplonychus rusticus.*

Alpha-terthienyl from roots of *Tagetes* species possessed all the desirable properties of a good insecticide/pesticide. It is fast acting, non-toxic, economic and property of degradation makes it more user friendly and safe (Manish et al. 2001). *Trichilia havanensis*-derived limonoids namely azadirone and 1,3+1,7-di-*O*-acetyl-havanensin did not show any toxic effect on two natural enemies *Chrysoperla carnea* and *Psyttalia concolor* (Pilar et al. 2006). Patil et al. (1997) reported that extracts of *Thuja occidentalis* and *Vitex negundo* did not significantly reduce egg hatching and spared the adults *C. carnea.* Water extract of *V. negundo* did not affect the hatchability and incubation period of *Rhynocoris marginatus* (Sahayaraj and Paulraj 1999a, b). Acetone extract of *Zingiber officinale* exhibited adverse effects on eggs, reduced predatory efficiency of larvae and induced poor rate of pupation to *Cheilomenes lunata* and *C. vicina* (Ofuya and Okuku 1994).

6 Effect on Honey Bees, Other Social Wasps and the Environment

Allophylus edulis extracts did not show any toxic effect against beneficial insect honey bee *Apis mellifera* (Lucia et al. 2009). Oil from *Allium sativum* was toxic to beneficial insects (Olkowski et al. 1995). Bees were moderately harmed by spraying *A. indica* extracts before flowering (Schmutterer and Hoist 1987). Neem products were found to be harmless to spiders, butterflies, ladybird beetles, wasps and bees (Saxena 1987; Schmutterer and Hoist 1987). Neem extracts had minimal toxicity on non-target organisms such as predators and pollinators (Naumann and Isman 1996) and degraded rapidly in the environment (Barrek et al. 2004). *A. indica* oil spray treatments were found to have no effect on adult honeybee populations (Melathopoulos et al. 2000). Neem seed oil (1 %) showed reduction in pollinator population 1 day after application (Harjindra Singh et al. 2010).

Azadirachtin, one of the more potent bioactive compounds from *A. indica*, was, in general, harmless to butterflies, bees, ladybirds and wasps (NRC 1992) and has very low mammalian toxicity and is relatively safe to beneficial insects (Gandhi et al. 1988). Azadirachtin was toxic to bee larvae, though less toxic to adults (Peng et al. 2000). Azadirachtin seems to be selective, non-mutagenic and readily degradable, with low toxicity to non-target and beneficial organisms, and causes minimal disruption to the ecosystem (Sundaram 1996). Azadirachtin can cause direct mortality, as it was determined for larvae of *Apis mellifera* and *Phormia terraenovae* (Rembold et al. 1981; Wilps 1987).

Commercial products from *A. indica* are reported to be harmless to natural enemies, pollinators and other non-target organisms (Ranga Rao et al. 2008; Singh and Singh 1996). *A. indica* formulations were found to be quite safe to spiders (Samiayyan and Chandrasekharan 1998). *A. indica*-based insecticides had negligible effects on beneficial insects and low environmental impacts (Schmutterer 1995; Haseeb et al. 2004; Greenberg et al. 2005; Isman 2006). *A. indica* preparations are said to be safe for bees; they do) affect the foraging behaviour and flight distances of bumble bees. The sublethal doses of azadirachtin affected the foraging distance of bumble bees (Karise et al. 2007). NeemAzal-T/S, a formulated product of *A. indica*, was proved to be harmless for the beneficial beetle (Hamd et al. 2005). When bees were caged in cotton field after spray, NeemAzal did not cause any mortality of honey bees lh after spray (Mann and Dhaliwal 2001).

An exclusive review by Boeke et al. (2004) expressed that *A. indica* provides many useful compounds that are used as pesticides and could be applied to protect stored seeds against insects. However, in addition to possible beneficial health effects, such as blood sugar-lowering properties and antiparasitic, anti-inflammatory, antiulcer and hepato-protective effects, also toxic effects are described. Also, they presented toxicological data from human and animal studies with oral administration of different neem-based preparations. The nonaqueous extracts appear to be the most toxic neem-based products, with an estimated safe dose (ESD) of 0.002 and 12.5 μg/kg bw/day. Less toxic are the unprocessed materials seed oil and the aqueous extracts (ESD 0.26 and 0.3 mg/kg bw/day and 2 μl/kg bw/day respectively). Most of the pure compounds show a relatively low toxicity (ESD azadirachtin 15 mg/kg bw/day). For all preparations, reversible effect on reproduction of both male and female mammals seems to be the most important toxic effect upon subacute or chronic exposure. From the available data, safety assessments for the various neem-derived preparations were made, and the outcomes are compared to ingestion of residues on food treated with neem preparations as insecticides. This leads to the conclusion that, if applied with care, use of neem-derived pesticides as an insecticide may be encouraged.

Clytostoma callistegioides, *Dolichandra cynanchoides*, *Dodonaea viscosa*, *Macfadyena unguis-cati*, *Phytolacca dioica*, *Prosopis juliflora*,

Salvia procurrens and *Salvia guaranitica* extracts did not show any toxic effects on the beneficial insect honey bee *Apis mellifera* (Lucia et al. 2009). *Chrysanthemum coronarium* whole plant extract was safe to beneficial natural enemies, humans and the environment (Mourad et al. 2008). Rotenone from *Derris elliptica* was toxic to bumble bees (Marletto et al. 2003). *Pongamia pinnata* (seed oil 1 %) showed reduction in pollinator population 1 day after application (Harjindra Singh et al. 2010).

7 Conclusions

From the foregoing, it is summarized that plant-derived insecticides show slight to moderate ill effects on parasitoids, predators and honeybees as obviously they also belong to the same class Insecta. But its innate biodegradability and low mammalian toxicity put them at height of 'safe/green' insecticides. In the case of parasitoids, certain guiding principles are suggested in accordance with multi-array activities of neem products in insects. Parasitoids are also susceptible, when they come in direct contact with plant origin insecticides including neem products. In such circumstances, blanket application of neem/any botanical product without understanding the behaviour of the parasitoid may adversely affect the beneficial capacity of the parasitoid. For example, the inundative release of the egg parasitoid *T. chilonis* should be resorted 3–4 days before/after neem product application. The external larval parasitoids are no exception to the ill effects, if they are in direct contact with neem products. To avoid this, for inundative releases, application of neem products may be followed by the release of the parasitoids and spraying may be avoided if the parasitoids are in larval stages in the field. Hence, pre-sampling is suggested to know the stage of the parasitoid, be it internal or external, for timing the application of botanical-based insecticides.

In the case of predatory insects and spiders, certain degree of selectivity is nevertheless apparent, as adult insects show no or relatively low sensitivity as in the case of earwigs, crickets, true bugs, beetles, lacewings and wasps. This can be explained by the fact that growth-disrupting compounds affect the first line juvenile instars of insects. The fecundity of neem-treated adult and predaceous parasitic insects and the fertility of their eggs are also not or only slightly affected by neem, in contrast to some phytophagous species. In some cases, the predation efficiency may be reduced. Nymphal/larval instars of beneficial insects are sensitive to neem products when topically treated, and reduction in food ingestion, delayed growth, difficulties in moulting, teretological and morphogenetic defects, reduced activity and increased mortality are normally observed in the laboratory. But far less drastic or even no effects are observed under semi-field or field conditions. This is partly due to the fast breakdown of the active principles under field conditions.

Neem and entomophages are twin gifts in IPM without endangering the agroecosystem. In fact, conservation biological control most commonly used the activity of native organisms. Hence, pre-sampling the parasitoids/predators is necessary in timing the application of botanical insecticides including neem products in order to avoid ill effects, if any. The 'integrated biological control' will include natural enemies vis-à-vis other botanical insecticides for organic production of agricultural products and to avoid the ill effects of synthetic chemical pesticides.

References

Abudulai M, Shepard BM, Mitchell PL (2004) Effects of neem (*Azadirachta indica* A. Juss) on predators of *Nezara viridula* (L.) (Hemiptera: Heteroptera: Pentatomidae). J Agric Urban Entomol 21:9–13

Adarkwah C, Daniel OO, Cornel A, Carmen B, Christoph R, Matthias S (2011) Integration of Calneem® oil and parasitoids to control *Cadra cautella* and *Corcyra cephalonica* in stored grain cereals. Phytoparasitica 39:223–233

Ahmad M, Ossiewatsch HR, Basedow T (2003) Effects of neem-treated aphids as food/hosts on their predators and parasitoids. J Appl Entomol 127:458–464

Alvarenga CD, Franca WM, Giustolin TA, Paranhos BA, Lopes GN, Cruz PL, Barbosa PR (2012) Toxicity of neem (*Azadirachta Indica*) seed cake to larvae of the mediterranean fruit fly, *Ceratitis Capitata* (Diptera:

Tephritidae), and its parasitoid, *Diachasmimorpha Longicaudata* (Hymenoptera: Braconidae). Fla Entomol 95:57–62

Anjum SM, Dildar G, Jalal M, Muhammad A, Sarfraz M (2007) Effect of various treatments of azadirachtin, spinosad and abamectin on the haemogram of *coccinella septempunctata* (coleoptera: Coccinellidae). Pak Entomol 29:151–164

Babu KS, Babu TR (2003) Effect of botanical insecticides on developmental stages of a larval parasitoid, *Bracon hebetor* Say. J Insect Sci 16:65–66

Babu GR, Rao GM, Rao PA (1998) Efficacy of neem oil and neem cake for the control of green leafhoppers, brown planthoppers and their effect on predators of brown planthoppers. Shashpa 5:91–94

Baitha A, Hameed SF, Singh R (2000) Effectiveness and economics of various treatments and their impact on spider population in rice ecosystem. Ann Plant Prot Sci 8:13–16

Balikai RA, Lingappa S (2004) Effect of natural plant products on predators of aphid *Melanaphis sacchari* (Zehntner) in sorghum. Indian J Entomol 66(1):57–59

Bandara KA, Kudagamage C (1996) Potential of neem for controlling Leaf-eating caterpillar complex on cabbage. In: Singh RP, Chari MS, Raheja AK, Kraus W (eds) Neem and environment, vol 1. Oxford/IBH Pub, New Delhi

Banken JAO, Stark JD (1997) Stage and age influence on the susceptibility of *Coccinella septempunctata* (Coleoptera: Coccinellidae) after direct exposure to Neemix, a neem insecticide. J Econ Entomol 90:1102–1105

Banken JAO, Stark JD (1998) Multiple routes of pesticide exposure and the risk of pesticides to biological controls: a study of neem and the seven-spot lady beetle, *Coccinella septempunctata* L. J Econ Entomol 91:1–6

Barbosa P, Gross P, Kemper J (1991) Influence of allelochemicals on tobacco hornworm and its parasitoid, *Cotesia congregata*. Ecology 72:1567–1575

Barrek S, Olivier P, Grenier-Loustalot MF (2004) Analysis of neem oils by LC–MS and degradation kinetics of azadirachtin-A in a controlled environment. Analy Biol Chem 378:753–763

Basappa H, Lingappa S (2004) Evaluation of various neem preparations against castor semilooper, *Achaea janata* Linn. (Lepidoptera: Noctuidae) under field conditions. J Res 32:25–28

Beckage NE, Metcalf JS, Nielson BD, Nesbit JD (1988) Disruptive effects of azadirachtin on development of Cotesia congregate in host tobacco hornworm larvae. Arch Insect Biochem Physiol 9:47–65

Beitzen HI, Hofmann R (1992) Versuchezur wirkung des azadireachtin haltigen, formulierten neemsamenextraktes, AZT-VR-NR aug *Hylobius obietis* L. *Lymantria monacha* L. and *Drino inconspicua*. Meig Z Pflan Pflnz 99:337–348

Bhardwaj V, Devi N, Raj D (2005) Toxicity of some insecticides and biopesticides to *Diadegma fenestralis* Himage and *Cotesia plutellae* (Kurdyumov), two parasitoids of diamondback moth. J Entomol Res 29:27–30

Boeke SJ, Boersma MG, Alink GM, Van Loon JJ, Van Huis A, Dicke M, Rietjens IM (2004) Safety evaluation of neem (*Azadirachta indica*) derived pesticides. J Ethnopharmacol 94:25–41

Boomathi N, Sivasubramanian P, Raguraman S, Premalatha K (2005) Safety of botanical and microbial pesticide mixtures to *Trichogramma chilonis* Ishii (Hymenoptera: Trichogrammatidae). J Biol Cont 19:23–28

Breithaupt J (1995) Effectiveness of *Trichogramma plasseyensis* Nagaraja to control the Asian Corn Stem Borer, *Ostrinia furnacalis* Guenée, in the lowlands of Papua New Guinea. Wajnberg, E. Colloques De L'Inra; Trichogramma and Other Egg Parasitoids 115–118

Burman AK, Basit A, Hazarika LK, Bhattacharya B (2003) Shaspa 10:175–180

Cano V, Gladstone SM (1994) Efecto de insecticida botanica, NIM-20, sobre el parasitismo por *Trichogramma pretiosum* an huevos del *Helicoverpa zea* en el cultivo de melon. Manecho Integrado de plagas 33:23–25

Chakraborti S (2004) Sustainable management of apical leaf curling in chilli. J Appl Zool Res 15:34–36

Chakraborti S, Chatterjee ML (1999) Bioefficacy of neem based pesticides against safflower aphid, *Dactynotus carthami* HRL and their impact on coccinellid predators. J Interacad 3:186–191

Chandra Mohan N, Nanjan K (1990) Effect of neem oil against diamond back moth in cabbage. Neem Newsl 7:35–36

Chandrababu A, Gautam RD, Garg AK (1997) Residual toxicity of some insecticides against the predatory beetle, *Brumoides suturalis* (Fab.). Ann Plant Prot Sci 5:131–136

Chari MS, Sreedhar U, Rao RSN, Reddy SAN (1996) Studies on compatibility of botanical and microbial insecticides to the natural enemies of *Spodoptera litura* F. Tobacco Res 22:32–35

Chari MS, Sreedhar U, Rao RSN (1997) Differential response of *Apanteles africanus* Cameron (Braconidae) and *Telenomus remus* Nixon (Scelionidae) to various pesticides on tobacco. J Insect Sci 10:181–183

Charleston DS, Rami K, Marcel D, Louise EMV (2005) Impact of botanical extracts derived from *Melia azedarach* and *Azadirachta indica* on populations of *Plutella xylostella* and its natural enemies: a Field test of laboratory findings. Biol Cont 33: 131–142

Chelliah S, Rajendran R (1984) Toxicity of insecticides to the predatory mirid bug, *Cyrtorhinus lividipennis* Reuter. Int Rice Res Newsl 9:15–16

Deepak KD, Choudhary AK (1998) Toxicity of some insecticides to *Tetrastichus pyrillae* Crawford (Hymenoptera: Eulophidae) an egg parasite of sugarcane leafhopper, *Pyrilla perpusilla* (Walker). Ann Plant Prot Sci 6:214–216

Deole SA, Bodhade SN, Mahajan LB, Deotale VY, Sharnagat BK (2000) Residual toxicity of some pesticides used in cotton pest management against a chrysopid, *(Chrysoperla carnea)*. J Soils Crops 10:279–291

Dhaliwal GS, Gill RS, Dilawari VK (1998) Management of insect pest complex of cabbage with neem based insecticides. In: Dhaliwal GS, Arora R, Randhawa NS, Dhawan AK (eds) Ecological agriculture and sustainable development. Indian Ecological Society and Centre for Research in Rural & Industrial Development, Chandigarh, pp 306–314

Dinesh KP, Sachin M, Veena BJ, Seetharama HG, Sreedharan K, Kumar PKV (2003) Evaluation of botanicals against mealy bug *Planococcus citri* Risso and its effect on parasitoid and attendant ant. J Coffee Res 31:139–152

Diraviam J, Viraktamath CA (1993) Toxicity of some insecticides to *Curinus coeruleus* Mulsant (Coleoptera: Coccinellidae), an introduced predator of the subabul psyllid. Entomology 18:77–79

Dobelin H (1997) Aphid control of cut anemones with Neem-Azal-T/S taking into account its effect on beneficials. In: Kleeberg H, Zebitz CPW (eds) Proceedings of 5th workshop on practice oriented results on the use and production of neem ingredients and pheromone, Wetzlar, Germany, pp 115–121

Drescher K, Madel G (1995) Auswirkungen eines Niem Extraktes auf den Pradator *Orius majusculus* Reuter (Heteroptera: Anthocoridae). Mitt Dtsch Ges Allg Angew Entomol 9:128

Eisenlohr K, Domange AL, Lenfant C, Sauphanor B (1992) Effects of NeemAzal-F on aphids and beneficial insects in peach orchards in France. In: Kleeberg H, Druck, Graphic (eds) Proceedings of workshop practice oriented results on the use and production of neem ingredient, Giessen, Germany, pp 27–40

Eyyüp MY, Başpınar H (2012) Effects of Neem on *Liriomyza trifolii* (Burgess) (Diptera: Agromyzidae) and its parasitoids on tomato greenhouse. Food Agric Environ 10:381–384

Facknath S (1999) Application of neem extract and intercropping for the control of cabbage pests in Mauritius, in *Azadirachta indica* A. Juss. In: Singh RP, Saxena RC (eds) *Azadirachta indica* A. Juss. Oxford &IBH Pub. Co. Pvt. Ltd., New Delhi, pp 165–176

Fernandez NJ, Palanginan EL, Soon LL, Botrella D (1992) Impact of neem on non-target organisms. In: Proceedings of final workshop on botanical pest control, IRRI, Los Banos, Philippines, pp 117–121

Forster P (1991) EinfluB Von Pflanzenschutzmitteln auf-larven and Adulte Von *Platypus dorsalis* (Pont) (Coleoptera: Carabidae) Und auf Adulte Von *Tachyporus hypnorum* (L.) (Coleoptera: Staphylinidae) in Labor-Halbfreiland Und Freilandversuchen. Doctoral thesis, University of Giessen, Germany

Gandhi M, Lal R, Sankaranarayanan A, Banerjee CK, Sharma PL (1988) Acute toxicity study of the oil from *Azadirachta indica* seed (neem oil). J Ethanopharm 23:39–51

Garg DK, Baranwal VK (1998) An integrated approach for pest management in basmati rice in Haryana. In: Dhaliwal GS, Arora R, Randhawa NS, Dhawan AK (eds) Ecological agriculture and sustainable development. Indian Ecological Society and Centre for Research in Rural & Industrial Development, Chandigarh, pp 306–314

Ghelani YH, Vyas HN, Jhala RC (2000) Toxicological impact of certain insecticides to *Rhynocoris fuscipes* Fabricius (Heteroptera: Reduviidae) a potential predator. J Appl Zool Res 11:54–57

Goldstein JA, Keil CB (1991) Prospects for integrated control of the Colorado potato beetle (Coleoptera: Chrysomelidae) using *Perillus bioculatus* (Hemiptera: Pentatomidae) and various pesticides. J Econ Entomol 84:1645–1651

Greenberg SM, Showler AT, Liu T (2005) Effects of neem-based insecticides on beet armyworm (Lepidoptera: Noctuidae). Insect Sci 12:17–23

Gunasena GH, Vinson SB, Williams HJ (1990) Effects of nicotine on growth, development, and survival of the tobacco budworm (Lepidoptera: Noctuidae) and the parasitoid *Campoletis sonorensis* (Hymenoptera: Ichneumonidae). J Econ Entomol 83:1777–1782

Hamd, EEA, Shafie HAF, Basedow T (2005) The different effects of two preparations of neem (*Azadirachta indica*) and of Sumicidin on the aphid predator *Hippodamia variegata* (Goeze) (Coleoptera: Coccinellidae). J Plant Dis Protect 112:580–585

Hamilton GC, Lashomb JH (1997) Effect of insecticides on two predators of the Colorado potato beetle, *Leptinotarsa decemlineata* (Coleoptera: Chrysomelidae). Fla Entomol 80:10–23

Hamilton GC, Lashomb JH, Patt J (1996) Impact of selected insecticides used in egg plant production on *Edovum puttleri* Grissell (Hymenoptera: Eulophidae). J Entomol Sci 31:432–439

Harjindra Singh R, Swaminathan N, Hussain T (2010) Influence of certain plant products on the insect pollinators of Coriander. J Biopest 3:208–211

Haseeb M, Liu TX, Jones WA (2004) Effects of selected insecticides on *Cotesia plutellae*, endoparasitoid of *Plutella xylostella*. Biocontrol 49:33–46

Hellpap C (1985) Populationsokologie und biologisch biotechnische Bekampfung von *Spodoptera* spp. in Nicaragua. Doctoral thesis, University of Frankfurt and Giessen, Germany

Heneidy AH, Barbosa P, Gross P (1988) Influence of dietary nicotine on the fall armyworm, *Spodoptera frugiperda* and its parasitoid, the ichneumonid wasp, *Hyposoter annulipes*. Entomol Exp Appl 46:227–232

Hermann P, Zebitz CPW, Kienzle J (1997) Effects of different NeemAzal formulations on larvae of the green lacewing *Chrysoperla carnea* Steph. (Neuroptera: Chrysopidae) in laboratory and semi-field. In: Kleeberg H, Zebitz CPW (eds) Practice orientated results on use and production of neem-ingredients and pheromones, Proceedings of the fifth workshop, Wetzlar, Germany, pp 183–188

Heyler NL, Gill G, Spredbury (1992) Pesticides chemistry an modern methods of residue analysis. GAU, Tests of Agrochemicals and Cultivars, 13

Hoelmer KA, Osborne LS, Yokumu RK (1990) Effects of neem extracts on beneficial insects in greenhouse culture. In: Locke JC, Lawson RH (eds) Proceedings of neem workshop on neem's potential in pest management programs USDA, Beltsville, MD, pp 100–105

Hohmann C, Silva FC, Denovaes T (2010) Selectivity of neem to *Trichogramma pretiosum* Riley and *Trichogrammatoidea annulata* De Santis (Hymenoptera: Trichogrammatidae). Neotrol Entomol 39:985–990

Holmes MS, Hassan E, Singh PP (1999) The contact and systemic action of neem seed extract against green peach aphid *Myzus persicae* Sulzer (Hemiptera: Aphididae). In: Singh RP, Saxena RC (eds) *Azadirachta indica* A Juss. Oxford & IBH Pub. Co. Pvt Ltd., New Delhi, pp 93–101

Hough GJ, Keil CB (1991) Prospects for integrated control of the Colorado potato beetle (Coleoptera: Chrysomelidae) using *Perillus bioculatus* (Heteroptera: Pentatomidae) and various pesticides. J Econ Entomol 84:1645–1651

Iannacone J, Lamas G (2003a) Efecto toxicológicos de extractos de molle (*Schinus molle*) lantana (*Lantana camara*) sobre *Chrysoperla externa* (Neuroptera: Chrysopidae), *Trichogramma pintoi* (Hymenoptera: Trichogrammatidae) *Copidosoma koehleri* (Hymenoptera: Encyrtidae). En el Perú Agric Téc 63:347–360

Iannacone J, Lamas G (2003b) Taxicological effects of neem, rotenone and cartap over three micorwasps parasitoid of agricultural pests in Peru. Boletín de Sanidad Vegetal Plagas 29:123–142

Imtiaz A, Naqvi SNH, Rahila T, Azmi MA, Yasmeen H, Asia A (1998) Efficacy of neem fractions (RB-a and RB-b) in comparison with *Melia* fraction (BB) against *Coccinella* sp. J Exp Zool 1:85–89

Ingawale UV, Mote UN, Tambe AB (2005) Field efficacy of natural and conventional insecticides against lucerne aphids and their impact on predators. J Maharashtra Agric Univ 30:46–49

Isman MB (2006) Botanical insecticides, deterrents, and repellents in modern agriculture and an increasingly regulated world. Ann Rev Entomol 51:45–66

Isman MB (2013) Botanical insecticides: a global perspective. In: American Chemical Society, 246th ACS national meeting and exposition, September 8-12, 2013, Indianapolis, Indiana. Picogram 84, p 105

Isman MB, Leatemia JA (2004) Toxicity and antifeedant activity of crude seed extracts of Annona squamosa (Annonaceae) against lepidopteran pests and natural enemies. Int J Trop Insect Sci 24:150–158

Jaastad G, Trandem N, Hovland B, Mogan S (2009) Effect of botanically derived pesticides on mirid pests and beneficials in apple. Crop Prot 28:309–313

Jakob G, Dickler E (1996) Auswirkungen von Niempraparaten auf den Apfelschalenwickler *Adoxophyes orana* (Lepidoptera: Tortricidae) und die ihn parasitierende Schlupfwespe *Colpoclypeus florus* Walk. (Hymenoptera: Eulophidae). In: Druck and Graphic (ed) Proceedings of 4th workshop practice oriented results on use and production of neem ingredients and pheromones, Giessen, Germany, pp 93–98

Jalali SK, Singh SP (2001) Residual toxicity of some insecticides to *Cheilomenes sexmaculata* (Fabricius). J Insect Sci 14:80–82

Jayaraj S (1991) An over view of botanical pest control research in TNGDNADU in Tamil Nadu, India. In: Proceedings of the botanical pest control project – Phase II, Midterm project review meeting, held on 28–31 July, at Dhaka, Bangladesh, International Rice Research Institute, Los Banos, Philippines, Abstract, pp 28–31

Jayaraj S, Bharathi M, Sundara Babu PC (1993) Integrated pest management. In: Randhawa NS, Parmar BS (eds) Neem research and development. Society of Pesticide Science, New Delhi, pp 154–167

Jhansi K, Sundara Babu PC (1987) Evaluation of effects of parasite, fungus and neem oil individually and in combination against spotted pod borer, *Maruca testulalis* (Geyer). Andhra Agric J 34:340–342

Joshi BG, Ramaprasad G, Sitaramaiah S (1982) Effect of neem seed kernel suspensions on *Telenomus remus*, an egg parasite of *Spodoptera litura*. Phytoparasitica 10:61–63

Jyothi M, Sannaveerappanavar VT (2003) Effect of aqueous neem seed kernel extract and neem oil emulsion on the parasitisation and survival of *Trichogramma chilonis* Ishii (Hymenoptera: Trichogrammatidae). In: Proceedings of symposium of biological control of Lepidopteran pests, Bangalore, India, 17–18 July 2002, pp 83–85

Kaethner M (1990) Wirkung von Niemsamenproduckten auf die Reproduktionsfahigkeit und Fitne von *Leptinotarsa decemlineata* Say, *Melolontha hippocastani* F. und *Melolontha melolontha* L. sowie Nebenwsirkungen auf die aphidophagen Nutzlinge *Coccinella septempunctata* L. und *Chrysoperla carnea* (Stephens). Doctoral thesis, University of Giessen, Germany

Kaethner M (1991) Keine Nebenwirkungen von Niempräparaten auf die aphidophagen Prädatoren *Chrysoperla carnea* (Steph.) und *Coccinella septempunctata* L. *Anz. Schädlingskd.* Pflanzenschutz Umweltschutz 64:97–99

Kain D, Agnello A (2002) Fruit ICM news. Newsl Exten Ohio State Univ 6:10–11

Kareem AA, Saxena RC, Malayba MT (1988) Effect of sequential neem (*Azadirachta indica*) treatment on green leafhopper (GLH), rice tungro virus (RTV) infection, predatory mirid and spiders in rice. Int Rice Res Newsl 13:37

Karim RANM, Chowdhury MMA, Haque MNM (1992) Current research on neem in rice in Bangladesh. In: Proceedings of final workshop on botanical pest control. IRRI, Los Banos, Philippines, pp 30–34

Karise R, Mand M, Koskor E, Bender A (2007) The sublethal effect of a botanical pesticide on the foraging behaviour of *B. terrestris*. Agronomy Research, 5:121–124

Kashio T (1983) Toxicity of certain insecticides and acaricides toward *Amblyseius eharai* Amitai & Swirski (Acarina: Phytoseiidae) [in Japanese]. Bull Fruit Tree Res Station Kuchinotsu 5:83–92

Khan MH, Ahmad N, Rashdi S, Ismail M, Rauf I, Tofique M (2013) Studies on the compatibility of neem oil with predator, *Chrysoperla carnea* for the management of aphids (Homoptera: Aphididae) in canola (*Brassica napus* L.). J Cereals Oilseeds 4:85–88

Klemm U, Schmutterer H (1993) Wirkungen von Niempräparaten auf die Kohl-motte *Plutella xylostella* L. und ihre natürlichen Feinde aus der Gattung Trichogramma. Z PflKrankh PflSchutz 100:113–128

Koodalingam A, Mullainadhan P, Arumugam M (2009) A antimosquito activity of aqueous kernel extract of soapnut *Sapindus emarginatus*: impact on various developmental stages of three vector mosquito species and non-target aquatic insects. Parasitol Res 105:1425–1434

Koul O, Wahab S (2004) Neem: today and in the new millennium. Kluwer Academic Publishers, Dordrecht

Krishnamoorthy A, Rajagopal D (1998) Effect of insecticides on the California red scale, *Aonidiella aurantii* (Maskell) and its natural enemies. Pest Manage Hortic Ecosyst 4:83–88

Krishniah NV, Kalode MB (1992) Environmental impact studies on neem. In: Proceedings of final workshop of IRRI-ADB-EWC on botanical pest control in rice based cropping systems. IRRI, Los Banos, Philippines, pp 50–53

Kumar P, Singh HP, Poehling H (2010) Effects of neem on adults of *Eretmocerus warrae* (Hym., Aphelinidae), a parasitoid of *Bemisia tabaci* (Hom., Aleyrodidae) in tropical horticulture systems. J Plant Dis Prot 117:273–277

Lakshmi VJ, Katti G, Krishnaiah NV, Lingaiah T (1998a) Laboratory evaluation of commercial neem formulations vis-à-vis insecticides against egg parasitoid, *Trichogramma japonicum* Ashmead (Hymenoptera; Trichogrammatidae). J Biol Cont 11:29–32

Lakshmi VJ, Katti G, Krishnaiah NV (1998b) Safety of neem formulations and insecticides to *Microvelia douglasi atrolineata* Bergroth (Heteroptera: Veliidae), a predator of planthoppers in rice ecosystems. J Biol Cont 11:33–36

Lakshmi VJ, Katti G, Krishnaiah NV, Kumat KM (1998c) Safety of neem formulations to *Cyrtorhinus lividipennis* Reuter, a predator of brown planthopper, *Nilaparvata lugens* (Stal) in rice crop. J Biol Cont 12:119–122

Lamb R, Saxena RC (1988) Effects of neem seed derivatives on rice leaf folders (Lepidoptera) and their natural enemies. In: Proceedings of final workshop of on botanical pest control of rice based cropping systems. IRRI, Los Banos, Philippines, p 47

Li KH, Xu X, Li YF, Meng QZ, Zhou LC (1986) Determination of toxicity of 29 chemicals to *Trichogramma japonicum* at various developmental stages. Nat Enemy Insect 8:187–194

Loke WH, Heng CK, Basirun N, Mardi AR (1992) Non-target effects of neem (*Azadirachta indica* A Juss) on *Apanteles plutellae* Kurdj. Cabbage, sawi and padi. In: Proceedings of 3rd international conference on plant protection in the tropics, vol 2, Malaysian Plant Protection Society (Malaysia), Pahang 20–23 Mar 1990, pp 108–110

Lowery DT, Isman MB (1996) Inhibition of aphid (Homoptera: Aphididae) reproduction by neem seed oil and azadirachtin. J Econ Entomol 89:602–607

Lucia C, Gonzalez-Coloma A, Gonzalez A, Diaz M, Santos E, Alonso-Paz E, Bassagoda MJ, Rossini C (2009) Screening of Uruguayan plants for deterrent activity against insects. Ind Crops Prod 29:235–240

Lyons B, Helson B, Bourchier R (1996) Effects of azadirachtin-based insecticides on the egg parasitoid, *Trichogramma minutum* Riley (Hymenoptera: Trichogrammatidae). Neem Newsl 13:50–51

Ma DL, Gordh G, Zalucki MP (2000) Toxicity of biorational insecticides to *Helicoverpa* spp. (Lepidoptera: Noctuidae) and predators in cotton field. Int J Pest Manage 46:237–240

Mamoon-ur–Rashid M, Khattak MK M, Abdullah K, Hussain S (2013) Toxic and residual activities of selected insecticides and neem oil against cotton mealy bug, *Phenacoccus solenopsis* Tinsley (Sternorrhyncha: Pseudococcidae) under laboratory and field conditions. Pak Entomol 33:151–155

Mani M, Krishnamoorthy A (1996) Response of the encyrtid parasitoid, *Tetracnemoidea indica* of the oriental mealybug, *Planococcus lilacinus* to different pesticides. Indian J Plant Prot 24:80–85

Mani M, Lakshmi VJ, Krishnamoorthy A (1997) Side effects of some pesticides on the adult longevity, progeny production and prey consumption of *Cryptolaemus montrouzieri* Mulsant (Coccinellidae: Coleoptera). Indian J Plant Prot 25:48–51

Manish N, Cherian B, Padh H (2001) Alpha-terthienyl: a plant derived new generation insecticide. Curr Sci 81:667–672

Mann GS, Dhaliwal GS (2001) Safety of Azadirachtin to honeybee, *Apis mellifera* Linnaeus, Foragers. Pest Res J 13:118–121

Mansour FA, Ascher KRS, Omari N (1986) Toxicity of neem (*Azadirachta indica*) seed kernel extracts prepared with different solvents on the spider, *Cheiracanthium mildei*. Phytoparasitica 14:73–76

Mansour FA, Ascher KRS, Abo MF (1993) Effects of Margosan-O™ and RD9, Repelin[R] on spiders and predacious mites. Phytoparasitica 21:205–211

Mansour FA, Ascher KRS, Abo MF (1997) Effects of Neemguard on phytophagous and predaceous mites and on spiders. Phytoparasitica 25:333–336

Mariappan V, Sethuraman K, Vidya F, Vidya KR, Karunakaran KR, Vidyasekaran P (1993) Management of pest and diseases of paddy crop with neem products. In: World neem conference, vol 9, Bangalore, India, pp 24–28

Markandeya V, Divakar BJ (1999) Effect of a neem formulation on four bioagents. Plant Prot Bull 51:28–29

Marletto F, Patetta A, Manno A (2003) Laboratory assessment of pesticide toxicity to bumblebees. Bull Insectol 56:155–158

Masood KK, Mamoon-ur-Rashid (2006) Evaluation of neem (*Azadirachta indica* A Juss) oil, neem seed water extract, baythroid TM against cotton bollworms and egg parasitoid, *Trichogramma chilonis*. Pak Entomol 28:5–9

Matter MM, Marei SS, Moadwad SM, El G (1993) The reaction of Aphis gossypii and its predator Coccinella undecimpunctata to some plant extracts. Bull Fac Agric Univ Cairo 44:417–432

McCloskey C, Arnoson JT, Donskov N, Chenier R, Kaminski J, Philogene BJR (1993) Third tropic level effects of azadirachtin. Can Entomol 125:163–165

Melathopoulos AP, Winston ML, Whitington R, Higo H, Le Doux M (2000) Field evaluation of neem and canola oil for the selective control of the honey bee (Hymenoptera: Apidae) mite parasites *Varroa jacobsoni* (Acari: Varroidae) and *Acarapis woodi* (Acari: Tarsonemidae). J Econ Entomol 93: 559–567

Men UB (1999) Braconid parasitoids from neem treated semilooper, *Hypomecis* sp. in Vidarbha, Maharashtra. Insect Environ 5:27

Men UB, Bhabad NS, Kandalkar HG (2002) Relative safety of some biopesticides and insecticides to Diaeretiella rapae McIntosh, a parasitoid of mustard aphid, *Lipaphis erysimi* Kaltenbach. Pest Manage Econ Zool 10:201–203

Michaud JP (2001) Responses of two ladybeetles to eight fungicides used in Florida citrus: implications for biological control. J Insect Sci 1:1–7

Michelakis SE, Vacante V (1997) The citrus leaf miner status in Greece. Bull Oil Seed Crop 20:81–82

Mineo N, Mineo G, Sinacori A (2000) Side effects of sprays with azadirachtin mixed with Biolid E(R) against the natural parasitoids of *Phyllocnistis citrella* Stainton (Lepidoptera: Gracillaridae). Bollettino di Zoologia Agraria e di Bachicoltura 32:157–162

Mishra NC, Mishra SN (1998) Impact of biopesticides on the useful insect pests and defenders of okra. Neem Newsle 15:18

Mitchell PL, Gupta R, Singh AK, Kumar P (2004) Behavioral and developmental effects of neem extracts on *Clavigralla scutellaris* (Hemiptera: Heteroptera: Coreidae) and its egg parasitoid, *Gryon fulviventre* (Hymenoptera: Scelionidae). J Econ Entomol 97:916–923

Mohan K (1989) Studies on the effect of neem products and vegetable oils against major pests of rice and safety to natural enemies. MSc (Ag.) thesis, Tamil Nadu Agricultural University, Coimbatore, India

Mohapatra RN, Patnaik NC, Mishra BK, Patnaik HP (1991) Comparative toxicity of some insecticides and neem oil against the brown planthopper, *Nilaparvata lugens* (Stal) and its predators. Orissa J Agric Res 4:239–240

Moser G (1994) Nutzlingskomplex urn *Aphis gossypii* Versuchmd Niems menkemwasser extrakten, an Beispeel von okra in der Dominikanischen Republik. Diploma Thesis, University of Giessen, Germany

Mourad L, Shonouda Salah O, Osama S, Amal A (2008) Insecticidal effect of *Chrysanthemum coronarium* L. flowers on the pest *Spodoptera littoralis* Boisd and its parasitoid *Microplitis rufiventris* Kok. with identifying the chemical composition. J Appl Sci 8:1859–1866

Nanda UK, Parija B, Pradhan NC, Nanda B, Dash DO (1996) Bioefficacy of neem derivatives against the insect pest complex of rice. In: Singh RP, Chari MS, Raheja AK, Kraus W (eds) Neem and environment, vol 1. Oxford & IBH Pub. Co. Ltd., New Delhi, pp 517–528

Nath L, Singh AK (2003) Safety of some plant extracts and a neem formulation to predators of the coriander aphid, *Hyadaphis coriandri* Das under field conditions. Pest Manage Econ Zool 11:159–162

National Research Council (1992) Neem: a tree for solving global problems. National Academic Press, Washington, DC, p 141

Naumann K, Isman MB (1996) Toxicity of neem (*Azadirachta indica* A. Juss.) seed extracts to larval honeybees and estimation of dangers from field application. Am Bee J 136:518–520

Nirmala R, Balasubramaniam G (1999) Residual toxicity effects of insecticides and neem based formulations on the predatory spiders of rice ecosystem. Neem Newsl 16:37–38

Obrycki JJ, Tauber MJ, Tingey WM (1986) Comparative toxicity of pesticides to *Edovum puttleri* (Hymenoptera: Eulophidae), an egg parasitoid of the Colorado potato beetle (Coleoptera: Chrysomelidae). J Econ Entomol 79:948–951

Ofuya TI, Okuku IE (1994) Insecticidal effect of some plant extracts on the cowpea aphid, *Aphis craccivora* Koch (Homoptera: Aphididae). Anzeiger für Schädlingskunde, Pflanzenschutz, Umweltschutz 67:127–129

Oi BY, Gordon G, Gimme W (2001) Effects of neem-fed prey on the predaceous insects, *Harmonia conformis* (Boisduval) (Coleoptera: Coccinellidae) and *Mallada signatus* (Schneider) (Neuroptera: Chrysopidae). Biol Cont 22:185–190

Olivella PE, Vogt H (1997) Seasonal occurrence, abundance, parasitism and leaf damage of leaf miner moths in apple orchards in South West Germany. Mitteilungen der Deutschen Gesellschaft fur Allgemeine und Angewandte Entomologie 11:611–617

Oliver JAI, Bringas YM (2000) Effects on the populations of the predator *Metacanthus tenellus* (Heteroptera: Berytidae) by the botanic insecticides rotenone and neem on tomato crop in Peru. Rev Colon Entomol 26:89–97

Olkowski W, Daar S, Olkowski H (1995) The gardener's guide to common sense pest control. The Taunton Press, Newtown

Osman MZ, Bradley J (1993) Effects of neem seed extracts on *Pholeastor* (*Apanteles*) *glomeratus* I. (Hymenoptera: Braconidae), a parasitoid of *Pieris brassicae* L. (Lepidoptera: Pieridae). J Appl Entomol 115:259–265

Oswald S (1989) Untersuchungen zur integrierten Bekamofung der Kokoswanze *pseudothreaptus wayi* Brown (Heteroptera: Coriedae) auf Sansibar. Doctoral thesis, University of Giessen, Germany

Patel MM, Yadav DN (1993) Toxicity of some botanical and chemical insecticide to *Menochilus sexmaculatus* (Coccinellidae) an important aphidophage and its parasite, *Tetrastichus coccinellae* (Eulophidae). In: World neem conference, Bangalore, India, p 24

Patel HM, Borad PK, Korat DM (2003) Bioefficacy of botanical materials against *Aphis* gossypii Glover infesting brinjal. In: Proceedings of national symposium on sustainable insect pest management, Chennai, 6–7 Feb 2003, pp 29–30

Patil RK, Rayar SG, Hiremath BH, Patil BR (1997) Effect of plant products on safflower aphid, Dactynotus carthami and on its natural predators, *Chrysoperla carnea*. J Oilseeds Res 14:35–38

Peng CYS, Trinh S, Lopez JE, Mussen EC, Hung A, Chuang R (2000) The effects of azadirachtin on the parasitic mite, *Varroa jacobsoni* and its host honey bee (Apis mellifera). J Apic Res 39:159–168

Perera DR, Armstrong G, Naylor REL, Senanayake N (2000) Response of *Chrysodeixis eriosoma* (Doubleday), *Plutella xylostella* L. and the parasitoid, *Cotesia plutellae* (Kurdjumov) to feeding deterrents. Trop Agric Res 12:186–198

Pilar M, Flor B, Manuel G, Rodríguez B, Díaz A, Huerta A, Zapata N, Viñuela E (2006). Effects of botanical insecticides on two natural enemies of importance in Spain: *Chrysoperla carnea* (Stephens) and *Psyttalia concolor* (Szépligeti). Pesticides and beneficial organisms, IOBC, WPRS Bulletin 29

Prasad GS, Logiswaran G (1998) Toxicity of different insecticides to the adult of *Cheilomenes sexmaculata* F. (Coccinellidae: Coleoptera). Insect Environ 3:117–118

Prasad D, Chand P, Haque MF (1987) Parasitization of *Heliothis armigera* (Hub.) in chickpea field as affected by pesticidal seed treatment. Res Dev Rep 4:236–238

Price JF, Schuster DJ (1991) Effects of natural and synthetic insecticides on sweetpotato whitefly *Bemisia tabaci* (Homoptera: Aleyrodidae) and its hymenopterous parasitoids. Flua Entomol 74:60–68

Punzo F (2005) Neem seed extract containing azadirachtin affects mortality, growth, and immunological function in the whipscorpion, *Mastigoproctus giganteus* (Lucas) (Arachnida, Uropygi). Bull Environ Cont Toxicol 75:684–690

Quang LD, Gee YL, Yong HC, Gyung JC, Kyoung SJ, Myung SP, Ho S, Young HH, Lim CH, Jin CK (2010) Insecticidal activities of crude extracts and phospholipids from *Chenopodium ficifolium* against melon and cotton aphid, *Aphis gossypii*. Crop Prot 29:1124–1129

Raguraman S (1987) Studies on the efficacy of neem products against rice insect pests. M.Sc. (Ag.) thesis, Tamil Nadu Agricultural University, Coimbatore, India

Raguraman S (2009) Neem and environment: integration of neem with bioagents in insect pest management. Madras Agric J 96:277–282

Raguraman S, Rajasekaran B (1996) Effect of neem products on insect pests of rice and predatory spider. Madras Agric J 83:510–515

Raguraman S, Singh RP (1998) Behavioural and physiological effects of neem (*Azadirachta indica*) seed kernel extracts on larval parasitoid, *Bracon hebetor*. J Chem Ecol 24:1241–1250

Raguraman S, Singh RP (1999) Biological effects of neem (*Azadirachta indica*) seed oil on eggs parasitoid, *Trichogramma chilonis*. J Econ Entomol 92:1274–1280

Ranga Rao GV, Visalakshi V, Suganthy M, Vasudeva Reddy P, Reddy YVR, Rameswar Rao V (2008) Relative toxicity of neem to natural enemies associated with the chickpea ecosystem: a case study. Int J Trop Insect Sci 27:229–235

Rao NB, Goud TR, Gour TB (2002) Toxicity of newer insecticides to egg parasitoid, *Trichogramma chilonis* Ishii. J Res 30:124–126

Rao MS, Shylaja M, Naik D (2003) Management of nematode induced wilt disease complex in tuberose (*Polianthes tuberosa* L.) cultivar prajwal using *Pochonia chlamydosporia* (*Verticillium chlamydosporium*) and *Trichoderma harzianum*. J Ornam Hortic 6:341–346

Reddy GVP, Guerro A (2000) Pheromone based integrated pest management to control the diamondback moth, *Plutella xylostella* in cabbage fields. Pest Manage Sci 56:882–888

Reddy PVR, Srikanth J (1996) Chemistry a modern methods of residue analysis. GAU, Anand, pp 46–62

Rembold H, Sharma GK, Czoppelt C (1981) Growth-regulating activity of azadirachtin in two holometabolous insects. In: Schmutterer H, Ascher KRS, Rembold H (eds) Natural pesticides from the neem tree (*Azadirachta indica A.* Juss), Proceedings of 1st international neem conference, Rottach- Egern, German Agency for Techn Coop Eschborn, Germany, pp 121–128

Roger C, Vincent C, Coderre D (1995) Mortality and predation efficiency of *Coleomegilla maculata lengi* Timb. (Coleoptera: Coccinellidae) following application of neem extracts (*Azadirachta indica* A. Juss., Meliaceae). J Appl Entomol 199:439–443

Rohan R (2000) The management of major insect pests *Bactrocera cucurbitae* and *Aulacophora* spp. in cucurbits under three intensive systems. BCPC Conf Pests Dis 3:981–985

Rowell B, Bunsong N, Satthaporn K, Phithamma S, Doungsa Ard C (2005) Hymenopteran parasitoids of diamondback moth (Lepidoptera: Ypeunomutidae) in Northern Thailand. J Econ Entomol 98:449–456

Saber M, Hejazi MJ, Hassan SA (2004) Effects of azadirachtin/Neemazal on different stages and adult life table parameters of *Trichogramma cacoeciae* (Hymenoptera: Trichogrammatidae). J Econ Entomol 97:905–910

Sahayaraj K, Karthikraja S (2003) Effect of biopesticides on *Rhynocoris marginatus* (Fab). J Biol Cont 17:43–45

Sahayaraj K, Paulraj MG (1999a) Effect of plant products on the eggs of *Rhynocoris marginatus* (Fab.) (Hemiptera: Reduviidae). Insect Environ 5:23–24

Sahayaraj K, Paulraj MG (1999b) Toxicity of some plant extracts against life stages of a reduviid predator, *Rhynocoris marginatus*. Indian J Entomol 28:342–344

Saikia P, Parameswaran S (2001) Contact toxicity of chemical and biopesticides against *Cnaphalocrocis medinalis* (Guenee) and *Trichogramma chilonis* Ishii. J Appl Zool Res 12:86–87

Saleem SA, Matter MM (1991) Relative effects of neem seed oil and Deenate on the cotton leaf worm. Spodoptera littoralis Boisd and the most prevalent predators in cotton fields in Menoufyia Governorate. Bull Fac Agric Univ Cairo 42:941–952

Samiayyan K, Chandrasekharan B (1998) Influence of botanicals on the spider populations of rice. Madras Agric J 85:479–480

Saminathan VR, Jayaraj S (2002) Exploring on effective botanical formulation 'for sustainable crop protection-a review. Vistas Entomol Res New Millenn 72–79

Sarode SV, Sonalkar VU (1999a) Ovicidal effect of some insecticides against *Chrysoperla carnea* (Stephens). Pestic Res J 11:97–98

Sarode SV, Sonalkar VU (1999b) Influence of different insecticides on parasitization of *Corcyra cephalonica* by *Trichogramma chilonis* Ishii. Pestic Res J 11:99–101

Saucke H (1995) Biological and integrated control of diamondback moth Plutella xylostella Lin. (Lepidoptera: Yponomeutidae) and other major pests in brassica crops, SPC/GTZ biological control project: final report 1992–1995, Papua New Guinea, Konedobu

Sauphanor B, Lenfant C, Sureau F (1995) Effects d'un extrait de neem (*Azadirachta indica* A. Juss.) sur le development de *Forficula auricularia* L. (Dermaptera). J Appl Entomol 119:215–219

Saxena RC (1987) Neem seed derivatives for management of rice insect pests-a review of recent studies. In: Schmutterer H, Ascher KRS, Rembold H (eds) Natural pesticides from neem tree and other tropical plants. GTZ, Eschborn, pp 81–93

Saxena RC (1989) Insecticides from neem tree. In: Arnason JT, Philogene BJR, Morand P (eds) Insecticides of plant origin. ACS symposium series 387. American Chemical Society, Washington, DC, pp 110–135

Saxena RC, Waldbauer GP, Liguido NJ, Puma BC (1981) Effects of neem seed oil on the rice leaf folder *Cnaphalocrosis medinalis*. In: Schmutterer H, Ascher KRS, Rembold H (eds) Natural pesticides from neem tree and other tropical plants. GTZ, Eschborn, pp 198–203

Saxena RC, Epino, PB, Cheng-Wen T, Puma BC (1984) Neem, chinaberry and custard apple: antifeedant and insecticidal effects of seed oils on leaf hopper and Planthopper pests of rice. In: Schmutterer H, Ascher KRS (eds) 2nd international neem conference, Rauischholzhausen, Germany, pp 403–412

Schauer M (1985) Die Wirkung von Nieminhaltsstoffen auf Blattlause und die Rubenblattwanze. Doctoral Thesis, Univ. of Giessen, Germany

Schicha E (1975) Predacious mites (Acarina: Phytoseiidae) on sprayed apple trees at Bathurst (NSW, Australia). J Aust Entmol Soc 14:217–219

Schicha E (1977) Phytophagous mites and their predators in an apple orchard at Bathurst (NSW, Australia). Zeitschristft fur Angewante Entomolgia 84:38–47

Schmidt G, Pesel E (1987) Studies on the sterilizing effects of neem extracts in ants, in natural pesticides from neem tree and other tropical plants. In: Schmutterer H, Ascher KRS, Rembold H (eds) Natural pesticides from neem tree and other tropical plants. GTZ, Eschborn, pp 361–373

Schmutterer H (1981) Ten years of neem research in federal Republic of Germany. In: Schmutterer H, Ascher KRS, Rembold H (eds) Natural pesticides from neem tree and other tropical plants. GTZ, Eschborn, pp 21–51

Schmutterer H (1992) Einfluß von Azadirachtin, einer azadirachtinfreien Fraktion eines alkoholischen Niemsamenextraktes und von formulierten Extrakten auf Verpuppung, Schlupf und Imagines der Kohlweißlingsbrackwespe *Apanteles glomeratus* (L.) (Hym., Braconidae). J Appl Entomol 113:79–87

Schmutterer H (1995) The neem tree *Azadirachta indica* A. Juss. and other meliaceous plants: sources of unique natural products for integrated pest management, medicine, industry and other purposes. VCH Publishers, Weinheim

Schmutterer H (1996) Side effects of neem products on insect pathogens, predators and parasitoids. Neem Newsl 13:54–55

Schmutterer H, Hoist H (1987) On the effects of the enriched and formulated neem seed kernel extract AZT-VR-K on *Apis mellifera* L. J Appl Entomol 103:208–213

Schmutterer H, Saxena RC, Von der Hyke J (1983) Morphogenetic effects of some partially purified fractions and methanolic extracts of neem seed on *Mythimna separata* (Walker) and *Cnaphalocrocis medinalis* (Guenee). Z Ang Entomol 95:230–237

Schneider A, Madel G (1992) Fekunditat und Vitalitat adultery Schlupfwespennach Exposition auf Niem (*Azadirachta indica*) behandelten Flachen. Mitt Dtsch Ges Allg Angew Entomol 8:273–278

Serra AC (1992) Untersuchungen zum Einsatz von Niem samenextrakten in Rahem integretter Ansaltze zur Bekampfung von Tomatenschadlingen in der Dominikanischen Repubik. Doctoral Thesis, Univ Giessen, Germany

Sharma HC, Leushner K, Gunasekhar D, Marthandamoorthi M, Bhaskariah K, Subramaniyam M, Sulthana N (1984) Insect antifeedants and growth inhibitors from *Azadirachta indica* and *Plumbago zeylanica*. In: Schmutterer H, Ascher KRS, Rembold H (eds) Natural pesticides from neem tree and other tropical plants. GTZ, Eschborn, pp 291–320

Sharma HC, Shankaram AVB, Nwanze KF, Singh PP (1999) Utilization of natural pesticides derived from neem and custard apple in integrated pest management. In: Singh RP, Saxena RC (eds) *Azadirachta indica* A. Juss. Oxford &IBH Pub. Co. Pvt. Ltd., New Delhi, pp 199–213

Shilke SS, Jadhav LD, Salunkhe GN (1990) Toxicity of some vegetable oils/extracts to *Chelonus blackburni* Cam. an egg-larval parasite of *Phthorimaea opercullela* (Zell.). Indian J Entomol 52:709–711

Shukla BC, Srivastava SK, Gupta R, (1988) Evaluation of neem oil for control of rice pests. In: Final workshop of IRRI-ADB-EWC project on botanical pest control in rice-based cropping system. International Rice Research Institute, Philippines

Silva FAC, Martinez SS (2004) Effects of neem seed oil aqueous solutions on survival and development of the predator *Cycloneda sanguinea* (L.) (Coleoptera: Coccinellidae). Neotrop Entomol 33:751–757

Simmonds MSJ, Manlove JD, Blaney WM, Khambay BPS (2000) Effect of botanical insecticides on the foraging and feeding behaviour of the coccinellid predator, *Cryptolaemus montrouzieri*. Phytoparasitica 28:99–107

Singh RP, Singh S (1996) Neem in agriculture: advantages and disadvantages. In: Lal OP (ed) Recent advances in Indian entomology. APC Press, New Delhi, pp 67–82

Singh RP, Singh Y, Singh SP (1985) Field evaluation of neem (*Azadirachta indica* A. Juss) seed kernel extracts against the pod borers of pigeon pea, (*Cajanus cajan* L.) Millsp. Indian J Entomol 47:111–112

Sithanantham S, Otieno ZN, Nyarko KA (1997) Preliminary observations on effects on neem sprays on some non target arthropods in cowpea ecosystem. In: Quatrieme conference Internationale sur les Ravageurs en Agriculture, Le Corum, Montpellier, France, pp 643–648

Sivagnaname N, Kalyanasundaram M (2004) Laboratory evaluation of methanolic extract of *Atlantia monophylla* (Family: Rutaceae) against immature stages of mosquitoes and non-target organisms. Mem Inst Oswaldo Cruz 99:115–118

Solis DR, Fagundes GG, Habib ME (2004) Tolerance of *Gryon gallardoi*, an egg parasitoid of *Leptoglossus zonatus* to *Azadirachta indica* seed oil extract. Rev Agric Piracicaba 79:328–334

Sontakke BK (1993) Field efficacy of insecticide alone and in combination with neem oil against insect pests and their predators in rice. Indian J Entomol 55:260–266

Sreenivasa AG, Patil BV (1998) Role of commercial neem products in cotton insect pest management. Neem Newsl 15:17

Srinivasa Babu K, Murthy MSN, Ramesh Babu T (1996) Effect of botanicals on certain parasitoids. In: Singh RP, Chari MS, Raheja AK, Kraus W (eds) Neem and environment, vol 2. Oxford & IBH Co. Pvt. Ltd., New Delhi, pp 1117–1126

Srinivasan G, Babu PCS (2000) Effect of neem products on predatory lacewing, *Chrysoperla carnea* Stephens (Chrysopidae: Neuroptera). Pestic Res J 12:123–126

Srivastava KP, Parmar BS (1985) Evaluation of neem oil emulsifiable concentrates against sorghum aphids. Neem Newsl 2:7

Srivastava M, Paul AVN, Rengasamy S, Kumar J, Parmar BS (1997) Effects of neem (*Azadirachta indica* A. Juss) seed kernel extracta on the larval parasitoid, *Bracon brevicornis* Wesm. (Hymenoptera: Braconidae). J Appl Entomol 121:51–57

Stansly PA, Liu TX (1997) Selectivity of insecticides to *Encarsia pergandiella* (Hymenoptera: Aphelinidae), an endoparasitoid of *Bemisia argentifolii* (Hemiptera: Aleyrodidae). Bull Entomol Res 87:525–531

Stark JD, Wennergren U (1995) Can population effects of pesticides is predicted from demographic toxicological studies? J Econ Entomol 88:1089–1096

Stark JD, Vargas RI, Wong TY (1990) Effects of neem seed extracts on tephritid fruit flies (Diptera: Tephritidae) and their parasitoids in Hawaii. In: Lock JC, Lawson RH (eds) Neem potential in pest management programme, USDA, ARS Baltomore, 86, pp 106–111

Stark JD, Wong TY, Vargas RI, Thalman RK (1992) Survival, longevity and reproduction of tephritid fruit-fly parasitoids (Hymenoptera: Braconidae) reared from fruit flies exposed to azadirachtin. J Econ Entomol 84:1125–1129

Stenseth C (1990) Whiteflies on ornamental plants in the greenhouse. Gartneryrket 80:16–18

Sundaram KMS (1996) Azadirachtin biopesticide: a review of studies conducted on its analytical chemistry, environmental behaviour and biological effects. J Environ Sci Health 31:913–948

Sundari MSN (1998) Residual toxicity of insecticides against *Cryptolaemus montrouzieri* Mulsant (Coleoptera: Coccinellidae). Insect Environ 4:45

Swaminathan RH, Hussain T (2010) Side effects of a few botanicals on the *Aphidophagous coccinellids*. J Biopest 3:081–084

Tadic M (1979) The toxicity of some synthetic insecticides to the Hymenoptera *Ooencyrtus kuvanae* How. and *Telenomus terebrans* Ratz. Zastita Bilja Beograd 30:205–210

Tedeschi R, Alma A, Tavella L (2001) Side-effects of three neem (*Azadirachta indica* A Juss) products on the predator *Macrolophus caliginosus* Wagner (Heteroptera: Miridae). J Appl Entomol 125:397–402

Teggelli GR, Biradar AP, Vastrad SM, Rajashaker DW (1998) Impact of pesticides on an endoparasitoid, *Campoletis chlorideae* Uchida. Adv Agric Res India 10:25–29

Thakur JN, Pawar AD (2000) Comparative toxicity of different insecticides against *Trichogramma chilonis* Ishii. J Biol Cont 14:51–53

Thirumurugan A, Koodalingam K (2005) Enhancing efficacy of *Trichogramma chilonis* against internode borer, *Chilo sacchariphagus indicus* in sugarcane. Cooperative Sugar 37:67–71

Tunca H, Kilinçer N, Ozkan C (2012) Side-effects of some botanical insecticides and extracts on the parasitoid, *Venturia canescens* (Grav.) (Hymenoptera: Ichneumonidae). Türk Entomol derg 36:205–214

Usman M, Inayatullah M, Usman A, Sohail K, Fahadsha HS (2012) Effect of egg parasitoid, *Trichogramma chilonis*, in combination with *Chrysoperla carnea* and neem seed extract against tomato fruitworm, *Helicoverpa armigera*. Sarhad J Agric 28:253–257

Vekaria MV, Patel GM (2000) Bioefficacy of botanicals and certain chemical insecticides and their combinations against mustard aphid, *Lipaphis erysimi*. Indian J Entomol 62:150–158

Velayuddan R, Senrayan R, Rajadurai S (1988) Parasitoid-host interactions with reference to *Anastatus ramakrishnae* (Mani) (Hymenoptera: Eupelmidae) in relation to pentatomid and coreid. Proc Indian Nat Sci Acad Biol Sci 54:145–153

Viggiani G (1973) Ricerche sulla entomofauna del nocciuolo. IX. Notizie sualcuni Omotteri (Ceresa bubalus (Fbrc.), Ledra aurita (L.) e Cicadella viridis (L.). Annalidella Facoltà di Scienze Agrarie della Università degli Studi di Napoli 7:153–160

Villanueva JA, Hoy MA (1998) Toxicity of pesticides to the citrus leaf miner and its parasitoid *Ageniaspis citricola* evaluated to assess their suitability for an IPM program in citrus nurseries. Biocontrol 43: 357–388

Vogt H (1993) Einsalz von Niempraparaten gagen *Adoxophyes arana* F.v. R und Untersuchungen zu Nebenwirkungen. In: Intern. Erfahrungsaust. uber For schungserg. *Zum Okologischen Obstbau,* LVWO Weinsberg, pp 51–55

Vogt H, Handel U, Vinuela E (1997) Field investigations on the efficacy of Neem Azal-T/S against *Dysaphis plantaginea* (Passerini) (Homoptera: Aphididae) and its effect on larvae of *Chrysoperla carnea* Stephens (Neuroptera: Chrysopidae). In: practice oriented results on use and production of neem - ingredients and pheromones, Proceedings of the fifth workshop, Wetzlar, Germany pp 105–114

Walters PJ (1976) Susceptibility of the *Stethorus* spp. (Coleoptera, Coccinellidae) to selected chemicals used in New South Wales apple orchards. J Aust Entomol Soc 15:49–52

Wilps H (1987) Growth and adult molting of larvae and pupae of the blowfly, Phormia terraenovae in relationship to azadirachtin concentrations. In: Schmutterer H, Ascher KRS, Rembold H (eds) Natural pesticides from neem tree and other tropical plants. GTZ, Eschborn, pp 299–314

Wu HL (1986) Determination of toxicity of some insecticides to *Lycosa pseudoannulata* and *Apanteles cypris*. Nat Enemies Insects 8:230–231

Yadav DN, Patel AR (1992) Effect of some botanical insecticides on oviposition of *Chrysopa scelestes* and their ovicidal action. In: Chari MS, Ramaprasad G (eds) Proceedings symposium on botanical pesticides in IPM, CTRI, Rajahmundry, India, pp 166–169

Retracted: Progress in the Development of Plant Biopesticides for the Control of Arthropods of Veterinary Importance

11

Srikant Ghosh and Reghu Ravindran

Abstract

Arthropods are infesting mammals, birds, and reptiles throughout the world and possess serious threat to economical livestock production industries. Besides causing direct effect on hosts through sucking of blood, irritation, damage of skin, retarded growth, and loss of production, they transmit a number of viral, rickettsial, bacterial, and protozoan diseases affecting wild, domestic animals and humans. For the control of these serious pests, synthetic chemicals are continuously used which possess inherent toxicities that endanger the health of farm operators, consumers, and the environment. Negative effects on human health led to resurgence in interest in botanical insecticides because of their minimal costs and ecological side effects. Botanicals affect only target pests and closely related organisms, are effective in very small quantities, decompose quickly, and provide residue-free food and safe environment to live. As an integrated pest management program, botanicals can greatly reduce the use of conventional insecticides or be used in rotation or in combination with other insecticides, thereby reducing the use of conventional insecticides and possibly mitigate the development of resistance in pest populations. Plant-based insecticides induce not only acute toxicity to pests but also deterrence and/or repellence which may contribute to overall efficacy against some pests that cause great economic losses as well as transmit diseases to

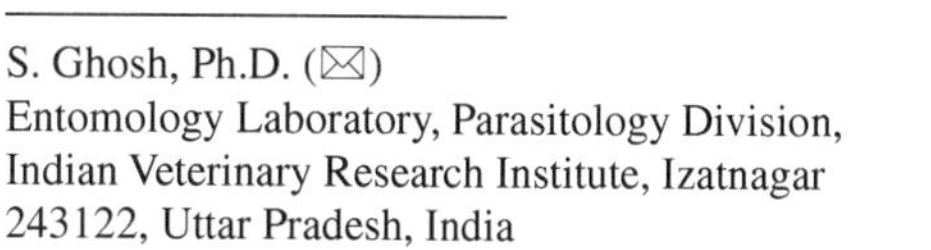

S. Ghosh, Ph.D. (✉)
Entomology Laboratory, Parasitology Division, Indian Veterinary Research Institute, Izatnagar 243122, Uttar Pradesh, India
e-mail: sghoshp@yahoo.co.in; sghoshtick@gmail.com

R. Ravindran, Ph.D.
Department of Parasitology, College of Veterinary and Animal Sciences, Pookode, Lakkidi, Wayanad, Kerala 673576, India

D. Singh (ed.), *Advances in Plant Biopesticides*,
DOI 10.1007/978-81-322-2006-0_11, © Springer India 2014

animals and humans. The present review is focused on progress in the development of plant-based insecticides effective against ticks, flies, lice, and mite infestations on milk- and meat-producing animals.

Keywords
Chemical insecticides • Resistance • Botanicals • Anti-flies • Lice • Mites • Ticks • Animal health and production

1 Introduction

Botanical pesticides have a proven track record and long use as simple extractives for pest control and have spun off important groups of synthetic pesticides with phytochemical leads such as pyrethroids and neonicotinoids. While botanicals are now a small part of the overall pesticide market due to replacement by synthetics, the new "green movement" has provided a favorable environment for the rebirth of botanical insecticides. Public concern over the use of chemical insecticides is growing. This has led to the large growth in organic agriculture where the industry self-regulates the use of products restricting synthetics but allowing some botanical pest control. Public resistance to adoption of genetically modified organisms is another factor favoring alternative control measures such as biopesticides, biocontrol, and other methodologies.

In reality, however, botanicals have certain advantages but an equal number of drawbacks in practical use. The advantages of botanical pesticides lie in their qualities such as rapid degradation, lack of persistence, or bioaccumulation in the environment, which are the major problems of synthetics. For example, DDT residues, still present in sandy soils in some countries decades after its use was discontinued, contaminated medicinal crops grown in these soils to levels which disallowed their export. On the other, botanical pesticides such as piperamides and alpha-terthienyl are degraded in the environment in hours or days. A long history of safe use of many herbal/natural products provided a confidence about their low risk, although this may not be implicit for those which will be developed in the future. The diversity and redundancy of phytochemicals in botanical extracts are also useful. Redundancy, which is the presence of numerous analogs of one compound, is known to increase the efficacy of extractives through analog synergism, reduce the rate of metabolism of the compounds, and prevent the evolution of pesticide resistance when selection occurs over several generations. From a research discovery point of view, the number of insect deterrents derived from plants seems endless as coadaptation appears to have produced a huge diversity of novel compounds across the plant kingdom and a remarkable redundancy of plant defenses within each plant species. Research activities have provided many applications of herbal pesticides as "behavior-modifying" antifeedants, "repellent/fumigant/insecticidal" essential oils, and/or innumerable novel modes of actions of large number of products to be developed in the future.

Despite many advantages, the market for botanical pesticides has a number of major challenges, and although there has been growth in the market, it has not grown in a comparable way like that of other botanical medicines in recent years. The Environmental Protection Agency (EPA) has fixed lesser registration requirements to a variety of traditionally used insecticide products. Hence, growth is expected in the fields of research, development, production, and marketing of botanical pesticides in developing/developed countries.

Flies, lice, mites, and ticks are causing severe problems in livestock industry directly reducing the efficiency of animals to enhance growth and production. Some of these arthropods are also transmitting many fatal diseases to animals (theileriosis, babesiosis, anaplasmosis, trypanosomiasis, ehrlichiosis, etc.) and to humans (CCHF, borreliosis, rickettsiosis, Rocky Mountain spotted fever, relapsing fever, and Colorado tick fever) (Ghosh et al. 2006; Azahaianambi and Ghosh 2007). Use of synthetic chemicals with insecticidal properties such as organochlorines, organophosphates, carbamates, pyrethroids, and macrocyclic lactones is the backbone to control arthropods all over the world. However, extensive and indiscriminate applications fostered not only several environmental and health concerns but also widespread development of resistance to almost all available chemicals (Nauen 2007; De la Fuente et al. 2007; Ghosh et al. 2007) and unwarranted toxic or lethal effects on nontarget organisms (Milam et al. 2000). Due to several undesirable and unsustainable features of chemical insecticides, integrated pest management (IPM) has been considered as the only sustainable option. Of the various components of IPM, strategic use of eco-friendly compounds of plant sources has been considered most potent, and literatures are flooded with information on insecticidal properties of plants grown in several countries (Murugan and Jeyabalan 1999; Amer and Mehlhorn 2006; Ravindran et al. 2012; Kaaya 2000; Ghosh et al. 2013).

An old review by Roark (1947) described around 1,200 plant species that have been listed in the literature having insecticidal potential. These studies have exposed an array of botanicals with a wide bioactive spectrum such as fungicides, nematicides, acaricides, insecticides, and carcinogenic inhibitors. Several investigators have resorted to explore plant resources to find alternate and eco-friendly compounds with potent insecticidal activity (Murugan and Jeyabalan 1999; Amer and Mehlhorn 2006). However, reviews especially focused on natural products showing activity against arthropods of veterinary medical importance are comparatively very limited. In the present chapter, a comprehensive account on botanical insecticides/acaricides useful for the control of arthropods affecting animals is provided.

1.1 Botanicals Against Flies

Plants produce numerous secondary compounds that serve as repellents, feeding deterrents, or toxicants to phytophagous insects (Levin 1976). Defensive phytochemicals are grouped into five broad categories: growth regulators, nitrogen compounds, phenolics, protease inhibitors, and terpenoids (Moore and Debboun 2007; Moore et al. 2007).

A repellent can generally be described as a substance that can be used "to cause movement away from stimuli to be repulsed" or "an agent of action as in any stimulus which elicits an avoiding reaction." Insects detect odors when volatile odor binds to odorant receptor (OR) proteins displayed on ciliated dendrites of specialized odor receptor neurons (ORNs) that are exposed to the external environment, often on the antennae and maxillary palps of insects. Some ORNs, such as OR83b that is important in olfaction, are blocked by synthetic repellent DEET (*N*,*N*-diethyl-3-methylbenzamide) (Ditzen et al. 2008).

An ideal repellent should provide protection for up to eight hours against an array of blood-feeding arthropods with a single treatment. All repellents exhibit some degree of volatility, and when applied, it allows for the production of a vapor layer, creating an unpleasant or offensive surface, smell, or taste to biting arthropods (Maibach et al. 1974). All repellent compounds have a relative vapor pressure which is directly correlated to vapor repellency. When vapor repellency is correlated with the boiling point of the chemical compound, optimal effective range falls somewhere between 230 and 260 °C, and so the compounds with lower boiling points have enough volatility to exert some vapor repellency, but not so much volatility that they evaporate away quickly (Brown and Hebert 1997). Therefore, naturally derived compounds with high vapor pressures will dissipate rather rapidly, whereas those with low vapor pressures will vaporize too slowly and may not supply enough volatile repellent compound to be effective (Paluch et al. 2010). A repellent's efficacy can be dramatically affected by sweating, abrasion of treated areas, heat, humidity, and getting treated areas wet or washed with water. In addition, environmental factors such as temperature, wind, and humidity affect repellent delivery systems, thereby influencing repellent effectiveness by impacting variability (Brown and Hebert 1997). A repellent's effectiveness is a combination of the relative vapor pressure (volatility) and delivery system (formulation).

Essential oils are generally known to have fumigant insecticidal properties, effects on behavior modification (attraction/repellency), and contact toxicity for different life stages (Koul et al. 2008). Natural oils are complexes of many biologically active constituents including terpenes, acyclic monoterpene alcohols, monocyclic alcohols, aliphatic aldehydes, aromatic phenols, monocyclic ketones, bicyclic monoterpene ketones, acids, and esters (Koul et al. 2008). The composition of oils from a particular plant species can be affected by the plant tissues used for extraction, cultivar variation, climatic and growth conditions, and the methods used for extraction and analysis. For this reason, there have been considerable efforts to examine the effects of individual components that are common to those essential oils known to have repellent properties (Isman 2000; Koul et al. 2008).

Sharma and Saxena (1974) evaluated the effects of a range of individual terpenoids on muscid flies and found a wide variety of effects. Some acted as attractants but had inhibitory effects on embryonic or larval development (eugenol and farnesol), whereas others repelled gravid females and inhibited embryonic/larval development. Fly responses to terpenoids were highly dose dependent and some were attractive at low concentrations but repellent at high ones. Larvicidal effects of the tested materials were modest at all doses. In another study, neem extracts and refined azadirachtin were found moderately toxic to larvae of the horn fly (*Haematobia irritans*), but doses required to control housefly larvae were deemed too high for practical application (Miller and Chamberlain 1989). Khan and Ahmed (2000) observed up to 85 % mortality of adult houseflies after exposure to neem extract, while Ezeonu et al. (2001) reported that extracts of sweet orange peels (*Citrus sinensis*) were effective as fumigants against adult flies.

Para-menthane-3,8-diol (PMD) – originating from the hydrodistillation of the essential oil from the leaves of eucalyptus (Curtis 1992) – was tested extensively for repellency and found mediocre to N,N-diethyl-3-methylbenzamide (DEET) (Schreck and Leonhardt 1991). These early studies used the formulations obtained from the Chinese producers that carried the active ingredient PMD in ethanol and the repellency was short lived. In reformulations, PMD concentration was increased to 50 % and ethanol was replaced with other carriers, and the repellency of the transformed PMD was found similar to repellents containing 50 % DEET (Trigg 1996). Further, in comparative study, Barnard and Xue (2004) ranked PMD as most effective after evaluating twelve commercially available repellent products, some of which contained up to 30 % DEET. In a very comprehensive study comparing various concentrations of PMD to DEET, researchers found virtually no difference in repellent performance during a 6-h field trial (Carroll and Loye 2006). The PMD has a lower vapor pressure than volatile monoterpenes found in most plant oils (Barasa et al. 2002) and provides very high level of protection from a broad range of insect vectors over several hours (Carroll and Loye 2006).

Among many well-known phytochemicals, azadirachtin (a tetranortriterpenoid compound) is one of the most extensively studied biological insecticides extracted from the seed kernel of the neem tree *Azadirachta indica* (Schmutterer 1990). This secondary metabolite is known to be effective against a wide range of insect pests as well as vectors (Sriwattanarungsee et al. 2008), and its efficacy against the targets has been proved by its ovicidal (Su and Mulla 1998), larvicidal, pupicidal (Coria et al. 2008), and growth inhibitory (Okumu et al. 2007) effects. Despite these merits, there are, however, at least two notable disadvantages with azadirachtin: it degrades rapidly upon exposure to sunlight, resulting in great loss of its insecticidal activity, and azadirachtin (0.5 and 1 ppm) in water suspension enhances the oviposition response of the mosquito *Culex tarsalis* and this enhancement was more pronounced in 1-week-old neem formulation (Su and Mulla 1999).

Palacios et al. (2009a, b) screened efficacy of essential oils of 21 medicinal and edible plants against muscid flies. Among the edible plants, essential oils from orange peel and eucalyptus leaves were the most toxic to flies. The limonene (92.5 %) and 1,8-cineole (56.9 %) were the principal components of these oils. Of the medicinal plants, essential oils high in pulegone, menthone, limonene, and 1,8-cineole were the most toxic. In a comprehensive survey, Pavela (2008) reported that essential oils of rosemary (*Rosmarinus officinalis*) and pennyroyal mint (*Mentha pulegium*) showed more activity against adult flies in both fumigant and contact toxicity assays. Essential oils of peppermint (*Mentha piperita*) and blue gum (*Eucalyptus globulus*) were the most effective of six plant extracts examined by Kumar et al. (2011) which showed both insecticidal and repellent properties. Application of an emulsifiable concentrate formulation of peppermint oil in field tests resulted in over 95 % control of flies (Kumar et al. 2011). The principal components of peppermint oil are menthone (20.9 %) and menthol (41.5 %) (Palacios et al. 2009b). As a part of an assessment of plants native to Chile, Urzua et al. (2010) found that essential oils from *Haplopappus foliosus* revealed good activity against adult flies and limonene was the most abundant component in the extract.

In a comprehensive study, Mansour et al. (2011) screened selected botanicals from agricultural wastes (*Opuntia vulgaris*, *Zea mays*, *Saccharum* spp., *Punica granatum*, *Citrus aurantifolia*), three weeds (*Cichorium intybus*, *Conyza aegyptiaca*, *Sonchus oleraceus*), three ornamental trees (*A. indica*, *Eucalyptus globulus*, *Salix safsaf*), and one agricultural crop (*Piper nigrum*) against *Musca*. The results of toxicity indicated the possible contact neurotoxic action of the active constituents of the plant species that is mainly related to the acetylcholinesterase and octopaminergic levels (Isman 2000) or the transformation of the alcohol present into the insect body into the corresponding esters by the active constituents (Tsao and Coats 1995). The volatile extracts of *C. aurantifolia* (lime) and *E. globulus* were found effective against muscid flies. It has long been recognized that the pesticidal activity of *Eucalyptus* refers to the compound 1,8-cineole which is abundantly present in the essential oil (Duke 2004). The ethanolic extracts of *C. aegyptiaca* ($LC_{50}=77.0$ ppm) and *P. nigrum* ($LC_{50}=50.1$ ppm) were found highly promising candidates as bioinsecticides.

Botanical extracts having characteristics of insect growth regulators (IGRs) can have a pronounced effect on the developmental period, growth, adult emergence, fecundity, fertility, and egg hatching resulting in effective control. The insect growth regulators are advantageous because they do not persist long in the environment due to their rapid biodegradation and low toxicity. The development of resistance to these substances has not yet been proved (Sheppard et al. 1992). Over 100 plant species contain bioactive substances related to phytoecdysones, phytojuvenoids, and anti-juvenile hormones, which act as IGRs (Varma and Dubey 1998). The use of ajugarins, isolated from *Ajuga remota* by Marcard et al. (1986), against mosquitoes is an example. Other examples include insect growth regulatory activity of crude petroleum ether–acetone extracts of 25 angiosperm plants on fly larvae (Neraliya and Srivastava 1996). The seed of *Peganum harmala* and leaves of *Acalypha indica*, *Santalum album*, *Griffonia simplicifolia*, and *Calotropis gigantea* were reported to have effect on larval–pupal and pupal–adult transformation stages of *Musca.* Seed extracts also induced some morphological abnormalities in larvae, pupae, and adult houseflies. The abnormalities included retardation of development of larvae, failure to emerge from the pupal case, and incomplete development of wings in adults that died 12 h after emergence (Bisseleua et al. 2008). Shaalan et al. (2005) compiled the list of plant species having effects on growth, development, reproduction, abnormalities, hatching rates, and fertility. Despite toxic effects and reductions in adult emergence, some phytochemicals such as those from *Annona squamosa* do not alter the larval developmental period. Zebitz (1984) reported the anti-ecdysteroid activity of neem seed kernel extract resulting in growth inhibition and prolonged developmental period. Comparable abnormalities have been recorded for *Musca* after treatment of the third larval instars with sesame, nigella, onion, diflubenzuron, and pyriproxyfen (Khater 2003). The abnormalities could be attributed to the metamorphosis inhibiting effect of plant oils, as a result of the disturbance of hormonal control, suggesting a type of insect growth-regulating activity (Hussien 1995).

Identification of phytochemicals with growth inhibition properties combined with a considerable capacity to reduce adult emergence is the desired end point of botanical insecticide research. Another desirable quality would be that a control agent induced IGR effects of less than the lethal dose so that recruitment could be reduced over time. Fortunately, adult emergence is commonly critically affected by phytochemicals, which are often capable of causing a combination of acute toxic and chronic growth effects. Hassan et al. (2012) reported IGR activities of *Ocimum tenuiflorum* and *Datura alba* leaf extracts after 72 h of treatment. A wide range of morphogenetic deformities were observed and recorded in different categories including larval–pupal mosaic, abnormal pupae, and pupal–adult mosaic according to stage of metamorphosis when death occurred due to abnormal growth and molting.

Abdel-Aal (1996) attributed such decrease in weight of treated fly to the decrease in total water content or decreased intensity of protein biosynthesis. Also, it may be due to the lack of proper sclerotization of the newly formed puparium or evaporation of body fluids.

Herbal products, NeemAzal T/S and Neemarin containing azadirachtin, were evaluated (Narladkar et al. 2006) for their efficacy against *Culicoides* larvae. The LC_{50} vales were 18.57 and 23.44 ppm, respectively. The oviposition deterrent and ovicidal effects were observed at 35 and 40 ppm. A herbal fly repellent formulation (AV/FRC/18) was found effective (Naraladker et al. 2011) for oviposition deterrence and ovicidal and larvicidal effects against *Culicoides* sp. in a farm in Maharashtra, India, without any adverse effects such as irritation, loss of production, or mortality of the animals. No residual effect was also noticed. Efficacy of a repellent lotion containing p-menthane-3,8-diol (16 %) and lemon grass oil (2 %) against *C. pachymerus* (Santamaría et al. 2012) was also investigated. Mean protection percentage of the repellent was 100 % up to 4 h and 99.5 % up to 5 h. The protection time was up to 6 h.

Khater et al. (2009) reported significant repellent properties of essential oils of camphor (*Cinnamomum camphora*), onion (*Allium cepa*), peppermint (*Mentha piperita*), chamomile (*Matricaria chamomilla*), and rosemary (*Rosmarinus officinalis*) against *Stomoxys calcitrans*, *Haematobia irritans*, and *Hippobosca equina* for 6 days posttreatment.

Besides the above extensively studied botanicals, there are a number of plant species having activities against flies. Some of the examples are listed in Table 11.1, but further detailed studies are required to establish the marketable potentiality of the plant species for the control of flies.

1.2 Botanicals Against Lice and Mites

Lice infestations in animals are common especially during winter season. All the life cycle stages of chewing lice (order Mallophaga) and sucking lice (order Anoplura) are seen on the host and cause skin irritation and of the skin and stimulate scratching, rubbing, and licking leading to restlessness, damage to hair coat, or fleece and hides and loss of milk production. Treatment may be warranted to reduce the damage to the hide (Radostitts et al. 2007).

While compiling ethnoveterinary uses of plant extracts as lousicides, Muhammad Ishtiaq Ch and Khan (2006) reported that *Adhatoda vasica* L. and *Cannabinus sativus* L. have the potential to kill lice infestations on animals. Robles (2004) reported lousicidal activity of tobacco (*Nicotiana tabacum*), tubli (*Derris philippinensis*), makabuhay (*Tinospora rumphii*), and neem (*Azadirachta indica*) at the concentrations of 10 %, 20 %, and 40 %, respectively, with 90 % mortality in vitro, whereas in vivo experimentation showed that

Table 11.1 Plants with their common names having fly repellent action

Family	Plant species	Common name	Part used	Animal species
Arecaceae L.	*Phoenix dactylifera*	Khajoor	–	Cattle, sheep
Asclepiadaceae (Ait.)	*Calotropis procera*	Aak	Aerial part	Cattle, goat
Lamiaceae	*Mentha arvensis* L.	Podina	Leaf	Sheep
Linaceae	*Linum usitatissimum* L.	Alsi	Leaf	Cattle, buffalo, sheep, goat
Meliaceae	*Azadirachta indica* A. Juss.	Neem	Seed, leaf	Do
Rutaceae	*Citrus medica* L.	Nimbu	Leaf	Sheep
Solanaceae	*Capsicum annuum* L.	Surkh mirch	Fruit	Goat
Solanaceae	*Nicotiana tabacum* L.	Tambaku	Leaf	Cattle, buffalo, sheep, goat
Zingiberaceae	*Curcuma longa* L.	Haldi	Root	Do
Araceae	*Arisaema flavum*	Soro ganda	Root	Do
Berberidaceae	*Berberis lyceum* L.	Sumbal	Do	Do
Dioscoreaceae	*Dioscorea deltoid*	Knis	Aerial parts	Do
Pinaceae	*Pinus roxburghii*	Cheer	Resin	Do
Podophyllaceae	*Podophyllum emodi*	Bankhakri	Aerial parts	Sheep
Rosaceae	*Prunus persica*	Arro	Leaf	Cattle, buffalo, sheep
Euphorbiaceae	*Ricinus communis*	Arind	Seed	Do
Polygonaceae	*Rumex hastatus*	Hola	Root	Do

only tobacco and makabuhay induced 45.9 % and 79.7 % reduction in infestations, respectively. Essential oils of pennyroyal, tea tree, and anise have potent insecticidal activity for killing lice and their eggs (Williamson 2007). The *Eucalyptus globulus* leaf oil-derived monoterpenoids were found to be highly toxic to eggs and to females (Yang et al. 2004). Khater et al. (2009) reported lousicidal properties of essential oils of camphor (*Cinnamomum camphora*), onion (*Allium cepa*), peppermint (*Mentha piperita*), chamomile (*Matricaria chamomilla*), and rosemary (*Rosmarinus officinalis*). All treated lice were killed after 0.5–2 min, and the number of lice infesting buffaloes was significantly reduced at 3, 6, 4, 6, and 9 days after treatment with camphor, peppermint, chamomile, onion, and d-phenothrin, respectively. Davidović et al. (2012) reported the efficacy of extracts of hellebore (*Helleborus* L., Ranunculaceae) and tobacco (*Nicotiana tabacum* L., Solanaceae) against cattle lice and mange infestations.

Disease caused by mite infestations is highly contagious to domestic and wild animals and is characterized by scratching and serous exudations, and infected animals usually cease to feed and can become severely debilitated (Bates 1999). As in the case of other arthropod infestations in animals, there are a good number of ethnoveterinary information available regarding the efficacy of the plant extracts against mites (Table 11.2). However, very few of these extracts were experimentally tested in vitro and in vivo against infestations. For example, Kim et al. (2004) screened 56 plant essential oils against poultry mite, *Dermanyssus gallinae*. Hundred percent mortality at 0.07 mg cm^{-2} was observed with essential oils from bay, cade, cinnamon, clove bud, coriander, horseradish, lime dis 5 F, mustard, pennyroyal, pimento berry, spearmint, thyme red, and thyme white oils. In fumigation tests with adults at 0.28 mg cm^{-2}, essential oils of cade, clove bud, coriander, horseradish, and mustard were more effective in closed containers than open ones, indicating acaricidal action of vapor phase.

Mägi et al. (2006) tested four medicinal plant extracts in 10 % ethanol solutions (hogweed, *Heracleum sosnowskyi*; Manden, mugwort, *Artemisia vulgaris* L.; tansy, *Tanacetum vulgare* L.; wormwood, *Artemisia absinthium* L.) and seven essential medicinal oils used in 1 % emulsions

Table 11.2 Plants with their local names have potential against lice and mite infestations in animals

Family	Plant species	Local name	Infestations/infection
Brassicaceae	*Brassica campestris*	Mustard oil	Pediculosis, mange
Cannabaceae	*Cannabis sativa*	Bhang	Pediculosis
Pinaceae	*Cedrus deodara*	Loo	Mange, pediculosis
	Zanthoxylum mays	Corn	Mange, pediculosis
Lythraceae	*Lawsonia inermis*	Mehndi/henna	Pediculosis
Meliaceae	*Melia azedarach* Linn.	Darenk	Do
Solanaceae	*Nicotiana plumbaginifolia* Viv.	Desi Tambaku	Do
Lamiaceae	*Ocimum sanctum* Linn.	Tulsi	Do
Pinaceae	*Pinus roxburghii* Sarg.	Chir	Do
Rosaceae	*Pyrus pashia* Buch.-Ham. ex D. Don	Kanth	Do
Sapindaceae	*Sapindus mukorossi* Gaertn.	Reetha	Do
Asclepiadaceae	*Tylophora hirsuta* Wight	Tripu	Lice and mites
Verbenaceae	*Vitex negundo* Linn.	Bana	Lice

(garlic, *Allium sativum* L.; black pepper, *Piper nigrum* L.; juniper, *Juniperus communis* L.; citronella grass, *Cymbopogon nardus* Rendle; pennyroyal, *Mentha pulegium* L.; eucalyptus, *Eucalyptus globulus* Labill.; tea tree, *Melaleuca alternifolia* Cheel) to control swine sarcoptic mange mites (*Sarcoptes scabiei* var. *suis*). All the preparations inhibited the development and were more or less lethal to mange mites, while tea tree and citronella volatile oil preparations proved to be the most effective. Among the ethanol-guided extracts, the most active extract was obtained from hogweed seeds in which 57–93 % of parasites died in 2–4 weeks of treatments of pigs. The extracts of local plants tansy and wormwood diminished the number of mites up to 44 % within the first week after treatment. The acaricidal activity of methanolic extracts from 40 medicinal plant species and steam distillate of *Cinnamomum camphora* against *D. gallinae* was evaluated. The steam distillate was found most effective in filter paper contact toxicity bioassay followed by extracts from *Asarum sieboldii* var. seoulens whole plant, *Eugenia caryophyllata* flower bud, and *Mentha arvensis var. piperascens* (Kim et al. 2007).

Psoroptic acariasis in rabbits is a global disease that causes considerable losses in the United States, Italy, Turkey, South Korea, India, and China. The parasite primarily inhabits the ears of rabbit (Ulutas et al. 2005; Singh et al. 2012). The *P. cuniculi* infestation can cause intense pruritus and the formation of crusts and scabs, which can completely cover the external ear canal and the internal surfaces of the pinna (Bates 1999). The potential acaricidal properties of an *Ailanthus altissima* bark extract were assessed against two common species of animal mites, *Psoroptes cuniculi* and *Sarcoptes scabiei* var. *cuniculi*, in vitro. The prepared extract exhibited significant acaricidal properties for both mite species. The extract at the concentrations of 1.0, 0.5, and 0.25 g/mL killed all tested *S. scabiei* within 7 h; however, only 1.0 and 0.5 g/mL of extract killed all treated *P. cuniculi* (Gu et al. 2013). Nong et al. (2012) evaluated in vivo efficacy of crofton weed (*Eupatorium adenophorum*) against *P. cuniculi*. Seven days after treatment the main clinical scores decreased from 3.32, 3.08, and 3.17 in the treated groups which were similar to treatment with ivermectin. Yong-Hua et al. (2008) reported anti-*Sarcoptes scabiei* var. *cuniculi* properties of *A. indica* with an LC_{50} value of 1.3 ppm for petroleum extract and 4.1 ppm for chloroform extract.

1.3 Botanicals Against Ticks

1.3.1 Repellents

Defensive phytochemicals are grouped into five broad categories: growth regulators, nitrogen compounds, phenolics, protease inhibitors, and terpenoids (Moore et al. 2007). The vast majority of phytochemicals that have been tested for repellency against ticks are terpenoids. A number of plants and essential oils from plants exhibit repellent properties against tick (Table 11.3).

Terpenoids

Terpenoids are a structurally diverse assembly of compounds that make up the largest group of secondary plant chemicals (Langenheim 1994). Terpenes are derived from units of isoprene and are classified sequentially as chains of isoprene (hemi-, mono-, sesqui-, di-, etc.) (Moore et al. 2007). Plant-derived terpenoids are repellent against several species of ticks. For example, Tunón et al. (2006) tested whole and fractioned compounds from the extract of southernwood, *Artemisia abrotanum* L., and the essential oil from the carnation flower, *Dianthus caryophyllus* L., against *I. ricinus* nymphs, and the monocyclic terpene eugenol isolated from both plants provided >90 % repellency, while the acyclic terpene alcohol, b-citronellol, isolated from carnation flower oil provided 84 % repellency. Dautel et al. (1999) reported repellent effect of essential oil of a number of plants including citronella, *Cymbopogon nardus* (L.) Rendle; peppermint, *Mentha piperita* L.; and lemon balm, *Melissa officinalis* L., against *I. ricinus* nymphs. Similarly, oils of citronella and lily of the valley, containing citronellol, have 83 and 67 % repellency against *I. ricinus* nymphs (Thorsell et al. 2006). They reported that 10 % clove oil, which contains high amounts of eugenol, provided 78 % repellency, while 10 % DEET provided 71 % repellency against *I. ricinus* nymphs for 8 h. Pållson et al. (2008) tested constituents in the essential oil from the flowers of aromatic tansy, *Tanacetum vulgare* L., against the nymphal *I. ricinus*. Extracts and oils of wormwood, *Artemisia absinthium* L., and marsh tea, *Rhododendron tomentosum* (Stokes), were tested against nymphal *I. ricinus* (Pållson et al. 2008) and found 78 % repellency of the extracts. The primary volatile compounds identified in *A. absinthium* and *R. tomentosum* were the terpenes, myrtenyl acetate (77.8 %), and (3Z)-hexanol (18.3 %), respectively.

Two terpenoids, callicarpenal and intermedeol, isolated from American beautyberry, *Callicarpa americana* L., and Japanese beautyberry, *C. japonica* Thunb., showed activity against ticks. Using a fingertip bioassay, Carroll et al. (2007) reported significant efficacy of intermedeol against nymphal *A. americanum* and *I. scapularis* isolated from *Callicarpa americana* L. and *C. japonica* Thunb., while callicarpenal isolated from the same plants provided 100 % repellency against *I. scapularis* (Carroll et al. 2007). Dietrich et al. (2006) isolated 14 compounds classified as monoterpenes, eremophilane sesquiterpenes, and eremophilane sesquiterpene derivatives from the essential oil of the heartwood of Alaskan cedar, and the repellency was compared with DEET against nymphal *I. scapularis* and was found comparative. Isolongifolenone, a sesquiterpene compound found in the South American tree, *Humiria balsamifera* St. (Aubl.), revealed 100 % efficacy to repel *I. scapularis* nymphs but 80 % against *A. americanum* (Zhang et al. 2009).

Several grasses have been suggested for use as anti-tick pastures. Thompson et al. (1978) and Mwangi et al. (1995a, b) compared recapture/climbing behavior of larvae of *R. (B.) microplus* and *R. appendiculatus* in pasture having molasses grass, *Melinis minutiflora* Beauv. and *Pennisetum clandestinum*, and reported significant reduction in recapture/climbing behavior of tested larvae. The mature gamba grass, *Andropogon gayanus* Kunth, having glandular trichomes was found effective against the larvae of *R. (B.) microplus* (Cruz-Vazquez and Fernández-Ruvalcaba 2000). Additionally, two tropical legumes, *Stylosanthes hamata* (L.) Taub. and *S. humilis* Kunth, with glandular trichomes exhibited acaricidal and repellent properties (Fernandez-Ruvalcaba et al. 1999). A number of African plants also showed tick repellent properties. For example, oil from wild basil, *Ocimum suave* Willd, was reported to have repellent activity against *R. appendiculatus*. Essential oil from the *Cleome monophylla* and *C. (Gynandropsis) gynandra* (L.) Brig. provided more than 90 % repellency of *R. appendiculatus* larvae (Ndungu et al. 1995; Lwande et al. 1999). Birkett et al. (2008) tested the hexane extract of resin of gum haggar, *Commiphora holtziana* Engl., against larvae of *R. (B.) microplus* and reported 5 h

Table 11.3 Plants that exhibit repellency against ticks, their taxonomic families, tick species repelled, and references

Scientific name	Common name	Family	Tick species	References
Andropogon gayanus	Gamba grass	Poaceae	*Rhipicephalus (Boophilus) microplus*	Thompson et al. (1978), Cruz-Vazquez and Fernández-Ruvalcaba (2000)
Cleome/Gynandropsis gynandra	African spider flower	Capparidaceae	*R. appendiculatus*	Malonza et al. (1992), Lwande et al. (1999)
Azadirachta indica	Neem tree	Meliaceae	*Ixodes ricinus*	Garboui et al. (2006)
Callicarpa americana	American beautyberry	Verbenaceae	*Amblyomma americanum, I. scapularis*	Carroll et al. (2007)
Callicarpa japonica	Japanese beautyberry	Verbenaceae	*A. americanum, I. scapularis*	Carroll et al. (2007)
Cleome monophylla	Spider plant	Capparidaceae	*R. appendiculatus*	Ndungu et al. (1995)
Artemisia abrotanum	Southernwood	Asteraceae	*I. ricinus*	Tunón et al. (2006)
Convallaria majalis	Lily of the valley	Liliaceae	*I. ricinus*	Thorsell et al. (2006)
Commiphora erythraea	Sweet myrrh	Burseraceae	*A. americanum, Dermacentor variabilis, I. scapularis*	Carroll et al. (1989)
Commiphora holtziana	Gum haggar	Burseraceae	*R. (B.) microplus*	Birkett et al. (2008)
Commiphora swynnertonii	-	Burseraceae	*R. appendiculatus*	Kaoneka et al. (2007)
Corymbia citriodora	Lemon-scented gum	Myrtaceae	*I. ricinus*	Trigg and Hill (1996), Gardulf et al. (2004), Jaenson et al. (2006)
Cymbopogon spp.	Citronella grass	Gramineae, Poaceae	*I. ricinus*	Thorsell et al. (2006)
Dianthus caryophyllus	Carnation	Caryophyllaceae	*I. ricinus*	Tunón et al. (2006)
Humiria balsamifera	Oloroso	Humiriaceae	*A. americanum, I. scapularis*	Zhang et al. (2009)
Lavandula angustifolia	Lavender	Lamiaceae	*Hyalomma marginatum rufipes, I. ricinus*	Jaenson et al. (2006), Mkolo and Magano (2007)
Lycopersicon hirsutum f. *glabratum*	Wild tomato	Solanaceae	*A. americanum, D. variabilis, I. scapularis, Ornithodoros parkeri*	Vanderherchen (2003), Witting-Bissinger et al. (2008), Bissinger et al. (2009a, b)
Melinis minutiflora	Molasses grass	Poaceae	*R. appendiculatus*	Thompson et al. (1978), Mwangi et al. (1995a, b)
Ocimum basilicum	Sweet basil	Lamiaceae	*I. ricinus*	Del Fabbro and Nazzi (2008)
Ocimum suave	Wild basil	Lamiaceae	*R. appendiculatus*	Mwangi et al. (1995a, 1995b)
Pelargonium graveolens	Geranium	Geraniaceae	*I. ricinus*	Jaenson et al. (2006)
Rhododendron tomentosum	Marsh tea	Ericaceae	*I. ricinus*	Jaenson et al. (2005)
Stylosanthes hamata	Caribbean stylo	Fabaceae	*R. (B.) microplus*	Muro Castrjón et al. (2003)
Stylosanthes humilis	Townsville stylo	Fabaceae	*R. (B.) microplus*	Do
Syzygium aromaticum	Clove	Myrtaceae	*I. ricinus*	Thorsell et al. (2006)
Tanacetum vulgare	Tansy	Asteraceae	*I. ricinus*	Pållson et al. (2008)

repellency period. Sesquiterpene hydrocarbons were assumed to be responsible for the repellency effect. Carroll et al. (1989) tested hexane extracts of gum resin from the shrub *Commiphora erythraea* Engler against *A. americanum*, *D. variabilis*, and *I. scapularis* and reported 100 % repellency against larvae and adults of *A. americanum.*

Although list has been enriched with information about repellent property of a number of plants and plant compounds, due to low yield and high cost of extraction of bioactive compounds, a few have been commercialized (Moore et al. 2007). A number of active ingredients commonly found in commercially available tick repellents are:

1. IR3535 or EBAAP (ethyl butylacetylaminopropionate): Staub et al. (2002) evaluated the effectiveness of EBAAP against *I. ricinus* in field tests and reported 41 % repellency.
2. PMD or para-menthane-3,8-diol: Trigg and Hill (1996) reported repellent activity against *I. ricinus* nymphs. In a field test, Gardulf et al. (2004) found significantly lower tick attachment on treated skin compared to untreated controls. Jaenson et al. (2006) tested oil of lemon eucalyptus and MyggA® Natural, a product similar to Citriodiol that contains 30 % oil of lemon eucalyptus with a minimum of 50 % PMD and small amounts of geranium, lavender, and rose extracts, against nymphal *I. ricinus*. Both products provided 100 % repellency. Field trials using cloth drags treated with MyggA® Natural or oil of lemon eucalyptus revealed 74 % and 85 % repellency, respectively. In another field study, it was observed that the blankets treated with two concentrations of MyggA® Natural (3.2 and 4.2 g/m^2) and the repellent RB86 (70 % neem oil containing azadirachtin) significantly reduced the number of *I. ricinus* nymphs collected by dragging compared to untreated blankets (Garboui et al. 2006).
3. Undecanone or 2-undecanone (methyl nonyl ketone): In tests using treated and untreated filter paper, BioUD with 7.75 % of 2-undecanone provided significantly greater mean percentage repellency against *A. americanum* and *I. scapularis* (Bissinger et al. 2009a). The BioUD provided 8-day repellency against *D. variabilis* on treated cotton cheesecloth (Witting-Bissinger et al. 2008) and an average of 93.2 % repellency against *A. americanum* over 7 weeks of testing (Bissinger et al. 2009b).
4. Dodecanoic acid or DDA: It occurs as the main compound in coconut and palm kernel oil. Schwantes et al. (2008) tested formulations of 10 % DDA against *I. ricinus* nymphs and reported a mean repellency of 86.5 %.

Laboratory bioassays were conducted to determine the activity of 15 chemical constituents isolated from the essential oil of Alaskan yellow cedar (AYC) against *I. scapularis* ticks. The repellent efficacy of nootkatone, from AYC, was comparable with DEET (Dietrich et al. 2006). In field trials for repellency, using treated cover, tick drags, and white cotton sheets, nootkatone was found more effective against deer ticks and lone star ticks than both Repel® brand Permanone® (0.05 % permethrin) and EcoSMART Organic Insect Repellent® (Schulze et al. 2011; Jordan et al. 2012). Scientists at CDC and Iowa State University demonstrated that nootkatone and the other compounds from AYC have a unique mode of action as compared to that of other known pesticides and repellents and considered them a potentially important alternative weapon against chemical-resistant arthropods (Tong et al. 2011).

Repellent activity against ticks is determined by a variety of factors, viz., the rate of evaporation, the importance of contact versus spatial repellency, the delivery rate to the receptor, and the potency of the compound to elicit repellent behavior. At the level of the sensilla, potency is affected by the delivery of the repellent to the receptor, affinity of the receptor protein for the repellent, degradation of the repellent in the sensilla, and potency (once in the receptor) of eliciting an effective repellent behavior. Conditions such as abrasion, humidity, temperature, and wind also affect the longevity of repellency (Moore et al. 2007). Formulation can play an important role in this equation. For example, an early field study showed that Indalone formulated as an emulsion provided 83 % repellency against ticks for 6 weeks compared to only 22 % provided by an aerosol formulation (Smith et al. 1954). However, formulation chemistry relative to tick repellents and repellents in general has been an understudied area in the scientific literature and might be a critical factor in repellent discovery and use in the future.

Screening of chemical libraries, the bioassay of different biological products from plants and animals, the development of structure–activity relationships, and serendipity have historically been critical factors in the research and development of repellents against vectors. The basic understanding of the mechanism of repellency from spatial versus contact to the molecular basis of odorant reception in ticks is minimal. The recent sequencing of the tick genome as well as new high-throughput DNA sequencing technologies, the ease for the de novo construction of transcriptomes from sample sizes as small as a single cell, and advances in bioinformatics are expected to strengthen our overall understanding of tick repellency at the molecular level. Although screening of chemical libraries and the examination of extracts from plants and animals will continue to be an important source for new compounds in the future, molecular and stereochemical modeling involving odorant transport, binding, and degradation of proteins as well as the development of in vitro and single cell receptor bioassays could be helpful to our basic understanding on the mechanism of repellency.

1.3.2 Larvicidal, Ovicidal, and Adulticidal

From an economic point of view, *R. (B.) microplus* (Acari, Ixodidae) is the main tick of the tropical and subtropical regions and one of the most important species throughout the world (Walker et al. 2003; Martins et al. 2006; Ghosh et al. 2007) (Fig. 11.1). Research activities on the use of plant extracts to control ticks have become intensified globally over the last decade. Even though a large number of investigations have proven the acaricidal activity of many plant extracts in the laboratory, follow-up studies are needed with the aim of validating this control strategy.

Based on available literature, the efficacy of plant extracts from 55 species belonging to 26 families has been evaluated against ticks (Table 11.4). Even though most of the studies were in vitro, compounds causing the acaricidal activity were identified in a few cases (Table 11.5). For example, Carroll et al. (1989) observed 80 % mortality with hexane extract of gum resin from the shrub *Commiphora erythraea* Engler against *A. americanum*, *D. variabilis*, and *I. scapularis*. Chungsamarnyart and Jiwajinda (1992) tested volatile oils steam-distilled from fresh and dried leaves of *Cymbopogon citratus* and *C. nardus* against adults and larvae of *R. (B.) microplus*. The activity was higher in extracts prepared from fresh leaves than dried leaves. The peel oils from *Citrus reticulata*, *C. suncris*, *C. maxima*, *C. sinensis*, and *C. hystrix* were tested against cattle tick, and the extracts from *C. reticulata* and *C. maxima* were found most effective (Chungsamarnyart and Jansawan 1996). Malonza et al. (1992) reported high levels of mortality of nymphs of *A. variegatum* and *R. appendiculatus* when treated with *Gynandropsis gynandra* extracts. Preliminary investigations on *M. azedarach* showed that the oily hexane and chloroform extracts obtained from ripe fruits affected larval mortality and reproduction of engorged females (Borges et al. 1994, 2003). Sousa et al. (2008) showed that the extracts from green fruits were 1.5 times more efficient than extracts from ripe fruits. However, lower efficiency was observed by Vivan (2005). Low interference with reproductive efficiency (32 %) was observed by Costa et al. (2008) when hydroalcoholic leaf extract from *A. indica* was used at 20 % concentration. The essential oil of the genus *Cymbopogon winterianus* Jowitt was tested against larvae and engorged females. Total inhibition of eclosion was observed at a concentration of 7.14 % and of egg conversion at 10 %. All the larvae died at concentrations between 5.5 and 7.14 %. The principal components of the essential oil, i.e., citronellal, geraniol, and citronellol, were tested against females, and the best results were observed for the first two components. However, the activity of the individual compounds remained inferior to that of the entire essential oil (Martins 2006). The essential oil of

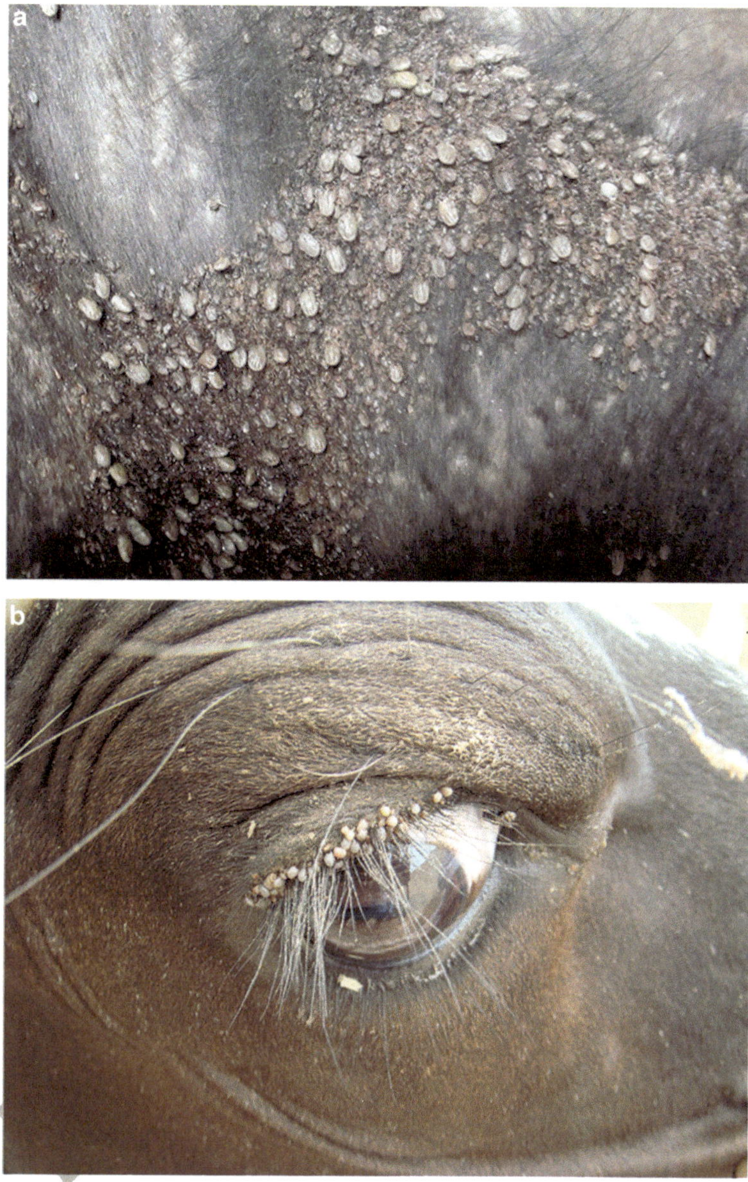

Fig. 11.1 (**a** and **b**) Crossbred cattle infested with ticks

Cymbopogon nardus caused 79 % inhibition of reproduction at 1 % concentration (Olivo et al. 2008). On the other hand, Costa et al. (2008) observed that this plant had low efficacy (17 %) against females using a hydroalcoholic extract from leaves at a concentration of 20 %. Alcoholic leaf extract of *C. citratus* at a concentration of 2 % showed efficacy of 18.4 % against engorged females (Broglio-Micheletti et al. 2009). The essential oil from the leaves of *Ageratum houstonianum* produced 100 % mortality of *R. lunulatus* on day 3 at 0.03 μL/g (Pamo et al. 2005). The hydrodistilled extracts of *Artemisia annua*, *A. vulgaris*, and *Ocimum kilimandscharicum* and oil seeds of *Pongamia glabra* were tested against *R. (B.) microplus.* The *O. kilimandscharicum* exhibited the highest efficacy of 98.34 % followed by *P. glabra* (96.67 %), *A. annua* (95 %), and *A. vulgaris*

Table 11.4 Showing the name of plants, family name, and parts of the plants tested positive against different tick species

Scientific name	Family name	Plant part(s) used (application form)	References
Acanthus pubescens (Thomson ex Oliv.) Engl.	Acanthaceae	Leaf/stem	Baerts and Lehmann (1989, 1991)
A. arboreus Forssk.		Leaf/root	Johns et al. (1995), Neuwinger (2000)
Rhinacanthus nasutus (L.) Kurz			Kamaraj et al. (2010)
Aloe latifolia Haw.	Aloeaceae	Leaf/juice	Watt and Breyer-Brandwijk (1962), Getahun (1976)
Aloe elgonica Bullock		Leaf/juice	Watt and Breyer-Brandwijk (1962), Getahun (1976)
Achyranthes aspera L.	Amaranthaceae		Zahir et al. (2009)
Rhus natalensis Bernh. ex Krauss var. *macrocarpa* (Schweinf.) Cufod.	Anacardiaceae	Aerial parts/root/seed	Janet (1986), Berger (1994), Hyde et al. (2011)
Lannea schimperi (Hochst. ex A. Rich.) Engl.	Anacardiaceae	Bark/root/leaf	Marin (1999), Jeruto et al. (2008)
Mangifera indica L.			Srivastava et al. (2008)
Annona senegalensis Pers. ssp. *senegalensis*	Annonaceae	Leaf/bark/root	Toyang et al. (1995)
Carissa edulis Vahl	Apocynaceae	Leaf/fruit/root	Ibrahim et al. (1999), Jawaid et al. (2011)
Acorus calamus Vach.	Acoraceae	Rhizome	Ghosh et al. (2011)
Microglossa pyrifolia (Lam.) O. Kuntze	Asteraceae	Whole plant-leaf-dust	ITDG and IIRR (1996)
Calea serrata Less.			Ribeiro et al. (2008a)
Chrysanthemum cinerariifolium (Trevir.) Vis.		Flower/leaf/stem/aerial parts	Cremlyn (1978)
Markhamia lutea (Benth.)	Bignoniaceae	Leaf/stem	Kayonga and Habiyaremye (1987), Rahmatullah et al. (2010), Jawaid et al. (2011)
Stereospermum kunthianum, Cham, Sandrine Petit		Root/bark	Malerich and Trauner (2003), Rahmatullah et al. (2010)
Jacaranda mimosifolia D. Don		Leaf/root/stem	Rojas et al. (2006), Rahmatullah et al. (2010), Jawaid et al. (2011)
Tamarindus indica L.	Caesalpiniaceae	Root/bark/leaf	Zemede and Mesfin (2001)
Senna didymobotrya (Fresn.) Irwin & Barneby		Root/bark/leaf	Mansingh and Williams (1998), UWMG (2003)
Cassia auriculata L.			Kamaraj et al.(2010)
C. alata L.		Leaves	Reghu Ravindran et al. (2012)
Maytenus senegalensis (Lam.) Exell.	Celastraceae	Fruit/leaf/stem/bark/root	El Tahir et al. (1998), El and Satti (1999), Bhatt et al. (2001), Kumar (2003)
Gynandropsis gynandra (L.) Briq.	Capparidaceae	Whole plant	Malonza et al. (1992), Dipeolu et al. (1992), Torto and Hassanali (1997)
Hypericum polyanthemum Klotzsch	Clusiaceae		Ribeiro et al. (2007)
Terminalia chebula Retz.	Combretaceae		Kamaraj et al. (2010)

(continued)

Table 11.4 (continued)

Scientific name	Family name	Plant part(s) used (application form)	References
Ipomoea batatas (L.) Lam.	Convolvulaceae	Leaf/stem/sap	Cevallos-Casals and Cisneros-Zevallos (2002)
Euclea divinorum Hiern.	Ebenaceae	Root/leaf/bar	Homer et al.(1992), Dagne et al. (1993), Hines and Eckman (1993), Lukwa et al. (2001)
Diospyros anisandra S. F. Blake			Rosado-aguilar et al. (2008)
Euphorbia tirucalli L.	Euphorbiaceae	Whole plant	Duke (1983a), Hines and Eckman (1993), Bowen and Hollinger (2002)
Bridelia scleroneura Mull.		Stem/root and/or bark	Anon (2001a), Schmidt (2003a)
Ricinus communis L.		Seed/stem/root	Mitchell and Ahmad (2006), Zahir et al. (2010), Ghosh et al. (2013)
Erythrococca bongensis Pax.		Stem/root/leaf	Kayonga and Habiyaremye (1987), Maundu et al. (2001), Schmidt (2003b)
Neoboutonia melleri (Mull. Arg.) Prain		Stem/root	Schmidt (2003b)
Macaranga kilimandscharica Pax.		Stem/root	Dewees (1995), Maundu et al. (2001)
Croton sylvaticus Hoschst. ex Krauss		Seed	Mwangi et al. (1998), Schmidt (2003b)
Hymenocardia acida Tul.		Bark/leaf	Irvine (1961), Marin (1999), Schmidt (2003b)
Sapium ellipticum Hochst. ex Krauss Pax		Root/bark	Samuelsson et al. (1992), Schmidt (2003b)
Phyllanthus ovalifolius Forssk.		Aerial parts	Schmidt (2003b)
Clutia abyssinica Jaub. & Spach		Aerial parts	Duke (2002), Schmidt (2003b)
Euphorbia candelabrum Kotschy.		Leaf/stem/sap	ICIPE Annual Report (1998/99)
Margaritaria discoidea (Baill.) G.L.Webster		Whole plant	Kaaya et al. (1995)
Manihot esculenta Crantz		Root/bark/stem	Schmidt (2003b)
Jatropha curcas Linn.		Leaf	Juliet et al. (2012)
Psorospermum febrifugum Spach	Guttiferae	Fruit/bark/stem/leaf	Irvine (1961), Cassady et al. (1990), Schmidt (2003c)
Harungana madagascariensis Lam. ex Poir.		Leaf/bark/stem	Woodland (1997), Irvine (1961), Schmidt (2003c)
Anisomeles malabarica (L.) R. Br	Lamiaceae		Zahir et al. (2009)
Hesperozygis ringens Benth.			Ribeiro et al.(2010)
Leucas aspera		Aerial part	Ravindran et al. (2011)
Cinnamomum zeylanicum Blume	Lauraceae		Álvarez et al. (2008)
Copaifera reticulata Ducke	Leguminosae		Fernandes and Freitas (2007)
Dahlstedtia pentaphylla (Taub) *Burk.*			Pereira and Famadas (2004)

(continued)

Table 11.4 (continued)

Scientific name	Family name	Plant part(s) used (application form)	References
Gloriosa superba L. ()	Liliaceae		Zahir et al. (2009)
Azadirachta indica A. Juss.	Meliaceae	Stem/bark/fruit/leaf/root	Williams (1993), Ndumu et al. (1999), Wilson and Mansingh (2002), Blair and Mansingh (2002), Abdel-Shafy and Zayed (2002)
Melia azedarach L.		Stem/bark/fruit leaf/root	Lindsay and Kaufman (1988), Cabral et al. (1996), Mansingh and Williams (1998), Borges et al. (2003)
Turraea holstii Gürke		Whole plant	Rajab et al. (1998)
Acacia sieberiana DC.	Mimosaceae	Root	Anon (2001a)
Acacia nilotica (L.) Bel.		Bark/root	Anon (2001b)
Artocarpus altilis Park.	Moraceae		Williams (1993)
Psidium guajava L.	Myrtaceae		Zahir et al. (2009)
Eucalyptus staigeriana F. Muell., *Eucalyptus citriodora* Hook.			Chagas et al. (2002)
Syzygium malaccense (L.) Merr. & Perry			Broglio-micheletti et al. (2009)
Lophira alata Banks ex Gaertn.	Ochnaceae	Whole plant	Anon (1977)
Ximenia americana L.	Olacaceae	Leaf/bark/root/fruit	Anon (2003)
Olea capensis L.		Leaf/bark/fruit	Hines and Eckman (1993), Stockbauer (2003)
Erythrina abyssinica Lam. ex DC.	Papilionaceae	Leaf/bark	Rulangaranga (1989), Hines and Eckman (1993), ITDG and IIRR (1996)
Rhynchosia resinosa (A. Rich.) Baker		Leaf/stem/root/bark	Matzigkeit (1990), William (1999)
Tephrosia vogelii Hook.f.		Leaf/stem/fruit/bud/bark/seed/root (whole plant)	Matzigkeit (1990), Toyang et al. (1995), Kambewa et al. (1997)
Mondia whitei (Hook.f.) Skeels	Periplocaceae	Root/stem	Okwemba (2002)
Pittosporum viridiflorum Sims.	Pittosporaceae	Stem bark/root/leaf	Coetzee et al. (1999), Berger et al. (2002), Seo (2002)
Zea mays L. kamayindi (Bukusu)	Poaceae	Leaf/stem	Lukwa et al. (2001), Ramos-Escudero et al. (2011)
Prunus africana (Hook.f.) Kalkman.	Rosaceae	Bark/leaf/root	Cunningham and Mbenkum (1993), Schippmann (2001)
Harrisonia abyssinica Oliv.	Simaroubaceae	Leaf/root/bark/stem	Johns et al. (1990), Kokwaro (1993), Fabry et al. (1996)

(continued)

Table 11.4 (continued)

Scientific name	Family name	Plant part(s) used (application form)	References
Nicotiana tabacum L.	Solanaceae	Leaf/stem-dust	Juliette De Ba Levy (1991), Dipeolu and Ndungu (1991), Berger (1994), Mwangi (1996), ITDG and IIRR (1996), Adoyo et al. (1997), Mansingh and Williams (1998), Adewusi and Afolayan (2010)
Physalis peruviana L.		Stem/root/fruit/leaf	Barbadine (2003), Wu et al. (2006), Franco et al. (2007), Arun and Asha (2007), Pardo et al. (2008), Kolar and Malbeck (2009),
Solanum incanum L.		Root, fruit juice	Hines and Eckman (1993), Lukwa et al. (2001), Adewusi and Afolayan (2010), Jawaid et al. (2011),
Capsicum frutescens L.		Fruit/leaf/stem	Stoll (1988), Berger (1994), Adoyo et al.(1997), Mansingh and Williams (1998),Regassaa (2000)
Withania somnifera (L.) Dunal		Stem/root/leaf	Singh and Kumar (1998), Van Der Hooft et al. (2005), Mirjalili et al. (2009)
Harrisonia abyssinica Oliv.	Simaroubaceae	Leaf/root/bark/stem	Watt and Breyer-Brandwijk (1962), Johns et al. (1990), Kokwaro (1993), Fabry et al. (1996)
*Grewia bicolo*r A. Juss.	Tiliaceae	Root/bark	Orwa et al. (2009)
Grewia trichocarpa Hochst. ex A. Rich.		Root	Brink (2009)
Steganotaenia araliacea Hochst.	Umbelliferae	Leaf/stem/bark	Rukangira (2003)
Rhoicissus revoilii Planch.	Vitaceae	Whole plant	ITDG (1996)
Rhoicissus tridentata (L.f.) Wild & R.B. Drumm.		Root or tuberous rootstock	Maundu et al. (2001), Michael (2003), Brookes and Katsoulis (2006), Gruber and O'Brien (2010)
Cyphostemma adenocaule (Steud. ex A. Rich.) Desc.		Whole plant	Polygenis-Bigendako (1990), Kokwaro (1993)
Vitex fischeri Gürke	Verbenaceae	Whole plant	Thijssen (2008)
Vitex doniana Sweet		Bark/root/leaf/fruit	Hines and Eckman (1993), Olusola et al. (1997), Rukangira (2001), Ky (2008)
Lantana camara L.		Whole plant	Adebayo et al. (1999)
Rotheca myricoides (Hochst.) Steane & Mabb.		Whole plant	Getahun (1976), Kebenei et al. (2004)
Clerodendrum myricoides (Hochst.) R. Br. ex Vatke		Leaf, juice, roots	Polygenis-Bigendako (1990), Kokwaro (1993), Tadesse (1994)
Clerodendrum myricoides (Hochst.) R. Br. ex Vatke		Leaf, juice, roots	Baerts and Lehmann (1991)

Table 11.5 Natural products tested against ticks and active compounds with acaricidal or insecticidal activity

Scientific name	Family	Reference	Natural products with acaricidal or insecticidal activity
Annona squamosa L.	Annonaceae	Magadum et al. (2009), Chungsamarnyart et al. (1990, 1991a)	Squamocin (Kawazu et al. 1989)
			Acetogenins (Hopp et al. 1998)
			Annotemoyin-1 (Parvin et al. 2003)
Annona muricata L.		Chungsamarnyart et al.(1991b), Broglio-micheletti et al.(2009)	Goniothalamicin (Alkofahi et al. 1988)
			Gigantetrocin A, annomontacin, bullatalicin (Alali et al. 1998)
			Squamocin (Guadaño et al. 2000)
Acorus calamus	Acoraceae	Ghosh et al. (2011)	Alpha-asarone (Ghosh et al. 2011)
Ricinus communis	Euphorbiaceae	Zahir et al. (2009)	Quercetin, gallic acid, flavone, kaempferol (Ghosh et al. 2013)
Hyptis verticillata Jacq.	Lamiaceae	Facey et al. (2005)	Cadina-4,10(15)-dien-3-one (Porter et al. 1995)
Cunila angustifolia Benth.		Apel et al.(2009)	α-Pinene, β-pinene, sabinene, menthofuran and 1,8-cineole (Apel et al. 2009)
C. spicata Benth.			
C. microcephala Benth.			
Tamarindus indicus L.	Leguminosae	Chungsamarnyart and Jansawan, (2001), Magadum et al. (2009)	Oxalic, malic, succinic, citric, and tartaric acids (Chungsamarnyart and Jansawan 2001)
Allium sativum L.	Liliaceae	Magadum et al.(2009)	Lectins (Bandyopadhyay et al. 2001)
Azadirachta indica A. Juss.	Meliaceae	Williams (1993), Valente et al. (2007), Srivastava et al. (2008), Costa et al. (2008), Magadum et al. (2009), Broglio-Micheletti et al. (2009, 2010)	Azadirachtin (Butterworth and Morgan 1971)
			Salanin (Meisner et al. 1981)
			Nimbin, nimbinin (Siddiqui et al. 1988)
			Meliatetraolenone, odoratone (Siddiqui et al. 2003)
Melia azedarach L.		Borges et al. (1994, 2003), Vivan (2005), Sousa et al. (2008)	Azedarachol (Nakatani et al. 1985)
			Trichilins, azedarachins (Nakatani et al. 1995)
			Meliacarpins (Bohnenstengel et al. 1999)
			Azadirachtin (Morgan and Thorton 1973)
Pimenta dioica (L.) Merr.	Myrtaceae	Brown et al.(1998)	Eugenol, methyleugenol (Brown et al. 1998)
Eucalyptus globules Labill.		Chagas et al.(2002), Magadum et al. (2009)	Terpenoids, δ-phenothrin, pyrethrum (Yang et al. 2004)
Piper aduncum L.	Piperaceae	Silva et al.(2009)	Dill apiol (Bernard et al. 1995; Silva et al. 2009)
P. mikanianum (Kunth) Steud., *P. xylosteoides* (Kunth) Steud., *P. amalago* L.		Ferraz et al. (2010)	Phenylpropanoids, monoterpene, and sesquiterpene hydrocarbons (Ferraz et al. 2010)
P. nigrum L.		Álvarez et al. (2008)	Piperine, pellitorine, pipercide (Miyakado et al. 1979)

(93.34 %). The egg masses and hatchability percentages were also affected (Stuti Vatsya et al. 2006).

In a comprehensive study, Karim (2006) reported comparative efficacy of aqueous and ethanolic extracts of *A. indica*, *Polygonum hydropiper*, *Annona reticulata*, *A. squamosa*, and *Cynodon dactylon* against *R. (B.) microplus* and 80–100 % efficacy was reported. The acaricidal activity of oleoresinous extract (oleoresin) from *Copaifera reticulata* was investigated against *R. (B.) microplus*, and LC_{50} and LC_{99} values were determined as 1,579 and 3,491 ppm, respectively (Fernandes et al. 2007). The essential oils of *Eucalyptus citriodora* and *E. staigeriana* killed 100 % of the treated larvae at a concentration of 10 %, while *E. globulus* had the same efficacy at 20 % concentration. Against engorged females, the maximum efficacy was observed at a concentration of 25 % for *E. citriodora*, 10 % for *E. globulus*, and 15 % for *E. staigeriana*. When the essential oils were formulated as concentrate emulsions, the effect was strengthened (Chagas et al. 2002). The hydroalcoholic extract of leaves of *Eucalyptus* sp. had an efficacy of 96 % against engorged females at a concentration of 10 % (Costa et al. 2008). From the same family, Broglio-Micheletti et al. (2009) tested the alcoholic extract of flowers of *Syzygium malaccense* at a concentration of 2 % and observed that the rate of interference with the engorged female reproduction was 59 %. Silva et al. (2009) tested different concentrations of hexane, ethyl acetate, and ethanolic extracts of *Piper aduncum* against larvae and adults of *R. (B.) microplus* and reported that LC_{50} of hexane extract was 9.3 mg/mL for larvae, while reproduction was reduced to 12.5–54.2 %. In the year 2007, Kheiradabi and Abyaneh (2007) reported up to 26.7 % mortality of adults of *R. (B.) annulatus* treated with flower extracts of *Matricaria chamomile*. The essential oil from leaves of *Hesperozygis ringens* was tested at 0.625–50 μL.mL^{-1} concentration (≈0.0625–5 %) inhibiting oviposition by 11.5–76.4 % with 95 % inhibition of larval hatchability at the highest concentration. Those concentrations were all lethal for 100 % of larvae. Pulegone was the main compound isolated from the essential oil (Ribeiro et al. 2010). In addition, essential oil from leaves of five *Cunila* species at concentrations of 2.5, 5, and 10 μL.mL^{-1} (≈0.25, 0.5, and 1 %) was tested, and *C. angustifolia* and *C. incana* caused 100 % mortality of larvae at the lowest concentration and *C. spicata* at a concentration of 5 μL.mL^{-1}. The active compounds identified in these plants were α-pinene, β-pinene, sabinene, menthofuran, and 1,8-cineole (Apel et al. 2009). Pereira and Famadas (2004) evaluated the action of the ethanolic root extract of *Dahlstedtia pentaphylla* (Leguminosae) against two strains of tick: one acaricide-sensitive Mozo strain and one from the field. The plant was less efficient against the field strain. The efficiency was close to 100 % at a concentration of 20 %: LC_{50} values are 1:34.94 mL (≈2.86 %) against engorged females and 1:231.337 mL (≈0.43 %) against larvae. The LC_{50} and LC_{99} of the oleoresinous extract of *Copaifera reticulata* (Leguminosae) against larvae were 1.579 ppm (≈0.16 %) and 3.491 ppm (≈0.35 %), respectively (Fernandes and Freitas 2007). Hexane, ethyl acetate, and ethanolic extracts from leaves of *Piper aduncum* (Piperaceae) were tested against engorged females in increasing, double concentrations from 5 to 100 mg.mL^{-1} (≈0.5–10 %). For all extracts, even at the highest concentration, the reproductive control was no higher than 62 %. Larval mortality was evaluated at concentrations of 1–20 mg.mL^{-1} (≈0.1–2 %) and was found to be 70.42, 40.5, and 17.2 % in the hexane, ethanol, and ethyl acetate extracts, respectively, at the highest concentration. Hydrodistillation of the hexane extract produced 6.8 % essential oil and 94.84 % sesquiterpene dill apiol, which caused 100 % larval mortality at 0.1 mg.mL^{-1} (≈0.01 %) (Silva et al. 2009). The essential oil of *Piper mikanianum* (LC_{50} 2.33 μL.mL^{-1}; ≈ 0.233 %) was more active against larvae than that of *Piper xylosteoides* (LC_{50} 6.15 μL.mL^{-1}; ≈ 0.615 %), while the oil of *Piper amalago* was inactive. The main compounds were phenylpropanoids, monoterpenes, and sesquiterpene hydrocarbons (Ferraz et al. 2010). The hexane and methanolic extracts from stems and leaves of *Hypericum polyanthemum* (Clusiaceae) were tested at concentrations of 6.25-50 mg.mL^{-1} (≈0.625–5 %). The effect against engorged females was low

(19.2 %) at the highest concentration of the hexane extract, but on other hand, it killed all larvae in all concentrations (Ribeiro et al. 2007). Similar effects were observed with the hexane extracts of stems and leaves of *Calea serrata* (Asteraceae) at the same concentrations (Ribeiro et al. 2008a). The LC_{50} and LC_{99} of the ethanolic extracts of stems of *Magonia pubescens* (Sapindaceae) were 365 ppm (≈0.036 %) and 4,000 ppm (≈0.4 %) against larvae (Fernandes et al. 2008). Over 95 % mortality of larvae was observed (Ribeiro et al. 2008b) with the essential oil of *Drimys brasiliensis* (Winteraceae) at concentrations ranging from 3.125 to 25 μL/mL (≈0.3125 to 2.5 %). The presence of sesquiterpenoids, cyclocolorenone, bicyclogermacrene, and alpha-gurjunene was confirmed in this oil. The ethanolic extract of seeds of *Annona muricata* L. (Annonaceae) at a concentration of 2 % had 100 % efficacy against engorged females (Broglio-Micheletti et al. 2009). Zahir et al. (2009) reported acaricidal activity of leaf ethyl acetate extract of *Achyranthes aspera* L., leaf methanolic extract of *A. malabarica*, flower methanolic extract of *Gloriosa superba*, and leaf methanolic extract of *R. communis* against *R. (B.) microplus*. Zahir et al. (2010) reported acaricidal activity of acetone and methanolic extracts of leaf of *Anisomeles malabarica*, methanolic extract of seed of *Gloriosa superba*, and methanolic extract of leaf of *Ricinus communis* against *Haemaphysalis bispinosa*.

Ghosh et al. (2011) screened 34 solvent-guided extracts prepared from 13 plants. Among the 34 extracts, 26 extracts showed no mortality within 72 h of application, while 12.0±4.9 % to 35.0±9.6 % mortality of treated ticks was recorded in other extracts. Of the effective extracts, the one prepared from rhizome of *Acorus calamus* proved highly efficacious and 100 % final mortality within 14 DPT. The LC_{85} value of the extract was determined as 11.26 %. Ravindran et al. (2011) reported acaricidal activity of crude ethanolic extract of aerial parts of *Leucas aspera* against *R. (B.) annulatus*. The percent adult mortality, inhibition of fecundity, and hatching of laid ova were studied at concentrations of 1.56, 3.13, 6.25, 12.5, 25, 50, and 100 mg/mL. Adult tick mortality was significant at the highest concentration tested. Inhibition of fecundity of treated groups differed significantly from control and was concentration dependent. The extract completely inhibited the eclosion of eggs from the treated ticks. Juliet et al. (2012) evaluated the effect of ethanolic extract of leaves of *Jatropha curcas* against *R. (B.) annulatus*. The extract at all concentrations tested (50–100 mg/mL) considerably blocked the hatchability of eggs when compared to control. However, the extract did not produce mortality of treated adults. The extract also did not significantly reduce the mass of eggs laid by the treated ticks. Ravindran et al. (2012) reported the acaricidal activity of ethanolic extracts of leaves of *Cassia alata* L. against *R. (B.) annulatus*. The highest mortality (45.8 %) and inhibition of fecundity (10.9 %) were observed at the highest concentration tested (100 mg/mL). The plant extract did not affect egg hatchability. A 95 % ethanolic extract of leaves of *Ricinus communis* was used to test the efficacy against reference acaricide-resistant lines by in vitro assay. The extract significantly affects the mortality rate of ticks in dose-dependent manner ranging from 35.0±5.0 to 95.0±5.0 % with an additional effect on reproductive physiology of ticks by inhibiting 36.4–63.1 % of oviposition. The leaf extract was found effective in killing 48.0, 56.7, and 60.0 % diazinon, deltamethrin, and multi-acaricide-resistant ticks, respectively. However, the cidal and oviposition limiting properties of the extract were separated when the extract was fractionated with hexane, chloroform, n-butanol, and water (Ghosh et al. 2013). Sunil et al. (2013) evaluated the acaricidal properties of crude ethanolic extract of *Cassia fistula* leaves against *R. (B.) annulatus*. The percentage of adult mortality, inhibition of fecundity, and hatching of ova laid were studied at different concentrations of the extract ranging from 50 to 100 mg/mL. The extract completely inhibited hatching of eggs at concentrations above 80 mg/mL of the extract. Mortality of adult engorged female ticks and inhibition of fecundity were concentration dependent. The LC_{50} value of extract against *R. (B.) annulatus* was 97.1 mg/mL.

Although a large number of plant extracts were reported to have acaricidal properties in vitro, very few of them were evaluated in vivo on tick-infested animals. Monthly sprays of ethanolic extracts of neem or weekly bathing in azadirachtin-rich aqueous 1:20 "Green Gold" controlled the bush tick and the cattle tick in Australia, but were less effective against the brown dog tick. Kalakumar et al. (2000) reported 60–75 % efficacy of custard seed oil extract on buffaloes and cattle infested naturally with *R. (B.) microplus*, *Hyalomma a. anatolicum*, and *R. haemaphysaloides*. In Jamaica, neem kernel extract controlled ticks on cattle and dogs. In Kenya, period for engorgement by larvae and nymphs of *Amblyomma variegatum* and larvae of *Rhipicephalus appendiculatus* was significantly prolonged due to the slow feeding on rabbit host sprayed with neem oil (Kaaya 2003). Neem treatment also led to a reduction in engorgement weight of larvae, nymphs, and adults of *A. variegatum*, *R. appendiculatus*, and *B. decoloratus* feeding on neem-treated rabbits, and fewer larvae and nymphs molted to the next developmental stage. Egg masses produced by neem-treated ticks weighed significantly less and their hatchability was adversely affected. Regardless of tick species, attachment by larvae was also significantly reduced on neem oil-treated rabbit. In trials conducted in pastures in Kenya, application of neem oil on cattle repelled all stages of *R. appendiculatus*, *B. decoloratus*, and *A. variegatum* (Kaaya 2003). The essential oil extracted from the leaves of *Ageratum houstonianum* was applied on animals infested with *R. annulatus*, and 95.1 % mortality on day 8 postapplication was recorded (Tedonkeng et al. 2005). Sivaramakrishnan et al. (1996) tested efficacy of a mixture of oils (*A. indica*, *Eucalyptus*, and *Pongamia*) on cattle and goats heavily infested with ticks. The neem and eucalyptus oil were found 92.2 % and 97.6 % effective, respectively.

The essential oil of *C. winterianus* was evaluated in two treatments: one through application of the crude oil on the animal's back and the other through aspersion of a solution oil: alcohol (1:10). A significant difference in the number of females was observed between the treated groups, 22–28 days after treatment (5.3–14.4 and 2.8–11.3 in the crude oil and aspersion groups, respectively) (Martins and Gonzalez 2007). Borges et al. (2005) reported efficacy of the hexane extract of ripe fruits of *M. azedarach* up to 63.6 %. Sousa et al. (2011) produced a concentrate emulsion of the hexane extract of *M. azedarach* and reported 89.0 % efficacy when treated at 0.5 %. There was greater action against larvae and adults than against nymphs. These in vivo results differed from in vitro observations (Borges et al. 1994, 2003; Sousa et al. 2008) in that the fruit extracts of the plant affected reproductive efficiency of females.

The efficacy of aqueous extracts from *A. indica* was compared with abamectin among artificially infested animals. One kilogram of leaves was mixed with 5 L of water and sprayed on the animals every week, for 4 weeks. Another group was treated once with abamectin. Similar tick counts were observed 15 and 30 days after treatment of both groups: 62.5 and 7.71 in the neem group and 50.5 and 16.0 in the abamectin group, respectively (Valente et al. 2007). Srivastava et al. (2008) compared the efficacy of ethanolic extract of neem seeds at a concentration of 8 % with cypermethrin, on experimentally infested animals. The mortality after five days of treatment in the neem group was 70.5 % while three days after the treatment with cypermethrin resulted in 92.4 % mortality. Considering the reproductive efficiency of surviving ticks, the efficacy was 68.32 and 80.48 % for the neem and cypermethrin groups, respectively. Similar results were observed by Magadum et al. (2009), using neem seeds at the same concentration as Srivastava et al. (2008). Pereira and Famadas (2006) tested the root extract of *D. pentaphylla* at concentrations of 1:10 and 1:20 on artificially infested bovines. The highest efficacies of 76 % were observed on the third and seventh days after treatment at 1:10 mL. The reproductive parameters of treated females were not affected by this extract. Olivo et al. (2008) evaluated the action of four increasing concentrations (1.25–5 %) of an aqueous extract of *Nicotiana tabacum* on naturally infested animals. The efficacies ranged from 62 % at a concentration of 5 % in the first week up to 77.5 %, 14 days after treatment, at the lowest concentration. By an in vivo experiment

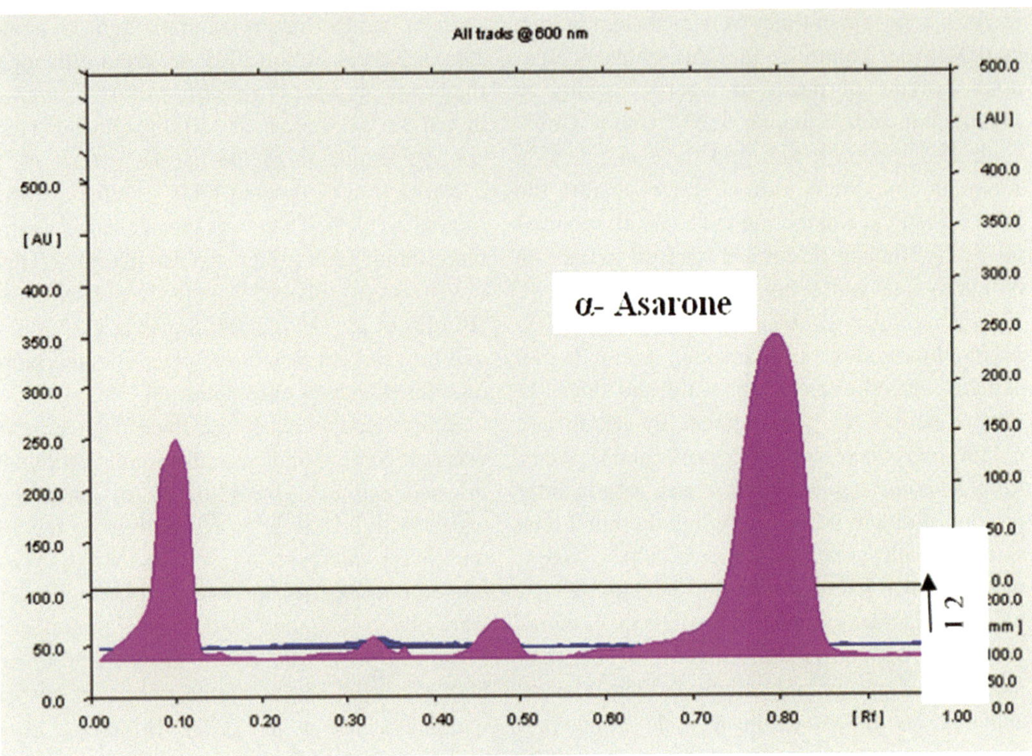

Fig. 11.2 HPTLC densitometric scan (at 600 nm) of *A. calamus* extracts (*1*) and reference compound α-asarone (*2*)

Ghosh et al. (2011) confirmed 42 % efficacy for the *A. calamus* extract. The extract was safe for topical application, and no adverse reaction was observed when animals were treated even at 50 % concentration, which was five times the concentration that was used in vivo. The α-asarone was identified as one of the active principles in *A. calamus* (Fig. 11.2). Ghosh et al. (2013) reported 59.9 % in vivo efficacy of leaf extract of *R. communis* after the first challenge; however, following the second challenge the efficacy was reduced to 48.5 %. The HPTLC fingerprinting profile of *R. communis* leaf extract showed presence of quercetin, gallic acid, flavone, and kaempferol which seemed to have synergistic acaricidal action (Fig. 11.3).

Taken together, these results do not point to any single component of plant extracts that standout as the critical element that accounts for activity against arthropods. Complex interactions may occur among major and minor constituents in an unforeseen manner that results in insecticidal/acaricidal activity. Similarly, mixtures of essential oils from different plants may have higher activity than individual extracts in ways that are difficult to predict. Judicious use of synergists could improve efficacy further. Further research on blends of essential oils and improved formulations and delivery system management of these botanicals could lead to substantial improvements in their performance for arthropod control.

2 Conclusion

A large number of different plant species representing different geographical areas around the world have been shown to possess phytochemicals that are capable of causing a range of acute and chronic toxic effects on different stages of arthropods. Any one of these effects taken alone is usually not impressive, but the combined ovi-

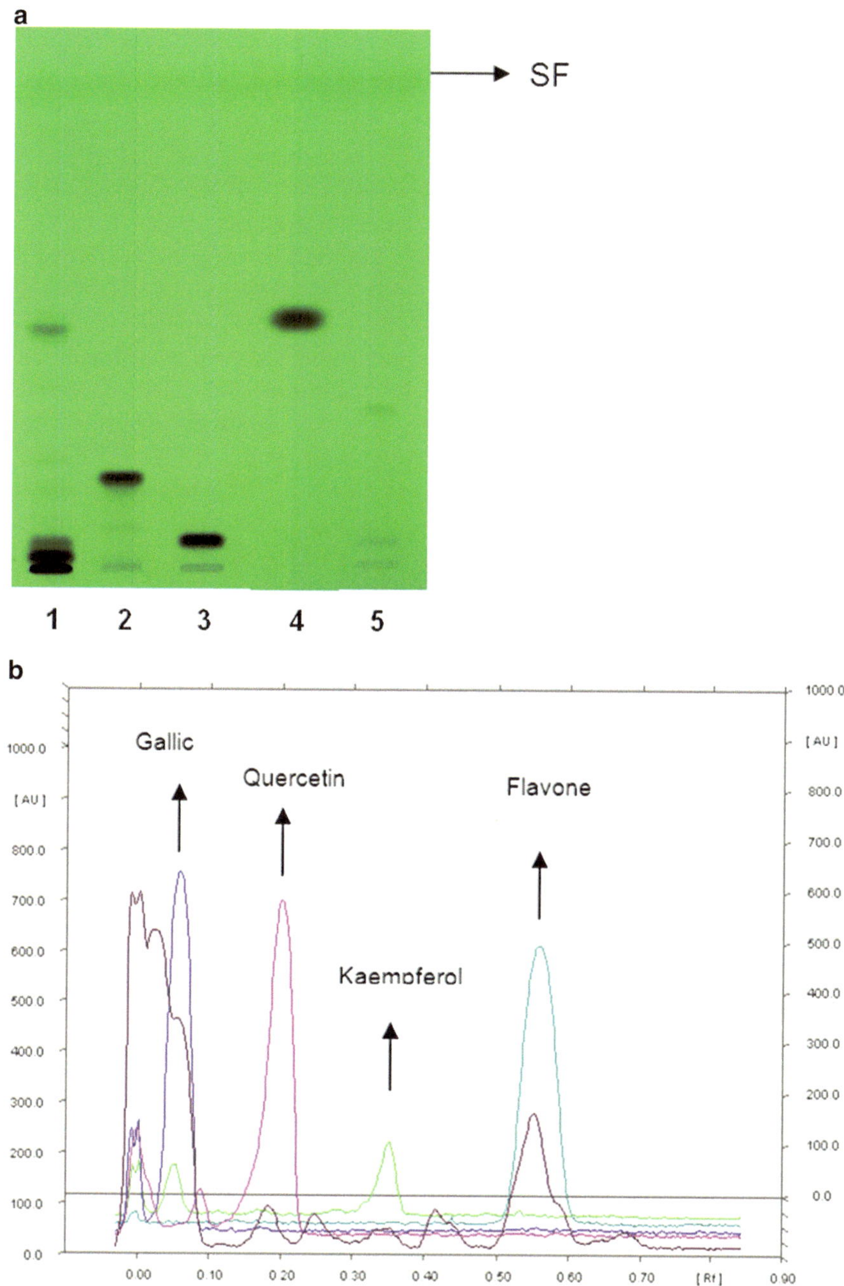

Fig. 11.3 (**a**) HPTLC fingerprint profile of *Ricinus communis* leaf extract under 254 nm (*1 R. communis* (leaf extract), *2* quercetin, *3* gallic acid, *4* flavone, *5* kaempferol) (**b**) HPTLC densitometric scan (at 254 nm) of sample (1) with reference compounds (2–5)

cidal, larvicidal, and IGR effects possessed by many phytochemicals can produce impressive results. The beneficial medicinal effects of plant materials typically result from active compounds present in the plant. These compounds are mostly secondary metabolites such as alkaloids, steroids, tannins, phenolic compounds, etc. The in vitro results of many of the herbal extracts are exemplary. However, the loss of efficiency of these extracts when tested on the animals is undoubtedly an important difficulty faced by most of the researchers. The large number of chemical compounds present in a single plant extract (Evans 1996) adds to this problem. The observed activity could be due to the effect of a single compound or combination of one or more. Several issues remain to be researched and improved upon including residual efficacy (degradation and persistence), preservation of nontarget organisms, development of resistance, and standardization of testing protocols (Mulla and Su 1999). The low persistence in the host body necessitates larger quantity of the plant extract or natural products for in vivo efficacy. For mass production of extracts, large quantities of plant materials are essential. Differences relating to edaphoclimatic conditions and the cultivation and conservation of plant materials for extract production may imply oscillation of the results (Heimerdinger et al. 2006). Storage of *M. azedarach* fruits for 5 months at room temperature caused a decrease in their acaricide effect (Sousa et al. 2008). Yakkundi et al. (1995) observed a 5 % reduction in azadirachtin after one month of storage of *A. indica* seeds and 35 % after 4 months. Likewise, Johnson et al. (1996) observed a decrease in azadirachtin and salanin after storage for 6 months. Even though the synthesis of chemical compounds is determined by the genetic characteristics of the plant, edaphoclimatic factors also interfere with this factor (Lapa et al. 1999).

It is also commonly stated that plant-based pesticides/repellents are better for the environment than synthetic molecules. While plant volatiles are naturally derived, distillation requires biomass energy, extraction commonly uses organic solvents that must be disposed of carefully, and growing the plants uses agrichemicals, such as fertilizers and pesticides (unless sourced from a sustainable and organic source). However, if carefully practiced, cash cropping of plants used for repellents provides a vital source of income for small-scale farmers in developing countries (Duke and DuCellier 1993) and can have beneficial environmental impact when planted in intercropping systems to prevent soil erosions (Zheng and He 1993). Therefore, it is important to carefully source of medicinal plants to avoid pitfalls associated with unsustainable cropping practices.

Very little has been researched and published on phytochemicals and their effects on nontarget organisms. From the conclusions of neem and tannin researches, it can be summarized that botanical derivatives are not free of risk and further field investigations are necessary. In India, Rao et al. (1995) applied *A. indica* to rice fields and found a marked reduction in the abundance of late instar culicine and anopheline larvae and pupae. No significant reduction of most nontarget organisms, including different aquatic insects, frog tadpoles, and plant spiders, was observed. Kreutzweiser et al. (2000) treated outdoor stream channels with different doses of a commercial azadirachtin formulation called Neemix 4.5, and no significant difference in abundance and species richness of aquatic insects was found between control and Neemix-treated stream channels at the low dose, but medium and high doses reduced the abundance and richness of aquatic insect communities compared with controls. These results suggest that neem is a risk to nontarget organisms at higher doses. Similarly, vegetable tannins have also been shown to exhibit biocidal properties among crustacean taxa representative of the nontarget fauna associated with Alpine mosquito breeding sites (Pautou et al. 2000).

Although there are a lot of encouraging in vitro laboratory results for the plant-based insecticides, in field trials most of them fail and do not fulfill the initial screening requirements of commercial houses. The onus is thus placed upon the researcher to carry out field trials for promising phytochemicals that are not picked up initially by insecticide manufacturers. It should be mentioned, however, that the high degree of biodegradation exhibited by most phytochemicals is what makes them eco-friendly and attractive as replacements of synthetic chemicals in the first place. Although the evaluation of phytochemicals is yet in its infancy and much research remains to further characterize promising agents and discover new agents, results described in this review suggest that botanical phytochemicals should be counted as future alternative to synthetic insecticides.

Acknowledgement The authors are thankful to World Bank-funded National Agricultural Innovation Project (code C2066) for funding the research project on development of phytoacaricides against economically important tick species. Sincere thanks are also due to Mr. Anil Kumar Sharma and Sachin Kumar for helping in compilation.

References

Abdel-Aal AE (1996) Biological, histological and physiological effects of some insect growth regulators on the greasy cutworm, *Agrotis ipsilon* (Lepidoptera: Noctuidae). M.Sc. thesis, Faculty of Science, Cairo University, Egypt

Abdel-Shafy S, Zayed AA (2002) *In vitro* acaricidal effect of plant extract of neem seed oil (*Azadirachta indica*) on egg, immature, and adult stages of *Hyalomma anatolicum excavatum* (Ixodoidea: Ixodidae). Vet Parasitol 106:89–96

Adebayo AT, Gbolade TA, Olaifa AA (1999) Comparative study of toxicity of some essential oils to larvae of three mosquito species. Nigerian J Nat Prod Med 3. http://www.inasp.info/ajol/journals/njnpm/vol3abs.html. 20 Oct 2003

Adewusi AE, Afolayan JA (2010) A review of natural products with hepatoprotective activity. J Med Plants Res 4:1318–1334

Adoyo F, Mukalama JB, Enyola M (1997) Kenya Woodfuel and Agroforestry Programme (KWAP), Busia, Kenya. Using Tithonia concoctions for termite control in Busia District, Kenya. ILEIA Newsletter 13, 24

Alali FQ, Kaakeh W, Bennet GW, McLaughlin JL (1998) Annonaceous acetogenins as natural pesticides: potent toxicity against insecticide-susceptible and resistant German cockroaches (Dictyoptera: Blattellidae). J Econ Entomol 91:641–649

Alkofahi A, Rupperecht JK, Smith DL, Chang CJ, McLaughlin JL (1988) Goniothalamicin and annonacin: bioactive acetogenins from *Goniothalamus giganteus* (Annonaceae). Cell Mol Life Sci 44:83–85

Álvarez V et al (2008) Control *in vitro* de garrapatas (*Boophilus microplus*; Acari: Ixodidae) mediante extractos vegetales. Rev Biol Trop 56:291–302

Amer A, Mehlhorn H (2006) Larvicidal effects of various essential oils against *Aedes, Anopheles and Culex* larvae (Diptera, Culicidae). Parasitol Res 99:466–472

Anon (1977) Posterior paralysis among sows kept on sawdust from the Azobe tree (Lophira alata var.procera). Tijdschr Diergeneesk 102, 582

Anon (2001a) Acacia sieberiana DC. http://www.ildis.org/LegumeWeb/6.00/taxa/373.shtml. 20 Oct 2003)

Anon (2001b) Termite Control HDRA – The Organic Organization. Ryton Organic Gardens, COVENTRY CV8 3LG, United Kingdom. www.hdra.org.uk. 27 Oct 2001

Anon (2003) On-farm technology testing. Sustainable agriculture extension manual. UNDP SANE (Sustainable Agriculture Networking and Extension) Program, Berkeley, CA

Apel MA, Ribeiro VL, Bordignon SA, Henriques AT, Von Poser G (2009) Chemical composition and toxicity of the essential oils from *Cunila* species (Lamiaceae) on the cattle tick *Rhipicephalus (Boophilus) microplus*. Parasitol Res 105:863–868

Arun M, Asha VV (2007) Preliminary studies on antihepatotoxic effect of *Physalis peruviana* Linn. (Solanaceae) against carbon tetrachloride induced acute liver injury in rats. J Ethnopharmacol 111:110–114

Azahaianambi P, Ghosh S (2007) Genetic transformation of arthropod vectors for control of vector borne diseases. Indian J Biotechnol 6:305–314

Baerts M, Lehmann J (1989) Guerisseurs et plantes medicinales de la region des cretes Zaire-Nil au Burundi. Musee royal de l'Afrique centrale, Tervuren, Belgique. Ann Sci Ecol 18:214

Baerts M, Lehmann J (1991) Plantes medicinales veterinaires de la region des cretesZaire-Nil au Burundi. Musee royal de l'Afrique centrale, Tervuren. Ann Sci Ecol 21:133

Bandyopadhyay S, Roy A, Das S (2001) Binding of garlic (*Allium sativum*) leaf lectin to the gut receptors of homopteran pests is correlated to its insecticidal activity. Plant Sci 161:1025–1033

Barasa SS, Ndiege IO, Lwande W, Hassanali A (2002) Repellent activities of stereoisomers of p-menthane-3,8-diols against *Anopheles gambiae* (Diptera: Culicidae). J Med Entomol 39:736–741

Barnard DR, Xue RD (2004) Laboratory evaluation of mosquito repellents against *Aedes albopictus, Culex nigripalpus*, and *Ochlerotatus triseriatus* (Diptera: Culicidae). J Med Entomol 41:726–730

Bates PG (1999) Inter- and intra-specific variation within the genus *Psoroptes* (Acari: Psoroptidae). Vet Parasitol 83:201–217

Berger A (1994) Using natural pesticides: current and future perspectives. A report for the plant protection improvement programme in Botswana, Zambia and Tanzania. Department of Entomology, Swedish University of Agricultural Sciences, P.O. Box 7044, S-750 07 Uppsala, Sweden. Number (ISBN, ISSN): ISSN: 1104-6422

Bernard CB, Krishanmurty HG, Chauret D, Durst T, Philogène BJ, Sánchez-Vindas P, Hasbun C, Poveda L, San Román L, Arnason JT (1995) Insecticidal defenses of *Piperaceae* from the Neotropics. J Chem Ecol 21:801–814

Bhatt DC, Nurani MA, Mitaliya KD, Baxi US (2001) Some ethnomedicinal plants of Barda Hills of Saurashtra, Gujarat, India. Adv Plant Sci 14(11):427–431

Birkett MA, Al Abassi S, Kröber T, Chamberlain K, Hooper AM, Guerin PM, Pettersson JA, Slade R, Wadhams LJ (2008) Antiectoparasitic activity of the gum resin, gum haggar from the East African plant, *Commiphora holtziana*. Phytochemistry 69:1710–1715

Bisseleua HBD, Gbewonyo SWK, Obeng- Ofori D (2008) Toxicity, growth regulatory and repellent activities of medicinal plant extracts on *Musca domestica* L. (Diptera: Muscidae). Afr J Biotechnol 7:4635–4642

Bissinger BW, Apperson CS, Sonenshine DE, Watson DW, Roe RM (2009a) Efficacy of the new repellent BioUD against three species of ixodid ticks. Exp Appl Acarol 48:239–250

Bissinger BW, Zhu J, Apperson CS, Sonenshine DE, Watson DW, Roe RM (2009b) Comparative efficacy of BioUD to other commercially available arthropod repellents against the ticks, *Amblyomma americanum* and *Dermacentor variabilis* on cotton cloth. Am J Trop Med Hyg 81:658–690

Blair M, Mansingh A (2002) Pesticidal potential of tropical plants111: acaricidal properties of few crude extracts and their fractions. In: Mansingh A, Young ED, Yee HT, Delgoda R, Robinson DE, Morrison YE, Lowe IH (eds) Proceedings of the first international symposium on Utilization of natural products in developing countries: trends and needs, held between 10 and 14 July 2000 at the Natural Products Institute, University of the West Indies, Mona, Kingston7, Jamaica, West Indies

Bohnenstengel FI, Wray V, Witte L, Srivastava RP, Proksch P (1999) Insecticidal meliacarpins (C-seco limonoids) from *Melia azedarach*. Phytochemistry 50:977–982

Borges LMF, Silva AC, Neves BP (1994) Teste "*in vitro*" de eficácia do cinamomo (*Melia azedarach*, L) sobre fêmeas ingurgitadas do *Boophilus microplus*, Can. (Acari: Ixodidae). Rev Patol Trop 23:175–179

Borges LMF et al (2003) *In vitro* efficacy of extracts of *Melia azedarach* against the tick *Boophilus microplus*. Med Vet Entomol 17:228–231

Borges LMF, Ferri PH, Silva WC, Silva WJ, Melo LS, Souza LAD, Soares SF, Faria KA, Gomes NA, Mori A, Silva NF (2005) Ação do extrato hexânico de frutos maduros de *Melia azedarach* (Meliaceae) sobre *Boophilus microplus* (Acari: Ixodidae) em bezerros infestados artificialmente. Rev Patol Trop 34:53–59

Bowen CR, Hollinger SE (2002) Alternative crops web site, 2 July 2002. Illinois State Water Survey, Champaign, IL. http://www.sws.uiuc.edu/data/altcrops/. 27 Sept 2003

Brink M (2009) *Grewia trichocarpa* Hochst. ex A.Rich. [Internet] record from protabase. In: Brink M, Achigan-Dako EG (eds) PROTA (Plant Resources of Tropical Africa/Ressources vegetales de l'Afrique tropicale), Wageningen, The Netherlands

Broglio-Micheletti SMF, Valente ECN, De Souza LA, Dias NS, Araujo AMN (2009) Extratos de plantas no controle de *Rhipicephalus* (*Boophilus*) *microplus* (Canestrini, 1887) (Acari: Ixodidae) em laboratório. Rev Bras Parasitol Vet 18:44–48

Broglio-Micheletti SMF, Dias NS, Valente ECN, De Souza LA, Lopes DOP, Santos JM (2010) Ação de extrato e óleo de nim no controle de *Rhipicephalus* (*Boophilus*) *microplus* (Canestrini, 1887) (Acari: Ixodidae) em laboratório. Rev Bras Parasitol Vet 19:46–50

Brookes KB, Katsoulis LC (2006) Bioactive components of *Rhoicissus tridentata:* a pregnancy-related traditional medicine. S Afr J Sci 102:267–272

Brown M, Hebert AA (1997) Insect repellents: an overview. J Am Acad Dermatol 36:243–249

Brown HA, Minott DA, Ingram CW, Williams LAD (1998) Biological activities of the extracts and constituents of *pimento*, *Pimenta dioica* L. against the Southern cattle tick, *Boophilus microplus*. Int J Trop Insect Sci 18:9–16

Butterworth JH, Morgan ED (1971) Investigation of the locust feeding inhibition of the seeds of the neem tree, *Azadirachta indica*. J Insect Physiol 17:969–977

Cabral MMO, Garcia ES, Rembold H, De Simone SG, Kelecon A (1996) Anti-moulting activity in Brazilian *Melia azedarach*. Mem Inst Oswaldo Cruz 91:117–118

Carroll SP, Loye J (2006) PMD, a registered botanical mosquito repellent with deet-like efficacy. J Am Mosq Control Assoc 22:507–514

Carroll JF, Maradufu A, Warthen JD Jr (1989) An extract of *Commiphora erythraea*: a repellent and toxicant against ticks. Entomol Exp Appl 53:111–116

Carroll JF, Cantrell CL, Klun JA, Kramer M (2007) Repellency of two compounds isolated from *Callicarpa americana* (Lamiaceae) against *Ixodes scapularis* and *Amblyomma americanum* ticks. Exp Appl Acarol 41:215–224

Cassady JM, Baird WM, Chang CJ (1990) Natural products as a source of potential cancer chemotherapeutic and chemopreventive agents. J Nat Prod 53:23–41

Cevallos-Casals BA, Cisneros-Zevallos LA (2002) Bioactive and functional properties of purple sweet potato (*Ipomoea batatas* (L.). Acta Hortic 583:195–203

Chagas ACS, Passos WM, Prates HT, Leite RC, Furlong J, Fortes ICP (2002) Efeito acaricida de óleos essenciais e concentrados emulsionáveis de *Eucalyptus* spp. em *Boophilus microplus*. Braz J Vet Res Anim Sci 39:247–253

Chungsamarnyart N, Jansawan W (1996) Acaricidal activity of peel oil of *Citrus* spp. on *Boophilus microplus*. Kasetsart J (Nat Sci) 30:112–117

Chungsamarnyart N, Jansawan W (2001) Effect of *Tamarindus indicus L.* against the *Boophilus microplus*. Kasetsart J (Nat Sci) 35:34–39

Chungsamarnyart N, Jiwajinda S (1992) Acaricidal activity of volatile oil from lemon and *citronella* grasses on tropical cattle ticks. Kasetsart J (Nat Sci) 26:46–51

Chungsamarnyart N, Jiwajinda S, Jansawan W (1990) Effects of plant crude-extracts on the cattle tick (*Boophilus microplus*) insecticidal action I. Kasetsart J (Nat Sci Suppl) 24:28–31

Chungsamarnyart N, Jiwajinda S, Jansawan W (1991a) Practical extraction of sugar apple seeds against tropical cattle ticks. Kasetsart J (Nat Sci Suppl) 25:101–105

Chungsamarnyart N, Jiwajinda S, Jansawan W (1991b) Acaricidal effect of plant crude-extracts on tropical cattle tick (*Boophilus microplus*). Kasetsart J (Nat Sci Suppl) 25:90–100

Coetzee C, Jefthas E, Reinten E (1999) Indigenous plant genetic resources of South Africa. In: Janick J (ed) Perspectives on new crops and new uses. ASHS Press, Alexandria, pp 160–163

Coria C, Almiron W, Valladares G, Carpinella C, Luduena F, Defago M, Palacios S (2008) Larvicide and oviposition deterrent effects of fruit and leaf extracts from *Melia azedarach* L. on *Aedes aegypti* (L.) (Diptera: Culicidae). Bioresour Technol 99:3066–3070

Costa FB, Vasconcelos PSD, Silva AMM, Brandao VM, Silva IAD, Teixeira WC, Guerra RMSN, Dos Santos ACG (2008) Eficácia de fitoterápicos em fêmeas ingurgitadas de *Boophilus microplus*, provenientes da mesorregião oeste do Maranhão, Brasil. Rev Bras Parasitol Vet 17:83–86

Cremlyn R (1978) Pesticides, preparation and mode of action. Wiley, New York

Cruz-Vazquez C, Fernández-Ruvalcaba M (2000) Anti-tick repellent effect of *Andropogon gayanus* grass on plots of different ages experimentally infested with *Boophilus microplus* larvae. Parasitol Día 24:88–91

Cunningham A, Mbenkum F (1993) Sustainability of Harvesting *Prunus africana* Bark in Cameroon. A medicinal plant in international trade. People and plants working paper, 2. UNESCO, Paris, France

Curtis CF (1992) Personal protection methods against vectors of disease. Rev Med Vet Entomol 80:543–553

Dagne E, Alemu M, Sterner O (1993) Flavonoids from *Euclea divinorum*. Bull Chem Soc Ethiopia 7:87–92

Dautel H, Kahl O, Siems K, Oppenrieder M, Müller-Kuhrt L, Hilker M (1999) A novel test system for detection of tick repellents. Entomol Exp Appl 91:431–441

Davidović V, Todorović Joksimović M, Stojanović B, Relić R (2012) Plant usage in protecting the farm animal health. Biotechnol Anim Husb 28(1):87–98

De Ba Levy J (1991) The complete herbal handbook for farm and stable. Faber and Faber Limited, London

De la Fuente J, Almázan C, Canales M, De la Lastra JM, Kocan KM, Willadsen P (2007) A ten-year review of commercial vaccine performance for control of tick infestations on cattle. Anim Health Res Rev 8:23–28

Del Fabbro S, Nazzi F (2008) Repellent effect of sweet basil compounds on *Ixodes ricinus* ticks. Exp Appl Acarol 45:219–228

Dewees AP (1995) Social and economic incentives for smallholder tree growing. A case study from Murang'a District, Central Kenya. Community Forestry Case Study Series. Food and Agriculture Organization of the United Nations, Via delle Terme di Caracalla, 00100 Rome, Italy. http://www.fao.org/docrep/U8995E/u8995e00.htm#Contents. 28 Sept 2003

Dietrich G, Dolan MC, Peralta-Cruz J, Schmidt J, Piesman J, Eisen RJ, Karchesy J (2006) Repellent activity of fractioned compounds from *Chamaecyparis nootkatensis* essential oil against nymphal *Ixodes scapularis* (Acari: Ixodidae). J Med Entomol 43:957–961

Dipeolu OO, Ndungu JN (1991) Acaricidal activity of Kupetaba, a ground mixture of natural products against *Rhipicephalus appendiculatus*. Vet Parasitol 38:327–338

Dipeolu OO, Mongi AO, Punya DK, Latif AA, Amoo OA, Odhiambo TR (1992) Current concepts and approach to control of livestock ticks in Africa. Discov Innov 4:35–44

Ditzen M, Pellegrino M, Vosshall LB (2008) Insect odorant receptors are molecular targets of the insect repellent deet. Science 319:1838–1842

Duke AJ (1983) *Euphorbia tirucalli* L. (Euphorbiaceae; Petroleum plant, Aveloz, Milkbush). In: Duke AJ (ed) Handbook of energy crops. Center for New Crops and Plant Products, Purdue University, West Lafayette

Duke AJ (2002) A mini-course in medical botany syllabus: African medicinal plants in world trade. Module 9: African. http://www.arsgrin.gov/duke/syllabus/index.html#warn. 06 Oct 2003

Duke JA (2004) Dr. Duke's phytochemical and ethnobotanical databases. Available online at: http://www.ars-grin.gov/duke

Duke JA, DuCellier JL (1993) CRC handbook of alternative cash crops. CRC Press, Boca Raton

El Tahir A, Satti GMH (1999) Antiplasmodial activity of selected Sudanese medicinal plants with emphasis on *Maytenus senegalensis* (Lam.) Exell. J Ethnopharmacol 64:227–233

El Tahir A, Ibrahim AM, Satti GMH, Theander TG, Kharazmi A, Khalid SA (1998) The potential antileishmanial activity of some Sudanese medicinal plants. Phytother Res 12:576–579

Evans WC (1996) The plant and animal kingdoms as sources of drugs. In: Trease and Evans' pharmacognosy. WB Saunders, London, pp 15–17

Ezeonu FC, Chidune GI, Ededi SC (2001) Insecticidal properties of volatile extracts of orange peels. Biores Technol 76:273–274

Fabry W, Okemo P, Aansorg R (1996) Activity of East African medicinal plants against *Helicobacter pylori*. Chemotherapy 42:315–317

Facey PC, Porter RB, Reese PB, Williams LA (2005) Biological activity and chemical composition of the essential oil from Jamaican *Hyptis verticillata* Jacq. J Agric Food Chem 53:4774–4777

Fernandes FF, Freitas EPS (2007) Acaricidal activity of oleoresinous extract from *Copaifera reticulata* (Leguminosae: Caesalpinioideae) against larvae of the southern cattle tick, *Rhipicephalus (Boophilus) microplus* (Acari: Ixodidae). Vet Parasitol 147:150–154

Fernandes FF, Bessa PAD, Freitas EPS (2008) Evaluation of activity of the crude ethanolic extract of *Magonia pubescens* St. Hil (Sapindaceae) against larvae of the cattle tick *Rhipicephalus (Boophilus) microplus* (Canestrini, 1887) (Acari: Ixodidae). Braz Arch Biol Technol 51:1147–1152

Fernandez-Ruvalcaba M, Cruz-Vazquez C, Soano-Vergara G, Garcia-Vazquez Z (1999) Anti-tick effects of *Stylosanthes humilis* and *Stylosanthes hamata* on plots experimentally infested with *Boophilus microplus* larvae in Morelos, Mexico. Exp Appl Acarol 23:171–175

Ferraz ABF, Balbino JM, Zini CA, Ribeiro VL, Bordignon SA, Von Poser G (2010) Acaricidal activity and chemical composition of the essential oil from three Piper species. Parasitol Res 107:243–248

Franco LA, Matiz GE, Calle J, Pinzon R, Ospina LF (2007) Antiinflammatory activity of extracts and fractions obtained from *Physalis peruviana* L. calyces. Biomedica 1:110–115

Garboui SS, Jaenson TGT, Pålsson K (2006) Repellency of MyggA Natural spray (para-menthane-3,8-diol) and RB86 (neem oil) against the tick *Ixodes ricinus* (Acari: Ixodidae) in the field of east-central Sweden. Appl Exp Acarol 40:271–277

Gardulf A, Wohlfart I, Gustafson R (2004) A prospective cross-over field trial shows protection of lemon eucalyptus extract against tick bites. J Med Entomol 41:1064–1067

Getahun A (1976) Some common medicinal and poisonous plants used in Ethiopian folk medicine. Faculty of Science, Addis Ababa University, Addis Ababa, p 63

Ghosh S, Azhahianambi P, De la Fluente J (2006) Control of ticks of ruminants with special emphasis on livestock farming system in India – present and future possibilities for integrated control: a review. Exp Appl Acarol 40:49–66

Ghosh S, Bansal GC, Gupta SC, Ray D, Khan MQ, Irshad H, Shahiduzzaman M, Seitzer U, Ahmed JS (2007) Status of tick distribution in Bangladesh, India and Pakistan. Parasitol Res 101(Suppl 2):207–216

Ghosh S, Sharma AK, Kumar S, Tewari SS, Rastogi S, Srivastava S, Singh M, Kumar R, Paul S, Ray DD, Rawat AKS (2011) In vitro and in vivo efficacy of Acorus calamus extract against *Rhipicephalus (Boophilus) microplus*. Parasitol Res 108(2):361–370

Ghosh S, Tiwari SS, Srivastava S, Sharma AK, Kumar S, Ray DD, Rawat AKS (2013) Acaricidal properties of *Ricinus communis* leaf extracts against organophosphate and pyrethroids resistant *Rhipicephalus (Boophilus) microplus*. Vet Parasitol 192:259–267

Gruber CW, O'Brien M (2010) Uterotonic plants and their bioactive constituents. Planta Med 77:207–220

Gu Xiaobin, Fang Chunlin, Yang Guangyou, Xie Yue, Nong Xiang, Zhu Junyang, Wang Shuxian, Peng Xuerong, Yan Qigui (2013) Acaricidal properties of an *Ailanthus altissima* bark extract against *Psoroptes cuniculi* and *Sarcoptes scabiei* var *cuniculi in vitro*. Exp Appl Acarol, Sept: 1–8

Guadaño A, Gutiérrez C, De La Peña E, Cortes D, González-Coloma A (2000) Insecticidal and mutagenic evaluation of two annonaceous acetogenins. J Nat Prod 63:773–776

Hassan MS, Qamar A, Khan I, Jacob P (2012) Larvicidal and IGR potential of *Ocimum tenuiflorum* and *Datura alba* leaf extracts against Malaria vector. Eur J Exp Biol 2:1370–1375

Heimerdinger A, Olivo CJ, Molento MB, Agnolim CA, Ziech MF, Scaravelli LFB, Skonieski FR, Both JF, Charao PS (2006) Extrato alcoólico de capim-cidreira (*Cymbopogon citratus*) no controle do *Boophilus microplus* em bovinos. Rev Bras Parasitol Vet 15:37–39

Hines DA, Eckman K (1993) Indigenous multipurpose trees of Tanzania: uses and economic benefits for people. In: FAO Miscellaneous/93/9 working paper, 1st edn. FAO, Rome/Ottawa, Ontario, Canada, ISBN 0-9697075-0-9. Cultural Survival Canada, Development Services Foundation of Tanzania and Tree Roots Canada. http://www.fao.org/docrep/X5327e/x5327e00.htm. 27 Sept 2003

Homer KA, Manji F, Beighton D (1992) Inhibition of peptidase and glycosidase activities of *Porphyromonas gingivalis, Bacteroides intermedius* and *Treponema denticola* by plant extracts. J Clin Periodontol 19:305–310

Hopp DC, Alali FQ, Gu ZM, McLaughlin JL (1998) Three new bioactive bis-adjacent THF-ring acetogenins from the bark of *Annona squamosa*. Bioorg Med Chem 6:569–575

Hussien KT (1995) Effect of some plant extracts in the control of a nonbiting muscoid fly. PhD thesis. Cairo University, Cairo, p 237

Hyde MA, Wursten BT, Ballings P (2011) Flora of Zimbabwe: cultivated plants: species information: *Rhus natalensis*. http://www.zimbabweflora.co.zw/cult/species.php?species id=136770. 08 Dec 2011

Ibrahim H, Abudurahman EM, Shock M, Ilyas N, Monday S, Chordima W (1999) Pharmacognostic studies of the stem of *Carissa edulis*. Nigerian J Nat Prod Med 3:31–38. http://www.inasp.info/ciajol/journals/njnpm/vol3abs.html. 20 Oct 2003

ICIPE Annual Report (1998/1999) In: Ng'eny-Mengech A, Sogo D, Mwanycky S, Ndung'u AC (eds) Sustainable methods for integrated tick management. The International Centre of Insect Physiology and Ecology (ICIPE), Animal Health Management (Ticks), ICIPE Science Press, Nairobi

Irvine FR (1961) Woody plants of Ghana with special reference to their uses. Oxford University Press, pp 1–868

Isman MB (2000) Plant essential oils for pest and disease management. Crop Protect 19:603–608

ITDG, IIRR (1996) Ethnoveterinary Medicine in Kenya. A field manual of traditional animal health care practice. Intermediate Technology Development Group (ITDG)/International Institute of Rural Reconstruction (IIRR), Nairobi, Kenya

Jaenson TGT, Pålsson K, Borg-Karlson AK (2005) Evaluation of extracts and oils of tick-repellent plants from Sweden. Med Vet Entomol 19:345–352

Jaenson TGT, Garboui S, Pålsson K (2006) Repellency of oils of lemon eucalyptus, geranium and lavender and the mosquito repellent MyggA Natural to *Ixodes ricinus* (Acari: Ixodidae) in the laboratory and field. J Med Entomol 43:731–736

Janet P (1986) Trees, plants and a rural community in the southern Sudan. An international Journal of Forestry and Forest Industries. FAO – Food and Agriculture Organization of the United Nations. Unasylva – No. 154 – Agroforestry– pathways 38:4

Jawaid T, Argal S, Singh S (2011) Botanicals and herbs: a traditional approach in treatment of epilepsy. J Pharmacol Res 4:1138–1140

Jeruto P, Lukhoba C, Ouma G, Otieno D, Mutai C (2008) An ethnobotanical study of medicinal plants used by the Nandi people in Kenya. J Ethnopharmacol 116:370–376

Johns T, Kokwaro JO, Kimanani EK (1990) Herbal remedies of the Luo of Siaya District, Kenya: establishing quantitative criteria for consensus. Econ Bot 44:369–381

Johns T, Faubert GM, Kokwaro JO, Mahunnah RLA, Kimanani EK (1995) Anti-giardial activity of gastrointestinal remedies of the Luo of East Africa. J Ethnopharmacol 46:17–23

Johnson S, Morgan ED, Peiris CN (1996) Development of the major triterpenoids and oil in the fruit and seeds of neem (*Azadirachta indica*). Ann Bot 78:383–388

Jordan RA, Schulze TL, Dolan MC (2012) Efficacy of plant-derived and synthetic compounds on clothing as repellents against *Ixodes scapularis* and *Amblyomma americanum* (Acari: Ixodidae). J Med Entomol 49:101–106

Juliet S, Ravindran R, Sunil AR, Ajith Kumar KG, Nair SN, Amithamol KK, Bandyapadhyay A, Rawat AKS, Ghosh S (2012) *Jatropha carcass* (Linn) leaf extract –a possible alternative for population control of *Rhipicephalus (Boophilus) annulatus*. Asian Pac J Trop Dis 2:225–229

Kaaya GP (2000) The potential for antitick plants as components of an integrated tick control strategy. Ann N Y Acad Sci USA 916:576–582

Kaaya GP (2003) Prospects for innovating tick control methods in Africa. Insect Sci Appl 23:59–67

Kaaya GP, Mwangi EN, Malonza MM (1995) Acaricidal activity of *Margaritaria discoidea* (Euphorbiaceae) plant extracts against the ticks *Rhipicephalus appendiculatus* and *Amblyomma variegatum* (Ixodidae). Int J Acarol 21:123–129

Kalakumar B, Kumar HSA, Kumar BA, Reddy KS (2000) Evaluation of custard seed oil and neem oil as acaricides. J Vet Parasitol 14:171–172

Kamaraj C, Rahuman AA, Bagavan A, Elango G, Rajakumar G, Zahir AA, Marimuthu S, Santhoshkumar T, Jayaseelan C (2010) Evaluation of medicinal plant extracts against blood-sucking parasites. Parasitol Res 106:1403–1412

Kambewa DMB, Mfitilodze WM, Huttner K, Wollny ABC, Phoya DKR (1997) The use of indigenous veterinary remedies in Malawi. In: Mathias-Mundy E, Rangnekar VD, McCorkle MC (eds) Ethnoveterinary medicine: alternatives for livestock development. Proceedings of an international conference held in Pune, India, on 4–6 Nov 1997. Volume 1: selected papers. File 3 of 9: Part 2: Validation of Ethnoveterinary Medicine. BAIF Development Research Foundation, Pune, India

Kaoneka B, Mollel M, Lyatuu F (2007) Leaf essential oil composition and tick repellency activity of *Commiphora swynnertonii* Burtt. J Biol Res Thessaloniki 8:213–216

Karim Md J (2006) Recent advances in disease control by natural products in animals and birds in Bangladesh. JIFS 3:15–20

Kawazu K, Alcantara JP, Kobayashi A (1989) Isolation and structure of neoannonin a novel insecticidal compound-from the seeds of *Annona squamosa*. Agric Biol Chem 53:2719–2722

Kayonga A, Habiyaremye FX (1987) Medecine traditionnelle et plantes medicinal esrwandaises. Contribution aux etudes ethnobotaniques de la florerwandaise. Prefecture de Gisenyi. Univ. Nat. Rwanda, Centre universitaire de recherche sur la pharmacopee et la medecine traditionnelle, CURPHAMETRA, inedit, Rwanda, p 121

Kebenei JS, Ndalut PK, Kiprono CP (2004) Larvicidal activity of myricoidine from *Clerodendrum myricoides*. Bull Chem Soc Ethiopia 18:225–227

Khan MF, Ahmed SM (2000) Toxicity of crude neem leaf extract against housefly (*Musca domestica* L.) adults as compared with DDVP, dichlorvos. Turk J Zool 24:219–223

Khater HF (2003) Biocontrol of some insects. PhD thesis, Zagazig University, Benha, pp 151

Khater HF, Mohamed RY, El-Madawy RS (2009) Lousicidal, ovicidal and repellent efficacy of some essential oils against lice and flies infesting water buffaloes in Egypt. Vet Parasitol 164:257–266

Kheirabadi- Pirali K, Abyaneh- Razzaghi M (2007) Biological activities of chamomile (*Matricaria chamomile)* flowers extract against the survival and egg laying of the cattle fever tick (Acari: Ixodidae). J Zhejiang Univ Sci B 8:603–696

Kim SI, Yi JH, Tak JH, Ahn YJ (2004) Acaricidal activity of plant essential oils against *Dermanyssus gallinae* (Acari: Dermanyssidae). Vet Parasitol 120:297–304

Kim SI, Na YE, Yi JH, Kim BS, Ahn YJ (2007) Contact and fumigant toxicity of oriental medicinal plant extracts against *Dermanyssus gallinae* (Acari: Dermanyssidae). Vet Parasitol 145:377–382

Kokwaro JO (1993) Medicinal plants of East Africa, 2nd edn. East African Literature Bureau, Kampala/Nairobi/Dar Es Salaam

Kolar J, Malbeck J (2009) Levels of the antioxidant melatonin in fruits of edible berry species. Planta Med 75:9

Koul O, Walia S, Dhaliwal GS (2008) Essential oils as green pesticides: potential and constraints. Biopest Int 4:63–84

Kreutzweiser DP, Capell SS, Scarr T (2000) A Community-level responses by stream insects to neem products containing azadirachtin. Environ Toxicol Chem 19:855–861

Kumar A (2003) Ayurvedic medicines: some potential plants for medicine from India. Energy Plantation Demonstration Project and Biotechnology Center, Department of Botany, University of Rajasthan, Jaipur, 302004 (Rajasthan) India

Kumar P, Mishra S, Malik A, Satya S (2011) Repellent, larvicidal and pupicidal properties of essential oils and their formulations against the housefly, *Musca domestica*. Med Vet Entomol 25:302–310

Ky KJM (2008) *Vitex doniana* Sweet. [Internet] Record from Protabase. In: Louppe D, Oteng-Amoako AA, Brink M (eds) PROTA (Plant Resources of Tropical Africa/Ressources vegetales de l'Afrique tropicale), Wageningen, Netherlands. http://database.prota.org/search.htm. 11 Dec 2011

Langenheim JH (1994) Higher plant terpenoids: a phytocentric overview of their ecological roles. J Chem Ecol 20:1223–1280

Lapa AJ et al (1999) Farmacologia e toxicologia de produtos naturais. In: Simões CM et al (ed) Farmacognosia: da planta ao medicamento. UFRGS, Porto Alegre, pp 181–196

Levin DA (1976) The chemical defenses of plants to pathogens and herbivores. Annu Rev Ecol Syst 7:121–159

Lindsay PJ, Kaufman WR (1988) The efficacy of azadirachtin on putative ecdysteroid – sensitive systems in the ixodid tick *Amblyomma americanum* L. J Insect Physiol 34:439–442

Lukwa N, Mutambu SL, Makaza N, Molgaard P, Furu P (2001) Perceptions about malaria transmission and control using anti-malaria plants in Mola, Kariba, Zimbabwe. Nigerian J Nat Prod Med 5, http://www.inasp.info/ajol/journals/njnpm/vol5abs.html. 27 Oct 2003

Lwande W, Ndakala AJ, Hassanali A, Moreka L, Nyandat N, Ndungu M, Amiani H, Gitu PM, Malonza MM, Punyua DK (1999) *Gynandropsis gynandra* essential oil and its constituents as tick (*Rhipicephalus appendiculatus*) repellents. Phytochemistry 50:401–405

Magadum S, Mondal DB, Ghosh S (2009) Comparative efficacy of *Annona squamosa* and *Azadirachta indica* extracts against *Boophilus microplus* Izatnagar isolate. Parasitol Res 105:1085–1091

Mägi E, Järvis T, Miller I (2006) Effects of different plant products against pig mange mites. Acta Vet Brno 2006(75):283–287

Maibach HL, Akers WA, Johnson HL (1974) Insects: topical insect repellents. Clin Pharmocol Ther 16:970–973

Malerich PJ, Trauner D (2003) Biomimetic synthesis of (±)-Pinnatal and (±)-Sterekunthal A. J Am Chem Soc 125:9554–9555

Malonza MM, Dipeolu OO, Amoo AO, Hassan SM (1992) Laboratory and field observations on anti-tick properties of the plant *Gynandropsis gynandra* (L.) Brig. Vet Parasitol 42:123–136

Mansingh A, Williams LAD (1998) Pesticidal potential of tropical plants – II. Acaricidal activity of crude extracts of several Jamaican plants. Insect Sci Appl 18:149–155

Mansour SA, Bakr RFA, Mohamed RI, Hasaneen NM (2011) Larvicidal activity of some botanical extracts, commercial insecticides and their binary mixtures against the housefly, *Musca domestica* L. Open Toxicol J 4:1–13

Marcard M, Zebitz CPW, Schmutterer H (1986) The effect of crude methanol extracts of *Ajuga* spp. on postembryonic development of different mosquito species. J Appl Entomol 101:146–154

Marin R (1999) Collection of medicinal plant samples from Africa and elsewhere. World botanical associates, http://www.worldbotanical.com/WBAMED2.HTM. 01 Oct 2003

Martins RM (2006) Estudio invitro de la acción acaricida del aceite esencial de la gramínea citronela de Java (*Cymbopogon winterianus* Jowitt) em la garrapata *Boophilus microplus*. Rev Bras Plantas Med 8:71–78

Martins RM, Gonzalez FHD (2007) Uso del aceite de citronela de Java (Cymbopogonwinterianus Jowitt) (Panicoidideae) como acaricida frente a la garrapata *Boophilus microplus* Canestrini (Acari: Ixodidae). Rev Bras Plantas Med 9:1–8

Martins JRS, Furlong J, Leite RC (2006) Controle de carrapatos. In: Barros-Battesti DMB, Arzua M, Bechara GH (Org) Carrapatos de importância médico-veterinário da Região Neotropical. Um guia ilustrado para identificação de espécies. Instituto Butantam, São Paulo, pp 145–153

Matzigkeit U (1990) Natural veterinary medicine. Ectoparasites in the tropics. Josef Margraf Verlag, Weikersheim

Maundu PM, Berger D, Ole Saitabau J, Nasieku C, Kipelian J, Mathenge M, Morimoto SG, Hoft Y (2001) Ethnobotany of the Loita Maasai: towards community management of the forest of the lost child-Experiences from the Loita ethnobotany project. People and plants working paper 8. UNESCO, Paris, France. http://www.rbgkew.org.uk/peopleplants/wp/wp8/index-html. 28 Sept 2003

Meisner J, Ascher KRS, Aly R, Warthen JD (1981) Response of *Spodoptera littoralis* (Boisd.) and *Eariasinsulana* (Boisd.) larvae to azadirachtin and salannin. Phytoparasitica 9:27–32

Michael HM (2003) Medicinal uses of succulents. http://www.succulents.co.za/Contact.shtml

Milam CD, Farris JL, Wilhide JD (2000) Evaluating mosquito control pesticides for effect on target and nontarget organisms. Arch Environ Contam Toxicol 39:324–328

Miller JA, Chamberlain WF (1989) Azadirachtin as a larvicide against the horn fly, stable fly and house fly. J Econ Entomol 82:1375–1378

Mirjalili MH, Moyano E, Bonfill M, Cusido RM, Palazon J (2009) Steroidal lactones from *Withania somnifera*, an ancient plant for novel medicine. Molecules 14:2373–2393

Mitchell SM, Ahmad MH (2006) A review of medicinal plant research at the University of the West Indies, Jamaica, 1948–2001. West Indian Med J 55:243–269

Mkolo MN, Magano MR (2007) Repellent effects of the essential oil of *Lavendula angustifolia* against adults of *Hyalomma marginatum rufipes*. J S Afr Vet Assoc 78:149–152

Moore SJ, Debboun M (2007) History of insect repellents. In: Debboun M, Frances S, Strickman D (eds) Insect repellents: principles, methods and uses. CRC Press, Boca Raton, pp 3–29

Moore SJ, Lenglet A, Hill N (2007) Plant-based insect repellents. In: Debboun M, Frances S, Strickman D (eds)

Insect repellents: insect repellents: principles, methods and uses. CRC Press, Boca Raton, pp 275–303

Morgan ED, Thorton MD (1973) Azadirachtin in the fruit of *Melia azedarach*. Phytochemistry 12:391–392

Muhammad Ishtiaq Ch, Khan MA (2006) Ethno veterinary medicinal uses of plants from Samahni Valley Dist. Bhimber (Azad Kashmir) Pakistan. Asian J Plant Sci 5:390–396

Mulla MS, Su T (1999) Activity and biological effects of neem products against arthropods of medical and veterinary importance. J Am Mosq Cont Assoc 15:133–152

Muro Castrjón F, Cruz-Vázquez C, Fernández-Ruvalcaba M, Molina-Torres J, Soria Cruz J, Ramos Parra M (2003) Repellence of *Boophilus microplus* larvae in *Stylosanthes humilis* and *Stylosanthes hamata* plants. Parasitol Latinoam 58:118–121

Murugan K, Jeyabalan D (1999) Effect of certain plant extracts against the mosquito, *Anopheles stephensi* Liston. Curr Sci 76:631–633

Mwangi EN (1996) The role of women in animal husbandry: the Kenyan perspective. In: Zessin HK (ed) Livestock production and diseases in the tropics: livestock production and human welfare. Proceedings of the VIII international conference of institutions of tropical veterinary medicine held from the 25–29 Sept 1995, Berlin, Germany. Volume II. Deutsche stiftung fuf Internationale Entwicklung. Zentralstelle fur Ernahrung und Landwirtschaft, Feldafing, Germany

Mwangi EN, Essuman S, Kaaya GP, Nyandat E, Munyinyi D, Kimondo M (1995a) Repellence of the tick *Rhipicephalus appendiculatus* by the grass *Melinis minutiflora*. Trop Anim Health Prod 27:211–216

Mwangi EN, Hassanali A, Essuman S, Myandat E, Moreka L, Kimondo M (1995b) Repellent and acaricidal properties of *Ocimum suave* against *Rhipicephalus appendiculatus* ticks. Exp Appl Acarol 19:11–18

Mwangi J, Thoithi GN, Addae-Mensah I, Achenbach H, Lwande W, Hassanali A (1998) Aromatic plants of Kenya III: volatile and some non-volatile constituents of *Croton sylvaticus* Hochst. East Central Afr J Pharm Sci 1:41–43

Nakatani M, Takao H, Miura I, Hase T (1985) Azedarachol, a steroid ester antifeedant from *Melia azedarach* var. Jpn Phytochem 24:1945–1948

Nakatani M, Huang RC, Okamura H, Iwagawa T, Tadera K, Naoki H (1995) Three new antifeeding meliacarpinins from Chinese *Melia azedarach* Linn. Tetrahedron 51:11731–11736

Naraladker BW, Khillare BS, Kalwaghe ST, Chghure GM, Ravikanth K, Maini S (2011) Effectiveness of phytogenic fly repellant product against dipteran flies. Int Res J Pharm Pharmacol 1:70–74

Narladkar BW, Deshpande PD, Vaniprasad V, Vadlamudi, Shivpuje PR, Deshpande AR (2006) Integrated management of *Culicoides* sp. of domesticated animals. J Vet Parasitol 20:125–128

Nauen R (2007) Insecticide resistance in disease vectors of public health importance. Pest Manag Sci 63:628–633

Ndumu PA, George JBD, Choudhury MK (1999) Toxicity of neem seed oil (*Azadirachta indica*) against the larvae of *Amblyomma variegatum* a three-host tick in cattle. Phytother Res 13:532–534

Ndungu M, Lwande W, Hassanali A, Moreka L, Chhabra SC (1995) *Cleome monophylla* essential oil and its constituents as tick (*Rhipicephalus appendiculatus*) and maize weevil (*Sitophilus zeamais*) repellents. Entomol Exp Appl 76:217–222

Neraliya S, Srivastava US (1996) Effect of plant extracts on post-embryonic development of the mosquito *Culex quinquefasciatus*. J Adv Zool 17:54–58

Neuwinger HD (2000) African traditional medicine: a dictionary of plant use and applications. Medpharm Scientific, Stuttgart

Nong X, Fang CL, Gu XB, Wang JH, Yang XB, Liu DY, Fu TF, Zhang Y, Zheng RH, Peng WP, Wang XR, Yang SXGY (2012) Acaricidal activity of extract from *Eupatorium adenophorum* against the *Psoroptes cuniculi* and *Sarcoptes scabiei in vitro*. Vet Parasitol 187:345–349

Okumu FO, Knols BGJ, Fillinger U (2007) Larvicidal effects of a neem (*Azadirachta indica*) oil formulation on the malaria vector *Anopheles gambiae*. Malar J 6:63

Okwemba A (2002) Local impotency herb to flavour yoghurt. An AWC feature. Horizon Magazine. http://www.nationaudio.com/News/DailyNation/. 17 Oct 2002

Olivo CJ, De Carvalho NM, Henrique J, da Silva S, Vogel FF, Massariol P, Meilnerz G, Agnolin C, Morel AF, Viau LV (2008) Óleo de citronela no controle do carrapato de bovinos. Ciência Rural Santa Maria 38:406–410

Olusola L, Zebulon SC, Okoye UF (1997) Effect of *Vitex doniana* stem bark on blood pressure. Nigerian J Nat Prod Med. http://www.inasp.info/ajol/journals/njnpm/vol1abs.html. 08 Nov 2003

Orwa C, Mutua A, Kindt R, Jamnadass R, Simons A (2009) Agroforestry database: a tree reference and selection guide version 4.0. http://www.worldagroforestry.org/af/treedb/. 11 Dec 2011

Palacios SM, Bertoni A, Rossi Y, Santander R, Urzúa A (2009a) Efficacy of essential oils from edible plants as insecticides against the house fly, *Musca domestica* L. Molecules 14:1938–1947

Palacios SA, Bertoni A, Rossi Y, Santander R, Urzua A (2009b) Efficacy of essential oils from edible plants as insecticides against the house fly, *Musca domestica* L. Molecules 14:1938–1947

Pållson K, Jaenson TGT, Backström P, Borg-Karlson AK (2008) Tick repellent substances in the essential oil of *Tanacetum vulgare*. J Med Entomol 45:88–93

Paluch GL, Bartholomay L, Coats J (2010) Mosquito repellents: chemical structure and olfaction. Pest Manage Sci 66:925–935

Pamo ET, Tendonkeng F, Kana JR, Payne VK, Boukila B, Lemoufouet J, Miegoue E, Nanda AS (2005) A study of the acaricidal properties of an essential oil extracted from the leaves of *Ageratum houstonianum*. Vet Parasitol 128:319–323

Pardo JM, Fontanilla MR, Ospina LF, Espinosa L (2008) Determining the pharmacological activity of *Physalis peruviana* fruit juice on rabbit eyes and fibroblast primary cultures. Inv Ophthalmol Visual Sci 7:3074–3079

Parvin S, Islam ME, Rahman MM, Haque ME (2003) Pesticidal activity of pure compound annotemoyin-1 Isolated from chloroform extract of the plant *Annona squamosa* Linn. against *Tribolium castaneum* (Herbst) (Coleoptera: Tenebrionidae). Pak J Biol Sci 6:1088–1091

Pautou MP, Rey D, David JP, Meyran JC (2000) Toxicity of vegetable tannins on crustacea associated with Alpine mosquito breeding sites. Ecotoxicol Environ Saf 47:323–332

Pavela R (2008) Insecticidal properties of several essential oils on the house fly (*Musca domestica* L.). Phytother Res 22:274–278

Pereira JR, Famadas KM (2004) Avaliação "*in vitro*" da eficiência do extrato da raiz do timbó (Dahlstedtia pentaphylla) (Leguminosae, Papilionoidae, Millettiedae) sobre *Boophilus microplus* (Canestrini, 1887) na região do Vale do Paraíba, São Paulo, Brasil. Arq Inst Biol 71:443–450

Pereira JR, Famadas KM (2006) The efficiency of extracts of *Dahlstedtia pentaphylla* (Leguminosae, Papilionoidae, Millettiedae) on *Boophilus microplus* (Canestrini, 1887) in artificially infested bovines. Vet Parasitol 142:192–195

Polygenis-Bigendako MJ (1990) Recherches ethnopharmacognosiques sur les plantes utilisees en medecine traditionnelle au Burundi occidental. These presentee en vue de l'obtention du grade de Docteur en sciences, annee acad. 1989–1990. Universite libre de Bruxelles, Faculty of Science Laboratory. De Botanique systematique et de Phytosociologie

Porter RBR, Reese PB, Williams LAD (1995) Acaricidal and insecticidal activities of cadina-4, 10 (15)-dien-3-one. Phytochemistry 3:735–738

Radostitts OM, Gay CC, Hinchcliff K, Constable PD (2007) Veterinary medicine, 10th edn. Reed Elsevier India Private Limited, India, pp 1596–1597

Rahmatullah M, Samarrai W, Jahan R, Rahman S, Sharmin N, Miajee UEMUZ, Chowdhury HM, Bari S, Jamal F, Bashar AMBA, Azad KA, Ahsan S (2010) An ethnomedicinal, pharmacological and phytochemical review of some Bignoniaceae family plants and a description of Bignoniaceae plants in folk medicinal uses in Bangladesh. Adv Natl Appl Sci 4:236–253

Rajab SM, Mulholland AD, Monkhe TV (1998) Triterpenoids from *Turraea holstii*. Phytochemistry 49:2585

Ramos-Escudero F, Mu Oz AM, Alvarado-Ort ZC, Alvarado N, Y Ez JA (2011) Purple corn (*Zea mays* L.) phenolic compounds profile and its assessment as an agent against oxidative stress in isolated mouse organs. J Med Food. http://www.ncbi.nlm.nih.gov/pubmed/22082063. 12 Dec 2011

Rao DR, Reuben R, Nagasampagi BA (1995) Development of combined use of neem (*Azadirachta indica*) and water management for the control of culicine mosquitoes in rice fields. Med Vet Entomol 9:25–33

Ravindran R, Juliet S, Sunil AR, Kumar KGA, Nair SN, Amithamol KK, Shynu M, Rawat AKS, Ghosh S (2011) Eclosion blocking effect of ethanolic extract of *Leucas aspera* (Lamiaceae) *on Rhipicephalus (Boophilus) annulatus*. Vet Parasitol 179:287–290

Ravindran R, Juliet S, Sunil AR, Ajith Kumar KG, Nair SN, Amithamol KK, Bandyopadhyay A, Rawat AKS, Ghosh S (2012) Acaricidal activity of *Cassia alata* against *Rhipicephalus (Boophilus) annulatus*. Exp Appl Acarol 56:69–74

Regassaa A (2000) The use of herbal preparations for tick control in western Ethiopia. J S Afr Vet Assoc 71:240–243

Ribeiro VL, Toigo E, Bordignon SA, Gonçalves K, Von Poser G (2007) Acaricidal properties of extracts from the aerial parts of *Hypericum polyanthemum* on the cattle tick *Boophilus microplus*. Vet Parasitol 147:199–203

Ribeiro VL, Avancini C, Gonçalves K, Toigo E, Von Poser G (2008a) Acaricidal activity of *Calea serrata* (Asteraceae) on *Boophilus microplus* and *Rhipicephalus sanguineus*. Vet Parasitol 151:351–354

Ribeiro VL, Rolim V, Bordignon S, Henriques AT, Dorneles GG, Limberger RP, Von Poser G (2008b) Chemical composition and larvicidal properties of the essential oils from *Drimys brasiliensis* Miers (Winteraceae) on the cattle tick *Rhipicephalus (Boophilus) microplus* and the brown dog tick *Rhipicephalus sanguineus*. Parasitol Res 102:531–535

Ribeiro VL, Rolim V, Bordignon S, Henriques AT, Dorneles GG, Limberger RP, Von Poser G (2010) Acaricidal properties of the essential oil from *Hesperozygis ringens* (Lamiaceae) on the cattle tick *Rhipicephalus (Boophilus) microplus*. Bioresour Technol 101:2506–2509

Roark RC (1947) Some promising insecticidal plants. Econ Bot 1:437 [Cited in: Curtis CF, Lines JD, Baolin L, Renz A. Natural and synthetic repellents. 75–79]

Robles AP (2004) Comparative efficacy of tobacco (*Nicotiana tobaccum*), makabuhay (*Tinosphora rumphi*), tubli (*Derris philippinensis*), and neem (*Azadirachta indica*) against carabao louse (*Haematopinus tuberculatus*). In: Maala CP, Landicho EF, Sarabia AS, Del Barrio AN, Ataby EP (eds) Abstract of researches on the Philippine water buffalo. Philippine CarabaoCenter, pp 139–140

Rojas J, Ochoa V, Ocampo S, Mũnoz J (2006) Screening for antimicrobial activity of ten medicinal plants used in Colombian folkloric medicine: a possible alternative in the treatment of non-nosocomial infections. BMC Compl Altern Med 6:2

Rosado-aguilar AJ, Caceres-Farfan M, Mendez-Gonzalez M, Borges-Argaez R, Rosado-Aguilar JA, Dorantes-Euan A, Rodrigues-Vivas RI, Garcia-Vazquez Z (2008) Actividad ixodicida de extractos crudos de *Diospyros anisandra* contra larvas de *Rhipicephalus (Boophilus) microplus* (Acari: ixodidae). Trop Subtrop Agroecosyst 8:297–301

Rukangira E (2001) The African herbal industry: constraints and challenges. Erboristeria Domani. http://www.conserveafrica.org/herbalindustry.rtf

Rukangira E (2003) Overview on the importance and the status of medicinal plants and traditional medicine in Africa. Sustainable Development International. http://www.sustdev.org/Features/04.09.01.shtml. 27 Oct 2003

Rulangaranga ZK (1989) Some important indigenous medicinal and aromatic plants in the wild flora of Tanzania mainland. Tropical Forestry Action Plan, working paper 24. Tanzania Ministry of Lands, Natural Resources and Tourism, Dar es Saalam, Tanzania

Samuelsson G, Faraha MH, Claeson P, Hagos M, Thulin M, Hedberg O, Warfa AM, Hassan AO, Elmi AH, Abdurahman AD, Elmi AS, Abdie YA, Alin MH (1992) Inventory of plants used in traditional medicine in Somalia. II. Plants of the families Combretaceae – Labiatae. J Ethnopharmacol 37:47–70

Santamaría E, Cabrera OL, Zipa Y, Pardo RH (2012) Field efficacy of repellent formulation containing para-menthane-3,8-diol and lemongrass against *Culicoides pachymerus* (Diptera: Ceratopogonidae) in Colombia. Biomedica 32:457–460

Schippmann U (2001) Medicinal plants significant trade. CITES Projekt S-109, Plants Committee Document PC9 9.1.3 (rev.). BFN Scripten – 39, BFN-German Federal Agency for Nature Conservation, pp 51–58

Schmidt JR (2003a) Agavaceae (Agave or Century Plant family). Plant Families Index, The Botanical Dermatology Database (BoDD), Index to Botanical Dermatology reviews. http://BoDD.cf.ac.uk/BoDDHomePage.html. 09 Oct 2003

Schmidt JR (2003b) Euphorbiaceae (Spurge family). Plant Families Index, The Botanical Dermatology Database (BoDD), Index to Botanical Dermatology Reviews

Schmidt JR (2003c) Leguminosae (Pea family). Plant families index, The Botanical Dermatology Database (BoDD), Index to botanical dermatology reviews. http://BoDD.cf.ac.uk/BoDDHomePage.html. 18 Oct 2003

Schmutterer H (1990) Properties and potential of natural pesticides from the neem tree, *Azadirachta indica*. Annu Rev Entomol 35:197–271

Schreck CE, Leonhardt BA (1991) Efficacy assessment of Quwenling: a mosquito repellent from China. J Am Mosq Contr Assoc 7:433–436

Schulze TL, Jordan RA, Dolan MC (2011) Experimental use of two standard tick collection methods to evaluate the relative effectiveness of several plant derived and synthetic repellents against *Ixodes scapularis* and *Ambloymma americanum* (Acari: Ixodidae). J Econ Entomol 104:2062–2067

Schwantes U, Dautel H, Jung G (2008) Prevention of infectious tick-borne diseases in humans: comparative studies of the repellency of different dodecanoic acid-formulations against *Ixodes ricinus* ticks (Acari: Ixodidae). Parasit Vectors 1:8–19

Seo Y (2002) A new bioactive triterpene saponin from *Pittosporum viridiflorum* from the Madagascar Rainforest. J Nat Prod 65:65–68

Shaalan EAS, Canyon D, Younes MWF, Hoda AW, Mansour AH (2005) A review of botanical phytochemicals with mosquitocidal potential. Environ Int 31:1149–1166

Sharma RN, Saxena KN (1974) Orientation and developmental inhibition in the housefly by certain terpenoids. J Med Entomol 11:617–621

Sheppard DC, Gaydon DM, Miller RW (1992) Resistance in house flies (Diptera: Muscidae) selected with 5.0 ppm feed-through cyromazine. J Agric Entomol 9:257–260

Siddiqui S, Siddiqui BS, Faizi S, Mahmood T (1988) Tetracyclic triterpenoids and their derivatives from *Azadirachta indica*. J Nat Prod 51:30–43

Siddiqui BS, Afshan F, Gulzar T, Sultana R, Naqvi SN, Tariq RM (2003) Tetracyclic triterpenoids from the leaves of *Azadirachta indica* and their insecticidal activities. Chem Pharm Bull 51:415–417

Silva WC, Martins JR, De Souza HE, Heinzen H, Cesio MV, Mato M, Albrecht F, De Azevedo JL, De Barros NM (2009) Toxicity of *Piper aduncum* L. (Piperales: Piperaceae) from the Amazon forest for the cattle tick *Rhipicephalus (Boophilus) microplus* (Acari: Ixodidae). Vet Parasitol 164:267–274

Singh S, Kumar S (1998) Withania somnifera. The Indian Ginseng Ashwagandha, Lucknow. ISBN 81-86943-91-9

Singh SK, Dimri U, Sharma MC, Swarup D, Kumar M, Tiwary R (2012) *Psoroptes cuniculi* induced oxidative imbalance in rabbits and its alleviation by using vitamins A, D3, E, and has adjunctive remedial. Trop Anim Health Prod 44:43–48

Sivaramakrishnan S, Kumar NS, Jeyabalan D, Babu R, Raja NS, Murugan K (1996) The effect of mixtures of neem, eucalyptus and pongamia oils on the mortality and biochemical profiles of the ticks, *Boophilus microplus* (Canestrini) (Acari: Ixodidae). Indian J Environ Toxicol 6:85–86

Smith CN, Cole MM, Gilbert IH, Gouck HK (1954) Field tests with tick repellents – 1949, 1950, and 1952. J Econ Entomol 47:13–19

Sousa LAD, Soares SF, Junior HBP, Ferri PH, Borges LMF (2008) Avaliação da eficácia de extratos oleosos de frutos verdes e maduros de cinamomo (*Melia azedarach*) sobre *Rhipicephalus (Boophilus) microplus* (Acari: Ixodidae). Rev Bras Parasitol Vet 17:36–40

Sousa LAD et al (2011) Potential synergistic effect of *Melia azedarach* fruit extract and *Beauveria bassiana* in the control of *Rhipicephalus (Boophilus) microplus* (Acari: Ixodidae) in cattle infestations. Vet Parasitol 175:320–324

Srivastava R, Ghosh S, Mandal DB, Azhahianambi P, Singhal PS, Pandey NN, Swarup D (2008) Efficacy of *Azadirachta indica* extracts against *Boophilus microplus*. Parasitol Res 104:149–153

Sriwattanarungsee S, Sukontason KL, Olson JK, Chailapakul O, Sukontason K (2008) Efficacy of neem extract against the blowfly and housefly. Parasitol Res 103:535–544

Staub D, Debrunner M, Amsler L, Steffen R (2002) Effectiveness of a repellent containing DEET and

EBAAP for preventing tick bites. Wilderness Environ Med 13:12–20

Stockbauer B (2003) Tree list for environmental nurseries. African Blackwood Conservation Project. http://www.blackwoodconservation.org/5-year program tree list.html. 18 Oct 2003

Stoll G (1988) Natural plant protection in the tropics. AGRECOL, Margraf Publishers Scientific Books, Weikersheim

Su T, Mulla MS (1998) Ovicidal activity of neem products (Azadirachtin) against *Culex tarsalis* and *Culex quinquefasciatus* (Diptera: Culicidae). J Am Mosq Contr Assoc 14:204–209

Su T, Mulla MS (1999) Effects of neem products containing azadirachtin on blood feeding, fecundity and survivorship of *Culex tarsalis* and *Culex quinquefasciatus* (Diptera: Culicidae). J Vector Ecol 24:202–215

Sunil AR, Amithamol KK, Juliet S, Nair SN, Kumar KGA, Soorya VC, Divya TM, Jyothimol G, Ghosh S, Ravindran R (2013) Acaricidal effect of *Cassia fistula* Linn. leaf ethanolic extract against *Rhipicephalus (Boophilus) annulatus*. Trop Biomed 30:231–237

Tadesse D (1994) Traditional use of some medicinal plants in Ethiopia. In: SeyaniJH, Chicuni AC (eds) Proceedings of the 13th plenary meeting of AETFAT, Zomba, Malawi, 2-11 Apr1991, vol. 1. National herbarium and Botanic Gardens of Malawi, Zomba, pp 273–293

Thijssen HJC (2008) *Vitex fischeri* Gurke. [Internet] Record from Protabase. In: Louppe D, Oteng-Amoako AA, Brink M (eds) PROTA (Plant Resources of Tropical Africa/Ressources vegetales de l'Afrique tropicale), Wageningen, The Netherlands

Thompson KC, Rao J, Romero T (1978) Antitick grasses as the basis for developing practical tropical tick control packages. Trop Anim Health Prod 10:179–182

Thorsell W, Mikiver A, Tunón H (2006) Repelling properties of some plant materials on the tick *Ixodes ricinus* L. Phytomedicine 13:132–134

Tong F, Gross AD, Dolan MC, Coats JR (2011) Effects of monoterpenoid insecticide/acaricide carvacrol, on [14C]- nicotine binding to the *Musca domestica* (house fly) nicotine acetylcholine receptor. Insect Biochem Mol Biol 65:957–960

Torto B, Hassanali A (1997) Progress in the search for anti-arthropod botanicals. Recent Res Dev Phytochem 1:475–488

Toyang JN, Nuwanyakpa M, Ndi C, Django S, Kinyuy CW (1995) Ethnoveterinary medicine practices in the Northwest Province of Cameroon. IK Monitor 3. http://www.nuffic.nl/ciran/ikdm/3-3/articles/toyang.html. 20 Oct 2003

Trigg JK (1996) Evaluation of a eucalyptus-based repellent against *Anopheles* spp. in Tanzania. J Am Mosq Contr Assoc 12:243–246

Trigg JK, Hill N (1996) Laboratory evaluation of a eucalyptus-based repellent against four biting arthropods. Phytother Res 10:313–316

Tsao R, Coats JR (1995) Starting from natural to make a better insecticides. Chentechnology 25:23–28

Tunón H, Thorsell W, Mikiver A, Malander I (2006) Arthropod repellency, especially tick (*Ixodes ricinus*), exerted by extract from *Artemisia abrotanum* and essential oil from flowers of *Dianthus caryophyllum*. Fitoterapia 77:257–261

Ulutas B, Voyvoda H, Bayramli G, Karagenc T (2005) Efficacy of topical administration of eprinomectin for treatment of ear mite infestation in six rabbits. Vet Dermatol 16:334–337

Urzua A, Santander R, Echeverría J, Cabezas N, Palacios SM, Rossi Y (2010) Insecticide properties of the essential oils from *Haplopappus foliosus* and *Bahia ambrosoides* against the house fly, *Musca domestica* L. J Chil Chem Soc 47:2945–2964

UWMG (2003) Uganda Wildlife Medicinal Garden (UWMG) A list of some of over 500 herbs in Uganda wildlife medicinal garden, with information about diseases they cure. http://www.uweczoo.org/Medicinalplants tree tags.pdf. 09 Nov 2003

Valente M, Barranco A, Sellaive-Villaroel AB (2007) Eficácia do extrato aquoso de *Azadirachta indica* no controle de *Boophilus microplus* em bovino. Arquivo Bras Med Vet Zootecnia 59:1341–1343

Van Der Hooft CS, Hoekstra A, Winter A, De Smet PA, Stricker BH (2005) Thyrotoxicosis following the use of Ashwagandha. Ned Tijdschr Geneeskd 149:2637–2638

Varma J, Dubey NK (1998) Prospectives of botanical and microbial products as pesticides of tomorrow. Cur Sci 76:172–179

Vatsya S, Yadav CL, Kumar RR, Banerjee PS (2006) *In vitro* acaricidal effect of some medicinal plants against *Boophilus microplus*. J Vet Parasitol 20:141–143

Vivan MP (2005) Uso do cinamomo (*Melia azedarach*) como alternativa aos agroquímicos no controle do carrapato bovino (*Boophilus microplus*). 72 f. Dissertação (Mestrado)-Universidade Federal de Santa Catarina, Florianópolis

Walker AR et al (2003) Ticks of domestic animals in Africa: a guide to identification of species. University of Edinburgh, Edinburgh

Watt JM, Breyer-Brandwijk MG (1962) The medicinal and poisonous plants of Southern and Eastern Africa, 2nd edn. E. and S. Livingstone Ltd., London

William NS (1999) Natural products drug discovery. http://chemistry.uah.edu/Faculty/setzer/natprod

Williams LAD (1993) Adverse effects of extracts of *Artocarpus altilis* Park and *Azadirachta indica* (A. Juss) on the reproductive physiology of the adult female tick, *Boophilus* (Canest.). Invertebr Reprod Dev 23:159–164

Williamson EM (2007) The medicinal use of essential oils and their components for treating lice and mite infestations. Nat Prod Commun 2:1303–1310

Wilson A, Mansingh A (2002) Pesticidal potential of tropical plants-VI. Toxic actions of ethanol extract of selected plants on *Cylas formicarius elegantulus* Summer and *Boophilus microplus* Canestrinii. In: Man Singh A, Young ED, Yee HT, Delgoda R, Robinson DE, Morrison YE, Lowe IH (eds) Proceedings of the first international symposium on utilization of natural products in developing countries: trends and needs, held between 10 and 14 July 2000 at the Natural Products Institute, University of the West Indies, Mona, Kingston 7, Jamaica, West Indies, pp 240–246

Witting-Bissinger BE, Stumpf CF, Donohue KV, Apperson CS, Roe RM (2008) Novel arthropod repellent, BioUD, is an efficacious alternative to DEET. J Med Entomol 45:891–898

Woodland DW (1997) Contemporary plant systematic, 2nd edn. Andrews University Press, Berrien Springs

Wu SJ, Tsai JY, Chang SP, Lin DL, Wang SS, Huang SN, Ng LT (2006) Supercritical carbon dioxide extract exhibits enhanced antioxidant and anti-inflammatory activities of *Physalis peruviana*. J Ethnopharmacol 108:407–413

Yakkundi SR, Thejavathi R, Ravindranath B (1995) Variation of Azadirachtin content during growth and storage of neem (*Azadirachta indica*) seeds. J Agric Food Chem 43:2517–2519

Yang YC, Choi HY, Choi WS, Clark JM, Ahn YJ (2004) Ovicidal and adulticidal activity of *Eucalyptus globulus* leaf oil terpenoids against Pediculus humanus capitis (Anoplura: Pediculidae). J Agric Food Chem 52:2507–2511

Yong-Hua Du, Ren-Yong Jia, Zhong-Qiong Yin, Zhong-Hui Pu, Jiao Chen,Fan Yang, Yu-Qun Zhang, Yang Lu (2008) Acaricidal activity of extracts of neem (*Azadirachta indica*) oil against the larvae of the rabbit mite *Sarcoptes scabiei* var. *cuniculi in vitro*. Vet Parast 157(1-2):144–148

Zahir AA, Rahuman AA, Kamaraj C, Bagavan A, Elango G, Sangaran A, Kumar BS (2009) Laboratory determination of efficacy of indigenous plant extracts for parasites control. Parasitol Res 105:453–461

Zahir AA, Rahuman AA, Bagavan A, Santhoshkumar T, Mohamed RR, Kamaraj C, Rajakumar G, Elango G, Jayaseelan C, Marimuthu S (2010) Evaluation of botanical extracts against *Haemaphysalis bispinosa* Neumann and *Hippobosca maculata* Leach. Parasitol Res 107:585–592

Zebitz CPW (1984) Effects of some crude and azadirachtin-enriched neem (*Azadirachta indica*) seed kernel extracts on larvae of *Aedes aegypti*. Aust J Entomol 39:208–211

Zhang A, Klun JA, Wang S, Carroll J, Debboun M (2009) Isolongifolenone: a novel sesquiterpene repellent of ticks and mosquitoes. J Med Entomol 46:100–106

Zheng H, He K (1993) In: IDRC (ed) Agroforestry systems in China. International Development Research Centre (IDRC)/Intercropping in rubber plantations and its economic benefits, Ottawa

Role of Plant Biopesticides in Managing Vectors of Communicable Diseases

12

Apiwat Tawatsin, Usavadee Thavara, Padet Siriyasatien, and Mir S. Mulla

Abstract

Some insects and arthropods frequently pose a serious risk to human health. They can cause painful bites and transmit some pathogens causing serious diseases such as malaria, dengue hemorrhagic fever, lymphatic filariasis, Lyme disease, river blindness, etc. Since the protective vaccines have not yet been available for most of these vector-borne diseases, vector controls are therefore the main strategies to prevent the diseases. Such plant-based products have been used in the control of insects of public health importance for centuries. However, the development and use of phytochemicals attracted considerable attention from researchers and industrial concerns in the last quarter of the century. Examples of major plant-based products used in pest control are pyrethrins, neem constituents, as well as many plant volatile essential oils for repelling hematophagous insects affording personal protection of humans from biting arthropods and noxious insects. The public has the perception that plant-based and other natural products are environmentally friendly and safer to use for vector control or apply to human skin as personal protectants than synthetic chemicals. Considerable advances have then been made in formulating phytochemicals to increase their efficacy, providing protection and acceptability in public health. In this chapter, we will dwell upon the research and development efforts leading to the development and use of plant essential oils for personal protection from anthropophilic insects and arthropods as well as the development and application of phytochemicals for the control of adult and preimaginal stages of disease vectors.

A. Tawatsin, Ph.D. (✉) • U. Thavara, Ph.D.
National Institute of Health, Department of Medical Sciences, Ministry of Public Health, Nonthaburi 11000, Thailand
e-mail: atawatsin@gmail.com

P. Siriyasatien, Ph.D.
Department of Parasitology, Faculty of Medicine, Chulalongkorn University, Bangkok 10330, Thailand

M.S. Mulla, Ph.D.
Department of Entomology, University of California, Riverside, CA 92521, USA

D. Singh (ed.), *Advances in Plant Biopesticides*,
DOI 10.1007/978-81-322-2006-0_12, © Springer India 2014

Keywords

Biopesticides • Repellents • Insects • Vectors • Human diseases

1 Vector-Borne Diseases

Some insects and arthropods frequently pose a serious risk to human health. Mosquitoes, for example, not only cause painful bites and nuisance but also transmit some pathogens causing serious human diseases such as malaria, yellow fever, dengue hemorrhagic fever (DHF), chikungunya fever, lymphatic filariasis, Japanese encephalitis, and West Nile encephalitis to humans (Bannister et al. 1996). Insects and other arthropods, such as blackflies, sand flies, tsetse flies, biting midges, ticks, chiggers, houseflies, and cockroaches, have been reported to transmit several pathogens to humans by biological or mechanical transmission. It has been estimated that more than half of the world's population, especially in tropical countries, is at risk from insect-borne diseases and at least 500 million people suffer from infections annually (WHO 1996). The diseases caused by various pathogens, such as bacteria, protozoans, nematodes, and viruses, are transmitted to human through the bite of the insect and arthropod vectors (Service 1993). The diseases, pathogens, and their vectors are summarized in Table 12.1. The search for effective vaccines against these diseases is still in progress. Vector control and personal protection from insect and arthropod bites are currently the most important measures to control vector-borne diseases.

2 Use of Plant Biopesticides to Manage Insect and Arthropod Vectors

Phytochemicals derived from extractions of various plant species have provided numerous activities against insects, including killing effects. Insecticides derived from plants have been developed as commercial products to control insects and arthropods that are human pests. The experimental plant extracts that were evaluated for their bioactivity against various insects and arthropods, such as mosquitoes, blackflies, biting midges, houseflies, cockroaches, ticks, chiggers, and land leeches, are briefly summarized in Table 12.2. A synopsis of results from some of these studies is presented below.

2.1 Mosquitoes

Mosquitoes are the most important insects among the vectors that transmit communicable diseases to humans. Malaria, for example, is one of the most serious diseases that threaten millions of human life annually, and it is transmitted by the bites of several species of *Anopheles* mosquitoes, such as *Anopheles gambiae*, *An. arabiensis*, *An. dirus* (Fig. 12.1), *An. minimus*, etc. Dengue hemorrhagic fever and chikungunya are transmitted by *Aedes aegypti* (Fig. 12.2) and *Ae. albopictus*. On the other hand, West Nile fever is transmitted by several species of *Culex* and *Aedes* mosquitoes. To minimize human-mosquito contact, efforts are directed toward prevention of mosquito bites. The current common protective measures include avoidance of high-risk exposure, such as outdoor activities during peak mosquito feeding periods, the application of mosquito repellents, the wearing of protective clothing, the use of anti-mosquito coils, the use of screens to discourage entrance of mosquitoes into houses, and the use of bed nets to prevent mosquito bites during sleep (Service 1993; WHO 1996). There are other strategies that could be applied to control mosquitoes, including immature and adult stages. To date, attempts have been made to investigate plant extracts having high-potential activities, such as repellent, oviposition deterrent, larvicide, and adulticide.

Table 12.1 Communicable diseases transmitted by various insects and arthropods

Diseases	Pathogens	Vectors
Malaria	*Plasmodium* protozoan (*P. vivax*, *P. falciparum*, *P. malariae*, *P. ovale*)	Several species of *Anopheles* mosquitoes
Dengue/dengue hemorrhagic fever	Dengue viruses (DEN-1, DEN-2, DEN-3, DEN-4)	*Ae. aegypti*, *Ae. albopictus*
Chikungunya	Chikungunya virus (CHIK-V)	*Ae. aegypti*, *Ae. albopictus*
Japanese encephalitis	JE virus (JEV)	*Cx. tritaeniorhynchus*, *Cx. gelidus*, *Cx. fuscocephala*
West Nile fever	West Nile virus (WNV)	Several species of *Culex* and *Aedes* mosquitoes
Lymphatic filariasis	Filarial worms (*Brugia malayi*, *Wuchereria bancrofti*)	Several species of *Mansonia* mosquitoes, *Ae. niveus*, *Cx. quinquefasciatus*
Leishmaniasis, sandfly fever	Several species of *Leishmania* parasites	Several species of *Phlebotomus* and *Lutzomyia* sand flies
River blindness	*Onchocerca volvulus*	Several species of *Simulium* blackflies
African trypanosomiasis (sleeping sickness)	*Trypanosoma* parasites (*T. brucei gambiense*, *T. brucei rhodesiense*)	Several species of *Glossina* tsetse flies
Chagas disease	*Trypanosoma cruzi*	Triatomine bugs
Diarrhea	Several species of diarrheal bacteria	Cockroaches, houseflies
Scrub typhus	*Orientia tsutsugamushi*	Several species of chiggers (larvae of mites)
Murine typhus (endemic typhus)	*Rickettsia typhi*	Several species of fleas
Epidemic typhus	*Rickettsia prowazekii*	*Amblyomma* ticks, body lice
Spotted fever	Several species of rickettsia bacteria	Several species of ticks and fleas
Lyme disease	*Borrelia* bacteria (*B. burgdorferi sensu stricto*, *B. afzelii*, *B. garinii*)	Deer ticks

2.1.1 Repellents

The use of repellents is a potentially practical and economical means of preventing of mosquito bites. It is important not only for the local people in the risk areas, especially the tropical countries, but also for the travelers who are vulnerable to the diseases from mosquito vectors when relaxing away from home. Although the most common mosquito repellents currently available in market containing deet (*N,N*-diethyl-3-methylbenzamide) have shown the excellent repellent efficacy against mosquitoes and other biting insects (Yap 1986; Coleman et al. 1993; Walker et al. 1996), toxicity problems after applications of deet varying from mild to severe reactions have been reported (Zadikoff 1979; Robbins and Cherniack 1986; Edwards and Johnson 1987; Qiu et al. 1998). To avoid these adverse effects, the attempts to investigate repellents derived from plant extracts in order to replace deet have been then conducted by many researchers. The development and use of locally available plant products showing repellent activity thus avails an alternative strategy for prevention of vector-borne diseases.

There are numerous publications containing information on plant-based repellents and protectants, and it will not be possible to list and review them all here. Three publications offer an enormous amount of information on research and development of personal protectants and repellents (Sukumar et al. 1991; Debboun et al. 2007; Moore et al. 2007). These publications have an impressive list of published works on this subject. A considerable amount of research has been carried out on the neem tree (*Azadirachta indica*) and its various products for the control of medically important insects, including reports on the usefulness of neem products as repellents, protectants, and deterrents (Mulla and Su 1999). The recent volume published on neem tree by the

Table 12.2 Reported bioactivity of plant extracts against various insects and arthropods transmitted diseases (alphabetically arranged by plant species)

Plant species (family)	Extracts used	Test insects/arthropods/organisms	Bioactivity	References
Achillea millefolium (Asteraceae)	Ethyl acetate extract	*Aedes* species	Repellent	Jaenson et al. (2006)
Ageratum conyzoides (Asteraceae)	Essential oil	*Ae. aegypti*	Repellent	Trongtokit et al. (2005)
Aglaia odorata (Meliaceae)	Essential oil	*Ae. aegypti, Ae. albopictus, An. dirus, Cx. quinquefasciatus*	Repellent, oviposition deterrent	Tawatsin et al. (2006a)
Allium sativum (Amaryllidaceae)	Essential oil	*Ae. aegypti*	Repellent	Trongtokit et al. (2005)
	Essential oil	*R. microplus*	Acaricide	Martinez-Velazquez et al. (2011)
Alpinia galanga (Zingiberaceae)	Essential oil	*Ae. aegypti, Ae. albopictus, An. dirus, Cx. quinquefasciatus*	Repellent, oviposition deterrent	Tawatsin et al. (2006a)
Ambrosia confertiflora (Asteraceae)	Ether extract	*Ae. aegypti*	Larvicide	Rodriguez et al. (2013)
Anethum graveolens (Apiaceae)	Essential oil	*Ae. aegypti*	Repellent	Choochote et al. (2007)
Apium graveolens (Apiaceae)	Ethanol extract	*Ae. aegypti*	Larvicide, adulticide, repellent	Choochote et al. (2004)
	Hexane extract	*Ae. aegypti*	Repellent	Tuetun et al. (2004)
	Hexane extract	*Aedes* species, *Anopheles* species, *Armigeres* species, *Culex* species, *Mansonia* species	Repellent	Tuetun et al. (2005)
	Essential oil	*Ae. aegypti*	Repellent	Trongtokit et al. (2005)
	Essential oil	*Ae. aegypti*	Adulticide	Chaiyasit et al. (2006)
	Essential oil	*Ae. aegypti, An. dirus*	Larvicide	Pitasawat et al. (2007)
	Hexane extract	*Aedes* species, *Anopheles* species, *Armigeres* species, *Culex* species, *Mansonia* species	Repellent	Tuetun et al. (2008, 2009)
Artemisia absinthium (Asteraceae)	Ethyl acetate extract, hexane	*I. ricinus*	Repellent	Jaenson et al. (2005)
Asarum heterotropoides (Aristolochiaceae)	Essential oil	*Ae. aegypti, Cx. pipiens, Oc. togoi*	Larvicide	Perumalsamy et al. (2009)
Azadirachta indica (Meliaceae)	Essential oil	*Anopheles* species, *Culex* species	Repellent	Das et al. (1999)
	Essential oil	*An. darlingi*	Repellent	Moore et al. (2002)
	Essential oil	*C. impunctatus*	Repellent	Blackwell et al. (2004)
	Essential oil	*I. ricinus*	Repellent	Garboui et al. (2006)
	Essential oil	*P. orientalis, P. bergeroti*	Repellent	Kebede et al. (2010)
	Essential oil	*Cx. quinquefasciatus*	Larvicide	Anjali et al. (2012)
	Ethanol extract	*M. domestica*	Adulticide	Mansour et al. (2012)
Carapa guianensis (Meliaceae)	Essential oil	*Ae. aegypti*	Larvicide	Silva et al. (2006)
Carum carvi (Apiaceae)	Essential oil	*Ae. aegypti*	Adulticide	Chaiyasit et al. (2006)
	Essential oil	*Ae. aegypti, An. dirus*	Larvicide	Pitasawat et al. (2007)
Catunaregam spathulifolia (Rubiaceae)	Water extract	*Haemadipsa* species	Repellent	Vongsombath et al. (2011)

(continued)

Table 12.2 (continued)

Plant species (family)	Extracts used	Test insects/arthropods/organisms	Bioactivity	References
Cedrus deodara (Pinaceae)	Essential oil	*An. stephensi*	Adulticide	Singh et al. (1984)
		M. domestica	Adulticide	Singh and Agrawal (1988)
Chamaecyparis nootkatensis (Cupressaceae)	Essential oil	*I. scapularis*	Repellent	Dietrich et al. (2006)
Chamaecyparis obtusa (Cupressaceae)	Methanol extract	*Ae. aegypti, Ochlerotatus togoi, Cx. pipiens pallens*	Larvicide	Jang et al. (2005)
Cichorium intybus (Asteraceae)	Ethanol extract	*M. domestica*	Adulticide	Mansour et al. (2012)
Cinnamomum camphora (Lauraceae)	Essential oil	*Ae. aegypti*	Repellent	Yang et al. (2004)
Cinnamomum cassia (Lauraceae)	Methanol extract	*Ae. aegypti*	Repellent	Yang et al. (2004)
	Essential oil	*Ae. aegypti*	Repellent	Chang et al. (2006)
	Essential oil	*Ae. albopictus, Ae. aegypti, Cx. pipiens*	Larvicide	Zhu et al. (2008)
Cinnamomum zeylanicum (Lauraceae)	Essential oil	*P. americana*	Repellent	Ahmad et al. (1995)
Citrullus colocynthis (Cucurbitaceae)	Petroleum ether extract	*Ae. aegypti, Cx. quinquefasciatus*	Larvicide	Rahuman and Venkatesan (2008)
Citrus aurantifolia (Rutaceae)	Essential oil	*Ae. albopictus*	Repellent	Das et al. (2003)
	Ethanol extract	*M. domestica*	Adulticide	Mansour et al. (2012)
Citrus bergamia (Rutaceae)	Essential oil	*Ae. aegypti, Cx. pipiens pallens*	Larvicide	Lee (2006)
Citrus hystrix (Rutaceae)	Essential oil	*Ae. aegypti, An. dirus, Cx. quinquefasciatus*	Repellent	Tawatsin et al. (2001)
	Essential oil	*Ae. aegypti*	Repellent	Trongtokit et al. (2005)
	Essential oil	*P. americana, B. germanica, N. rhombifolia*	Repellent	
	Essential oil	*Ae. aegypti*	Larvicide	Sutthanont et al. (2010)
Citrus reticulata (Rutaceae)	Essential oil	*Ae. aegypti*	Larvicide	Sutthanont et al. (2010)
Clausena excavata (Rutaceae)	Essential oil	*Ae. aegypti, Ae. albopictus*	Repellent	Cheng et al. (2009)
Commiphora myrrha (Burseraceae)	Essential oil	*Ae. aegypti, Cx. pipiens pallens*	Larvicide	Lee (2006)
Conyza aegyptiaca (Compositae)	Ethanol extract	*M. domestica*	Adulticide	Mansour et al. (2012)
Cordia curassavica (Boraginaceae)	Water extract	*Ae. aegypti*	Larvicide	Mohammed and Chadee (2007)
Croton pseudopulchellus (Euphorbiaceae)	Essential oil	*An. gambiae*	Repellent	Odalo et al. (2005)
Croton roxburghii (Euphorbiaceae)	Essential oil	*Aedes* species, *Armigeres* species, *Culex* species	Repellent	Vongsombath et al. (2012)
Curcuma aeruginosa (Zingiberaceae)	Ethanol extract	*Ae. togoi, Ar. subalbatus, Cx. quinquefasciatus, Cx. tritaeniorhynchus*	Repellent	Pitasawat et al. (2003)

(continued)

Table 12.2 (continued)

Plant species (family)	Extracts used	Test insects/arthropods/organisms	Bioactivity	References
Curcuma aromatica (Zingiberaceae)	Essential oil	*Anopheles* species, *Culex* species	Repellent	Das et al. (1999)
	Ethanol extract	*Ae. togoi, Ar. subalbatus,*	Repellent	Pitasawat et al. (2003)
	Essential oil, hexane extract	*Cx. quinquefasciatus, Cx. tritaeniorhynchus, Ae. aegypti*	Repellent, larvicide, adulticide	Choochote et al. (2005)
Curcuma longa (Zingiberaceae)	Essential oil	*P. americana*	Repellent	Ahmad et al. (1995)
	Essential oil	*Ae. aegypti, Ae. albopictus, An. dirus,*	Repellent	Tawatsin et al. (2001)
		Cx. quinquefasciatus, Simulium species, *Haemadipsa* species	Oviposition deterrent	Tawatsin et al. (2006a, b)
	Essential oil	*Ae. albopictus, Ae. aegypti, Cx. pipiens*	Larvicide	Zhu et al. (2008)
Curcuma xanthorrhiza (Zingiberaceae)	Ethanol extract	*Ae. togoi, Ar. subalbatus, Cx. quinquefasciatus, Cx. tritaeniorhynchus*	Repellent	Pitasawat et al. (2003)
Curcuma zedoaria (Zingiberaceae)	Essential oil	*Ae. aegypti*	Adulticide	Chaiyasit et al. (2006)
	Essential oil	*Ae. aegypti, An. dirus*	Larvicide	Pitasawat et al. (2007) and Champakaew et al. (2007)
Cymbopogon citratus (Poaceae)	Essential oil	*P. americana*	Repellent	Ahmad et al. (1995)
	Essential oil	*Ae. aegypti*	Larvicide	Cavalcanti et al. (2004)
Cymbopogon nardus (Poaceae)	Essential oil	*Ae. aegypti*	Repellent	Trongtokit et al. (2005)
Cymbopogon winterianus (Poaceae)	Essential oil	*Ae. aegypti, An. dirus, Cx. quinquefasciatus*	Repellent	Tawatsin et al. (2001)
Cyperus rotundus (Cyperaceae)	Essential oil	*B. germanica*	Adulticide	Chang et al. (2012)
Daucus carota (Apiaceae)	Ethanol, hexane extract	*Cx. annulirostris*	Larvicide	Shaalan et al. (2006)
Eleutherococcus trifoliatus (Araliaceae)	Essential oil	*Ae. aegypti, Ae. albopictus, An. dirus, Cx. quinquefasciatus*	Repellent, oviposition deterrent	Tawatsin et al. (2006a)
Endostemon tereticaulis (Lamiaceae)	Essential oil	*An. gambiae*	Repellent	Odalo et al. (2005)
Eucalyptus citriodora or *Corymbia citriodora* (Myrtaceae)	PMD	*An. darlingi*	Repellent	Moore et al. (2002)
	PMD	*Haemadipsa* species	Repellent	Kirton (2005)
	PMD	*I. ricinus*	Repellent	Garboui et al. (2006)
	PMD	*L. carteri*	Repellent	Carroll and Loye (2006a)
	PMD	*Ae. aegypti, Ae. vexans, Oc. melanimon, Oc. increpitus*	Repellent	Carroll and Loye (2006b)
	Essential oil	*Ae. albopictus, Ae. aegypti, Cx. pipiens*	Larvicide	Zhu et al. (2008)
	Leave (direct burning and thermal expulsion)	*An. arabiensis, An. pharoensis*	Repellent	Dugassa et al. (2009)
Eucalyptus grandis (Myrtaceae)	Essential oil	*Ae. aegypti*	Larvicide	Lucia et al. (2007)

(continued)

Table 12.2 (continued)

Plant species (family)	Extracts used	Test insects/arthropods/organisms	Bioactivity	References
Eucalyptus globulus (Myrtaceae)	Essential oil	*L. imphalum*	Repellent	Eamsobhana et al. (2009)
	Essential oil	*M. domestica*	Repellent, larvicide, pupicide	Kumar et al. (2011)
	Ethanol extract	*M. domestica*	Adulticide	Mansour et al. (2012)
Eugenia caryophyllata (Myrtaceae)	Essential oil	*S. calcitrans*	Repellent	Hieu et al. (2010)
Foeniculum vulgare (Apiaceae)	Essential oil	*Ae. aegypti*, *An. dirus*	Larvicide	Pitasawat et al. (2007)
Hedychium coronarium (Zingiberaceae)	Essential oil	*Ae. aegypti*, *Ae. albopictus*, *An. dirus*, *Cx. quinquefasciatus*	Repellent, oviposition deterrent	Tawatsin et al. (2006a)
Houttuynia cordata (Saururaceae)	Essential oil	*Ae. aegypti*, *Ae. albopictus*, *An. dirus*, *Cx. quinquefasciatus*	Repellent, oviposition deterrent	Tawatsin et al. (2006a)
Hyptis suaveolens (Lamiaceae)	Ethyl acetate extract	*Ae. aegypti*, *Aedes* species	Repellent	Jaenson et al. (2006)
	Essential oil	*Aedes* species, *Armigeres* species, *Culex* species	Repellent	Vongsombath et al. (2012)
Illicium verum (Schisandraceae)	Ethanol extract	*Cx. quinquefasciatus*	Larvicide	Pitasawat et al. (1998)
	Essential oil	*Ae. aegypti*	Adulticide	Chaiyasit et al. (2006)
Kaempferia galanga (Zingiberaceae)	Ethanol extract	*Cx. quinquefasciatus*	Larvicide	Pitasawat et al. (1998)
	Hexane extract	*Ae. aegypti*, *Ar. subalbatus*, *Ma. uniformis*,	Larvicide, adulticide	Choochote et al. (1999)
		Anopheles species, *Culex* species	Repellent	
	Essential oil	*Ae. aegypti*	Repellent	Choochote et al. (2007)
	Essential oil	*Ae. aegypti*	Larvicidal	Sutthanont et al. (2010)
Khaya senegalensis (Meliaceae)	Acetone, ethanol, hexane, methanol extract	*Cx. annulirostris*	Larvicide	Shaalan et al. (2006)
Levisticum officinale (Apiaceae)	Essential oil	*S. calcitrans*	Repellent	Hieu et al. (2010)
Lippia graveolens (Verbenaceae)	Essential oil	*R. microplus*	Repellent	Martinez-Velazquez et al. (2011)
Litsea cubeba (Lauraceae)	Essential oil	*Ae. aegypti*, *Ae. albopictus*, *An. dirus*, *Cx. quinquefasciatus*	Repellent, oviposition deterrent	Tawatsin et al. (2006a)
	Essential oil	*Aedes* species, *Armigeres* species, *Culex* species		
Luetzelburgia auriculata (Fabaceae)	Ethanol extract	*Ae. aegypti*	Ovicide, larvicide, pupicide	Souza et al. (2011)
Manglietia garrettii (Magnoliaceae)	Essential oil	*Ae. aegypti*, *Ae. albopictus*, *An. dirus*, *Cx. quinquefasciatus*	Repellent, oviposition deterrent	Tawatsin et al. (2006a)
Melaleuca alternifolia (Myrtaceae)	Essential oil	*L. imphalum*	Repellent	Eamsobhana et al. (2009)

(continued)

Table 12.2 (continued)

Plant species (family)	Extracts used	Test insects/arthropods/organisms	Bioactivity	References
Melaleuca cajuputi (Myrtaceae)	Essential oil	*Ae. aegypti*, *Ae. albopictus*, *An. dirus*, *Cx. quinquefasciatus*	Repellent, oviposition deterrent	Tawatsin et al. (2006a)
Melaleuca ericifolia (Myrtaceae)	Essential oil	*Ae. vigilax*, *V. carmenti*, *M. vetustissima*, *C. ornatus*, *C. immaculatus*	Repellent	Greive et al. (2010)
Melia azedarach (Meliaceae)	Essential oil	*P. orientalis*, *P. bergeroti*	Repellent	Kebede et al. (2010)
Mentha piperita (Lamiaceae)	Essential oil	*M. domestica*	Repellent, larvicide, pupicide	Kumar et al. (2011)
Mentha spicata (Lamiaceae)	Essential oil	*Ae. aegypti*	Repellent	Trongtokit et al. (2005)
Momordica charantia (Cucurbitaceae)	Methanol extract	*Ae. aegypti*, *Cx. quinquefasciatus*	Larvicide	Rahuman and Venkatesan (2008)
Murraya paniculata (Rutaceae)	Essential oil	*Ae. aegypti*, *Ae. albopictus*, *An. dirus*, *Cx. quinquefasciatus*	Repellent, oviposition deterrent	Tawatsin et al. (2006a)
Myracrodruon urundeuva (Anacardiaceae)	Ethanol extract	*Ae. aegypti*	Ovicide, larvicide, pupicide	Souza et al. (2011)
Myrica gale (Myricaceae)	Essential oil,	*I. ricinus*	Repellent	Jaenson et al. (2005)
	Ethyl acetate extract	*Ae. aegypti*, *Aedes* species	Repellent	Jaenson et al. (2006)
Myristica fragrans (Myristicaceae)	Essential oil	*Ae. aegypti*, *Ae. albopictus*, *An. dirus*, *Cx. quinquefasciatus*	Repellent, oviposition deterrent	Tawatsin et al. (2006a)
Nardostachys chinensis (Valerianaceae)	Methanol extract	*Ae. aegypti*	Repellent	Yang et al. (2004)
Nepeta cataria (Lamiaceae)	Essential oil	*Ae. albopictus*, *Ae. aegypti*, *Cx. pipiens pallens*	Repellent	Zhu et al. (2006)
	Essential oil	*Ae. intrudens*, *S. decorum*	Repellent	Spero et al. (2008)
Ocimum americanum (Labiatae)	Essential oil	*Ae. aegypti*, *An. dirus*, *Cx. quinquefasciatus*	Repellent	Tawatsin et al. (2001)
	Essential oil	*Ae. aegypti*	Larvicide	Cavalcanti et al. (2004)
Ocimum basilicum (Labiatae)	Essential oil	*Ae. aegypti*, *An. minimus*, *Cx. quinquefasciatus*	Repellent	Phasomkusolsil and Soonwera (2010)
Ocimum fischeri (Labiatae)	Essential oil	*An. gambiae*	Repellent	Odalo et al. (2005)
Ocimum forskolei (Labiatae)	Essential oil	*An. gambiae*	Repellent	Odalo et al. (2005)
Ocimum gratissimum (Labiatae)	Essential oil	*Ae. aegypti*	Larvicide	Cavalcanti et al. (2004)
	Essential oil	*S. damnosum*	Repellent	Usip et al. (2006)
Ocimum sanctum (Labiatae)	Essential oil	*Ae. aegypti*, *Ae. albopictus*, *An. dirus*, *Cx. quinquefasciatus*	Repellent, oviposition deterrent	Tawatsin et al. (2006a)
Paeonia suffruticosa (Paeoniaceae)	Methanol extract	*Ae. aegypti*	Repellent	Yang et al. (2004)
Pandanus odorus (Pandanaceae)	Essential oil	*P. americana*	Repellent	Ahmad et al. (1995)
Pelargonium graveolens (Geraniaceae)	Essential oil	*Ae. aegypti*	Repellent	Trongtokit et al. (2005)
	Essential oil	*L. imphalum*	Repellent	Eamsobhana et al. (2009)

(continued)

Table 12.2 (continued)

Plant species (family)	Extracts used	Test insects/arthropods/organisms	Bioactivity	References
Pemphis acidula (Lythraceae)	Methanol, benzene, acetone extract	*Cx. quinquefasciatus, Ae. aegypti*	Larvicide, ovicide, repellent	
Pimenta racemosa (Myrtaceae)	Essential oil	*Ae. aegypti, Cx. pipiens pallens*	Larvicide	Lee (2006)
Pinus longifolia (Pinaceae)	Essential oil	*An. culicifacies* *Cx. quinquefasciatus*	Repellent	Ansari et al. (2005)
Piper aduncum (Piperaceae)	Essential oil	*Ae. albopictus*	Repellent	Misni et al. (2009)
Piper betle (Piperaceae)	Essential oil	*Ae. aegypti, Ae. albopictus, An. dirus, Cx. quinquefasciatus*	Repellent, oviposition deterrent	Tawatsin et al. (2006a)
Piper longum (Piperaceae)	Essential oil	*Ae. aegypti*	Adulticide	Chaiyasit et al. (2006)
	Ethanol extract	*Ae. aegypti*	Larvicide	Chaithong et al. (2006)
	Ethanol extract	*St. aegypti*	Adulticide	Choochote et al. (2006)
Piper nigrum (Piperaceae)	Essential oil	*Ae. aegypti, Ae. albopictus, An. dirus, Cx. quinquefasciatus*	Repellent, oviposition deterrent	Tawatsin et al. (2006a)
	Ethanol extract	*M. domestica*	Adulticide	Mansour et al. (2012)
Piper retrofractum (Piperaceae)	Water extract	*Ae. aegypti, Cx. quinquefasciatus*	Larvicide	Chansang et al. (2005)
Piptadenia moniliformis (Fabaceae)	Ethanol extract	*Ae. aegypti*	Ovicide, larvicide, pupicide	Souza et al. (2011)
Piper ribesoides (Piperaceae)	Ethanol extract	*Ae. aegypti*	Larvicide	Chaithong et al. (2006)
	Ethanol extract	*St. aegypti*	Adulticide	Choochote et al. (2006)
Piper sarmentosum (Piperaceae)	Ethanol extract	*Ae. aegypti*	Larvicide	Chaithong et al. (2006)
	Ethanol extract	*St. aegypti*	Adulticide	Choochote et al. (2006)
Plectranthus longipes (Lamiaceae)	Essential oil	*An. gambiae*	Repellent	Odalo et al. (2005)
Pogostemon cablin (Lamiaceae)	Essential oil	*Ae. aegypti*	Repellent	Trongtokit et al. (2005)
		S. calcitrans	Repellent	Hieu et al. (2010)
Psidium guajava (Myrtaceae)	Essential oil	*Ae. aegypti, Ae. albopictus, An. dirus, Cx. quinquefasciatus, Simulium* species, *Haemadipsa* species	Repellent, oviposition deterrent	Tawatsin et al. (2006a, b)
Punica granatum (Lythraceae)	Ethanol extract	*M. domestica*	Adulticide	Mansour et al. (2012)
Rosmarinus officinalis (Lamiaceae)	Essential oil	*R. microplus*	Repellent	Martinez-Velazquez et al. (2011)
Rhododendron tomentosum (Ericaceae)	Essential oil	*I. ricinus*	Repellent	Jaenson et al (2005)
	Ethyl acetate extract	*Ae. aegypti, Aedes* species	Repellent	Jaenson et al. (2006)

(continued)

Table 12.2 (continued)

Plant species (family)	Extracts used	Test insects/arthropods/organisms	Bioactivity	References
Ruta chalepensis (Rutaceae)	Ether extract, methanol extract	*Ae. aegypti*	Larvicide	Rodriguez et al. (2013)
Salix safsaf (Salicaceae)	Ethanol extract	*M. domestica*	Adulticide	Mansour et al. (2012)
Santalum album (Santalaceae)	Essential oil	*Ae. albopictus*, *Ae. aegypti*, *Cx. pipiens*	Larvicide	Zhu et al. (2008)
Santalum (Santalaceae)	Essential oil	*Cx. molestus*, *Ae. camptorhynchus*	Larvicide	Spafford et al. (2007)
Sapindus rarak (Sapindaceae)	Water extract	*Haemadipsa* species	Repellent	Vongsombath et al. (2011)
Schefflera leucantha (Araliaceae)	Essential oil	*Ae. aegypti*, *Ae. albopictus*, *An. dirus*, *Cx. quinquefasciatus*	Repellent, oviposition deterrent	Tawatsin et al. (2006a)
Sonchus oleraceus (Asteraceae)	Ethanol extract	*M. domestica*	Adulticide	Mansour et al. (2012)
Spilanthes acmella (Compositae)	Ethanol extract	*Cx. quinquefasciatus*	Larvicide	Pitasawat et al. (1998)
	Essential oil	*Ae. aegypti*	Repellent	Trongtokit et al. (2005)
Syzygium aromaticum (Myrtaceae)	Essential oil	*P. americana*	Repellent	Ahmad et al. (1995)
	Essential oil	*Ae. aegypti*	Repellent	Trongtokit et al. (2005)
	Essential oil	*L. imphalum*	Repellent	Eamsobhana et al. (2009)
	Essential oil	*Ae. aegypti*	Larvicide	Sutthanont et al. (2010)
Thymus vulgaris (Lamiaceae)	Essential oil	*Ae. albopictus*, *Ae. aegypti*, *Cx. pipiens pallens*	Repellent	Zhu et al. (2006)
	Essential oil	*S. calcitrans*	Repellent	Hieu et al. (2010)
	Ether extract	*Ae. aegypti*	Larvicide	Rodriguez et al. (2013)
Vernonia elaeagnifolia (Compositae)	Water extract	*Haemadipsa* species	Repellent	Vongsombath et al. (2011)
Vitex trifolia (Verbenaceae)	Essential oil	*Ae. aegypti*, *Ae. albopictus*, *An. dirus*, *Cx. quinquefasciatus*	Repellent, oviposition deterrent	Tawatsin et al. (2006a)
Zanthoxylum armatum (Rutaceae)	Essential oil	*Anopheles* species, *Culex* species	Repellent	Das et al. (1999)
	Seed oil	*Ae. aegypti*	Repellent	Kwon et al. (2011)
Zanthoxylum fagara (Rutaceae)	Ether extract	*Ae. aegypti*	Larvicide	Rodriguez et al. (2013)
Zanthoxylum limonella (Rutaceae)	Essential oil	*Ae. albopictus*	Repellent	Das et al. (2003)
	Essential oil	*Ae. aegypti*	Repellent	Trongtokit et al. (2005)
	Essential oil	*Ae. aegypti*, *An. dirus*	Larvicide	Pitasawat et al. (2007)
Zanthoxylum piperitum (Rutaceae)	Essential oil	*Ae. aegypti*	Repellent	Choochote et al. (2007) and Kamsuk et al. (2007)
Zea mays (Poaceae)	Ethanol extract	*M. domestica*	Adulticide	Mansour et al. (2012)
Zingiber cassumunar (Zingiberaceae)	Essential oil	*L. imphalum*	Repellent	Eamsobhana et al. (2009)
	Essential oil	*Ae. aegypti*, *An. minimus*, *Cx. quinquefasciatus*	Repellent	Phasomkusolsil and Soonwera (2010)

(continued)

Table 12.2 (continued)

Plant species (family)	Extracts used	Test insects/arthropods/organisms	Bioactivity	References
Zingiber officinale (Zingiberaceae)	Essential oil	*P. americana*	Repellent	Ahmad et al. (1995)
	Essential oil	*Ae. aegypti*, *Ae. albopictus*, *An. dirus*, *Cx. quinquefasciatus*, *Simulium* species *Haemadipsa* species	Repellent, oviposition deterrent	Tawatsin et al. (2006a, b)
Zingiber zerumbet (Zingiberaceae)	Essential oil	*Ae. aegypti*	Larvicide	Sutthanont et al. (2010)

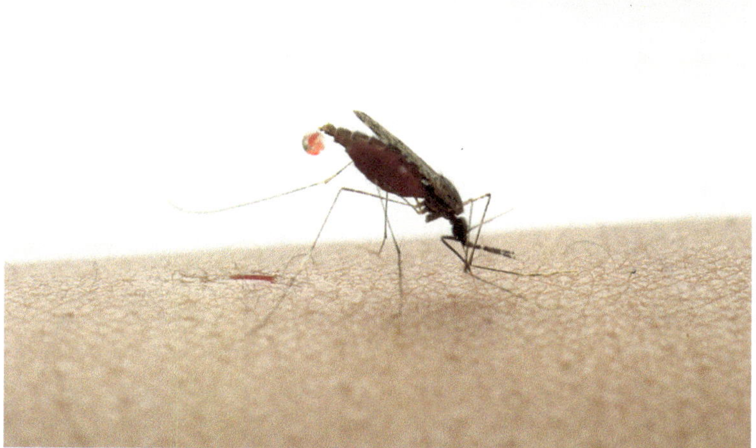

Fig. 12.1 *Anopheles dirus*

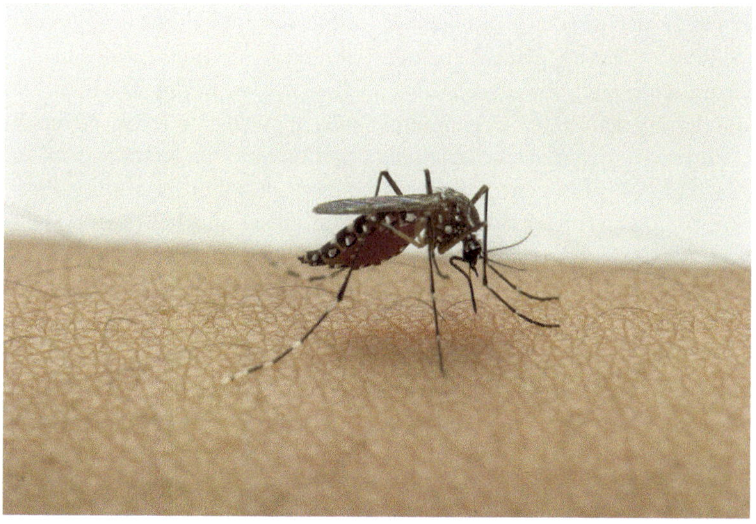

Fig. 12.2 *Aedes aegypti*

Neem Foundation (Schmuterer 2002) contains valuable information on the neem tree and the unique products derived from this tree, which are used in pest control and many industrial, medicinal, and agricultural enterprises. The neem tree (*A. indica*) has attracted a great deal of attention as a source of agricultural biopesticides. In India, neem tree parts have been used for thousands of years as insect repellents and protectants. In recent years, materials from this plant have been developed for the control of medical insects. The neem tree has about 35 bioactive compounds which show toxicity, causing sterility, decreased fecundity, as well as repellency and deterrence. Proper formulations can increase toxicity and repellency of neem and other plant products. Bioactivity is related to concentration of bioactive principles and the extent of release and absorption (Isman 1999; Mulla and Su 1999; Schmuterer 2002). Bioactive materials from neem tree have multiple modes of action, such as repellency, deterrence, sterility, growth regulation, and toxicity (Isman 1999; Mulla and Su 1999; Schmuterer 2002).

Essential oils are extracted from particular parts of plants by steam distillation, and mosquito repellents derived from plant extracts are mostly essential oils. Evidently, essential oils are known to evaporate at room temperature over other oils. As per reviews of literature on this aspects is concerned, several essential oil-bearing plants have been identified as source of potential mosquito repellents. These plants include pyrethrum (*Chrysanthemum cinerariaefolium*), sweet almond (*Prunus amygdalus*), geranium (*Pelargonium graveolens*), hemlock (*Conium maculatum*) (Macnay 1939; Rutledge and Gupta 1996), neem (*A. indica*) (Ansari and Razdan 1995; Sharma et al. 1995), turmeric (*Curcuma longa*) (Tawatsin et al. 2001), aromatic turmeric (*C. aromatica*) (Pitasawat et al. 2003), kaffir lime (*Citrus hystrix*) (Tawatsin et al. 2001), citronella grass (*Cymbopogon nardus*) (Dethier 1948; Tyaki et al. 1998), hairy basil (*Ocimum americanum*) (Chokechaijaroenporn et al. 1994; Palsson and Jaenson 1999; Tawatsin et al. 2001), eucalyptus (*Eucalyptus globulus*) (Trigg 1996a), clove (*Syzygium aromaticum*), makaen (*Zanthoxylum limonella*) (Trongtokit et al. 2004), thyme (*Thymus vulgaris*) (Choi et al. 2002; Park et al. 2005), and *Melaleuca ericifolia* (Greive et al. 2010).

Ocimum species (*Basilicum* in particular) contains a number of repellent compounds which also act as mosquito larvicides. Species of *Hyptis, Mentha,* and *Thymus* also contain insect-repellent principles, and they are used in various ways by the local communities. *Tagetes* species have exhibited larvicidal and insecticidal properties, and their essential oils act as repellents in some species and not others. *Artemisia* species have both essential oils which act as repellent and other principles acting as larvicides. Lemon eucalyptus extracts from *Eucalyptus citriodora* or *Corymbia citriodora* (from China) have been the subject of many tests. The essential oil extract was noted to have repellency to mosquitoes, slightly better than most *Eucalyptus* species. The compound *p*-menthane-3,8-diol (PMD) was identified as a by-product. This material was proven to be highly repellent, equaling the repellency of deet (commercial repellent). The Chinese name of this repellent is Quwenling, "effective repellent of mosquitoes."

Moore et al. (2007) tabulate the protection time for essential oils in extracts of 37 plants. The protection time has been found increased corresponding to concentration of essential oils. Most of the protection times for 50 % concentration were from 0 to 80 min. The longest protection time were 88 min for *Z. limonella* and 60 min for each of *Apium graveolens*, *C. nardus*, and *Pogostemon cablin*. The rest of the plant essential oils showed low levels of repellency. The same authors provide a comparison in protection time among 8 commercial plant-based products with Skinsensation (7 % deet). Six commercial products were inferior to Skinsensation, while 2 commercial plant products (Bite Blocker and Repel) proved superior to deet. Bite Blocker contains oil of coconut, geranium, and soybean (Consep Inc., Bend Oregon), and Repel contains oil of lemon eucalyptus and PMD (Wisconsin Pharmacal Co, Jackson WI). As mentioned earlier, appropriate formulations will greatly enhance the effectiveness of plant essential oils as insect repellents. Gerberg and Novak (2007) list 50 plants and the names of repellent products derived from them. Apparently, there are other plants and products used locally that have not been commercialized.

The repellent chemicals found in different parts of plants have been identified in some plant extracts. These constituents not only act as repellents but also as feeding deterrents, toxicants, growth regulators, etc. The major groups of chemical substances identified are categorized as alkaloids, phenols, terpenoids, and others. Different plant species and their parts yield different groups of chemicals and in varying quantities. Some plants may have one or a few phytochemicals possessing repellent properties, while others may have quite a few principles. The citronella group of grasses originating in India has widespread distribution. These plants have varying amounts of repellent chemicals (mostly terpenoids), but the most abundant repellent chemicals contained are citronella, citronellol, and geraniol. The grass *C. nardus* or citronella is the most common species used in commercial blending of repellent products. Several other species of *Cymbopogon* are used in commercially available plant products.

Citronella oil has long been known and widely used in India, Sri Lanka, and Java; unfortunately, its early history is unrecorded (Weiss 1997). However, citronella oil extracted from *C. nardus* (formerly *Andropogon nardus*) has been broadly recognized as a mosquito repellent since 1882 (Dethier 1948). Citronella oil, which is available commercially, usually comes from *C. nardus* and *C. winterianus*; they are usually defined as Ceylon and Java types, respectively (Weiss 1997). *C. nardus* seems to have been studied as a mosquito repellent more than *C. winterianus*. In India, Osmani et al. (1972) demonstrated that citronella oil extracted from *C. nardus* was as effective as dimethyl phthalate against the mosquito *Culex quinquefasciatus* (formerly *Cx. pipiens fatigans*) in the field. The oil provided almost complete protection against *Anopheles culicifacies* and *Cx. quinquefasciatus* under field conditions (Ansari and Razdan 1995), and it provided complete protection against *Ae. aegypti*, *An. stephensi*, and *Cx. quinquefasciatus* for as long as 8–10 h under laboratory and field conditions (Tyaki et al. 1998).

In Thailand, all citronella oils formulated as topical mosquito repellents have been extracted from *C. nardus*. Jaruwichitratana et al. (1988) showed that 14 % citronella cream repelled *Culex* mosquitoes for 1 h under field conditions, whereas Wasuwat et al. (1990) found that 14 % citronella cream provided repellency against *Ae. aegypti* for two hours under laboratory conditions. Sree-iam (1991) showed that 15–30 % citronella oil was likely to be the effective concentration range to be formulated as a topical repellent in various forms (i.e., solution, gel, and emulsion) with protection times of 2.0–3.7 h against *Ae. aegypti* under laboratory conditions. On the other hand, Suwonkerd and Tantrarongroj (1994) revealed that repellencies of 1.25, 2.5, and 5 % citronella cream against *An. minimus* under laboratory conditions were 2 h, whereas that of 10 % formulation provided at least 4 h. It is important to note that these studies were conducted on human skin-based methods (Fig. 12.3).

The plant essential oil possesses enormous potential for its bioactivities against various arthropods including storage pests (Singh et al. 1989). The essential oils of turmeric, especially obtained from *C. longa*, were widely studied for repellent/insecticidal efficacies against many arthropods, such as stored-grain pests (Su et al. 1982; Jilani and Su 1983; Jilani et al. 1988; Jilani and Saxena 1990), mites (Gulati and Mathur 1995), and cockroaches (Ahmad et al. 1995). However, very few evaluations of this plant have been conducted against mosquitoes (Tawatsin et al. 2001). Ross (1999) reported that root essential oil of turmeric extracted by petroleum ether was active against *Ae. aegypti*, but the effective concentration and protection time were not reported. Tawatsin et al. (2001) demonstrated that the repellent containing turmeric volatile oil with addition of 5 % vanillin provided repellency against *Ae. aegypti* for about 4 h and against *An. dirus* and *Cx. quinquefasciatus* for 8 h under laboratory conditions.

For basil, a study in Tanzania showed that juices squeezed from the green leaves and inflorescences of *O. americanum* and *O. suave* rubbed

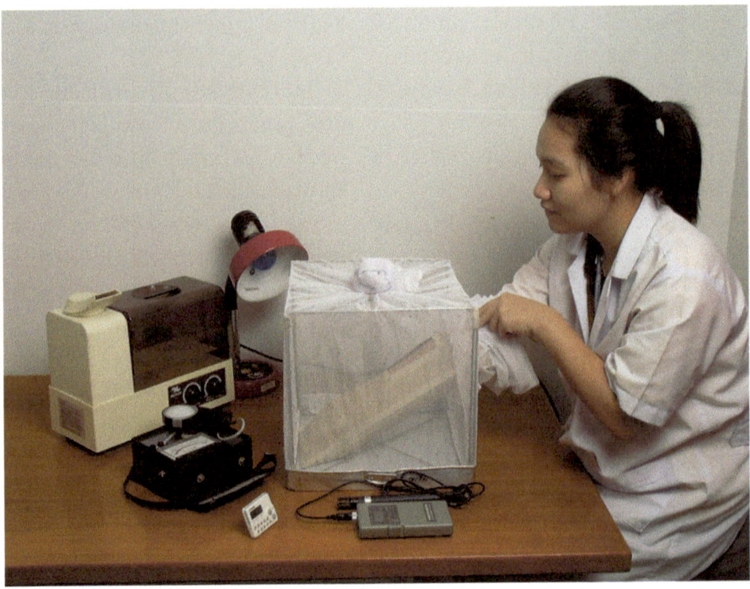

Fig. 12.3 Repellent testing against mosquitoes in laboratory

on a leg of a volunteer could reduce biting of *An. gambiae* by about 50 % (White 1973). *Ocimum* volatile oils obtained from *O. gratissimum*, *O. tenuiflorum*, *O. basilicum*, and *O. americanum* had repellency of 2.25, 1.75, 1.25, and 0.25 h, respectively, against *Ae. aegypti*, using laboratory mouse skin (Chokechaijaroenporn et al. 1994). On the other hand, in a study in Guinea Bissau, West Africa, fresh *O. canum* (syn. *O. americanum*) could reduce biting by anopheline mosquitoes (mostly *An. gambiae* and *An. pharoensis*) by about 63.6 % under field conditions between 2,000 and 2,200 h (Palsson and Jaenson 1999).

Trongtokit et al. (2005) carried out an intensive investigation on mosquito repellent in 38 essential oils obtained by steam distillation under laboratory conditions using human arm inserted into a cage and found that the oils of *S. aromaticum* (clove) and *Z. limonella* (makaen) provided the longest period of complete repellency for 2–4 h. Subsequently, the gel products containing 20 % clove oil or 10 % clove plus 10 % makaen oil were formulated, and they exhibited a high degree of repellencies for 5 h under both laboratory and field conditions (Trongtokit et al. 2004).

Attempts to increase the efficacy of mosquito repellents by formulation with various materials (e.g., perfume fixatives, vanillin, polymers) have been carried out by many researchers. Khan et al. (1975a) combined seven aromatic nitrogen compounds (commercial perfume fixatives) with four mosquito repellents (deet, ethyl hexanediol, dimethyl phthalate, and Indalone®) to investigate the effects on prolonged repellency; four compounds (Tibetene®, Ambrette, Givambrol, and musk xylol) could significantly increase repellency of deet against *Ae. aegypti*, whereas all compounds showed no effect with ethyl hexanediol, dimethyl phthalate, and Indalone®. Khan et al. (1975b) also studied the effect of the addition of vanillin (4-hydroxy-3-methoxybenzaldehyde) to seven mosquito repellents (deet, ethyl hexanediol, dimethyl phthalate, Indalone®, triethylene glycol monohexyl ether, triethylene glycol monoheptyl ether, and triethylene glycol ethylhexyl ether) and found that vanillin could significantly prolong repellency against *Ae. aegypti* of all test repellents. It was suggested that vanillin was more potent than were the previously studied perfume fixatives, as they were more specific than vanillin and were effective only when used with deet (Khan et al. 1975b). Vanillin prolonged repellency of repellent by reducing its evaporation rate (Khan et al. 1975b).

Tawatsin et al. (2001) tested volatile oils from plants against three mosquito species determining medium protection time. Essential oils from turmeric (*C. longa*), citronella (*C. winterianus*), and hairy basil (*O. americanum*) formulated with vanillin as additive gave essentially the same protection time (6–8 h) as deet for *Ae. aegypti*, *An. dirus*, and *Cx. quinquefasciatus*. In another field study, the extracts of citronella oil (10–13 %), eucalyptus oil (15 %), and tea tree oil (5 %) provided protection (93–100 %) for 4 h from night-biting mosquitoes (*Culex* species and *Mansonia* species) (Thavara et al. 2002). Again, the extent of protection and duration were similar to the standard chemical repellent deet.

In a comprehensive study, Tawatsin et al. (2006a) extracted and formulated essential oils from 18 plant species and compared them for protection time with deet and IR3535 (ethyl butylacetylaminopropionate) against 4 species of mosquitoes (*Ae. aegypti*, *Ae. albopictus*, *An. dirus*, and *Cx. quinquefasciatus*). Chemical composition of the oils for each plant species was also determined. The oil yield from the 18 plants ranged from 0.04 to 0.80 %. The oils were extracted from leaves, flowers, or rhizomes, depending on the plant. The plant-based products afforded low to medium protection time (0.3–2.8 h) against *Ae. aegypti* as compared to deet (7.5 h) and IR 3535 (6.7 h). It seems that *Ae. aegypti* was not sensitive to these plant extracts. On the other hand, all plant extracts with the exception of *Litsea cubeba*, *Aglaia odorata*, *Myristica fragrans*, *Psidium guajava*, *Murraya paniculata*, and *Zingiber officinale* provided at least 7 h protection against *Ae. albopictus*, equaling that of deet or IR3535. All 18 plant extracts gave high levels of protection time (7–8 h) against *An. dirus*, equaling that of deet or IR3535. Only four plant species yielded around 6 h protection time against *Cx. quinquefasciatus*, and the remaining four species yielded protection time of 7–8 h, comparable to the time afforded by the two synthetic chemicals. The plants giving 7.8–8 h protection against this mosquito were *Piper nigrum* and *C. longa*. It is possible that formulation manipulation might increase activity and duration of plants giving less protection using unformulated material. But this would not be necessary, as the number of active plants currently known is adequate. A field study with four plant-based repellents was carried out against night-biting (*Culex* and *Mansonia*) and day-biting (*Ae. aegypti* and *Ae. albopictus*) mosquitoes (Tawatsin et al. 2006b). The four plant oils, fingerroot, *Boesenbergia rotunda* (10 %); guava, *P. guajava* (10 %); turmeric, *C. longa* (10 %); and a commercial product Repel Care 9.5 % (turmeric oil 5 % and *E. citriodora* oil 4.5 %), were compared with deet. All repellents provided 100 % protection for 9 h. The same products tested on human volunteers against day-biting mosquitoes gave almost 100 % protection for 5 h and provided 76–93 % protection for 8 h. In these tests, the efficacies of the plant extracts were equal to that of deet against both day- and night-biting mosquitoes.

Celery (*A. graveolens*) is an outstanding plant which possesses various kinds of bioactivities against mosquitoes, including repellent, larvicidal, and adulticidal effects (Choochote et al. 2004; Tuetun et al. 2004, 2005, 2008, 2009; Trongtokit et al. 2005; Chaiyasit et al. 2006; Pitasawat et al. 2007). However, the repellent activity of celery seed extract seems to be an extraordinary effect over other activities. The hexane extract of *A. graveolens* seed exhibited strong repellent activity against *Ae. aegypti* in laboratory and a broad range of mosquito species belonging to various genera, such as *Ae. gardnerii*, *Ae. lineatopennis*, *Armigeres subalbatus*, *Cx. tritaeniorhynchus*, *Cx. vishnui*, *Cx. quinquefasciatus*, and *Mansonia uniformis* (Tuetun et al. 2004, 2005). Subsequently, the best formulation of the hexane extract of *A. graveolens* seed called G10 was developed and evaluated for repellent activity under laboratory and field conditions. It is noteworthy that G10 showed excellent repellency against *Ae. aegypti* in laboratory (Tuetun et al. 2008) and against a wide range of mosquito species belonging to various genera, including *Aedes*, *Anopheles*, *Armigeres*, *Culex*, and *Mansonia* in the field (Tuetun et al. 2009). It is expected that G10 will be commercial available in Thailand in the near future.

In Australia, the essential oil extracted from *M. ericifolia* was found to be an effective repellent against various bloodsucking insects, including mosquitoes and biting midges (Greive et al. 2010). It was revealed that the three formulations (alcohol-based spray, emulsion, and gel) of repellents containing essential oil of *M. ericifolia* (5 %) as active ingredient were highly effective against *Ae. vigilax* and *Verrallina carmenti* mosquitoes in the field and they also were of comparable repellency and not significantly different to a commercial repellent containing 69.75 g/L deet (Greive et al. 2010).

2.1.2 Mosquito Coils

The most common devices sold for personal protection are mosquito coils and they are used in large quantities on a worldwide basis. It has been estimated that the number of mosquito coils sold worldwide reaches to 29 billion, where 95 % are used in Asia. Mosquito coils are used primarily by low-income communities, while vaporizing mats and aerosols are used by the middle- and high-income sectors. Mulla et al. (2001) noted that in mosquito-infested areas of Thailand, the residents spent about US$12.50–25 per residence per annum for protection from biting mosquitoes. Similar levels of expenditures were reported for some communities in India (Snehalatha et al. 2003). In most of the coils, the active ingredients consist of synthetic pyrethroids. The use of plant materials in mosquito coils is then attracting greater attention. In actual practice, many people in low-income communities burn plant parts to achieve protection from biting insects. This practice serves the same purpose as a manufactured mosquito coil sold on the market. There have been numerous studies on the efficacy of mosquito coils based on the active ingredients consisting of synthetic pyrethroids. In contrast, there are very few studies published on the efficacy of mosquito coils containing plant materials. In one study, Tawatsin et al. (2002) incorporated plant materials into the coil and tested them in the field against night-biting mosquitoes. In total nine plant species were evaluated and each plant part was galanga (*Alpinia galanga*) rhizomes, fingerroot (*B. rotunda* or *B. pandurata*) rhizomes, turmeric (*C. longa*) rhizomes, cardamom (*Elettaria cardamomum*) rhizomes, neem (*A. indica*) leaves, Siamese cassia (*Cassia siamea*) leaves, citronella grass (*C. nardus*) leaves, eucalyptus (*E. citriodora*) leaves, and Siam weed (*Eupatorium odoratum*) leaves. The plant material constituted 25 % of the coil and a blank coil was also run along with the test coils. The mosquitoes caught in this study included 12 species belonging to 5 genera: *Aedes*, *Anopheles*, *Armigeres*, *Culex*, and *Mansonia*. All nine plant materials tested individually provided protection of 50–71 % in 3 h of testing. Citronella grass and eucalyptus leaves are highly potential active ingredients to be used in mosquito coil formulations, while Siamese cassia leaves, galanga rhizomes, and neem leaves also express fairly potentiality. According to the results obtained from this study, it is not recommended to use fingerroot, turmeric, cardamom, and Siam weed for mosquito coil formulations since their repellency was less than 60 %. By comparison, coils containing synthetic pyrethroids (0.3 % D-allethrin) showed repellency of 84–86 %. It seems that further studies are needed to increase the efficacy of plant-based mosquito coils through formulation technology.

2.1.3 Oviposition Deterrent

The studies on oviposition deterrent activity of chemical compounds and insect repellents have been carried out against mosquito vectors, whereas those of plant extracts are scarce. Xue et al. (2001, 2003) reported oviposition deterrent effects of deet and several repellent compounds, such as AI3-37220, AI3-35765, AI3-54995, AI3-55051, etc., against *Ae. albopictus* under laboratory as well as field conditions. Xue et al. (2006) also pointed out the oviposition deterrent effectiveness (76–100 % repellency) against *Ae. albopictus* of 21 commercial insect-repellent products (at 0.1 % concentration), including 12 botanical, 6 deet based, and 3 synthetic organics. As for the plant extracts, Mehra and Hiradhar (2002) revealed that the crude acetone extract of *Cuscuta hyalina* was effective oviposition deterrent against *Cx. quinquefasciatus* at the concentration of 80 ppm. The oviposition deterrent effects of 18 essential oils and the two chemical repellents deet and IR3535 (at 0.01 % concentration) against *Ae. aegypti* were carried out by Tawatsin et al. (2006a) in Thailand. All essential oils exhibited oviposition deterrent activity against the mosquitoes with various degrees of repellency, ranging from 16.6 to 94.7 %, whereas deet and IR3535 provided no repellency. Relatively high oviposition deterrencies were obtained from *C. longa* (94.7 %), *S. leucantha* (91.6 %), *Z. officinale* (90.1 %), *V. trifolia* (89.1 %), *M. cajuputi* (87.9 %), *H. coronarium* (87.5 %), *P. guajava* (87.1 %), *M. garrettii* (86.1 %), and *H. cordata* (85 %). Further studies are needed to formulate the active essential oils intended for treatment of water-storage containers, the most common breeding sites of *Ae. aegypti* in Thailand. As a matter of fact the oviposition avoidance of insecticide-treated water-storage containers by gravid female mosquitoes would reduce levels of larval populations (Moore 1977). The active essential oils that possess oviposition deterrent activity and probably include larvicidal effects against *Ae. aegypti* would be the most interesting plant-based products for control of mosquitoes.

2.1.4 Larvicides

There are numerous studies on larvicidal activity of various plant species from many countries around the globe against several mosquito vectors, such as *Ae. aegypti*, *Ae. albopictus*, *Aedes togoi*, *An. arabiensis*, *An. dirus*, *Cx. pipiens pallens*, and *Cx. quinquefasciatus*. Although it will not be able to list and review them all, some selected studies are revealed here. Batra et al. (1998) reported that 5 % oil-water emulsion of neem (*A. indica*) completely inhibited emergence of the immature stages of *Ae. aegypti*, *An. stephensi*, and *Cx. quinquefasciatus* under laboratory conditions, when applied at the dosages from 0.1 to 0.4 ml/L. This study reveals the effect of neem oil as the interference of the hormones controlling metamorphosis and molting of larvae. The neem oil emulsion also showed up to 100 % reduction of larvae and pupae of *Ae. aegypti* larvae in small stagnant water habitats; however, neem oil also exhibited adverse killing effect on some beneficial predators, such as *Gambusia affinis* and *Anisops* (Batra et al. 1998). Moreover, neem products possessed not only larvicidal activity against mosquito larvae but also could have marked effects on blood feeding, fecundity, and survivorship of the adult mosquitoes when applied at the sublethal dosages (Su and Mulla 1999). Recently, Anjali et al. (2012) disclosed the effect of droplet size on the larvicidal activity of neem oil nanoemulsion. It was found that the neem oil nanoemulsion with the smallest droplet size of about 31.03 nm was more effective in controlling *Cx. quinquefasciatus* larvae as compared with the larger droplet sizes (Anjali et al. 2012). This obviously implies that the larvicidal activity of essential oils against mosquito larvae could be increased by the reduced size and uniform spreading of the fine droplet emulsion. The nanoemulsion technology then should be applied to other essential oils to improve larvicidal activity.

The ethanolic extracts of ten plant species were tested for larvicidal activity against *Cx. quinquefasciatus* larvae under laboratory conditions in Thailand. Among these, the ethanolic extracts from three plants, *Kaempferia galanga*, *Illicium verum*, and *Spilanthes acmella*, showed a high degree of larvicidal effect against the larvae of *Cx. quinquefasciatus* with LC_{50} of about 50.54, 54.11, and 61.43 ppm, respectively (Pitasawat et al. 1998). In addition, the hexane fraction of *K. galanga* also demonstrated a high larvicidal activity against *Cx. quinquefasciatus* larvae with LC_{50} of 42.33 ppm (Choochote et al. 1999). Essential oils extracted from five plant species, *A. graveolens*, *Carum carvi*, *C. zedoaria*, *Foeniculum vulgare*, and *Z. limonella*, were investigated for larvicidal activity against larvae of *Ae. aegypti* and *An. dirus* in laboratory. It was found that all essential oils were effective against *Ae. aegypti* larvae with LC_{50} ranging from 24.61 to 54.62 ppm and *An. dirus* larvae with LC_{50} ranging between 29.69 and 40.23 ppm (Pitasawat et al. 2007). In particular, the essential oil of *Z. limonella* was the most effective against *Ae. aegypti* larvae, while *C. zedoaria* oil was the most effective against *An. dirus* larvae. On the other hand, water extracts (lyophilized and crude extracts) of *Piper retrofractum* also showed relatively high larvicidal activity against larvae of *Ae.*

aegypti and *Cx. quinquefasciatus* with LC_{50} of about 6.3–79 ppm and 4.33–135 ppm, respectively (Chansang et al. 2005). It was found that the extracts derived from the ripe fruits were more effective against the test larvae than those obtained from the unripe ones. However, the larvicidal activities of these extracts were lost upon aging and temperature of the storage of the extracts (Chansang et al. 2005). This study, hence, reveals some weak points of biopesticides when applied for controlling of mosquito larvae.

In Brazil, essential oils distilled from nine species of local plants were evaluated for larvicidal effect against *Ae. aegypti* larvae in laboratory. It was revealed that two essential oils derived from *Ocimum gratissimum* and *O. americanum* demonstrated high level of larvicidal activity with LC_{50} of about 60 and 67 ppm, respectively (Cavalcanti et al. 2004). A study carried out in the USA revealed that some botanical-based mosquito repellents also have larvicidal activity (Zhu et al. 2008). In this study, four essential oils extracted from *C. longa*, *E. citriodora*, *Santalum album*, and *Cinnamomum cassia* were evaluated for larvicidal activity against *Ae. aegypti*, *Ae. albopictus*, and *Cx. pipiens* larvae under laboratory conditions, and the results showed that *S. album* oil (at the dosage 0.2 mg/ml) was the most effective oil against three species with LT_{50} of about 1.06, 1.82, and 1.55 h, respectively. Perumalsamy et al. (2009) disclosed that the essential oil extracted from *Asarum heterotropoides* possessed larvicidal activity against larvae of *Cx. pipiens pallens*, *Ae. aegypti*, and *Oc. togoi* with LC_{50} of about 21.07, 23.82, and 27.64 ppm, respectively. In addition, the *A. heterotropoides* was identified for chemical constituents, and it was found that safrole was the most toxic constituent to the larvae of *Cx. p. pallens* (LC_{50} 8.22 ppm) and *Ae. aegypti* (LC_{50} 2.6 ppm), while terpinolene was most effective to *Oc. togoi* larvae (LC_{50} 1.6 ppm) (Perumalsamy et al. 2009). Recently, leaves and seeds of *Tribulus terrestris* was extracted with methanol and examined for larvicidal activity against *An. arabiensis* larvae in laboratory (El-Sheikh et al. 2012). The results showed that the methanolic seed extract provided over three times better larvicidal effect against the larvae than did the leaf extract, with LC50 of about 36.5 and 123.1 ppm, respectively.

Very recently, Rodriguez et al. (2013) revealed the larvicidal and cytotoxicity activities of 11 native plants from Mexico against *Ae. aegypti* larvae under laboratory conditions. Among those plants extracted, it was found that the ether extracts of *Ruta chalepensis* (LC_{50} 1.8 μg/ml), *T. vulgaris* (LC_{50} 4.45 μg/ml), and *Z. fagara* (LC_{50} 75.1 μg/ml) were significantly effective against *Ae. aegypti* larvae with low toxicity to Vero cells and these extracts could be developed as excellent biopesticides for controlling mosquito larvae in the field.

2.1.5 Adulticides

There are a few studies of adulticidal activity of plant extracts against insect vectors. Singh et al. (1984) demonstrated that essential oil extracted from *Cedrus deodara* provided knocking down 50 % of adult *An. stephensi* at the concentration of 0.4452 %. Later on himachalol and ß-himachalene were identified as insecticidal principles from *C. deodara* oil which is one of the cheapest plant essential oils (Singh and Agrawal 1988). In a study conducted in Thailand, two extracts of *Curcuma aromatica*, hexane-extracted fraction and essential oil, were investigated for adulticidal activity against female *Ae. aegypti* by topical application under laboratory conditions (Choochote et al. 2005). The results revealed that both extracts showed a promising adulticidal effect against the mosquitoes while the hexane extract of *C. aromatica* (LC_{50} 1.60 μg/mg female) was slightly more effective than the essential oil (LC_{50} 2.86 μg/mg female). Later, Choochote et al. (2006) found that the ethanolic extracts of three *Piper* species, *P. longum*, *P. ribesoides*, and *P. sarmentosum*, provided better adulticidal effect against *Stegomyia aegypti* than did the extracts of *C. aromatica*. The adulticidal activity (LD_{50} values) obtained from *P. sarmentosum*, *P. ribesoides*, and *P. longum* were 0.14, 0.15, and 0.26 μg/female, respectively (Choochote et al. 2006). Essential oils extracted from five plant species, *A. graveolens*, *Carum carvi*, *C. zedoaria*, *P. longum*, and *I. verum*, were also evaluated for adulticidal efficacy against

Ae. aegypti by topical application in laboratory (Chaiyasit et al. 2006). It was found that all five essential oils exhibited adulticidal activity against both laboratory and natural field strains of *Ae. aegypti* with an LC_{50} ranging between 5.44 and 8.83 μg/mg female. These promising plant extracts mentioned above are alternatives in developing mosquito adulticides as an effective measure used in controlling mosquito vectors. However, they are required formulation technology for using with ultralow volume (ULV) application.

2.2 Blackflies

Blackflies, most species belong to the genus *Simulium*, are medically important insects that transmit several diseases, including river blindness in many areas, such as Africa and Central and South America. They are also common nuisance pests for humans around the world, especially in the USA where many US states have to have programs to suppress populations of blackflies in particular places. The common strategy to prevent biting of blackflies relies mainly upon application of synthetic repellents on skin, mostly containing deet as active ingredient. However, some researches and use of plant-based repellents against blackflies have been reported from various parts of the world. One study carried out in the Cote d'Ivoire (Africa) revealed that cocoa oil, which is usually used in some rural areas to prevent biting insects provided excellent repellency against *Simulium damnosum* (Pitroipa et al. 2002). Later, repellents formulated from coconut (*Cocos nucifera*), palm nut (*Elaeis guineensis*), and gobi (*Carapa procera*) also showed high repellency against *Simulium* bites in the field tests conducted in the Cote d'Ivoire (Sylla et al. 2003). In Nigeria, Usip et al. (2006) demonstrated that the volatile oil (essential oil) derived from *O. gratissimum* could reduce the biting rate of *S. damnosum*, ranging from 79 to 90 % for at least 3 h after application. The repellent efficacy of extracts from *O. gratissimum* against blackflies was confirmed by later study in Nigeria by Sam-Wobo et al. (2011). It was disclosed that the repellents formulated from ethanolic extracts of *O. gratissimum* roots and *Pistia hyptis* leaves were effective against *S. damnosum* with approximately 78 % repellency. In Thailand, essential oils extracted from *B. rotunda*, *P. guajava*, and *C. longa* and one commercial plant-based repellent (containing turmeric oil 5 % and eucalyptus oil 4.5 %) were tested on human volunteers against blackflies (*S. nigrogilvum* 99 % of the flies collected). All four plant-based repellents gave 100 % protection from the bites of blackflies for 9 h, protection decreasing to 93 and 90 % after 10 and 11 h of application of the repellents (Tawatsin et al. 2006b). The extent and duration of protection was the same for deet and the plant products. Essential oil derived from catmint, *Nepeta cataria*, was also evaluated for repellency against blackflies. It was found that the repellents containing the dehydrogenated catmint oil (15 %) were extremely effective against blackflies, primarily *S. decorum* with complete protection for 7.5 h in field tests in the USA (Spero et al. 2008). Very recently, a study on plant-based repellents was carried out against blackflies in India (Hazarika et al. 2013). The results showed that the repellents containing essential oils extracted from *Homalomena aromatica*, *Vitex negundo*, and *Ageratum conizoides* provided a high degree of repellency against various blackfly species (*S. himalayense*, *S. barraudi*, *S. kupari*, *S. indicum*, and *S. rufibasis*) for at least 5 h. The studies of plant-based repellents against blackflies mentioned above point out the high-potential use as effective repellents to prevent biting of blackflies.

2.3 Biting Midges

Biting midges are bloodsucking insects belonging into many genera, such as *Culicoides*, *Leptoconops*, and *Forcipomyia*. Although most of the biting midges cannot transmit any diseases to humans, they are accepted as serious nuisance pests deterring human life in many parts of the world, for example, the UK, Australia, and the USA. Avoiding the infested areas, wearing protective clothing and applying insect repellent on skin seem to be the best strategies to protect humans from these biting insects (Fradin and Day 2002). A plant-based insect repellent, containing *p*-menthane-3,8-diol (PMD: derived from essential oil of lemon eucalyptus, *E. citriodora* or *C. citriodora*) as active ingredient, showed complete repellency at least 5 h against *Culicoides variipennis* in laboratory testing (Trigg and Hill 1996) and for up to 7 h against *Culicoides impunctatus* in the field trials in Scotland (UK), and it also demonstrated a high degree of protection of about 99.5 % between 8 and 10 h after an application as compared with 97 % protection obtained from deet (Trigg 1996b). In another study conducted in California (USA), two PMD formulations (spray and lotion) also exhibited extremely high protective level against *Leptoconops carteri* biting midges in the field and half of the PMD-treated volunteers received no bites during 6 h of continuous exposure per treatment (Carroll and Loye 2006a). On the other hand, essential oil of neem derived from *A. indica* was evaluated for repellent and antifeedant activity against *C. impunctatus* under laboratory conditions. It was revealed that *C. impunctatus* females were repelled by neem oil (≥1 %) in a Y-tube olfactometer testing, and the proportion of blood feeding of the wild-caught parous females of *C. impunctatus* was significantly reduced by topical application of neem oil (≥1 %) employing a membrane feeder technique (Blackwell et al. 2004). Essential oils extracted from various Australian plants, such as *Backhousia anisata*, *B. citriodora*, *Callitris columellaris*, *C. glaucophylla*, *Eremophila mitchellii*, *Leptospermum liversidgei*, *L. petersonii*, *M. ericifolia*, *M. linariifolia*, and *M. uncinata*, were prepared in a simple cream base and evaluated for repellency under laboratory conditions, using *Ae. aegypti* as the screened insects (Greive et al. 2010). Based on laboratory results, essential oil from *M. ericifolia* was then selected to be further studied for repellency against *Culicoides* biting midges in the field. Interestingly, the three formulations (alcohol-based spray, emulsion, and gel) of repellents containing essential oil of *M. ericifolia* (5 %) as active ingredient were highly effective against *C. ornatus* and *C. immaculatus*, and they also were of comparable repellency and not significantly different to deet in repelling the biting midges (Greive et al. 2010).

2.4 Housefly and Bush Fly

Housefly, *Musca domestica*, is one of the most potential flying insects that mechanically transmit various diarrheal pathogens to human foods; it is also an important nuisance pest of human leisure both outdoor and indoor, especially in the summer time. Most of the control strategies for housefly rely mainly on synthetic chemicals applied as adulticides and larvicides. There have been very few researches on plant-based insecticides or repellents against housefly until recently. Singh and Singh (1991) studied the repellent and insecticidal properties of 31 plant essential oils, and they revealed that essential oils of *O. gratissimum*, *T. serpyllum*, *I. verum*, *M. fragrans*, and *C. amada* showed 100 % repellency (knockdown) and essential oils of *A. calamus* and *T. serpyllum* showed about 40 % of mortality at 2 % concentration in acetone as compared to malathion, deet, DMP, thymol, and pyrethrum oleoresin containing 2 % of pyrethrin. This study explored the possibility of utilization of essential oils at sweet shops, meat shops, and in areas prone to housefly menace in Asia and other parts of world. Maganga et al. (1996) revealed that pine oil containing myrcene, *p*-cymene, γ-terpinene, and linalool as major constituents possessed repellent and feeding deterrent activities against *M. domestica* under laboratory conditions. It was found that pine oil could repel 95 % of the houseflies away from the stimulus at distance of about 6 mm and the oil at low volume of treatment (10 μL) could also inhibit any feeding by the flies even after 24 h after treatment. Recently, attempts were made to investigate repellent and adulticidal activities of 13 plant species against *M. domestica* under laboratory conditions (Mansour et al. 2012). The ethanolic extracts of *A. indica*, *Cichorium intybus*, *Conyza aegyptiaca*, and *Sonchus oleraceus* showed at least 90 % of repellent and antifeedant activities against the houseflies. It was also found that the ethanolic extracts of *C. intybus*, *C. aegyptiaca*, and *S. oleraceus* when applied in sugar bait exhibited equal or better insecticidal activity (based on LD_{50} value) against *M. domestica* as compared with the synthetic chemical insecticides, such as deltamethrin, chlorpyrifos, methomyl, and flufenoxuron (Mansour et al. 2012). On the other hand, another study was carried out to investigate field repellency of essential oil against bush fly, *Musca vetustissima*, in Australia (Greive et al. 2010). Actually, bush fly is a nonbiting insect but it is recognized as mechanical vector that transmits enteric diseases and eye infection (Hughes 1970). It was found that three formulations (alcohol-based spray, emulsion, and gel) of repellents containing essential oil of *M. ericifolia* (5 %) as active ingredient showed equal repellency against the bush fly as the commercial repellent containing deet (Greive et al. 2010). These studies clearly demonstrate the potential use of plant extracts to be formulated as effective repellents or insecticides against housefly and bush fly.

2.5 Cockroaches

Cockroaches are important urban pests around the globe. They are also recognized as insects that potentially carry and transmit many pathogens, such as bacteria, virus, fungi, protozoa, and helminths (Cochran 1982). Attempts have been made to investigate some repellents derived from plants that could repel cockroaches. In Malaysia, essential oils extracted from six plant species, *C. longa*, *Z. officinale*, *C. zeylanicum*, *S. aromaticum*, *C. citratus*, and *Pandanus odorus*, were evaluated for repellency against *Periplaneta americana* under laboratory conditions (Ahmad et al. 1995). It was found that the essential oils of the first four species exhibited complete repellency (100 %) against the cockroach at the concentration of about 12 ppm, while the essential oils of *C. citratus* and *P. odorus* provided 93 % and 57 % repellency, respectively. Generally, *C. longa*, *Z. officinale*, *C. zeylanicum*, and *S. aromaticum* are local plants and abundant in Southeast Asia; the development of cockroach repellent from essential oils of these plants is likely possible. The essential oil of catnip (*N. cataria*), two purified isomers of nepetalactone, and deet were tested for repellency against adult male German cockroaches by Peterson et al. (2002), and it was found that the E,Z-nepetalactone was significantly more active than equivalent doses of deet, the essential oils, or Z,E-nepetalactone. Recently, seven plant-based repellents derived from essential oils of *B. rotunda*, *C. hystrix*, *C. longa*, *L. cubeba*, *P. nigrum*, *P. guajava*, and *Z. officinale* were tested against the cockroaches *P. americana*, *Blatella germanica*, and *Neostylopyga rhombifolia* in the laboratory in Thailand (Thavara et al. 2007). All seven products reduced cockroach populations drastically, and *N. rhombifolia* was less susceptible than the other two species. Among the seven plants, kaffir lime (*C. hystrix*) was the most effective giving the highest repellency (87–100 %), higher than that of naphthalene used as a standard. *C. hystrix* essential oil was formulated as repellent (20 %) for field tests. In three village tests in Phitsanulok Province (Thailand), treatments reduced cockroaches by 50–70 % in 9–12 days posttreatment. Similarly, in two areas in Bangkok, kaffir lime formulation (20 %) treatment caused 60–80 % reduction in cockroaches. This level of reduction is equivalent to that of some insecticidal treatments. This shows a high potential of *C. hystrix* essential oil to be commercially formulated as effective repellent to repel cockroaches. The insecticidal and repellent properties of volatile constituents of essential oils were evaluated against the American cockroach (*P. americana*) in laboratory (Ngoh et al. 1998). It is interesting to note that the benzene derivatives (eugenol, methyl eugenol, isoeugenol, safrole, and isosafrole) were better toxicants and repellents to the cockroaches than the monoterpenes (limonene, cineole, and *p*-cymene).

2.6 Ticks

Ticks are medically important arthropods that transmit various pathogens and cause many diseases to humans and animals, such as *Borrelia burgdorferi* (Lyme disease), *Francisella tularensis* (tularemia), and *Coxiella burnetii* (Q fever). These diseases can be protected by applications of repellents or acaricides to prevent and control tick vectors. Both repellents and acaricides against ticks usually contain synthetic chemical as active ingredients, whereas those derived from plants are really scarce. Trigg and Hill (1996) demonstrated that PMD when applied at the dosage 0.3 μL/cm^2 could reduce significantly attachment and feeding of the deer tick (*Ixodes ricinus*) on rabbit ears in laboratory. In addition, an average of 77.5 % mortality in nymphs on the treated ears was also occurred as compared with 11.6 % on the untreated ears. This study reveals the repellent and acaricidal effects of PMD and the possibility to be used as a personal protection against ticks. Evaluations of plant extracts were carried out in Sweden to investigate repellency against host-seeking deer tick (*I. ricinus*). In this study, the essential oil extracted from *Rhododendron tomentosum* (10 % in acetone) provided 95 % repellency against *I. ricinus*, whereas the ethyl acetate extracts of *R. tomentosum* and *Artemisia absinthium* exhibited more than 70 % repellency against the ticks (Jaenson et al. 2005). This reveals the potential use of essential oils or extracts from *R. tomentosum* and *A. absinthium* as plant-based tick repellents. As for the botanical acaricide, it is interesting to note that one commercial product, Eco-Exempt IC2 (IC2), containing essential oils from rosemary (10 %) and peppermint (2 %) as active ingredients demonstrated excellent acaricidal activities against *I. scapularis* in field trials carried out in oak-pine forest in southern Maine, USA (Elias et al. 2013). It was found that IC2 was effective as bifenthrin (a synthetic pyrethroid served as positive reference) in controlling nymphs through the rest of nymphal season and reducing larvae through 14 months post-spray. In addition, IC2 could control adult ticks for 6 months post-spray during the fall adult season as compared with 12 months for bifenthrin. No long-term impacts on some nontarget arthropods, such as Coleoptera, Hymenoptera, and Collembola, were observed in the areas treated with IC2 as all numbers rebounded by 20 days post-spray. However, IC2 was phytotoxic to the leafy portions of some plants but they could recover by the next growing season. This is an effective plant-based acaricide, which is commercially available in the market now.

2.7 Chiggers

Chiggers are larval stage of trombiculid mites which is the only parasitic stage of the mite's life cycle. They feed on various hosts, such as mammals, birds, reptiles, amphibians, and humans (accidental hosts). Several species of *Leptotrombidium* chiggers are vectors of scrub typhus, an important disease caused by rickettsia, in many areas of Asia. In Thailand, essential oils derived from 13 plant species were evaluated for repellency against chiggers of *Leptotrombidium imphalum* under laboratory conditions (Eamsobhana et al. 2009). Among these, only four essential oils from *S. aromaticum* (5 % in ethanol), *M. alternifolia* (40 % in ethanol), *E. globulus* (undiluted oil), and *Z. cassumunar* (undiluted oil) demonstrated an excellent result (100 % repellency) against *L. imphalum* chiggers. Therefore, the essential oil from *S. aromaticum* could be commercially formulated as an effective repellent against chiggers at the low concentration, and it could be an alternative repellent to prevent scrub typhus elsewhere.

2.8 Land Leeches

Land leeches are severe pests of man and animals in national parks, wilderness, and wetlands. They could be quite a deterrent to outdoor activity and enjoyment of wilderness and open spaces. In Malaysia, Kirton (2005) revealed that a commercial spray containing 40 % (w/w) Citriodiol, an extract of the leaves of *C. citriodora* (or *E. citriodora*) and primarily consists of PMD, showed excellent repellency against land leeches, mostly *Haemadipsa sylvestris*, when applied on the footwear and trouser legs (tucked into socks). Later, three plant-based repellents derived from essential oil of *B. rotunda* (10 %), *P. guajava* (10 %), and *C. longa* (10 %) and one commercial repellent (containing turmeric oil 5 % and eucalyptus oil 4.5 %) as mentioned previously by Tawatsin et al. (2006b) and deet (10 %) were tested on human volunteers against land leeches in national park in Thailand. The volunteers walked for 10 min along the trail in each hour interval through the walking trail and collected attached land leeches on both legs (one treated with repellent and the other untreated as control). The leeches were in the genus *Haemadipsa*. All four plant-based repellents and deet provided 100 % protection from leeches for 8 h. It was also found that the leeches in contact with repellent-treated surfaces shrined and died within minutes. Not only were the products repellent to leeches but also highly toxic to them. The leeches were highly sensitive to plant-based repellents, and these can be used as sprays for clearing leech infestations in trails and wilderness areas. Recently, water extracts obtained from three plant species (*Sapindus rarak*, *Vernonia elaeagnifolia*, and *Catunaregam spathulifolia*) were applied on cotton stockings and tested for repellency against terrestrial bloodsucking leeches in Lao PDR (Vongsombath et al. 2011). It was found that *S. rarak* demonstrated a high level of repellency (82.6 %), while *V. elaeagnifolia* (63 %) and *C. spathulifolia* (62.6 %) provided moderate repellency. *Haemadipsa trimaculosa and H. sylvestris* were the predominant species of leeches found in this study. More studies are needed on prevention and control of land leeches.

3 Conclusion

It can be concluded that plant biopesticides have a multiple mode of actions against bloodsucking insects and arthropods. They can be used for repelling and control of larval and adult vectors, mostly mosquitoes. The essential oils extracted from plants offer a rich source of products that can be used in preparations to repel and protect humans form biting insects and arthropods. They can be employed in lotions and aerosols properly formulated for personal protection. The repellents derived from essential oils and other extracts of various plant species have demonstrated excellent repellency against mosquitoes, blackflies, biting midges, ticks, chiggers, land leeches, houseflies, and cockroaches. This is one of the best eco-friendly strategies to prevent diseases transmitted by insects and other arthropods when the protective vaccines have not yet been available for particular diseases. Evidently, PMD derived from the lemon eucalyptus (*E. citriodora*) seems to be the most promising active ingredient to be formulated and used as effective repellents against various hematophagous insects and arthropods, and it has been already commercialized in the market. On the other hand, celery (*A. graveolens*) and neem (*A. indica*) extracts also provide various kinds of bioactivities, such as repellent, larvicide, and adulticide, against many species of vector mosquitoes. The extracts obtained from these two plants could be developed and formulated as commercial repellents or other forms of biopesticides against mosquito vectors. However, further research is needed to identify and develop more potential active principles and formulations.

References

Ahmad FBH, Mackeen MM, Ali AM, Mashirun SR, Yaacob MM (1995) Repellency of essential oils against the domiciliary cockroach, *Periplaneta americana*. Insect Sci Appl 16:391–393

Anjali CH, Sharma Y, Mukherjee A, Chandrasekaran N (2012) Neem oil (Azadirachta indica) nanoemulsion – a potent larvicidal agent against *Culex quinquefasciatus*. Pest Manag Sci 68:158–163

Ansari MA, Razdan RK (1995) Relative efficacy of various oils in repelling mosquitoes. Indian J Malariol 32:104–111

Ansari MA, Mittal PK, Razdan RK, Sreehari U (2005) Larvicidal and mosquito repellent activities of Pine (*Pinus longifolia*, Family: Pinaceae). J Vect Borne Dis 42:95–99

Bannister BA, Begg NT, Gillepie SH (1996) Infectious disease. Blackwell, London

Batra CP, Mittal PK, Adak T, Sharma VP (1998) Efficacy of neem oil-water emulsion against mosquito immatures. Indian J Malariol 35:15–21

Blackwell A, Evans KA, Strang RHC, Cole M (2004) Toward development of neem-based repellents against the Scottish Highland biting midges *Culicoides impunctatus*. Med Vet Entomol 18:449–452

Carroll SP, Loye J (2006a) Field test of a lemon eucalyptus repellent against *Leptoconops* biting midges. J Am Mosq Control Assoc 22:483–485

Carroll SP, Loye J (2006b) PMD, a registered botanical mosquito repellent with deet-like efficacy. J Am Mosq Control Assoc 22:507–514

Cavalcanti ESB, De Morais SM, Lima MAA, Santana EWP (2004) Larvicidal activity of essential oil from Brazilian plants against *Aedes aegypti* L. Mem Inst Oswaldo Cruz, Rio de Janeiro 99:541–544

Chaithong U, Choochote W, Kamsuk K, Jitpakdi A, Tippawangkosol P, Chaiyasit D, Champakaew D, Tuetun B, Pitasawat B (2006) Larvicidal effect of pepper plants on *Aedes aegypti* (L.) (Diptera: Culicidae). J Vector Ecol 31:138–144

Chaiyasit D, Choochote W, Rattanachanpichai E, Chaithong U, Chaiwong P, Jitpakdi A, Tippawangkosol P, Riyong D, Pitasawat B (2006) Essential oils as potential adulticides against two populations of *Aedes aegypti*, the laboratory and natural field strains, in Chiang Mai province, northern Thailand. Parasitol Res 99:715–721

Champakaew D, Choochote W, Pongpaibul Y, Chaithong U, Jitpakdi A, Tuetun B, Pitasawat B (2007) Larvicidal efficacy and biological stability of a botanical natural Product, Zedoary oil-impregnated sand granules, against *Aedes aegypti* (Diptera: Culicidae). Parasitol Res 100:729–737

Chang KS, Tak JH, Kim SI, Lee WJ, Ahn YJ (2006) Repellency of *Cinnamomum cassia* bark compounds and cream containing cassia oil to *Aedes aegypti* (Diptera: Culicidae) under laboratory and indoor conditions. Pest Manag Sci 62:1032–1038

Chang KS, Shin EH, Park C, Ahn YJ (2012) Contact and fumigant toxicity of *Cyperus rotundus* steam distillate constituents and related compounds to insecticide-susceptible and -resistant *Blattella germanica*. J Med Entomol 49:631–639

Chansang U, Zahiri NS, Bansiddhi J, Boonruad T, Thongsrirak P, Mingmuang J, Benjapong N, Mulla MS (2005) Mosquito larvicidal activity of aqueous extracts of long pepper (*Piper retrofractum* Vahl) from Thailand. J Vector Ecol 30:195–200

Cheng SS, Chang HT, Lin CY, Chen PS, Huang CG, Chen WJ, Chang ST (2009) Insecticidal activities of leaf and twig essential oils from *Clausena excavata* against *Aedes aegypti* and *Aedes albopictus* larvae. Pest Manag Sci 65:339–343

Choi WS, Park BS, Ku SK, Lee SE (2002) Repellent activities of essential oils and monoterpenes against *Culex pipiens pallens* Coquillett. J Am Mosq Control Assoc 18:348–351

Chokechaijaroenporn O, Bunyapraphatsara N, Kongchuensin S (1994) Mosquito repellent activities of *Ocimum* volatile oils. Phytomedicine 1:135–139

Choochote W, Kanjanapothi D, Panthong A, Taesotikul T, Jitpakdi A, Chaithong U, Pitasawat B (1999) Larvicidal, adulticidal and repellent effects of *Kaempferia galanga*. Southeast Asian J Trop Med Public Health 30:470–476

Choochote W, Tuetun B, Kanjanapothi D, Rattanachanpichai E, Chaithong U, Chaiwong P, Jitpakdi A, Tippawangkosol P, Riyong D, Pitasawat B (2004) Potential of crude seed extract of celery, *Apium graveolens* L., against the mosquito *Aedes aegypti* (L.) (Diptera: Culicidae). J Vector Ecol 29:340–346

Choochote W, Chaiyasit D, Kanjanapothi D, Rattanachanpichai E, Jitpakdi A, Tuetun B, Pitasawat B (2005) Chemical composition and anti-mosquito potential of rhizome extract and volatile oil derived from *Curcuma aromatica* against *Aedes aegypti* (Diptera: Culicidae). J Vector Ecol 30:302–309

Choochote W, Chaithong U, Kamsuk K, Rattanachanpichai E, Jitpakdi A, Tippawangkosol P, Chaiyasit D, Champakaew D, Tuetun B, Pitasawat B (2006) Adulticidal activity against *Stegomyia aegypti* (Diptera: Culicidae) of three *Piper spp*. Rev Inst Med Trop Sao Paulo 48:33–37

Choochote W, Chaithong U, Kamsuk K, Jitpakdi A, Tippawangkosol P, Tuetun B, Champakaew D, Pitasawat B (2007) Repellent activity of selected essential oils against *Aedes aegypti*. Fitoterapia 78:359–364

Cochran DG (1982) Cochroach: biology and control. WHO/CDS/WHOPES/99.3

Coleman RE, Robert LL, Roberts LW, Glass JA, Seeley DC, Laughinghouse A, Perkins PV, Wirtz RA (1993) Laboratory evaluation of repellents against four anopheline mosquitoes (Diptera: Culicidae) and two phlebotomine sand flies (Diptera: Psychodidae). J Med Entomol 30:499–502

Das NG, Nath DR, Baruah I, Talukdar PK, Das SC (1999) Field evaluation of herbal mosquito repellents. J Commun Dis 31:241–245

Das NG, Baruah I, Talukdar PK, Das SC (2003) Evaluation of botanicals as repellents against Mosquitoes. J Vect Borne Dis 40:49–53

Debboun M, Frances SP, Strickman D (2007) Insect repellent, principles, methods and uses. CRC Press, Boca Raton

Dethier VG (1948) Chemical insect attractants and repellents. H.K. Lewis & Co.Ltd, London

Dietrich G, Dolan MC, Peralta-Cruz J, Schmidt J, Piesman J, Eisen RJ, Karchesy JJ (2006) Repellent activity of fractioned compounds from *Chamaecyparis nootkatensis* essential oils against nymphal *Ixodes scapularis* (Acari: Ixodidae). J Med Entomol 43:957–961

Dugassa S, Medhin G, Balkew M, Seyoum A, Gebre-Michael T (2009) Field investigation on the repellent activity of some aromatic plants by traditional means against *Anopheles arabiensis* and *An. pharoensis* (Diptera: Culicidae) around Koka, central Ethiopia. Acta Trop 112:38–42

Eamsobhana P, Yoolek A, Kongkaew W, Lerdthusnee K, Khlaimanee N, Parsartvit A, Malainual N, Yong HS (2009) Laboratory evaluation of aromatic essential oils from thirteen plant species as candidate repellents against *Leptotrombidium* chiggers (Acari: Trombiculidae), the vector of scrub typhus. Exp Appl Acarol 47:257–262

Edwards DL, Johnson CE (1987) Insect-repellent-induced toxic encephalopathy in a child. Clin Pharm 6:496–498

Elias SP, Lubelczyk CB, Rand PW, Staples JK, Amand TWS, Stubbs CS, Lacombe EH, Smith LB, Smith RP Jr (2013) Effect of a botanical acaricide on Ixodes scapularis (Acari: Ixodidae) and nontarget arthropods. J Med Entomol 50:126–136

El-Sheikh TMY, Bosly HAM, Shalaby NM (2012) Insecticidal and repellent activities of methanolic extract of *Tribulus terrestris* L. (Zygophyllaceae) against the malaria vector *Anopheles arabiensis* (Diptera: Culicidae). Egypt Acad J Biol 5:13–22

Fradin MS, Day JF (2002) Comparative efficacy of insect repellents against mosquito bites. N Engl J Med 347:13–18

Garboui SS, Jaenson TGT, Palsson K (2006) Repellency of MyggA® natural spray (*para*- menthane-3,8-diol) and RB86 (neem oil) against the tick *Ixodes ricinus* (Acari: Ixodidae) in the field in east-central Sweden. Exp Appl Acarol 40:271–277

Gerberg LJ, Novak RJ (2007) Considerations on the use of botanically-derived repellent products. In: Debboun M, Frances SP, Strickman D (eds) Insect repellents, principles, methods and uses. CRC Press, Boca Raton

Greive KA, Staton JA, Miller PF, Peters BA, Oppenheim VMJ (2010) Development of *Melaleuca* oils as effective natural-based personal insect repellents. Aus J Entomol 49:40–48

Gulati R, Mathur S (1995) Effect of eucalyptus and mentha leaves and curcuma rhizomes on *Tyrophagus putrescentiae* (Schrank) (Acarina: Acaridae). Exp Appl Acarol 19:511–518

Hazarika S, Dhiman S, Rabha B, Bhola RK, Singh L (2013) Repellent activity of some essential oils against *Simulium* species in India. J Insect Sci 12:1–9

Hieu TT, Kim SI, Lee SG, Ahn YJ (2010) Repellency to *Stomoxys calcitrans* (Diptera: Muscidae) of plant essential oils alone or in combination with *Calophyllum inophyllum* nut oil. J Med Entomol 47:575–580

Hughes RD (1970) The seasonal distribution of bushfly (*Musca vetustissima* Walker) in South-East Australia. J Animal Ecol 39:691–706

Isman MB (1999) Neem and related natural products. In: Hall FR, Menn JJ (eds) Biopesticides, use and delivery. Humana Press, Ottawa

Jaenson TGT, Palsson K, Borg-Karlson AK (2005) Evaluation of extracts and oils of tick- repellent plants from Sweden. Med Vet Entomol 19:345–352

Jaenson TGT, Palsson K, Borg-Karlson AK (2006) Evaluation of extracts and oils of mosquito (Diptera: Culicidae) repellent plants from Sweden and Guinea-Bissau. J Med Entomol 43:113–119

Jang YS, Jeon JH, Lee HS (2005) Mosquito larvicidal activity of active constituent derived from *Chamaecyparis obtusa* leaves against 3 mosquito species. J Am Mosq Control Assoc 21:400–403

Jaruwichitratana S, Tanyasittisuntorn P, Timpatanapong P (1988) Comparison of citronella cream and placebo cream in protection of mosquito biting. J Rama Med 11:94–97

Jilani G, Saxena RC (1990) Repellent and feeding deterrent effects of turmeric oil, sweetflag oil, neem oil, and a neem-based insecticide against lesser grain borer (Coleoptera: Bostrychidae). J Econ Entomol 83: 629–634

Jilani G, Su HCF (1983) Laboratory studies on several plant materials as insect repellents for protection of cereal grains. J Econ Entomol 76:154–157

Jilani G, Saxena RC, Deuda BP (1988) Repellent and growth-inhibiting effects of turmeric oil, sweetflag, neem oil, and "Margosan-O" on red flour beetle (Coleoptera: Tenebrionidae). J Econ Entomol 81: 1226–1230

Kamsuk K, Choochote W, Chaithong U, Jitpakdi A, Tippawangkosol P, Riyong D, Pitasawat B (2007) Effectiveness of *Zanthoxylum piperitum*-derived essential oils as an alternative repellent under laboratory and field applications. Parasitol Res 100: 339–345

Kebede Y, Gebre-Mitchael T, Balkew M (2010) Laboratory and field evaluation of neem (*Azadirachta indica* A. Juss) and Chinaberry (*Melia azedarach* L.) oils as repellents against *Phlebotomus orientalis* and *P. bergeroti* (Diptera: Psychodidae) in Ethiopia. Acta Trop 113:145–150

Khan AA, Maibach HI, Skidmore DL (1975a) Addition of perfume fixatives to mosquito repellents to increase protection time. Mosq News 35:23–26

Khan AA, Maibach HI, Skidmore DL (1975b) Addition of vanillin to mosquito repellents to increase protection time. Mosq News 35:223–225

Kirton LG (2005) Laboratory and field tests of the effectiveness of the lemon-eucalyptus extract, Citriodiol, as a repellent against land leeches of the genus *Haemadipsa* (Haemadipsidae). Ann Trop Med Parasitol 99:695–714

Kumar P, Mishra S, Malik A, Satya S (2011) Repellent, larvicidal and pupicidal properties of essential oils and their formulations against the housefly, *Musca domestica*. Med Vet Entomol 25:302–310

Kwon HW, Kim SI, Chang KS, Clark JM, Ahn YJ (2011) Enhanced repellency of binary mixtures of *Zanthoxylum armatum* seed oil, vanillin, and their aerosols to mosquitoes under laboratory and field conditions. J Med Entomol 48:61–66

Lee HS (2006) Mosquito larvicidal activity of aromatic medicinal plant oils against Aedes aegypti and Culex pipiens pallens. J Am Mosq Control Assoc 22:292–295

Lucia A, Audino PG, Seccacini E, Licastro S, Zebra E, Masuh H (2007) Larvicidal effect of *Eucalyptus grandis* essential oil and turpentine and their major components on *Aedes aegypti* larvae. J Am Mosq Control Assoc 23:299–303

Macnay GC (1939) Studies on repellents for biting flies. Can Entomolog 71:38–44

Maganga ME, Gries G, Gries R (1996) Repellency of various oil and pine oil constituents to house flies (Diptera: Muscidae). Environ Entomol 25: 1182–1187

Mansour SA, Bakr RFA, Hamouda LS, Mohamed RI (2012) Adulticidal activity of some botanical extracts, commercial insecticides and their binary mixtures against the housefly, *Musca domestica* L. Egypt Acad Biol Sci 5:151–167

Martinez-Velazquez M, Rosario-Cruz R, Castillo-Herrera G, Flores-Fernandez JM, Alvarez AH, Lugo-Cervantes E (2011) Acaricidal effect of essential oils from *Lippia graveolens* (Lamiales: Verbenaceae), *Rosmarinus officinalis* (Lamiales: Lamiaceae), and *Allium sativum* (Liliales: Liliaceae) against *Rhipicephalus* (*Boophilus*) *microplus* (Acari: Ixodidae). J Med Entomol 48:822–827

Mehra BK, Hiradhar PK (2002) *Cuscuta hyalina* Roth., an insect development inhibitor against common house mosquito *Culex quinquefasciatus* Say. J Environ Biol 23:335–359

Misni N, Sulaiman S, Othman H, Omar B (2009) Repellency of essential oil of *Piper aduncum* against *Aedes albopictus* in the laboratory. J Am Mosq Control Assoc 25:442–447

Mohammed A, Chadee DD (2007) An evaluation of some Trinidadian plant extracts against larvae of *Aedes aegypti* mosquitoes. J Am Mosq Control Assoc 23:172–176

Moore CG (1977) Insecticide avoidance by ovipositing *Aedes aegypti*. Mosq News 43:291–293

Moore SJ, Lenglet A, Hill N (2002) Field evaluation of three plant-based insect repellents against malaria vectors in Vaca Diez province, the Bolivian Amazon. J Am Mosq Control Assoc 2002:107–110

Moore SJ, Lenglet A, Hill N (2007) Plant-based insect repellents. In: Debboun M, Frances SP, Strickman D (eds) Insect repellents, principles, methods and uses. CRC Press, Boca Raton

Mulla MS, Su T (1999) Activity and biological effects or neem products against arthropods of medical and veterinary importance. J Am Mosq Control Assoc 15:133–152

Mulla MS, Thavara U, Tawatsin A, Kong-Ngamsuk W, Chompoosri J (2001) Mosquito burden and impact and the poor: measures and costs for personal protection in some communities in Thailand. J Am Mosq Control Assoc 17:153–159

Ngoh SP, Choo LEW, Pang FY, Huang Y, Kini MR, Ho SH (1998) Insecticidal and repellent properties of nine volatile constituents of essential oils against the American cockroach, *Periplaneta americana* (L.). Pestic Sci 54:261–268

Odalo JO, Omolo MO, Melebo H, Angira J, Njeru PM, Ndiege IO, Hassanali A (2005) Repellency of some plants from the Kenyan coast against *Anopheles gambiae*. Acta Trop 95:210–218

Osmani Z, Anees I, Naidu MB (1972) Insect repellent cream from essential oils. Pesticides 6:19–21

Palsson K, Jaenson TGT (1999) Plant products used as mosquito repellents in Guinea Bissau, West Africa. Acta Trop 72:39–52

Park BS, Choi WS, Kim JH, Kim KH, Lee SE (2005) Monoterpenes from thyme (*Thymus vulgaris*) as potential mosquito repellents. J Am Mosq Control Assoc 21:80–83

Perumalsamy H, Kim NJ, Ahn YJ (2009) Larvicidal activity of compounds isolated from *Asarum heterotropoides* against *Culex pipiens pallens*, *Aedes aegypti* and *Ochlerotatus togoi* (Diptera: Culicidae). J Med Entomol 46:1420–1423

Peterson CJ, Nemetz LT, Jones LM, Coat JR (2002) Behavioral activity of catnip (Lamiaceae) essential oil compounds to the German cockroach (Blattodea: Blattellidae). J Econ Entomol 95:377–380

Phasomkusolsil S, Soonwera M (2010) Insect repellent activity of medicinal plant oils against *Aedes aegypti* (Linn.), *Anopheles minimus* (Theobald) and *Culex quinquefasciatus* Say based on protection time and biting rate. Southeast Asian J Trop Med Public Health 41:83–840

Pitasawat B, Choochote W, Kanjanapothi D, Panthong A, Jitpakdi A, Chaithong U (1998) screening for larvicidal activity of ten carminative plants. Southeast Asian J Trop Med Public Health 29:660–662

Pitasawat B, Choochote W, Tuetun B, Tippawangkosol P, Kanjanapothi D, Jitpakdi A, Riyong D (2003) Repellency of aromatic turmeric *Curcuma aromatica* under laboratory and field conditions. J Vector Ecol 28:234–240

Pitasawat B, Champakaew D, Choochote W, Jitpakdi A, Chaithong U, Kanjanapothi D, Rattanachanpichai E, Tippawangkosol P, Riyong D, Tuetun B, Chaiyasit D (2007) Aromatic plant-derived essential oil: an alternative larvicide for mosquito control. Fitoterapia 78:205–210

Pitroipa X, Sankara D, Konan L, Sylla M, Doannio JM, Traore S (2002) Evaluation of cocoa oil for individual protection against *Simulium damnosum* s.l. Med Trop 62:511–516

Qiu H, Jun HW, McCall JW (1998) Pharmacokinetics, formulation, and safety of insect repellent N, N-diethyl-3-methylbenzamide (deet): a review. J Am Mosq Control Assoc 14:12–27

Rahuman AA, Venkatesan P (2008) Larvicidal efficacy of five cucurbitaceous plant leaf extracts against mosquito species. Parasitol Res 103:133–139

Robbins PJ, Cherniack MG (1986) Review of the biodistribution and toxicity of the insect repellent N, N-diethyl-m-toluamide (deet). J Toxico Environ Health 18:503–525

Rodriguez YT, Estrada FRM, Suarez AEF, Torres NW, Aranda RS (2013) Larvicidal and cytotoxicity activities of extracts from 11 native plants from Northeastern Mexico. J Med Entomol 50:310–313

Ross IA (1999) Medicinal plants of the world. Humana Press Inc., New Jersey

Rutledge LC, Gupta RK (1996) Reanalysis of the C.G. Macnay mosquito repellent data. J Vector Ecol 21:132–135

Sam-Wobo SO, Adeleke MA, Mafiana CF, Surakat OH (2011) Comparative repellent activities of some plant extracts against *Simulium damnosum* complex. Vector Borne Zoonotic Dis 11:1201–1204

Schmuterer H (2002) The neem tree (*Azadirachta indica*) and other meliaceous plants: Sources of unique natural products for integrated pest management, medicine, industry and other purposes. Neem Foundation, Mumbai

Service MW (1993) Mosquitoes (*Culicidae*). In: Lane RP, Crosskey RW (eds) Medical insects and arachnids. Chapman & Hall, London

Shaalan EA, Canyon DV, Younes MWF, Abdel-Wahab H, Mansour AH (2006) Efficacy of eight larvicidal botanical extractions from *Khaya senegalensis* and *Daucus carota* against *Culex annulirostris*. J Am Mosq Control Assoc 2006:433–436

Sharma SK, Dua VK, Sharma VP (1995) Field studies on the mosquito repellent action of neem oil. Southeast Asian J Trop Med Public Health 26:180–182

Silva OS, Prophiro JS, Nogared JC, Kanis L, Emerick S, Blazius RD, Romao PRT (2006) Larvicidal effect of andiroba oil, *Carapa guianensis* (Meliaceae), against *Aedes aegypti*. J Am Mosq Control Assoc 22:699–701

Singh D, Agrawal SK (1988) Himachalol and β-Himachalene: insecticidal principles of Himalayan cedarwood oil. J Chem Ecol 14:1145–1151

Singh D, Singh AK (1991) Repellent and insecticidal properties of essential oils against housefly, *Musca domestica* L. Insect Sci Appl 12:487–491

Singh D, Rao SM, Tripathi AK (1984) Cedarwood oil as a potential insecticidal agent against Mosquitoes. Naturwissenschaften 71:265–266

Singh D, Siddiqui MS, Sharma S (1989) Reproduction retardant and fumigant properties in essential oils against rice weevil (Coleoptera: Curculionidae) in stored wheat. J Econ Entomol 83:727–733

Snehalatha KS, Ramaiah KD, Vijay Kumar KN, Das PK (2003) The mosquito problem and type and costs of personal protection measures used in rural and urban communities in Pondicherry region, South India. Acta Trop 88:3–9

Souza TM, Farias DF, Soares BM, Viana MP, Lima GPG, Machado LKA, Morais SM, Carvalho AFU (2011) Toxicity of Brazilian plant seed extracts to two strains of *Aedes aegypti* (Diptera: Culicidae) and nontarget animals. J Med Entomol 48:846–851

Spafford H, Jardine A, Carver S, Tarala K, Van Wess M, Weinstein P (2007) Laboratory determination of efficacy of a *Santalum spicatum* extract for mosquito control. J Am Mosq Control Assoc 23:304–311

Spero NC, Gonzalez YI, Scialdone MA, Hallahan DL (2008) Repellency of hydrogenated catmint oil formulations to black flies and mosquitoes in the field. J Med Entomol 45:1080–1086

Sree-iam S (1991) The development of mosquito-repellent formulations containing citronella oil and Plai oil (*Zingiber cassumunar* Roxb.). Dissertation, Mahidol University

Su T, Mulla MS (1999) Effects of neem products containing azadirachtin on blood feeding, fecundity and survivorship of *Culex tarsalis* and *Culex quinquefasciatus* (Diptera: Culicidae). J Vector Ecol 24:202–215

Su HCF, Hoevat R, Jilani G (1982) Isolation, purification, and characterization of insect repellents from *Curcuma longa* L. J Agr Food Chem 30:290–292

Sukumar K, Perich MJ, Boobar LR (1991) Botanical derivatives in mosquito control. A review. J Am Mosq Control Assoc 7:210–237

Sutthanont N, Choochote W, Tuetun B, Junkum A, Jitpakdi A, Chaithong U, Riyong D, Pitasawat B (2010) Chemical composition and larvicidal activity of edible plant-derived essential oils against the pyrethroid-susceptible and -resistant strains of *Aedes aegypti* (Diptera: Culicidae). J Vector Ecol 35:106–115

Suwonkerd W, Tantrarongroj K (1994) Efficacy of essential oil against mosquito biting. Com Dis J 20:4–11

Sylla M, Konan L, Doannino JM, Traore S (2003) Evaluation of the efficacy of coconut (*Cocos nucifera*), palm nut (*Eleais guineensis*) and gobi (*Carapa procera*) lotions and creams in individual protection against Simulium damnosum s.i. bites in Cote d'Ivoire. Bull Soc Pathol Exot 96:104–109

Tawatsin A, Wratten SD, Scott RR, Thavara U, Techadamrongsin Y (2001) Repellency of volatile oils from plants against three mosquito vectors. J Vector Ecol 26:76–82

Tawatsin A, Thavara U, Chompoosri J (2002) Field evaluation of mosquito coils derived from plants against night-biting mosquitoes in Thailand. In: Mulla MS (ed) Proceedings of the Third International Conference on Biopesticides 3. Compact Print, Bangkok

Tawatsin A, Asavadachanukorn P, Thavara U, Wingsinkongman P, Bansidhi J, Boonruad T, Chavalittumrong P, Soonthornchareonnon N, Komalamisra N, Mulla MS (2006a) Repellency of essential oil extracted from plants in Thailand against four mosquito vectors (Diptera: Culicidae) and oviposition deterrent effects against *Aedes aegypti* (Diptera: Culicidae). Southeast Asian J Trop Med Public Health 37:915–932

Tawatsin A, Thavara U, Chansang U, Chavalittumrong P, Boonruad T, Wongsinkongman P, Bansidhi J, Mulla MS (2006b) Field evaluation of deet, Repel Care®, and three plant- based essential oils repellents against mosquitoes, black flies (Diptera: Simuliidae), and land leeches (Arhynchobdellida: Haemadipsidae) in Thailand. J Am Mosq Control Assoc 22:306–313

Thavara U, Tawatsin A, Chompoosri J (2002) Phytochemicals as repellents against mosquitoes in Thailand. In: Mulla MS (ed) Proceedings of third international conference on Biopesticides 3. Compact Print, Bangkok

Thavara U, Tawatsin A, Bhakdeenuan P, Wongsinkongman P, Boonruad T, Bansiddhi J, Chavalittumrong P, Komalamisra N, Siriyasatien P, Mulla MS (2007) Repellent activity of essential oils against cockroaches in Thailand. Southeast Asian J Trop Med Public Health 38:663–673

Trigg JK (1996a) Evaluation of a eucalyptus-based repellent against *Anopheles* spp. In Tanzania. J Am Mosq Control Assoc 12:243–246

Trigg JK (1996b) Evaluation of a eucalyptus-based repellent against *Culicoides impunctatus* (Diptera: Ceratopogonidae) in Scotland. J Am Mosq Control Assoc 12:329–330

Trigg JK, Hill N (1996) Laboratory evaluation of a eucalyptus-based repellent against four biting arthropods. Phytother Res 10:313–316

Trongtokit Y, Rongsriyam Y, Komalamisra N, Krisadaphong P, Apiwathanasorn C (2004) Laboratory and field trial of developing medicinal local Thai plant products against four species of mosquito vectors. Southeast Asian J Trop Med Public Health 35:325–333

Trongtokit Y, Rongsriyam Y, Komalamisra N, Apiwathnasorn C (2005) Comparative efficacy of 38 essential oils against mosquito bites. Phytother Res 19:303–309

Tuetun B, Choochote W, Rattanachanpichai E, Chaithong U, Jitpakdi A, Tippawangkosol P, Riyong D, Pitasawat B (2004) Mosquito repellency of the seeds of celery (Apium graveolens L.). Ann Trop Med Parasitol 98:407–417

Tuetun B, Choochote W, Kanjanapothi D, Rattanachanpichai E, Chaithong U, Chaiyong P, Jitpakdi A, Tippawangkosol P, Riyong D, Pitasawat B (2005) Repellent properties of celery, *Apium graveolens* L., compared with commercial repellents against mosquitoes under laboratory and field conditions. Trop Med Int Health 10:1190–1198

Tuetun B, Choochote W, Pongpaibul Y, Junkum A, Kanjanapothi D, Chaithong U, Jitpakdi A, Riyong D, Pitasawat B (2008) Celery-based topical repellents as a potential natural alternative for personal protection against mosquitoes. Parasitol Res 104:107–115

Tuetun B, Choochote W, Pongpaibul Y, Junkum A, Kanjanapothi D, Chaithong U, Jitpakdi A, Riyong D, Wannasan A, Pitasawat B (2009) Field evaluation of G10, a celery (*Apium graveolens*)-based topical repellent, against mosquitoes (Diptera: Culicidae) in Chiang Mai province, northern Thailand. Parasitol Res 104:515–521

Tyaki BK, Shahi AK, Kaul BL (1998) Evaluation of repellent activities of *Cymbopogon* essential oils against mosquito vectors of malaria, filariasis and dengue fever in India. Phytomedicine 5:324–329

Usip LPE, Opara KN, Ibanga ES, Atting IA (2006) Longitudinal evaluation of repellent activity of *Ocimum gratissimum* (Labiatae) volatile oil against *Simulium damnosum*. Mem Inst Oswaldo Cruz, Rio de Janeiro 101:201–205

Vongsombath C, Boer HJ, Palsson K (2011) Keeping leeches at bay: field evaluation of plant-derived extracts against terrestrial blood-sucking leeches (Haemadipsidae) in Lao PDR. Acta Trop 119:178–182

Vongsombath C, Palsson K, Bjork L, Borg-Karlson A, Jaenson TGT (2012) Mosquito (Diptera: Culicidae) repellency field tests of essential oils from plants traditionally used in Laos. J Med Entomol 49:1398–1404

Walker TW, Robert LL, Copeland RA, Githeko AK, Wirtz RA, Githure JI, Klein TA (1996) Field evaluation of arthropod repellents, deet and piperidine compound, AI3-37220, against *Anopheles arabiensis* in western Kenya. J Am Mosq Control Assoc 12:172–176

Wasuwat S, Sunthonthanasart T, Jarikasem S, Putsri N, Phanrakwong A, Janthorn S (1990) Mosquito repellent efficacy of citronella cream. J Sci Tech 5:62–68

Weiss EA (1997) Essential oil crops. CAB International, Wallingford

White GB (1973) The insect repellent value of *Ocimum* spp. (Labiatae): traditional anti-mosquito plants. Afr Med J 50:248–252

World Health Organization (WHO) (1996) Report of the WHO informal consultation on the evaluation and testing of insecticides. World Health Organization, Geneva

Xue RD, Barnard DR, Ali A (2001) Laboratory and field evaluation of insect repellents as oviposition deterrents against mosquito *Aedes albopictus*. Med Vet Entomol 15:126–131

Xue RD, Barnard DR, Ali A (2003) Laboratory evaluation of 18 repellent compounds as oviposition deterrents of *Aedes albopictus* and as larvicides of *Aedes aegypti*, *Anopheles quadrimaculatus* and *Culex quinquefasciatus*. J Am Mosq Control Assoc 19:397–403

Xue RD, Barnard DR, Ali A (2006) Laboratory evaluation of 21 insect repellents as larvicides and as oviposition deterrents of *Aedes albopictus* (Diptera: Culicidae). J Am Mosq Control Assoc 22:126–130

Yang YC, Lee EH, Lee HS, Lee DK, Ahn YJ (2004) Repellency of aromatic medicinal plant extracts and a steam distillate to *Aedes aegypti*. J Am Mosq Control Assoc 20:146–149

Yap HH (1986) Effectiveness of soap formulations containing deet and permethrin as personal protection against outdoor mosquitoes in Malaysia. J Am Mosq Control Assoc 2:63–67

Zadikoff CM (1979) Toxic encephalopathy associated with the use of insect repellent. J Ped 95:140–142

Zhu J, Zeng X, Yanma LT, Qian K, Han Y, Xue S, Tucker B, Schultz G, Coats J, Rowley W, Zhang A (2006) Adult repellency and larvicidal activity of five plant essential oils against mosquitoes. J Am Mosq Control Assoc 22:515–522

Zhu J, Zeng X, O'neal M, Schultz G, Tucker B, Coats J, Bartholomay L, Xue RD (2008) Mosquito larvicidal activity of botanical-based mosquito repellents. J Am Mosq Control Assoc 24:161–168

13 Management of Mite Pests in Honeybee Colonies Through Botanicals

Dwijendra Singh

Abstract

Honeybee mite species have been recorded to hamper the honey production in colonies and adversely affect the normal foraging potential of honeybees that helps in cross-pollination of most of the arable flowering plant species the world over. Among various honeybee mite species, the most common are parasitic and tracheal mites whose outbreak has been found to destroy honeybee colony even to the extent of 100 % in different continents. Due to residue problem, development of resistance, and adverse effects on non-target organism of synthetic acaricides, efforts have been made after the 1980s to investigate various botanicals and phytochemicals as natural products for managing mostly *Varroa jacobsoni* Oud., *Varroa destructor*, and *Acarapis woodi* (Rennie) in both laboratory and field conditions in colonies of *Apis mellifera* L. The most studied phytochemicals, thymol and menthol, have been screened extensively for their bioactivity in managing honeybee mites, which are synthesized by various plant floras especially from genera *Thymus* and *Mentha*, respectively. However, other botanicals and phytochemicals have also shown promising results in managing the mite pests in honeybee colonies, and various commercial products have also been developed based on thymol alone or blended with other natural products. Apart from evaluation of edible plant products, the biologically active plant species, namely, medicinal and aromatic plants which synthesize high content of biologically active phytochemicals, may be evaluated to develop novel and safe acaricides in the future to manage different species of mite pests in honeybee colonies.

D. Singh, Ph.D. (✉)
Microbial Technology & Entomology Department,
Crop Protection Division, CSIR-Central Institute
of Medicinal and Aromatic Plants, P.O. CIMAP,
Lucknow 226 015, Uttar Pradesh, India
e-mail: dsinghlko@gmail.com

D. Singh (ed.), *Advances in Plant Biopesticides*,
DOI 10.1007/978-81-322-2006-0_13, © Springer India 2014

Keywords
Honeybees • Mite • *Varroa* Mite • Pests of honeybees • Pest Management • Essential oil • Botanical pesticide • Beekeeping • Beehive

1 Introduction

The mite pests adversely influence the honey productivity in beekeeping colonies globally. So far, 258 species of arthropods have been recorded associated with beehives of which 160 species of mites have been recorded in honeybee nests. Among several species of mites, *Acarapis woodi* (Rennie), *A. externus* Morgenthaler, *A. dorsalis* Morgenthaler, *Pyemotes ventricosus* (Newp.), *Varroa jacobsoni* Oud., and *Tropilaelaps clareae* Delfinado and Baker are known as major pests of honeybee. Based on activity of mite species, two types of mites are found to hamper the production of honey, namely, parasitic and nonparasitic mites that attack honeybees (Grobov 1979; Mladan et al. 1986; Qamar 2000).

The *Varroa* mite, *V. jacobsoni*, is the most serious ectoparasitic pest affecting honeybees, *Apis mellifera* L., worldwide (Elzen et al. 2001a, b). The hive of the honeybee is a suitable habitat for diverse mites (Acari), including nonparasitic, omnivorous, pollen-feeding species and parasites. The biology and damage of the three main pest species *A. woodi*, *V. jacobsoni*, and *T. clareae* have been reviewed along with detection and control methods in past literature (Sammataro et al. 2000). But utilization of botanicals and their phytochemicals to control mite pests in honeybee colonies causing reduction in honey production review is limited.

The parasitic mites *V. jacobsoni, V. destructor* and the tracheal mite *A. woodi* are widely distributed, and they attack the most industrious honeybee colonies of *Apis mellifera* worldwide. The *Varroa* mite alone is reported to destroy 100 % *A. mellifera* colony in the western regions (Dejong 1990). However, Calderone and his co-workers (1997) have estimated about 30–90 % of colony losses. Several synthetic acaricides developed on the basis of petrochemical derivatives – namely, amitraz, cymiazole, fluvalinate, flumethrin, clofentezine, chlorobenzilate, and coumaphos – have been evaluated but only few of them have been found effective to control the honeybee mite species (Ferrer-Dufol et al. 1991). Milani (1995) reported that *Varroa* mite has developed resistance to fluvalinate. Besides the development of resistance in parasitic mite through acaricide, the synthetic acaricides may contaminate the honey and wax products with these pesticides (De Ruijter 1995; Wallner 1995). Synthetic acaricides may also leave residue in honeybee products, quickly develop resistance in mite pests, and destroy other life-forms in honeybee colonies.

Natural products derived from various plant species to manage the mite pests could be a better bioresource tool because they are eco-friendly, have little chances of developing resistance in mite pests, have the least residue problem as compared to synthetic acaricides, and are least hazardous to other life-forms of honeybees in colonies. Under the survey of scientific literature in the area of botanicals and phytochemicals to manage honeybee mite species, it has been observed that plant products containing secondary metabolites might be a potential tool to develop novel and safe acaricide for managing the major honeybee mite pests (Qamar 2000).

2 Parasitic Mite Management

2.1 *Varroa jacobsoni*

Among other parasitic mites, *V. jacobsoni* is found to damage *A. mellifera* colonies, sometimes to the tune of 100 %. Vegetable oils and plant products containing secondary metabolites have been reported to manage the parasitic mite pests during beekeeping. Since *A. mellifera* is one of the most industrious species of bees for honey production, bioactive materials from botanicals have been evaluated to manage the parasitic mite in honeybee colonies.

Thymol is found in various plant products and vegetative oils which have been found to be one of the potential plant molecules to control the parasitic mite as compared to amitraz – a synthetic acaricide (Marchetti and Barbattini 1984). Thymol tested as mite fumigant against *V. jacobsoni* in the colony of *A. mellifera* showed 81 % mortality (Lodesani et al. 1990). However, thymol powder used at the rate of 0.5 g/ hive killed 97 % of *Varroa* mite (Chiesa 1991). The application of thymol on top of the hive showed significant reduction in controlling the incidence of *Varroa* mites (Calderone and Spivak 1995). Mladan and co-workers (1986) have also suggested the use of thymol for higher activity against *Varroa* mite. Aerosol prepared based on thymol (5–15 g/l air), camphor (50–150 ug/l of air), and menthol (20–60 g/l of air) gave complete mortality of the parasitic mite (Imdorf et al. 1995). Thymol blended with the oils of *Chenopodium* and fir showed the highest bioactivity against *Varroa* mite. When thymol powder (1–6 g/hive) was applied in honeybee colonies, 92.6 % mortality was recorded (Higes and Llorente 1997). Among thymol with cineol, citronellal, or linalool mixed in ratio of 1:1 (w/w), thymol and cineol showed 56.4 % of mite mortality in the honeybee colonies (Calderone et al. 1997). The spray prepared from menthol 3.7–4 % showed miticidal effect against *Varroa* mite in colonies of *A. mellifera* (Steen 1992). A formulated product "APILIFE" containing 74 % thymol, 16 % eucalyptus oil, and 3 % each of camphor and menthol controlled 75 % of *Varroa* mite when hive plates were impregnated (Gregorc and Jeleng 1996). Treatment with thymol ($n=24$) resulted in an average mite mortality of 75.4 ± 5.79 %, significantly less than that attained with Apistan (tau-fluvalinate) or formic acid. The addition of essential oils did not affect treatment efficacy of either formic acid or thymol (Calderone 2000).

The essential oil obtained from *Heterotheca latifolia* showed potential acaricidal activity against *V. jacobsoni* in the laboratory. Even high concentration (5 % *H. latifolia*) in distilled water and emulsifier did not show bee toxicity (Ruffinengo et al. 2002). Celery extracts and fresh celery juice were mixed at different concentrations, and the mixtures were applied to parasitized bees with *V. jacobsoni*; the fresh celery juice had a significant influence on *V. jacobsoni* mortality, but 50 % fresh celery juice was found effective and safe to honeybees in laboratory studies (Yan et al. 2006).

2.2 *Varroa destructor*

Essential oils obtained from plant species have shown excellent acaricidal activities against major mite pests in honeybee colony (Ellis 2001; Umpierrez et al. 2011). Plant essential oils of the family Labiatae have shown promising results in controlling the parasitic mite in honeybee colonies (Colin 1990). Linalool is found in several aromatic plant species. Linalool (1 %) mixed with sugar syrup demonstrated total control of the *Varroa* mite (Prigli et al. 1991). The powder of *Tussilago farfara* at 2–4 g/hive caused a knock-down effect on the parasitic varroa mite (Palmeri 1995). The menthol obtained from essential oil of plant species of the family Labiatae and applied to control the parasitic mite was found to be effective against *Varroa* mite at 30–32 °C than controlling the tracheal mite at 25 °C. However, the application of menthol at 20 °C temperature was not effective (Hoopingarner and Zabik 1992).

Thymol and thymol blended with other essential oils or essential oil components offer a promising exception. During controlling *V. destructor* in honeybee colonies, residues of thymol found in honey collected from the beehives ranged from 0.021 to 0.288 for thymol in dust, from 0.119 to 0.311 for thymol solution in sugar syrup, and from 0.041 to 0.277 mg Kg^{-1} for thymol solution in alcohol (Emsen and Dodologlu 2009). These formulations controlled *V. destructor* mites to the tune of 90 %, and their residues in honey were found low even after long-term treatments (Hu et al. 2005). Studies have shown that relying solely on a single treatment with essential oil components is not sufficient generally to maintain mite population below the economic injury level (Imdorf et al. 1999). A thymol-based miticide – Apiguard – was found effective in controlling the *Varroa* and tracheal mites in field studies. Honey collected from control colonies had no detectable levels of thymol (Mattila and Otis 1999). Similarly, other thymol-based biomiticides, namely, Thymovar® and BeeVital®, controlled significantly the *V. destructor* to 96.91 and 88.66 %, respectively, and these biopesticides did not pose mortality of the queen, brood, and adult honeybee *A. mellifera* (Akyol and Yeninar 2008). Commercial formulations, namely, APIGUARD containing thymol and Api Life containing thymol, eucalyptol, menthol, and camphor based on natural products, have been utilized to control the varroa (*V. destructor*) in Friuli-Venezia Giulia in Italy in the year 2005 (Greatti 2005).

The attractive-repellent test against the *V. destructor* mite and honeybee revealed that ethanolic botanical extract of *Baccharis flabellate* stimulated the olfactory stimulus of mites and was found to be a repellent (Damiani et al. 2011). Among leaves and stem extract of *Swietenia mahogany* and *S. macrophylla* tested against *Varroa* mite in honeybee colony, *S. mahogany* bark and *S. macrophylla* leaves gave 100 and 95 % reduction after 48 h of treatment in mite when applied at 500 ppm concentration (El Zalabani et al. 2012).

The topical application of formic acid and marjoram gave a 97.6 % mortality of *Varroa* species. Honeybee mites were found to be controlled by the application of a formulation blended with sulfur, garlic, and pepper (Greatti and Barbatinin 1996). Natural oil extracted from *Eucalyptus camaldulensis* killed 83 % of the total fallen mites in honeybee colonies (Rezk and Gadelhak 1997). The smoke extracts prepared based on a volatile compound of 2,6-dimethoxyphenol showed significant contact activity in causing *Varroa* to dislodge from adult honeybees (Elzen et al. 2001a, b).

Among botanical oils neem, thymol, and canola examined for control of parasitic mites in honeybees colonies in British Columbia, neem oil spray (5 % solution) killed 90±6 % of varroa mites. Colonies treated with the thymol oil spray had a significantly lower tracheal mite infestation (1.3±7.5 %) at the end of the treatment period than the untreated group (23.3±6.0 %). However, neem and thymol oil spray treatments were found detrimental to bees too (Whittington et al. 2000). Peng and co-workers (2000) found that feeding host larvae of honeybees with azadirachtin derived from *Azadirachta indica* significantly reduced the fecundity of mother varroa mites ($P<0.001$). However, azadirachtin is an already known plant-based insecticide on which detailed studies are needed to further evaluate its efficacies against mites and its safety to honeybee's adult workers, queens, and drones. Plant extracts – neem oil, mixture (neem oil, garlic oil, and tobacco oil), and tobacco oil – have shown encouraging results in managing *V. destructor* (Qayyoum et al. 2013). Melathopoulos and co-workers (2000a, b) have reported that neem oil applied topically to infested bees in the laboratory proved highly effective against both *Varroa* and *Acarapis* species. However, honeybees were also found deterred from feeding on sucrose syrup containing >0.01 mg/ml of neem-aza, and bee mortality was found less than 10 % in 48 h after treatment in the laboratory.

Plant essential oils are attractive chemicals for the control of *Varroa* mites on honeybees because they are perceived as "natural" compounds that will not contaminate hive products. Laboratory experiments revealed that mites were killed by origanum, a thymol mixture, clove, bay, and tea tree (*Melaleuca*). Origanum, the thymol mixture, cineole (eucalyptol), and the commercial product Bee Calm, all lodged varroa (Sammataro et al. 1998). As another potential ecological tool, superfine-ground, plain white sugar dusting on infested honeybee was found effective to control varroasis (Fakhimzadeh 2000, 2001). The grapefruit leaf-burning residue containing phenolic 3-methylphenol was most active in causing *Varroa* to detach from bees in the laboratory indirect and direct exposure bioassays (Elzen et al. 2001a, b).

With a view to replace the synthetic chemicals in controlling the *Varroa* mite in *A. mellifera*, a mixture containing dosages of thymol (15 g), menthol (1 g), eucalyptus oil (3 ml), and citronella oil (1 ml) gave efficacy of 89.71–90.20 % in infested hives (Bacandritsos et al. 2004). The experimental data showed superiority of the mixture of thymol obtained from *Thymus vulgaris*, and the extract of *Artemisia cina* caused an obvious reduction in the rate of mite infestation either in sealed brood cells or on adult bees (Elbassiouny et al. 2006). A 10 ml pyrethrum extract containing 4 % pyrethrin and 16 % piperonyl butoxide as a synergist sprayed in the hive resulted in a 95 % kill of *Varroa* mites. However, it is suggested that residual analysis of pyrethrin should be a prerequisite before its recommendation for commercial use (Nijhuis et al. 1987).

3 Tracheal Mite Management

The tracheal mite, *A. woodi*, is one of the major honeybee mite pests, which also cause serious threat to honeybee colony. Different isomers of natural and synthetic menthol have been used to control the tracheal mite (Wilson et al. 1989). Menthol obtained from plant species of the family Labiatae has been found to control the tracheal mite *A. woodi*. Application of menthol as a solid cake reduced the population of *A. woodi* from 25.3 to 1.6 % during a 12-month period. Another experimental result revealed that the use of 50 g pack of solid menthol placed on the bottom board of the hive resulted in 82 % reduction in *A. woodi* population (Cox et al. 1988; Morrison 1988; Moffett et al. 1989). Menthol with vegetable oil also decreased the population of *A. woodi* (Delaplane 1992). When 30–60 g menthol was placed on foam strips, mite population was found reduced but brood mortality and high concentration of residues in honey were recorded (Morrison 1988; Nelson et al. 1993).

Oils of clove, eucalyptus, peppermint, and marjoram extract resulted in reductions of 46–100 % of the population of *A. woodi* in 15 days (Abu-Zaid et al. 1990). Tracheal mites were found to be deterred after application of 60 g/hive of *Artemisia* powder (Abu-Zaid and Salem 1991). One parts of the groundnut, sunflower, soybean and canola (rapeseed) oil with two parts of sugar (w/w) reduced the mite prevalence (Calderone and Shimanuki 1995). The population of tracheal mites was found reduced by application of oil patties and sugar with or without antibiotic (Terramycin) in honeybee hives (Sammataro et al. 1994). Liu (1995) have reported that neem-based pesticide – MORGOSAN 03 ml/l containing 3,000 ppm azadirachtin – controlled the adult tracheal mites and also acted as ovicidal to mite eggs. The treated colony with neem-based pesticide was also found to have collected more pollen grains as well as higher production of honey. Citronellal and clove oil were recorded to decrease the population of *A. woodi* 78.2 and 63.4 %, respectively. The bee mortality was also found to be lower after application of these plant products (Calderone et al. 1991).

Most of the earlier researchers have given focus to manage the parasitic mite *Varroa jacobsoni* and the tracheal mite *Acarapis woodi* and recently against *Varroa destructor* through the use of secondary metabolites obtained from plant species with special reference to thymol and menthol. Most efforts were taken to manage the *Varroa* mites during the 1980s–1990s. No attention was made to discover plant products/secondary metabolites as tools of honeybees before 1980. It is also concluded that only few plant species and plant oil/molecules have been investigated to manage the honeybee mite. Evidently, the use of synthetic acaricides may pose a great threat to the brood and other hive products because of the development of resistance in mites and contamination of honey and bee waxes. Therefore, the use of essential oils and their major constituents may be a basis to investigate more potential bioresource as plant acaricide to use against the honeybee mites. Since most of the secondary metabolites from plant species have shown potential activity against honeybee mite management, the medicinal and aromatic plant species which are known for synthesizing high amount of biologically active secondary metabolites might be screened in the future to investigate more potential and safe botanical acaricides. However, the edible plant materials such as wheat flour may also be useful for mite management (Logilo and Pinessi 1991). Besides, a comprehensive review till 2000 on mite management with special reference to utilization of plant secondary metabolites and focus on apiculture strategies have already been available elsewhere (Singh et al. 2001; Singh 2007). The methods of application of botanicals and their natural products against mites in honeybee colony; their adverse effects on the brood, collection of pollen quality/quantity, and honey productivity; and residue and resistance problem in honeybee mite management are intriguing factors in discovering new tools for the management of honeybee mite pests in honeybee colonies.

References

Abu-Zaid MI, Salem MM (1991) Evaluation of certain doses of wormwood as bioactive agent against the acarine mite, *Acarapis woodi* (Rennie). Bull Entomol Soc Egypt Econ Ser 17:121–125

Abu-Zaid MI, Mazzeed MM, Salem MM (1990) Natural bioactive substances for controlling *Acarapis woodi* (Rennie). Bull Entomol Soc Egypt Econ Ser 16:283–287

Akyol E, Yeninar H (2008) Controlling *Varroa destructor* (Acari: Varroidae) in honeybee *Apis mellifera* (Hymenoptera: Apidae) colonies by using Thymovar® and Beevital®. Ital J Anim Sci 7:237–242

Bacandritsos N, Bourbos E, Anastassiadou H (2004) Possibility of treating acariasis caused by *Varroa jacobsoni* with natural substances. Veterinaraski Glasnik 58:359–368

Calderone NW (2000) Effective fall treatment of *Varroa jacobsoni* (Acari: Varroidae) with a new formulation of formic acid in colonies of *Apis mellifera* (Hymenoptera: Apidae) in the Northeastern United States. J Econ Entomol 93:1065–1075

Calderone NW, Shimanuki H (1995) Evaluation of four seed derived oils as control for *Acarapis woodi* (Acari: Tarsonemidae) in colonies of *Apis mellifera* (Hymenoptera: Apidae). J Econ Entomol 88:805–809

Calderone NW, Spivak M (1995) Plant extracts for the control of the parasitic mite *Varroa jacobsoni* (Acari: Varroidae) in colonies of the western honeybee (Hymenoptera: Apidae). J Econ Entomol 88:1211–1215

Calderone NW, Bruce WA, Allen-Wardell G, Shimanuki H (1991) Evaluation of botanical compound for the control of the honeybee tracheal mite, *Acarapis woodi*. Am Bee J 131:589–599

Calderone NW, Wilson WT, Spivak M (1997) Plant extracts for the control of the parasitic mite *Varroa jacobsoni* (Acari: Varroidae) and *Acarapis woodi* (Acari: Tarsenomidae) in colonies of *Apis mellifera* (Hymenoptera: Apidae). J Econ Entomol 90:1080–1086

Chiesa F (1991) Effective control of varroatosis using powdered thymol. Apidologie 22:135–145

Colin ME (1990) Essential oils of Labiatae for controlling honey bee varroasis. J Appl Entomol 110:19–25

Cox RL, Moffett JO, Ellis M, Wilson WT (1988) Long-term beneficial effects of menthol treatment on honeybee colonies infested with tracheal mites *Acarapsis woodi*. Am Bee J 128:801

Damiani N, Gende LB, Maggi MD, Palacios S, Marcangeli JA, Eguaras MJ et al. (2011) Repellent and acaricidal effects of botanical extracts on *Varroa destructor*. Parasitol Res 108:79–86

De Ruijter A (1995) Issues in the control of *Varroa* infestation. In: Matheson A (ed) New perspectives on Varroa. IBRA, Cardiff, pp 24–26

Dejong D (1990) Mites: Varroa and other parasites of brood. In: Morse RA, Nowogrodzki R (eds) Honey bee pest, predators and diseases, 2nd edn. Cornell University Press, Ithaca

Delaplane KS (1992) Controlling tracheal mites (Acari: Tarsenomidae) in colonies of honeybee (Hymenoptera: Apidae) with vegetable oil and menthol. J Econ Entomol 85:2114–2118

El Zalabani SM, Hesham I, Ei-Askary et al (2012) Acaricidal activity of *Swietenia mahogany* and *Swietenia macrophylla* ethanolic extracts against *Varroa destructor* in honeybee colonies. Exp Parasitol 130:166–170

Elbassiouny AM, Abdel-Megeed MI, El-Shaarawi MO, Mohammed GM (2006) Effect of colony management and plant extraction on varroa mites attacking honeybee colonies. Ann Agric Sci (Cairo) 51:559–672

Ellis JD Jr (2001) The future of varroa control: integrating current treatments with the latest advancements. Am Bee J 141:127–131

Elzen PJ, Elzen GW, Stipanovic RD (2001a) Biological activity of grapefruit leaf burning residue extract and isolated compounds on *Varroa jacobsoni*. Am Bee J 141:369–371

Elzen PJ, Stipanovic RD, Rivera R (2001b) Activity of two preparations of natural smoke products on the behavior of *Varroa jacobsoni* Oud. Am Bee J 141:289–291

Emsen B, Dodologlu A (2009) The effects of using different organic compounds against honey bee mite (*Varroa destructor* Anderson and Trueman) on colony developments of honey bee (*Apis mellifera* L.) and residue levels in honey. J Anim Vet Adv 8:1004–1009

Fakhimzadeh K (2000) Potential of super-fine ground, plain white sugar dusting as an ecological tool for the control of varroasis in the honey bee (*Apis mellifera*). Am Bee J 140:487–491

Fakhimzadeh K (2001) The effects of powdered sugar varroa control treatments on *Apis mellifera* colony development. J Agric Res 40:105–109

Ferrer-Dufol M, Vinuales MAI, Sanchez-Adedo C (1991) Comparative tests of fluvalinate and flumethrin to control *Varroa jacobsoni* Oudemans. J Apic Res 30:103–106

Greatti M (2005) Control of varroa in Friuli Venezia Giulia. Notiziario ERSA 18:43–47

Greatti M, Barbatinin R (1996) Efficacy of a product containing sulphur, garlic, and pepper against *Varroa jacobsoni* and its tolerance by *Apis mellifera*. Selezione Veterinaria 1:9–20

Gregorc E, Jeleng J (1996) Control of *Varroa jacobsoni* Oud. in honey bee colonies using Apilife-var. Zbornk Veterinarrske Fakultete, Univerza Ljulbljna 33:231–235

Grobov OF (1979) Pathology of honey bee. Veterinaria, Moscow 6:69–71

Higes M, Llorente J (1997) Thymol, tests of efficacy in the control of varroasis in honey production colonies. Vida Apicola 81:14–17

Hoopingarner RA, Zabik MJ (1992) Tracheal mite (*Acarapsis woodi*) temperature dependent mortality with saturated menthol vapour. Am Bee J 132:809–810

Hu F, Zhu W, Li Y (2005) Use of essential oils for controlling *Varroa destructor* in honey bee colonies. Chin Bull Entomol 42:375–378

Imdorf A, Kilchenman V, Bogdanov S, Bachofen B, Beretta C (1995) Toxic effects of thymol, camphor, menthol and eucalyptol on *Varroa jacobsoni* Oud. and *Apis mellifera* L. in a laboratory test. Apidologie 26:27–31

Imdorf A, Bognadov S, Ochova RI, Calderone NW (1999) Use of essential oil for the control of *Varroa jacobsoni* Oud. in honeybee colonies. Apidologie 30:209–228

Liu TP (1995) Controlling tracheal mites in colonies of honeybee with neem (Margosan-O) and flumethrin (Bayvarol). Am Bee J 135:562–566

Lodesani M, Bergomi S, Pellacani A, Carpana E, Rabitti T (1990) A comparative study on the efficacy of some products for controlling Varroa and determination of the residue. Apicoltura 6:105–130

Logilo G, Pinessi E (1991) Use of wheat flour for the ecological control of varroasis. Apicoltore Modemo 82:185–192

Marchetti S, Barbattini R (1984) Comparative effectiveness of treatments used to control *Varroa jacobsoni* Oud. Apidologie 15:363–378

Mattila HR, Otis GW (1999) Trails of Apiguard, a thymol-based miticide. Part 1. Efficacy for control of parasitic mites and residues in honey. Am Bee J 139:947–952

Melathopoulos AP, Winston ML, Whittington R, Higo H, LeDoux M (2000a) Field evaluation of neem and canola oil for the selective control of the honeybee (Hymenoptera: Apidae) mite parasites *Varroa jacobsoni* (Acari: Varroidae) and *Acarapis woodi* (Acari: Tarsonemidae). J Econ Entomol 93:559–567

Melathopoulos AP, Winston ML, Whittington R, Smith T, Lindberg C, Mukai A, Moore M (2000b) Comparative laboratory toxicity of neem pesticides to honey bees (Hymenoptera: Apidae), their mite parasites *Varroa jacobsoni* (Acari: Varroidae) and *Acarapis woodi* (Acari: Tarsonemidae). J Econ Entomol 93:199–209

Milani N (1995) The resistance of *Varroa jacobsoni* Oud. to pyrethroids: A laboratory assay. Apidologie 26:361–440

Mladan V, Erski-Biljic M, Jakic D, Simic M (1986) Application of some preparation for the control of Varroa diseases. 2 Veterinarski Glasnik 40:509–514

Moffett JO, Cox RL, Ellis M, Rivera R, Wilson WT, Carduso-Tamez D, Vargas CJ (1989) Menthol reaches winter population of tracheal mites, *Acarapsis woodi* in honey bee from Mexico and Nebraska. Southwest Entomol 14:57–65

Morrison J (1988) Menthol clears bees breathing tubes. Apicultural Research, USDA 36:11

Nelson D, Sporns P, Kristiansen P, Mills P, Li M (1993) Effectiveness and residue levels of three methods of menthol application to honey bee colonies for the control of tracheal mites. Apidologie 24:549–556

Nijhuis H, Enss K, Vob S, Tietgen I, Tietgen W (1987) The use of pyrethrum-extract for control of bee mites, *Varroa jacobsoni* in apiculture. Pyreth Post 16:118–119

Palmeri V (1995) Studies on the efficacy of powders of *Tussilago farfara* against *Varroa jacobsoni*. Apicoltore Moderno 86:67–75

Peng CYS, Trinh S, Lopez JE, Mussen EC, Hung A, Chuang R (2000) The effects of azadirachtin on the parasitic mite, *Varroa jacobsoni* and its host honey bee (*Apis mellifera*). J Apic Res 39:159–168

Prigli M, Suhayda J, Beckessy G (1991) Control of *Varroa jacobsoni* on honey bees with terpenes. International Patent Application WO 91 07 875, 16 pp

Qamar M (2000) Evaluation of some natural products for controlling mites of *Varroa* sp. parasitizing honeybees in Sokoto. Shashpa 7:49–52

Qayyoum MA, Khan BS, Bashir MH (2013) Efficacy of plant extracts against honey bee mite, *Varroa destructor* (Acari: Varroidae). World J Zool 8:212–216

Rezk HA, Gadelhak G (1997) Impact of four plant oil extracts on the control of the parasitic mite, *Varroa jacobsoni* Odu. (Acari: Prostigmata). Alexendria J Agric Res 42:105–113

Ruffinengo SR, Eguaras MJ, Cora D, Rodriguez E, Bedascarrasbure E, Bailac PN, Ponzi MI (2002) Biological activity of *Heterotheca latifolia* oil against *Varroa jacobsoni*. J Essent Oil Res 14:462–464

Sammataro D, Cobey S, Smith BH, Needham CR (1994) Controlling tracheal mites (Acari: Tarsenomidae) in honey bees (Hymenoptera: Apidae) with vegetable oil. J Econ Entomol 87:910–916

Sammataro D, D-Hoffman G, Needham G, Wardell G (1998) Some volatile plant oils as potential control agents for *varroa* mites (Acari: Varroidae) in honeybee colonies (Hymenoptera: Apidae). Am Bee J 138:681–685

Sammataro D, Gerson U, Needham G (2000) Parasitic mites of honeybees: life history, implications, and impact. Annu Rev Entomol 45:519–548

Singh D (2007) Apiculture in India. Curr Sci 92:1335–1336

Singh Dwijendra, Niyoliya NK, Kumar S (2001) Advances in the use of plant products in honeybee mite management. In: Proceedings of the seventh international conference on tropical bees: management and diversity & fifth Asian Apicultural Association conference held during 19–25 March 2000 at Chiang Mai, Thailand organized by IBRA and AAA, Chiang Mai, pp 237–242

Steen J Van Der (1992) The effects of mixture of etheric oils on *Varroa jacobsoni* infesting honeybee colony. Apidologie 23:383–389

Umpierrez ML, Santos E, Gonzalez A, Rossini C et al. (2011) Plant essential oils as potential control agents of varroatosis. Phytochem Rev 10:227–244

Wallner K (1995) The use of varrocides and their influence on the quality of bee production. Am Bee J 135:817–831

Whittington R, Winston ML, Melathopoulos A, Higo HA (2000) Evaluation of the botanical oils neem, thymol and canola sprayed to control *Varroa jacobsoni* Oud. (Acari: Varroidae) and *Acarapis woodi* (Acari: Tarsonemidae) in colonies of honey bees (*Apis mellifera* L., Hymenoptera: Apidae). Am Bee J 140:567–572

Wilson WT, Pettis JS, Collins AM (1989) Efficacy of different isomers of menthol against the honeybee tracheal mite. Am Bee J 129:826

Yan WY, Jiao YL, Huang K, Zeng Z (2006) Studies on the effect of celery extracts on *Varroa jacobsoni*. Acta Agriculturae Universitatis Jiangxiensis 28:912–914

14 Nanotechnology and Plant Biopesticides: An Overview

K. Sahayaraj

Abstract

The increasing worldwide population and demand for higher food stocks require application of modern techniques for manifold agricultural production as well as minimizing losses during cultivation of crops, transportation, and storage. Various methods have been involved and used to combat insect pests since ancient period. During the nineteenth century, inorganic, botanical, and natural pesticides have been mostly exploited resulting in reduced losses in agricultural yield. Synthesis of new chemicals based on discovery of new structure elucidation and biological activities of various compounds have been invented globally. Further, the severe adverse effects of the chemical pesticides on the environment mainly due to development of resistance, residue problem, and harmful effect on beneficial arthropods have been recorded in the past that led to unrest among public which resulted quick actions for stricter regulations and legislation for reducing their wider use. Since the resistance problem in pests have been rapidly found in mostly all the classes of arthropods, the advance approach to increase the efficiency of pesticides is now known to be the application of nanotechnologies in modern formulation technologies. In this chapter we have discussed in detail about the possibility of application of nanotechnology in improving the bioactivities of plant biopesticides against pests of agriculture and storage.

Keywords

Nanomaterials • Bioreduction • Biopesticides • Pest management

K. Sahayaraj, Ph.D. (✉)
Crop Protection Research Centre, St. Xavier's College,
Palayamkottai 627002, Tamil Nadu, India
e-mail: Ksraj48@gmail.com

D. Singh (ed.), *Advances in Plant Biopesticides*,
DOI 10.1007/978-81-322-2006-0_14, © Springer India 2014

1 Introduction

Insect pests, diseases, and weeds cause considerable damage to potential agricultural production worldwide. Insects fed on crop plants, forest trees, medicinal plants, weeds, etc. They also feed food and other stored products causing huge amount of loss and also deterioration of food quality. In the context of crop protection, conventionally synthetic pesticides of different class have been used widely beyond geographical boundaries globally. The World Health Organization (WHO) has already estimated the human death to be about 20,000 people by pesticide poisoning in the Third World annually. Synthetic pesticides developed on the basis of petrochemical derivatives are reported to be accumulating in soil and groundwater where they have been threatening the health of the entire ecosystems. Pest outbreak, insect resistance against pesticides, pest resurgence, health hazards to human beings, as well as domestic and wild animals are considered as important adverse impacts of pesticides (Casida and Quistad 2000).

Nanotechnology is a new emerging area that deals with materials having the size of 10^{-9} or one million part of meter. Both button-up and top-up approaches have been followed using physical and chemical methods. There are enormous side effects put forth by many scientists. Many industrial people have been utilizing these particles in a useful manner. Similarly, agriculturists have been utilizing this technology for excellent plant-growth stimulation (Misra et al. 2013), genetic improvement of crops, plant disease diagnosis, postharvest management and food biotechnology (Prasann 2007; Mousavi and Rezaei 2011), disease and stress resistance (Brecht et al. 2003), plant disease control (Hae-Jun et al. 2006), and delivery of DNA and chemicals into the plant (Wang et al. 2002). The tomato seedlings growing in carbon nanotubes (CNT) showed enhanced growth due to enhanced water uptake which caused by penetration of CNT as vehicle to deliver desired molecules either nutrient or biocides into the seeds during germination. The role of nanotechnology in general (Rai and Ingle 2012; Margulis-Goshen and Magdassi 2013) – solid and liquid formulations of nanoparticles (Goswami et al. 2010), novaluron nanoparticles (Elek et al. 2010), pesticide delivery by nanosuspensions (Chin et al. 2011), and fungi-based nanomaterials (Kashyap et al. 2013) in particular – was proposed by many scientists for agriculture pest management. Further, scientific publication and patent right related to the perspectives of nanotechnology in plant production and protection have been highlighted (Goodsell 2004; Sinha et al. 2009; Gogos et al. 2012).

As an alternative, botanicals have been recommended for crop production and protection. Further, botanicals have certain advantages and certainly having demerits. However, botanical insecticides are not widely used in conventional crop production but organic crop producers in industrialized countries recognize them. Botanical pesticides are extracted from stems, seeds, roots, leaves, and flower heads of different plant species. Flowers of pyrethrum are used to extract pyrethrin (*Tanacetum cinerariifolium* Trev. Schultz Bip.), a botanical insecticide with a long history of safe and effective use in pest management. The bioactive substance of pyrethrin is a mixture of six active components (pyrethrins I and II, cinerins I and II, and jasmolins I and II) (Grdiša et al. 2013). The insecticidal potential of pyrethrin was recognized decades ago, and dried and ground flowers have traditionally been used worldwide for agricultural pests (Casida. 1980; Dubey et al. 2010; Prota et al. 2013), storage (Campos and Phillips 2013), and household pest management (Lu et al. 2013).

Indian lilac or neem tree or margosa tree (*Azadirachta indica* A. Juss, *Melia azadirachta* L., or *Antelaea azadirachta* L.) (Meliaceae) or neem (*Azadirachta indica* A. Juss. = *Melia azadirachta* L.) a most common tree in India. Partially or complete inhibition of fecundity (Sahayaraj 1998; Ghazawi et al. 2007; Silva et al. 2013) by neem and/or sometimes egg hatchability (Khan et al. 2007), reduction of the life span (Dutta et al. 2013), oviposition repellence against females (Barati et al. 2013), direct ovicidal effects, antifeedant effects against young ones and adults (Ikeura et al. 2014), formation of permanent larvae (Andrade-Coelho et al. 2009), insect growth regulator effects at molting between one stage to another (Viñuela et al. 2001) and especially in the prepupal stage and analogous lesions during the emergence of adults (Schmutterer et al. 1988) and act as pesticide.

Rotenone (*Lonchocarpus utilis* A. C. Sm.) has been shown to be an effective control agent of many pest species, including orthopterans (Sandoval-Mojica et al. 2011), coleopterans (Sandoval-Mojica et al. 2011), aphids (Fei Yi et al. 2012), mites (Satta et al. 2008), and moths (Wang et al. 2007). Quassia (*Quassia amara* L.) (Psota et al. 2010), sabadilla (*Schoenocaulon officinale* Schltdl. & Cham.), and ryania (R*yania speciosa* Vahl.) (Regnault-Roger 2012) have been utilized worldwide as botanicals for pest management. Recently development of technical research capabilities, particularly relating to the botanical-based nanoparticles have not been used so far in the pest management program. Hence, this review is focused on plant biopesticide strategies available for the management of insect pests and role of nanomaterials and nanotechnology as modern approaches for the management of insect pests in agriculture.

2 Nanotechnology

Nanotechnology employs nanoparticles (NPs) having a dimension of 100 nm or less or 10 and 1,000 nm. Metallic nanoparticles are mostly prepared from noble metals such as gold, silver, and platinum. Metal oxidase materials (TiO2, ZnO, AgO, and MgO), ceramics, silicates, magnetic materials, semiconductor quantum dots (QDs), lipids, polymers, dendrites, and emulsions have been used to make NPs. Nanomaterials are of both organic and inorganic origin. They are synthesized by physical, chemical, and biological methods. Nanopesticides, nanomicrobicides, and nanobiopesticides are being used efficiency in agriculture (Owolade et al. 2008).

The field of nanotechnology is one of the most active areas of research in modern material sciences. Nanotechnology is a field that is developing day by day, making an impact in all spheres of human life and creating a growing sense of excitement in the life sciences especially biomedical devices and biotechnology (De Villiers and Hoisington 2011). The use of nanoparticles is gaining impetus in the present century as they posses defined chemical, optical, and mechanical properties.

2.1 Application of Nanotechnology

Nanotechnology is exploited for many applications in the chemical, agricultural, medical, biological, and cosmetics industries and many more. However, the use of (nanoparticles) nanotechnology in agriculture is still in its infancy stage. In the beginning, mesoporous silica nanoparticles (MSNs) for biocide (imidacloprid) (Popat et al. 2012), porous hollow silica nanoparticles (PHSNs) for water-soluble pesticide validamycin (Liu et al. 2006). Insecticidal effect of nanostructured alumina was evacuated against two major insect pests in stored food such as *Sitophilus oryzae* L. and *Rhyzopertha dominica* (F.). Results reveal that nanostructured alumina caused significant mortality after 3 days of continuous exposure to treated wheat. Further, 9 days after treatment, the LD_{50} ranged from 127 to 235 mg kg^{-1} (Stadler et al 2010). At 15–30 nm size range, the diatomaceous earth (DE) was used to design amorphous nanosized hydrophilic, hydrophobic, and lipophilic silica (silica nanoparticle – SNP), and its entomotoxicity was tested against rice weevil *Sitophilus oryzae.* Results showed SNP was found to be highly effective against *Sitophilus oryzae* causing >90 % mortality, indicating the effectiveness of SNP to control stored-product insect pests (Debnath et al. 2011). Nanosilica which prepared from silica has been used as nano-pesticide. The exoskeleton of insects is composed of lipid which is used as a water barrier to protect the insects and thereby prevent death from desiccation. But nanosilica of 3–5 nm gets absorbed into the exoskeleton (cuticle) lipids by physisorption and thereby causes death of insects Barik et al. (2008).

Synthetic pesticides released using nanoparticles efficiently in order to increase their efficiency, better uniformity of coverage for highly active compounds and less exposure to workers, relative to compounds solubilized in organic solvents. For instance, polymeric stabilizers bifenthrin nanoparticles have been proposed for the slow release of bifenthrin nanoparticles (Liu et al. 2008) and increase pesticide loading from 50 to 91 %. Previously, Liu et al. (2006) also released validamycin, another pesticide using porous hollow silica nanoparticles (PHSNs).

Other materials have also been proposed in insect pest management (Joseph and Morrison 2006; Scott 2007; Scrinis and Lyons 2007; Bhattacharyya 2009; Bhattacharyya et al 2010; Harper 2010) as well as plant protection (Ghormade et al. 2011). Nanosilica may be useful against stored grain, household pests, animal parasites, fungal organisms, worms, etc. (Ghormade et al. 2011). Further, many studies confirmed that metal nanoparticles are effective against plants pathogens, insects, and pests. Ghormade et al. (2011) proposed the following potential applications of nanotechnology in agriculture: delivery of nanocides–pesticides encapsulated in nanomaterials for controlled release; stabilization of biopesticides with nanomaterials; slow release of nanomaterial-assisted fertilizers, biofertilizers, and micronutrients for efficient use; and field applications of agrochemicals and nanomaterial-assisted delivery of genetic material for crop improvement.

2.2 Nanoencapsulation and Feasible Method of Nanotechnology

Encapsulation is a prime technique that has been followed by people for pesticide release. Nanoencapsulation is a process through which a chemical such as an insecticide is slowly but efficiently released to a particular host plant for insect pest control. Nanoencapsulation with nanoparticles in the form of pesticides allows for proper absorption of the chemical into the plants unlike the case of larger particles (Scrinis and Lyons 2007). Initially Allan et al. (1971) tried to release an insecticide through a polymeric encapsulation, but they have not succeeded in their efforts. Later, oil-in-water nanoemulsions for pesticide formulations and release (Wang et al. 2007) has been proposed. But, botanicals or their bioactive principles or plant oil has been utilized in nanoencapsulation techniques for pest management program. For example, to release *Moringa oleifera* (synonym: *Moringa pterygosperma*) Lamm. (Moringaceae) extract has released utilized cashew gum as surfactant (Paula et al. 2012). Previously, Mugisha-Kamatenesi et al. (2008) recorded insecticidal activity of *M. oleifera* leaf against many pests including banana weevil, bean fly, cereal stem borers, pod feeders, grain moth, rodents, moths, termites, birds, aphids, and cutworms at Uganda. But not much work has been undertaken. The proposed study has opened an avenue for the utilization of *M. oleifera* leaf in pest management. In another study, Lai et al. (2006) encapsulated *Artemisia arborescens* L. essential oil into solid lipid nanoparticles (SLN) and formulated two different SLN using high-pressure homogenization technique. These SLN was tested against adult *Bemisia tabaci*. Results reveal that both *Artemisia arborescens* (LC100 = 0.1 mg/cm^2) and its SLN (0.5–1.5 %) showed similar impacts.

Essential oils are particularly abundant in some families of plants – conifers, Rutaceae, Umbelliferae, Myrtaceae, and Labiatae – and are often utilized (Ho et al. 1996; Hori 1996; Yang et al. 2010; Zhao et al. 2013) in many economically important pest and disease management (Isman 2006). As a recent advancement, garlic essential oil (Yang et al. 2009) against *Tribolium castaneum*, neem seed oil (Kulkarni et al. 1999), and *Lippia sidoides* essential oil (Paula et al. 2011) were released utilizing polyethylene glycol, alginate–glutaraldehyde, and chitosan–angico gum as encapsulating agents, respectively. Further, N-(octadecanol-1-glycidyl ether)-O-sulfate chitosan with octadecanol glycidyl ether was used for rotenone (Lao et al. 2010), while carboxymethyl chitosan–ricinoleic acid for azadirachtin (Feng and Peng 2012).

Pesticidal plants and microbes possessing the capacity to reduction of metal ions (silver and gold) liner moderately or rapidly at ambient condition in a novel and in triggering in biogenesis of nanoparticles. This kind of bioreduced nanoparticle can be utilized in biointensive integrated pest and disease management. Recently, a good number of works have been done with regard to the plant- as well as microbe-assisted reduction of metallic nanoparticles, showing lots of scope in this multidisciplinary field. Further, pesticidal terpenoids, flavonoids, tannins, and entomotoxic proteins (Rajakumar and Rahuman 2011) were also identified as reducing and copping in biogenesis of nanoparticles.

2.3 Biopesticides and Nanotechnology

Indiscriminate use of pesticides and fertilizers causes environmental pollution, increases pathogen and pest resistance, and results in loss of biodiversity. Reduced soil biodiversity and nitrogen fixation contribute to the bioaccumulation of pesticides, destruction of the habitats of many animals, and health hazards to human beings and domestic and wild animals. Along with chemicals, physical methods, biological methods, and mechanical methods have been individually utilized or in combination (integrated pest management – IPM) for the eradication of pest and diseases as a result to protect on crops and income the production. Later, an improved concept, biointensive integrated pest management (BIPM) was introduced proportion of biological origination (or biopesticides) has increased. Bionano materials by virtual of biological along with nanomaterials related properties, as potential agro-biotechnological application for alternative of synthetic pesticides and also a part of BIPM. Keeping the concept in mind, this chapter is high lightly both nanotechnology and biopesticides in an overall prospective manner. Plant extracts can be utilized as plant-based pesticides (Fig. 14.1a) as well as for bioreduced nanomaterial synthesis (Fig. 14.1b).

Biopesticides are certain types of pesticides derived from natural materials as animals, plants, bacteria, and certain minerals. These include fungi such as *Beauveria* sp., bacteria such as *Bacillus* sp., neem extract, and pheromones. Similarly, canola oil and baking soda have pesticide applications and are considered as biopesticides. Among them, both plants and microbes have been utilized for the synthesis of nanomaterials. For instance, nanoparticles of silver, nickel, cobalt, zinc, and copper have also been synthesized inside the live plants.

2.3.1 Advantages of Biopesticides

Biopesticides in general have a narrow target range and a very specific mode of action; are slow acting; have relatively critical application times; suppress, rather than eliminate, a pest population; have limited field persistence and a short shelf life; are safer to humans and the environment than conventional pesticide; present no residue problems; do not persist in the environment; present a relatively low risk to beneficial predators and parasites (nontarget organisms); usually break down rapidly in the environment; are easily metabolized by animals receiving sublethal doses; have a broad spectrum of activity; and are easy to process and use. Considering these advantages, biopesticides can be utilized for the synthesis of nanomaterials and utilized in agriculture for the betterment of human beings and also to our nature.

2.3.2 Applications of Nanotechnology in Agriculture

Nanotechnology is either directly or indirectly used in agriculture for the delivery of nanocides– pesticides encapsulated with nanomaterials (NMs), slow release of nanomaterial-assisted fertilizers, stabilization of biopesticides with NMs, biofertilizers and micronutrients for efficient use, field application of agrochemicals, NM-assisted delivery of genetic materials (DNA/double-stranded RNA) using gold (Ramanathan et al. 2009), starch (Liu et al. 2008), and chitosan (Zhang et al. 2010) and nanosensor for pesticides like carbofuran (Guo et al. 2009) with gold nanoparticles, DDT (Lisa et al. 2009) with gold particles; dimethoate (Guo et al. 2009) with zirconium oxide (Lisa et al. 2009), iron oxide (Ghormade et al. 2011), organophosphate (Wang et al. 2009), paraoxon with silica (Ramanathan et al. 2009), and pyrethroid with iron oxide (Kaushik et al. 2009) detection.

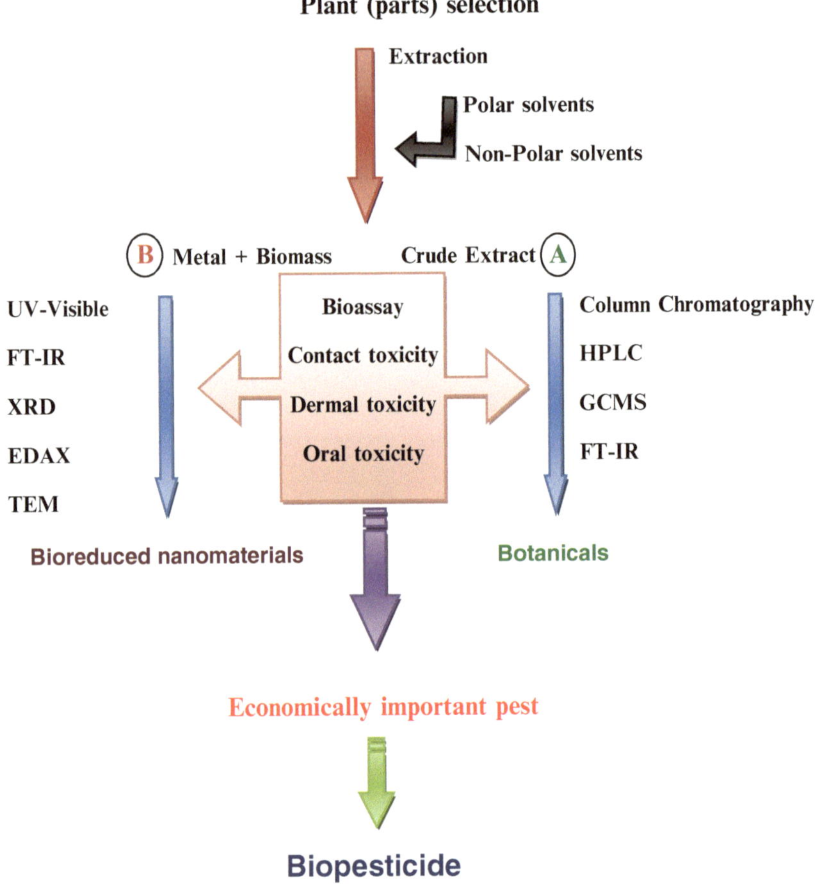

Fig. 14.1 Schematic representation of biopesticide: botanicals-A and bioreduced metal nanomaterial-B for pest management

2.3.3 What Made Nanoparticles for Plant Protection?

NPs possess insecticidal property due to their characteristics like extraordinary strength; more chemical reactivity; possessing high electrical conductivity and optical properties; distinct physical, chemical, and biological properties associated with the atomic strength; specific maintenance of size and shape; size–depth qualities; high surface-to-volume ratio as well as high stability; particles arranged into ordered layer due to hydrogen bonding, dipole forces, hydrophilic and hydrophobic interaction, surface tension, and gravity; they are released slowly at the same time efficient to the host plant, which deter the insect attack; higher mobility and lower toxicity; highly pest specific and effective; and higher self-assembly, stability, specificity, encapsulation, and biocompatibility (Wang et al. 2009).

2.3.4 Nanomaterials in Insect Pest Management

By considering the abovementioned benefits and environmental friendliness of traditional methods, which leading to the development of new and modern approaches for management of insect pest is needs. Keeping this novel idea in mind, application of nanomaterials in pest management is a potential approach. Plants of both terrestrial as well as marine (Fig. 14.2) have been utilized for the synthesis of nanomaterials. Bionano materials should not be agglutination capacity (Fig. 14.3a, b) in prime importance for

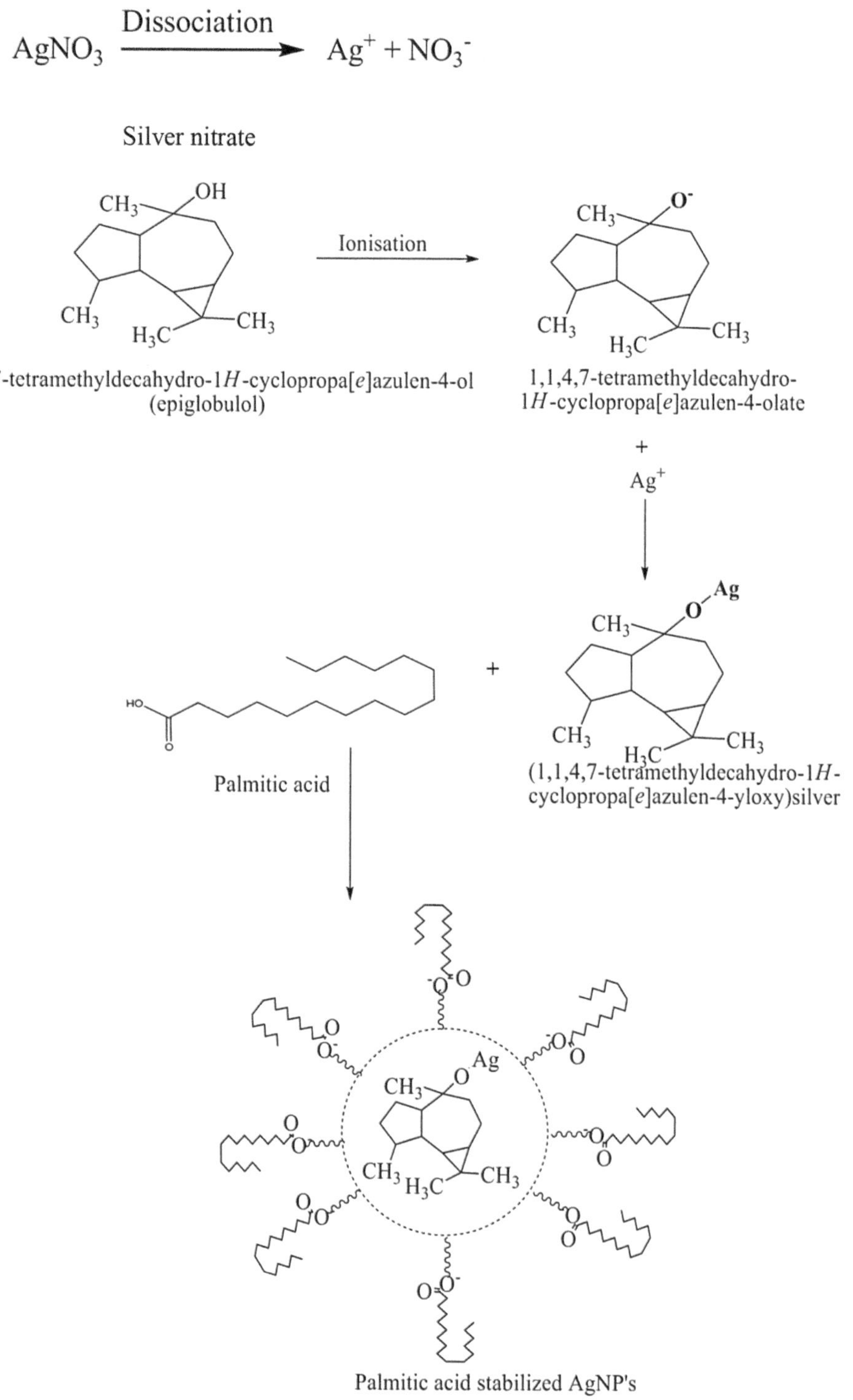

$$AgNO_3 \xrightarrow{\text{Dissociation}} Ag^+ + NO_3^-$$

Silver nitrate

Ionisation

1,1,4,7-tetramethyldecahydro-1*H*-cyclopropa[*e*]azulen-4-ol (epiglobulol)

1,1,4,7-tetramethyldecahydro-1*H*-cyclopropa[*e*]azulen-4-olate

+ Ag^+

Palmitic acid

(1,1,4,7-tetramethyldecahydro-1*H*-cyclopropa[*e*]azulen-4-yloxy)silver

Palmitic acid stabilized AgNP's

Fig. 14.2 Possible chemical reaction involved in the synthesis of silver nanoparticles by epiglobulol and palmitic acid present in ethyl acetate extract identified by GC-MS in *Caulerpa scalpelliformis*

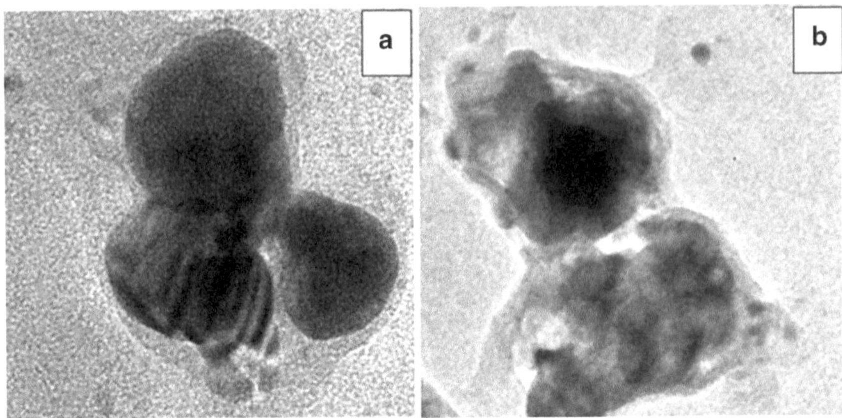

Fig. 14.3 Agglutinations of *Padina pavonica* silver nanomaterials (**a**) or silver nanoparticles with silver biomass (**b**)

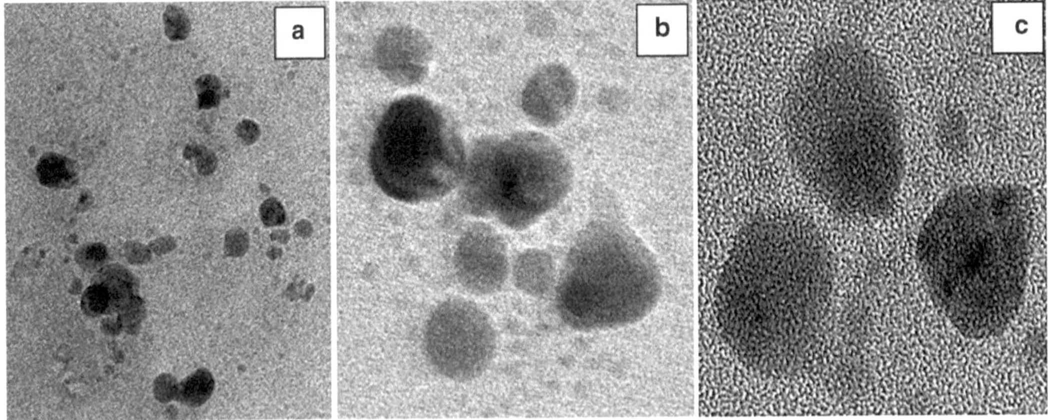

Fig. 14.4 Bioreduced silver nanomaterials of *Padina pavonica* (**a**), PG (**b**) and *Vernonia* (**c**)

their utility in pest management. The nanomaterials are of different size and shapes (Fig. 14.4a–c).

Literature available on this topic brings to close conclusion that only few researchers all over the world are working in this area, and hence, there is a pressing need to apply nanotechnology, and thus, it warrants detailed study. The research studies carried out related to management of insect pest have been reviewed here, which can be used as an efficient and potential modern approach. Bio-based nanoparticles can be used as novel pesticides and drug carriers.

Although there are a number of studies on the toxic effect of nanoparticles or biogenetic nanoparticle agent on plant pathogenic bacteria, fungi, and viruses, very few research has been carried out to investigate the toxicity effect of nanoparticles on insects. Nanotechnology opens an avenue for efficient delivery of both synthetic chemical pesticides and biopesticides with nano-sized preparation or NM-based agrochemical formulation (Ghormade et al. 2011). Further, Bailey et al. (2010) emphasized that the integration of biomolecules (enzymes, primary and secondary metabolites, whole cells) with nanostructure leads to synthesis of new bionano materials which have enormous applications including in agriculture, for instance, the impact of silver and zinc nanoparticle against *Aphis nerii* Boyer de Fonscolombe (Hemiptera: Aphididae)

(Rouhani et al. 2012). *Tinospora cordifolia* leaf extract reduced AgNPs and showed pediculicidal and larvicidal activity (Jayaseelan et al. 2011).

Various natural sources like the shell wall of phytoplankton, epidermis of vegetables, burnt pretreated rice hulls and straw at thermoelectric plants, and volcanic soil are considered as the best sources of *Amorphous nanosilica*. It displayed promising potential as a biopesticide. They also proposed the mechanism of how the *Amorphous nanosilica* cause insect death, i.e., the silica NPs were physio-sorbed by the cuticular lipids disrupting the protective barrier and thereby causing death of insects purely by physical means. Later, Debnath et al. (2011) reported that silica NPs caused 100 % mortality in *Sitophilus oryzae*. Furthermore, surface-charged modified hydrophobic silica NPs (3–5 nm) were successfully used to control a range of agricultural insect pests and animal ectoparasites of veterinary importance (Ulrichs et al. 2006). Insect mortality due to silica nanoparticle treatment was obtained at dose rates almost comparable with those of commercially available DE formulations ranging from 500–5,000 mg kg^{-1} (Subramanyam and Roesli 2000).

Few studies show nanoencapsulation preparations being effective against pests and vectors. Nanoencapsulation is a process by which a chemical is slowly but efficiently released to the particular host for insect pest control. Nanoparticles loaded with generalist essential oil are effective against *Tribolium castaneum* Hurst (Yang et al. 2009). In another study (Chakravarthy et al. 2012a), DNA-tagged gold nanoparticles against *Spodoptera litura* (Fab.). Very recently, Chakravarthy et al. (2012b) utilize Cds, Ag, and TiO_2 particles against *S. litura.* They further proposed that tagging of gold nanoparticles increased the mortality of the pest while the day of exposure increased (645.75, 377.21, and 256.31 ppm for third, fourth, and fifth day after exposure, respectively). Nanotechnological methods were proposed for the management of *Helicoverpa armigera* Hübner (Vinutha et al. 2013). Since essential oil of *Artemisia arborescens* was stable (Lai et al. 2006), it was incorporated with solid lipid NPs and developed an emulsion to improve stability and insecticidal activity (Lai et al. 2006). Another form of nanomaterials used biopesticides in an amorphous nanosilica. It is obtained from different natural sources such as phytoplankton's shell wall, epidermis of vegetables, burnt pretreated rice mills and straw, and volcanic soil. Primarily, the silica nanoparticles were transformed as surface-charged modified hydrophobic Si NPs and successfully utilized to control many agricultural pests and animal ectoparasites (Ulrichs et al. 2006). DNA-tagged gold nanoparticles were used as pesticide against *Spodoptera litura* (Chakravarthy et al. 2012a) and would therefore be useful components for an integrated pest management strategy (Table 14.1).

Table 14.1 Nanoparticle and bioreduced nanoparticle activities against various insect pests with citations

Nanoparticles	Pest name	Activities	Citation
Ag^+	*Aphis nerii*	Insecticidal	Rouhani et al. (2012)
Ag–Zn	*Aphis nerii*	Insecticidal	Rouhani et al. (2012)
Cds	*Spodoptera litura*	Insecticidal	Chakravarthy et al. (2012a)
Ag	*Spodoptera litura*	Insecticidal	Chakravarthy et al. (2012a)
Tio_2	*Spodoptera litura*	Insecticidal	Chakravarthy et al. (2012a)
Al_2O_3	*Sitophilus oryzae*	Repellency	Sabbour (2012)
Tio_2	*Sitophilus oryzae*	Insecticidal Ovipositional	Debnath et al. (2011)
Silica	*Sitophilus oryzae*	Insecticidal	Rajakumar and Rahuman (2011)
Nanosilica	Insect	Pesticidal	Barik et al. (2008)
Eclipta prostrata	*Culex quinquefasciatus*	Larvicidal	Rajakumar and Rahuman (2011)
Leaf-based $AgNo_3$	*Anopheles subpictus*	Pediculocidal	Rajakumar and Rahuman (2011)
Garlic oil + polyethylene glycol	*Tribolium castaneum*	Pesticidal	Yang et al. (2009)

2.3.5 Storage Pest Management

Using the melt-dispersion method, a polyethylene glycol (PEG) coated nanoparticles loaded with garlic essential oil was prepared, and their insecticidal activity against adult *Tribolium castaneum* has been evaluated. The control efficacy against adult *T. castaneum* remained over 80 % after 5 months, as against free garlic essential oil at a similar concentration (640 mg/kg) which was only 11 % (Yang et al. 2009). The author proposed that the higher insecticidal activity was presumably due to the slow and persistent release of the active components from the nanoparticles. Goswami et al. (2010) reported that the applications of different surface functionalized hydrophilic nanoparticles, silica (SNP), aluminum oxide (ANP), zinc oxide (ZNP), and titanium dioxide (TNP) nanoparticle were tested against rice weevil *S. oryzae*. Stadler et al. (2010) successfully tested nano-alumina against two stored grain pests, *Sitophilus oryzae* Linn. and *Rhyzopertha dominica* (Fab.); results are encouraging. During the same period, Teodoro et al. (2010) also recorded the insecticidal activity of nanostructured alumina against the same pests. Results revealed that continuous exposure of *S. oryzae* and *R. dominica* to nanostructured alumina-treated wheat significantly reduced the population after 3 days. NPs caused 100 % mortality in *Sitophilus oryzae* (Debnath et al. 2011). Silver nanoparticles (Ag NPs) were synthesized by using aqueous leaf extracts of *Euphorbia prostrate* and tested against adults of *Sitophilus oryzae* L. The LD_{50} values of aqueous extract, $AgNO_3$ solution, and synthesized Ag NPs were 213.32, 247.90, and 44.69 mg/kg^{-1}, respectively (Abduz Zahir et al. 2012).

2.3.6 Possible Mechanism of Nanomaterials on Insects

In mosquito, the mortality is mainly due to recapturing of midgut as a result discharge of larval inner contents low to the death (Zhang et al. 2010). Initially, it was proposed by Asharani et al. (2007) that Ag NPs bind with chromosome and are capable of including cell proliferation in the cell lines of zebra fish. It shows the nonmaterial's are having the capacity to enter cells and cause cellular damage (Hussain et al. 2005; Ji et al. 2007). In another study it was observed that nanotube filled with aluminosilicate easily stick to plant surface, while in graders' of nanotube have the ability to stick to the exoskeleton of insect pest and subsequently enter the body and influence certain physiological functions leads to the death. Ulrichs et al. (2006) proposed that since nanoparticles easily circulated in the neurological as well as biochemical changes of the animal.

2.3.7 Nanoparticles for Diseases Management

Plants

Fungi, bacteria, and viruses are important pathogens of vegetables, fruits, cereals, pulses, and other crops. As an alternative to synthetic chemicals, the use of nanoparticles as antimicrobial agents has become more common as technology advances. Until now, limited research provided some evidence of the application of nanoparticles for the control of plant diseases caused by pathogenic phytopathogens. In the present section, the antimicrobial activity of nanoparticles of silver, silicon, titanium dioxide, magnesium, iron, and zinc and its oxide has been discussed briefly.

Another important potential application of nanotechnology is in the management of plant diseases. Titanium dioxide-based nanoparticles greatly affect disease development and subsequently increase the yield of edible cowpea. Hydrophobic Si NPs were successfully applied as a thin film on cereal seeds, which decreased the fungal growth and increased germination (Robinson and Salejova-Zadrazilova 2010). Further, it was also proposed by Kim et al. (2006) that Ag NPs can be used for the infesting various plant pathogens. Again during 2007, Kim and his coworkers tested antifungal activity of silver ions and nanoparticles against plant pathogenic fungi, *Bipolaris sorokiniana* and *Magnaporthe grisea*, both under laboratory, and also on ryegrass, *Lolium perenne*. Results revealed that silver ions and NPs reduced colony formation of spores. Silver NPs reduced spore formation, reduced the number of germinating fragments and sprout length, and reduced the growth of fungal hyphae. Jo (2009) tested various forms of silver ions and nanoparticles (SNP) on two plant pathogenic fungi, *Bipolaris sorokiniana* and *Magnaporthe grisea*, under growth chamber. SNP significantly reduced these two fungal diseases on perennial ryegrass *Lolium perenne*. The number of spores formed by *Fusarium culmorum* mycelia increased in the culture after contact with silver nanoparticles with a 2.5 ppm solution of silver nanoparticles greatly reduced the number of mycelial growth, germinating fragments, and sprout length relative to the control (Marek et al 2010). Application of 100 ppm silver nanoparticles showed the highest inhibition rate for both before and after the outbreak of disease on cucumbers and pumpkins at field level (Kabir et al. 2010). Silica NPs were successfully applied as a thin film on seeds to decrease fungal growth and boost cereal germination (Robinson and Salejova-Zadrazilova 2010). Previously, Park et al. (2006) reported that nanosized silica–silver controlled *Botrytis cinerea, Rhizoctonia solani, Colletotrichum gloeosporioides, Magnaporthe grisea*, and *Pythium ultimum* within 3 days of spraying. Nanosized silver colloidal solution of WA-CV-WA13B, WA-AT-WB13R, and WA-PR-WB13R (AgNPs) at concentrations of 10, 25, 50, and 100 ppm against *Alternaria alternata, Alternaria brassicicola, Alternaria solani, Botrytis cinerea, Cladosporium cucumerinum, Corynespora cassiicola, Cylindrocarpon destructans, Didymella bryoniae, Fusarium* spp., *Glomerella cingulata, Monosporascus cannonballus, Pythium aphanidermatum, Pythium spinosum*, and *Stemphylium lycopersici.* The results indicated that AgNPs possess antifungal properties against these plant pathogens at 90–100 %. Among the AgNPs, WA-CV-WB13R AgNPs caused maximum inhibition of most fungi (Kim et al. 2012).

Nanoparticles, particularly titanium dioxide, help produce new pesticides, because it decreased the disease incidence of cowpea (says Owolade et al. 2008). Phytopathogenic fungi such as *Rhizoctonia solani, Sclerotinia sclerotiorum*, and *S. minor* hyphal growth (12 %, 36 %, and 41 % for *Rhizoctonia solani, Sclerotinia sclerotiorum*, and *S. minor*, respectively) and sclerotial germination grow inhibited by 7 ppm silver nanoparticles.

Antifungal activities of zinc oxide nanoparticles (ZnO NPs) at different concentrations (0, 3, 6, and 12 mmol L^{-1}) and their mode of action against two postharvest pathogenic fungi, namely, *Botrytis cinerea* and *Penicillium expansum*, were investigated (Lili et al. 2011). Growth of *F. graminearum* was significantly inhibited by inclusion of the ZnO nanoparticles in a mung bean broth agar and in sand. Further, the ZnO NPs did not prevent metabolites from a biocontrol bacterium, *Pseudomonas chlororaphis* O6, from inhibiting *Fusarium* growth; no synergism was observed in the mung bean agar (Dimkpa et al. 2013). Results revealed that ZnO NPs at concentrations greater than 3 mmol L^{-1} significantly inhibit the growth of *B. cinerea* and *P. expansum* by preventing the development of conidiophores and conidia, which eventually led to the death of fungal hyphae. Nanoparticles of magnesium, iron, and zinc significantly inhibit the germination of spores of *Penicillium notatum, Aspergillus niger*, and *Nigrospora oryzae* Berk. Among the three nanoparticles, the nano-MgO at highest concentration was found most effective in reducing the spore germination, followed by nano-FeO and nano-ZnO at the same concentration, respectively (Wani et al. 2012).

Nanocomposite is another form of nanotechnological tool which can be utilized in agriculture either for pest management or for phytopathogens management. For instance, Cioffi et al. (2005) tested the antifungal activity of polymer-based copper nanocomposite against plant pathogenic fungi.

Animal

Dreadful NPV disease of silkworm was easily managed by plant-stem-derived silica by enhancing the survival rate from 0 to 30 % by SiO_2 through inhibiting sporulation and growth of fungi (Goswami et al. 2010). Further, they recorded the solid and liquid formulations of the silver nanoparticles (SNP), aluminum oxide (ANP), zinc oxide, and titanium dioxide used for the management of rice weevil and grasserie disease in *Bombyx mori* caused by *Sitophilus oryzae* and baculovirus (*B. mori* nuclear polyhedrosis virus – BmNPV), respectively (Goswami et al. 2010).

It was reported that hydrophilic SNP was most effective on the first day. On day 2, more than 90 % mortality was obtained with SNP and ANP. After 7 days of exposure, 95 and 86 % mortality were reported with hydrophilic and hydrophobic SNP and nearly 70 % of the insects were killed when the rice was treated with lipophilic SNP.

2.4 Merits and Demerits

1. Application of amorphous nanosilica did not alter either respiration or photosynthesis of many horticultural and crop plants (Ghormade et al. 2011).
2. Use of amorphous as nanosilica as nano-biopesticides is considered safe for humans by WHO.
3. Uses of these tiny molecules are more efficient utilization of active agents, resource saving, safety, etc.
4. Bioreduced nanomaterials reduced the phytotoxicity by lowering the mobility of the active agent in soil, hence reducing its residue in the food chain.

2.5 Conclusion and Future Focus

The abovementioned review shows that instead of synthetic pesticides, farmers are advised to utilize biopesticides and botanical-based nanomaterials for pest and disease management to protect our environment, our fellow human beings, and our domestic animals, as our natural resources, like plants, have degradable and low-cost secondary metabolites. Laboratories, as well as field experiments, are imperative to utilize nanomaterial-based insecticides to eradicate devastating pests and disease. Further, encapsulation of botanicals with nanomaterials could be encouraged in the future for their preparation.

References

Abduz Zahir A, Bagavan A, Kamaraj C, Elango G, Rahuman AA (2012) Efficacy of plant-mediated synthesized silver nanoparticles against *Sitophilus oryzae*. J Biopest 5(Supplementary):95–102

Allan GG, Chopra CS, Neogi AN, Wilkins RM (1971) Design and synthesis of controlled release pesticide-polymer combinations. Nature 234:349–351

Andrade-Coelho CA, Souza NA, Gouveia C, Silva VC, Gonzalez MS, Rangel EF (2009) Effect of fruit and leaves of Meliaceae plants (*Azadirachta indica* and *Melia azedarach*) on the development of *Lutzomyia longipalpis* larvae (Diptera: Psychodidae: Phlebotominae) under experimental conditions. J Med Entomol 46:1125–1130

Asharani PV, Nair G, Zhiyuan H, Manoor P, Valiyaveetti S (2007) Potential health impacts of silver nanoparticles. Abstracts of papers, 234th ACS National Meeting, Boston, MA, USA, 19–23 August 2007, TOXI-099

Bailey KL, Boyetchko SM, Langle T (2010) Social and economic drivers shaping the future of biological control: a Canadian perspective on the factors affecting the development and use of microbial biopesticides. Biol Control 52:221–229

Barati R, Golmohammadi G, Ghajarie H, Zarabi M, Mansouri R (2013) The effects of some botanical insecticides and pymetrozine on life table parameters of silver leaf whitefly Bemisia tabaci Gennadius (Hemiptera: Aleyrodidae). Pesticidi i fitomedicina 28(1):47–55

Barik TK, Sahu B, Swain V (2008) Nano-silica from medicine to pest control. J Parlog Res 103(1):253–258

Bhattacharyya A (2009) Nanoparticles from drug delivery to insect pest control. Akshar 1(1):1–7

Bhattacharyya A, Bhaumik A, Rani PU, Mandal S, Epidi TT (2010) Nano-particles-a recent approach to insect pest control. Afr J Biotech 9:3489–3493

Brecht M, Datnoff L, Nagata R, Kucharek T (2003) The role of silicon in suppressing tray leaf spot development in St. Augustine grass. Publication in University of Florida, pp 1–4

Campos M, Phillips TW (2013) Laboratory evaluation of attract-and-kill formulations against the Indianmeal moth, *Plodia interpunctella* (Hübner) (Lepidoptera: Pyralidae). Retrieved from http://krex.ksu.edu

Casida JE (1980) Pyrethrum flowers and pyrethroid insecticides. Environ Health Perspect 34:189

Casida JE, Quistad GB (2000) Insecticide targets: learning to keep up with resistance and changing concepts of safety. Agric Chem Biotechnol 43:185–191

Chakravarthy AK, Bhattacharyya A, Shashank PR, Timothy T, Doddabasappa EB, Swapan Mandal K (2012a) DNA-tagged nano gold: a new tool for the control of the armyworm, *Spodoptera litura* Fab. (Lepidoptera Noctuidae). Afr J Biotechnol 11(38): 9295–9301

Chakravarthy AK, Chandrashekharaiah B, Subhash K, Bhattacharya A, Dhanabala K, Gurunatha K, Ramesh P (2012b) Bio efficacy of inorganic nanoparticles CdS, Nano-Ag and Nano-Tio2 against *Spodoptera litura* (Fabricius) (Lepidoptera: Noctuidae). Curr Biotica 6(3):271–281

Chin CP, Wu HS, Wang SS (2011) New approach to pesticide delivery using nanosuspensions: research and applications. Ind Eng Chem Res 50:7637–7643

Cioffi N, Torsi L, Ditaranto N, Tantillo G, Ghibelli L, Sabbatini L, Traversa E (2005) Copper nanoparticle/polymer composites with antifungal and bacteriostatic properties. Chem Mater 17(21):5255–5262

Debnath N, Das S, Seth D, Chandra R, Bhattacharya SC, Goswami A (2011) Entomotoxic effect of silica nanoparticles against *Sitophilus oryzae* (L.). Pest Sci 84:99–105

De Villiers SM, Hoisington DA (2011) The trends and future of biotechnology crops for insect pest control. Afr J Biotechnol 10(23):4677–4681

Dimkpa CO, McLean JE, Britt DW, Anderson AJ (2013) Antifungal activity of ZnO nanoparticles and their interactive effect with a biocontrol bacterium on growth antagonism of the plant pathogen *Fusarium graminearum*. BioMetals. doi:10.1007/s10534-013-9667-6

Dubey NK, Shukla R, Kumar A, Singh P, Prakash B (2010) Prospects of botanical pesticides in sustainable agriculture. Curr Sci 98(4):479–480

Dutta P, Reddy SGE, Borthaku BK (2013) Effect of neem kernal aqueous extract (NKAE) in Tea Mosquito Bug, *Helopeltis theivora* (Waterhouse, 1886) (Heteroptera: Miridae). Munis Entomol Zool 8(1):213–218

Elek N, Hoffman R, Raviv U (2010) Novaluron nanoparticles: formation and potential use in controlling agricultural insect pests. Colloid Surf A 372(1–3):66–72

Fei Yi, Chunhua Zou, Qiongbo Hu, Meiying Hu (2012) The joint action of Destruxins and botanical insecticides (Rotenone, Azadirachtin and Paeonolum) Against the Cotton Aphid, *Aphis gossypii* Glover. Molecules 17:7533–7542. doi:10.3390/molecules17067533

Feng BH, Peng LF (2012) Synthesis and characterization of carboxymethyl chitosan carrying Ricinoleic Functions as an emulsifier for azadirachtin. Carbohydr Polym 88:576–582

Ghazawi NA, El-Shranoubi ED, El-Shazly MM, Rahman KA (2007) Effects of azadirachtin on mortality rate and reproductive system of the grasshopper Heteracris littoralis Ramb. (Orthoptera: Acrididae). J Orthoptera Res 16(1):57–65

Ghormade V, Deshpande MV, Paknikar KM (2011) Perspectives for nano-biotechnology enabled protection and nutrition of plants. Biotechnol Adv 29(6):792–803

Gogos A, Knauer K, Bucheli TD (2012) Nanomaterials in plant protection and fertilization: current state, foreseen applications, and research priorities. J Agric Food Chem 60(39):9781–9792

Goodsell DS (2004) Bionanotechnology: lessons from nature. Wiley-Liss, Hoboken

Goswami A, Roy I, Sengupta S, Debnath N (2010) Novel applications of solid and liquid formulations of nanoparticles against insect pests and pathogens. Thin Solid Films 519(3):1252–1257

Grdiša M, Babić S, Periša M, Carović-Stanko K, Kolak I, Liber Z, Satovic Z (2013) Chemical diversity of the natural populations of Dalmatian Pyrethrum (*Tanacetum cinerariifolium* (Trevir.) Sch. Bip.) in Croatia. Chem Biodivers 10(3):460–472

Guo YR, Liu SH, Gui WJ, Zhu GN (2009) Gold immune chromatographic assay for simultaneous detection of carbofuran and triazophos in water samples. Anal Biochem 389:32–39

Hallberg K (2010) Towards a responsible research in nanoscience and nanotechnol. Pasi Nano-Bio 1–47

Harper S (2010) New approaches needed to gauge safety of nanotech-based pesticides, Researchers Urge. Published in Phys Chem 4(33):2010–2012

He L, Liu Y, Mustapha A, Lin M (2011) Antifungal activity of zinc oxide nanoparticles against *Botrytis cinerea* and *Penicillium expansum*. Microbiol Res 166(3):207–215

Ho SH, Koh L, Huang Y, Sim KY (1996) The oil of garlic, *Allium sativum* L. (Amaryllidaceae), as a potential grain protectant against *Tribolium castaneum* (Herbst) and *Sitophilus zeamais* Motsch. Postharvest Biol Technol 9:41–48

Hori M (1996) Settling inhibition and insecticidal activity of garlic and onion oils against *Myzus persicae* (Sulzer) (Homoptera: Aphididae). Appl Entomol Zool 31:605–612

Hussain SM, Hess KL, Gearhart JM, Geiss KT, Schlager JJ (2005) In vitro toxicity of nanoparticles in BRL 3A rat liver cells. Toxicol In Vitro 19:975–983

Ikeura H (2014) Use of plant volatile for plant pathogens and pests management. In: Sahayaraj K (ed) Basic and applied aspects of biopesticides. Springer, India, pp 181–192

Isman MB (2006) Botanical insecticides, deterrents, and repellents in modern agriculture and an increasingly regulated world. Annu Rev Entomol 51:45–66

Jayaseelan C, Rahuman AA, Rajakumar G, Vishnu AK, Santhosh kumar T, Marimuthu S, Bagavan A, Kamaraj C, Zahir AA, Elango G (2011) Synthesis of pediculocidal and larvicidal silver nanoparticles by leaf extract from heartleaf moonseed plant, *Tinospora cordifolia* Miers. Parasitol Res 109:185–194

Ji JH, Jung JH, Kim SS, Yoon JU, Park JD, Choi BS, Chung YH, Kwon IH, Jeong J, Han BS, Shin JH, Sung JH, Song KS, Yu IJ (2007) Twenty-eight-day inhalation toxicity study of silver nanoparticles in Sprague-Dawley rats, Inhal. Toxicology 19:857–887

Jo YK (2009) Antifungal activity of silver ions and nanoparticles on phytopathogenic fungi. Plant Dis 93(10):1037–1043

Joseph T, Morrison M (2006). Nanotechnology in agriculture and food a nanoforum report. www.nanoforum.org

Kabir L, Sang-Woo Kim, Jin Hee Jung, Yun Seok Kim, Kyoung Su Kim, Youn Su Lee (2010) Inhibition effects of silver nanoparticles against powdery mildews on cucumber and pumpkin. Microbiolgy 39(1):26–32

Kashyap PL, Kumar S, Srivastava AK, Sharma AK (2013) Myconanotechnology in agriculture: a perspective. World J Microbiol Biotechnol 29:191–207

Kaushik A, Solanki PR, Ansarib AA, Malhotra BD, Ahmad S (2009) Iron oxide-chitosan hybrid nanobiocomposite based nucleic acid sensor for pyrethroid detection. J Biochem Eng 46:132–140

Khan M, Hossain MA, Islam MS (2007) Effects of neem leaf dust and a commercial formulation of a neem compound on the longevity, fecundity and ovarian development of the melon fly, *Bactrocera cucurbitae* (Coquillett) and the oriental fruit fly, *Bactrocera dorsalis* (Hendel) (Diptera: Tephritidae). Pak J Biol Sci 10(20):3656–3661

Kim J, Grate JW, Wang P (2006) Nanostructures for enzyme stabilization. Chem Eng Sci 61:1017–1026

Kim TH, Jiang HL, Jere D, Parka IK, Cho MH, Nah JW (2007) Chemical modification of chitosan as a gene carrier in vitro and in vivo. Prog Polym Sci 32:726–753

Kim SW, Jin Hee Jung, Kabir Lamsal, Yun Seok Kim, Ji Seon Min, Youn Su Lee (2012) Antifungal effects of silver nanoparticles (AgNPs) against various plant pathogenic fungi. Mycobiology 40(1):53–58

Kulkarni AR, Soppimath KS, Aminabhavi TM, Dave AM, Mehta MH (1999) Application of sodium alginate beads crosslinked with glutaraldehyde for controlled release of pesticide. Polymers News 2:285–286

Lai F, Wissing SA, Müller RH, Fadda AM (2006) Artemisia arborescens L essential oil-loaded solid lipid nanoparticles for potential agricultural application: preparation and characterization. AAPS PharmSciTech 7(1):E10–E18

Lao SB, Zhang ZX, Xu HH, Jiang GB (2010) Novel amphiphilic chitosan derivatives: synthesis, and an increasingly regulated world. Annu Rev Entomol 51:45–46

Lisa M, Chouhan RS, Vinayaka AC, Manonmani HK, Thakur MS (2009) Gold nanoparticles based dipstick immune-assay for the rapid detection of dichlorodiphenyl-trichloroethane: an organochlorine pesticide. Biosens Bioelectron 25:224–227

Liu F, Wen LX, Li ZZ, Yu W, Sun HY, Chen JF (2006) Porous hollow silica nanoparticles as controlled delivery system for water-soluble pesticide. Mater Res Bull 41:2268–2275

Liu Y, Tong Z, Prudhomme RK (2008) Stabilized polymeric nanoparticles for controlled and efficient release of bifenthrin. Pest Manage Sci 64:808–812

Lu C, Adamkiewicz G, Attfield KR, Kapp M, Spengler JD, Tao L, Xie SH (2013) Household pesticide contamination from indoor pest control applications in urban low-income public housing dwellings: a community-based participatory research. Environ Sci Technol 47(4):2018–2025

Marek JK, Magdalena K, Anna G (2010) The effect of silver nanoparticles on phytopathogenic spores of *Fusarium culmorum*. Can J Microbiol 56(3):247–253

Margulis-Goshen K, Magdassi S (2013) Nanotechnology: an advanced approach to the development of potent insecticides. In: Advanced technologies for managing insect pests. Springer, Dordrecht, pp 295–314

Misra AN, Misral M, Singh R (2013) Nanotechnology in agriculture and food industry. Int J Pure Appl Sci Technol 16(2):1–9

Mousavi SR, Rezaei M (2011) Nanotechnology in agriculture and food production. J Appl Environ Biol Sci 1:414–419

Mugisha-Kamatenesi M, Deng AL, Ogendo JO, Omolo EO, Mihale MJ, Otim M, Buyungo JP, Bett PK (2008) Indigenous knowledge of field insect pests and their management around lake Victoria basin in Uganda. Afr J Environ Sci Technol 2(8):342–348

Owolade OF, Ogunleti DO, Adenekan MO (2008) Titanium dioxide affects disease development and yield of edible cowpea. Agric Food Chem 7(50): 2942–2947

Park H, Kim SH, Kim HJ, Choi SH (2006) A new composition of nanosized silica-silver for control of various plant diseases. Plant Pathol J 22:295–302

Paula HCB (2011) Preparation and characterization of chitosan/cashew gum beads loaded with *Lippia sidoides* essential oil. Mater Sci Eng C 31(2):173–178

Paula HCB, Rodrigues MLL, Ribeiro WLC, Stadler AS, Paula RCM, Abreu FOMS (2012) Protective effect of cashew gum nanoparticles on natural Larvicide from *Moringa oleifera* seeds. J Appl Polym Sci 124:1778–1784

Popat A, Liu J, Hu Q, Kennedy M, Peters B, Lu GQM, Qiao SZ (2012) Adsorption and release of biocides with mesoporous silica nanoparticles. Nanoscale 4(3):970–975

Prasann BM (2007) Nanotechnology in agriculture. ICAR National Fellow, Division of Genetics, IARI, New Delhi

Prota N, Bouwmeester HJ, Jongsma MA (2013) Comparative antifeedant activities of polygodial and pyrethrins against whiteflies (*Bemisia tabaci*) and aphids (Myzus persicae). Pest Manage Sci 70(4):682–688

Psota V, Ouředníčková J, Falta V (2010) Control of Hoplocampa testudinea using the extract from Quassia amara in organic apple growing. Hort Sci (Prague) 3:139–144

Rai M, Ingle A (2012) Role of nanotechnology in agriculture with special reference to management of insect pests. Appl Microbiol Biotechnol 94:287–293

Rajakumar G, Rahuman A (2011) Larvicidal activity of synthesized silver nanoparticles using *Eclipta prostrata* leaf extract against filariasis and malaria vectors. Acta Tropica 118:196–203

Ramanathan M, Luckarift HR, Sarsenova A, Wild JR, Ramanculov ER, Olsen EV (2009) Lysozyme-mediated formation of protein–silica nano-composites for biosensing applications. Colloids Surf B Biointerfaces 73:58–64

Regnault-Roger C, Vincent C, Arnason JT (2012) Essential oils in insect control: low-risk products in a high-stakes world. Annu Rev Entomol 57:405–424

Robinson DKR, Salejova-Zadrazilova G (2010) Nanotechnologies for nutrient and biocide delivery in agricultural production. Working paper version

Rouhani M, Amin Samih M, Kalantari S (2012) Insecticide effect of silver and zinc nanoparticles against *Aphis nerii* Boyer De Fonscolombe (Hemiptera: Aphididae). Chilean JAR 72(4):590–594

Sabbour MM, Alipour HR, Abbasi S, Ghavidel S, Farahani HF, Sabaghi M, Rahman MK (2012) Entomotoxicity assay of two Nanoparticle Materials 1-(Al2O3 and TiO2) Against Sitophilus oryzae Under Laboratory and Store Conditions in Egypt. JNAS Journal-2012-1-4/103-108

Sahayaraj K (1998) Antifeedant effect of some plant extracts on the Asian armyworm, Spodoptera litura (Fabricius). Curr Sci 74:523–525

Sandoval-Mojica AF, Capinera JL (2011) Antifeedant effect of commercial chemicals and plant extracts against Schistocerca americana (Orthoptera: Acrididae) and Diaprepes abbreviatus (Coleoptera: Curculionidae). Pest Manag Sci 67(7):860–868

Satta A, Floris I, Caboni P, Cabras P, Eguaras M, Velis G (2008) New experimental data on use of rotenone as an acaricide for control of Varroa destructor in honey bee colonies. J Econ Entomol 101(4):1075–1080

Schmutterer H (1988) Potential of azadirachtin-containing pesticides for integrated pest control in developing and industrialized countries. J Insect Physiol 34(7):713–719

Scott NR (2007) Nanotechnology opportunities in agriculture and food systems. Biological and Environmental Engineering, Cornell University NSF Nanoscale Science & Engineering Grantees conference, Arlington, VA, 5 December 2007

Scrinis G, Lyons K (2007) The emerging nano-corporate paradigm: nanotechnology and the transformation of nature, food and Agri food systems. Int J Sociol Food Agric 15(2):1–23

Silva MA, Bezerra-Silva GCD, Vendramim JD, Mastrangelo T (2013) Sublethal effect of neem extract on mediterranean fruit fly adults. Rev Bra de Fruti 35(1):93–101

Sinha S, Pan I, Chanda P, Sen SK (2009) Nanoparticles fabrication using ambient biological resources. J Appl Biosci 19:1113–1130

Stadler T, Buteler M, Weaver DK (2010) Novel use of nanostructured alumina as an insecticide. Pest Manage Sci 66:577–579. doi:10.1002/ps.1915

Subramanyam B, Roesli R (2000) Inert dusts. In: Alternatives to pesticides in stored-product IPM. Springer US, Boston, pp 321–380

Teodoro S, Micaela B, David KW (2010) Novel use of nano-structured alumina as an insecticide. Pest Manag Sci 66(6):577–579

Ulrichs C, Krause F, Rocksch T, Goswami A, Mewis I (2006) Electrostatic application of inert silica dust based insecticides onto plant surfaces. Comm Agric Appl Bio Sci 71:171–178

Viñuela E, Medina MP, Schneider MI, González M, Budia F, Adan A, Del EStal P (2001) Comparison of side effect of spinosad, tebufenozide and azadirachtin on the predators *Chrysoperla cornea* and *Podisus maculiventris* and the parasitoids Opius concolor Hyposoter didymator under laboratory conditions. IOBC/WPRS Bull 24:25–34

Vinutha JS, Bhagat D, Bakthavatsalam N (2013) Nanotechnology in the management of polyphagous pest *Helicoverpa armigera*. J Acad Indus Res 1(10):606–608

Wang YA, Li JJ, Chen HY, Peng XG (2002) Stabilization of inorganic nanocrystals by organic dendrons. J Am Chem Soc 124:2293–2298

Wang L, Li X, Zhang G, Dong J, Eastoe J (2007) Oil-in-water nanoemulsions for pesticide formulations. J Colloid Interface Sci 314(1):230–235

Wang H, Wang J, Choi D, Tang Z, Wu H, Lin Y (2009) EQCM immunoassay for phosphorylated acetylcholinesterase as a biomarker for organophosphate exposures based on selective zirconia adsorption and enzyme-catalytic precipitation. Biosens Bioelectron 24:2377–2383

Wani AH, Amin M, Shahnaz M, Shah MA (2012) Antimycotic activity of nanoparticles of MgO, FeO and ZnO on some pathogenic fungi. Int J Manuf Mater Mech Eng 2(4):59–70

Yang FL, Li XG, Zhu Z, Lei CL (2009) Structural characterization of nanoparticles loaded with garlic essential oil and their insecticidal activity against *T. castaneum* (Herbst) (Coleoptera: Tenebrionidae). J Agric Food Chem 57(21):10156–10162

Yang FL, Zhu F, Lei CL (2010) Garlic essential oil and its major component as fumigants for controlling *Tribolium castaneum* (Herbst) in chambers filled with stored grain. J Pest Sci 83(3):311–317

Zhang X, Zhang J, Zhu KY (2010) Chitosan/double-stranded RNA nanoparticle-mediated RNA interference to silence chitin synthase genes through larval feeding in the African malaria mosquito (*Anopheles gambiae*). Insect Mol Biol 19:683–693

Zhao NN, Zhang H, Zhang XC, Luan XB, Zhou C, Liu QZ, Shi WP, Liu Z (2013) Evaluation of acute toxicity of essential oil of garlic (*Allium sativum*) and its selected major constituent compounds against over wintering *Cacopsylla chinensis* (Hemiptera: Psyllidae). J Econ Entomol 106(3):1349–1354

Phytochemical Pesticides

15

Suresh Walia, Supradip Saha, and Virendra S. Rana

Abstract

Many phytochemical pesticides exhibiting broad spectrum of activity against pests and diseases have long been considered as attractive alternative to synthetic chemical pesticides as they are biodegradable, target specific, and pose no or less hazard to the environment or to human health. Although a large number of studies suggest that plant-based materials do affect arthropod pests, vectors, and other pathogens, yet only a handful of botanicals are currently used in agriculture, and there are few prospects for commercial development of new botanical products. Several factors appear to limit the success of botanicals, most notably regulatory barriers and the availability of competing products of microbial origin that are cost-effective and relatively safe compared with their predecessors. In the context of agricultural pest management, botanical pesticides are best suited for use in organic food production and can play a much greater role in the production and post-harvest protection of food and food products in developing countries. There is thus a need to organize natural sources, develop quality control, adopt standardization strategies, and modify regulatory mechanisms.

Keywords

Phytochemicals • Essential oils • Limonoids • Saponins • Rocaglamides • Ryanodines • Isobutylamides • Polyol esters • Acetogenins • Light-activated botanicals

S. Walia, Ph.D. (✉) • S. Saha, Ph.D.
V.S. Rana, Ph.D.
Division of Agricultural Chemicals, Indian Agricultural Research Institute,
New Delhi 110 012, India
e-mail: suresh_walia@yahoo.com

D. Singh (ed.), *Advances in Plant Biopesticides*,
DOI 10.1007/978-81-322-2006-0_15, © Springer India 2014

1 Introduction

Intensification of agriculture through expansion of irrigation facilities, introduction of high-yielding varieties, and application of increased amounts of agrochemicals has been continued. Besides these, the cultural practices have been modified to achieve maximum productivity/ha from the available land. However, along with various technological achievements, severe outbreaks of insect pests, diseases, and weeds in agricultural crops have also increased. Many hitherto unknown insect pests and diseases have assumed serious status, and some of them have developed resistance to one or more groups of pesticides. In addition, pesticides have contaminated different components of our environment and posed the potential health hazard to consumers. Therefore, the future pest problems will have to be tackled in an environmentally benign manner as a part of a sustainable crop production technology.

Plant biodiversity has provided an excellent source of biologically active materials for use in traditional crop protection. They can be as simple as crude or purified extracts of plant parts or plant synthesized secondary metabolites that are toxic to herbivores and pathogens and so are believed to act as defense compounds. Over the years, more than 6,000 species of plants have been screened, and more than 2,500 plant species belonging to 235 families (Saxena 1998) were found to possess biological activity against various categories of pests. It is thus likely that novel and potent molecules that can be used for pest suppression still remain to be discovered from many plant species for further development and utilization.

Although noted for the complexity of their chemical structures and biosynthetic pathways, phytochemicals have been investigated for their chemical properties extensively since the 1850s. Recognition of the biological properties of large number of phytochemicals has fueled the current focus for the search of new insecticides, herbicides, and insect behavior-modifying chemicals. Many of these compounds have been shown to have important adaptive significance in protection against herbivory (Croteau et al. 2000). Most of the compounds have been established as insect antifeedants (Koul 2005). Although plant secondary metabolites mediate a wide variety of complex interactions, some of these are released into the environment, mostly volatile or essential oil constituents, alcohol and aldehydes, ketones, esters, aromatic phenols, mono- and sesquiterpenes, etc., and others are produced in the plant for defense. In fact, during the last few decades, literature has been flooded with umpteen studies and comprehensive reviews (Arnason et al. 2004; Champagne et al. 1992; Copping and Duke 2007; Dev and Koul 1997; Harborne 1977; Isman 2005, 2006; Isman and Machial 2006; Koul and Dhaliwal 2001; Koul et al. 2005a, b; Nguefack et al. 2007; Omar et al. 2007; Parmar and Walia 2001; Reitz et al. 2008; Rosell et al. 2008; Saxena 1987). Processes for insect repellent and insecticides products of plant origin have also been patented (Kumar et al. 1995, 1999; Singh et al. 1998, 2000).

2 Vegetable Oils, Fatty Acids, and Soaps

Vegetable oils and their by-products are considered to be quite safe to humans and other living beings. Mixing food grains, pulses, and oilseeds with locally available plant oils is widely practiced by the traditional farmers for protecting grains against insect infestation (Pereirra 1983). A number of plant by-products have been found to be protectants against a number of stored grain insect pests (Hill and Schoonhoven 1981; Pandey et al. 1981; Doharey et al. 1990; Ayyangar and Rao 1989; Dixit and Saxena 1990). These have been traditionally used in many Asian and African countries as contact insecticides to protect grain especially legumes against storage insects (Pandey et al. 1981; Grainge and Ahmed 1988). The vegetable oils have been found useful for controlling whiteflies, young scales, mites, and many other plant pests present during the growing season. Oils have also proven useful in managing some plant diseases. Studies have revealed that the crude oils of rice bran, cottonseed, and palm kernel were most effective at rates of 1–6 g/kg against *Callosobruchus maculatus*. While the cottonseed oil provided good control of *B. tabaci* on cotton (Broza et al. 1988), cottonseed and neem oils at 0.5 and 1.0 % dose were found to show similar effectiveness as 0.05 % fenpropathrin exhibits against both adults and nymphs of *B. tabaci* (Puri et al. 1991, 1994). Further, the emergence of *C. chinensis* adult was completely prevented by the application of 0.5 % neem oil and 7.5 % karanj oil for up to 100 days after treatment without any adverse effect of the oils on seed germination (Khaire et al. 1992). Groundnut and other vegetable oils at 0.3 % (w/w) provided full protection of green gram against *C. maculatus* (Verma and Pandey 1978; Pandey et al. 1981). Complete control of *Sitophilus granarius* was achieved for 60 days by mixing wheat grain with cottonseed, soybean, maize, or peanut oil at 10 g/kg seeds (Qi and Burkholder 1981). Plant-derived oils have been shown to be effective against a number of insects and mite pests in home garden (Butler and Henneberry 1991).

The edible oils presumably act by interfering with normal respiration by plugging the orifices, called spiracles, resulting in suffocation of the insect. In the case of eggs, the oils possibly caused coagulation of the protoplasm by penetrating through microphylls. The application of oils to termites and scale insects resulted in the reduction of the interfacial tension between solvent and water leading to water loss from the insects. With the addition of 1 % glycerol monooleate, the insects lost twice as much weight as those treated with oil alone. Further, crude vegetable oils and less viscous oils were more active than pure and viscous oil (Shaaya and Ikan 1979; Calderon 1979). Since vegetable oils are generally the mixture of mono-, di-, or triglycerides of mainly oleic, stearic, and/or palmitic acids, their insect control activity has been attributed to ester linkages as corresponding hydrolytic product glycerol is devoid of any activity.

Among the straight-chain fatty acids, C_9 to C_{11} acids were found to be most effective in preventing oviposition of *C. maculatus* and causing mortality to *S. oryzae* adults. Pelargonic (C_9), capric (C_{10}), and undecanoic (C_{11}) acids were the most effective in preventing oviposition. At concentration ranging from 0.4 to 1.6 g/kg, these compounds exhibited strong insect repellent activity. Interestingly, conversion of fatty acids into corresponding methyl esters resulted in total loss of the activity. While capric, lauric, and myristic acids reportedly exhibited mosquitocidal activity at LC_{50} 0.0004–0.0014 % against *Aedes triseriatus* (Lalonde et al. 1979), the higher fatty acids such as palmitoleic acid, oleic acid, and linoleic acid were moderately active against *Sitophilus oryzae* and *Tribolium castaneum* (Deshpande et al. 1974). The salts of fatty acids generally known as soaps have been used as insecticides for the control of insect pests of plants (Weinzierl 2000). The most common soaps are made of the potassium salts of fatty acids, which are known to disrupt the structure and permeability of cell membranes of the insect. The cell contents get leaked from the damaged cells leading to quick cell death. A wide range of small- and soft-bodied insect species such as aphids, mealybugs, thrips, and spider mites and whiteflies are sensitive to soap. Some large-bodied insects, such as caterpillars, leafhoppers, and beetles, are also susceptible to soap. Generally, insecticidal soaps have a minimal impact on beneficial insects such as ladybird beetle larvae, parasitic wasps, and honeybees, but it can be quite harmful to soft-bodied predators. The selective action of soaps, their low mammalian toxicity, biodegradability, and high degree of safety to humans are their major advantages. Effects of soap are rapid, usually resulting in death of susceptible insects within a few minutes after exposure. Resistance to insecticidal soaps is less likely to develop as quickly as to the more traditional pesticides. Soaps, oils and neem products have been used as biorational pesticides for IPM in tomato (Schuster and Stansly 2000).

3 Leaf Waxes and Exudates

Leaf glandular trichomes and the exudates such as cuticular waxes produced by them play a significant role in determining resistance and susceptibility to infestation by insects in these plants (Shepherd et al. 1999a, b). These exudates produce a microcrystalline layer of waxy material comprising of a number of secondary plant metabolites such as glycolipids, glycerolipids, as well as free fatty acids/esters and terpenes. These materials, besides providing in-built plant resistance to invading pests, are active against certain phytophagous insects and pathogens. The presence of high level of long-chain alcohols such as hexacosanol (C_{26}) in cabbage wax and triacontanol (C_{30}) in alfalfa has been associated with resistance to larvae of the diamondback moth *Plutella xylostella*. High levels of α-amaryl alkanoates and cycloartenol alkanoates in epicuticular waxes in plants act as defense systems against insects. Abundance of cycloartenol alkanoates in raspberry, *Rubus idaeus* L., is considered as significant factor in resistance against *Amphorophora idaei*. Interestingly, aphid-derived triacylglycerol found only in the leaf waxes of aphid-infested plants serves as an index of aphid infestation. Antixenotic resistance of various *Brassica* species to turnip root fly, *Delia floralis*, has been reportedly linked to the presence of wax esters (Shepherd et al. 1999a, b). Glycerolipids in combination with glycolipids such as glucose and sucrose esters make *Nicotiana benthamiana* resistant to attack by the hornworm, *Manduca sexta* (Matsuzaki et al. 1992). In addition, there are a number of secondary metabolites that provide defense to plants against pests. Therefore, a general approach has been to use botanical products against pests in the form of various extracts containing a group of active ingredients of diverse chemical nature or the isolated cuticular waxes and exudates. Such compounds induce various types of inhibitions in the developmental processes of pests.

4 Essential Oils

Essential oils are mainly obtained by hydrodistillation or steam distillation from aromatic and medicinal plants and are used for medicinal, fragrance, and flavoring purposes in the cosmetic, perfumery confectionary, and pharmaceutical industries and in aromatherapy. The oils are mainly composed of complex mixtures of monoterpenes and sesquiterpenes and their oxygenated hydrocarbons. A number of plant species have been traditionally used for protection of stored commodities, especially in the Mediterranean region and in southern Asia, but interest in the oils was renewed with emerging demonstration of their fumigant and contact insecticidal activities to a wide range of pests in the 1990s (Koul et al. 2008; Isman 2000).

Other essential oils such as lemongrass (*Cymbopogon flexuosus*), eucalyptus (*Eucalyptus globulus*), rosemary (*Rosmarinus officinalis*), vetiver (*Vetiveria zizanioides*), clove (*Eugenia caryophyllus*), and thyme (*Thymus vulgaris*) are known for their pest control properties. Peppermint (*Mentha piperita*) is reported to repel ants, mice, flies, lice, and moths, while pennyroyal (*Mentha pulegium*) wards off fleas, ants, lice, mosquitoes, ticks, and moths. Spearmint (*Mentha spicata*) and basil (*Ocimum basilicum*) are also effective against flies. Some other essential oil-bearing plants, namely, wormwood (*Artemisia vulgaris*), cajeput (*Melaleuca leucadendron*), geranium (*Pelargonium roseum*), lavender (*Lavandula angustifolia*), peppermint (*Mentha piperita*), and red cedarwood (*Juniperus virginiana*), are also effective against various pests. The volatile oil constituents of *Mentha species* are highly effective against stored grain pests, namely, *Callosobruchus maculatus* and *Tribolium castaneum* (Tripathi et al. 2000). Essential oils from cinnamon (*Cinnamomum zeylanicum*), lemongrass (*Cymbopogon citratus*), lavender (*Lavandula angustifolia*), tansy (*Tanacetum vulgare*), *Rabdosia melissoides*, calamus (*Acorus calamus*), clove (*Eugenia caryophyllata*), basil (*Ocimum basilicum*), wintergreen (*Gaultheria procumbens*), jeera (*Cuminum cyminum*), kala jeera (*Bunium persicum*), ajwain (*Trachyspermum ammi*), saunf (*Foeniculum vulgare*), musk (*Abelmoschus moschatus*), cedarwood (*Cedrus deodara*), and other plant species are also known for their varied pest control properties. Citronella (*Cymbopogon nardus*) oil has been used for over 50 years both as an insect repellent and an animal repellent. Combining few drops each of citronella, lemon (*Citrus limon*), rose (*Rosa damascena*), lavender, and basil essential oils with 1 l of distilled water is effective to ward off indoor insect pests. The larvicidal activity of citronella oil has been mainly attributed to its major monoterpenic constituent citronellal (Zaridah et al. 2003).

Vetiver (*Vetiveria zizanioides*) oil has been historically known to protect clothes and other valuable materials placed in closets, drawers, and chests from damaging insect pests. Lavender oil derived from *L. angustifolia* is also known for its pest control properties. *Eucalyptus* (*Eucalyptus globulus*, *E. citriodora*) essential oil exhibits significant pest control properties due to the presence of 1,8-cineole and other essential oil constituents. The insect repellent activity of thyme oil is due to thymol and other phenolic constituents.

The essential oil of catnip (*Nepeta cateria*) is highly effective for repelling mosquitoes, bees, and other flying insects. It repels mosquitoes ten times more than DEET. It is particularly effective against *Aedes aegypti* mosquito known to carrying yellow fever virus. The most active constituent in catnip has been identified as nepetalactone. Dill oil obtained from dill plant (*Anethum sowa*) as by-product of dill industry is also a rich source of carvone. The other major constituent of *A. sowa*, namely, dillapiole is well known for its insecticide synergistic properties. It also occurs to the extent of about 40–60 % in *Anethum graveolens* seed oil and more than 51 % in spearmint oil (*Mentha spicata*). The turmeric (*Curcuma longa*) oil from leaves was rich in α-phellandrene (70 %) and exhibited significant growth inhibition and larval mortality against *S. obliqua* (Agarwal et al. 1999). The leaf oil showed good ovicidal and nymphicidal activity against *D. koenigii* and moderate knockdown effect against *Tribolium castaneum*. Curcumin and ginger oil at 0.2 % concentration showed 86 % inhibition of the mycelial growth of the test fungus *R. solani*. Essential oil extract of *C. longa* has also been found effective against various pests (Agarwal and Walia 2003).

Antifungal activity of certain essential oils or their components have also been assessed and found most effective against *Botrytis cinerea* (Wilson et al. 1997), *Monilinia fructicola* (Tsao and Zhou 2000), *Rhizoctonia solani*, *Fusarium moniliforme* and *Sclerotinia sclerotiorum* (Muller et al. 1995), *F. oxysporum* (Bowers and Locke 2000), *Aspergillus niger* (Paster et al. 1995), *A. flavus* (Montser and Carvajal 1998), *Penicillium digitatum* (Daferera et al. 2000), *F. solani*, *R. solani*, *Pythium ultimum*, *Colletotrichum lindemuthianum* (Zambonelli et al. 1996), *Alternaria padwickii*, *Bipolaris oryzae*, and peanut fungi (Krishna et al. 2007; Nguefack et al. 2007). Tomato seedlings transplanted in soil treated with 700 ml/l of palmarosa oil and 700 ml/l of lemongrass oil were free from bacterial wilt, and 100 % of plants in thymol treatments were free of *R. solanacearum* (Pradhanang et al. 2003).

Essential oils also showed herbicidal activity. Dudai et al. (1999) found that application of essential oil from *Cymbopogon citratus* (lemongrass) in soil inhibited germination of both mono- and dicotyledonous plant species. The citronella oil is reported to have preemergence herbicide activity (Clay et al. 2005). The essential oils and mixture and blends have been found to control weeds and grasses (Brentwood 2003). Citronella allelochemicals at a high dose largely killed foliage of some tree species within 1 day of application but most species regrew strongly. *Senecio jacobaea* was the most susceptible species, with good control 2 months after application of the higher dose (Clay et al. 2005), and volatile cineoles are also allelopathic to weedy plant species (Romagni et al. 2000). Thus ability of these natural plant products to kill or reduce weed populations represents an alternative to the use of toxic weedicide.

The rapid action of essential oils against some pests is indicative of a neurotoxic mode of action, and there is evidence for interference with the neuromodulator octopamine (Kostyukovsky et al. 2002) by some oils and with GABA-gated chloride channels by others (Priestley et al. 2003). Some of the purified terpenoid constituents of essential oils are moderately toxic to mammals, but, with few exceptions, the oils themselves or products based on oils are mostly nontoxic to mammals, birds, and fish (Stroh et al. 1998). Owing to their volatility, essential oils have limited persistence under field conditions. Recent evidence for an octopaminergic mode of action for certain monoterpenoids (Bischof and Enan 2004; Kostyukovsky et al. 2002), combined with their relative chemical simplicity, may yet find these natural products useful as lead structures for the discovery of new neurotoxic insecticides with good mammalian selectivity. Cedarwood oil, himachalol, and β-himachalene have been found to possess insecticidal properties (Singh et al. 1984; Singh and Agrawal 1988; Singh and Rao 1986; Singh 1996). Essential oils have also been reported to show reproduction retardant, fumigant, repellent, and insecticidal activities (Singh et al. 1989; Singh and Singh 1991).

4.1 Essential Oil Constituents

Essential oil constituents are primarily lipophilic compounds that act as toxins, feeding deterrents, and oviposition deterrents to a wide variety of insect pests. Mono- and sesquiterpenes are formed by the condensation of isopentenyl pyrophosphate units. These are believed to aid plants in chemical defense against phytophagous insects, fungi, and bacteria. Insecticidal properties of several monoterpenoids to the housefly, red flour beetle, and southern corn rootworm have been reported (Rice and Coats 1994). Although many monoterpenoids have insecticidal properties, the degree of toxicity of different compounds to one species differs considerably. Cornelius et al. (1997) evaluated toxicity of monoterpenoids against *Coptotermes formosanus*, a subterranean termite of which eugenol was found most effective as termiticide. It was also effective as a fumigant and as feeding deterrent. Similarly, limonene found in the essential oil of various *Citrus* leaves and fruit peels exhibited significant insect control properties (Karr and Coats 1988).

Lichtenstein et al. (1974) reported that carvone isolated from aerial part of dill plants (*Anethum graveolens* L.) exhibits considerable insecticidal activity against *Drosophila* and *Aedes* spp. Lee et al. (1997) evaluated acute toxicity of 34 naturally occurring monoterpenoids against three insects species. The study indicated that citronellic acid and thymol were most toxic against housefly, while citronellol and thujone were most effective against the western corn rootworm. Tripathi et al. (2003) reported that L-carvone completely suppressed the egg hatching of *T. castaneum* at concentration 7.22 mg/cm^2. It also suppressed larval and adult survival (Ouden et al. 1993). Hierro et al. (2004) reported the action of different monoterpenic compounds against *Anisakis simplex* larvae and found that geraniol, citronellol, citral, carvacrol, and cuminaldehyde were active at 12.5 μg/ml.

Some essential oil constituents exhibit insect attractant, repellent, and deterrent properties. Eugenol from cloves (*Eugenia caryophyllus*), 1,8-cineole from *Eucalyptus globulus*, citronellal from lemongrass *Cymbopogon nardus*, pulegone from *Mentha pulegium*, and thymol and carvacrol from *Thymus vulgaris* were among the most active constituents, whereas D-limonene and L-limonene were repellent and attractant. α-Pinene has also been reported to exhibit antifeedant and growth inhibitory activities to the adults and larvae of *T. castaneum*. Limonene, one of the major constituents of *Citrus* oils, has been reported to inhibit the development and growth of German cockroach, *Blattella germanica*. Similarly, eugenol, methyl eugenol, isoeugenol, safrole, and isosafrole are better toxicants and repellents than monoterpenes limonene, cineole, and p-cymene. The essential oil from root of sweet flag, *Acorus calamus*, is also known for its insecticidal and antigonadal actions associated with its most abundant constituent β-asarone. Cinmethylin, a herbicidal analogue of cineole, has been sold in Europe and Asia. The triketone herbicides are derivatives of the plant-produced phytotoxin, leptospermone (Pimental 2005). Composition and insect repellent activity of the essential oil of marigold (*T. erecta*) flower have been recently reported (Ray et al. 2000). Thymol and carvacrol are definitely active against most fungal species tested (Kurita et al. 1981; Muller et al. 1995; Tsao and Zhou 2000). The mechanism of action of these compounds against fungi is unknown but may be related to their general ability to dissolve or otherwise disrupt the integrity of cell walls and membranes (Isman and

Machial 2006). Many monoterpenes from plant sources have been evaluated as feeding deterrents against insects (Koul 1982; Koul et al. 2008). Sesquiterpenes are an important source of insect antifeedants, insecticidal, and antifeedant agents (Cooper-Driver and LeQuesne 1987; Eigenbrode et al. 1994; Hough-Goldstein 1990; Isman et al. 1989; Ivie and Witzel 1982; Rodriguez 1985; Rossiter et al. 1986; Tsunao et al. 1993). Tricyclic silphinene, a sesquiterpene, namely, 11-β-acetoxy-5α-(angeloyloxy)-silphinen-3-one and its two hydrolytic products, 11-β-hydroxy-5α-(angeloyloxy)-silphinen-3-one and 11-β,5α-dihydroxysilphinen-3-one reported from *Senecio palmensis* (Asteraceae), possessed strong antifeedant activity against Colorado potato beetle, *Leptinotarsa decemlineata* (Say) (Mullin et al. 1997). Feeding deterrents sesquiterpene lactone, angelate argophyllin-A, and 3-*O*-methyl niveusin-A were isolated from inflorescences of cultivated sunflower. α-Cyperone, a sesquiterpene isolated from the *Cyperus rotundus* (nut grass) tubers, showed insecticidal activity against diamondback moth, *Plutella xylostella* (Linn) (Thebtaranonth et al. 1995). Poligodial, warburganal, and muzigadial are among some of the potential drimane sesquiterpenes having anti-insect and antifungal properties (Jansen and Groot 1991). Inhibition of feeding in monophagous as well as polyphagous insects has been attributed to enol and α,β-unsaturated aldehyde group(s) in these molecules (Kubo and Nakanishi 1978). The biological activity is primarily due to their ability to form adducts with amino groups rather than sulfhydryl group of the receptors (Caprioli et al. 1987). Kauranoid alcohols have been reported from the important medicinal plant *Croton lacciferus* commonly found in Sri Lanka and India (Ratnayake et al. 1988). These compounds are moderately insecticidal against *Aphis craccivora* Koch (Ratnayake et al. 1992). Costunolide and parthenolide, the sesquiterpene lactones isolated from the fruits of *Magnolia salicifolia*, were toxic to *Aedes aegypti* inducing absolute mortality within 24 h at 15 ppm (Kelm et al. 1997). Conifer resin, a synergistic blend of monoterpene olefins with antiherbivore and antipathogen activity, and diterpene acids that are toxic and deterrent to herbivores have good potential in agricultural pest control (Gershenzon and Dudareva 2007). Also, menthol-containing formulation inhibits adzuki bean beetle, *Callosobruchus chinensis* L. (Coleoptera; Bruchidae), population in pulse grain storage (Singh and Mehta 2010).

5 Clerodane Diterpenes

Clerodane diterpenes, 3,13*E*-clerodane-15-oic acid, 4,13*E*-clerodane-15-oic acid, 18-oxo-3,13*E*-clerodane-15-oic acid, and 2-oxo-3,13*E*-clerodane-15-oic acid, from Nigerian plant *Detarium microcarpum* were found to be feeding deterrents (Lajide et al. 1995a), particularly against workers of the subterranean termite, *Reticulitermes speratus*. The exceptionally hard wood of a Nigerian plant *Xylopia aethiopica* also withstands attack from termites and other insects destructive to wooden structures, which led to the isolation of ent-kauranes, (−) kaur-16-en-19-oic acid having strong termite antifeedant activity against workers of *Reticulitermes speratus* Kolbe (Lajide et al. 1995b). Several natural neo-clerodane diterpenoids isolated from *Linaria saxatilis* and some semisynthetic derivatives were tested against several insect species with different feeding adaptations. The antifeedant tests showed that the oliphagous *Leptinotarsa decemlineata* was the most sensitive insect, followed by the aphid *Myzus persicae*. The polyphagous *Spodoptera littoralis* was not deterred by these diterpenoids; however, following oral administration, some of these compounds did have postingestive antifeedant effects on this insect. In general terms, the antifeedant effects of these compounds were species-dependent and more selective than their toxic/postingestive effects. The study of their structure-activity relationships showed that both the decalin moiety and the chain at C-9 determined their bioactivity. Furthermore, the presence of a 4,18-epoxy/diol moiety was an important feature for both the antifeedant and the toxic/postingestive effects (Gonzalez-Coloma et al. 2000). On the whole, the study of neo-clerodane diterpenoids from structural elucidation to biological activity has been extensively discussed recently (Coll and Tandrón 2007), and many diterpenoids have been reported as both insecticidal and feeding deterrents against various insect species and discussed comprehensively (Dev and Koul 1997; Koul 2005).

Neo-clerodane diterpenes were a promising group of compounds that affect the feeding behavior of insect pests. These were isolated from various species of *Teucrium*, *Ajuga*, and *Scutellaria* (family Labiatae) that also inhibit feeding in Lepidopteran larvae. Approximately 150 neo-clerodanes have been isolated (Merritt and Ley 1992; Ortego et al. 1995), and among these, eriocephalin and teucvin were quite effective along with ajugarins isolated from *Ajuga remota* (Kubo et al. 1982, 1983; Pickett et al. 1987). Jodrellin-B, also reported from *Scutellaria woronowii*, was the most active compound in this series, and scutalpin-C from *Scutellaria alpina javalambrensis* was very active against *S. littoralis* (Munoz et al. 1997). It has been shown that saturation of the dihydrofuran ring and addition of the tigloyl ester function at C-1 results in decreased activity (Anderson et al. 1989; Cole et al. 1990). Both clerodane and neo-clerodane group of diterpenoids are well known for their insecticidal (Camps and Coll 1993) and antifeedant activities (Belles et al. 1985; Cole et al. 1990).

6 Conventional Botanicals: Nicotine, Rotenones, Sabadilla, and *Pyrethrum*

Besides traditional use of botanicals, their commercial use began in the nineteenth century with the introduction of nicotine from *Nicotiana tabacum*, rotenone from *Lonchocarpus* sp., derris dust from *Derris elliptica*, and pyrethrum from *Chrysanthemum cinerariaefolium*. Nicotine, an alkaloid obtained from *Nicotiana tabacum*, *N. rustica*, and *N. glutinosa*, is another well-established botanical insecticide. Nicotine analogues like nor-nicotine and anabasine also possess insecticidal properties. Nicotine is active against piercing-sucking insects such as aphids, leafhoppers, whiteflies, thrips, and mites (Dhaliwal and Koul 2007). However, due to high mammalian toxicity and detrimental effect on human health, its use as an insecticide has decreased considerably.

Rotenones, the first-generation botanical pesticides, have been extensively used in the past to control household and agricultural pests. Its use, however, had to be dispensed with due to high fish and/or mammalian toxicity. Rotenone is a naturally occurring chemical with insecticidal, acaricidal, and piscicidal properties. It is obtained from the roots of *Lonchocarpus* or *Derris*. Although rotenone is the primary constituent in insecticides containing these preparations, a second isoflavone, deguelin, also possesses similar biological properties (Dayan et al. 2009; Caboni et al. 2004). The use of *Derris* root powder was first patented in 1912. It is a selective, nonspecific insecticide, used in home gardens for insect control, for lice and tick control on pets, and for fish eradications as part of water body management. Rotenone is not very soluble in water and is used either as a dust or in an oil or kerosene solution. It exerts its toxic action by acting as a general inhibitor of cellular respiration.

Sabadilla alkaloid derived from sabadilla (*Schoenocaulon officinale* A. Gray) and a number of *Veratrum* species generally referred to as *Veratrum* alkaloids are also known for their insect control properties. The insecticidal activity of sabadilla comes from the alkaloid fraction, which constitutes 3–6 % of the extract. Two most important lipophilic alkaloids in the extract have been identified as veratridine and cevadine, the former being more insecticidal. The major effects of sabadilla poisoning include muscle rigor in mammals and paralysis in insects. Its mode of action is similar to that of the pyrethroids and acts through disruption of nerve cell membranes causing loss of nerve function, an increase in the duration of the action potential, repetitive firing, and a depolarization of the nerve membrane potential due to effects on the sodium channel. Sabadilla alkaloids are labile and break down rapidly in sunlight. These are less toxic to mammals than most other insecticides and are therefore safe to use.

Pyrethrum, the most widely used botanical insecticide is extracted from the flowers of *Chrysanthemum cinerariaefolium* (pyrethrum). It is highly effective against houseflies, mosquitoes, fleas, lice, and many other indoor arthropod pests. The toxins, namely, pyrethrins, cinerins, and jasmolins, have some unusual insecticidal properties, most striking being the immediate knockdown or paralysis on contact which causes most flying insects to drop almost immediately upon exposure (Casida and Quistad 1995). These compounds act both on the central nervous system and in the peripheral nervous system causing repetitive discharges, followed by convulsions. Pyrethrins have low toxicity to vertebrates and have wide acceptance worldwide. Like most other natural pesticides, pyrethrins are labile, have limited stability under field conditions, and are rapidly degraded by sunlight and heat. These are generally formulated with synergists such as piperonyl butoxide (PBO) to inhibit detoxification and improve insect mortality. Natural pyrethrins are considered as the best example of products manipulated in the laboratory to discover highly effective synthetic pyrethroid group of insecticides.

Thus, successful use of traditional botanicals has aroused further interest on exploring plant biodiversity for new bioactive phytochemicals and extractives as possible source of pest control agents. Some of the recent developments are described as under.

7 Isobutylamides

A large number of unsaturated isobutylamides have been isolated from fruits, stems, and leaves of various species of the genus *Piper* (Piperaceae), which are known to have diverse insecticidal actions. These include *P. nigrum*, *P. acutisleginum*, *P. khasiana*, *P. longum*, *P. pedicellosum*, and *P. thomsoni* (Parmar and Walia 2001). Screening of other species of the genus points to numerous other potential sources of natural insecticides, such as *P. retrofractum* from Thailand, *P. guineense* from West Africa, and *P. tuberculatum* from Central America (Isman 2001). Some of the active compounds include piperlonguminine, piperine, pipericide, dihydropipericide, and pellitorine. Recently pellitorine and 4,5-dihydropiperlonguminine were extracted from the seeds of *Piper tuberculatum* Jacq., (Piperaceae) in yields of 6.10 and 4.45 %, respectively. The acute toxicity to the velvet bean caterpillar, *Anticarsia gemmatalis* (Hubner), of these compounds was determined. The LD_{50} and LD_{90} values were 31.3 and 104.5 μg/insect, respectively, for pellitorine and 122.3 and 381.0 μg/insect for 4,5-dihydropiperlonguminine (Navickiene et al. 2007). Being neurotoxic, these amides show both knockdown and lethal action against pyrethroid susceptible and resistant insects. They are extremely unstable molecules but are toxic to a range of insect pests. The information on their environmental or mammalian toxicity is scanty, primarily because they are not yet commercially available. Scott et al. (2003) demonstrated that the amides present in the *Piper* plants have higher toxicity when they are combined in binary, tertiary, and quaternary mixtures, and it is also suggested that seed extracts of piper plants may be more powerful than the isolated compounds (Navickiene et al. 2007).

According to Scott et al. (2003), the piperamides found in *Piper* species are bifunctional, as isobutyl amide functionality is combined with a methylenedioxyphenyl (MDP) moiety. In addition, the piperamides showed dual biological activities, being neurotoxic and also inhibitors of cytochrome P450 enzymes. Fortunately, the risk to human health is much reduced because the active components have had a safe history as food additives and spices (Scott et al. 2003). Miranda et al. (2003) evaluated the susceptibility of *Apis mellifera* L. to pellitorine and found LD_{10} values of 39.1 ng active ingredient/larva, and if LD_{50} of pellitorine is compared with the velvet bean caterpillar (31.3 μg/insect), a value is 1,000 times higher than the LD_{10} for honeybee larvae. Thus, the honeybee larvae were shown to be highly susceptible to pellitorine. Further, evaluation of the effects of piperamides on other nontarget organisms, such as the pests' natural enemies, needs to be carried out.

8 Limonoids

Chemistry and use of neem in pest control is well documented (Isman 2006; Koul 1996, 2008; Koul and Wahab 2004; Randhawa and Parmar 2007; Schmutterer 2002). Within limonoid class, neem (*Azadirachta indica* A. Juss.) is an excellent source of bioactive limonoids. Neem plant constitutes number of bioactive compounds of which tetranortriterpenoids, C-secomeliacins, or limonoids are more predominant. Some of the potential compounds of this group include salannin, desacetyl salannin (salannin group), nimbin, desacetyl nimbin, 4-epinimbin (nimbin group), and azadirachtinoids (azadirachtin group). Azadirachtin, a tetranortriterpenoid (C_{26}-terpenoid), has received the utmost attention as bioactive molecule because of its relative abundance in neem seed kernel. It is a powerful insect antifeedant and growth-regulating substance and exhibits considerable promise as an insecticide (Schmutterer 1990). It can control 413 insect species from several orders (Singh and Saxena 1999). It inhibits feeding and growth of insects throughout a wide variety of insect taxa including Lepidoptera (Anarson et al. 1985; Barnby and Klocke 1987; Koul et al. 1987), Diptera (Miller and Chamberlain 1989), Orthoptera (Ascher et al. 1989; Champagne et al. 1989; Mordue et al. 1985; Rao and Subrahmanyam 1986), Hemiptera (Dorn et al. 1986; Garcia and Rembold 1984; Koul 1984; Redfern et al. 1982), and Hymenoptera (Rembold et al. 1984). The highest IGR activity is associated with 3-detigloylazadirachtin-B ($LC_{50,}$ 0.08 ppm) followed by hydrogenated products, namely, dihydroazadirachtin-B and dihydroazadirachtin-A, and the ethanol adducts 23-α-ethoxy-22,23-dihydroazadirachtin-A which were equally effective in exhibiting LC_{50} of 0.74 and 0.52 ppm, respectively (Yamasaki and Klocke 1987). The antifeedant effects of azadirachtin-A are well known (Ascher 1993; Jacobson 1989; Mordue and Blackwell 1993; Schmutterer 1990). The discovery of antifeedant activity of neem seed kernel extract gave the momentum to investigate the plant thoroughly (Pradhan et al. 1962). Besides antifeedant, repellent, and insect growth regulatory activity, azadirachtin has properties like a partial reduction or complete inhibition of fecundity and/or egg hatchability, reduction of the life span of adults, insect sterilization, oviposition deterrence, ovicidal, formation of deformed larvae, and molts between larval or nymphal instars giving rise to larval-pupal, nymphal-pupal, nymphal-adult, and pupal-adult intermediates (Ascher 1993). Azadirachtin is a common example of a natural plant defense chemical affecting feeding, through chemoreception (primary antifeedancy), that consists in the blockage of the input from receptors that normally respond to phagostimulants or from stimulation of specific deterrent cells or both and through a reduction in food intake due to toxic effects if consumed (secondary antifeedancy), where food intake is reduced after application of azadirachtin in ways which bypass the mouthpart chemoreceptors (Mordue and Blackwell 1993). Growth regulatory effects was attributed to a disruption of endocrine events as the downregulation of hemolymph ecdysteroid level through the blockage of release of PTTH, prothoracicotropic hormone, from the brain-corpus cardiacum complex or to a delay in the appearance of the last ecdysteroid peak showing a complete molt inhibition. These groups of meliacins have effect on insect reproduction, leading to disruption of ovarian development, fecundity, and fertility and prevention of oviposition (Dorn et al. 1986). Female *S. exempta* when emerged from larvae topically treated with 0.01 and 0.1 μg azadirachtin exhibited reduced fecundity due to failure of many oocytes to mature (Tanzubil and McCaffrey 1990).

Dozens of studies have demonstrated the toxicity of neem oils, oil cakes, extracts, leaves, roots, or root exudates against many species of nematodes (Akhtar and Mahmood 1996; Akhtar 1998; Akhtar and Alam 1991; Egunjobi and Afolami 1976). Devakumar and Goswami (1992) evaluated the nematicidal activity of eight C-secomeliacins including azadirachtins and found their LC_{50} ranging between 55 and 157 ppm.

Azadirachtin is highly photolabile, either breaking down or isomerizing under sunlight, which led to significant reduction in bioactivity (Barnby et al. 1989). Short environmental persistence is the major problem limiting its use in agriculture. Initially, research thrust was given toward the bioactivity of neem extracts against agriculturally important phytophagous pests. Once the potential of the extracts is proven, researches were targeted toward extraction of active molecule, protocol of azadirachtin extraction, and its characterization. Huge potential of the molecule led to synthesis by 22 years of effort by Ley and his group at Imperial College, University of London, and later at Cambridge University, the synthesis of azadirachtin was completed in 71 steps (Jauch 2008). The synthesis was finally completed in what is described as a relay method.

Biotechnological interventions have also been made for the production of insecticides from field-grown plants (Akula et al. 2003; Prakash et al. 2005). Prakash and Srivastava (2007) have successfully developed scale-up procedure of *Azadirachta indica* suspension culture for azadirachtin production in stirred tank bioreactor with two different impellers. The study recommended the feasibility of large-scale azadirachtin production in stirred tank bioreactors.

Limonoids from the *Melia azedarach* (chinaberry Meliaceae), and related species, in general showed different types of activities against insect pests (Champagne et al. 1992). *Melia* extracts possess insecticidal, acaricidal, and fungicidal properties, and some of the limonoids isolated are 21-α-acetoxymelianol (Ntalli et al. 2010), meliantriol, melianone, melianol (Lavie and Jain 1967), meliacin, meliacarpin (Li 1999), and meliartenin (Carpinella et al. 2003). The chinaberry fruit powder has been demonstrated as biofumigant to control *M. incognita*-infested soil (EC_{50}, 0.34 % w/w). The activity against root-knot nematode was attributed to the defatted methanol extract at EC_{50}, 0.916 % w/w, which paralyzed half the population tested. The study further revealed that the nematicidal activity does not lay only on in its limonoids' contents comprising of 3-α-tigloylmelianol, melianone, 21-β-acetoxy-melianone, methyl kulonate, etc. but in its organic acids, aldehydes, and alcohols as well. The foremost nematicidal principle, exhibiting activity similar to commercial nematicide fosthiazate, has been identified as furfurale (Ntalli et al. 2010). The compounds like meliatoxins have been recorded as potential compounds for pest control (Koul et al. 2002; Macleod et al. 1990). Similarly, toosendanin, an antifeedant limonoid from the bark of the trees *Melia toosendan* and *M. azadirachta* (Meliaceae), has been subjected to considerable research as a botanical pesticide (Chiu 1989; Chen et al. 1995; Koul et al. 2002). Vertebrate selectivity of this compound is very favorable (LD_{50} mice = 10 g/kg) (Isman 1994). Studies conducted on the antifeedant activity of the fruit extract against a variety of herbivore and granivorous insects through choice and no-choice tests revealed that meliartenin and its interchangeable isomer 12-hydroxyamoorastatin inhibited feeding of *Epilachna paenulata* Germ. Larvae (ED_{50} value of 0.80 μg/cm). The activity was comparable to toosendanin (Carpinella et al. 2003). Azadirachtin was found to influence the head protein and metamorphosis development in *Helicoverpa armigera* Hub. (Neoliya et al. 2005, 2007; Neoliya and Singh 2005).

Citrus species also contain the bitterness causative compounds, which belong to limonoid group; limonin is one of the potential known compounds. A few other *Citrus* limonoids, including nomilin, nomilinic acid, ichangin, and obacunoic acid, are also bitter. Among these, limonin and nomilin are known to deter feeding in Lepidopterans and Coleopterans with variable efficacies (Champagne et al. 1992). It appears that furan, and epoxide groups played a major role in the activity of these compounds. A possible role of C-7 is implied by the modest activity of the 7-hydroxylated de-epoxy system (Bentley et al. 1988). For instance, highly reduced activity of deoxyepilimonol against limonin demonstrates the above conclusion. In certain cases, the cyclohexenone A ring and the *α*-hydroxy enone group in the B ring appear to be important for antifeedant activity. Also, the absence of 14–15 epoxides may not drastically reduce antifeedant activity (Govindachari et al. 1995). Some structural-activity relationships have also been drawn by preparing some semisynthetic derivatives of *Citrus* limonoids, suggesting the potential of functional groups for the activity (Ruberto et al. 2002).

Chemical investigation of the diethyl ether extract of the stem bark of *Khaya ivorensis* A Chev (Meliaceae) afforded ten limonoids of angolensates, ring D-opened limonoids, and mexicanolides. Methyl angolensate and 1,3,7-trideacetylkhivorin displayed the highest antifungal activity against *Botrytis cinerea* and inhibited 62.8 % and 64.0 % mycelial growth at 1,000 mg/l and 73.3 % and 68.6 % mycelial growth at 1,500 mg/l. 3,7-Dideacetylkhivorin showed stronger antifungal and antibacterial activities than methyl 6-hydroxyangolensate against all of the test fungi and bacteria except *Penicillium expansum* Link and was reported for the first time (Abdelgaleil et al. 2005).

Antifeedant activity of a mixture of limonoids 1,7-di-*O*-acetylhavanensin and 3,7-di-*O*-*acetylhavanensin* isolated from seeds of *Trichilia havanensis* (Meliaceae) and *neo*-clerodane diterpene, scutecyprol A, isolated from *Scutellaria valdiviana* (Lamiaceae) to fifth instar larvae of the beet armyworm, *Spodoptera exigua*, has been investigated. Choice and no-choice feeding assays, nutritional tests, and post-treatment studies indicated that scutecyprol A acts as an insect feeding deterrent against *S. exigua*, whereas the antifeedant activity of mixture is likely associated with a toxic mode of action. The mixture of limonoids significantly increased glutathione S-transferases during the treatment and post-treatment periods, whereas esterases were inhibited during the treatment period. On the contrary, scutecyprol A did not have any significant effect on any of the enzymatic processes. Hence, the metabolic response of *S. exigua* larvae to the ingestion of the secondary metabolites tested depends on their mode of action (Caballero et al. 2008) suggesting that mixture of limonoids may play a role in pest control and resistance management. Thus, *M. azedarach* fruit extract and its active principle have interesting potential for use in pest control programs.

9 Saponins and Sapogenins

Plant saponins show a diverse range of biological activities. Irrespective of differences in their carbohydrate and aglycone structure, saponins as a class are characterized by their bitter taste, foaming in aqueous solution, and hemolytic activity. The ability of the saponins to reduce surface tension has been utilized in making emulsion stabilizers with results comparable with fatty acid salts. Saponins are divided into three major groups, triterpenic, basic steroidal, and steroidal alkaloid saponins. The fungistatic action of several saponins and saponin fractions was tested on 15 species of plant pathogenic fungi. The study revealed that *Sclerotinia fructicola*, *C. purpurea*, *Trichothecium roseum*, and *Pyricularia oryzae* were most sensitive to the saponins, while *Botrytis allii*, *Alternaria solani*, *Coniophora cerebella*, *R. solani*, and *Fusarium* species were resistant. Some of the steroidal saponins such as asparanins I and B isolated from *Asparagus adscendens* (Roxb.) and triterpenic saponins isolated from *Albizia chinensis* and *Acacia concinna* pods exhibited nematicidal activity against the root-knot nematode, *Meloidogyne incognita* (Meher et al. 1988). Interestingly, saponins with more sugar units in the glycon moiety were less active than those with fewer sugar units. Medicagenic acid (XII) and several of its glycosides are present in the roots of alfalfa (*Medicago sativa* L.) and exhibit fungicidal properties. Alfalfa meal has been shown to control avocado root rot caused by *Phytophthora cinnamomi* Rands and several other plant pathogens (Oleszek et al. 1992). Saponins of this type are believed to exert their effects by interacting with membrane sterols, proteins, and phospholipids. Bioassay-directed fractionation led to the isolation of five bioactive oleanolic acid saponins from the roots of *Viguiera decurrens* (Marquina et al. 2001). Of these, saponin-IV with two sugar substitution in two sites showed significant insecticidal activity (LC_{50} value 80 μg ml^{-1}) against *Epilachna varivestis*. Tomatine and solanine, the steroidal alkaloid glycosides from *Solanum tuberosum* and *Lycopersicon esculentum*, and their aglycone tomatidine and solanidine exhibit antifeedant properties (Dahlman and Hibbs 1967). These compounds are attracting renewed interest because of their arthropod resistance and interesting insecticidal and nematicidal action. Triterpenic saponins extracted from *Diploknema butyracea* (*Madhuca indica*) and *Sapindus mukorossi* (soap nut) have been reported to exhibit insect feeding deterrent and insect regulatory activity (Saha et al. 2010a). These compounds also exhibit plant growth stimulant activity (Saha et al. 2010b). Structure-biological activity relationships in triterpenic saponins and the relative activity of protobassic acid and its derivatives against plant pathogenic fungi and synergistic/potentiation interaction between nematostatic constituents from *Azadirachta indica* (neem) and phytochemicals from *D. butyracea* and *S. mukorossi* have been well established (Saha et al. 2010c).

10 Naphthoquinones

The biological activity of substituted-1,4-naphthoquinones was first reported by Fieser et al. (1948). Lapachol obtained from the wood extract of *Tabebuia serratifolia* (Bignoniaceae) showed antifungal activity (Velasquez et al. 2004). The compound was found to be more active (LC_{50}=20.8 ppm) against the larvae of *Aedes aegypti* than the amine derivatives (LC_{50}=242.6–899.4 ppm) obtained from an ethanolic bark extract of *T. serratifolia* (Oliviera et al. 2002). Two active principles from the Chilean plant, *Calceolaria andina* (Scrophulariaceae), related to the familiar garden "slipper" plants have been identified as hydroxynaphthoquinone and its acetate, designated as BTG 505 and BTG 504, which are effective against a range of commercially important pests including the tobacco whitefly, *Bemisia tabaci*; aphids; and the two-spotted spider mite, *Tetranychus urticae* (Khambay et al. 1999). These compounds have been characterized as 2-(1, 1-dimethylprop-2-enyl)-3-hydroxy-1,4-naphthoquinone and the corresponding acetate, 2-acetoxy-3-(1,1-dimethylprop-2-enyl)-1, 4-naphthoquinone. Their activities against 29 pest species and 9 beneficial species of arthropod from a total of 11 orders have been determined. No cross-resistance has been observed for strains resistant to established classes of insecticide. The mammalian toxicity of this compound is also low. These compounds offer opportunities both as lead structures for analogue synthesis (Khambay et al. 1997a) and as new botanical pesticides (Khambay et al. 1997b; Khambay et al. 1999) exhibiting low mammalian toxicity unlike other naphthoquinones. The primary mode of action in insects is by inhibition of complex III of the mitochondrial respiratory chain (Khambay and Jewess 2000). The insecticidal and fungicidal properties of dunnione (a known naphthoquinone) with natural BTG 505 have been compared (Khambay et al. 2003). Although dunnione showed practically no activity against the housefly, *Musca domestica*, the whitefly *B. tabaci*, the beetle *Phaedon cochleariae*, or the spider mite *T. urticae* unlike BTG 504 and BTG 505, dunnione has an unusually broad spectrum of antifungal activity. The mode of action of dunnione is primarily through initiation of redox cycling, whereas BTG 505 acts by inhibiting mitochondrial complex III (Khambay et al. 2003).

11 Rocaglamides

The genus *Aglaia* consisting of some 130 species widely distributed in the Indo-Malaysian region (Nugroho et al. 1999) has attracted considerable attention in the past decade as a possible source of bioactive natural products. Phytochemical investigations of *Aglaia* have revealed the presence of rocaglamides (Ishibashi et al. 1993; Proksch et al. 2001), aglains (Bacher et al. 1999), bisamides (Brader et al. 1998), triterpenes (Weber et al. 2000), and lignans (Wang et al. 2004) with interesting biological activities. Rocaglamide derivatives feature a cyclopentatetrahydrobenzofuran skeleton and have been shown to have strong insecticidal activity of which some derivatives showed activity comparable to azadirachtin against neonate larvae of *Spodoptera littoralis*, *Ostrinia* species, and the gram pod borer, *Helicoverpa armigera* (Brader et al. 1998; Gussregan et al. 1999; Koul et al. 2004a; Nugroho et al. 1997a, b, 1999). The insecticidal mode of action of rocaglamides results from the inhibition of protein synthesis, explaining the long time-to-death in treated insects (Satasook et al. 1993). The insecticidal activity of rocaglamides can be attributed to the presence of the furan ring system since the closely related aglains, possessing a pyran ring, are devoid of insecticidal activity

(Nugroho et al. 1999). Acylation of the OH group at C-1 caused a reduction of insecticidal activity in neonate larvae of *S. littoralis* compared with other rocaglamide derivatives with a hydroxyl substituent isolated from the twigs of *A. duperreana* (Nugroho et al. 1997a). There is a decline in insecticidal activity for rocaglamide derivatives featuring an unsubstituted C-2 in contrast with analogues possessing an amide or carboxylic substituent at this position. A similar trend has been noted in other rocaglamide derivatives isolated from *A. odorata* (Gussregan et al. 1999; Nugroho et al. 1999) and *A. elliptica* (Nugroho et al. 1997b). Substitution of a hydroxy group with the methoxy group at C-8b resulted in a complete loss of activity in compounds that were isolated from roots of *A. duperreana* (Chaidir et al. 1999), showing the importance of the OH group at C-8b. Among the various botanicals isolated from *Aglaia odorata*, *A. elliptica*, and A. *duperreana* (Meliaceae), rocaglamide is the most effective (EC_{50}=0.8 ppm). It is slightly more potent than azadirachtin (EC_{50}=1.0 ppm) against some insect species (Janprasert et al. 1993).

12 Polyol Esters

Polyol esters are known for their diverse pest control and other biological properties. Some of these products have been comprehensively discussed (Dev and Koul 1997; Koul 2005). The powdered root bark of Chinese bittersweet (*Celastrus angulatus*) is traditionally used in China to protect plants from insect damage. An insect antifeedant celangulin with two possible structures 1 and 2 has been reported from this plant. This non-alkaloidal sesquiterpene polyol ester has a dihydroagarofuran skeleton with seven hydroxyl functions, five of which are acylated, one benzoylated, and one free (Wakabayashi et al. 1988). Insecticidal alkaloids with α, α-dehydroagarofuran skeleton such as wilfordine from *Tripterygium wilfordii* (Yamada et al. 1978) and wilforine alkaloid from *Maytenus rigida* as insect antifeedant (Delle-Monache et al. 1984) have also been reported. Some terpenes particularly the spirocaracolitones isolated from Rutales were effective antifeedants (Omar et al. 2007). The antifeedant and insecticidal activity of these polyol esters against *Pieris rapae* and *Ostrinia furnacalis* (Guenee) has been attributed to the ester moieties attached to the decalin portion of the molecule (Koul and Walia 2009).

From the seed oil of *Euonymus bungeanus*, three sesquiterpene polyol esters, namely, 6-1,2-diacetoxy-1β,2β,9α-tri(β-furancarbonyloxy)-4α-hydroxy-β-dihydroagarofuran, 6α-1,2-diacetoxy-1β,9α-di(β-furancarbonyloxy)-4α-hydroxy-2β-2-methylbutanoyloxy-1β-2-methyl butanoyl-β-dihydroagarofuran, and 6α-1,2-diacetoxy-2β,9α-di(β-furancarbonyloxy)-4α-hydroxy-1β-2-methylbutanoyl-1β-dihydroagarofuran, have been reported (Tu et al. 1990). Two compounds, 6α-1,2-diacetoxy-1β,9α-di(β-furancarbonyloxy)-4α-hydroxy-2β-2-methylbutanoyl oxy-1β-2-methyl butanoyl-β-dihydroagarofuran and 6α-1,2-diacetoxy-2β,9α-di(β-furancarbonyloxy)-4α-hydroxy-1β-2-methylbutanoyl-1β-dihydroagarofuran, were antifeedants against *Pieris rapae*. The antifeedant and insecticidal activities of these polyol esters have been attributed to the ester moieties attached to the decalin portion of the molecule. Veratridine and cevadine, the bioactive alkaloidal constituents of sabadilla (*Schoenocaulon officinale*), are among the most potent natural product insecticides. The significance of acyl moiety in these compounds was emphasized when protoveratrine A and B were found to be active (Ujvari et al. 1991). The insecticidal activity was however significantly reduced when these esters were hydrolyzed to corresponding veracevine and ryanodol suggesting the importance of ester functions in the bioactive alkaloid or terpene polyester molecules.

Plant glucose and sucrose esters occur naturally in leaf cuticular waxes of wild *tobacco Nicotiana gossei* and several other solanaceous plants. These esters are composed of lower fatty acids (C-2 to C-10) and have been found to be toxic to the tobacco aphid, *Myzus nicotianae*, and the greenhouse whiteflies, *Trialeurodes vaporariorum* and *Bemisia argentifolii*. Sucrose and glucose esters being thick and sticky have an apparent role in the entrapment of insects that become physically tapped in the sticky trichome exudates (Burke et al. 1987). If present in sufficiently large quantities, these esters immobilize the small and tender insects like aphids and whiteflies, causing mortality. Phytochemical investigations of a large number of *Nicotiana* species have resulted in the isolation of a variety of glucose esters (Matsuzaki et al. 1989a, b). The esters isolated from *N. bigelovii* was identified as 2,3,4-tri-*O*-acyl-α,β-D-glucopyranose. The fatty acid moieties located at 2, 3, and 4 positions consisted of acetic acid, methylpropionic acid, 2-methylbutanoic acid, 3-methylpentanoic acid, and/or 4-methylpentanoic acid. Glucose esters of 3-nitropropanoic acid, karakin, coronarian, and cibarian, isolated from *Coronilla varia*, *Lotus pedunculatus*, and few other plants, were found toxic to larvae of *Costelytra zealandica* (white grass grub). At 1 % dose, 70, 55, and 75 % mortality were observed for coronarian, cibarian, and karakin, respectively (Hutchins et al. 1984). Acyl sugars deterred the green peach aphid *Myzus persicae* (Sulzer) from feeding on *Solanum berthaultii* (Lapointe and Tingey 1984; Neal et al. 1990) and on *Lycopersicon pennellii* (Correll), a wild relative of the cultivated tomato (Goffreda et al. 1989). The triacylated glucose esters extracted from *N. bigelovii* and *N. miersii*, tetraacylated glucose esters from *N. miersii*, and three sucrose esters isolated from *N. umbratica* caused severe root damage and inhibited shoot growth of barnyard grass, *Echinochloa crusgalli* above 5×10^{-3} M concentration (Matsuzaki et al. 1989a, b, 1991). Severson et al. (1985) reported the occurrence of a series of sucrose esters in the cuticular waxes of the tobacco leaves in which only the glucose moiety was substituted at the C-6 position with an acetate group and at C-2, C-3, and C-4 positions with either methylvaleryl, isovaleryl, isobutyryl, methylcaproyl, butyryl, valeryl, caproyl, or propionyl moieties. The major bioactive constituent was identified as 6-*O*-acetyl-2,3,4-tri-*O*-(β-methylvaleryl)-sucrose (Severson et al. 1985; Einolf and Chan 1984). Bioassay-guided approach led to the isolation of two new sucrose esters from *N. glutinosa* (Matsuzaki et al. 1988), which differed from the ones reported by Severson et al. (1984) in the position of acetyl moiety (C-3′) on the fructose ring and other acyl moieties in the glucose ring. These were identified as (2,3,4-tri-*O*-acyl)-α-d-glucopyranosyl-(3-*O*-acetyl)-β-d-fructofuranoside and (2,3,4-tri-*O*-acyl)-α-D-glucopyranosyl-β-d-fructofuranoside. In another significant study, three sucrose esters were isolated from the surface lipids of leaves of *N. cavicola* (Ohya et al. 1994). The common features found in all three sucrose esters were the presence of one acetyl residue at fructose ring and free hydroxyl groups at 2 and 3 positions of glucose ring. These sugar esters were different from those reported from *N. tabacum* and *N. glutinosa*, *N. umbratica*, and *N. benthamiana* in terms of their ester positions and fatty acid composition.

The presence of glucose and sucrose esters has also been reported from leaves of tomato (Burke et al. 1987), wild tomato and potato species (King et al. 1987, 1988; Neal et al. 1990), and petunia (Kays et al. 1994). The insecticidal sucrose esters in *Nicotiana gossei*, a wild relative of tobacco, show considerable plant toxicity against the greenhouse whitefly (Neal et al. 1994; Severson et al. 1985). Insecticidally active sugar esters isolated from the leaf surface of *N. gossei* have been identified as 2,3-di-*O*-acyl-6′-*O*-acetyl sucrose and 2,3-di-*O*-acyl-1′,6′-diacxetylsucrose (Buta et al. 1993) and the two glucose esters as 1-*O*-acetyl-2,3-di-*O*-acylglucose and 2,3-di-*O*-acylgucose (Severson et al. 1994). The acyl group in these compounds is mainly 5-methylhexanoyl-5-methoxyheptanoyl. Sucrose esters isolates from *N. plumbaginifolia*, *N. gossei*, *N. palmerii*, *N. cavicola*, *N. amplexicaulis*, and *N. langsdorfii* showed potent antibacterial activity (Chortyk et al. 1993). The apparent structure-activity relationship studies revealed that high carbon-chain acyl groups were associated with increased activity.

Sugar esters reportedly disrupt the integrity of cellular membranes and uncouple oxidative phosphorylation, similar to the action of insecticidal soaps. According to Puterka and Severson (1995), sugar esters disrupt the structure of the insect cuticle. It has been stated that leaf surface moisture and ambient relative humidity affects the efficacy of *Nicotiana gossei* sugar esters. For example, the application of hygroscopic materials such as humectants at the site of application improve the toxicity of natural sugar esters from *N. gossei* and other *Nicotiana* species as well as certain synthetic sugars against tobacco aphids (Xia et al. 1997a, b). Certain sugar esters have also been synthesized (Chortyk and Nottingham 1995; Chortyk et al. 1996) for possible use against whitefly and other soft-bodied arthropod pests damaging agricultural products.

13 Acetogenins

The tropical plant family Annonaceae despite of its relatively large size is chemically one of the least explored. The discovery of uvaricin in 1982, the first of the acetogenins, as an in vivo active antileukemic (P-388) agent, invigorated wide interest in this family. Annonaceous acetogenins are a class of polyketide natural products found in plants of the family Annonaceae. These are derived from C-32/C-34 fatty acids that are combined with an isopropanol unit. They are usually characterized by a long aliphatic chain bearing a terminal methyl-substituted α,β-unsaturated γ-lactone ring (sometimes rearranged to a keto-lactone) with one, two, or three tetrahydrofuran (THF) rings located along with the hydrocarbon chain studded with a number of hydroxyl, acetoxyl, ketones, epoxides, and/or double bonds function. Bioactive acetogenins such as annonins and related compounds, namely, squamocin, asimicin, annonacins, and cohibinsin, occur widely in twigs and branches, unripe fruits, and seeds of several *Annona* (custard apple) species (Annonaceae). Entire group of annonaceous acetogenins has been patented as pesticide in which asimicin was claimed as a structurally defined pesticidal acetogenin (Parmar and Walia 2001). Johnson et al. (2000) have isolated hundreds of acetogenins from the Annonaceae and found their value as insecticides. Isman (2006) has demonstrated that crude ethanolic extracts or even aqueous extracts of seeds from *A. squamosa* collected at several sites in eastern Indonesia were effective against the diamondback moth, *Plutella xylostella*. Natural products from the plants of the family Annonaceae, collectively called Annonaceous acetogenins, are potent inhibitors of the NADH-ubiquinone reductase (Complex I) activity of mammalian mitochondria. Comparison of some acetogenins with rotenone and piericidin indicated that rolliniastatin-1 and rolliniastatin-2 are more powerful than piericidin in terms of both their inhibitory constant and the protein dependence of their titer in bovine submitochondrial particles. Squamocin and otivarin, contrary to the other acetogenins, behave qualitatively like rotenone (Esposti et al. 1994). Acetogenins are slow stomach poisons, particularly effective against chewing insects such as lepidopterans and the Colorado potato beetle, *Leptinotarsa decemlineata*. Annonaceous acetogenins and/or their crude extracts can thus be employed as safe, effective, economical, and environmentally friendly pesticides with an emphasis on the home garden, ornamental, greenhouse, and produce markets, pending regulatory approval.

14 Light-Activated Botanicals

Toxicity of some of the phytochemicals is increased following their exposure to light radiations. Such light-activated phototoxins like substituted acetylenes, thiophenes, acetylenic thiophenes, quinones, furanocoumarins, and related compounds exhibit significant pest control properties. For example, oil of the desert plant *Artemisia monosperma* has been reported to contain a phototoxin 3-methyl-3-phenyl-1,4-pentadiyne which under light-induced conditions has been found to be as active as DDT against the housefly *Musca domestica* and cotton leafworm *S. littoralis* larvae. Random screening of *A. pontica* yielded an acetylenic epoxide, namely, ponticaepoxide, which when applied to the mosquito larvae under UV light exhibited LC_{50} of 1.47 ppm (Marchant and Cooper 1987). α-Terthienyl (αT) is found in abundance in the floral, foliar, and root extracts of *Tagetes minuta*.

It has been demonstrated that irradiation of α-terthienyl with near UV light generates a reactive singlet oxygen species responsible for enhanced nematicidal activity (Chan et al. 1975; Bakker et al. 1979). Under nonirradiated conditions, it exhibited low toxicity to *Aedes* mosquito larvae, and the activity was substantially increased upon irradiation by near UV light. An acetylenic thiophene 5-(3-buten-1ynyl)-2,2-bithienyl isolated from *Tagetes* roots not only exhibited nematicidal activity but also showed insecticidal activity against several herbivorous insects such as *M. sexta* and *P. rapae* and mosquitoes like *Aedes aegypti*. Under photosensitizing conditions, α-T is more toxic (LC_{50}=20 ppb) to *A. aegypti* larvae than malathion (LC_{50}=62 ppb) and less toxic than chlorpyrifos (LC_{50}=1 ppb) and temephos (LC_{50}=3 ppb). Matricaria ester and *cis*-dehydromatricaria ester, the two polyacetylenic compounds, have been found to be ovicidal to freshly laid eggs of the fruit fly. Under the influence of UV light, its activity was dramatically enhanced. Similarly 2-(non-1-en-3,5,7-trinyl) furan exhibited excellent mosquitocidal activity against *Aedes aegypti* larvae exhibiting LC_{50} of 0.079 ppm under UV light (Uhlenbroek and Bijloo 1959; Kawazu et al. 1977).

Photoactivated natural toxins generally operate by two modes of action. Phototoxin first absorbs light and generates activated species of oxygen. In one action, the excitation energy is transferred to molecular oxygen to produce highly reactive singlet oxygen superoxide, hydroperoxide, or hydroxyl radicals through electron transfer mechanisms, which ultimately damage important biomolecules (Arnason et al. 1983). The other mode of action is photogenotoxic and cause damage independently of oxygen by reacting directly with DNA (Towers 1984; Berenbaum 1987). Since the mode of action of phototoxins is quite different from the conventional synthetic pesticides, there is no likelihood of cross-resistance to conventional larvicides such as malathion (Hasspieler et al. 1988). However, light-induced toxicity to vertebrates and its possible effects on nontarget organisms need further study before these are considered as an alternative to current mosquito larvicides.

15 Miscellaneous Phytochemical Pesticides

Antifungal activity of phytochemicals such as cinnamaldehyde against *Verticillium fungicola*, *Rhizoctonia* sp., *Pythium* sp., *Sclerotinia homoeocarpa*, and *Fusarium moniliforme* and a combination of L-glutamic acid and γ-aminobutyric acid against powdery mildew explains the potential of plant products in fungal pathogen control (Copping and Duke 2007). Antifungal activities of natural substances from *Eucalyptus dalrympleana*, *E. globulus*, *E. gunnii*, and *E. urnigera* were evaluated against postharvest pathogens of kiwifruits, *Botrytis cinerea*, *Botryosphaeria dothidea*, and *Diaporthe actinidiae*. Eucalyptus-derived phenolics such as gallic acid were found to be effective in checking mycelial growth and spore germination of *B. cinerea* at relatively high concentrations. The results suggest that gallic acid can be a safer and more acceptable alternative to current synthetic fungicides controlling soft rot decay of kiwifruit during postharvest storage (Oh et al. 2008). Three different sesquiterpene lactones, namely, hydroxyachillin from *Artemisia lanata*, parthenolide from *Magnolia grandiflora*, and dehydrocostus lactone and costunolide from *Costus* resin oil, have been reported to possess fungicidal activity (Wedge et al. 2000).

Plant secondary metabolites are also a potential source for new herbicides (Devine et al. 1993; Duke 1990; Copping 1996; Vyvyan 2002) and offer new modes of action, more specific interactions with weeds, and less harmful to the environmental. For example, mesotrione (Duke et al. 2000; Mitchell et al. 2001) was basically derived from a natural compound leptospermone obtained from the roots of *Callistemon citrinus*. Juglone from walnut trees has been found effective against redroot pigweed, velvet leaf and barnyard grass. Dhurrin from sorghum, gallic acid from spurge, trimethylxanthine from coffee, and cinch from *Eucalyptus* are some other important plant products having potential herbicide activity (Narwal et al. 1999). Aqueous extract of kava roots showed high allelopathic potential and has suppressed germination and growth of lettuce, radish, barnyard grass, and monochoria.

Out of the nine kava root lactones, six compounds namely desmethoxyyangonin, kavain, 7,8-dihydrokavain, yangonin, methysticin, and dihydromethysticin showed great herbicidal and antifungal activities. The growth of lettuce and barnyard grass was significantly inhibited at 1–10 ppm, and four plant fungi, namely, *Colletotrichum gloeosporioides*, *Fusarium solani*, *Fusarium oxysporum*, and *Trichoderma viride*, were significantly inhibited at 10–50 ppm. This study suggested that kava lactones may be useful for the development of bioactive herbicides and fungicides (Xuan et al. 2006). Similarly, allelochemicals from *Lantana camara* against aquatic weeds *Eichhornia crassipes* and *Microcystis aeruginosa* have been shown to inhibit their growth, and the compounds responsible belong to pentacyclic terpenoids, lantadenes A and B (Kong et al. 2006). In general, it is observed that the mixture of plant extracts with compounds appears to be more toxic than individual compound due to synergistic, potentiating, and/or additive effects of the constituents (Hummelbruner and Isman 2001; Koul et al. 2003, 2004b, c). Such mixtures show interactions for longer duration of time (Akhtar and Isman 2003; Chockalingam et al. 1990). Many medicinal and aromatic plants and their active compounds have been found to possess antifeedant, repellant, insecticidal, and insect growth regulatory activities and are used in integrated pest management (Tripathi et al. 1990; Singh and Mehta 1998; Singh et al. 2001, 2003; Raguraman and Singh 1997; Tripathi et al. 1987; Tripathi and Singh 1994; Singh 1996, 1998, 2005, 2006; Anshul et al. 2013). Himachalol and β-himachalene derived from the cheapest essential oil of Cedrus deodara Roxb. have been found as insecticidal principles against housefly and adzuki bean beetles (Singh and Agrawal 1988).

16 Commercial Potential

Among the traditional botanicals, pyrethrum, neem, rotenone, sabadilla, ryania, and nicotine are registered products for use in different countries. Several azadirachtin-based insecticides are sold in the United States, and pyrethrum, rotenone, and nicotine are registered products in Canada. In Europe, pyrethrum, neem, rotenone, and nicotine are allowed; however, there is variation in registration protocols in different European countries. In Asia, India leads in the use of botanicals where a number of products are registered under provisional registration. Among the latest commercial botanicals, neem-based products seem to have good future. However, apart from neem, some more botanicals have demonstrated their antifeedant efficacy in the field. Application of polygodial or methyl salicylate has shown that aphid populations are reduced with concomitant increases in yields of winter wheat and in one case comparable to that achieved with the pyrethroid insecticide cypermethrin (Pickett et al. 1997).

In the United States, commercial development of insecticides based on plant essential oils has been greatly facilitated by exemption from registration for certain oils commonly used in processed foods and beverages (Quarles 1996). Several essential oil-based insecticides, fungicides, and herbicides using rosemary oil, clove oil, and thyme oil as active ingredients have been developed particularly for control of greenhouse pests and diseases and for control of domestic and veterinary pests (Koul et al. 2008). EcoSMART Technologies has introduced insecticides containing eugenol and 2-phenethyl propionate aimed at controlling crawling and flying insects, under the brand name EcoPCO®. Hexa-Hydroxyl®, a synergistic blend of plant oils effective against a broad spectrum of pests, has also been developed. An insecticide/acaricide containing rosemary oil as the active ingredient has also been introduced for use on horticultural crops under the name EcoTrol™. Another rosemary oil-based fungicide is sold under the name Sporan™, while a formulation of clove oil sold as Matran™ is used for weed control. Several smaller companies in the United States and the United Kingdom have developed garlic oil- and mint oil-based pest control products for home and garden use. Menthol has been approved for use in North America for the control of tracheal mites in beehives, and a product produced in Italy (Apilife VAR™) containing thymol, cineole, menthol, and camphor is used to control *Varroa* mites in honeybees. Israel start-up Botanocap is developing a slow-release technology for essential oils to make relatively environmentally friendly pesticides. The company has developed a patented technology for the gradual release of essential oils and natural components. It possesses patents on capturing essential oils in capsules to achieve the delayed release effect.

Additional challenges to commercialize plant-based pesticides include nonavailability of sufficient quantities of plant material, standardization and refinement of pesticide products, and regulatory approval (Isman 2005). In addition, as the chemical profile of plant species can vary naturally depending on geographic, genetic, climatic, annual, or seasonal factors, pesticide manufacturers must take additional steps to ensure that their products will perform consistently. Finally, once all of these issues are addressed, regulatory approval will continue to be a barrier until commercial products satisfy the regulatory systems. If used indiscriminately, these may also result in the development of resistance to insect pests and diseases. Regulatory mechanism remains the most formidable barrier for commercialization. In many jurisdictions, no distinction is made between synthetic pesticides and biopesticides, including botanicals. Regulatory mechanism therefore needs to be developed for safety, quality control, and development of biosafety standards for botanicals.

17 Conclusion

Natural products have historically been used in numerous applications for the control of pests infesting agricultural crops. However, inadequate availability of raw materials, problems in large-scale production, poor shelf life, diminished residual toxicity under field conditions, and non-availability of reliable standards are some of the constraints cited as major impediments to their commercialization. Aside from the inherent chemical complexity, natural products have insufficient potency or inadequate specificity to act as leads for synthesis. Simply because a compound is a natural product does not automatically ensure that it is safe as most of the toxic mammalian poisons are natural products and many of them are plant products. Before recommending for pest control, such products must be assessed for their environment safety and toxicity to mammals and nontarget organisms. So far, crude natural product mixtures have been used as natural pesticidal preparations. In the present regulatory climate, it is mandatory to establish the chemical profile of bioactive constituents in such products. Concerns have also been raised about the possible development of resistance in insects that feed on such plants. Several pertinent issues relating to human and environment safety, cost-effective production, stabilization for improved field persistence, enhancement of pesticidal spectrum for sustained efficacy, and development of new and improved techniques for their analysis and quality control require serious attention. The potential of synthetic modification to improve upon a natural product or a lead molecule has been demonstrated successfully in the past with the discovery of synthetic analogues with improved activity and enhanced stability. Such fascinating developments shall continue to emphasize the importance of natural products as important sources for developing new agricultural chemicals for use in pest control.

References

Abdelgaleil SAM, Hashinaga F, Nakatani M (2005) Antifungal activity of limonoids from *Khaya ivorensis*. Pest Manag Sci 61:186–190

Agarwal M, Walia S (2003) Pest control potential of phytochemicals derived from *Curcuma longa* and *Zingiber officinale*. In: Dureja P, Saxena DB, Kumar J, Singh SB, Gopal M, Tanwar RS (eds) Proceedings of international conference on pesticides, environment, food security, society of pesticide science, New Delhi, pp 110–119

Agarwal M, Walia S, Dhingra S (1999) Pest control properties of turmeric leaf oil against *Spilosoma obliqua. Dysdercus Koenigii* and *Tribolium castaneum*. In: Proceedings of 2nd all India people's technology congress, Calcutta, pp 1–7

Akhtar M (1998) Biological control of plant-parasitic nematodes by neem products in agricultural soil. Appl Soil Ecol 7:219–223

Akhtar M, Alam MM (1991) Integrated control of plant-parasitic nematodes on potato with organic amendments, nematicide and mixed cropping with mustard. Nematol Mediterr 19:169–171

Akhtar Y, Isman MB (2003) Larval exposure to oviposition deterrents after subsequent oviposition behavior in generalist, *Trichoplusia ni* and specialist, *Plutella xylostella* moths. J Chem Ecol 29:1853–1870

Akhtar M, Mahmood I (1996) Control of plant-parasitic nematodes with organic and inorganic amendments in agricultural soil. Appl Soil Ecol 4:243–247

Akula C, Akula A, Drew R (2003) Somatic embryogenesis in clonal neem, azadirachta indica a. juss. and analysis for in vitro azadirachtin production. In Vitro Cell Dev Biol Plant 39:304–310

Anarson JT, Philgone BJR, Donskov N, Hudon N, Mc Dougall C, Fortier G, Morand P, Gardner D, Lambert J, Morris C, Nozzolillo C (1985) Antifeedant and insecticidal properties of azadirachtin to the European corn borer, *Ostrinia nubilalis*. Entomol Exp Appl 66:29–34

Anderson JC, Blaney WM, Cole MDS, Fellows LE, Ley SV, Sheppard RN (1989) The structure of two new clerodane diterpenoid potent insect antifeedants *from Scutellaria woronowii*, Jodrellin A and B. Tetrahedron Lett 30:3737–3740

Anshul N, Bhakuni RS, Gaur R, Singh D (2013) Isomeric flavonoids of *Artemisia annua* (Asterales: Asteraceae) as insect growth inhibitors against *Helicoverpa armigera* (Lepidoptera: Noctuidae). Fla Entomol 96(3):897–903

Arnason JT, Towers GHN, Philogene BJR, Lambert JDH (1983) The role of natural photosensitizers in plant resistance to insects. In: Hedin PA (ed) Plant resistance to insects. ACS symposium series 208. American Chemical Society, Washington, DC, pp 139–151

Arnason JT, Guillet G, Durst T (2004) Phytochemical diversity of insect defenses in tropical and temperate plant families. In: Carde RT, Miller GJ (eds) Advances in insect chemical ecology. Cambridge University Press, Cambridge, pp 1–20

Ascher KRS (1993) Nonconventional insecticidal effects of pesticides available from the neem tree, *Azadirachta indica*. Arch Insect Biochem Physiol 22:433–449

Ascher KRS, Streloke M, Schmidt GH, Warthen JD Jr (1989) The antifeedant effects of neem seed kernel extract and azadirachtin on nymphs of *Eyprepocnemis plorans*. Phytoparasitica 17:167–174

Ayyangar GSC, Rao PJ (1989) Azadirachtin effects on consumption and utilization of food and midgut enzymes of *Spodoptera litura* (Fab.). Ind J Entomol 51:373–376

Bacher M, Brader G, Hofer O, Greger H (1999) Oximes from seeds of *Atalantia ceylanica*. Phytochemistry 50:991–994

Bakker J, Gommers FJ, Nieuwenhuis I, Wynberg H (1979) Photoactivation of the nematicidal compound alpha-terthienyl from roots of marigolds (*Tagetes* species). A possible singlet oxygen role. J Biol Chem 254:1841–1844

Barnby MA, Klocke JA (1987) Effects of azadirachtin on the nutrition and development of the tobacco budworm, *Heliothis virescens*. J Insect Physiol 33:69–75

Barnby MA, Klocke JA, Darlington MV, Yamasaki RB (1989) Uptake, metabolism, and excretion of injected tritiated 22, 33-dihydroazadirachtin in last instar larvae of *Heliothis virescens*. Entomol Exp Appl 52:1–6

Belles X, Camps F, Coll J, Dollars PM (1985) Insect antifeedant activity of clerodane diterpenoids against larvae of *Spodoptera littoralis* (Lepidoptera). J Chem Ecol 11:1439–1445

Bentley MD, Rajab MS, Alford AR, Mendel MJ, Hassanali A (1988) Structure- activity studies of modified citrus limonoids as antifeedants for Colorado potato beetle larvae, *Leptinotarsa decemlineata*. Entomol Exp et Appl 49:189–193

Berenbaum M (1987) Charge of the light brigade: phototoxicity as a defense against insects. In: Heitz JK, Downum KR (eds) Light- activated pesticides. ACS symposium series 339. American Chemical Society, Washington, DC, pp 206–216

Bischof LJ, Enan EE (2004) Cloning, expression and functional analysis of an octopamine receptor from *Periplaneta americana*. Insect Biochem Mol Biol 34:511–521

Bowers JH, Locke LC (2000) Effect of botanical extracts on the population density of *Fusarium oxysporum* in soil and control of Fusarium wilt in the green house. Plant Dis 84:300–305

Brader G, Vajrodays S, Greger H, Bacher M, Kalchhauser H, Hofer O (1998) Bisamides, lignans, triterpenes, and insecticidal cyclopenta[b]benzofurans from *Aglaia* species. J Nat Prod 61:1482–1490

Brentwood TN (2003) Herbicidal compositions containing plant essential oils and mixtures or blends thereof. US patent 6506707

Broza M, Butler GD, Henneberry TJ (1988) Cotton seed oil for the control of *Bemisia tabaci* on cotton. In: Proceedings of Beltwide cotton production research conference. National Cotton Council of America, Memphis, TN, p 301

Burke BA, Goldsby G, Mudd JB (1987) Polar epicuticular lipids of lycopersicon pennellii. Phytochemistry 26:2567–2571

Buta JG, Lusby WR, Neal JW, Waters RM, Pittarelli GW (1993) Sucrose esters from *Nicotiana gossei* active against the greenhouse whitefly *Trialeurodes vaporariorum*. Phytochemistry 32:859–864

Butler GD Jr, Henneberry TJ (1991) Effect of oil spray on sweetpotato whitefly and phytotoxicity on watermelons, squash and cucumbers. Southwest Entomol 16:63–72

Caballero C, Lopez-Olguin J, Ruiz M, Ortego F, Castanera P (2008) Antifeedant activity and effects of terpenoids on detoxication enzymes of the beet armyworm, *Spodoptera exigua* (Hübner). Span J Agric Res 6:177–184

Caboni P, Sherer TB, Zhang N, Taylor G, Na HM, Greenamyre JT, Casida JE (2004) Rotenone, deguelin, their metabolites, and the rat model of Parkinson's disease. Chem Res Toxicol 17(11):1540–1548

Calderon M (1979) Mixing chickpea with paraffin oil to prevent *Callosobruchus maculatus* (F.) infestation. Progress report for the year 1978–79. Stored Product

Division, Special publication no. 140, Ministry of Agriculture, Israel

Camps F, Coll J (1993) Insect allelochemicals from Ajuga plants. Phytochemistry 32:1361–1370

Caprioli V, Cimino G, Colle R, Gavagnin M, Sodano G, Spinella A (1987) Insect antifeedant activity and hot taste for humans of selected natural and synthetic 1,4-dialdehydes. J Nat Prod 50:146–151

Carpinella MC, Defago MT, Valladares G, Palacios SM (2003) Antifeedant and insecticide properties of a limonoid from *Melia azedarach* (Meliaceae) with potential use for pest management. J Agric Food Chem 51(2):369–374

Casida JE, Quistad GB (1995) Pyrethrum flowers: production, chemistry, toxicology and uses. Oxford University Press, Oxford

Chaidir JH, Nugroho BW, Bohnenstengel FI, Wray V, Witte L, Hung PD, Kiet LC, Proksch P (1999) New insecticidal rocaglamide derivatives from flowers of *Aglaia duperreana* (Meliaceae). Phytochemistry 52:837–842

Champagne DE, Isman MB, Towers GHN (1989) Insecticidal activity of phytochemicals and extracts of the Meliaceae. In: Arnason J et al (eds) Insecticides of plant origin. ACS symposium series 387, New York, pp 95–109

Champagne DE, Koul O, Isman MB, Scudder GGE, Towers GHN (1992) Biological activities of limonoids from the Rutales. Phytochemistry 31:377–394

Chan GFQ, Towers GH, Mitchell JC (1975) Ultraviolet-mediated antibiotic activity of thiophene compounds of *Tagetes*. Phytochemistry 14:2295–2296

Chen W, Isman MB, Chiu SF (1995) Antifeedant and growth inhibitory effects of the limonoid toosendanin and *Melia toosendan* extracts on the variegated cutworm, *Peridroma saucia* (Lep., Noctuidae). J Appl Entomol 119:367–370

Chiu SF (1989) Recent advances in research on botanical insecticides in China. In: Arnason JT, Philogene BJR, Morand P (eds) Insecticides of plant origin. American Chemical Society symposium series 387, Washington, DC, pp 69–77

Chockalingam S, Thenmozhi S, Sundari MSN (1990) Larvicidal activity of different products against mosquito larvae. J Environ Biol 11:101–104

Chortyk OT, Nottingham SF (1995). Natural sugar esters as potent whitefly insecticides. In: 209th national meeting. American Chemical Society, 2–6 April 1995, Anaheim, CA

Chortyk OT, Severson RF, Cutler HC, Sisson VA (1993) Antibiotic activities of sugar esters isolated from selected Nicotiana species. Biosci Biotechnol Biochem 57:1355–1356

Chortyk OT, Pomonis JG, Johnson AW (1996) Syntheses and characterizations of insecticidal sucrose esters. J Agric Food Chem 44:1551–1557

Clay DV, Dixon FL, Willoughby I (2005) Natural products as herbicides for tree establishment. Forestry 78(1):1–9

Cole MD, Anderson JC, Blaney WM, Fellows LE, Ley SV, Sheppard RN et al (1990) Neoclerodane insect antifeedants from *Scutellaria galericulata*. Phytochemistry 29:1793–1796

Coll J, Tandrón YA (2007) Neo-Clerodane diterpenoids from *Ajuga*: structural elucidation and biological activity. Phytochem Rev 7:25–49

Cooper-Driver GA, LeQuesne PW (1987) Diterpenoids as insect antifeedants and growth inhibitors: role in *Solidago* species. In: Waller GR (ed) Allelochemicals: role in agriculture and forestry. ACS symposium series 330. American Chemical Society, Washington, DC, pp 534–550

Copping LG (1996) Crop protection agents from nature: natural products and analogues. Critical reports on applied chemistry, vol 35. Royal Society of Chemistry, Thomas Graham House, Cambridge

Copping LG, Duke SO (2007) Natural products that have been used commercially as crop protection agents. Pest Manag Sci 63:524–554

Cornelius ML, Grace KJ, Yates JR (1997) Toxicity of monoterpenoids and other natural products to the Formosan subterranean termite (Isoptera: Rhinotermitidae). J Econ Entomol 90:320–325

Croteau R, Kutchan TM, Lewis NG (2000) Natural products (secondary metabolites). In: Buchanan B, Gruissem W, Jones R (eds) Biochemistry and molecular biology of plants. Am Soc Plant Physiol:1250–1318

Daferera DJ, Ziogas BN, Polissiou MG (2000) GC-MS analysis of essential oils from some Greek aromatic plants and their fungi toxicity on *Penicillium digitatum*. J Agric Food Chem 48:2576–2581

Dahlman DL, Hibbs ET (1967) Response of *Empoasca fabae* (Cicadellidae:Homopter) to tomatine, solanine, leptine I, tomatidine, solanidine and demissidine. Ann Entomol Soc Am 60:732–740

Dayan FE, Cantrell CL, Duke SO (2009) Natural products in crop protection. Bioorg Med Chem 17:4022–4034

Delle-Monache F, Maring Bettolo GB, Bernays EA (1984) Isolation of insect antifeedant alkaloids from *Maytenus rigida* (Celastraceae). Z Angew Entomol 97:406–414

Deshpande RS, Adhikari PR, Tipnis HP (1974) Stored grain pest control agents from *Nigella sativa* and *Pogostemon heyneanus*. Bull Grain Technol 12:232

Dev S, Koul O (1997) Insecticides of plant origin. Harwood Academic Publishers Gmbh, Amsterdam

Devakumar C, Goswami BK (1992) Nematicidal principles from neem (Azadirachta indica A. Juss) Part III. Isolation and bioassay of some neem meliacins. Pestic Res J 4:81–86

Devine M, Duke SO, Fedtke C (1993) Naturally occurring chemicals as herbicides. In: Devine MD, Duke SO, Fedtke C (eds) Physiology of herbicide action. Prentice Hall, Englewood Cliffs, pp 395–424

Dhaliwal GS, Koul O (2007) Biopesticides and pest management: conventional and biotechnological approaches. Kalyani Publishers, New Delhi

Dixit OP, Saxena RC (1990) Insecticidal action of *Premma integrifolia* against *Callosobruchus chinensis*. Pesticides 24:29–31

Doharey RB, Katiyar RN, Singh KM (1990) Ecotoxicological studies on pulse beetles infesting green gram. Bull Grain Technol 28(2):116–119

Dorn A, Rademacher JM, Sehn E (1986) Effects of azadirachtin on the moulting cycle, endocrine system and ovaries in last-instar larvae of the milkweed bug, *Oncopeltus fasciatus*. J Insect Physiol 32:321–328

Dudai N, Poljakoff-Mayber A, Mayer AM, Putievsky E, Lerner HR (1999) Essential oils as allelochemicals and their potential use as bioherbicides. J Chem Ecol 25:1079–1089

Duke SO (1990) Natural pesticides from plants. In: Janick J, Simon JE (eds) Advances in new crops. Timber Press, Portland, pp 511–517

Duke SO, Dayan FE, Rimando AM (2000) Natural products and herbicide discovery. In: Cobb HS, Kirkwood RC (eds) Herbicides and their mechanisms of action. Sheffield Academic Press, Sheffield, pp 105–133

Egunjobi OA, Afolami SO (1976) Effect of neem (*Azadirachta indica*) leaf extracts on populations of *Pratylenchus brachyurus* and on the growth and yield of maize. Nematologica 22:125–132

Eigenbrode SD, Thumble JT, Millar JG, White KK (1994) Topical toxicity of tomato sesquiterpenes to the beat armyworm and the role of these compounds in resistance derived from an accession of *Lycopersicon hirsutum* f. typicum. J Agric Food Chem 42:807–810

Einolf WN, Chan WG (1984) Estimation of sucrose esters in tobacco by direct chemical ionization mass spectrometry. J Agric Food Chem 32(4):785–789

Esposti MD, Ghelli A, Ratta M, Cortes D, Estornell E (1994) Natural substances (acetogenins) from the family Annonaceae are powerful inhibitors of mitochondrial NADH dehydrogenase (complex I). Biochem J 30(1):161–167

Fieser LF, Berliner E, Bondhus FJ, Chang FC, Dauben WG, Ettlinger MG, Fawaz G, Fields M, Fieser M, Heidelberger C, Heymann H, Seligman AM, Vaughan WR, Wilson E, Wu M, Leffler MT, Hamlin KE, Hathaway RJ, Matson EJ, Moore EE, Moore MB, Rapala RT, Zaugg HE (1948) Naphthoquinone antimalarials. J Am Chem Soc 70:3151–3244

Garcia ES, Rembold H (1984) Effects of azadirachtin on ecdysis of *Rhodnius prolixus*. J Insect Physiol 30:939–941

Gershenzon J, Dudareva N (2007) The function of terpene natural products in the natural world. Nat Chem Biol 3:408–414

Goffreda JC, Mutschler MA, Ave DA, Tengey WM, Steffens JC (1989) Aphid deterrence by glucose esters in glandular trichome exudates of the wild tomato, *Lycopersicon pennellii*. J Chem Ecol 15:2135–2147

Gonzalez-Coloma A, Gutierrez C, Del Corral JMM, Gordaliza M, De La Puente ML, San Feliciano A (2000) Structure and species-dependent insecticidal effects of neo-clerodane diterpenes. J Agric Food Chem 48:3677–3681

Govindachari TR, Narasimhan NS, Suresh G, Partho PD, Gopalakrishnan G, Krishna Kumari GN (1995) Structure-related insect antifeedant and growth regulating activities of some limonoids. J Chem Ecol 21:1585–1600

Grainge M, Ahmed S (1988) Handbook of plants with pest-control properties. Wiley, New York

Gussregan B, Puhr M, Nugroho BW, Wray V, Witte I, Proksch P (1999) New insecticidal rocaglamide derivative from flower of *Aglaia odorata*. Z Naturforsch 52:339–343

Harborne JB (1977) Introduction to ecological biochemistry. Academic Press, New York

Hasspieler B, Arnason JT, Downe AER (1988) Toxicity, localization and elimination of the phototoxin alpha-terthienyl in mosquito larvae. J Am Mosq Control Assoc 4:479–484

Hierro I, Valero A, Perez P, Gonzalez P, Cabo MM, Montilla MP, Navarro MC (2004) Action of different monoterpenic compounds against *Anisakis simplex* s.l. L3 larvae. Phytomedicine 11:77–82

Hill J, Schoonhoven AV (1981) Effectiveness of vegetable oil fractions in controlling the Mexican bean weevil. J Econ Entomol 74:478–479

Hough-Goldstein JA (1990) Antifeedant effects of common herbs on the Colorado potato beetle (Coleoptera: Chrysomelidae). Environ Entomol 19:234–238

Hummelbruner LA, Isman MB (2001) Acute, sublethal, antifeedant and synergistic effects of monoterpenoid essential oil compounds on the tobacco cutworm, *Spodoptera litura* (Lepidoptera:Noctuidae). J Agric Food Chem 49:715–720

Hutchins RFN, Sutherland ORW, Gnanasunderam C, Greenfield WJ, Williams EM, Wright HJ (1984) Toxicity of nitro compounds from *Lotus pedunculatus* to grass grub (*Costelytra zealandica*) (Coleoptera: Scarabaeidae). J Chem Ecol 10:81–93

Ishibashi F, Satasook C, Isman MB, Towers GHN (1993) Insecticidal 1*H*-cyclopentatetrahydro[*b*]benzofurans from *Aglaia odorata* (Lour.) (Meliaceae). Phytochemistry 32:307–310

Isman MB (1994) Botanical insecticides and antifeedants: new sources and perspectives. Pestic Res J 6:11–19

Isman MB (2000) Plant essential oils for pest and disease management. Crop Prot 19:603–608

Isman MB (2001) Biopesticides based on phytochemicals. In: Koul O, Dhaliwal GS (eds) Phytochemical biopesticides. Harwood Academic Publishers, Amsterdam, pp 1–12

Isman MB (2005) Problems and opportunities for the commercialization of insecticides. In: Regnault-Roger C, Philogene BJR, Vincent R (eds) Biopesticides of plant origin. Lavoisier, Paris, pp 283–291

Isman MB (2006) Botanical insecticides, deterrents, and repellents in modern agriculture and an increasingly regulated world. Ann Rev Entomol 51:45–66

Isman MB, Machial CM (2006) Pesticides based on plant essential oils: from traditional practice to commercialization. In: Rai M, Carpinella MC (eds) Naturally occurring bioactive compounds. Elsevier BV, Amsterdam, pp 29–44

Isman MB, Brard NL, Nawrot J, Harmatha J (1989) Antifeedant and growth inhibitory effects of bakkenolide-A and other sesquiterpene lactones on the variegated cutworm, *Peridroma saucia* Hubner (Lep., Noctuidae). J Appl Entomol 107:524–529

Ivie GW, Witzel DA (1982) Sesquiterpene lactones: structure, biological action and toxicological significance. In: Keeler RF, Tu AT (eds) Handbook of natural toxins. Marcel Dekker Inc., New York, pp 543–584

Jacobson M (1989) Pharmacology and toxicology of neem. In: Jacobson M (ed) Phytochemical pesticides, vol 1. The neem tree. CRC Press, Boca Raton, pp 133–153

Janprasert J, Satasook C, Sukumalanand P, Champagne DE, Isman MB, Wiriyachitra P, Towers GHN (1993) Rocaglamide, a natural benzofuran insecticide from *Aglaia odorata*. Phytochemistry 32:67–69

Jansen BJM, Groot A (1991) The occurrence and biological activity of drimane sesquiterpenoids. Nat Prod Rep 8:309–318

Jauch J (2008) Total synthesis of azadirachtin – finally completed after 22 years. Angew Chem Int Ed 47:34–37

Johnson HA, Oberlies NH, Alali FQ, McLaughlin JE (2000) Thwarting resistance: annonaceous acetogenins as new pesticidal and antitumor agents. In: Cutler SJ, Cutler JG (eds) Biologically active natural products: pharmaceuticals. CRC Press, Boca Raton, pp 173–183

Karr LL, Coats JR (1988) Insecticidal properties of *d-limonene*. J Pestic Sci 13:287–290

Kawazu K, Ariwa M, Kii Y (1977) An ovicidal substance cis dehydro matricaria ester from solidago altissima. Agric Biol Chem 41(1):223–224

Kays SJ, Severson RF, Nottingham SF, Chalfant RB, Chortyk O (1994) Possible biopesticide from *Petunia* for the control of sweetpotato *whitefly* (Bemisia tabaci) on vegetable crop. Proc Fla State Hortic Soc 107:163–167

Kelm MA, Nair MG, Schutzki RA (1997) Mosquitocidal compounds from *Magnola salicifolia*. Int J Pharmcognosy 35:84–90

Khaire VM, Kachare BV, Mote UN (1992) Efficacy of different vegetable oils as grain protectants against pulse beetle, Callosobruchus chinensis L. in increasing storability of pigeonpea. J Stored Prod Res 28:153–156

Khambay BPS, Jewess P (2000) The potential of natural naphthoquinones as the basis for a new class of pest control agents-an overview of research at IACR-Rothamsted. Crop Prot 19:597–601

Khambay BPS, Batty D, Beddy DG, Denholm I, Cahill MR (1997a) A new group of plant derived naphthoquinone pesticides. Pestic Sci 50:291–296

Khambay BPS, Brddle DG, Simmonds MSJ (1997b) In pursuit of insecticidal compounds from plants. In: Wrigley S, Hayes M, Thomas R, Chrystal E (eds) Phytochemical diversity-a source of new industrial products. Royal Society of Chemistry, Cambridge, pp 158–169

Khambay BPS, Batty D, Cahill MR, Denholm I, Mead-Briggs M, Vinall S, Niemeyer HM, Simmonds MSJ (1999) Isolation, characterization, and biological activity of naphthoquinones from *Calceolaria andina* L. J Agric Food Chem 47:770–775

Khambay BPS, Batty D, Jewess PJ, Bateman GL, Hollomon DW (2003) Mode of action and pesticidal activity of the natural product dunnione and of some analogues. Pest Manag Sci 59:174–182

King RR, Singh RP, Bouicher A (1987) Variation in sucrose esters from the type B glandular trichomes of certain wild potato species. Am Potato J 64:529–534

King RR, Calhoun LA, Singh RP (1988) 3,4-di-*O*- and 2,3,4-tri-*O*-acylated glucose esters from the glandular trichomes of non-tuberous *Solanum species*. Phytochemistry 27:3765–3768

Kong CH, Wang P, Zhang CX, Zhang MX, Hu F (2006) Herbicidal potential of allelochemicals from *Lantana camara* against *Eichhornia crassipes* and the alga *Microcystis aeruginosa*. Weed Res 46:290–295

Kostyukovsky M, Rafaeli A, Gileadi C, Demchenko N, Shaaya E (2002) Activation of octopaminergic receptors by essential oil constituents isolated from aromatic plants: possible mode of action against insect pests. Pest Manag Sci 58(11):1101–1106

Koul O (1982) Insect feeding deterrents in plants. Ind Rev Life Sci 2:97–125

Koul O (1984) Azadirachtin-I. Interaction with the development of red cotton bugs. Entomol Exp Appl 36:85–88

Koul O (1996) Mode of azadirachtin action. In: Randhawa NS, Parmar BS (eds) Neem. New Age International Publishers Ltd., New Delhi, pp 160–170

Koul O (2005) Insect antifeedants. CRC Press, Boca Raton

Koul O (2008) Phytochemicals and insect control: an antifeedant approach. Crit Rev Plant Sci 27:1–24

Koul O, Dhaliwal GS (2001) Phytochemical biopesticides. Harwood Academic Publishers, Amsterdam

Koul O, Wahab S (2004) Neem: today and in the new millennium. Kluwer Academic Publishers, Dordrecht

Koul O, Walia S (2009) Comparing impacts of plant extracts and pure allelochemicals and implications for pest control. CAB Rev: Perspect Agric Vet Sci Nutr Nat Resour 4:1–30

Koul O, Amanai K, Ohtaki K (1987) Effect of azadirachtin on the endocrine events of *Bombyx mori*. J Insect Physiol 33:103–108

Koul O, Multani JS, Singh G, Wahab S (2002) Bioefficacy of toosendanin from *Melia dubia* (syn. *M. azedarach*) against gram pod borer, *Helicoverpa armigera* (Hubner). Curr Sci 83:1387–1391

Koul O, Multani JS, Singh G, Daniewski WM, Berlozecki S (2003) 6ß-hydroxygedunin from *Azadirachta indica*. Its potentiation effects with some non-azadirachtin limonoids in neem against lepidopteran larvae. J Agric Food Chem 51:2937–2942

Koul O, Multani JS, Goomber S, Daniewski WM, Berlozecki S (2004a) Activity of some non-azadirachtin limonoids from *Azadirachta indica* against lepidopteran larvae. Aust J Entomol 43:189–195

Koul O, Singh G, Singh R, Singh J, Daniewski WM, Berlozecki S (2004b) Bioefficacy and mode of action of some limonoids of salannin group from *Azadirachta indica* A. Juss and their role in a multicomponent system against lepidopteran larvae. J Biosci 29:409–416

Koul O, Kaur H, Goomber S, Wahab S (2004c) Bioefficacy of rocaglamide from *Aglaia elaeagnoidea* (syn. *A. roxburghiana*) against gram pod borer, *Helicoverpa armigera* (Hubner). J Appl Entomol 128:177–1781

Koul O, Singh G, Singh R, Multani JS (2005a) Bioefficacy and mode-of-action of aglaroxin A from *Aglaia elaeagnoidea* (syn. *A. roxburghiana*) against *Helicoverpa armigera* and *Spodoptera litura*. Entomol Exp Appl 114:197–204

Koul O, Singh G, Singh R, Singh J (2005b) Bioefficacy and mode-of-action of aglaroxin B and aglaroxin C from *Aglaia elaeagnoidea* (syn. *A. roxburghiana*) against *Helicoverpa armigera* and *Spodoptera litura*. Biopest Int 1:54–64

Koul O, Walia S, Dhaliwal GS (2008) Essential oils as green pesticides: potential and constraints. Biopest Int 4:63–84

Krishna KG, Pande R, Harsha S (2007) Evaluation of essential oils and their components for broad-spectrum antifungal activity and control of late leaf spot and crown rot disease in peanut. Plant Dis 91:375–379

Kubo I, Nakanishi K (1978) Some terpenoid insect antifeedants from tropical plants. In: Geissbuhler H (ed) Advances in pesticide science, part 2. Pergamon Press, Oxford, pp 284–294

Kubo I, Klocke JA, Miura I, Fukuyama Y (1982) Structure of ajugarin-IV. J Chem Soc Chem Commun 11:618–619

Kubo I, Fukuyama Y, Chapya A (1983) Structure of ajugarin-IV. Chem Lett 10:223

Kumar S, Singh D, Mehta SS, Srivastava AK (1995) A process for the preparation of pest repellent tablet. Application no. 2441 Del 1995 dated 27-03-97, Granted Indian Patent No. 184803 dated 27-04-2001

Kumar S, Tyagi BR, Bahl JR, Khanuja SPS, Shasany AK, Shukla RS, Sattar A, Singh D, Haseeb A, Singh VP, Ram P, Singh K, Singh S, Singh SP, Patra NK, Alam M, Naqvi AA, Ram M, Agarwal KK, Singh K (1999) Mint plant *Mentha arvensis*, Himalaya. US Patent No. 10935 dated 6/01/99, Application No. 08/855, 768

Kurita N, Miyaji M, Kurane R, Trakahara Y (1981) Antifungal activity of components of essential oils. Agric Biol Chem 45:945–952

Lajide L, Escoubas P, Mizutani J (1995a) Termite antifeedant activity in *Detarium microcarpum*. Phytochemistry 40:1101–1104

Lajide L, Escoubas P, Mizutani J (1995b) Termite antifeedant activity in *Xylopia aethiopica*. Phytochemistry 40:1105–1112

Lalonde RT, Morris CD, Wong CF, Gardner LC, Exkert DJ, King DR, Zimmerman RH (1979) Response of Aedes triseriatus larvae to fatty acids of *Cladophora glomerata*. J Chem Ecol 5:371–382

Lapointe SL, Tingey WM (1984) Feeding response of the green peach aphid (Homoptera: Aphididae) to potato glandular trichomes. J Econ Entomol 77:386–389

Lavie D, Jain MK (1967) Tetranortriterpenoids from *Melia azadirachta* L. Chem Commun 6:278–280

Lee S, Tsao R, Peterson C, Coats JR (1997) Insecticidal activity of monoterpenoids to western corn rootworm (Coleoptera: Chrysomelidae), twospotted spider mite (Acari: Tetranychidae), and house fly (Diptera: Muscidae). J Econ Entomol 90(4):883–892

Li X (1999) Recent studies on insecticidal activities of limonoids from meliaceaous plants. Entomol Sinica 6:283–288

Lichtenstein EP, Liang TT, Schulz KR, Schnoes HK, Carter GT (1974) Insecticidal and synergistic components isolated from dill plants. J Agric Food Chem 22:658–664

Macleod JK, Moeller PDR, Molinski TF, Koul O (1990) Antifeedant activity and ^{13}C NMR spectral assignments of the meliatoxins. J Chem Ecol 16:2511–2518

Marchant YY, Cooper GK (1987) Structure and function relationships in Polyacetylene photoactivity. In: Heitz JR, Downum KR (eds) Light-activated pesticides. ACS symposium series, pp 241–254

Marquina S, Maldonado N, Garduño-Ramírez ML, Aranda E, Villarreal ML, Navarro V, Bye R, Delgado G, Alvarez L (2001) Bioactive oleanolic acid saponins and other constituents from the roots of Viguiera decurrens. Phytochemistry 56:93–97

Matsuzaki T, Koseki K, Koiwai A (1988) Germination and growth inhibition of surface lipids from nicotiana species and identification of sucrose esters. Agric Biol Chem 52(8):1889–1897

Matsuzaki T, Shinozaki Y, Suhara S, Ninomiya M, Shigematsu H, Koiwai A (1989a) Isolation of glycolipids from the surface lipids of *Nicotiana bigelovii* and their distribution in *Nicotiana* species. Agric Biol Chem 53:3079–3082

Matsuzaki T, Shinozaki Y, Suhara S, Shigematsu H, Koiwai A (1989b) Isolation and characterization of tetra- and triacylglucose from surface lipids of *Nicotiana miersii*. Agric Biol Chem 53:3343–3345

Matsuzaki T, Fukamiya N, Okano M, Fujita T, Tagahara K, Lee KH (1991) Picrasinoside H, a new quassinoid glucoside, and related compounds from the stem wood of Picrasma ailanthoides. J Nat Prod 54(3):844–848

Matsuzaki T, Shinozaki Y, Hagimori M, Tobita T, Shigematsu H, Koiwai A (1992) Novel glycerolipids and glycolipids from the surface lipids of *Nicotiana benthamiana*. Biosci Biotechnol Biochem 56:1565–1569

Meher HC, Walia S, Sethi CL (1988) Effect of steroidal and triterpenic saponins on the mobility of juveniles of Meloidogyne incognita. Indian J Nematol 18:244–247

Merritt AT, Ley SV (1992) Clerodane diterpenoids. Nat Prod Rep 9:243–287

Miller DA, Chamberlain WF (1989) Azadirachtin as larvicide against the hornfly, stable fly and house fly. J Econ Entomol 82:1375–1378

Miranda JEM, Navickiene HMD, Nogueira-Couto RH, De Bortoli S, Kato MJ, Bolzani VS et al (2003) Susceptibility of *Apis mellifera* (Hymenoptera: Apidae) to pellitorine, an amide isolated from *Piper tuberculatum* (Piperaceae). Apidologie 34:409–415

Mitchell G, Bartlett DW, Fraser TEM, Hawkes TR, Holt DC, Townson JK, Wichert RA (2001) Mesotrione: a new selective herbicide for use in maize. Pest Manag Sci 57:20–28

Montser BR, Carvajal M (1998) Control of *Aspergillus flavus* in maize with plant essential oils and their components. J Food Prot 61:616–619

Mordue AJ, Blackwell A (1993) Azadirachtin : an update. J Insect Physiol 39:903–924

Mordue AJ, Cottee PK, Evans KA (1985) Azadirachtin: its effect on gut mortality, growth and moulting in *Locusta*. Physiol Entomol 10:431–437

Muller RF, Berger B, Yegen O (1995) Chemical composition and fungi toxic properties to phytopathogenic fungi of essential oils of selected aromatic plants growing wild in Turkey. J Agric Food Chem 43:2262–2266

Mullin CA, Gonzalez-Coloma A, Gutierrez C, Reima M, Eichenseer H, Hollister B et al (1997) Antifeedant effects of some novel terpenoids on chrysomelid beetles: comparison with alkaloids on an alkaloid-adapted and non-adapted species. J Chem Ecol 23:1851–1866

Munoz DM, Dela Torre MC, Rodriguez B, Simmonds MSJ, Blaney WM (1997) Neo-Clerodane insect antifeedants from *Scutellaria alpina* sub sp. Javalambrensis. Phytochemistry 44:593–597

Narwal SS, Hoagland RE, Dilday RH, Reigosa MJ (1999) Allelopathy in ecological agriculture and forestry. Kluwer Academic Publishers, Dordrecht

Navickiene HMD, Miranda JE, Bortoli SA, Kato MJ, Bolzani VS, Furlan M (2007) Toxicity of extracts and isobutyl amides from *Piper tuberculatum*: potent compounds with potential for the control of the velvetbean caterpillar, *Anticarsia gemmatalis*. Pest Manag Sci 63:399–403

Neal JJ, Tingey WM, Steffens JC (1990) Sucrose esters of carboxylic acids in glandular trichomes of *Solanum berthaultii* deter settling and probing by green peach aphid. J Chem Ecol 16:487–497

Neal JW Jr, Buta JG, Pittarelli GW, Lusby WR, Bentz JA (1994) Novel sucrose esters from *Nicotiana gossei*: effective biorationals against selected horticultural insect pests. J Econ Entomol 87:1600–1607

Neoliya NK, Singh D (2005) Effect of some neem based formulations on the metamorphic development of gram pod borer, *Helicoverpa armigera* Hub. (Lepidoptera: Noctuidae). In: Proceedings of ICIPE 2002 held at NBRI, Lucknow, India, Plant response to environmental stress, pp 459–463

Neoliya NK, Singh D, Sangawan RS (2005) Azadirachtin influences total head protein content of *Helicoverpa armigera* Hub. Larvae. Curr Sci 88(12):1889–1890

Neoliya NK, Singh D, Sangwan RS (2007) Azadirachtin based insecticides induce alteration in *Helicoverpa armigera* Hub. head polypeptides. Curr Sci 92(1):94–99

Nguefack J, Nguikwie SK, Fotio D, Dongmo B, Zollo PH, Amvam LV, Nkengfack AE, Poll L (2007) Fungicidal potential of essential oils and fractions from *Cymbopogon citratus*, *Ocimum gratissimum* and *Thymus vulgaris* to control *Alternaria padwickii* and *Bipolaris oryzae*, two seed borne fungi of rice. J Essent Oil Res 19:581–587

Ntalli NG, Menkissoglu-Spiroudi U, Giannakou IO (2010) Nematicidal activity of powder and extracts of *Melia azedarach* fruits against *Meloidogyne incognita*. Ann Appl Biol 156:309–317

Nugroho BW, Edrada RA, Gussregen B, Wray V, Witte L, Bringmann G, Proksch P (1997a) Insecticidal rocaglamide derivatives from *Aglaia Duppereana*. Phytochemistry 44:1455–1461

Nugroho BW, Gussregen B, Wray V, Witte L, Bringmann G, Gehling M, Proksch P (1997b) Insecticidal rocaglamide derivatives from *Aglaia elliptica* and *A. harmsiana*. Phytochemistry 45:1579–1585

Nugroho BW, Edrada RA, Wray V, Witte L, Bringmann G, Proksch P (1999) An insecticidal rocaglamide derivative and related compounds from *Aglaia odorata*. Phytochemistry 51:367–376

Oh S, Kim JA, Jeon H, Park JC, Koh YJ, Hur H, Hur J (2008) Antifungal activity of *Eucalyptus*-derived phenolics against postharvest pathogens of kiwifruits. Plant Pathol J 24:322–327

Ohya I, Shinozaki Y, Tobita T, Takahashi H, Matsuzaki T, Koiwai A (1994) Sucrose esters from the surface lipids of *Nicotiana cavicola*. Phytochemistry 37:143–145

Oleszek W, Jurzysta M, Gorski PM (1992) Alfalfa saponins the allelopathic agents. In: Rizvi SJH, Rizvi V (eds) Allelopathy: basic and applied aspects. Chapman & Hall, London

Oliviera MF, Lemos TG, De Mattos MC, Segundo TA, Santiago GM, Braz-Filho R (2002) New enamine derivatives of lapachol and biological activity. An Acad Bras Cienc 74:211–221

Omar S, Marcotte M, Fields P, Sanchez PE, Poveda L, Mata R et al (2007) Antifeedant activities of terpenoids isolated from tropical Rutales. J Stored Prod Res 43:92–96

Ortego F, Rodriguez B, Castanera P (1995) Effects of neo-Clerodane diterpenes from *Teucrium* on feeding behavior of Colorado potato beetle larvae. J Chem Ecol 21:1375–1386

Ouden HD, Den VJH, Alkema DPW, Dev JJ, Derks PSM (1993) Experiments with volatile oil substances in slow release formulations causing repellency for oviposition by the cabbage root fly, Phorbia brassicae Bche (Diptera, Anthomydae). J Appl Entomol 115:307–312

Pandey GP, Doharey RB, Verma BK (1981) Efficacy of some vegetable oil for protecting greengram against the attack of *Callosobruchus maculatus* (F.). Ind J Agric Sci 51:910–912

Parmar BS, Walia S (2001) Prospects and problems of phytochemical biopesticides In: Koul O, Dhaliwal GS (eds) Phytochemical biopesticides. Hardwood Academic Publishers, Amsterdam, pp 133–210

Paster N, Menasherou M, Ravid U, Juven B (1995) Antifungal activity of oregano and thyme essential oils applied as fumigants against fungi attacking stored grain. J Food Prot 58:81–85

Pereirra J (1983) The effectiveness of six vegetable oils as protectants of cowpea and bambaera groundnuts against infestations by *Callosobruchus maculatus*. J Stored Prod Res 19:57–62

Pickett JA, Dawson GW, Griffith DC, Hassanali A, Merritt LA, Mudd A et al (1987) Development of plant derived antifeedants for crop protection. In: Greenhalgh R, Roberts TR (eds) Pesticide science and biotechnology. Blackwell Scientific Publications, Oxford, pp 125–128

Pickett JA, Wadhams LJ, Woodcock CM (1997) Developing sustainable pest control from chemical ecology. Agric Ecosyst Environ 64:149–156

Pimental D (2005) Encyclopedia of pest management, CRC Press, Boca Raton, FL. powder and extracts of Melia azedarach fruits against Meloidogyne incognita. Ann Appl Biol 156(2):309–317

Pradhan S, Jotwani MG, Rail BK (1962) The neem seed deterrent to locusts. Ind Farm 12:711

Pradhanang PM, Momol MT, Olson SM, Jones JB (2003) Effects of plant essential oils on *Ralstonia solanacearum* population density and bacterial wilt incidence in tomato. Plant Dis 87:423–427

Prakash G, Srivastava AK (2007) Azadirachtin production in stirred tank reactors by Azadirachta indica suspension culture. Process Biochem 42:93–97

Prakash G, Emmannuel CJSK, Srivastava AK (2005) Variability of azadirachtin in *Azadirachta indica* (neem) and batch kinetics studies of cell suspension culture. Biotech Bioproc Eng 10:198–204

Priestley CM, Williamson EM, Wafford KA, Sattelle DB (2003) Thymol, a constituent of thyme essential oil, is a positive allosteric modulator of human GABA receptors and a homo-oligomeric GABA receptor from *Drosophila melanogaster*. Br J Pharmacol 140(8):1363–1372

Proksch P, Edrada R, Ebel R, Bohnenstengel IF, Nugroho WB (2001) Chemistry and biological activity of rocaglamide derivatives and related compounds in *Aglaia* species (Meliaceae). Curr Org Chem 5:923–938

Puri SN, Butler GD, Henneberry JJ (1991) Plant derived oils and soap solutions as control for the whitefly on cotton. J Appl Zool Res 2:1–5

Puri SN, Bhosle BB, Elyas M, Butler GD, Henneberry JJ (1994) Detergents and plant derived oils for control of the sweetpotato whitefly on cotton. Crop Prot 13:45–48

Puterka GJ, Severson RF (1995) Activity of sugar esters isolated from leaf trichomes of *Nicotiana gossei* to pear psylla (Homoptera: Psyllidae). J Econ Entomol 88:615–619

Qi Y, Burkholder WE (1981) Protection of stored wheat from the granary weevil by vegetable oils. J Econ Entomol 74:502–505

Quarles W (1996) EPA exempts least-toxic pesticides. IPM Pract 18:16–79

Raguraman S, Singh D (1997) Biopotentials of *Azadirachta indica* and *Cedrus deodara* oils on *Callosobruchus chinensis* Linn. Int J Pharm (USA) 35:344–348

Randhawa NS, Parmar BS (eds) (2007) Neem, 2nd edn. New Age International Pvt Ltd. Publishers, New Delhi

Rao PJ, Subrahmanyam B (1986) Azadirachtin induced changes in development, food utilization and haemolymph constituents of *Schistocerca gregaria* Forsk. Z Angewandte Chemie Int Edition 102:217–224

Ratnayake BM, Wimalasiri WR, Macleod JK (1988) Ent-kauranes and oleananes from *Croton lacciferus*. Phytochemistry 27:869–871

Ratnayake S, Rupprecht SK, Potter WM, Mclaughlin JL (1992) Evaluation of various parts of the paw paw tree, *Asimina triloba* (Annonaceae) as commercial sources of the pesticidal annonacious acetogenins. J Econ Entomol 85:2353–2356

Ray DP, Walia S, Dureja P, Singh RP (2000) Composition and repellent activity of the essential oil of marigold (Tagetes erecta) flower. Indian Perfum 44:267–270

Redfern RE, Kelly TJ, Borokovec AB, Hayes DK (1982) Ecdysteroid titres and moulting aberrations in last stage *Oncopeltus fasciatus* nymphs treated with insect growth regulators. Pestic Biochem Phys 18:351–356

Reitz SR, Maiorino G, Olson S, Sprenkel R, Crescenzi A, Momol MT (2008) Interesting plant essential oils and kaolin for the sustainable management of thrips and tomato spotted wilt on tomato. Plant Dis 92:878–886

Rembold H, Forster H, Czoppelt CH, Rao PJ, Sieber KP (1984) The azadirachtins, a group of insect growth regulators from the neem tree. In: Schmutterer H, Ascher KRS (eds) Natural pesticides from the neem tree and other tropical plants proceedings of the second international neem conference, 25–28 May 1983. Rauischholzhausen, Eschborn, pp 153–162

Rice PJ, Coats JR (1994) Structural requirements for monoterpenoid activity against insects, Chapter 8. In: Hedin PA (ed) Bioregulators for crop protection and pest control. American Chemical Society, Washington, DC, pp 92–108

Rodriguez E (1985) Insect feeding deterrents from semi arid and arid land plants. In: Hedin PA (ed) Bioregulators for pest control. ACS symposium series 276. American Chemical Society, Washington, DC, pp 447–453

Romagni JG, Allen SN, Dayan FE (2000) Allelopathic effects of volatile cineoles on two weedy plant species. J Chem Ecol 26:303–313

Rosell G, Quero C, Coll J, Guerrero A (2008) Biorational insecticides in pest management. J Pestic Sci 33:103–121

Rossiter M, Gershenzon J, Mabry TJ (1986) Behavioural and growth responses of specialist herbivores, *Homoeosoma electellum* to major terpenoid of its host *Helianthus* spp. J Chem Ecol 12:1505–1521

Ruberto G, Renda A, Tringali C, Napoli EM, Simmonds MSJ (2002) Citrus limonoids and their semisynthetic derivatives as antifeedant agents against *Spodoptera frugiperda* larvae. A structure-activity relationship study. J Agric Food Chem 50:6766–6774

Saha S, Walia S, Kumar J, Dhingra S, Parmar BS (2010a) Screening for feeding deterrent and insect growth regulatory activity of triterpenic saponins from *Diploknema butyracea* and *Sapindus mukorossi*. J Agric Food Chem 58:434–440

Saha S, Walia S, Kumar J, Parmar BS (2010b) Plant growth regulatory activity of triterpenic saponins. J Appl Bot Food Qual 83(2):189–195

Saha S, Walia S, Kumar J, Parmar BS (2010c) Structure-biological activity relationships in triterpenic saponins: the relative activity of protobassic acid and its derivatives thereof against plant pathogenic fungi. Pest Manag Sci 66:825–831

Satasook G, Isman MB, Wiriyachita P (1993) Activity of rocaglamide, an insecticidal natural product, against the variegated cut worm, *Peridroma saucia*, (Lepidoptera: Noctuidae). Pestic Sci 36:53–58

Saxena RC (1987) Antifeedants in tropical pest management. Insect Sci Appl 18:731–736

Saxena RC (1998) Botanical pest control. In: Dhaliwal GS, Heinrichs EA (eds) Critical issues in insect pest management. Commonwealth Publishers, New Delhi, pp 155–179

Schmutterer H (1990) Properties and potential of natural pesticides from the neem tree, *Azadirachta indica*. Ann Rev Entomol 35:271–292

Schmutterer H (ed) (2002) The neem tree: source of unique natural products for integrated pest management, medicine, industry and other purposes, 2nd edn. Neem Foundation, Mumbai

Schuster DJ, Stansly PA (2000) Response of two lacewing species to biorational and broad-spectrum insecticides. Phytoparasitica 28:297–304

Scott IM, Jensen H, Scott JG, Isman MB, Arnason JT, Philogène BJR (2003) Botanical insecticides for controlling agricultural pests: piperamides and the Colorado potato beetle *Leptinotarsa decemlineata* Say (Coleoptera: Chrysomelidae). Arch Insect Biochem Physiol 54:212–225

Severson RF, Arrendale RF, Chortyk OT, Johnson AW, Jackson DM, Gwynn GR, Chaplin JF, Stephenson MG (1984) Quantitation of the major cuticular components from green leaf of different tobacco types. J Agric Food Chem 32:566–570

Severson RF, Arrendale RF, Chortyk OT, Green CR, Thome FA, Stewart JL, Johnson AW (1985) Isolation and characterization of the sucrose esters of the cuticular waxes of green tobacco leaf. J Agric Food Chem 33:870–875

Severson RF, Chortyk OT, Stephenson MG, Akey DM, Neal JW Jr, Pittarelli GW et al (1994) Characterization of natural pesticides from *Nicotiana gossei*. In: Hedin PA (ed) Bioregulators for crop protection and pest control. ACS symposium series 557. American Chemical Society, Washington, DC, pp 109–121

Shaaya E, Ikan R (1979) Insect control using natural products. In Gessbuhler H (ed) Advances in pesticide science. Pergamon Press, Oxford, Part 2, pp 303–306

Shepherd T, Robertson GW, Griffiths DW, Birch ANE (1999a) Epicuticular wax composition in relation to aphid infestation and resistance in red raspberry *Rubus idaeus*. Phytochemistry 52:1239–1254

Shepherd T, Robertson GW, Griffiths DW, Birch ANE (1999b) Epicuticular wax ester and triacyl glycerol composition in relation to aphid infestation and resistance in red raspberry *R. idaeus*. Phytochemistry 52:1255–1277

Singh D (1996) Medicinal and aromatic plants in insect pest management. In: Narwal SS, Tauro P (eds) Allelopathy in pest management for sustainable agriculture. Scientific Publisher, Jodhpur, pp 125–136

Singh D (1998) Non-conventional medicinal and aromatic plants as source of botanical insecticides. In: Proceedings of Ist national symposium for pest management in horticultural crops, Bangalore, pp 195–199

Singh D (2005) Potential of insect growth regulators in *Helicoverpa* management. In: Saxena H, Rai AB, Ahmed R, Gupta S (eds) Recent advances in Helicoverpa management. Indian Society of Pulse Research/Indian Institute of Pulse Research, Kanpur, pp 241–249

Singh D (2006) Status and prospects of integrated pest management strategies in selected crops- medicinal plants. In: Amerika Singh, Sharma OP, Garg DK (eds) Integrated pest management: principles and applications, vol 2. Applications. CBS Publishers and Distributors, New Delhi, pp 615–636

Singh D, Agrawal SK (1988) Himachalol and β-himachalene: insecticidal principles of Himalayan cedarwood oil. J Chem Ecol 14:1145–1151

Singh D, Mehta SS (1998) Screening of plant material for repellent and insecticidal properties against pulse beetle (*Callosobruchus chinensis*) and house fly (*Musca domestica*). J Med Arom Plant Sci 20:397–400

Singh D, Mehta SS (2010) Menthol containing formulation inhibits Adzuki Bean Beetle, *Callosobruchus chinensis* L. (Coleoptera; Bruchidae) population in pulse grain storage. J Biopestic 3(3):523–530

Singh D, Rao SM (1986) Effect of cedarwood oil on reproduction of *Dysdercus koenigii* (F.). Curr Sci 55:422–423

Singh RP, Saxena RC (1999) *Azadirachta indica* A Juss. Oxford & IBH Publishing Pvt. Ltd., Delhi, 330 pp

Singh D, Singh AK (1991) Repellent and insecticidal properties in essential oils against housefly, *Musca domestica* L. Int J Trop Insect Sci 12(4):487–491

Singh D, Rao SM, Tripathi AK (1984) Cedarwood oil as a potential insecticidal agent against mosquitoes. Naturwissenschaften 71:265–266

Singh D, Siddiqui MS, Sharma S (1989) Reproduction retardant and fumigant properties in essential oils against rice weevil (Coleoptera: Curculionidae) in stored wheat. J Econ Entomol (USA) 83:727–733

Singh D, Mehta SS, Ram RD, Kumar S (1998) A novel formulation useful as pest repellent for stored grains. Granted Indian Patent No. 189730 dated 27-01-2004 Application No. 1974/Del/98 dated 10-07-98

Singh D, Mehta SS, Kumar N, Shukla YN, Mishra M, Kumar S (2000) A process for preparation of fraction having insecticidal properties from *Catharanthus roseus*. (NF 183/98)

Singh D, Mehta SS, Kumar S (2001) Bio-potential of some botanical tablet formulations in controlling the population of *Callosobruchus chinensis* L. and *Tribolium castaneum* Herbst. In: Proceedings of 2nd international symposium on biopesticides (Toxicology, safety, evaluation, production and proper use) held at Wuhan, China during 26–30 October 1998, pp 95–99

Singh D, Mehta SS, Neoliya NK, Shukla YN, Mishra M (2003) New possible insect growth regulators from *Catharanthus roseus*. Curr Sci 84(9):1184–1186

Stroh J, Wan MT, Isman MB, Moul DJ (1998) Evaluation of the acute toxicity to juvenile pacific coho salmon and rainbow trout of some plant essential oils, a formulated product, and the carrier. Bull Environ Contam Toxicol 60:923–930

Tanzubil PB, McCaffrey AR (1990) Effects of azadirachtin in the African armyworm (*Spodoptera exempta*). Entomol Exp Appl 57:115–121

Thebtaranonth C, Thebtaranonth Y, Wanauppathamkul S, Yuthavong Y (1995) Antimalarial sesquiterpenes from tubers of *Cyperus rotundus*: structure of 10,12-peroxycalamenene, a sesquiterpene endoperoxide. Phytochemistry 40:125–128

Towers GHN (1984) Interactions of light with phytochemicals in some natural and novel systems. Can J Bot 62:2900–2911

Tripathi AK, Singh D (1994) Screening of natural products for insect anti-feedant activity Part I. Plant extracts. Indian J Entomol 56(2):129–133

Tripathi AK, Rao SM, Singh D, Chakravarty RB, Bhakuni DS (1987) Antifeedant activity of plant extracts against *Spilosoma obliqua* Walker. Curr Sci 56:607–609

Tripathi AK, Singh D, Jain DC (1990) Persistency of Tylophorine as an insect antifeedant against *Spilosoma obliqua* Walker. Phytother Res (UK) 4:405–406

Tripathi AK, Prajapati V, Aggarwal KK, Kumar S, Prajapti V, Kumar S, Kukreja AK, Dwivedi S, Singh AK (2000) Effect of volatile oil constituents of *Mentha species* against stored grain pests, *Callosobruchus maculatus* and *Tribolium castaneum*. J Med Arom Plant Sci 22:549–556

Tripathi AK, Prajapati V, Kumar S (2003) Bioactivities of l-carvone, d-carvone, and dihydrocarvone toward three stored product beetles. J Econ Entomol 96:1594–1601

Tsao R, Zhou T (2000) Antifungal activity of monoterpenoids against postharvest pathogens *Botrytis cinerea* and *Monilinia fructicola*. J Essent Oil Res 12:113–121

Tsunao MK, Hassanali A, Jondiko IJO, Tocco B (1993) Mutangin, a dihydroagarofuranoid sesquiterpene insect antifeedant from *Elaeodendron buchananii*. Phytochemistry 34:665–667

Tu YK, Zhou DGW, Chen YZ, Pan XF (1990) Bioactive sesquiterpene polyol esters from *Euonymus bungeanus*. J Nat Prod 53:603–608

Uhlenbroek JH, Bijloo JD (1959) Investigations on nematicides: II. Structure of a second nematicidal principle isolated from Tagetes roots. Recueil des travaux chimiques des Pays-Bas 78:382–390

Ujvari I, Eya BK, Grendell RL, Toia RF, Casida JE (1991) Insecticidal activity of various 3-acyl and other derivatives of veracevine relative to the veratrum alkaloids veratridine and cevadine. J Agric Food Chem 39:1875–1881

Velasquez J, Rojas LB, Usubillaga A (2004) Antifungal activity of naphthoquinone from *Tabebuia serratifolia* (Vahl, Nicholson). CIEN 12:64–69

Verma BK, Pandey GP (1978) Treatment of stored greengram seed with edible oils for the protection from *Callosobruchus maculatus* (F.). Ind J Agric Sci 48:72–75

Vyvyan JR (2002) Allelochemicals as leads for new herbicides and agrochemicals. Tetrahedron 58:1631–1646

Wakabayashi N, Wu WJ, Waters RM (1988) Celangulin: a non-alkaloidal insect antifeedant from Chinese bittersweet *Celastrus angulatus*. J Nat Prod 51:537–542

Wang L, Huang J, You M, Liu B (2004) Time dose mortality modeling and virulence indices for six strains of *Verticillium lecanii* against sweetpotato whitefly, *Bemisia tabaci* (Gennadius). J Appl Entomol 128:494–500

Weber S, Puripattanavong J, Brecht V, Frahm AW (2000) Phytochemical investigation of *Aglaia rubiginosa*. J Nat Prod 63:636–642

Wedge DE, Galindo JCG, Macias FA (2000) Fungicidal toxicity of natural and synthetic sesquiterpene lactone analogues. Phytochemistry 53:747–757

Weinzierl RA (2000) Botanical insecticides, soaps, and oils. In: Rechcigl JE, Rechcigl NA (eds) Biological and biotechnological control of insect pests. Lewis Publisher, Boca Raton, USA, pp 101–121

Wilson CL, Solar JM, Ghaouth A, Wisniewski ME (1997) Rapid evaluation of plant extracts and essential oils for antifungal activity in *Botrytis cinerea*. Plant Dis 81:204–210

Xia Y, Johnson AW, Chortyk OT (1997a) Effect of leaf surface moisture and relative humidity on the efficacy of sugar esters from *Nicotiana gossei* against the tobacco aphid (Homoptera: Aphididae). J Econ Entomol 90:1010–1014

Xia Y, Johnson AW, Chortyk OT (1997b) Enhanced toxicity of sugar esters to the tobacco aphid using humectants. J Econ Entomol 90:1015–1021

Xuan TD, Elzaawely AA, Fukuta M, Tawata S (2006) Herbicidal and fungicidal activities of lactones in Kava (*Piper methysticum*). J Agric Food Chem 54:720–725

Yamada K, Shizuri Y, Hirata Y (1978) Isolation and structure of a new alkaloidal alatimine and an insecticidal alkaloid. Tetrahedron 34:1915–1925

Yamasaki RB, Klocke JA (1987) Structure-bioactivity relationships of azadirachtin, a potent insect control agent. J Agric Food Chem 35:467–471

Zambonelli A, Aulerjo AZ, Bianchi A, Albasini A (1996) Effects of essential oils on phytopathogenic fungi *in vitro*. J Phytopathol 144:491–494

Zaridah MZ, Nor Azah MA, Abu Said A, Mohd. Faridz ZP (2003) Larvicidal properties of citronellal and *Cymbopogon nardus* essential oils from two different localities. Trop Biomed 20:169–174

Mode of Action of Plant-Derived Natural Insecticides

16

Thimmappa Shivanandappa
and Yallappa Rajashekar

Abstract

Most of the chemical insecticides are neurotoxic, acting on targets in the central nervous system such as the membrane ion channels (DDT, pyrethroids), the enzyme acetylcholinesterase (organophosphate, carbamate), and the receptors of neurotransmitters (avermectins, neonicotinoids). The recently introduced diamide group of insecticides target the novel ryanodine receptor in the nervous system. Since pests continue to evolve resistance to compounds currently in use, new compounds with new modes of action are needed. Natural products could be a promising source for novel pest control agents. The origin of many of the important insecticide classes is traceable to a natural source as in the case of pyrethroids, avermectins, spinosads, and neonicotinoids. Although insect control agents acting on targets other than the nervous system such as insect growth regulators (e.g., azadirachtin, JH analogues, ecdysone antagonists) have been developed, due to their lack of contact toxicity, they are not quite successful, but find a place in the integrated pest management. Recent progress in understanding the biology of insect olfaction and taste offers new strategies for developing selective pest control agents. Decalesides, recently discovered natural insecticides, represent a new class of plant-derived insecticides targeting the tarsal gustatory receptors. In this chapter, we focus on the toxicity and mode of action of natural insecticides.

T. Shivanandappa, Ph.D. (✉)
Department of Zoology, University of Mysore,
Manasagangotri, Mysore 570006, Karnataka, India
e-mail: tshivanandappa@yahoo.com

Y. Rajashekar, Ph.D.
Animal Bioresources Programme, Institute of
Bioresources and Sustainable Development (IBSD),
Department of Biotechnology, Government of India,
Takyelpat, Imphal, Manipur 795 001, India

D. Singh (ed.), *Advances in Plant Biopesticides*,
DOI 10.1007/978-81-322-2006-0_16, © Springer India 2014

Keywords
Pest control • Plant-derived insecticides • Mode of action • Neurotoxicity • Insect growth regulators • Insect olfactory and gustatory receptors

1 Introduction

The serious problems of insecticide resistance in pests and the contamination of the biosphere associated with large-scale use of synthetic pesticides have led to the search for eco-friendly pesticides with greater selectivity to the pests. This awareness has created a worldwide interest in the development of alternative strategies, including the discovery of newer insecticides (Dayan et al. 2009; Rajashekar et al. 2012a, b; Miresmailli and Isman 2014). However, newer insecticides will have to meet entirely different standards. They must be pest-specific, non-phytotoxic, nontoxic to mammals, eco-friendly, less prone to pest resistance, relatively less expensive, and locally available (Hermawan et al. 1997). This has necessitated a reexamination of the century-old practices of protecting stored products using plant derivatives (Talukder 2006). Plant-derived pest control agents are more readily biodegradable, less likely to contaminate the environment, and relatively less toxic to mammals. Therefore, researchers are seeking newer naturally occurring insecticides that meet the present-day regulatory requirements (Talukder and Howse 1995; Talukder et al. 2004; Isman 2008; Yao et al. 2008; Rajashekar et al. 2010).

Since ancient times, there have been efforts to protect the agricultural harvest against pests. The Egyptian and Indian farmers used traditional methods such as mixing the stored grain with fire ash (Varma and Dubey 1999). The ancient Romans used false hellebore (*Veratrum album*) as a rodenticide, and the Chinese are credited with discovering the insecticidal properties of *Derris* species, whereas pyrethrum was used as an insecticide in Persia and Dalmatia (Ahmed and Grainge 1986). In many other parts of the world, locally available plants are currently in wide use to protect stored products against damage caused by insect infestation (Golob and Gudrups 1999; Akthar et al. 2008; Tripathi et al. 2009). In northern Cameroon, cowpeas are traditionally mixed with sieved ash after threshing and the mixture put into mud granary or a clay jar (Wolfson et al. 1991). In Eastern Africa, leaves of the wild shrub *Ocimum suave* and the cloves of *Eugenia aromatica* are traditionally used as stored grain protectants (Powel 1989). Owusu (2001) suggested the natural and cheaper methods for the control of stored-product pests of cereals, with traditionally useful Ghanaian plant materials. In some South Asian countries, food grains such as rice and wheat are traditionally stored by mixing with 2 % turmeric powder (Saxena et al. 1988). The use of oils in stored-products pest control is also an ancient practice. Botanical insecticides such as pyrethrum, derris, nicotine, oil of citronella, and other plant extracts have been used for centuries (Park et al. 2003; Sim et al. 2006). Talukder (2006) has listed 43 plant species as insect repellents, 21 plants as insect feeding deterrents, 47 plants as insect toxicants, 37 plants as grain protectants, 27 plants as insect reproduction inhibitors, and 7 plants as insect growth and development inhibitors. Several essential oils isolated from the Myrtaceae family were reported as a rich source of natural insecticides (Ebadollahi 2013). The organic extracts of *Calotropis procera* and *Annona squamosa* are potential alternatives to chemical insecticides (Begum et al. 2013).

2 Toxicity and Mode of Action of Insecticides

A complete understanding of the mode of action of an insecticide requires knowledge of how it affects a specific target site within an organism. The target site is usually a critical protein or enzyme in insects. Although most insecticides have multiple biological effects, toxicity is usually attributed to a single major effect.

Table 16.1 Biochemical sites of action of natural insecticides

Common name	Class of insecticide	Targeted system	Mode of action
Abamectin	Avermectin	Nervous system	Chloride channel activator
Azadirachtin	Botanical from neem oil	Growth and development/ metabolic processes	Prothoracicotropic hormone (PTTH) inhibitor; phagostimulant disruptor
Bacillus thuringiensis	Microbial	Metabolic processes	Insect midgut membrane disruptor
Cinnamaldehyde	Botanical	Energy production	Exact mode of action not well understood; possibly interference with glucose uptake or utilization
Decalesides I and II	Botanical (natural trisaccharides)	Nervous system	Inhibition of sodium pump (?)
Emamectin benzoate	Avermectin	GABA-gated chloride channels	Chloride channel activators
Pyrethrins I and II	Botanical (pyrethrum)	Nervous system	Sodium channel modulator
Rotenone	Botanical	Mitochondrial electron transport system	Electron transport inhibitor—site 1
Ryanodine	Botanical	Calcium channels (ryanodine receptor)	Activation
Spinosad	Spinosyn	Nervous system	Nicotinic acetylcholine receptor agonist (mimic)

Insecticides can be classified according to their mode of entry into insect: (1) stomach poisons, (2) contact poisons, and (3) fumigants. However, another way insecticides can be classified is by their mode of action. Most insecticides affect one of the five biological systems in insects. These include (1) the nervous system, (2) production of energy, (3) endocrine system, (4) the development of the integument (cuticle), and (5) water balance (Table 16.1).

2.1 Insecticides That Affect the Nervous System

Most of classical insecticides fit into these categories, namely, pyrethroids, organophosphates, and carbamates, as they adversely affect the nervous system (Fig. 16.1).

2.2 Voltage-Gated Sodium Channels

Several classes of insecticides are axonic poisons which interfere with axonal conduction of nerve impulse. In the nervous system, there are two types of voltage-gated channels that contribute to the action potential, namely, the sodium and potassium channels. It is believed that DDT and pyrethroid insecticides act on sodium channels, which causes a delay in sodium channel closing, resulting in excessive neuroexcitation, finally leading to loss of control of the coordinated movement, paralysis, and death (Matsumura 1985; Bloomquist 1999; Nauen 2006; Du et al. 2013; Von Stein et al. 2013).

2.3 Calcium Channels

Calcium channels are present in nerve terminals and muscles. Under normal conditions, muscle contraction is mediated by calcium channels. Depolarization caused by an action potential in the motor nerve terminal activates calcium channels which triggers an influx of Ca^{2+} ions. These ions stimulate the release of the chemical transmitter, glutamate, which diffuses across the synaptic cleft and binds to a receptor-operated ion channel, resulting in the influx of sodium and calcium ions. This influx subsequently activates the sarcoplasmic reticulum calcium channels in the muscle to release calcium ions into the protein filaments, resulting in muscle contraction. Studies have

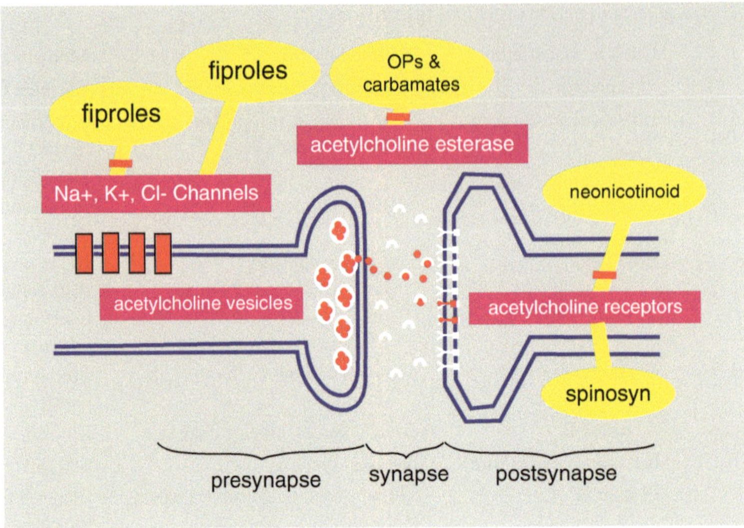

Fig. 16.1 Schematic diagram of insecticides that affect the nervous system (The above figure does not belong to Shivanandappa and Rajashekar. It is an extract figure from Internet)

shown that the active component of *Ryania*, ryanodine, activates the calcium channels in the sarcoplasmic reticulum of the skeletal muscle. The activated calcium channels then release an excess of calcium ions in the protein filaments and induce skeletal muscle contraction and paralysis in insects (Bloomquist 1999; Jeanguenat 2013).

2.4 Inhibition of Acetylcholinesterase

Organophosphate and carbamate insecticides act by interfering with the synaptic transmission of the nerve impulse in the insect nervous system. Organophosphate insecticides are generally regarded as irreversible inhibitors of the enzyme acetylcholinesterase (AChE). Because of the inability of phosphorylated AChE to hydrolyze acetylcholine, the concentration of acetylcholine in the synapse builds up, and excessive neuroexcitation occurs because of the prolonged binding of acetylcholine to its postsynaptic receptor. The signs of intoxication include restlessness, hyperexcitability, tremors, convulsions, paralysis, and, finally death.

The mode of action of the carbamate insecticides is similar to that of the organophosphates on the CNS, and the symptoms of intoxication are similar to those with the organophosphates. Decarbamylation of acetylcholinesterase is rapid, typically in minutes, and, therefore, the carbamate insecticides are regarded as reversible acetylcholinesterase inhibitors (Matsumura 1985; Von Osten et al. 2004; Casida and Durkin 2013; Cespedes et al. 2013).

2.5 GABA-Gated Chloride Channels

Avermectins, derived from streptomyces, are axonic poisons acting on the gamma-aminobutyric acid (GABA)-gated chloride channel in the nervous system (Abalis et al. 1986; Lahm et al. 2013). Avermectins block the chloride channel causing nerve hyperexcitation of the system, resulting in tremors and uncoordinated movement and leading to paralysis and death (Nauen 2006).

2.6 Nicotinic Acetylcholine Receptor

Nicotinic acetylcholine receptors in the insect nervous system are present on post and presynaptic nerve terminals and the cell bodies of interneurons, motor neurons, and sensory neurons (Jeschke and

Nauen 2005; Green et al. 2013). The insecticides nicotine, neonicotinoids, and spinosyns all mimic acetylcholine by acting on the acetylcholine receptor as agonists which causes an influx of sodium ions and the generation of action potentials; under normal physiological conditions, the synaptic action of acetylcholine is terminated by the enzyme acetylcholinesterase, which rapidly hydrolyzes the neurotransmitter. Since these insecticides are not hydrolyzed or destroyed by AChE, the persistent activation leads to an overstimulation of cholinergic synapses, resulting in hyperexcitation, convulsion, paralysis, and finally death of the insect (Matsuda et al. 2001; Green et al. 2013).

2.7 Insecticides That Inhibit Energy Production

Only a handful of chemicals that inhibit the production of energy are currently in use as insecticides. The well-known energy production-inhibiting insecticides are hydramethylnon (inhibitor of cytochrome bc_1), hydrogen cyanide, and phosphine (inhibitor of cytochrome oxidase). These chemicals bind to the protein, cytochrome, in the electron transport system of the mitochondria, which blocks the production of ATP (Karami-Mohajeri and Abdollahi 2013). Insects killed by these chemicals die on their feet. They essentially "run out of gas" (Hollinshaus 1987).

Many new chemicals are being developed for use as energy production inhibitors. Chemicals in this class are pyrrole, thiourea, quinazoline, and sulfuryl fluorides which show promise as pesticides that inhibit energy production.

2.8 Insecticides That Affect the Endocrine System

The chemicals typically referred to as insect growth regulators (IGRs) act on the endocrine system of insects. IGRs are somewhat specific for insect groups, cause slow death of the insects, have low mammalian toxicity, and show nonpersistence in the environment. Most of the currently registered IGRs mimic the juvenile hormone produced in the insect brain. Juvenile hormone acts to keep insects in the immature state. When sufficient growth has occurred, juvenile hormone production ceases triggering the molt to the adult stage (Birnbaum 2013). IGR chemicals such as hydroprene, methoprene, and pyriproxyfen are juvenile hormone mimics (Staal et al. 1985; Ishaaya 2001). Insects treated with these chemicals are unable to molt successfully to the adult stage and cannot reproduce normally.

2.9 Insecticides That Inhibit Cuticle Production

These chemicals, known as chitin synthesis inhibitors or CSIs, are often grouped with the insect growth regulators (IGRs). These chemicals inhibit the production of chitin. Chitin is a major component of the insect exoskeleton. Insect poisoned with CSIs are unable to synthesize new cuticle, thereby preventing them from molting successfully to the next stage (Oberlander and Smagghe 2001; Meyer et al. 2013). The most notable chemicals being used as a CSIs are the benzoylphenyl ureas. This class of insecticides includes lufenuron, which is a systemic insecticide used for flea control, diflubenzuron which is used against fly larvae in manure (Ishaaya and Casida 1974; Leighton et al. 1981), and cyromazine which does not inhibit the biosynthesis of chitin, but is a molting disruptor.

2.10 Insecticides Affecting Water Balance

Insecticides with this mode of action include boric acid, diatomaceous earth, and sorptive dusts. Insects have a thin wax covering their body that helps to prevent water loss from the cuticular surface. Silica aerogels (sorptive dusts) and diatomaceous earth are very effective at absorbing oils. Therefore, when an insect is in contact with these chemicals, the protective waxy covering on the insect is absorbed by the chemical resulting in rapid water loss from the cuticle and eventually death from desiccation (Ware and Whitacre 2004). Unfortunately, insects that live in environments with high relative humidity or that have ready access to a water source show increased tolerance to silica aerogels and diatomaceous earth (Rigsby et al. 2013).

3 Some Important Natural Insecticides with Mammalian Safety

The botanical insecticides that have primarily been used and are commercially available include pyrethrin, rotenone, sabadilla, ryania, nicotine, and azadirachtin (Table 16.2, Fig. 16.2) (El-Wakeil 2013).

3.1 Ryania

The active components of ryania are derived from the roots and woody stems of the plant *Ryania speciosa*, native to Trinidad (Pepper and Carruth 1945; Charles 1954). Ryania works as both contact and stomach poison. It has the longest residual activity among the botanical insecticides. Ryania has a unique mode of action and affects muscles by binding to the calcium channels in the sarcoplasmic reticulum which causes calcium ion flow into the cells and death that follows very rapidly. Ryania is effectively synergized by piperonyl butoxide (PBO) and is reported to be most effective in hot water. Ryania works best on caterpillars (codling moth, corn earworm). However, it has been also found effective on a wide range of insects and mites, including potato beetle, lace bugs, aphids, and squash bug.

Altriset is the first termiticide product featuring an active ingredient from the anthranilic diamide class of chemistry. The class was inspired by research into the insecticide properties of a natural substance found in the bark of trees and shrubs of the genus *Ryania*. Calteryx, the active ingredient in Altriset, is a synthetic compound that affects the ryanodine receptors in the insect muscle fiber. These receptors regulate the flow of calcium into the cell cytoplasm to control muscle contraction. Altriset binds to the ryanodine and causes it to remain open, resulting in a depletion of calcium ions that disrupts muscle contraction. Altriset was designed to target ryanodine receptors of specific insects such as termites (Fig. 16.3) (Cordova et al. 2006; Neoh et al. 2012).

Ryania has low mammalian toxicity, with an LD_{50} of 750 mg/kg bw (Table 16.3). It is moderately toxic to mammals by ingestion and only slightly toxic by dermal exposure. Ingestion of large doses causes weakness, deep and slow respiration, vomiting, diarrhea, and tremors, sometimes followed by convulsions, coma, and death. Purified ryanodine is approximately 700 times more toxic than the crude ground or powdered wood and causes poisoning symptoms similar to those of synthetic organophosphate insecticides. Depending on exposure, symptoms of poisoning may include sweating, headache, twitching, muscle cramps, mental confusion, tightness in chest, blurred vision, vomiting, evacuation of bowels and bladder, convulsions, respiratory collapse, coma, and death.

3.2 Rotenone ($C_{23}H_{228}O_6$)

Rotenone is derived from the roots of the two legume plants, namely, *Lonchocarpus* sp. and *Derris* sp. originally obtained from the East Indies, Malaya, and South America. Rotenone is one of the toxic botanical insecticides with an LD_{50} of 350 mg/kg bw to mammals (Table 16.3). In fact, rotenone is more toxic than both carbaryl and malathion, two commonly used synthetic insecticides. Also, rotenone is extremely toxic to fish (Isman 2006). This botanical insecticide works both as contact and stomach poison. Rotenone is slower acting than most other botanical insecticides, taking several days to kill pests; rotenone shows broad spectrum of activity on many insects and mite pests, including leaf-feeding beetles, caterpillars, lice, mosquitoes, ticks, fleas, and fire ants. It degrades rapidly in air and sunlight. Rotenone blocks NADH dehydrogenase (complex 1) in the respiratory chain, blocking the flow of electrons from NADH to coenzyme Q. Therefore, no ATP is formed from NADH, but two ATPs are formed per $FADH_2$ (Fig. 16.4). The

Table 16.2 List of insecticidal active principles of plants

Active principle	Plant species	Insect toxicity	Insect species	References
Anonaine	*Annona reticulata*	Contact	*C. chinensis*	Oliver-Bever (1986)
Azadirachtin	*Azadirachta indica*	Contact/IGR	Stored grain pests, aphids	Morgan (2009)
E-anethole	*Foeniculum vulgare*	Contact	*S. oryzae*, *C. chinensis*	Kim and Ahn (2001)
β-Asarone	*Acorus calamus*	Contact	Stored grain pests	Baxter et al. (1960)
Z-Asarone	*Acorus calamus*; *Acorus gramineus*	Contact	*S. zeamais*	Yao et al. (2008)
Bornyl acetate	*Chamaecyparis obtusa*	Contact	*S. oryzae*	Park et al. (2003)
Camphor	*Ocimum kilimandacharicum*	Contact	*S. oryzae*	Obeng-Ofori et al. (1998)
(+)-3-Carene	*Baccharis salicifolia*	Contact	*T. castaneum*	Obeng-Ofori et al. (1998)
Carvacrol	*Thujopsis dolabrata*	Contact/fumigant	*S. oryzae*, *C. chinensis*	Ahn et al. (1998)
Carvone	*Carum carvi*	contact	*S. oryzae*, *R. dominica*	Afifi et al. (1989)
Caryophyllene oxide	*Origanum vulgare*	Contact/fumigant	*T. castaneum*	Kim et al. (2010)
1,8 Cineole	*Eucalyptus*	Contact/fumigant	*R. dominica* *T. castaneum*	Prates et al. (1998)
Cinnamaldehyde	*Cinnamomum aromaticum*	Contact	*T. castaneum*, *S. zeamais*	Park et al. (2000)
p-Cymene (cymol)	*Chenopodium ambrosioides*	Contact	*S. zeamais* *B. germanica*	Tapondjou et al. (2002)
Decalesides I and II	*Decalepis hamiltonii*	Contact	*Stored grain and household insects*	Rajashekar et al. (2012b)
Dioctyl hexanedioate	*Conyza dioscoridis*	Contact	*T. castaneum*, *S. granaries*	Peterson et al. (1989)
Eugenol	*Citrus*	Fumigant	*S. oryzae*	Prates et al. (1998)
Estragole	*Foeniculum vulgare*	Contact	*S. oryzae* *L. serricorne*	Kim and Ahn (2001)
(+)- fenchone	*Foeniculum vulgare*	Contact	*S. oryzae* *L. serricorne*	Kim and Ahn (2001)
Geraniol	*Pelargonium graveolens*	Fumigant	*T. confusum*, *T. castaneum*, *S. oryzae*	Stamopoulos et al. (2007)
Hexadecane	*Chenopodium ambrosioides*	Contact	*T. castaneum*, *S. granaries*	Peterson et al. (1989)
Hexadecanoic acid	*Convolvulus arvensis*	Contact	*T. castaneum*, *S. granaries*	Peterson et al. (1989)
Linalool	*Ocimum canum* Sims.	Fumigant	*S. oryzae*, *R. dominica*	Weaver et al. (1991) and Kim et al. (2010)
Limonene	*Citrus*	Contact	*T. castaneum*	Park et al. (2003)
(−)- limonene	*Baccharis salicifolia*	Contact/fumigant	*T. castaneum*	Garcia et al. (2005)
Nicotine	*Nicotiana tabacum*	Contact	Mites, aphids, thrips, leafhopper	Tiwari et al. (1995)
Pyrethrins I and II	*Tanacetum cinerariaefolium*	Contact; stomach poison	Stored grain pests, crop pests	Tattersfield et al. (1929)
β-Pinene	*Baccharis salicifolia*	Contact	*T. castaneum*	Garcia et al. (2005)

(continued)

Table 16.2 (continued)

Active principle	Plant species	Insect toxicity	Insect species	References
α-Pinene	*Baccharis salicifolia*	Fumigant	*T. castaneum*	Garcia et al. (2005)
Plumbagin	*Plumbago europaea*	Fumigant	*M. domestica*	Gujar (1990)
Rotenone	*Lonchocarpus* sp.	Contact; stomach poison	Crop pests, lace bugs, *S. oryzae*	Isman (2006)
Ryania	*Ryania speciosa*	Contact; stomach poison	Potato beetle, aphids, lace bugs, stored grain pests	Jefferies et al. (1992)
Sabadilla	*Schoenocaulon officinale*	Contact; stomach poison	Stinks, thrips, squash bugs, leafhoppers, caterpillars	Hare and Morse (1997)
Spinosyns A and D	*Saccharopolyspora spinosa*	Stomach poison	Stored grain pests	Sparks et al. (2001)
Verbenone	*Artemisia* spp.	Fumigant	Stored grain and household insects	Koul et al. (2008)

nature of the binding site is still unclear. Rotenone is a powerful inhibitor of cellular respiration, the process that converts nutrient compounds into energy at the cellular level. In insects, rotenone exerts its toxic effects primarily on nerve and muscle cells, causing rapid cessation of feeding. Death occurs several hours to a few days after exposure. Because of its high toxicity to fish, it is often used as a fish poison (piscicide) in water management programs. It is effectively synergized by PBO or MGK 264 (Yamamoto 1970; Rattan 2010; Casida and Durkin 2013).

Although rotenone is a potent cellular toxin, mammals detoxify ingested rotenone efficiently via liver enzymes. As with pyrethrins, rotenone is more toxic by inhalation than by ingestion. Exposure to high doses may cause nausea, vomiting, muscle tremor, and rapid breathing. Very high doses may cause convulsions followed by death from respiratory paralysis and circulatory collapse. Direct contact with rotenone may be irritating to skin and mucous membranes. Treatment of poisoning is symptomatic. Chronic exposure to rotenone may lead to liver and kidney damage. Although some rodent testing has shown that chronic dietary exposure to rotenone may induce tumor formation, the most recent US EPA registration standard considers rotenone to be noncarcinogenic. Rotenone is one of the more acutely toxic botanicals. In comparison, pure, unformulated rotenone is more toxic than pure carbaryl (Sevin®) or malathion, two commonly used synthetic insecticides. In the form of a 1 % dust, rotenone poses roughly the same acute hazard as the commonly available 5 % Sevin dust. Commercial rotenone products have presented little hazard to man over many decades. Neither fatalities nor systemic poisonings in humans have been reported in relation to ordinary use.

3.3 Pyrethrin/Pyrethrum

(Pyrethrin I: $C_{21}H_{28}O_3$, pyrethrin II: $C_{22}H_{28}O_5$)

Pyrethrins I and II are derived from the seeds or flower of *Tanacetum cinerariaefolium* (Isman 2006; Copping and Duke 2007) which is grown in Africa, Ecuador, and Kenya. Pyrethrin is one of the oldest household insecticides still available and is fast acting, providing almost immediate "knockdown" of insects following an application. It is both contact and stomach poison. The plant material has a very short residual activity, degrading rapidly under sunlight, air, and moisture, which means that frequent applications may be required. Pyrethrin can be used up until harvest, as there is no waiting interval required between initial application and harvest of food crops (Casida and Quistad 1995).

Pyrethrin kills insects by disrupting the sodium and potassium ion-exchange process in insect nerves and interrupting the normal transmission

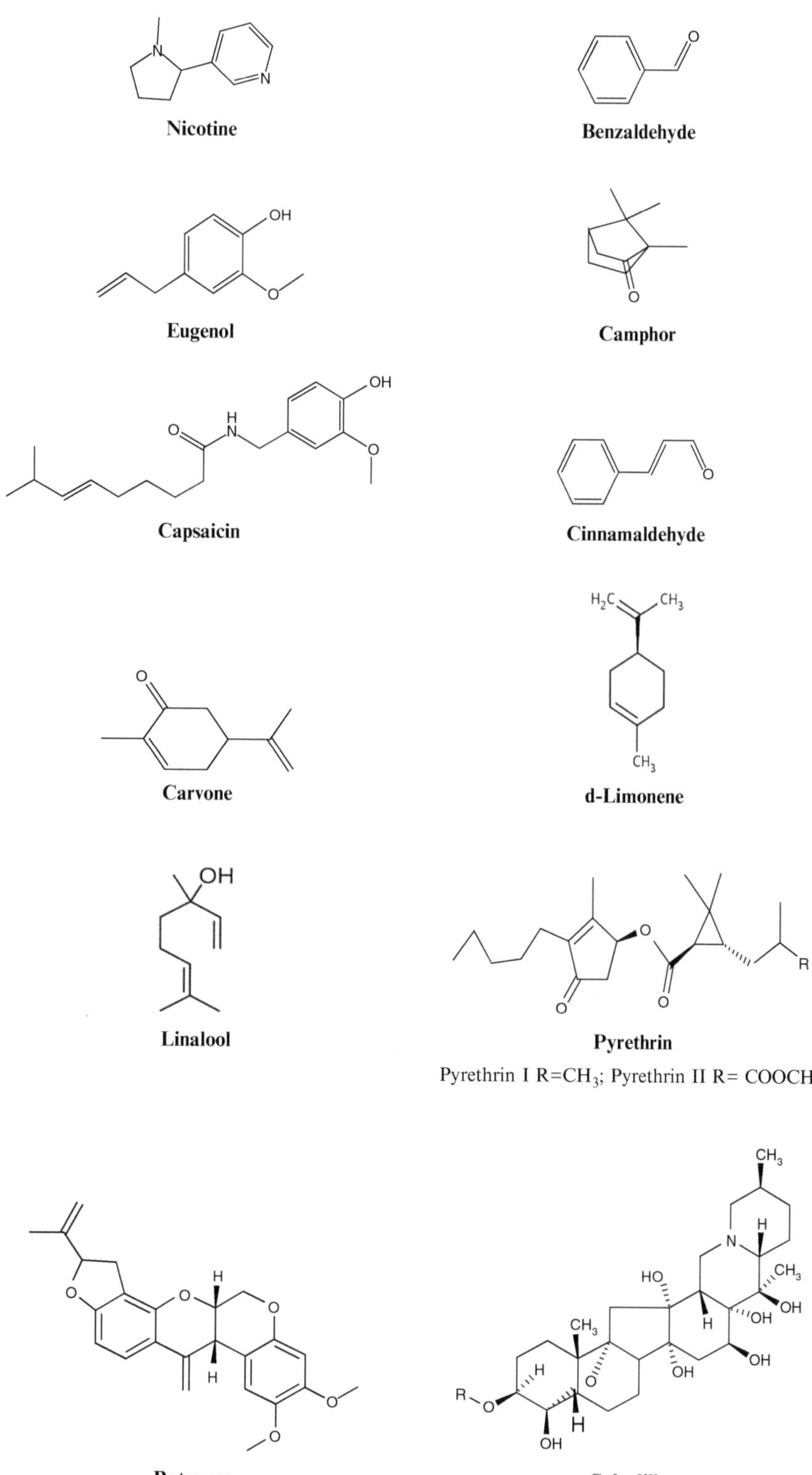

Fig. 16.2 Molecular structures of some important natural insecticides

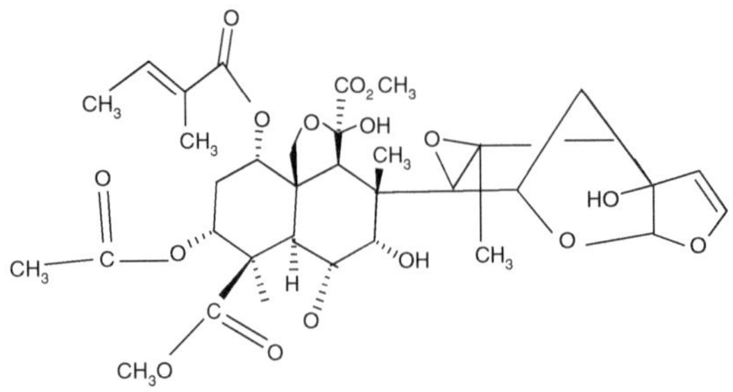

Azadirachtin

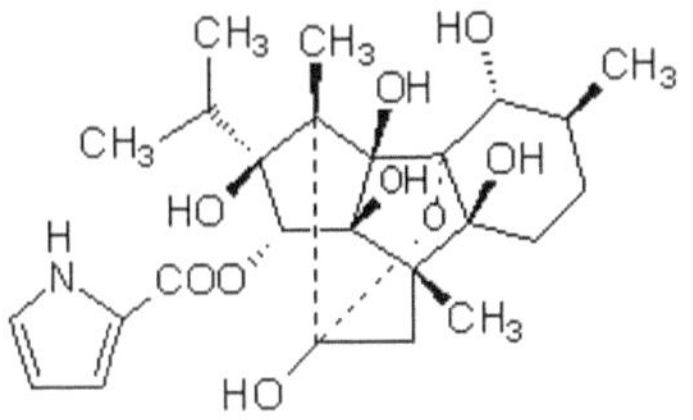

Ryanodine

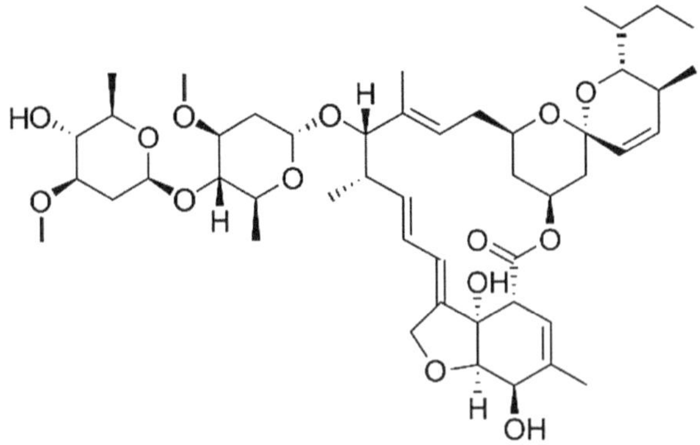

Avermectins

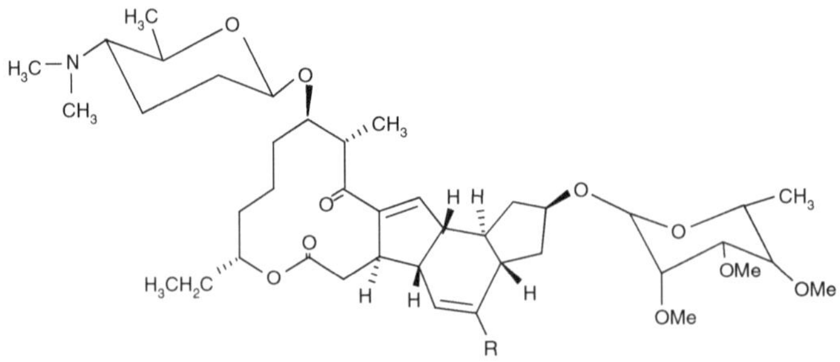

Spinosyn

Spinosyn A;R=H

Spinosyn D;R=CH_3

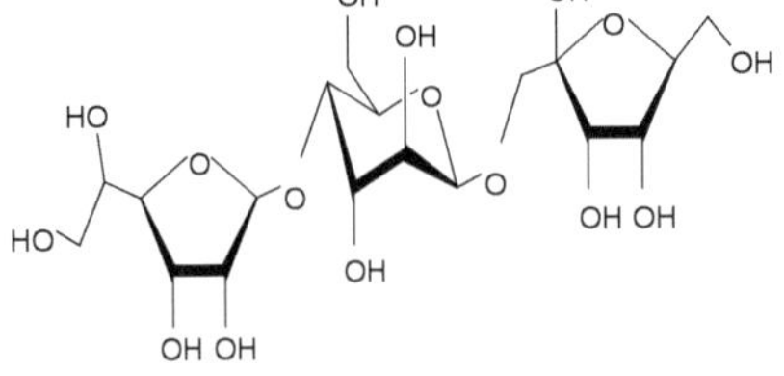

Decaleside I

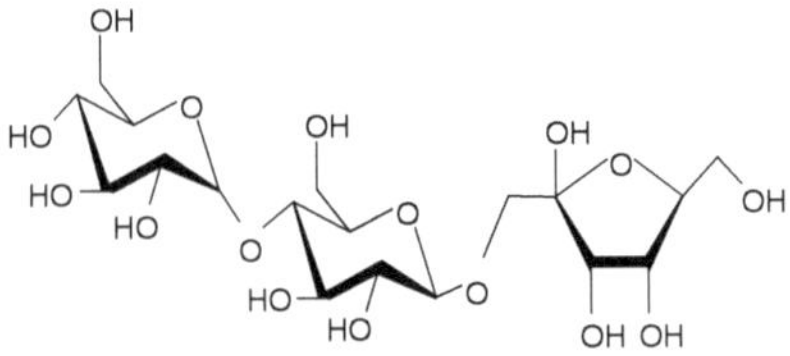

Decaleside II

Fig. 16.2 (continued)

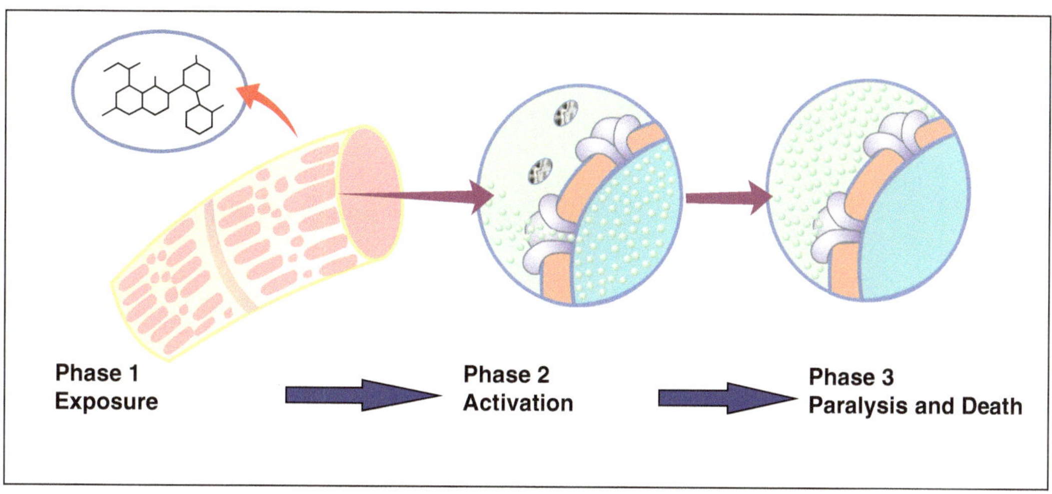

Fig. 16.3 Schematic diagram of mode of action of Ryania (The above figure does not belong to Shivanandappa and Rajashekar. It is an extract figure from Internet)

Table 16.3 Insecticidal activity and mammalian toxicity of some natural insecticides

Natural insecticides	Insect toxicity[a]	Mammalian toxicity (Oral rat LD_{50} (mg/kg bw))
Anethole	C, S	2,090
β-Asarone	C, S	275
Azadirachtin	IGR, R	13,000
Carvacrol	C	810
Cineole (1–8)	C, F	2,480
Cinnamaldehyde	C	1,160
Cuminic aldehyde	C, S	1,390
Eugenol	C, F	500
Nicotine	C	50
Pyrethrins I and II	C, S	1,200
Rotenone	S	350
Ryania	C, S	750
Sabadilla	C, S	5,000
Spinosad	C	3,738

[a]*C* contact, *S* stomach poison, *F* fumigant, *IGR* insect growth regulator, *R* repellent (Golob and Gudrups 1999)

of nerve impulses. Pyrethrin shows insecticidal activity on wide range of insects and mites, including flies, fleas, beetles, and spider mites (Casida and Quistad 1998; Rattan 2010). Pyrethrins exhibit low mammalian toxicity and few cases of human poisonings have ever been reported. The LD_{50} of pyrethrin is 1,200–1,500 mg/kg bw (Table 16.3). Cats, however, are highly susceptible to poisoning by pyrethrins, and care must be taken to follow label directions closely when using products containing pyrethrins to treat cats for fleas.

When ingested, pyrethrins are not readily absorbed from the digestive tract and rapidly hydrolyzed under the acidic conditions of the gut and in the liver. Pyrethrins are more toxic to mammals by inhalation than by ingestion due to direct route to the bloodstream. Exposure to high doses may cause nausea, vomiting, diarrhea, headaches, and other nervous disturbances. Repeated contact with crude pyrethrum dusts may cause skin irritation or allergic reactions. The allergens that cause these reactions are not present in products containing refined pyrethrins. Tests indicate that chronic exposure to pyrethrins does not cause genetic mutations or birth defects. There is no single antidote for acute pyrethrin poisoning. Treatment of poisoning is symptomatic, i.e., the various symptoms of poisoning are treated individually as they occur because there is no way to counteract the source of the poisoning directly.

3.4 Nicotine

Nicotine, which is derived from *Nicotiana tabacum*, is highly toxic to mammals among the botanical insecticides with an LD_{50} between 50

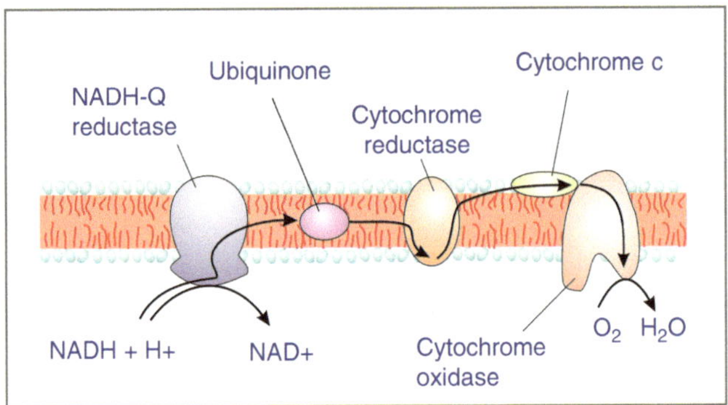

Fig. 16.4 Schematic diagram of electron transport affected by rotenone (The above figure does not belong to Shivanandappa and Rajashekar. It is an extract figure from Internet)

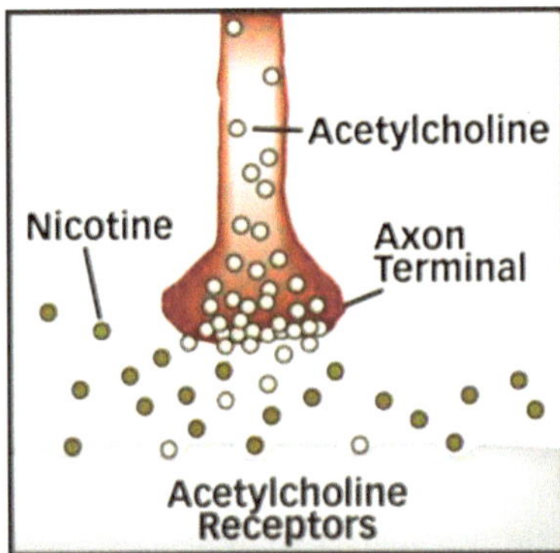

Fig. 16.5 Schematic diagram of mode of action of nicotine (Source: Lindstrom 1997)

and 60 mg/kg bw (Yamomoto 1999; Isman 2006). It is extremely harmful to humans. Nicotine, a fast-acting nerve toxin, works as a contact poison. Nicotine is most effective on soft-bodied insects and mites, including aphids, thrips, leafhoppers, and spider mites. Many caterpillars are resistant to nicotine. It kills insects (and humans) through action on nicotinic cholinergic receptors at the nerve synapses (junctures), causing uncontrolled nerve firing, and by mimicking acetylcholine (ACh) at the nerve-muscle junctions in the central nervous system (Fig. 16.5) (Ujavary 1999; Rattan 2010). Despite the fact that smokers regularly inhale small quantities of nicotine in tobacco smoke, nicotine in pure form is extremely toxic to mammals and is considered a Class I (most dangerous) poison. Nicotine is particularly hazardous because it penetrates the skin, eyes, and mucous membranes readily; both inhalation and dermal contact may result in death. Ingestion is slightly less hazardous due to the effective detoxifying action of the liver.

Symptoms of nicotine poisoning are extreme nausea, vomiting, excess salivation, evacuation of bowels and bladder, mental confusion, tremors, convulsions, and finally death by respiratory failure and circulatory collapse. Poisoning occurs very rapidly and is often fatal. Treatment for nicotine poisoning is symptomatic, and only immediate treatment, including prolonged artificial respiration, may save a victim of nicotine poisoning. Nicotine has been responsible for numerous serious poisonings and accidental deaths because of its rapid penetration of skin and mucous membranes and because of the concentrated form in which it is used. Certain plant types, such as roses, may be harmed or injured by nicotine sprays.

3.5 Azadirachtin

Azadirachtin is derived from the well-known neem tree, *Azadirachta indica*, grown in India and Africa (Isman 2006). Azadirachtin has an extremely low mammalian toxicity and is the least toxic among the botanical insecticides, with an LD_{50} of 13,000 mg/kg bw. Azadirachtin is considered a contact poison; however, it has "some" systemic activity in plants when applied to the foliage. The material is generally nontoxic to beneficial insects and mites. Azadirachtin has broad mode of action, as a feeding deterrent, insect growth regulator, repellent, and sterilant; it may also inhibit oviposition (Rembold 1989; Isman 2006). Recently Qiao et al. (2013) reported that Azadirachtin can interfere with the insect's central nervous system via inhibition of excitatory cholinergic transmission and partly blocking the calcium channel.

Azadirachtin is active on a broad range of insects, including stored grain pests, aphids, caterpillars, and mealybugs (Morgan 2009). As a repellent, neem prevents insects from initiating feeding. As a feeding deterrent, it causes insects to stop feeding either immediately after the first "taste" (due to the presence of deterrent taste factors) or at some point soon after ingesting the food (due to secondary hormonal or physiological effects of the deterrent substance). As a growth regulator, neem is thought to disrupt normal development interfering with chitin synthesis. Susceptibility to the various effects of neem differs by species.

3.6 Sabadilla

Sabadilla is derived from the seeds of plant *Schoenocaulon officinale*, which is grown in Venezuela. Sabadilla is one of the least toxic registered botanical insecticides, with mammalian LD_{50} of 5,000 mg/kg bw. Sabadilla works as a contact and a stomach poison. Similar to other botanical insecticides, it has minimal residual activity and degrades rapidly in sunlight and moisture (rainfall). Sabadilla works by affecting nerve cell membranes, causing loss of nerve function, paralysis, and death (Copping and Duke 2007; Rattan 2010). Sabadilla, in the form of dusts made from ground seeds, is one of the least toxic botanicals. Purified veratrine alkaloids are quite toxic, however, and are considered on par with the most toxic synthetic insecticides. Sabadilla can be severely irritating to skin and mucous membranes and has a powerful sneeze-inducing effect when inhaled. Ingestion of small amounts may cause headaches, severe nausea, vomiting, diarrhea, cramps, and reduced circulation. Ingestion of very high doses may cause convulsions, cardiac paralysis, and respiratory failure. Sabadilla alkaloids can be absorbed through the skin or mucous membranes. Systemic poisoning by sabadilla preparations used as insecticide has been very rare or nonexistent (Duke et al. 2010). It is effective against caterpillars, leafhoppers, thrips, stink, and squash bugs.

3.7 Neonicotinoids

Neonicotinoids are a new class of insecticides related to nicotine. The name literally means "new nicotine-like insecticides." Neonicotinoids being used in agriculture are imidacloprid, acetamiprid, and thiamethoxam. Like nicotine, the neonicotinoids act on neurotransmitter receptors in the nerve synapse. They are much more toxic to invertebrates, like insects, than to mammals, birds, and other higher organisms. One thing that has made neonicotinoid insecticides popular in pest control is their water solubility, which allows them to be applied to soil and be taken up by plants. Efforts have been made to develop nicotinyl insecticides with high affinity to the insect nicotinic acetylcholine receptor (nAChR), resulting in the development of a new group of insecticides (Elbert et al. 1996). Neonicotinoids act through contact and ingestion and result in the cessation of feeding within several hours of contact and death shortly after. Neonicotinoids do not inhibit cholinesterase or interfere with sodium channels. Therefore, its mode of action is different from those of organophosphate, carbamate, and pyrethroid compounds. It appears that neonicotinoid acts as an agonist of insect nicotinic acetylcholine receptors in a manner different from other insecticides. Neonicotinoids interact with nicotinic acetylcholine receptors at the central and peripheral nervous system, resulting in excitation and paralysis, followed by death (Fig. 16.6). These compounds interact with nAChR in a structure-activity relationship (Tomizawa et al. 1995; Tomizawa and Casida 2005), resulting in excitation and paralysis, followed by death. Their selectivity results from a higher affinity to the insect nAChR than to that of vertebrates, in contrast to the original nicotine. Hence, it has been suggested that these new compounds are called neonicotinoids.

Comparative toxicity was carried out, under laboratory conditions, of two neonicotinoids, acetamiprid and imidacloprid, against the whitefly

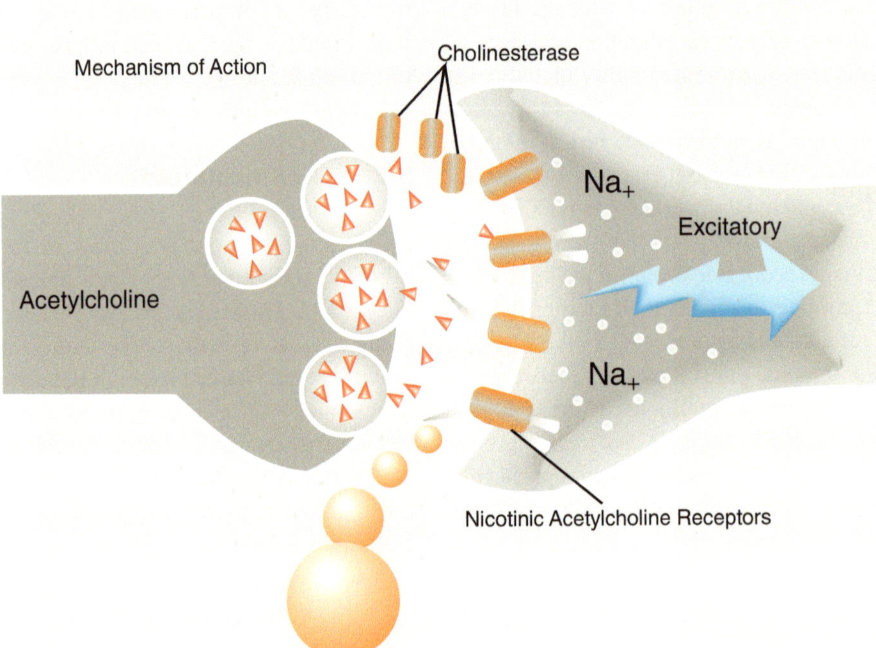

Fig. 16.6 Schematic diagram of mode of action of spinosad (The above figure does not belong to Shivanandappa and Rajashekar. It is an extract figure from Internet)

Bemisia tabaci (Gennadius), using foliar and systemic applications on cotton seedlings (Horowitz et al. 1998). The ovicidal and nymphicidal activities of foliar application of acetamiprid on cotton seedlings were, according to LC_{50} and LC_{90} values, 10- and 18-fold more potent than imidacloprid. However, when applied to soil, imidacloprid was more potent than acetamiprid. In other assays, it was found that the comparative toxicity of the two neonicotinoids could be different when assayed on different host plants. The translaminar activity of imidacloprid on cabbage leaves against *Myzus persicae* was superior to that of acetamiprid, while against *Aphis gossypii* on cotton, the activity of imidacloprid was inferior to that of acetamiprid. The neonicotinoids act specifically on sucking pests and have a mild or no effect on parasitoids and predators, and as such, they fit well in various IPM programs (Ishaaya 2001).

3.8 Avermectins

Avermectins, which are macrocyclic lactones, are derived from the actinomycetes *Streptomyces avermitilis* (Putter et al. 1981). This molecule is most effective against agricultural pests with lethal concentration of 90 % (LC_{90}) in the range of 0.02 ppm for mites and less toxic to stored-product pests. It is effective on internal parasites of domestic animals (Mrozik et al. 1989; Lasota and Dybas 1991). Avermectins bind with high affinity to glutamate-gated channels which occur in invertebrate nerve and muscle cells, causing an increase in the permeability of the cell membrane to chloride ions with hyperpolarization of the nerve or muscle cell leading to paralysis and death of the pest either directly or by causing the worms to starve. At least one study, however, seems to suggest a depolarizing rather than hyperpolarizing role for avermectins on the glutamate-gated chloride channel (Pemberton et al. 2001; Rattan 2010). However, in either case, the end result is the deactivation of the channel by manipulation of chloride levels. Selective activity of compounds of this class is attributable to the facts that some mammals do not have glutamate-gated chloride channels and that avermectins have low affinity for mammalian ligand-gated chloride channels. Visible activity such as feeding or egg laying stops shortly after exposure, though death may not occur for several days (Deng et al. 1991). Technical avermectin is quite toxic, with an oral rat LD_{50} of 30 mg/kg bw.

Abamectin/avermectin has been shown to cause pupil dilation, mild skin irritation, vomiting, convulsions and/or tremors, and coma in laboratory animals. Because it is a nerve poison, it can also cause nervous system depression in mammals at very high doses. A study in rats given 0.40 mg/kg/day of abamectin showed decreased lactation, increased stillbirths, and an increased likelihood of producing unhealthy offspring, demonstrating a strong chance of similar effects in humans at high enough doses. EPA reviewed toxicological data from the manufacturer in connection with a 1987 petition for establishment of a tolerance in citrus oil and citrus pulp. EPA's reviewers found that avermectin does not cause birth defects in rats and rabbits, but can cause cleft palate in mice. The calculated "lowest effect level (LEL)" for the latter effect was quite low at 0.10 mg/kg/day. EPA reviewers stated that "studies on mutagenicity demonstrated an overall negative potential" (ETN 1996). Abamectin is also very toxic to fish and aquatic invertebrates (FCH 2000).

3.9 Spinosads

Spinosad is a mixture of spinosyn A and spinosyn D, originally isolated from the soil actinomycetes, *Saccharopolyspora spinosa*, used as bioinsecticide (Copping and Menn 2000). Spinosad is recommended for the control of a very wide range of caterpillars, leaf miners, and foliage-feeding beetles. Spinosad may be used on crops (including cotton), vegetables, fruit trees, turf, vines, and ornamentals, and it is under development for use on livestock animals and in livestock premises. The actual mode of action of spinosad has not been established. In insects, the poisoning symptoms of spinosad are muscle contractions due to postural changes, typically elevation of the body and straightening of the legs (Salgado et al. 1998). After many hours of excitation, the movements and fine tremors finally cease and the ensuing paralysis is due to neuromuscular fatigue. In electrophysiological studies with cockroach neurons, the spinosad was shown to activate nAChRs and action could be blocked by the selective nicotinic antagonist α-bungarotoxin. Spinosad could also prolong the action of ACh. Furthermore, the spinosad affected GABA receptors in isolated insect neurons (Salgado 1997; Orr et al. 2009; Rattan 2010). Spinosyns have a novel mode of action, primarily targeting binding sites on nicotinic acetylcholine receptors that are distinct from those of other insecticides leading to disruption of acetylcholine neurotransmission (Sparks et al. 2001; Snyder et al. 2007) (Fig. 16.6). Spinosad has a very low mammalian toxicity and is considered nontoxic to mammals but slightly toxic to fish. Spinosad is rapidly degraded on soil surfaces by photolysis and, below the soil surface, by soil microorganisms (Saunders and Bert 1997).

3.10 Limonene

Limonene is extracted from citrus oils. The oral LD_{50} is reported to be greater than 5,000 mg/kg bw. Linalool is a closely related material that is also an extract from orange and other citrus fruit peels. Citrus oil extracts have been combined with insecticidal soap for use as contact poisons against aphids and mites. Limonene and linalool are contact poisons (nerve toxins) (Kostyukovsky et al. 2002). They have low oral and dermal toxicities. Both the compounds evaporate readily from treated surfaces and have no residual effect.

3.11 Decaleside

(Decaleside I: $C_{21}H_{28}O_3$, Decaleside II: $C_{22}H_{28}O_5$)

Decalesides I and II, the natural insecticides isolated and characterized from the roots of *Decalepis hamiltonii*, are novel trisaccharides that are toxic to a variety of insect species by contact exposure (Rajashekar et al. 2012b). The insect toxicity of decalesides was comparable to that of chemical insecticides based on the LC_{50} values, and the rapidity of action even surpassed that of chemical insecticides as evident from the time course and knockdown effect in contact

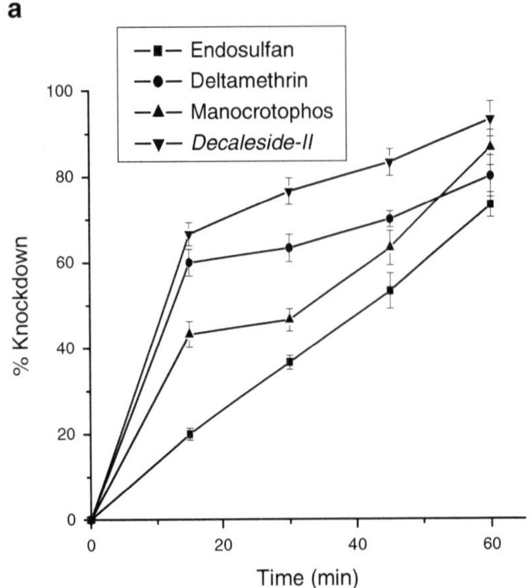

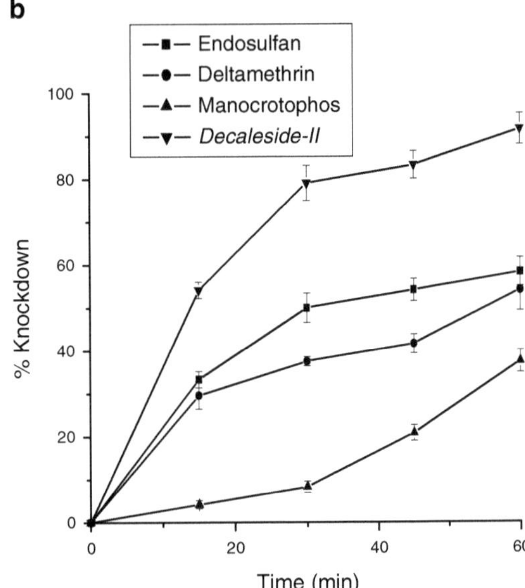

Fig. 16.7 Comparison of the insecticidal action (time course) of *Decaleside II* with that of chemical insecticides in (**a**) *Musca domestica* (housefly) and (**b**) *Blattella germanica* exposure at LC_{50} by contact bioassay ($n = 4$, error bars, s.e.m.) (Source: Rajashekar et al. 2012b)

bioassay (Fig. 16.7). Although several natural compounds have been reported to exhibit contact toxicity, there is no comparative study of the toxicity of a natural compound with that of synthetic insecticides on insects in a contact bioassay. The nature of insect toxicity of decaleside in the bioassay, based on the symptoms, indicated hyperactivity as seen in the increased movements followed by knockdown with legs showing twitching movements indicating a neuromuscular effect. The neurotoxic symptoms were similar to that of pyrethroids. However, the lack of toxicity of decalesides orally or by topical application on the abdomen was intriguing.

Since decalesides I and II are trisaccharides, it was of interest whether the mode of action involves the chemical sensilla in the tarsi in which gustatory receptors are located. In order to test this hypothesis, experiments were done by the authors (Rajashekar et al. 2012b) in which the lower part of the legs (tarsi) of insects (houseflies and cockroaches) was surgically ablated and then exposed to decaleside-treated surface in the contact bioassay. The surgical ablation of tarsi did not cause mortality of the insects nor their movement drastically in the bioassay. Further, a less invasive method such as masking the sensilla in the tarsi by application of molten wax was done to see if it blocks the insecticidal effect. In both of these cases, toxicity was abolished by tarsal ablation as well as wax treatment. Direct application of decaleside II to the tarsi of the first pair of legs was effective in killing the cockroaches, whereas wax application protected against the toxic action. This compelling evidence demonstrated the requirement of the exposure by contact of tarsi to decaleside for the toxic action and, therefore, implicates the gustatory (sugar) receptors in the insecticidal action (Fig. 16.8). Further, to test the possible involvement of gustatory (sugar) receptors, simple experiments were performed in which we studied the effect of sugars (mono-, di-, and trisaccharides) on the insect toxicity of decaleside. Experiments in which cockroaches were exposed to decaleside-treated paper with or without various sugars showed that sugars protected against the toxicity but not an amino acid (Fig. 16.9). Among the sugars, maltotriose, a trisaccharide, was most effective in rescuing the toxicity of decaleside and the effect was dose

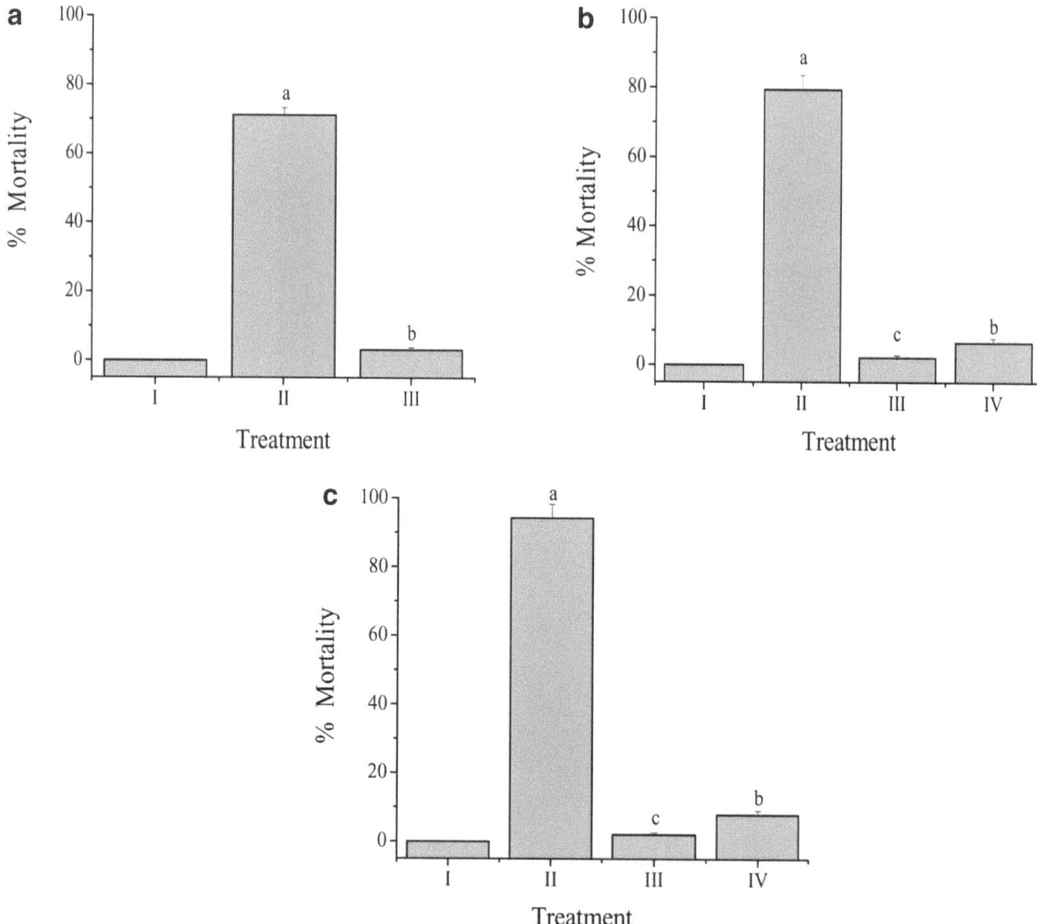

Fig. 16.8 Experimental demonstration of the tarsi-mediated contact toxicity of *Decaleside II* in the German cockroach. (**a**) Effect of tarsal ablation on the insecticidal activity of *Decaleside II* in German cockroach by contact bioassay. (**I**) Intact control (solvent), (**II**) intact control + *Decaleside II*, (**III**) tars *II* ablated + *Decaleside II* ($n = 4$, error bars, s.e.m.). One-way ANOVA, $^{***}P < 0.001$. (**b**) Effect of wax application on tarsi, on the toxicity of *Decaleside II* in German cockroach, by contact bioassay. (**I**) Intact control (solvent), (**II**) intact control + *Decaleside II*, (**III**) wax treated (solvent control), (**IV**) wax treated + *Decaleside II* ($n = 4$, error bars, s.e.m.). One-way ANOVA, $^{***}P < 0.001$. (**c**) Effect of wax application on the tarsi, on the toxicity of *Decaleside II* in the German cockroach, by topical application on the leg. (**I**) Control (solvent only), (**II**) untreated + *Decaleside II* (1 mg/insect), (**III**) wax treated on tarsi (+solvent), (**IV**) with wax treated tarsi + *Decaleside II* (1 mg/insect) ($n = 4$, error bars, s.e.m.). One-way ANOVA. Values denoted by different alphabets differ significantly ($^{***}P < 0.001$)

dependent. Further, when decaleside was subjected to chemical and enzymatic hydrolysis, insecticidal activity was lost. These experiments clearly demonstrated the involvement of the gustatory (sugar) receptors located in the tarsi for the insecticidal action of Decaleside. Since decalesides I and II being natural trisaccharides are detected by the gustatory receptors in the tarsi, the insecticidal action requires contact with the tarsi as shown experimentally. Abolition of the toxicity of decaleside by hydrolysis indicates that only the intact trisaccharide molecules exhibit the insecticidal activity. However, other natural trisaccharides such as raffinose, melezitose, maltotriose, and other sugars were not toxic to insects, but protected against the toxic action of Decalesides I and II, indicating the common site of action involving the gustatory receptors (Fig. 16.10).

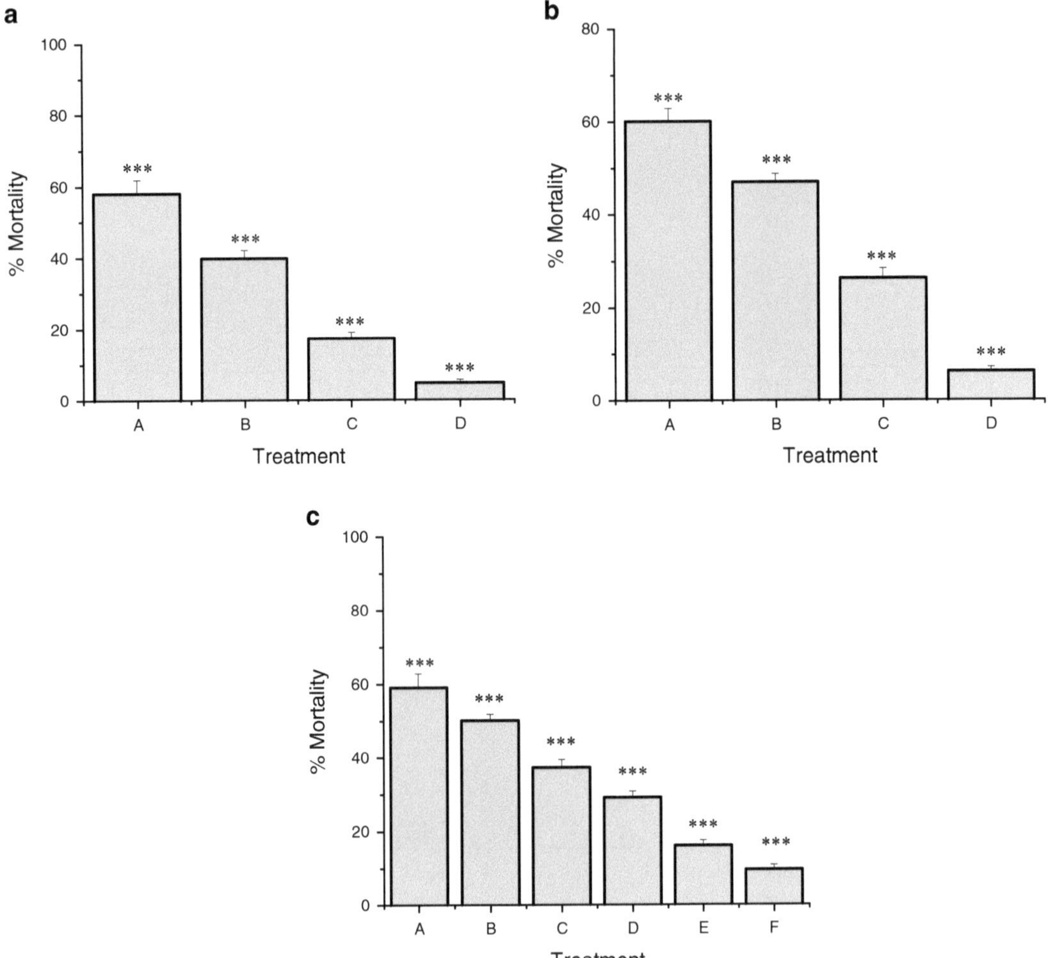

Fig. 16.9 Effect of sugars on the insecticidal activity of *Decaleside II* in German cockroach exposure at LC_{50} (0.07 mg/cm^2) by contact bioassay. (**a**) Monosaccharide (glucose). A. *Decaleside II*, B. *Decaleside II*+glucose (1:1), C. *Decaleside II*+glucose (1:2), and D. *Decaleside II*+glucose (1:3). Ratio in molar concentration (n=4, error bars, s.e.m.). One-way ANOVA, $^{***}P<0.001$. (**b**) Disaccharide (Trehalose). A. *Decaleside II*, B. *Decaleside II*+trehalose (1:1), C. *Decaleside II*+trehalose (1:2), and D. *Decaleside II*+trehalose (1:3). Ratio in molar concentration (n=4, error bars, s.e.m.). One-way ANOVA, $^{***}P<0.001$. (**c**) Trisaccharide (maltotriose). A. *Decaleside II*, B. *Decaleside II*+maltotriose (1:0.1), C. *Decaleside II*+maltotriose (1:0.25), D. *Decaleside II*+maltotriose (1:0.5), E. *Decaleside II*+maltotriose (1:0.75), and F. *Decaleside II*+maltotriose (1:1). Ratio in molar concentration (n=4, error bars, s.e.m.). One-way ANOVA, ***values significantly different from the control ($^{***}P<0.001$). (Source: Rajashekar et al. 2012b)

The precise biochemical or molecular target for the action of decaleside is not known yet. However, preliminary evidence from the authors' laboratory (our unpublished work) suggests the involvement of sodium pump in the mode of action. Electrophysiological evidence is needed to support this hypothesis.

Future investigations will throw more light in these areas and, no doubt, will lead to exciting novel mechanisms in the signal transduction process involving the gustatory receptors.

4 Conclusion

Conventional pest management has been significantly influenced by bioactive natural products that are used directly or, in a derived form, as

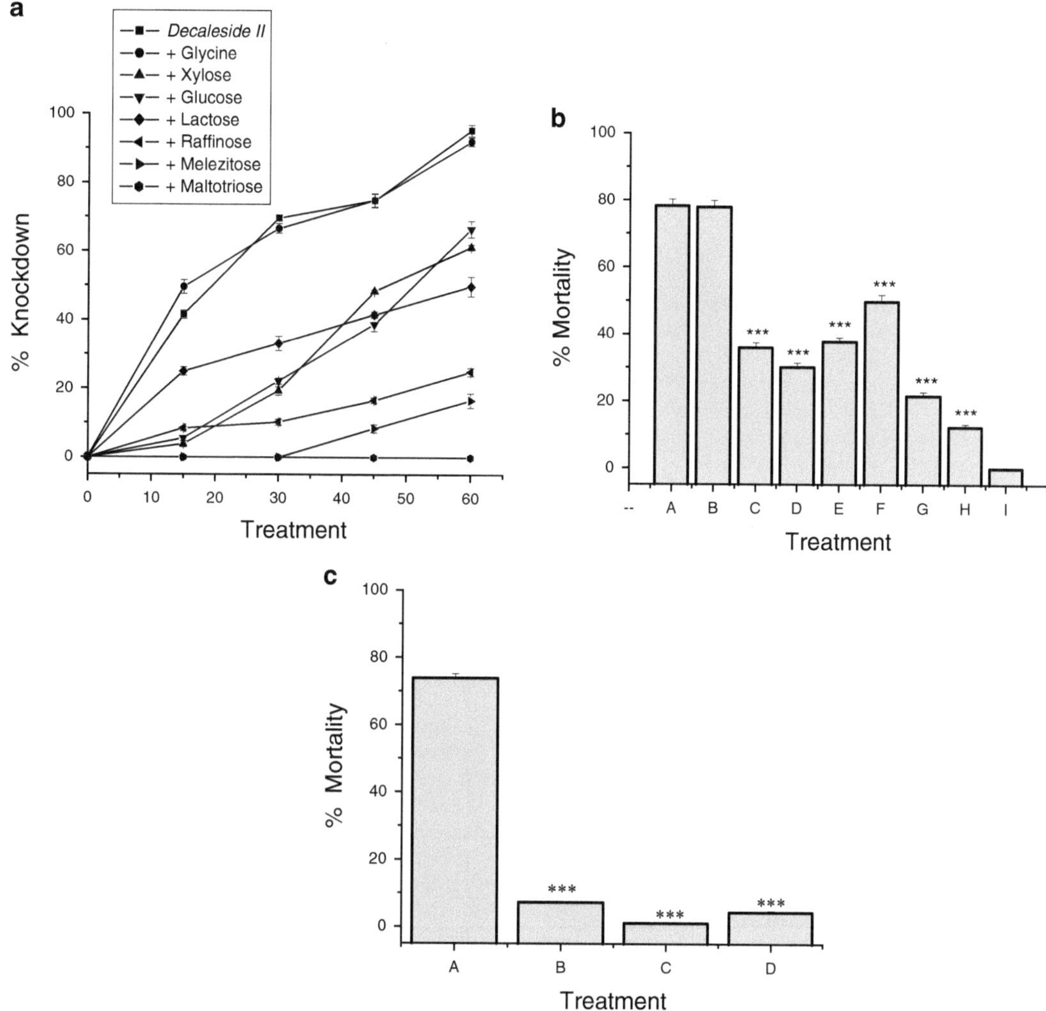

Fig. 16.10 Effect of sugars on the insecticidal activity of *Decaleside II* in the German cockroach (**a**) Knockdown at equimolar (LC_{50}) concentration of *Decaleside II*+Sugars ($n=4$, error bars, s.e.m). (**b**) Mortality at equimolar concentration: **A**. *Decaleside II*, **B**. *Decaleside II*+Glycine, **C**. *Decaleside II*+Glucose, **D**. *Decaleside II*+Xylose, **E**. *Decaleside II*+Lactose, **F**. *Decaleside II*+Trehalose, **G**. *Decaleside II*+Raffinose, **H**. *Decaleside II*+Melezitose, **I**. *Decaleside II*+Maltotriose ($n=4$, error bars, s.e.m.). One-way ANOVA, $^{***}P<0.001$. (**c**) Effect of hydrolysis of *Decaleside II* on the insecticidal activity. (**A**) Control (without hydrolysis), (**B**) acid hydrolysis, (**C**) enzymatic hydrolysis by β-galactosidase, (**D**) enzymatic hydrolysis by α-glucosidase ($n=4$, error bars, s.e.m.). One-way ANOVA, $^{***}P<0.001$ (Source: Rajashekar et al. 2012b)

pesticides. Bio-based pesticides are commonly used as alternatives to synthetic compounds in organic agriculture and stored grain protection. These past successes and the current public's concern over the impact of synthetic pesticides on the environment ensure a continued interest in searching nature for eco-friendly pest management agents. There is great scope for the discovery of novel molecules showing promise as newer biopesticides from natural sources considering the tremendous biodiversity in many parts of the world. Parallelly, new knowledge of insect biology will open up newer targets for devising insecticides that will eventually pave the way for combating the phenomenon of resistance and achieving selectivity in pest control agents.

References

Abalis IM, Eldefawi AT, Eldefawi ME (1986) Action of avermectin B_{1a} on the γ-aminobutyric $acid_A$ receptor and chloride channels in rat brain. J Biochem Toxicol 1:69

Afifi FA, Salem M, Hekal AM (1989) Insecticidal properties of the extracts of lupin seed and caraway fruits against some stored product insects. Ann Agric Sci 34(1):401–414

Ahmed S, Grainage M (1986) Potential of the Neem Tree (*Azadirachta indica*) for pest control and rural development. Econ Bot 40(2):201–209

Ahn YJ, Lee SB, Lee HS, Kim GH (1998) Insecticidal and acaricidal activity of carvacrol and β-thujaplicine derived from *Thujopsis dolabrata* var. hondai sawdust. J Chem Ecol 24:81–90

Akhtar Y, Yeoung YR, Isman MB (2008) Comparative bioactivity of selected extracts from Meliaceae and some commercial botanical insecticides against two noctuid caterpillars, *Trichoplusia* ni and *Pseudaletia*. Phytochem Rev 7:77–88

Baxter RM, Dandiya PC, Kandal SI (1960) Separation of the hypnotic potentiating principles from the essential oil of *Acorus Calamus* L. of Indian origin by liquid gas chromatography. Nature 185:466

Begum N, Sharma B, Pandey RS (2013) *Calotropis procera* and *Annona squamosa*: potential alternatives to chemical pesticides. Br J Appl Sci Technol 3(2):254–267

Birnbaum LS (2013) When environmental chemicals act like uncontrolled medicine. Trends Endocrinol Metab 24(7):321–323

Bloomquist JR (1999) Insecticides: chemistries and characteristics. In: Radcliff EB, Hutchison (eds) Radcliffes IPM World Text book. University of Minnesota, St. Paul. URL: http://ipmworld.umn.edu

Casida JE, Durkin KA (2013) Neuroactive insecticides: targets, selectivity, resistance and secondary effects. Annu Rev Entomol 58:99–117

Casida JE, Quistad GB (1995) Pyrethrum flowers: production, chemistry, toxicology and uses. Oxford University Press, Oxford

Casida JE, Quistad GB (1998) Golden age of insecticide research: past, present, or future? Annu Rev Entomol 43:1–16

Cespedes CL, Manoz E, Salazar JR, Yamaguchi L, Werner E, Alarcon J, Kubo I (2013) Inhibition of cholinesterase activity by extracts, fractions and compounds from *Calceolaria talcana* and *C. integrifolia* (Calceolariaceae: Scrophulariaceae). Food Chem Toxicol 62:919–926

Charles LJ (1954) Toxicity of the botanical insecticide *Ryania speciosa* to *Culex pipiens fatigans* Wied. Bull Entomol Res 45:403–410

Copping LG, Duke SO (2007) Natural products that have been used commercially as crop protection agents. Pest Manag Sci 63:524–554

Copping LG, Menn JJ (2000) Biopesticides: a review of their action, applications and efficacy. Pest Manag Sci 56:651–676

Cordova D, Benner EA, Sacher MD, Rauh JJ, Sopa JS, Lahm GP, Selby TP, Stevenson TM, Flexner L, Gutteridge S, Rhoades DF, Wu L, Smith RM, Tao Y (2006) Anthranilic diamides: a new class of insecticides with a novel mode of action, ryanodine receptor activation. Pestic Biochem Physiol 84(3):196–214

Dayan FE, Cantrell CL, Duke SO (2009) Natural products in crop protection. Bioorg Med Chem 17:4022–4034

Deng Y, Palmer CJ, Casida JE (1991) Housefly brain γ-aminobutyric acid-gated chloride channel: target for multiple classes of insecticides. Pestic Biochem Phys 41(1):60–65

Du Y, Nomura Y, Satar G, Hu Z, Nauen R, He SY, Zhorov BS, Dong K (2013) Molecular evidence for dual pyrethroid-receptor sites on a mosquito sodium channel. Proc Natl Acad Sci U S A 110(29):11785–111790

Duke SO, Cantrell CL, Meepagala KM, Wedge DE, Tabanca N, Schrader KK (2010) Natural toxins for use in pest management. Toxins 2(8):1943–1962

Ebadollahi A (2013) Essential oils isolated from myrtaceae family as natural insecticides. Annu Rev Res Biol 3(3):148–173

Elbert A, Nauen R, Cahill M, Devonshire AL, Scarr A, Sone S, Steffens R (1996) Resistance management for chloronicotinyl insecticides using imidacloprid as an example. Pflanzenschutz Nachr Bayer 49:5–54.

El-Wakeil NE (2013) Botanical pesticides and their mode of action. Gesunde Oflanzen 65(4):125–149

Extension Toxicology Network (ETN) (1996) Pesticide Information Profiles: Abamectin. http://ace.orst.edu/cgi-bin/mfs/01/pips/abamecti.htm

Farm Chemicals Handbook 2000 (FCT) (2000) Meister Publishing Co., Willoughby

Garcia M, Donadel OJ, Ardanaz CE, Tonn CE, Sosa ME (2005) Toxic and repellent effects of *Baccharis salicifolia* essential oil on *Tribolium castaneum*. Pest Manag Sci 61:612–618

Golob P, Gudrups I (1999) The use of spices and medicinals as bioactive protectants for grains, FAO agricultural sciences bulletin no. 137. FAO, Rome

Green BT, Welch KD, Panter KE, Lee S (2013) Plant toxins that affect nicotinic acetylcholine receptors: a review. Chem Res Toxicol 26(8):1129–1138

Gujar GT (1990) Plumbagin, a naturally occurring naphthoquinone. Its pharmacological and insecticidal activity. Fitoterapia LXI(5):387–394

Hare DJ, Morse JG (1997) Toxicity, persistence and potency of sabadilla alkaloid formulations to Citrus thrips (Thysanoptera: Thripidae). J Econ Entomol 90(2):326–332

Hermawan W, Nakajima S, Tsukuda R, Fujisaki K, Nakasuji F (1997) Isolation of an antifeedant compound from *Andrographis paniculata* (Acanthaceae) against the diamondback, *Plutella xylostella* (Lepidoptera: Yponomeutidae). Appl Entomol Zool 32:551–559

Hollinshaus JG (1987) Inhibition of mitochondrial electron transport by hydramethylnon: a new amidinohydrazone insecticides. Pestic Biochem Phys 27:61–70.

Horowitz AR, Mendelson Z, Weintraub PG, Ishaaya I (1998) Comparative toxicity of foliar and systemic application of acetamiprid and imidacloprid against the cotton white fly Bemisiatabaci (Hemiptera: Aleurodidae). Bull Entomol Res 88:437–442.

Ishaaya I (2001) Biochemical process related to insecticide action. In: Ishaaya I (ed) Biochemical sites of insecticide action and resistance. Springer, Berlin, pp 1–16

Ishaaya I, Casida JE (1974) Dietary TH 6040 alters composition and enzyme activity of house fly larval cuticle. Pestic Biochem Phys 4:484–490.

Isman MB (2006) Botanical insecticides, deterrents and repellents in modern agriculture and an in increasingly regulated world. Annu Rev Entomol 51:45–56

Isman MB (2008) Botanical insecticides: for richer, for poorer. Pest Manag Sci 64:8–11

Jeanguenat A (2013) The story of a new insecticidal chemistry class: the diamides. Pest Manag Sci 69(1):7–14

Jefferies PR, Toia RF, Brannigan B, Pessah I, Casida JE (1992) Ryania insecticide: analysis and biological activity of natural ryanoids. J Agric Food Chem 40(1):142–146

Jeschke P, Nauen R (2005) Neonicotinoid insecticides. In: Gilbert LI, Iatrou K, Gill SS (eds) Comprehensive molecular insect biology, vol 5. Elsevier, London

Karmi-Mohajeri S, Abdollahi M (2013) Mitochondrial dysfunction and organophosphorus compounds. Toxicol Appl Pharmacol 270(1):39–44

Kim DH, Ahn YJ (2001) Contact and fumigant activities of constituents of *Foeniculum vulgare* fruit against three coleopteran stored-product insects. Pest Manag Sci 57(3):301–306

Kim S-II, Yoon J-S, Jung JW, Hong K-B, Ahn YJ, Kwon HW (2010) Toxicity and repellency of *Origanum* essential oils and its components against *Tribolium castaneum* (Coleoptera: Tenebrionidae). J Asia Pac Entomol 13(4):369–373

Kostyukovsky M, Rafaeli A, Gileadi C, Demchenko N, Shaaya E (2002) Activation of octopaminergic receptors by essential oil constituents isolated from aromatic plants: possible mode of action against insect pests. Pest Manag Sci 58(11):1101–1106

Koul O, Walia S, Dhaliwal GS (2008) Essential oils as green pesticides: potential and constraints. Biopestic Int 4(1):63–88

Lahm GP, Cordova D, Barry JD, Pahutski TF, Smith BK, Long JK, Benner EA, Holyoke CW, Joraski K, Xu M, Schroeder ME, Wagerle T, Mahaffey MJ, Smith RM, Tong M-H (2013) 4-Azolyphenyl isoxazoline insecticides acting at the GABA gated chloride channel. Bioorg Med Chem Lett 23(10):3001–3006

Lasota JA, Dybas RA (1991) Avermectins, a novel class of compounds: implications for use in arthropod pest control. Annu Rev Entomol 36:91–117

Leighton T, Marks E, Leighton F (1981) Pesticides: insecticides and fungicides are chitin synthesis inhibitors. Science 213:905

Lindstrom J (1997) Nicotinic acetylcholine receptors in health and disease. Mol Neurobiol 15(2):193–222

Matsuda K, Buckingham SD, Kleier D, Rauh JJ, Grauso M, Sattele DB (2001) Neonicotinoids: insecticides acting on insect nicotinic acetylcholine receptors. Trends Pharmacol Sci 22(11):573–580

Matsumura F (1985) Toxicology of insecticides, 2nd edn. Plenum Press, New York

Meyer F, Flotenmeyer M, Moussian B (2013) The sulfonylurea receptor Sur is dispensable for chitin synthesis in *Drosophila melanogaster* embryos. Pest Manag Sci 69(10):1136–1140

Miresmailli S, Isman MB (2014) Botanical insecticides inspired by plant-herbivore chemical interactions. Trends Plant Sci 19(1):29–35

Morgan DE (2009) Azadirachtin, a scientific gold mine. Bioorg Med Chem 17:4096–4105

Mrozik H, Eskola P, Linno BO, Lusi A, Shih TL, Tischler M, Waksmunski FS, Wyvratt MJ, Hilton NJ, Anderson TE (1989) Discovery of novel avermectins with unprecedented insecticidal activity. Cell Mol Life Sci 45(3):315–316

Nauen R (2006) Insecticide mode of action: return of the ryanodine receptor. Pest Manag Sci 62:690–692

Neoh K-B, Hu J, Yeoh B-H, Lee C-Y (2012) Toxicity and horizontal transfer of chlorantraniliprole against the Asian subterranean termite *Coptotermes gestroi* (Wasmann): effects of donor: recipient ratio, exposure duration and soil type. Pest Manag Sci 68(5):749–756

Obeng-Ofori D, Reichmuth CH, Bekele AJ, Hassanali A (1998) Toxicity and protectant potential of camphor, a major component of essential oil of *Ocimum kilmanscharicum* against four- stored product beetles. Int J Pest Manag 44:203–209

Oberlander H, Smagghe G (2001) Imaginal discs and tissue cultures as targets for insecticide action. In: Ishaaya I (ed) Biochemical sites of insecticide action and resistance. Springer, Berlin

Oliver-Bever B (1986) Medicinal plants in tropical West Africa. Cambridge University Press, Cambridge

Orr N, Shaffner AJ, Richey K, Crouse GD (2009) Novel mode of action of spinosad: receptor binding studies demonstrating lack of interaction with known insecticidal target sites. Pestic Biochem Physiol 95(1):1–5

Owusu EO (2001) Effect of some Ghanaian plant components on control of two stored-product insect pests of cereals. J Stored Prod Res 37:85–91

Park Il-K, Lee HS, Lee SG, Park JD, Ahn YJ (2000) Insecticidal and fumigant activities of *Cinnamomum cassia* bark-derived materials against *Mechoris ursulus* (Coleoptera: Attelabidae). J Agric Food Chem 48(6):2528–2531

Park C, Kim SI, Ahn YJ (2003) Insecticidal activity of asarones identified in *Acorus gramineus* rhizome against three coleopteran stored product insects. J Stored Prod Res 39:333–342

Pepper BP, Carruth LA (1945) A new plant insecticide for control of the European corn borer. J Econ Entomol 38:59–66

Peterson GS, Kandil MA, Abdallah MD, Fraq AAA (1989) Isolation and characterization of biologically-active compounds from some plant extracts. Pestic Sci 25:337–342

Pemberton DJ, Franks CJ, Walker RJ, Holden-Dye L (2001) Characterization of Glutamate-gated chloride channels in the pharynx of wild-type and mutant Caenorhabditis elegans Delineates the role of the sub unit Glu Cl-α2 in the function of the native receptor. Mol Pharmocol 59(5):1037–1045.

Powel RG (1989) Higher plants as a source of new insecticidal compounds. Pestic Sci 27:228–229

Prates HT, Santos JP, Waquil JM, Fabris JD, Oliveira AB, Foster JE (1998) Insecticidal activity of monoterpenes against *Rhyzopertha dominica* (F.) and *Tribolium castaneum* (Herbst). J Stored Prod Res 34(4):243–249

Putter I, Mac Connell JG, Preiser FA, Haidri AA, Ristich SS, Dybas RA (1981) Avermectins: novel insecticides, acaricides and nematicides from a soil micro organism. Cell Mol Life Sci 37(9):963–964

Qiao J, Zou X, Lai D, Yan Y, Wang Q, Li W, Deng S, Xu H (2013) Azadirachtin blocks the calcium channel and modulates the cholinergic miniature synaptic current in the central nervous system of *Drosophila*. Pest Manag Sci. doi:10.1002/ps.3644

Rajashekar Y, Gunasekaran N, Shivanandappa T (2010) Insecticidal activity of the root extract of *Decalepis hamiltonii* against stored-product insect pests and its application in grain protection. J Food Sci Technol 43(3):310–314

Rajashekar Y, Bakthavatsalm N, Shivanandappa T (2012a) Botanicals as grain protectants. Pysche 2012(2012):Article ID 646740, 13 pp

Rajashekar Y, Rao LJM, Shivanandappa T (2012b) Decaleside: a new class of natural insecticides targeting tarsal gustatory sites. Naturwissenschaften 99(10): 843–852

Rattan MS (2010) Mechanism of action of insecticidal secondary metabolites of plant origin. Crop Prot 29(9):913–920

Rembold H (1989) Azadirachtins: their structure and mode of action. In: Arnason JT, Philogene BJR, Morand P (eds) Insecticides of plant origin, ACS symposium series. American Chemical Society, Washington, DC, pp 150–163

Rigsby CM, Cipollini D, Amstutz EM, Smith TJ, Yoder JA (2013) Water conservation features of Ova of *Agrilus planipennis* (Coleoptera: Buprestidae). Environ Entomol 42(2):363–369

Salgado VL (1997) The mode of action of spinosad and other insect control products. Down Earth 52(1):35–44

Salgado VL, Sheets JJ, Watson GB, Schmidt AL (1998) Studies on the mode of action of spinosad: insect symptoms and physiological correlates. Pestic Biochem Physiol 60:91–102

Saunders DG, Bret BL (1997) Fate of spinosad in the environment. Down Earth 52:21–28

Saxena RC, Jillani G, Kareem AA (1988) Effects of neem on stored grain insects. In: Jacobson M (ed) Focus on phytochemical pesticides, volume 1: the neem tree. CRC Press, Boca Raton

Sim MJ, Choi DR, Ahn YJ (2006) Vapour phase toxicity of plant essential oils to *Cadra cautella* (Lepidoptera: Pyralidae). J Econ Entomol 99:593–598

Snyder DE, Meyer J, Zimmermann AG, Qiao M, Gissendanner SJ, Cruthers CR, Slone RL, Young DR (2007) Preliminary studies on the effectiveness of the novel pulicide, spinosad, for the treatment and control of fleas on dogs. Vet Parasitol 150(4):345–351

Sparks TC, Crouse GD, Durst G (2001) Natural products as insecticides: the biology, biochemistry and quantitative structure-activity relationships of spinosyns and spinosoids. Pest Manag Sci 57:896–905

Staal GB, Henrick CA, Grant DL, Moss DW, Johnson MC, Rudolph RR, Donahue WA (1985) Cockroach control with juvenoids. In: Hedin PA (ed) Bioregulators for pest control, Symposium series 276. American Chemical Society, Washington, DC, pp 201–258

Stamopoulas DC, Damos P, Karangianidae G (2007) Bioactivity of five monoterpenoid vapours to *Tribolium confusum* (du Val) (Coleoptera: Tenebrionidae). J Stored Prod Res 43(4):571–577

Talukder FA (2006) Plant products as potential stored product insect management agents- a mini review. Emir J Agric Sci 18:17–32

Talukder FA, Howse PE (1995) Laboratory evaluation of toxic and repellent properties of the pitharaj tree, *Aphanamixis polstachya* Wall & Parker, against *Sitophilus oryzae* (L.). Int J Pest Manag 40:274–279

Talukder FA, Islam MS, Hossain MS, Rahman MA, Alam MN (2004) Toxicity effects of botanicals and synthetic insecticides on *Tribolium castaneum* (Herbst) and *Rhyzopertha dominica* (F). Bangladesh J Environ Sci 10(2):365–371

Tapondjou LA, Adler A, Bonda H, Fontem DA (2002) Efficacy of powder and essential oil from *Chenopodium ambrosioides* leaves as post-harvest grain protectant against six-stored product beetles. J Stored Prod Res 38:395–402

Tattersfield F, Hobson RP, Gimingham CT (1929) Pyrethrin I and II: their insecticidal valve and estimation in Pyrethrum (*Chrysanthemum cinebariaefolium*). I. J Agric Sci 19:266–296

Tiwari A, Kumar ML, Saxena RC (1995) Effect of *Nicotiana tabacum* on *Tribolium castaneum*. Pharm Biol 33(4):348–350

Tomizawa M, Casida JE (2005) Neonicotinoid insecticide toxicology: mechanisms of selective action. Annu Rev Pharmacol Toxicol 45:247–268

Tomizawa M et al (1995) Pharmacological characteristics of insect nicotinic acetylcholinereceptor with its ion channel and the comparison of the effect of nicotinoids and neoni-cotinoids. J Pest Sci 20:57–64

Tripathi AK, Upadhyay S, Bhuiyan M, Bhattacharya PR (2009) A review on prospects of essential oils as biopesticides in insect-pest management. J Pharmacogn Phytother 1(5):52–63

Ujvary I (1999) Nicotine and other alkaloids. In: Yamomoto I, Casida JE (eds) Nicotinoid insecticides and the nicotinic acetylcholine receptor. Springer, Tokyo, pp 29–70. ISBN 4-431-70213-x

Varma J, Dubey NK (1999) Perspective of botanical and microbial products as pesticides of tomorrow. Curr Sci 76:172–179

Von Osten JR, Tinoco-Ojanguren R, Soares AMUM, Guilhermino L (2004) Effect of pesticide exposure on Acetylcholinesterase activity in subsistence farmers from Campeche, Mexico. Arch Environ Health 59(8):418–425

Von Stein RT, Silver KS, Soderlund DM (2013) Indoxacarb, metaflumizone, and other sodium channel inhibitor insecticides: mechanism and site of action on mammalian voltage-gated sodium channels. Pestic Biochem Physiol 106(3):101–112

Ware GW, Whitacre DM (2004) The pesticide book, 6th edn. MeisterPro Information Resources, Willoughby

Weaver DK, Dunkel FV, Ntezurubanza L, Jakson LL, Stock DK (1991) Efficacy of linalool, a major component of freshly milled *Ocimum canum* Sims. (Lamiaceae) for protection against post harvest damage by certain stored-product Coleoptera. J Stored Prod Res 27:213–220

Wolfson JL, Shade RE, Mentzer PE, Murdock LL (1991) Efficacy of ash for controlling infestations of *calloso-bruchus maculatus* (F.) (Coleoptera: Bruchidae) in stored cowpea. J Stored Prod Res 27(4):239–243

Yamamoto I (1970) Mode of action of pyrethroids, nicotinoids and rotenoids. Annu Rev Entomol 15:257–272

Yamamoto I (1999) Nicotine to nicotinoids. In: Yamomoto I, Casida JE (eds) Nicotinoid insecticides and the nicotinic acetylcholine receptor. Springer, Tokyo, pp 3–28. ISBN 4-431-70213-x

Yao Y, Cai W, Yang C, Xue D, Huang Y (2008) Isolation and characterization of insecticidal activity of (Z)-asarone from *Acorus calamus* L. Insect Sci 15:229–236

Limitation of Plant Biopesticides

17

Roman Pavela

Abstract

The seeking of new alternatives of synthetic insecticides for the safe environment and health has become an important issue of scientific research which may enable us to obtain safe foods. Although botanical insecticides (BIs) can never entirely replace the amounts of produced synthetic insecticides, they may significantly contribute to seeking the solution of problems associated with application of synthetic pesticides. Three most important arguments support the use of BIs: environmental safety, low or no toxicity for vertebrates and prevention of resistance development. The above-mentioned assets of BIs make us believe in the need of increasing the society-wide efforts leading to further expansion of practical use of these products. However, despite these assets, several limiting factors are associated with BIs that prevent their wider use or restrict their practical applications. This chapter is an effort to critically summarise the advantages and disadvantages of botanical insecticides including major factors that pose limitations to their practical use.

Keywords

Botanical insecticides • Plant extracts • Biopesticides

1 Introduction

Pest control is known as old as agriculture, as there has always been a need to keep crops free from pests. In order to maximise food production, it is advantageous to protect crops from competing species of plants, as well as from herbivores competing with humans. The efforts to minimise agricultural losses caused by insects have motivated gradual development of various techniques aimed at reducing the

R. Pavela, Ph.D. (✉)
Crop Research Institute,
Drnovska 507, 161 06 Praha 6, Czech Republic
e-mail: pavela@vurv.cz

D. Singh (ed.), *Advances in Plant Biopesticides*,
DOI 10.1007/978-81-322-2006-0_17, © Springer India 2014

occurrence of pests. Manual collection of pests and mechanical destruction of infested plants were probably the oldest methods being uneconomical and impractical for modern agriculture. However, efforts of pest control using various inorganic and organic compounds soon emerged.

Chemical pesticides date back 4,500 years, when the Sumerians used sulphur compounds as insecticides. The ancient Chinese, similarly to the Romans, mined arsenic and its first medicinal use can be traced back to 200 B.C. in Shen Nong Ban Cao Jing, the first traditional Chinese medicine book (Liu et al. 2008). However, by 900 A.D., arsenic was being used to control garden pests. Paris green, or copper (II) acetoarsenite, is a rodenticide and insecticide made from copper (II) acetate and arsenic trioxide (Richardson 2000). Its use in the United States as an insecticide dates from 1867, when it was effectively used against the Colorado potato beetle, *Leptinotarsa decemlineata* (Say) (Casagrande 1987). Both arsenic and sulphur have been used as pesticides to the current day. For example, arsenic trioxide and orthoarsenic acid are registered for use as wood protectants against termites, borers and fungal wood rots. By the fifteenth century, heavy metals such as mercury and lead were also being widely used as insecticides and fungicides (Ware and Whitacre 2004) and continued to be used until the mid-twentieth century.

Besides inorganic compounds, various plant extracts, decocts or only crushed plants were also used. The first predecessors of today's sophisticated botanical insecticides (BIs) were thus created. It has been known from preserved written materials that the use of BIs dates back at least 3,000 years. Use of dust from *Chrysanthemum cinerariifolium* and *C. coccineum* flowers was most common in ancient Europe. The use of dust from flowers of these plants does date back far in the history and can be best documented by the decree of King Xerxes I who ordered (at around 470 BC) that the dust from the flowers of *C. cinerariifolium* should be used against louses and fleas. We also know that Vergil (at around 25 BC) described the use of the mixture of nitre and amurca of olives as a good drenching solution for seeds, which improved their germination capacity and growth of the plants. The first record reference of natural substance was that of Marco Polo in 1300 A.D. mentioning use of oil for controlling mange (*Sarcoptes scabiei* var. *cameli*) in camels (Randhawa 1980).

A note of the use of *Derris* sp. extract against pests was preserved from 1649; tobacco extract was commonly used as early as in 1690 as an insecticide, and soap solution was recommended in 1787 to destroy diseases and pests in plants (Dubey 2011; Randhawa 1980).

BIs were used in Europe until the end of World War II when they were replaced with much cheaper synthetic pesticides (e.g. organochlorines, organophosphates, carbamates, and synthetic pyrethroids). Nevertheless, their indiscriminate use has resulted in the development of resistance by pests (insects, weeds, etc.), resurgence and outbreak of new pests, toxicity to nontarget organisms and hazardous effects on the environment endangering the sustainability of ecosystems (Jeyasankar and Jesudasan 2005).

In the recent years the EU has employed a fundamental reform of the Common Agricultural Policy (CAP) highlighting the respect to the environmental, food safety and animal welfare standards, imposing farmlands' cross-compliance with good agricultural and environmental conditions (Schillhorn van Veen 1999). Due to environmental side effects and health concerns, many synthetic carbamate, organophosphate and organophthalide pesticides have been banned (EEC 1991) or are being under evaluation (European Commission 2009). At present times, possible ways have been sought to reduce the consumption of synthetic pesticides while preserving the quality of production and high yields. Biopesticides offer one of such alternatives of plant protection. The use of biopesticides is associated with benefits and drawbacks, similarly as other methods. This chapter is an effort to critically summarise the advantages and disadvantages of botanical insecticides including

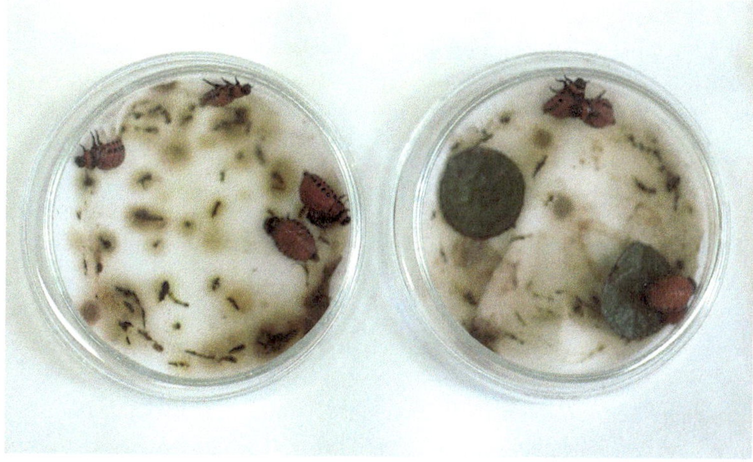

Fig. 17.1 No-choice test with 1 % extracts obtained from seeds of *Leuzea carthamoides* against larvae of *Leptinotarsa decemlineata* after 48 h application

Fig. 17.2 Efficiency of potential botanical insecticides based on extract of *Leuzea carthamoides* and control against larvae of *Leptinotarsa decemlineata* in potato field after 15 days of application

major factors that pose limitations to their practical use.

2 Most Commonly Used Botanical Insecticides

BIs, as they are understood today, are any products that contain secondary plant metabolites as their active substances. Timely application of these products can prevent damage that could be otherwise caused by insect pests. The application of BIs may either kill the pest (the mechanism of action of such BIs causes acute or chronic toxicity) (Pavela 2008, 2009a; Rattan 2010) or modify the behaviour of the pest to such an extent that the pest cannot cause any further damage (the mechanism of action of such BIs shows repellent, antiovipositional or antifeedant effects Figs. 17.1 and 17.2) (Singh and Mehta 1998; Pavela 2007; Koul 2008). BIs are produced by farmers themselves in some regions (especially in countries with a profound tradition of the use of such products such as in India, China and some African countries); the effect of such farmers' products then depends on experience of the grower, preparation method and application itself. In addition, several dozens of commercial products are available today, which involve various formulations and are usually made of historically verified plants.

Quite a lot of information is available about the most common BIs (Isman 1997, 2000, 2006,

2008; Dubey 2011; Koul 2008); in spite of that, below we shall remember the most important information at least.

2.1 Pyrethrum

Pyrethrum is an extract from dried flowers of *Chrysanthemum cinerariifolium* (Trevir.) Vis. (syn. *Pyrethrum cinerariifolium* Trevir., *Tanacetum cinerariifolium* (Trevir.) Schultz. Bip.) (Asteraceae) or *Chrysanthemum coccineum* Willd. (Syn. *C. roseum* Adams, *Pyrethrum roseum* (Adams) M.B., *Pyrethrum carneum* M.B.), and the active principles "pyrethrins" may reach a 3 % content.

Chemical components of the extract are sesquiterpenes, triterpenes and sterols, flavonoids, *n*-alkanes and fatty acids. The "pyrethrins" are meroterpenes (mixed biosynthesis: a terpene-derived unit is attached to a non-terpene moiety), esters of the chrysanthemic or pyrethric acid with ketocyclopentene alcohols: pyrethrolone (pyrethrins I and II), cinerolone (cinerins I and II) and jasmolone (jasmolin I and II). Often available at 25–50 % concentration, pyrethrin I and II are present in greatest amounts (Koul 2008). Pyrethrins show neurotoxic action by blocking voltage-gated sodium channels, prolonging their opening and thereby causing a rapid knockdown effect and death. Synthetic analogues (pyrethroids) have been developed to overcome low stability to UV light without compromising biodegradability and to maintain insecticidal potency while minimising fish toxicity. The structure of modern pyrethroids differs quite a lot from that of model pyrethrins as well as their molecular mode of action (Rattan 2010).

2.2 Nicotinoids

Nicotine, 3-(1-methyl-2-pyrrolidinyl) pyridine, an alkaloid commonly found in the nightshade family (Solanaceae), was the first widely used botanical insecticide. It is naturally extracted from the leaves of tobacco, *Nicotiana tabacum* L. and *N. rustica* L., and where it constitutes approximately 0.6–3.0 % of the dry weight of tobacco (Yamamoto and Casida 1999).

Botanical insecticides based on nicotine are a rapidly acting neurotoxin which penetrates the insect's central nervous system and targets the nicotine acetylcholine receptor (NAChR), directly at the synapse. It mimics acetylcholine, which is a ligand to the NAChR (Rattan 2010). The effect of nicotine in insects is to produce stimulation at low concentrations and blockage at higher concentrations. It thus causes a sequential behaviour in intoxicated insects with twitching, followed by convulsions and paralysis (Perry et al. 1998).

Nicotine is most effective on soft-bodied insects and mites, including aphids, thrips, leafhoppers and spider mites. Commercial tobacco contains other alkaloids such as nornicotine (found specially in *Nicotiana glutinosa* L. and *Nicotiana sylvestris* Speg. Comes) and anabasine (*Nicotiana glauca* being its most important source) and a range of other "minor" related alkaloids (Rosell et al. 2008). Nicotine was first synthesised by Pictet and Crepieux in 1893. It was used as an insecticide (albeit, as synthetic nicotine sulphate) until the second half of the twentieth century. Unfortunately, nicotine is also highly toxic to mammals (e.g. rat and mouse oral $LD_{50}=24$ and 50 mg kg^{-1}, respectively) (Yamamoto and Casida 1999). Related to nicotine in action and partly in structure, the highly insecticidal neonicotinoids originated from screening novel synthetic chemicals. This newest major class of neuroactive insecticides is increasingly used because of their outstanding potency and systemic action for crop protection against pests (Rosell et al. 2008).

2.3 Rotenoids

Rotenone is an isoflavonoid molecule, and the major constituent of insecticidal, acaricidal, and piscicidal cubé resin. It is commonly available as a dust containing 1–5 % of active ingredients from rhizomes or roots of the tropical plants genera *Lonchocarpus*, *Derris* or *Tephrosia* (Fabaceae), or extracts (as resins) usually with up to 45 % total rotenoids. The four major active ingredients of cubé resin are rotenone, deguelin, rotenolone and tephrosin, totalling 77 wt% (Isman 2008).

Rotenone was first isolated from *Lonchocarpus nicou* (Aublet) DC by Geoffroy in 1895 (LaForge et al. 1933). Roark (1932) published a detailed review of derris and its insecticidal properties, covering the period 1747–1931. The first reported use of rotenone-containing plant parts was in the 1800s from Southeast Asia (Indonesia and Malesia), South America and tropical Africa (LaForge et al. 1933) as an aid for catching fish, by means of paralysing them. Rotenone is also produced by plants in the genera *Pachyrhizus* and *Piscidia* (Fabaceae) in North America, *Mundulea* (Fabaceae) in Africa, *Duboisia* (Solanaceae) in Australia or *Verbascum* (Scrophulariaceae) in Europe where it has also been used as a pesticide by indigenous peoples (Betarbet and Greenamyre 2008). The most commonly used source of commercial rotenone is tuba, *Derris elliptica* Benth. Rotenone is obtained from the crystallisation of an extract of root/tuber with organic solvents, and produces its insecticidal action by inhibiting respiratory metabolism, specifically by inhibiting complex 1 of the mitochondrial electron transport chain (Li et al. 2003). It requires ingestion to be an effective poison, although there is evidence it can also penetrate the integument (Perry et al. 1998). Behavioural characteristics of derris intoxication in insects include paralysis of mouthparts and subsequent starvation and inactivity, followed by knockdown, paralysis and generally slow death.

Rotenone has been used as an insecticide and acaricide in dust and bait formulations in horticultural crops against caterpillars, aphids, thrips and spider mites.

2.4 Azadirachtin

Among the natural products, one of the most promising natural compounds is azadirachtin, an active compound extracted from the *Azadirachta indica* A. Juss (neem) tree (Meliaceae). All parts of the *A. indica* tree possess insecticidal activity but seed kernel is the most effective. It has a multitude of pesticidal active ingredients which are together called "triterpene or limnoids". The four best limnoids compounds are: azadirachtin, salannin, meliantriol and nimbin. Azadirachtin itself is a group of compounds such as azadirachtin A–G. Of these, azadirachtin-A (Aza A) is the most plentiful and biologically active one which has shown antifeedant, growth inhibitory, growth regulatory and toxic effects against a number of insect pests and it is generally Aza A that is used for commercial insecticides (Salehzadeh et al. 2003; Anuradha and Annadurai 2008) (Fig. 17.3).

A. indica has certain distinct advantages over most of the other commercially used plants as natural pesticides. Azadirachtin is active against 550 insect species, mostly relating to orders Coleoptera, Dictyoptera, Diptera, Heteroptera, Homoptera, Isoptera, Lepidoptera, Orthoptera, Siphonaptera and Thysanoptera, is the main component responsible for (Mordue and Blackwell 1993). While many subsistence farmers in countries such as India still use crude neem extracts, a number of commercial extracts based on azadirachtin have been developed, in particular NeemAzal T/S (Trifolio-M GmbH Germany) and AzaMax® (Organic Crop Protectants, Lilyfield NSW, Australia). Isman (2004) reported that neem products accounted for one-third of botanical insecticides used in California in 2003. Nevertheless, as Isman (2006) points out, despite all of its apparent advantages, neem has not lived

Fig. 17.3 *Azadirachta indica* – fruits

Fig. 17.4 Plant of cucumber after application of new botanical pesticides based on oil obtained from seeds of *Pongamia pinnata* (*left*) and control plant (*right*) damaged by pests (*Tetranychus urticae*)

up to the publicity and hyperbole about it and is yet to become a mainstream, commercial insecticide.

Azadirachtin is one of the most efficient natural substances with an insecticidal effect. Its efficacy has been found high even at very low dosages, and in addition, it is received very well by the plants also through the root system or through vascular bundles, in which case azadirachtin is built in the vacuoles and provides efficient protection of the plant against phytophagous pests (Pavela et al. 2004) (Fig. 17.4).

2.5 Annonaceous Acetogenins

Tattersfield and Potter (1940) were the first to report detailed investigations to assess the insecticidal properties of extracts of seeds from *Annona* spp., in particular *Annona reticulata* L. and *A. squamosa* L. (Magnoliales: Annonaceae), although McIndoo and Sievers (1924) had much earlier listed five species of *Annona* with insecticidal properties. Pesticide activity of Annonaceae species was attributed to their content of isoquinoline alkaloids. However, the insecticidal activity was traced to a newly isolated type of compound, displaying a long-chain fatty acid structure (usually C-32 or C-34, such as for asimicin). Since then, well over 400 of them have been reported. One major group displays a 4-methyl-2-substituted butenolide (L1, as in asimicin) as g-lactone head. Less common moieties within this group are 4-hydroxy-L1, 4-(hydroxymethylene) butenolide, 3-hydroxybutenolide and 3-methoxy-4-methylenebutenolide. The second major skeletal type displays a 2-(2-oxopropyl)-4-substituted butenolide (L2) with the corresponding butenolide as less common function. The most common heterocyclic fragments are built out of tetrahydrofuran rings [mono-THF, adjacent bis-THF (as in asimicin), nonadjacent bis-THF, adjacent tris-THF] (Esposti et al. 1994; Rosell et al. 2008).

Less common heterocycles are tetrahydropyran or epoxy rings. These compounds are potent inhibitors of complex I (NADH:ubiquinone oxidoreductase) in mitochondrial electron transport systems and slow-acting stomach poisons, a mode of action identical to that of rotenone (Rattan 2010).

Table 17.1 Botanical insecticides approved for use in specific countries

Country	*Chrysanthemum cinerariifolium*	*Derris* spp.	*Nicotiana tabacum*	*Azadirachta indica*	Others
Australia	X	X	–	X	Citrus oils
New Zealand	X	X	–	X	
India	X	X	X	X	Ryania, pongam oil, specified essential oils
Philippines	X	–	–	–	
Hungary	X	–	–	X	Quassia
Denmark	X	X	–	X	Lemongrass, clove, eucalyptus oils
Germany	X	–	–	X	Specified essential oils
Czech	X	–	X	X	Pongam oil, quassia, specified essential oils
Netherlands	X	–	–	X	
Spain	X	–	–	X	Pongam oil, quassia, specified essential oils
United Kingdom	X	X	X	–	Specified essential oils
South Africa	X	–	–	–	Specified essential oils
Brazil	X	X	–	X	Garlic
United States	X	X	X	X	Specified essential oils, ryania, sabadilla
Canada	X	X	X	–	Specified essential oils
Mexico	X	X	–	X	Garlic, capsicum

Source: (Isman 2006) and personal communications
X = in country are registered this botanical insecticide, and – = is not known to use this botanical insecticide

2.6 Essential Oils

An important group of plant secondary metabolites are the essential oils. They are the active principles responsible for odours of plants. Plant essential oils (or their constituents) have been valued as insecticides, owing to their broad-spectrum activity. Direct toxicity, oviposition and feeding deterrence, repellency or attraction appear to result from interaction with the insect nervous system, either by acetylcholinesterase inhibition or antagonism of the octopamine receptors. Essential oils contain liquid, more or less volatile, compounds of many classes of organic substances – acyclic and isocyclic hydrocarbons and their oxygenated derivatives – and in some cases contain S and/or N (Bakkali et al. 2008).

Applications to crop protection (stored product pests), mosquito repellency (citronella oil), control of domestic pests (cockroaches, ants, fleas, etc.), *Varroa* mite control, as aphicides and acaricides (cinnamon oil) and urban insect control (eugenol-based products from basil or clove) have been reported (Regnault-Roger et al. 2005) A range of insecticide products may be also blends of plant oils, such as rosemary, clove, peanut oil, thyme, lemon grass, cinnamon or pennyroyal as components (Isman 2000). Other used botanical insecticides approved for use in specific countries are presented in Table 17.1.

3 Factors that Restrict the Use of Botanical Insecticides

In spite of present legislative and social-economic changes that lead to an effort of gradually reducing the consumption of synthetic pesticides to a necessary minimum, their use may have a negative impact on nontarget organisms (bird, fish and honeybee) in ecosystems, in rates much higher than, for example, fungicides or herbicides (Casida 2012). Moreover, synthetic pesticide residues in foods may pose a threat to human health and especially to the health of children (Paoletti and Pimentel 2000; Roberts et al. 2012). Although BIs can never entirely replace the amounts of produced synthetic insecticides, they may significantly contribute to seeking the solution of problems associated with application of synthetic pesticides. Three most important arguments support the use of BIs:

1. *Environmental safety*: Products based on plant extracts are generally considered as safe for

the environment and nontarget organisms (Dubey 2011). This assumption is based on the fact that most of these products contain secondary plant metabolites, which are subject to relatively rapid degradation in nature and thus do not burden the environment with their residues (Isman 2006; Koul 2008). Active substances of BIs are quite natural and easily degradable in natural ecosystems through common degradation processes.

2. *Low or no toxicity for vertebrates*: Biological efficacy of substances contained in BIs is based on substances of defensive nature that are synthesised by the plants through antibiosis or antixenosis (Tiffin 2000). These substances are synthesised by plants as specialised molecules that have a negative impact only on pathogens and insects feeding on their tissues (Dubey 2011). These substances (in commonly applied doses) are harmless and often rather beneficial for homeothermic animals including man, considering their medicinal effects, which have been scientifically documented (Bruneton 2001).
3. *Prevention of resistance development*: The combination of behavioural and physiological actions of botanical pesticides deters the development of resistance. This is common phenomenon in natural ecosystem where herbivorous insects are controlled by plant allelochemicals (Ratan 2010).

The above-mentioned assets of BIs make us believe in the need of increasing the society-wide efforts leading to further expansion of practical use of these products. However, despite these assets, several limiting factors are associated with BIs that prevent their wider use or restrict their practical applications.

3.1 Factors Restricting Efficacy

The multiple mechanisms of action (see Table 17.2) give phytochemicals unique properties that make it very useful in today's agricultural industry (Rattan 2010). Some phytochemicals work as contact poisons, causing acute toxicity (e.g. pyrethrins, rotenone, monoterpenes). These substances usually block or inhibit some important physiological processes in insects (Rattan 2010), and therefore their mechanism of action is not selective, and upon application, they may have a negative impact also on nontarget organisms, particularly predators and parasitoids found at the place of application at the given time (Simmonds et al. 2000, 2002). Application of these BIs should be directed only against those insect species where there is no principal risk of affecting nontarget organisms (e.g. pests in enclosed areas, storage pests, vectors and troublesome insects). However, types of substances were successfully obtained in recent decades that are selective, given that they act as feeding poisons causing chronic toxicity (e.g. azadirachtin, nimbilin, 20-hydroxyecdysone) or affecting the behaviour of insects (they have repellent, antiovipositional or antifeedant effects). Although highly selective and friendly for nontarget organisms, such products often provide a delayed or insufficient effect only against some development stages of the pest (Simmonds et al. 2002; Isman 2006, 2008; Dubey 2011).

Another factor that restricts efficacy is a relatively short time of persistence of the effect of phytochemicals, which is caused either by rapid biodegradation in the environment (e.g. azadirachtin, pyrethrin) or rapid release in the surrounding environment (e.g. monoterpenes). However, stability in the environment and prolonged persistence time have been studied recently and in some cases even addressed in various formulations of the BIs (Regnault-Roger et al. 2005; Isman 2006; Pavela 2012).

3.2 Factors Restricting the Dose

The application dose of BIs is a limiting factor considering the total costs of application. Some phytochemicals are qualitatively significant and provide their effect already in low application doses or concentrations (pyrethrum, azadirachtin); other phytochemicals are quantitatively significant and their efficacy is manifested only upon higher doses or concentrations (e.g. polyphenols, some terpenoids). The overall consumption of phytochemicals necessary to achieve the desired

Table 17.2 Mechanism of action of some botanical insecticides

System	Mechanism of action	E.g. compound	Plant source
Cholinergic system	Inhibition of acetylcholinesterase (AChE)	Essential oils	*Mentha* spp., *Lavandula* spp.
	Cholinergic acetylcholine nicotinic receptor agonist/antagonist	Nicotine	*Nicotiana* spp., *Delphinium* spp.,
			Haloxylon salicornicum
GABA system	GABA-gated chloride channel	Thymol, silphinenes	*Thymus vulgaris*
Mitochondrial system	Sodium and potassium ion exchange disruption	Pyrethrin	*Chrysanthemum cinerariifolium*
	Inhibitor of cellular respiration (mitochondrial complex I electron transport inhibitor or METI)	Rotenone	*Lonchocarpus* spp.
	Affect calcium channels	Ryanodine	*Ryania* spp.
	Affect nerve cell membrane action	Sabadilla	*Schoenocaulon officinale*
Octopaminergic system	Block octopamine receptors by working through tyramine receptors cascade	Essential oils	*Cedrus* spp., *Pinus* spp., *Citronella* spp., *Eucalyptus* spp.
Miscellaneous	Hormonal balance disruption	Azadirachtin	*Azadirachta indica*

Source: (Rattan 2010)

effect is also a limiting factor in commercialisation of new BIs (Mordue and Blackwell 1993; Bakkali et al. 2008; Dubey 2011).

Applications of some phytochemicals through the root system or injections in conductive tissues therefore seem very promising. In this case, the consumption of the BIs and thus also the price of their application are much lower than those of common application (Pavela et al. 2005). For example, the plants integrate azadirachtin in their own tissues, and thus a single application of BIs can protect the plants for their entire vegetative period (Pavela et al. 2004).

3.3 Factors Restricting Availability

Although the production of BIs has been increasing worldwide, their current market supply is short. Therefore, international research has been supported to seek novel phytochemicals suitable for the development of BIs. Although as many as several dozens of new, highly effective phytochemicals have been found, which are perspective for the development of new BIs (Singh et al. 1989; Singh and Metha 1998; Pavela 2008, 2009a, b, 2011a, b; Chermenskaya et al. 2010; Bakkali et al. 2008), only very few commercial products have been marketed. Isman (1997, 2000, 2004) postulated reasons for this lack of success: in particular the need for sustainable supply of the botanical resource, standardisation of the chemically complex extracts and regulatory approvals. As a result, there are a range of matters to be considered in the development of an interesting discovery into a successful commercial product.

Particularly, the high costs related to the registration of new BIs limit commercialisation of new products. These costs must be reflected by manufacturers in the final price of their products, which then become competitive for synthetic insecticides only with difficulty. Higher prices of these products are given especially by their limited production and thus by the need to grow a sufficient amount of plant materials in the required quality and quantity of the active substances.

4 Current Research

Plants, as long-lived stationary organisms, must resist attackers over their lifetime, so they produce and exude constituents of the secondary metabolism, playing an important role in their defence mechanisms. In fact, the phytochemicals' research has its roots in allelochemistry, involving the complex chemical-mediated interactions between a plant and other organisms in its environment (Rossel et al. 2008).

It has been estimated that approximately 300,000 plant species exist on our planet, while every species is able to synthesise even several dozens of biologically active substances many of which are unique. The purpose of research is thus to find phytochemicals with a novel and sufficiently effective mechanism of action that could be either obtained directly from the plants or that could become a model for the synthesis of their analogues.

Reports of assessment of a wide range of potential pesticides from endemic plants have been reported for a number of countries, including India (Rao 1957; Prakash and Rao 1997), China (Yu et al. 2005), Kyrgyzstan (Chermenskaya et al. 2010), Laos (De Boer et al. 2010) and Eurasia (Pavela 2008, 2009a, b, 2011a, b). In addition, there have been reports of botanical insecticide use in various crops; e.g. rice, wheat, pulses and vegetables (Prakash et al. 2008; Pavela 2012), stored products (Rao 1957; Dubey 2011; Yankanchi and Gadacher 2010) and for veterinary use (George et al. 2008). The results of present research indicate a number of active substances, particularly among terpenoids, phenols and polyphenols (Dubey 2011). Not only substances with direct insecticidal activity (Rattan 2010) can be found among them but also those that modify specific behaviour of insects, and thus those where selectivity to nontarget organisms can be assumed (Koul 2008; Dudey 2011). Such substances are highly perspective in terms of practical use (Figs. 17.5 and 17.6).

5 The Future of Botanical Insecticides

BIs have accompanied the man worldwide since time immemorial. In some regions, the use of plant extracts has a deep-rooted tradition, not impaired by synthetic pesticides as much as in Europe, for example. BIs probably cannot be expected to completely replace the inexpensive, synthetically produced insecticides. However, their application anywhere where the benefits of these products can be used to the full extent would be advisable. Such benefits include particularly their safety for the environment and health. They would thus find their place especially in the protection of those crops or agricultural products where application of synthetic insecticides can be expected to be associated with a high risk of integration of hazardous residues (thus, e.g. in the growing of fruits, vegetables or in the storage of fruits and agricultural commodities) (Jeyasankar and Jesudasan 2005; Isman 2006; Pavela 2012).

Another application of BIs can be found in human and animal protection against medically important vectors or against troublesome insects. Whenever a direct contact of humans or animals with insecticides occurs, it is very important to make sure that the risk of such an application is minimal (Pavela 2008, 2009a, b).

6 Conclusion

BIs have long been touted as attractive alternatives to synthetic chemical pesticides for pest management because botanicals reputedly pose little threat to the environment and to human health. The body of scientific literature documenting bioactivity of plant derivatives to pests continues to expand rapidly, yet only a handful of botanicals are currently used in agriculture in the industrialised world, and there are few prospects for commercial development of new botanical products. Several factors appear to limit the success of botanicals, most notably regulatory barriers and the availability of competing products (newer synthetics, fermentation products, microbials) that are cost-effective and relatively safe compared with their predecessors (Isman 2006). Botanical pesticides presently play only a minor role in crop protection; increasingly stringent regulatory requirements in many jurisdictions have prevented all but a handful of botanical products from reaching the marketplace in North America and Europe in the past 20 years (Isman 2008).

Fig. 17.5 Structures of major active ingredients

Pyrethrin I, R = CH_3
Pyrethrin II, R = CO_2CH_3

Nicotine

Rotenone

Quassin

Azadirachtin A

Ryanodine

Fig. 17.6 Harvest of flowers of *Chrysanthemum cinerariifolium* for producing botanical insecticides

Therefore, clear rules must be designed for marketing BIs, and these rules should be more favourable given the positive aspects of these products. Although BIs represent only a small segment (approximately 1 %) of the international market (Rosell et al. 2008) including products based on microorganisms, they can be viewed as an important alternative of plant protection, and their use should be supported, particularly due to significantly lower risks for the environmental and for our health.

Acknowledgements Financial support for this work was provided by the Ministry of Education, Youth and Sports (No. LH11133).

References

Anuradha A, Annadurai RS (2008) Biochemical and molecular evidence of azadirachtin binding to insect actins. Curr Sci 95:1588–1593

Bakkali F, Averbecka S, Averbecka D, Idaomar M (2008) Biological effects of essential oils – a review. Food Chem Toxicol 46:446–475. doi:10.1016/j.fct.2007.09.106

Betarbet R, Greenamyre JT (2008) Complex I inhibition, rotenone and Parkinson's disease. In: Nas R, Przedlovski S (eds) Parkinson's disease- molecular and therapeutic insights from model systems. Elsevier, New York, pp 745–749

Bruneton J (2001) Pharmacognosy, phytochemistry, medicinal plants, 2nd edn., repr. Paris [u.a.]: Lavoisier Tec, France

Casagrande RA (1987) The Colorado potato beetle: 125 years of mismanagement. Bull Entomol Soc Am 33:142–150

Casida JE (2012) The greening of pesticide–environment interactions: some personal observations. Environ Health Perspect 120:487–493. doi:10.1289/ehp.1104405

Chermenskaya TD, Stepanycheva EA, Shchenikova AV, Chakaeva AS (2010) Insectoacaricidal and deterrent activities of extracts of Kyrgyzstan plants against three agricultural pests. Ind Crop Prod 32:157–163. doi:10.1016/j.indcrop.2010.04.009

De Boer H, Vongsombath C, Pålsson K, Björk L, Jaenson TG (2010) Botanical repellents and pesticides traditionally used against hematophagous invertebrates in Lao People's Democratic Republic: a comparative study of plants used in 66 villages. J Med Entomol 47:400–414

Dubey NK (2011) Natural products in pest management. CAB International, London

EEC (1991) Council Directive 91/414/EEC of 15 July 1991 concerning the placing of plant protection products on the market. Office for Official Publications of the European Communities, OJ No L 230, Brusseuls, 19.8.1991, p 194

Esposti MD, Ghelli A, Ratta M, Cortes D, Estornell E (1994) Natural substances (acetogenins) from the family Annonaceae are powerful inhibitors of mitochondrial NADH dehydrogenase (Complex I). Biochem J 301:161–167

European Commission (2009) Council Regulation (EC) no. 1107/2009 of 21 October 2009 concerning the placing of plant protection products on the market and repealing Council Directives 79/117/EEC and 91/414/EEC. Off J Europ Union 52:1–51. doi:10.3000/17252555.L_2009.309.eng

George DR, Guy JH, Arkle S, Harrington D, De Luna C, Okello EJ, Shiel RS, Port G, Sparaganoa OA (2008) Use of plant-derived products to control arthropods of veterinary importance: a review. Anim Biodivers Emerg Dis Ann N Y Acad Sci 1149:23–26

Isman MB (1997) Neem and other botanical insecticides: barriers to commercialization. Phytoparasitica 25:339–344

Isman MB (2000) Plant essential oils for pest & disease management. Crop Prot 19:603–608. doi:10.1016/S0261-2194(00)00079-X

Isman MB (2004) Factors limiting commercial success of neem insecticides in North America and Western Europe. In: Koul O, Wahab S (eds) Neem: today and in the new millennium. Kluwer Academic Publishers, Dordrecht

Isman MB (2006) Botanical insecticides, deterrents, and repellents in modern agriculture and an increasingly regulated world. Annu Rev Entomol 51:45–66. doi:10.1146/annurev.ento.51.110104.151146

Isman MB (2008) Botanical insecticides: for richer, for poorer. Pest Manag Sci 64:8–11

Jeyasankar A, Jesudasan RWA (2005) Insecticidal properties of novel botanicals against a few lepidopteran pests. Pestology 29:42–44

Koul O (2008) Phytochemicals and insect control: an antifeedant approach. Crit Rev Plant Sci 27:1–24

LaForge FB, Haller HL, Smith LE (1933) The determination of the structure of rotenone. Chem Rev 12:181–213

Li N, Ragheb K, Lawler G, Sturgis J, Rajwa B, Melendez JA, Robinson JP (2003) Mitochondrial complex I inhibitor rotenone induces apoptosis through enhancing mitochondrial reactive oxygen species production. J Biol Chem 10:8516–8525

Liu J, Lu Y, Wu Q, Goyer RA, Waalkes MP (2008) Mineral arsenicals in traditional medicines: orpiment, realgar, and arsenolite. J Pharmacol Exper Ther 326:363–368

McIndoo NC, Sievers AF (1924) Plants tested for insecticidal value or reported to possess insecticidal properties, USDA Bulletin 1201. United States Department of Agriculture, Washington, DC

Mordue (Luntz) AJ, Blackwell A (1993) Azadirachtin: an update. J Insect Physiol 39:903–924

Paoletti MG, Pimentel D (2000) Environmental risks of pesticides versus genetic engineering for agricultural pest control. J Agric Environ Ethics 12:279–303

Pavela R (2007) Lethal and sublethal effects of thyme oil (*Thymus vulgaris* L.) on the house fly (*Musca domestica* Lin.). J Essent Oil Bear Plants 10:346–356

Pavela R (2008) Larvicidal effects of various Euro-Asiatic plants against Culex quinquefasciatus Say larvae (Diptera: Culicidae). Parasitol Res 102:555–559

Pavela R (2009a) Larvicidal effects of some Euro-Asiatic plants against *Culex quinquefasciatus* Say larvae (Diptera: Culicidae). Parasitol Res 105:887–892

Pavela R (2009b) Larvicidal property of essential oils against *Culex quinquefasciatus* Say (Diptera: Culicidae). Ind Crop Prod 30:311–315

Pavela R (2011a) Screening of Eurasian plants for insecticidal and growth inhibition activity against *Spodoptera littoralis* larvae. Afr J Agric Res 6:2895–2907

Pavela R (2011b) Insecticidal and repellent activity of selected essential oils against pollen beetle, *Meligethes aeneus* (Fabricius) adults. Ind Crop Prod 34:888–892. doi:10.1016/j.indcrop.2011.02.014

Pavela R (2012) Efficacy of three newly developed botanical insecticides based on pongam oil against *Plutella xylostella* L. larvae. J Biopestic 5:62–70

Pavela R, Barnet M, Kocourek F (2004) Effect of azadirachtin applied systemically through roots of plants on the mortality, development and fecundity of the cabbage aphid (*Brevicoryne brassicae*). Ptytoparasitica 32:286–294. doi:10.1007/BF02979823

Pavela R, Harmatha J, Barnet M, Vokac K (2005) Systemic effects of phytoecdysteroids on the cabbage aphid *Brevicoryne brassicae* (Sternorrhyncha: Aphididae). Eur J Entomol 102:647–653

Perry AS, Yamamoto I, Ishaaya I, Perry RY (1998) Insecticides in agriculture and environment. Springer, Berlin

Prakash A, Rao J (1997) Botanical pesticides in agriculture. CRC Lewis Publications, Boca Raton

Prakash A, Rao J, Nandagopal V (2008) Future of botanical pesticides in rice, wheat, pulses and vegetables pest management. J Biopestic 1:154–169

Randhawa MS (1980) History of agriculture in India, vol 1. Indian Council of Agricultural Research, New Delhi

Rao DS (1957) The insecticidal property of petals of several common plants of India. Econ Bot 11:274–276

Rattan RS (2010) Mechanism of action of insecticidal secondary metabolites of plant origin. Crop Prot 29:913–920

Regnault-Roger C, Philogène BJ, Vincent C (2005) Biopesticides of plant origin. Lavoisier Publishing, Paris

Richardson HW (2000) Copper compounds. Ullmann's Encyclopedia of industrial chemistry. Published online only doi:10.1002/14356007.a07_567

Roberts JR, Karr CJ, Council on Environmental Health (2012) Pesticide exposure in children. Pediatrics 130:E1765–E1788. doi:10.1542/peds.2012-2758

Rosell G, Quero C, Coll J, Guerrero A (2008) Biorational insecticides in pest management. J Pestic Sci 33:103–121. doi:10.1584/jpestics.R08-01

Salehzadeh A, Akhkha A, Cushley W, Adams RLP, Kusel JR, Strang RHC (2003) The antimitotic effect of the neem terpenoid azadirachtin on cultured insect cells. Insect Biochem Mol Biol 33:681–689

Schillhorn van Veen TW (1999) Agricultural policy and sustainable livestock development. Int J Parasitol 29:7–15

Simmonds MSJ, Manlove JD, Blaney WM, Khambay BPS (2000) Effect of botanical insecticides on the foraging and feeding behavior of the coccinellid predator *Cryptolaemus montrouzieri*. Phytoparasitica 28:99–107. doi:10.1007/BF02981738

Simmonds MSJ, Manlove JD, Blaney WM, Khambay BPS (2002) Effects of selected botanical insecticides on the behaviour and mortality of the glasshouse whitefly Trialeurodes vaporariorum and the parasitoid Encarsia formosa. Entomol Exp Appl 102:39–47. doi:10.1046/j.1570-7458.2002.00923.x

Singh D, Mehta SS (1998) Screening of plant material for repellent and insecticidal properties against pulse beetle (*Callosobruchus chinensis*) and house fly (*Musca domestica*). J Med Arom Plant Sci 20:397–400

Singh D, Siddiqui MS, Sharma S (1989) Reproduction retardant and fumigant properties in essential oils against rice weevil (Coleoptera: Curculionidae) in stored wheat. J Econ Entomol 83:723–733

Tattersfield F, Potter C (1940) The insecticidal properties of certain species of Annona and of an Indian strain of *Mundulea sericea* ("Supli"). Appl Biol 27:262–272

Tiffin P (2000) Are tolerance, avoidance, and antibiosis evolutionarily and ecologically equivalent responses of plants to herbivores? Am Nat 155:128–138

Ware G, Whitacre D (2004) The pesticide book, 6th edn. Meister Media, Willoughby

Yamamoto I, Casida JE (eds) (1999) Nicotinoid insecticides and the nicotinic acetylcholine receptor. Springer, Tokyo

Yankanchi SR, Gadacher AH (2010) Grain protectant efficacy of certain plant extracts against rice weevil, *Sitophilus oryzae* L. (Coleoptera: Curculionidae). J Biopestic 3:511–513

Yu XP, Lu ZX, Chen JM, Xu HX, Zheng XS, Chen LZ, Zhang JF, Shentu XP (2005) Review on research and development of botanical pesticides in China. Acta Agric Zhejiangensis 17:42–48 (in Chinese)

Production and Consumption of Biopesticides

18

Errol Hassan and Ayhan Gökçe

Abstract

The intensive use of synthetic pesticides in pest control activities can cause resistance and therefore resurgence of target pests. Undesirable effects on the environment, including reduction in natural enemies (predators and parasitoids) and beneficial insects, are also possible. A major concern is the effects of synthetic pesticides on human health. In the last few decades, biopesticides have emerged as a potential alternative to synthetic insecticides. Currently, biopesticides share only a small portion of global pesticide market, but growth is faster in this area than in synthetic insecticides. This growth is mainly driven by a rising interest in the demand for organic agricultural products that is most pronounced in western countries. This review will discuss biopesticide history, categories, advantages, disadvantages, conventional and nonconventional extraction technology, and consumption.

Keywords

Essential oils • Environment friendly • Biopesticide history • Extraction technology • Biopesticide global consumption • Biopesticides advantages • Pest management • Biopesticide categories • Botanical pesticides • Natural products • Friendly to natural enemies

E. Hassan, Ph.D. (✉)
School of Agriculture and Food Sciences,
The University of Queensland Gatton,
Gatton, QLD 4343, Australia
e-mail: e.hassan@uq.edu.au

A. Gökçe, Ph.D.
Plant Production and Technologies Department,
Faculty of Agricultural Sciences and Technologies,
University of Nigde, Nigde, Turkey

D. Singh (ed.), *Advances in Plant Biopesticides*,
DOI 10.1007/978-81-322-2006-0_18, © Springer India 2014

1 Introduction

Synthetic pesticides are most commonly used for controlling insect pests, diseases, and weeds. Many farmers prefer synthetic pesticides over other pest control products because of their efficacy and rapidity of action. However, repeated application of synthetic pesticides can result in pest resistance. According to Pedigo (1989), insect pest populations may develop resistance after exposure to an insecticide for several generations, but this is dependent on the insects' life cycle. Insects with a shorter life cycle are more likely to develop resistance to insecticides. Repeated use of synthetic insecticides has frequently disrupted natural biological control systems and led to resurgence of target insect species. This has been due to causing undesirable effects on nontarget organisms, such as important insect predators and (Epstein et al. 2000; Van Hamburg and Guest 1997; Papachristos and Milonas 2008) parasitoids (Borgemeister et al. 1993). Beneficial insects such as earthworms (Reddy and Rao 2008) and pollinators (Nderitu et al. 2007; Loucif-Ayad et al. 2008) have also been affected by application of synthetic pesticides to the detriment of crop species. These effects on nontarget insect species occur mainly because many synthetic insecticides have a broad spectrum of action and long residual activity.

In addition, the large-scale use of chemical pesticides in agriculture may cause adverse effects on humans (Hayes and Laws 1991; Lopez et al. 2005). Besides degrading slowly, many synthetic pesticides are also fat soluble which means they can accumulate in animals, plants, and humans. Consequently, the prolonged use and application of synthetic chemical insecticides has detrimental impacts on the environment and can negatively affect human health (Nayak and Chhibber 2002; Wheeler 2002; Lopez et al. 2005).

Insect pests are notorious for their ability to adapt to control methods. Thus, implementation of an integrated pest management (IPM) system is the best strategy for controlling them. A combination of strategies including crop rotation, cultivar resistance, and biological control helps to keep insect pest populations below economic thresholds. Biopesticides which generally have slow-acting activity when applied independently can be integrated into an IPM system with other techniques to enhance their efficacies (Bailey et al. 2010).

2 History of Biopesticides

For thousands of years, humans have been using various crop protection products to control insects, diseases, and weeds that harm or destroy food crops. Before the invention of synthetic insecticides in the twentieth century, the most likely primary source of pesticides was of natural origin. The first generation of pesticides was probably derived from inorganic materials (sulfur, copper, mercury, arsenic, etc.). The Sumerians were the first recorded using sulfur compounds to kill insects and mites around 2500 BC (Dent 2000). The oldest documentation of the use of plant-based pesticides was found in India and is dated as 2000 BC (Ignacimuthu 2012). In 1200 BC, the Chinese applied mercury and arsenical compounds to control body lice (Dent 2000). At this time, the Chinese also applied insecticides of plant origin for seed treatments (Dent 2000). History records that farmers in the seventeenth century used nicotine in an effort to control plum beetles (Lopez et al. 2011). More recently, extensive uses of botanical insecticides were recorded between the late 1800s and 1940s (Henn and Weinzierl 1989).

The use of botanical insecticides dropped sharply in commercial agriculture following the introduction of synthetic insecticides in the mid-1940s which were less expensive, more effective, and more persistent than their botanical predecessors (Henn and Weinzierl 1989). At the time of World War II, one of the first organochlorine insecticides to be used widely in insect pest control was dichlorodiphenyltrichloroethane, commonly known as DDT. Between 1945 and the 1970s, pyrethrins were used only for household and industrial sprays, whereas nicotine was used for greenhouse/orchards, and rotenone was in limited use for home gardens (Henn and Weinzierl 1989).

The interest in using pesticides of plant origin for crop protection was renewed when the potential negative effects of synthetic insecticides on target and nontarget organisms were revealed. The publication of a book entitled *Silent Spring* by Rachel Carson in 1962 became the most defining moment that led to the banning of a number of synthetic insecticides that were mainly organochlorines such as DDT (McKinlay et al. 2012). It is not surprising that, in the time since this publication, there has been a marked increase in the study of pesticides of plant origin. A rapid increase in the number of studies relating to the development of insecticides of natural, and particularly organic, origin was observed during the 1990s. Studies of plant ecology have led to the identification of various biochemical compounds that act as insect repellents and insect antifeedants. Among the most widely studied plant products are neem oil/extract (Schmutterer 1995) and essential oils (Isman 2000).

There are currently a considerable number of plant-based crop protection products that are available commercially (Cloyd et al. 2009). In addition to plant materials that have a long history of traditional use such as neem, rotenone, and pyrethrum, more recently developed botanical pesticides contain vegetable oils and/or essential oils. Although botanical pesticides are generally lower in efficacy than the synthetic ones, their demands are expected to increase. Organic agricultural products continue to gain popularity, especially in developed countries, indicating that food safety is a major concern for consumers and this has driven the development of biopesticides.

3 Category of Biopesticides

The US Environmental Protection Agency (EPA) categorizes biopesticides into biochemical pesticides, microbial pesticides, and plants containing added genetic material. However, the most common categories of biopesticides are botanicals and microbial pesticides. Botanical pesticides refer to pesticides of plant origin, whereas microbial pesticides refer to pesticides that include microorganisms such as bacteria, fungi, and viruses.

3.1 Microbial Pesticides

Plant insects and pathogens are often naturally infected (for insects) or antagonized (for pathogens) by various kinds of microorganisms. Some common fungi in nature such as *Trichoderma* spp. have the ability to antagonize plant pathogens either through parasitization, antibiosis, or competition, while other fungi, such as *Beauveria bassiana*, act as entomopathogens against a number of insect species. These two fungi are used commercially in plant nurseries, forestry plantations, and field crops for insect pest control (Quarles 2011). A report suggested that *Beauveria bassiana* accounted for about one third of the 171 mycoinsecticides and mycoacaricides available commercially (Faria and Wraight 2007). A number of fungi have also been developed for weed control although only a few of them are available commercially (Chutia et al. 2007).

Other microbial pesticides include those products based on bacteria. These products are developed for the management and control of plant diseases, nematodes, insects, and weeds. The most well-known and widely used microbial pesticide is *Bacillus thuringiensis*, or Bt. It produces insecticidal proteins that are harmful to the larvae of lepidoptera, coleoptera, and diptera. The proteins cause insect pests to stop feeding and finally die. Bt has been proven to be so effective that it has been in commercial use in the agricultural sector for more than 50 years. It was reported that the majority of biopesticides sales involved *B. thuringiensis* (Sanchis and Bourguet 2008).

There are a number of virus formulations available for insect control (Quarles 2011). The most common viruses in use are granulosis virus (GV) and nuclear polyhedrosis virus (NPV). Both are baculoviruses, a family of large rod-shaped viruses. These viruses are active against a number of insect pests. To be active, these viruses must be ingested by target insects. When a susceptible caterpillar, for example, ingests the viruses, the microcapsule dissolves and the virus begins to infect the cells lining the midgut. After a significant buildup of the virus inside the insect's body, symptoms will be noticed such as refusal to eat, and in later developmental stages the cuticle

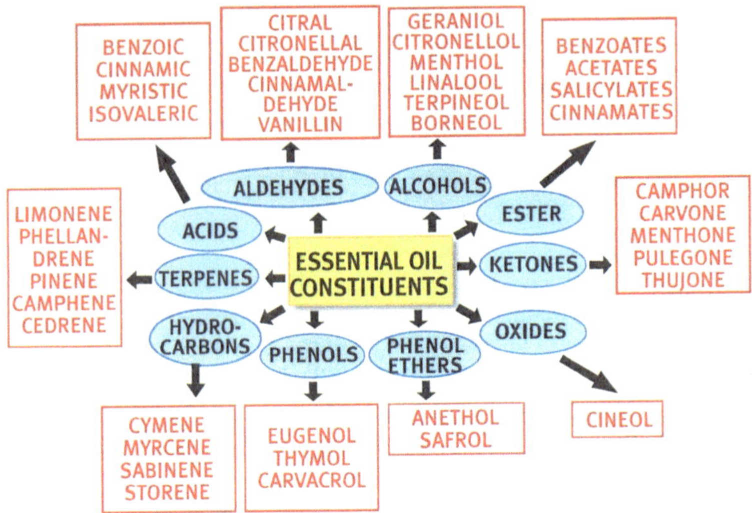

Fig. 18.1 Diverse chemical groups present in essential oils adapted from (Handa 2008)

ruptures easily. It will take 3–8 days for a virus to cause death. An example of a commercially successful viral insecticide is the granulovirus of the codling moth *Cydia pomonella*.

3.2 Botanical Pesticides

In addition to producing primary metabolites such as carbohydrates, proteins, and lipids, plants are known to produce secondary metabolites. A plant secondary metabolite is not directly involved in normal growth, development, and reproduction, but it usually performs an ecological role to govern the relation between plants and other organisms. Some compounds may have bitter taste that act to repel or deter herbivorous insects while other compounds can be toxic. There are three major classes of plant secondary metabolites: terpenes, phenolics, and alkaloids. Azadirachtins, the most active compounds of neem extract/oil, are triterpenoids. Terpenes are made up of carbon and hydrogen. Oxidation of terpenes results in terpenoids. Monoterpenoids and sesquterpenoids are the primary components of plant essential oils. Figure 18.1 shows the different chemical groups found in plants.

Plant secondary metabolites can be extracted as valuable sources of botanical pesticides. Grainge et al (1984) provided a list of plants that have pest control properties such as attractant, antifeedant, repellent, growth inhibitor, and insecticidal. Among those plants, 1,053 are insecticidal, 384 are antifeedants, 297 are repellents, and 32 are growth inhibitors (Ignacimuthu 2012). Of the many plants that have been identified as having insect control properties, only about 30 have commercial potential, and these include neem (*Azadirachta indica* A. Juss), tobacco (*Nicotiana tabacum* L.), pyrethrum (*Tanacetum cinerariaefolium* (Trevir.) Sch.), sabadilla (*Sabdilla officinarum* Brandt Ratzeb), derris (*Derris chinensis* Benth), basil (*Ocimum tenuiflorum* L.), jatropha (*Jatropha curcas* L.), spearmint (*Mentha spicata* L. and *M. arvensis*), garlic (*Allium sativum* L.), and eucalyptus (*Eucalyptus globulus* Labill) (Ignacimuthu 2012).

One of the most widely used phytochemicals is azadirachtin which is derived from the neem tree (*Azadirachta indica*). This bioactive compound affects the reproductive and digestive processes of a number of important pests. Azadirachtin's structure is similar to the natural insect-molting hormone ecdysone. Azadirachtin interrupts the ability of the insect pest to undergo metamorphosis (Hassan 1999; Bhuiyan, et al. 2001) and can cause sterility of emerged adults. Immature insects that ingest azadirachtin may molt prematurely or die. Deformation of adults often occurs due to hormonal disturbance.

4 Advantages and Disadvantages of Biopesticides

4.1 Advantages

When exposed to sunlight, plant-based compounds breakdown easily into nontoxic substances within hours or days. Therefore, food contamination by biopesticides is less likely to happen. Contamination of soil and groundwater is a rare case except for in rotenone-based botanical insecticides.

Insect resistance to biopesticides is less likely to occur because the complexity of mixtures in the plant extracts and oils can act to disrupt the selection process (Berenbaum and Zangerl 1996; Scott et al. 2003). A number of plant products are fast acting, thereby preventing insect feeding and ultimately avoiding further crop damage. Furthermore, because many plant-based biopesticides act on the insect's gut enzymes and breakdown faster in the environment, they can be selectively applied to insect pests and are therefore safer to nontarget organisms (Hedin and Hollingworth 1997; Krishnamoorthy and Atmakuri 1999; Schmutterer 1997).

Biopesticides typically have multiple modes of action and do not rely on a single target site for efficacy. This situation causes insects to have a low tolerance to botanical insecticides as their detoxification enzymes can have difficulty in metabolizing the mixture (Berenbaum and Zangerl 1996; Scott et al. 2003). For example, azadirachtin (from neem) has been in long-term use as an insecticide for controlling crop and storage pests with no evidence of tolerance in target organisms found to date (Kraiss and Cullen 2008; Rharrabe et al. 2008). The responsible and safe use of biopesticides has the potential to extend the effective field life of products by curtailing the development of resistant insect pest populations.

Biopesticides provide growers with valuable tools to manage insect pests in an environmentally friendly manner. Generally, biopesticides have been shown to have limited adverse effects on the environment, flora and fauna, and the wider public (Brown 1978; Georghiou and Saito 1983; Hayes and Laws 1991; Lopez et al. 2005). Other positive aspects of using biopesticides are their ability to help maintain beneficial insect populations, their nonhazardous application, and their effective use in resistance management programs.

Today's consumers are more aware and knowledgeable about their health and the foods that they eat. To meet the growing demand for producing high-quality fruit and vegetables year-round, growers are mindful that consumers view produce grown with less or no synthetic chemical inputs as safer to eat, healthier, and friendlier to the environment. Consequently, organic farming has expanded greatly in the USA, Europe, and other countries.

Most biopesticides worldwide are exempt from residue limits on fresh and processed foods. When growers use biopesticides, either alone or together with reduced application of conventional pesticides, consumer exposure to regulated pesticide residues is reduced. This provides benefits to food producers, grocery retailers, and consumers. Essentially, these factors provide a solid foundation for the provision of wholesome food and vegetables to meet the supply and demand of today's sophisticated consumers.

Insecticides of plant origin include different compounds that provide various modes of action that can be targeted toward specific insect pests. Biopesticides generally do not act or harm against beneficial insect populations. It is important to protect beneficial insects and animals as they help pollinate plants and can be natural predators of the insects infesting the crop. For example, bees, ladybird beetles, butterflies, birds, and parasitic wasps fall into this category. Maintaining natural enemy populations against insect pests, biopesticides play a significant role in integrated pest management (IPM) programs. This approach in combination with cultural and biological controls keeps beneficial insects healthy and helps keep insect pest populations below economic threshold.

4.2 Disadvantages

Rapid degradation of most botanical insecticides can be considered as an advantage, but consequently more frequent application is required. Commercial formulations of botanical insecticides are often more expensive than the synthetics mainly due to the high cost of raw materials (labor intensive in collecting neem seeds) and extraction processes in order to optimize the quantity of active compounds. Therefore, investigation of the insecticidal properties of cheap and readily available natural products such as vegetable oils is strongly encouraged. It is important to note that crop protection products must be readily available with assured continuity of the supply before they are adopted on a broad scale. The quantity of plant secondary metabolites often varies between seasons and locations so it is not easy to standardize the quality of botanical pesticides.

5 Extraction Techniques for Biopesticides

Plant extraction is the separation or isolation of particular chemicals from plant tissue using selective solvents in a range of procedures. Extraction techniques separate the soluble plant metabolites and leave behind the insoluble cellular material. Extraction is a critical step in exploring bioactivity of phytochemicals. Plants produce complex mixtures of metabolites including alkaloids, glycosides, terpenoids, flavonoids, and lignans. Various techniques are available to further isolate fractions or individual compounds from the crude extracts. Essential oils are plant products usually obtained by distillation or other extraction methods. Figure 18.2 shows plant organs containing essential oils.

The quality of a plant extract is determined by the plant parts, extraction solvents, extraction techniques (extraction technology), and type of equipment used. Figure 18.3 shows special structure of the plant tissue where essential oils are usually available. A number of methods are available for phytochemical extraction, and the

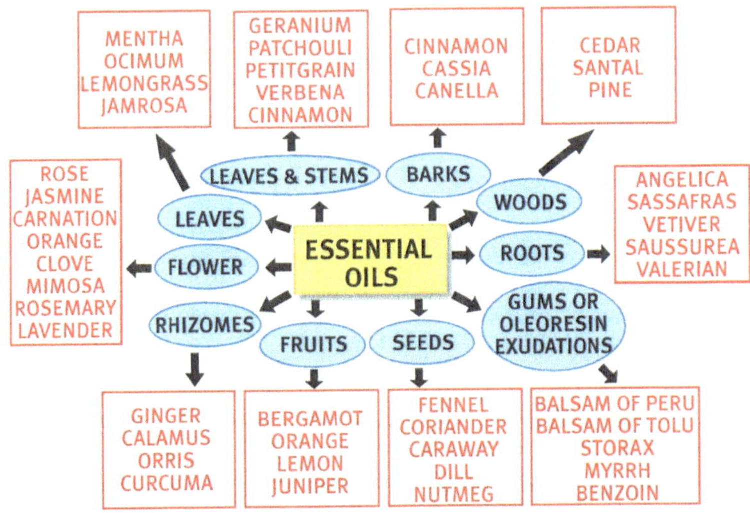

Fig. 18.2 Plant organs containing essential oils (Handa 2008)

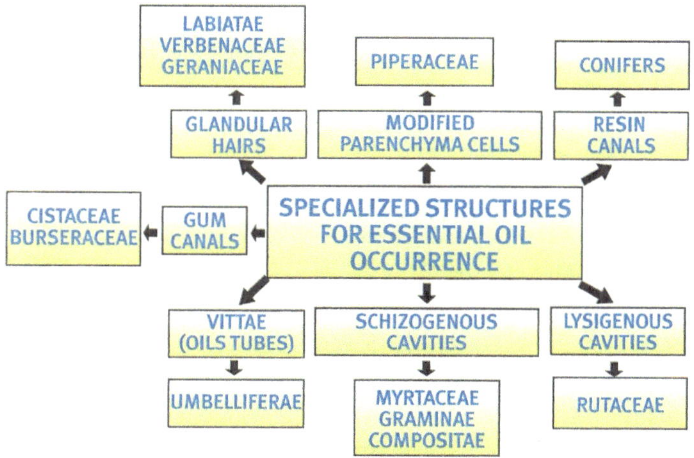

Fig. 18.3 Family-specific plant tissues responsible for producing essential oils (Handa 2008)

choice among them is determined by the physicochemical properties and stability of the phytoconstituents to be obtained. Figure 18.4 shows general methods for producing essential oils from plant materials. For the extraction of volatile compounds such as essential oils, the simplest methods are hydro-distillation and steam distillation, while for nonvolatile compounds, they are cold fat extraction, expression, maceration, and solvent extraction. More advanced extraction technologies are available including supercritical fluid extraction.

5.1 Maceration

Maceration is an inexpensive way to extract essential oils and other bioactive compounds. This extraction technique involves several steps (Azmir et al. 2013; Handa 2008): Firstly, plant materials are ground into small particles to increase surface areas for optimal mixing with solvents and then an appropriate solvent is added in a closed vessel. The vessel is kept at room temperature for at least three days or until the soluble matter has dissolved (constant shaking is often needed). Following this, the liquid is strained off and the marc (solid material) is pressed and then the combined liquid is filtered. To increase solvent efficiency, gentle heat may be used during the extraction process if this is not destructive to the bioactive compounds (Handa 2008).

There are several factors that need to be considered when performing extractions with maceration techniques (Singh 2008):

1. Continuous hot extraction should be avoided when phytochemicals are susceptible to high temperature.
2. Solvent selection depends on the solubility of the phytochemicals.
3. Solvent can be recovered under reduced pressure to minimize evaporation temperatures so that thermolabile compounds can be retained.
4. For a large-scale extraction, it is very important to improve the efficiency of extraction to minimize solvent requirements.

Circulatory extraction is an example of an improved maceration procedure (Singh 2008). In circulatory extraction, the solvent is circulated continuously through plant materials by pumping it from the bottom of the vessel (through an outlet) and then redistributing it over the surface of the plant materials.

5.2 Hydro-distillation

Hydro-distillation is a common method to isolate essential oils from plant materials. Three types of hydro-distillation are available: water distillation,

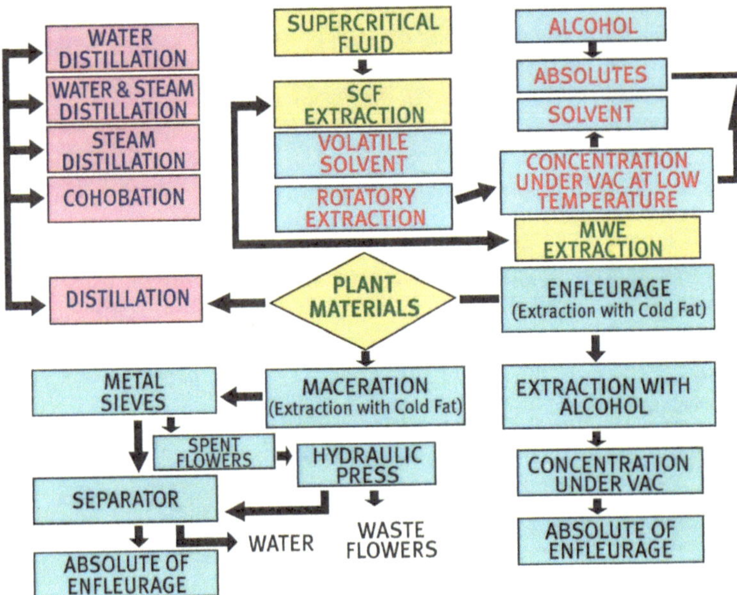

Fig. 18.4 Methods of extraction of essential oils from plants (Handa 2008)

water and steam distillation, and direct steam distillation (Handa 2008).

5.2.1 Water Distillation

In water distillation, the plant material is placed inside a still compartment and immersed in water that is then brought to boil by heating. This process frees plant biochemicals, particularly volatile compounds (essential oils), from the plant tissue. The vapor is then condensed by indirect cooling with water. The distillate is then filtered through a separator to separate oil from distillate water. The advantages of this type of distillation are as follows: it can be used for finely powdered material or plant materials that normally form lumps when contacted with steam, and the still compartments are inexpensive, easy to build, and suitable for field operations. The disadvantages of water distillation are as follows: it is impossible to extract all of the essential oils and constituent chemicals may be hydrolyzed (certain esters) or polymerized (aldehydes). This process is higher in cost than other distillation techniques as a larger volume of material must be processed to extract a similar volume of essential oils making this technique time and resource intensive. Therefore, water distillation is used only when other hydro-distillation techniques (water and steam distillation and direct steam distillation) are unsuitable.

5.2.2 Water and Steam Distillation

The equipment used for water and steam distillation is similar to that used for water distillation. The difference is that the plant material is placed above the boiling water on a perforated grid in water and steam distillation. When compared with water distillation, this method produces a higher yield of oil with less hydrolysis and polymerization while being faster and more energy efficient. However, this distillation technique reduces the capacity of the still.

5.2.3 Direct Steam Distillation

In direct steam distillation, the steam used to distil plant materials is generated from a boiler placed outside the still. This distillation method is advantageous because the released steam can easily be controlled and the temperature generated will not exceed 100 °C so thermal degradation of phytochemicals is less likely to occur. The disadvantage is that the equipment required is more expensive than the other two distillation techniques. However, as it can accommodate a large-scale oil production, this distillation technique is most widely utilized.

5.3 Expression

Expression refers to a physical process, usually achieved by a combination of crushing and pressing, that forces the plant material to release its oil. This method works well for plant materials that contain high quantity of oil such as seeds. Neem oil and many vegetable oils can be extracted by this method. Citrus or lemon essential oils can also be extracted from fruit peel using expression. The expression method often generates heat through friction that may damage the quality of oil.

5.4 Cold Pressing

Cold pressing is a process of expression in which temperatures are kept as low as possible to maintain the natural structure of the oil and therefore preserve quality. There is still no global agreement in the standard maximum temperature of cold pressing. Some suggest that it should be no more than 27 °C, but the others have a standard of no more than 49 °C. In this extraction method, stainless steel presses are commonly used for pressing and grinding.

5.5 Hot Continuous Extraction (Soxhlet)

Soxhlet extraction has been widely used to isolate phytochemicals. The general procedures are described in Azmir et al. (2013) and Handa (2008) as follows: (1) a small amount of powdered plant material is placed in a thimble (a porous bag), which is placed in a designated chamber of the soxhlet apparatus; (2) solvent is placed in a glass flask at the bottom of soxhlet apparatus that is heated; (3) the vapor that is produced is then condensed so that the solvent drips into thimble; (4) when the chamber of the thimble becomes full, the solvent flows back into the solvent flask; and (5) this process continues until the plant material is fully extracted.

5.6 Microwave-Assisted Extraction (MAE)

In microwave-assisted extraction, the following process occurs (Pangarkar 2008): (1) there is a direct transfer of heat to plant material that causes instantaneous heating of moisture in the solid; (2) the heated moisture evaporates and creates a high vapor pressure; and (3) the vapor pressure breaks cell walls causing them to release oils. The advantages of MAE are reduced heat degradation, reduced processing cost, reduced extraction time, much lower energy usage, and much lower solvent usage (Pangarkar 2008).

5.7 Supercritical Fluid Extraction (SFE)

Supercritical fluid extraction (SFE) is a new important alternative to conventional extraction methods such as hydro-distillation and organic solvent extraction. This new extraction method does not cause hydrolysis or degradation of thermolabile compounds as often occurs with hydro-distillation or conventional solvent extraction. With SFE, there is no need to remove expensive solvent via evaporation.

A supercritical fluid refers to any substance in a stage where distinct liquid and gas phases do not exist. This occurs when the substance's temperature and pressure are above its critical point. The most commonly used supercritical fluid is CO_2. The critical temperature of CO_2 is 31 °C and the critical pressure is 73.825 bars.

Supercritical fluid extraction results in extracts with low viscosities and high diffusiveness that are easily recovered as the supercritical fluid can return to gas phase by simply depressurizing, leaving little or no solvent residue. The main disadvantage is that the investment cost for SFE is higher than the traditional atmospheric extraction techniques (Martin et al. 2012). Another disadvantage is that CO_2 has low polarity and is more effective for extraction of nonpolar substances. However, this limitation has been addressed by the use of chemical modifiers that can increase polarity (Azmir et al. 2013). The common modifiers are alcohols (1–10 % of the supercritical fluid), which allow the mixtures to dissolve into more polar compounds (Martin et al. 2012). SFE is applicable for the extraction of numerous important phytochemicals such as pyrethrins, azadirachtins, rotenone, and also essential oils (Martin et al. 2012).

5.8 Pressurized Liquid Extraction (PLE)

Pressurized liquid extraction or PLE is a new extraction method conducted under elevated temperature and pressure. This method was first reported by Richter et al. (1996). PLE is less time intensive and requires less solvent when compared with traditional solvent extraction techniques (Richter et al. 1996). PLE also provides a potential alternative to supercritical fluid extraction for the extraction of polar compounds (Kaufmann and Christen 2002).

A study by Sae-Yun et al. (2006) revealed that PLE had higher extraction efficiency in extracting rotenone (from *Derris elliptica* Benth and *Derris malaccensis* Prain) compared with the conventional maceration method. Time and solvent consumption during extraction was reduced from 72 h and 10 ml/g of dried sample using maceration to only 30 min and 3 ml/g of dried sample via PLE.

5.9 Enzyme-Assisted Extraction (EAE)

Enzyme-assisted extraction (EAE) can be used to extract oils from plant parts such as seeds. Cellulase, amylase, and pectinase are among the enzymes that are most frequently used for oil extraction (Rosenthal et al. 1996). Two common approaches are available for this extraction method, namely, enzyme-assisted aqueous extraction (EAAE) and enzyme-assisted cold pressing (EACP) (Latif and Anwar 2009). In EAAE, the enzymes help to degrade the seed cell walls and rupturing the polysaccharide-protein colloid; however, in EACP, the enzymes only facilitate the hydrolysis (Latif and Anwar 2009). Little is known on the use of enzyme-assisted extraction for phytochemicals with pest control properties. However, these extraction methods are expected to have a role in the near future since many plant vegetable and essential oils are currently used as biopesticides. Several studies revealed that enzyme-assisted extraction methods can increase the yield of oil with only little change on the quality (Latif and Anwar 2009; Sowbhagya et al. 2009).

6 Production of Biopesticides

Agriculture sectors face many obstacles in producing profitable crops. One of the primary obstacles is pest pressure in agroecosystems. High pest control standards in crop production have led the industry to rely on conventional pesticides that result in increased chemical residues at harvest, increased development of resistance, and consequently, increased cost to growers. The Food Quality Protection Act (1996) encouraged the reduction in the use of conventional pesticides in food production for human consumption and also encouraged farmers to use more environmentally friendly pesticides, such as biopesticides.

The global biopesticides market has been growing since the mid-1990s and is expected to reach $3.2 billion by 2017 (Global Biopesticides Market – Trends and Forecasts (2012–2017)) and Thakore (2006). In this growth, organic farming and chemical-free crops are the key factors encouraging international companies to add more biopesticides to their product ranges. This trend is reflected in the recent increase in biopesticide product registrations in the USA from 175 registered in 1998 to 452 in 2009 (Regnault-Roger 2012). In Canada, 86 products were registered for controlling agricultural and domestic pests in 2010 (Agriculture and Agri-Food Canada 2010). In recent years, more biopesticides have been approved compared to conventional pesticides by authorities. For instance, the US Environmental Protection Agency approved eight biopesticides in 2011 of which half were microbial and the other half botanical (Agropages.com 2012). In China, insecticidal microbial biopesticides are the leading group with 13 registrations, followed by 6 fungicides and 2 nematicides (Agrow 2013; Zhang et al. 2011).

Microbial pesticides include bacteria, fungi, viruses, protozoa, and nematodes and are the leading category in production and consumption of biopesticides. The major companies that dominate the microbial biopesticides market include: AG Biotech Australia Pty Ltd., AgraQuest Inc., Bayer Crop Protection Gmbh, BioWorks Inc., BionTech Inc., Certis USA LLC, Greeneem, Isagro SpA, Kumiai Chemical Industry Co. Ltd., Marrone Bio Innovations, Koppert B.V., Prophyta Biologischer Pflanzenschutz GmbH, San Jacinto Environmental Supplies, Sumitomo Chemical Co. Ltd., Syngenta International AG, Troy Biosciences Inc., and Valent Biosciences Corporation. Microbial pesticides accounted for approximately 63 % of the global biopesticide market with *Bacillus thuringiensis* and its subspecies being the most widely used (marketsandmarkets.com 2012). The USA, China, Russia, and India are the leading producers of microbial pesticides (Leng et al. 2011). *Bacillus subtilis*, *Trichoderma gamsii*, and *Trichoderma harzianum* are the most popular microbial biopesticides in the USA, and their market share significantly increased in recent years (Agropages.com 2013a). In China, a total of 23 registered microbial pesticides (9 bacteria, 7 fungi, and 7 viruses) are sold, and these are manufactured by 134 different companies (Agrow 2013). Chinese biopesticides companies' cumulative sales were 6.2 billion yuan, and this amount accounted for 12 % of the total pesticide sales in 2010 (Agropages.com 2011). While in India, the total registered biopesticide products numbered 12. Leng et al. (2011) assert that there are 410 Indian companies manufacturing biopesticides, with 130 of these operated by private sector (Leng et al. 2011).

Several botanical biopesticides, such as pyrethrins, azadirachtin, rotenone, and essential oils, are already widely used in specialty crop production, particularly for organic production. These compounds are usually considered relatively safe for humans and the environment. The unique mode of action of botanical biopesticides promises potential for their use in rotation with other reduced-risk pesticides. Microbial biopesticides are dominating the biopesticide market, but new active compounds are expected to contribute to the growth of botanical biopesticide sales.

Pyrethrum, azadirachtin (neem), rotenone, and essential oils are four major types of botanical biopesticide that dominate the world biopesticide market. Kenya and Australia are the leading pyrethrum-producing countries, with Kenya alone producing 70 % of pyrethrum world production (Cavoski et al. 2011). Pyrethrum is also grown in Tanzania and Ecuador but their productions are small quantities (Attia et al. 2013). There are many formulations of pyrethrum used in agricultural pest management, public health, and structural pest control (Table 18.1). Pyrethrum sales account for 80 % of total global botanical insecticides (Isman 2005).

Azadirachtin is the active compound of neem oil that is extracted from the seeds of the neem tree (*Azadirachta indica*). The neem tree is native to India and Burma and has also been planted widely in Asia, Africa, and Australia. India has the largest neem tree plantations with 20 million rootstocks and accounting for 60 % of the entire neem tree population (Agropages.com 2013b). The total neem oil production of India is approximately 2.5 lakh (Indian measurement) tonnes. However, this number may rise to 7 lakh tonnes with some development programs. India exported US$ 5.73 million neem-based products in 2012 to the leading importers USA, Italy, Japan, and Spain (Agropages.com 2013b).

Traditionally, aromatic plants were used in pest control. Mint, thyme, clove, and rosemary oils and their constituents are very effective against stored product insects and some important plant diseases and weeds. Tables 18.2 and 18.3 show lists of selected, commercially available essential

Table 18.1 Examples of commercially available botanical insecticides and nematicides[a]

Active ingredient	Plant species	Company	Product name
Azadirachtin	*Azadirachta indica*	ThermoTrilogy	Neemix 90 EC, Neemazid, Trilogy 90 EC, Triact 90 EC, Bioneem, Margosan-O, Azalin Align, Turplex, Bollwhim
		Fortune	Fortune Aza, Fortune Biotech
		Agridyne	Azatin
		Karapur Agro	Neem Suraksha, Proneem, Neem Wave, Aza Technical
		Trifolio-M	NeemAzal
		Krishi Rasayan	Kayneem
		Rallis	Neemolin
		Consep	Surefire Neemachtin
		Stanes and PBT International	Nimbecidine
		EID Parry and Andermatt	NeemAzad
		Cyclo	Neememulsion
		Rallis	Neemitox, Neemolin
		RPG	Blockade
		Agro Logistics	Neem Cake Agroneem
		Gowan	AZA-Direct
		PBI/Gordon	EI-783, EI-791, Azatrol EC
		JB Chemicals	Jawan
		Biostadt	Neemactin
		Agri Life	Margosom
		Tagros	Trineem
		Biocontrol Network and Safer	Bioneem
		Biocontrol Network	NeemPlus Liquid
		Organica	K+ Neem
		SOM Phytopharma	AquaNeem
		Vipesco	Vineem
Nicotine	*Nicotiana tabacum*	United Phosphorus Ltd	Stalwart
		Hortichem	No-Fid
		Vitax	XL-All Nicotine
		Dow AgroSciences	Nicotine 40 % Shreds
		Bonide	Tobacco Dust
		United Phosphorus Ltd	Nico Soap
Plant pelargonic and related fatty acids		Dow AgroSciences	M-Pede, De-Moss, Scythe I, Akzo-Nobel, Thinex

(continued)

Table 18.1 (continued)

Active ingredient	Plant species	Company	Product name
Pyrethrins	*Chrysanthemum cinerariaefolium*	Prentiss	Prentox Pyrethrum Extract, ExciteR
		Syngenta	Alfadex
		MGK	Pyrocide, Evergreen, Premium Pyganic 175, Pyganic Crop Protection EC 1.4, Pyganic Crop Protection EC 5.0
		Frunol Delicia	Milon
		Agropharm	Pycon
		Consep	CheckOut
		Kemio	Hash
		Wilbur-Ellis	Py-rin Growers Spray
		Diatect	Diatect II, Diatect III, Diatect V
		Bonide	Bonide Liquid Rotenone-Pyrethrin Spray, Garden Dust, Earth Friendly Fruit Tree Spray/Dust
		Woodstream	Safer Yard
			Garden Insect Killer
		Natural Animal Health Products	Ecozone Pyrethrum Insect Powder
Rotenone	*Derris* spp., *Lonchocarpus* spp., *Tephrosia* spp.	Tifa	Chem Sect, Cube Root, Chem-Fish, Rotenone Extract
		Vipesco	Vironone
		Penick	PBNox
		Wright Webb	Pyrellin
		Prentiss	Prenfish, Noxfish, Nusyn-Noxfish, Synpren Fish
		Bonide	Rotenone 5 %, Bonide Liquid Rotenone-Pyrethrin Spray, Garden Dust, Rotenone-Copper Dust
Sabadilla	*Schoenocaulon officinale*	Dunhill Chemical	Veratran D, Veratran
Ryania	*Ryania speciosa*	AgriSystems International	Natur Gro R-50, Natur Gro Triple Plus
		Dunhill Chemical	Ryan50
Starch syrup		Kyoyu Agri	YE-621
Capsaicin	*Capsicum frutescens*	Soil Technologies	Armorex, Nemastroy, Valoram
		Bonide	Bonide Hot Pepper Wax
		Champon	Dazitol
		Hot Pepper Wax	Hot Pepper Wax
		Biocontrol Network	Hot Pepper Wax Animal Repellent

(continued)

Table 18.1 (continued)

Active ingredient	Plant species	Company	Product name
Cinnamaldehyde	*Cassia tora*	Monterey	Vertigo
		Proguard	Cinnacure
Phenethyl propionate	*Mentha piperita*	Spectrum	Bag-a-Bug Japanese Beetle Trap, Japanese Beetle Combo Bait
		Bioganic	Ecozap WaspHornet Insecticide, Ecozap Crawling and Flying Insecticide, Bioganic Flying Insect Killer
		EcoSmart	EcoExempt HC, Matran, Ecopco D Dust Insecticide, Ecopco AC, Ecopco Jet
		Arbico	Ecosafe
		Trece	Japanese Beetle Bait II, Trece Japanese Beetle Trap
		Woodstream	Ringer Japanese Beetle Bait
		Suterra	Surefire Japanese Beetle Trap
		Vipesco	Vizubon-D
Garlic extract	*Allium sativum*	ECOSpray Ltd	Nemguard
Saponin			Nema-Q
Plant-derived porphyrin derivative	*Quassia amara*		Nemastop

[a]Data adapted from Cooping and Duke (2007) and Dayan et al. (2009)

oil-based fungicides and herbicides. Definitive lists are difficult to compile as aromatic plants are grown all over the world and there is no accurate total harvest of these crops.

7 Consumption of Biopesticides

Biopesticides contribute only a small percentage of the total pesticide market in the world; however, it is expected their use will rise faster than synthetic insecticides. In 2011, the global biopesticide market was valued at US$1.3 billion (Marketsandmarkets.com 2012), whereas the total pesticide market (biopesticides and synthetic pesticides) was valued at US$37.5 billion (BCC Research 2012). The global market value of biopesticides is predicted to reach US$2.1 billion in 2012 and US$3.7 billion in 2017, giving a compounded annual growth rate (CAGR) of 12 % (BCC Research 2012). On the other hand, the global synthetic pesticide market value was predicted to reach US$44 billion in 2012 and US$61.5 billion in 2017, giving a CAGR of only 7 % (BCC Research 2012). Increasing demand for biopesticides from the USA and European countries is expected to continue because of market growth in organic vineyards, vegetables, and tree crops. For example, Table 18.4 shows the biopesticides usage in California from 2012 to 2013. It is interesting to see that the use of vegetable oils increased every year. This growth is reflected in the world with areas devoted to organic farming estimated to be at around 22 million hectares with 7.287 million hectares in Europe (Regnault-Roger 2012).

Table 18.2 Commercially available botanical herbicides[a]

Active ingredient	Plant species	Company	Product name
Clove oil	*Syzygium aromaticum*	EcoSmart	Matran EC
		St. Gabriel Laboratories	BurnOut II concentrate
		St. Gabriel Laboratories	Poison Ivy Defoliant
Fatty acids	Several plant species	Monterey	M-Pede
Pelargonic acid	Geraniaceae family members	Amvac	Hinder
		Neudorff	Quik-RTU
		Nufarm	Neo-Fat
		Russell	Naturell WK herbicide
		Otsuka	Oleate
		Koppert	Savona
		Neudorff	Neu 1128
		Dow Agrosciences	Scythe I
Phenethyl propionate	*Mentha piperita*	EcoSmart	EcoSmart HC
		EcoExempt	EcoExempt HC
		Bioganic	Bioganic Weed and Grass Killer
Pine oil	*Pinus* spp.	Certified Organics Ltd	Interceptor
Citronella oil	*Cymbopogon* spp.	Barrier Biotech Ltd	Barrier H

[a]The data in the table gathered from Cooping and Duke (2007) and Dayan et al. (2009)

Table 18.3 Commercially available botanical fungicides and bactericides

Active ingredient	Plant species	Company	Product
Cinnamaldehyde	*Cassia tora*	Monterey	Vertigo
		Proguard	Cinnacure
L-Glutamic acid plus γ-aminobutyric acid	Many plants	Auxein Corporation	AuxiGro
Jojoba oil	*Simmondsia californica*	IJO Products	E-Race
	S. chinensis		Eco E-Race
		Soil Technologies	Permatrol
Laminarin	*Laminaria digitata*	Goemar	Iodus 40
Milsana	*Reynoutria sachalinensis*	KHH BioScience	Milsana
Pink plume poppy extract	*Macleaya cordata*	Camas	Qwel
Giant knotweed extract	*Reynoutria sachalinensis*	Marrone Bio Innovations	Regalia
Tea tree oil	*Melaleuca alternifolia*	SIA "Biomor Latvija"	Timorex
Clove oil	*Syzygium aromaticum*	Xeda International	Bioxeda
Fenugreek seed powder	*Trigonella foenum-graecum*		Stimulia

Data adapted from Cooping and Duke (2007) and Dayan et al. (2009)

In the USA and Europe, the *Cydia pomonella* granulovirus is applied to combat codling moth on apples (Chandler et al. 2011). For example, in Washington State, the biggest apple producer in the USA, it is used on 13 % of the apple crop (Chandler et al. 2011). In Brazil, 4 million ha (approximately 35 %) of soybean crops have been sprayed with *Nucleopolyhedrovirus* in order to control the soybean caterpillar *Anticarsia gemmatalis* in the mid-1990s (Moscardi 1999). Also in Brazil, the entomopathogenic fungus *Metarhizium anisopliae* is used against spittlebugs in approximately 750,000 ha of sugarcane and 250,000 ha of grassland annually (Lomer et al. 2001).

Table 18.4 The total usage of biopesticide in California between 2002–2011

	Total usage (%)									
Active ingredient	2002	2003	2004	2005	2006	2007	2008	2009	2010	2011
Acetic acid	–	–	–	–	–	–	–	0.01	0.12	–
Agrobacterium radiobacter	0.01	0.02	0.02	–	0.03	0.06	–	0.01	0.01	–
Azadirachtin	0,10	0.11	0.29	0,10	0.20	0.24	0.21	0.23	0.13	0.13
Bacillus pumilus	–	–	–	0.26	0.48	0.74	0.77	0.64	0.48	0.47
Bacillus sphaericus	0.25	0.81	1.41	2.52	3.83	2.13	2.04	1.66	0.92	0.63
Bacillus subtilis	1.01	1.38	1.67	1.04	1.45	1.83	1.59	1.48	1.5	1.53
Bacillus thuringiensis	7.37	10.58	15.59	17.24	20.61	27.96	27.05	27.56	21.8	16.15
Beauveria bassiana	0.06	0.06	0.09	0.06	0.05	0.07	0.05	0.03	0.03	0.04
Buffalo gourd root powder	–	–	–	–	–	0.01	0.03	–	–	–
Canola oil	–	–	–	–	–	–	–	–	0.01	–
Castor oil	0.03	0.10	0.04	0.01	–	–	–	–	–	–
Chenopodium ambrosiodes near *ambrosiodes*	–	–	–	–	–	–	–	1.86	0.73	0.49
Neem oil	16.09	4.83	8.65	8.65	8.15	11.68	9.97	9.71	8.16	4.39
Coniothyrium minitans	0.01	0.01	0.02	–	–	–	–	0.01	0.01	0.01
Gamma-aminobutyric acid	7.08	0.54	0.86	0.64	0.36	0.20	0.09	0.02	–	–
Garlic	0.04	0.02	0.02	0.01	0.01	0.01	0.02	0.00	0.30	–
Lavandulyl senecioate	–	–	–	–	–	–	–	0.04	0.03	0.38
Limonene	4.64	2.26	1.44	3.39	2.77	7.27	4.33	5.16	3.97	3.78
Linalool	0.01	0.02	0.02	0.01	0.01	0.01	0.01	0.01	0.08	0.01
Margosa oil	–	–	–	–	–	–	–	–	0.04	0.48
Menthol	–	–	–	0.01	–	–	–	–	–	–
Myrothecium verrucaria	1.93	3.75	3.98	2.06	2.11	3.16	2.27	2.13	1.61	1.73
Oil of citronella	–	–	–	–	–	–	–	–	–	0.02
Oil of jojoba	0.07	0.11	0.30	0.26	0.81	0.76	1.15	0.31	0.29	0.08
Oil of lemon eucalyptus	–	–	–	–	–	–	–	–	–	–
Paecilomyces lilacinus	–	–	–	–	–	–	–	–	0.02	0.03
Pseudomonas fluorescens	0.07	0.16	0.09	0.07	0.08	0.06	0.04	0.03	0.02	0.02
Quillaja	–	–	–	–	0.01	0.03	0.11	0.04	0.05	0.07
Reynoutria sachalinensis	–	–	–	–	–	–	–	0.02	0.63	0.92
S-abscisic acid	–	–	–	–	–	–	–	0.01	0.06	0.12
Sesame oil	–	–	–	–	–	0.09	0.05	0.08	0.09	0.08
Soybean oil	1.77	2.63	5.02	3.41	5.94	1.55	1.14	2.59	1.68	1.50
Sucrose octanoate	–	–	–	–	–	–	0.16	0.37	0.08	0.01
Thyme	–	–	–	–	0.01	0.05	0.06	0.07	0.09	0.04
Trichoderma harzianum	–	–	–	–	–	–	–	–	0.04	0.01
Vegetable oil	5.89	11.23	24.80	15.41	21.66	16.24	25.72	17.92	22.75	32.17
Yucca schidigera	–	–	–	–	–	–	–	0.01	0.03	0.10

Data was obtained from the Summary of Pesticide Use Report Data 2011 published by California Department of Pesticide Regulation

Biopesticides decompose in nature without bioaccumulation. Additionally, as these products often exist in the agroecosystem with few negative effects on nontarget organisms, they are regarded as environmentally friendly. Successful implementation of biopesticides will result in residue reduction, thereby producing a potentially safer product for the consumer. With the development and implementation of biopesticide control strategies, farmers and growers can decrease their reliance on the use of conventional pesticides against insect pests while at the same time, decreasing insect resistance to conventional pesticides.

8 Biopesticide Solutions

In essence, biopesticides are an innovative and safe solution for crop protection and management. The global biopesticide market is expanding rapidly in response to a demand for more environmentally friendly products. Biopesticide products exist on the market for the management of a wide range of important agricultural pests and diseases.

For agricultural growers, using cost-effective production methods is only part of the equation to harvesting and storing crops. The grower's income is determined on crop quality and yield. In an effort to achieve this output, biopesticide products can help growers improve crop quality and yield under challenging conditions. Microbial and biochemical products can be used in weed management and crop protection against pathogens, harmful insects, and numerous plant diseases that may otherwise divert or restrict a crop's access to valuable resources such as water, sunlight, or nutrients. Therefore, biopesticides promote crop health therefore increasing its salability. With organic farming systems becoming more popular due to consumer demand, biopesticide products provide growers with a means by which to produce higher-quality crops and larger yields that conform to consumer requirements.

9 Conclusion

World agricultural sectors face many obstacles, and pest pressure is particularly significant due to the evolution of resistance to synthetic pesticides. Many growers still rely on the use of conventional pesticides to protect their crops for financial reasons. However, due to their negative effects, the use of conventional pesticides has been strictly regulated. The use of alternatives such as biopesticides provides benefits as they are safer for the environment. Although the global market value is relatively small compared to that of synthetic pesticides, the growth rate in the biopesticides market value is encouraging. Growth in the agricultural sectors of affluent countries, in which the use of synthetic pesticides is controversial due to concerns related to human health, can only lead to increased demand for biopesticides.

References

Agriculture and Agri-Food Canada (2010) Directory of biopesticides for agricultural crops in OECD countries. https://www5.agr.gc.ca/MPDD-CPM.search-recherche.do?lang=eng

Agropages.com (2011) China pesticide sales amount to 51.8 billion Yuan in 2010. http://news.agropages.com/News/NewsDetail---3589.htm

Agropages.com (2012) New pesticides registered or launched in 2011 (Insecticides and others). http://news.agropages.com/News/NewsDetail---3865.htm

Agropages.com (2013a) North America biopesticides market watch. http://news.agropages.com/News/NewsDetail---9975.htm

Agropages.com (2013b) Neem-based pesticide market growing at 7–9 % annually in India. http://news.agropages.com/News/NewsDetail---10438.htm

Agrow (2013) Biopesticides. http://www.agrow.com

Attia S, Grissa KL, Lognay G, Bitume E, Hance T, Mailleux AC (2013) A review of the major biological approaches to control the worldwide pest *Tetranychus urticae* (Acari: Tetranychidae) with special reference to natural pesticides. J Pest Sci 86:361–386

Azmir J, Omar AKM, Zaidul ISM, Rahman MM, Sharif KM, Mohamed A, Sahena F, Jahurul MHA, Ghafoor K, Norulaini NAN (2013) Techniques for extraction of bioactive compounds from plant materials: a review. J Food Eng 117(4):426–436

Bailey A, Chandler D, Grant WP, Greaves J, Prince G, Tatchell M (2010) Pest management with biopesticides. In: Bailey A, Chandler D, Grant WP, Greaves J, Prince G, Tatchell M (eds) Biopesticides: pest management and regulation. CABI, Wallingford, pp 71–130

BCC Research (2012) Global markets for biopesticides. http://www.bccresearch.com/pressroom/report/code/CHM029D

Berenbaum MR, Zangerl AR (1996) Phytochemical diversity – adaptation or random variation? In: Romeo JT, Saunders JA, Barbosa P (eds) Phytochemical diversity and redundancy in ecological interactions, vol 30. Plenum Press, New York, pp 1–24

Bhuiyan MKR, Hassan E, Isman MB (2001) Growth inhibitory and lethal effect of some botanical insecticides and potential synergy by dillapiole in *Spodoptera litura* (Fab.) (Lepidoptera: Noctuidae). J Plant Dis Prot 108(1):82–88

Borgemeister C, Poehling HM, Dinter A, Holler C (1993) Effects of insecticides on life-history parameters of Aphid parasitoid *Aphidius rhopalosiphi* Hym, Aphidiidae. Entomophaga 38(2):245–255

Brown AWA (1978) Ecology of pesticides. Wiley, New York

Cavoski I, Caboni P, Miano T (2011) Natural pesticides and future perspectives. In: Stoytcheva M (ed) Pesticides in the modern world – pesticides use and management. InTech. ISBN:978-953-307-459-7, doi:10.5772/17550. Available from: http://www.intechopen.com/books/pesticides-in-the-modern-world-pesticides-use-and-management/natural-pesticides-and-future-perspectives

Chandler D, Bailey AS, Tatchell GM, Davidson G, Greaves J, Grant WP (2011) The development, regulation and use of biopesticides for integrated pest management. Philos Trans R Soc B 366:1987–1998

Chutia M, Mahanta JJ, Bhattacharyya N, Bhuyan M, Boruah P, Sarma TC (2007) Microbial herbicides for weed management: prospects, progress and constraints. Plant Pathol J 6(3):210–218

Cloyd RA, Galle CL, Keith SR, Kalscheur NA, Kemp KE (2009) Effect of commercially available plant-derived essential oil products on arthropod pests. J Econ Entomol 102(4):1567–1579

Cooping LG, Duke SO (2007) Natural products that have been used commercially as crop protection agents. Pest Manag Sci 63:524–544

Dayan FE, Cantrell CL, Duke SO (2009) Natural products in crop protection. Bioorg Med Chem 17:4022–4034

Dent D (2000) Insect pest management. CABI Publishing, Wallingford

Epstein DL, Zack RS, Brunner JF, Gut L, Brown JJ (2000) Effects of broad-spectrum insecticides on epigeal arthropod biodiversity in Pacific Northwest apple orchards. Environ Entomol 29(2):340–348

Faria MR, Wraight SP (2007) Mycoinsecticides and mycoacaricides: a comprehensive list with worldwide coverage and international classification of formulation types. Biol Control 43(3):237–256

Georghiou GP, Saito T (1983) Pest resistance to pesticides. Plenum Press, New York

Grainge M, Ahmed S, Mitchell WC, Hylin JW (1984) Plant species reportedly possessing pest-control properties-A database. Resource Systems Institute, East-west Center, Honolulu

Handa SS (2008) An overview of extraction techniques for medicinal and aromatic plants. In: Handa SS, Khanuja SPS, Longo G, Rakesh DD (eds) Extraction technologies for medicinal and aromatic plants. ICS-UNINDO, Trieste, pp 21–54

Hassan E (1999) The insecticidal effects of neem seed kernel extract on eggs and larvae of *Helicoverpa armigera* (Hubner). J Plant Dis Protect 106(5):523–529

Hayes J, Laws E (1991) Handbook of pesticide toxicology, vol 1. Academic, San Diego

Hedin PA, Hollingworth RM (1997) New applications for phytochemical pest-control agents. In: Hedin PA, Hollingworth RM, Masler EP, Miyamoto J, Thompson DG (eds) Phytochemicals for pest control, vol 658. American Chemical Society, Washington, DC, pp 1–12. ISBN 658

Henn T, Weinzierl R (1989) Botanical insecticides and insecticidal soaps. Circular 1296:1–18

Ignacimuthu S (2012) Insect pest control: using plant resources. Alpha Science International, Oxford

Isman MB (2000) Plant essential oils for pest and disease management. Crop Prot 19(8–10):603–608

Isman MB (2005) Problems and opportunities for the commercialization of botanical insecticides. In: Regnault-Roger C, Philogene BJR, Vincent C (eds) Biopesticides of plant origin. Lavaisier, Paris, pp 283–291

Kaufmann B, Christen P (2002) Recent extraction techniques for natural products: microwave-assisted extraction and pressurised solvent extraction. Phytochem Anal 13(2):105–113

Kraiss H, Cullen EM (2008) Insect growth regulator effects of azadirachtin and neem oil on survivorship, development and fecundity of *Aphis glycines* (Homoptera: Aphididae) and its predator, *Harmonia axyridis* (Coleoptera: Coccinellidae). Pest Manag Sci 64(6):660–668

Krishnamoorthy K, Atmakuri R (1999) Residues and persistence of azadirachtin in cotton lint seed oil and soil. Pestic Res J 11(2):181–188

Latif S, Anwar F (2009) Physicochemical studies of hemp (*Cannabis sativa*) seed oil using enzyme-assisted cold-pressing. Eur J Lipid Sci Technol 111(10): 1042–1048

Leng P, Zhang PG, Zhao M (2011) Applications and development trends in biopesticides. Afr J Biotechnol 10(86):19864–19873

Lomer CJ, Bateman RP, Johnson DL, Langewald J, Thomas M (2001) Biological control of locusts and grasshoppers. Annu Rev Entomol 46:667–702

Lopez O, Fernandez-Bolanos JG, Gil MV (2005) New trends in pest control: the search for greener insecticides. Green Chem 7(6):431–442

Lopez O, Fernandez-Bolanos JG, Gil MV (2011) Classical insecticides: past, present and future. In: Lopez O, Fernandez-Bolanos JG (eds) Green trends in insect control. The Royal Society of Chemistry, Cambridge

Loucif-Ayad W, Aribi N, Soltani N (2008) Evaluation of secondary effects of some Acaricides on *Apis mellifera intermissa* (Hymenoptera: Apidae): acetylcholinesterase and glutathione S-transferase activities. Eur J Sci Res 21(4):642–649

Marketsandmarkets.com (2012) Global biopesticides market – trends & forecasts. http://www.marketsandmarkets.com/Market-Reports/biopesticides-267.html

Martin L, Marques JL, Gonzalez-Coloma A, Mainar AM, Palavra AMF, Urieta JS (2012) Supercritical methodologies applied to the production of biopesticides: a review. Phytochem Rev 11(4):413–431

McKinlay R, Dassyne J, Djamgoz M, Plant JA, Voulvoulis N (2012) Agricultural pesticides and chemical fertilisers. In: Plant JA, Voulvoulis N, Ragnarsdottir KV (eds) Pollutants, human health and the environment: a risk based approach, 1st edn. Wiley, West Sussex, pp 181–206

Moscardi F (1999) Assessment of the application of baculoviruses for control of Lepidoptera. Annu Rev Entomol 44:257–289

Nayak SK, Chhibber RC (2002) Evaluation of relative toxicity of endosulfan and methyl chlorpyriphos against *Spodoptera litura* (Fabricius). Shashpa 9(2):191–193

Nderitu J, Kasina M, Nyamasyo G, Oronje ML (2007) Effects of insecticide applications on sunflower (*Helianthus annuus* L.) pollination in Eastern Kenya. World J Agric Sci 3(6):731–734

Pangarkar VG (2008) Microdistillation, thermomicrodistillation and molecular distillation techniques. In: Handa SS, Khanuja SPS, Longo G, Rakesh DD (eds) Extraction technologies for medicinal and aromatic plants. ICS-UNINDO, Trieste, pp 129–143

Papachristos DP, Milonas PG (2008) Adverse effects of soil applied insecticides on the predatory coccinellid *Hippodamia undecimnotata* (Coleoptera: Coccinellidae). Biol Control 47(1):77–81

Pedigo LP (1989) Entomology and pest management. Macmillan, New York

Quarles W (2011) New biopesticides for IPM and organic production. IPM Pract 33(7/8):1–20

Reddy NC, Rao JV (2008) Biological response of earthworm, *Eisenia foetida* (Savigny) to an organophosphorus pesticide, profenofos. Ecotoxicol Environ Saf 71(2):574–582

Regnault-Roger C (2012) Trends for commercialization of biocontrol agent (biopesticide) products. In: Merillon JM, Ramawat KG (eds) Plant defence: biological control. Dordrecht, New York, pp 139–160

Rharrabe K, Amri H, Bouayad N, Sayah F (2008) Effects of azadirachtin on post-embryonic development, energy reserves and (alpha-amylase activity of *Plodia interpunctella* Hubner (Lepidoptera: Pyralidae). J Stored Prod Res 44(3):290–294

Richter BE, Jones BA, Ezzell JL, Porter NL, Avdalovic N, Pohl C (1996) Accelerated solvent extraction: a technique for sample preparation. Anal Chem 68(6):1033–1039

Rosenthal A, Pyle DL, Niranjan K (1996) Aqueous and enzymatic processes for edible oil extraction. Enzyme Microb Technol 19(6):402–420

Sae-Yun A, Ovatlarnporn C, Itharat A, Wiwattanapatapee R (2006) Extraction of rotenone from *Derris elliptica* and *Derris malaccensis* by pressurized liquid extraction compared with maceration. J Chromatogr A 1125(2):172–176

Sanchis V, Bourguet D (2008) Bacillus thuringiensis: applications in agriculture and insect resistance management. A Rev Agron Sustain Dev 28(1):11–20

Schmutterer H (1995) The neem tree: source of unique natural products for integrated pest management, medicine, industry and other purposes. VCH, Weinheim

Schmutterer H (1997) Side-effects of neem (*Azadirachta indica*) products on insect pathogens and natural enemies of spider mites and insects. J Appl Entomol-Zeitschrift Fur Angewandte Entomologie 121(2):121–128

Scott IM, Jensen H, Scott JG, Isman MB, Arnason JT, Philogene BJR (2003) Botanical insecticides for controlling agricultural pests: piperamides and the Colorado potato beetle *Leptinotarsa decemlineata* say (Coleoptera: Chrysomelidae). Arch Insect Biochem Physiol 54(4):212–225

Singh J (2008) Maceration, percolation and infusion techniques for the extraction of medicinal and aromatic plants. In: Handa SS, Khanuja SPS, Longo G, Rakesh DD (eds) Extraction technologies for medicinal and aromatic plants. ICS UNINDO, Trieste, pp 67–82

Sowbhagya HB, Purnima KT, Florence SP, Rao AGA, Srinivas P (2009) Evaluation of enzyme-assisted extraction on quality of garlic volatile oil. Food Chem 113(4):1234–1238

Thakore Y (2006) The biopesticide market for global agricultural use. Ind Biotechnol 2:194–208

Van Hamburg H, Guest PJ (1997) The impact of insecticides on beneficial arthropods in cotton agroecosystems in South Africa. Arch Environ Contam Toxicol 32(1):63–68

Wheeler WB (2002) Role of research and regulation in 50 years of pest management in agriculture – prepared for the 50th anniversary of the journal of agricultural and food chemistry. J Agric Food Chem 50(15):4151–4155

Zhang W, Jiang F, Ou J (2011) Global pesticide consumption and pollution: with China as a focus. Proc Int Acad Ecol Environ Sci 1(2):125–144

Retracted: Formulation, Registration, and Quality Regulation of Plant Biopesticides

19

D.K. Hazra, P.K. Patanjali, and S.K. Raza

Abstract

The environmental hazards resulting from the intensive use of synthetic pesticides have made it imperative to consider alternative approaches for sustainable agricultural development. Biopesticides of plant origin have been found to be a better alternative, and the research and development during the past 30 years have led to a perfect understanding of the interactions between plants and pests. These new crop protection agents of plant origin and their place in integrated pest biocontrol have been recognized, and a lot of R&D activities are going on to develop user and environment-friendly formulations with commercial viability. In this chapter, attempts have been made to discuss the existing biocontrol agents, their uses in crop protection, the search for new supply sources, and current and future commercial developments.

Keywords

Antifeedant • Repellant • Botanical pesticide • Pyrethrum • *Azadirachta indica* • Plant essential oils • Botanical pesticide formulation • Registration • Quality regulation of plant biopesticides

1 Introduction

Over the next 20 years, crop production will have to increase significantly to meet the needs of a rising human population. This has to be done without damaging the other public goods—environment and social that farming brings. There will be no "silver bullet" solution to the impending food production challenge. Rather, a series of innovations must be developed to meet the different needs of farmers according to their local circumstances (Bastiaans et al. 2008).

D.K. Hazra, Ph.D. (✉)
AINP on Pesticide Residues, Department of Agricultural Chemicals, Bidhan Chandra Krishi Viswavidyalaya, Mohanpur, Nadia, West Bengal, India
e-mail: dipakipft@gmail.com

P.K. Patanjali, Ph.D. • S.K. Raza, Ph.D.
Institute of Pesticide Formulation Technology (IPFT), Gurgaon 122016, India

D. Singh (ed.), *Advances in Plant Biopesticides*,
DOI 10.1007/978-81-322-2006-0_19, © Springer India 2014

One way to increase food availability is to improve the management of pests. There are estimated to be around 67,000 different crop pest species including plant pathogens, weeds, invertebrates, and some vertebrate species, and together they cause about a 40 % reduction in the world's crop yield. Crop losses caused by pests undermine food security alongside other constraints, such as inclement weather, poor soils, and farmers' limited access to technical knowledge (Speranza et al. 2008). Since the 1960s, pest management in the industrialized countries has been based around the intensive use of synthetic chemical pesticides. Alongside advances in plant varieties, mechanization, irrigation, and crop nutrition, they have helped increase crop yields by nearly 70 % in Europe and 100 % in the United States (Pretty 2008). However, the use of synthetic pesticides is becoming significantly more difficult owing to a number of interacting factors:

- The injudicious use of broad-spectrum pesticides can damage human health and the environment (MEA 2005). Some of the "older" chemical compounds have caused serious health problems in agricultural workers and others because of inadequate controls during manufacture, handling, and application.
- Excessive and injudicious prophylactic use of pesticides can result in management failure through pest resurgence, secondary pest problems, or the development of heritable resistance (Van Emden and Service 2004). Worldwide, over 500 species of arthropod pests have been found resistant to one or more insecticides (Hajek 2004), while there are close to 200 species of herbicide-resistant weeds (Heap 2010).
- Pesticide products based on "old" chemistry are being withdrawn because of the new health and safety legislation (PSD 2008a, b). However, the rate at which new, safer chemicals are being made available is very low. This is caused by a fall in the discovery rate of new active molecules and the increasing costs of registration (Thacker 2002).
- Further pressures on pesticide use arise from concerns expressed by consumers and pressure groups about the safety of pesticide residues in food. These concerns are voiced despite the fact that pesticides are among the most heavily regulated of all chemicals.

Over the past 50 years, crop protection has relied heavily on synthetic chemical pesticides, but their availability is now declining due to the new legislation and the evolution of resistance in pest populations, residue problem, and destructions to beneficial organisms. Therefore, alternative pest management tactics are needed. Biopesticides are pest-management agents based on living microorganisms or natural products (Gelernter 2005). They have proven potential for pest management, and they are being used across the world. However, they are regulated by systems designed originally for chemical pesticides that have created market entry barriers by imposing burdensome costs on the biopesticide industry. There are also significant technical barriers in making biopesticides more effective (Marrone 2007). In the European Union, a greater emphasis on integrated pest management (IPM) as part of agricultural policy may lead to innovations in the way that biopesticides are regulated (Moser et al. 2008). There are also new opportunities for developing biopesticides in IPM by combining ecological science with post-genomics technologies. The new biopesticide products that will result from this research will bring with them new regulatory and economic challenges that must be addressed through joint working between social and natural scientists, policy makers, and industries (Morse et al. 2002).

2 Current Botanicals in Use

At present there are four major types of botanical products used for insect control (pyrethrum, rotenone, neem, and essential oils), along with three others in limited use (ryania, nicotine, and sabadilla). Additional plant extracts and oils (e.g., garlic oil, *Capsicum oleoresin*) see limited (low volume) regional use in various countries, but these are not in common use as far as India is concerned. In discussing the extent to which each of the more important botanical insecticides is used, often the data are referred published annually by the State of California's Department of Pesticide Regulation (CDPR 2005). Although not necessarily representative of uses in other jurisdictions, total pesticide use (in terms of active ingredient applied) in California is estimated to the tune of more than 175 million pounds, or 80,000 tonnes, in 2003. This amount represents approximately 6 % of the global pesticide use, of which 91 % was used for agriculture. Pesticide use data in California are reported by active ingredient and by crop or other use and as such are perhaps the world's most accurate and detailed records of pesticide use available to the general public.

2.1 Pyrethrum

Pyrethrum refers to the oleoresin extracted from the dried flowers of the pyrethrum daisy, *Tanacetum cinerariaefolium* (Asteraceae). The flowers are grounded to fine powder and then extracted with hexane or a similar nonpolar solvent; removal of the solvent yields an orange-colored liquid that contains the active principle (Casida and Quistad 1995; Glynne-Jones 2001). These are three esters of chrysanthemic acid and three esters of pyrethric acid. Among the six esters, those incorporating the alcohol pyrethrolone, namely, pyrethrins I (Fig. 19.1) and II, are the most abundant and account for most of the insecticidal activity. Technical grade pyrethrum, the resin used in formulating commercial insecticides, typically contains from 20 to 25 % pyrethrins (Casida and Quistad 1995). The insecticidal action of the pyrethrins is characterized by a rapid knockdown effect, particularly in flying insects, and hyperactivity and convulsions in most insects. These symptoms are results of the neurotoxic action of the pyrethrins, which block voltage-gated sodium channels in nerve axons. As such, the mechanism of action of pyrethrins is qualitatively similar to that of DDT and many synthetic organochlorine insecticides. In purity, pyrethrins are moderately toxic to mammals (rat oral acute LD_{50} values range from 350 to 500 mg kg^{-1}), but technical-grade pyrethrum is considerably less toxic (1,500 mg kg^{-1}) (Casida and Quistad 1995). Pyrethrins are especially labile in the presence of the UV component of sunlight, a fact that has greatly limited their use outdoors. A recent study indicated that the half-lives of pyrethrins on field-grown tomato and bell pepper fruits were 2 h or less (Antonious 2004). This problem created the impetus for the development of synthetic derivatives (pyrethroids) that are more stable in sunlight. The modern pyrethroids, developed in the 1970s and 1980s, have been highly successful and represent one of the rare examples of synthetic pesticide chemistry based on a natural product model.

However, note that the modern pyrethroids bear little structural resemblance to the natural pyrethrins, and their molecular mechanism of action differs as well. The data generated on use of pyrethrum from California in 2003 clearly demonstrate the dominance of this material among botanicals: Pyrethrum accounted for 74 % of all botanicals used that year, but only 27 % of that amount was used in agriculture (≈800 kg). Major uses of pyrethrum in California are for structural pest control, in public health, and for treatment of animal premises. Pyrethrum is the predominant botanical in use, perhaps accounting for 80 % of the global botanical insecticide market (CDPR 2005; Isman 2005). For many years, world production of pyrethrum was led by Kenya, with lesser quantities produced in Tanzania and Ecuador. In the past 5 years, Botanical Resources Australia, with plantings in Tasmania, has become the second largest producer in the world (~30 % of the world's production at present). Pyrethrum produced in Tasmania is qualitatively similar to that produced in the East.

2.2 Neem

Two types of botanical insecticides can be obtained from the seeds of the Indian neem tree, *Azadirachta indica* (Meliaceae) and neem oil (Schmutterer 2002). Neem oil, obtained by cold-pressing seeds, can be effective against soft-bodied insects and mites but is also useful in the management of phytopathogens. Apart from the physical effects of neem oil on pests and fungi, disulfides in the oil likely contribute to the bioactivity of this material. More highly valued than neem oil are medium-polarity extracts of the seed residue after the removal of the oil, as these extracts contain the complex triterpene azadirachtin (Fig. 19.2). Neem seeds actually contain more than a dozen azadirachtin analogs, but the

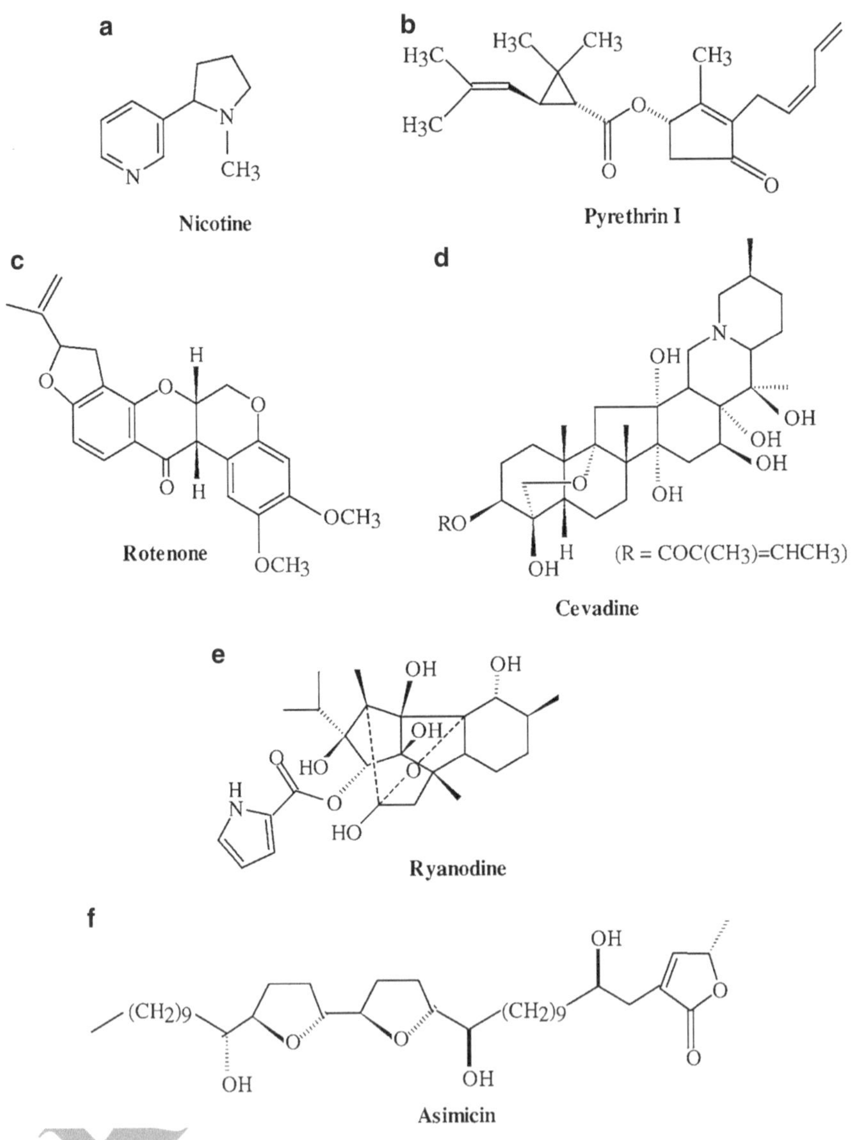

Fig. 19.1 Active constituents of some botanical insecticides from various plant sources discussed in this review. (**a**) Nicotine, (**b**) pyrethrin I, (**c**) rotenone, (**d**) cevadine, (**e**) ryanodine, and (**f**) asimicin

major anti-insect bioactive form is azadirachtin, and the remaining minor analogs likely contribute little to overall efficacy of the extract (Rembold and Mwangi 2002). Seed extracts include considerable quantities of other triterpenoids, notably salannin, nimbin, and derivatives thereof. The role of these other natural substances has been controversial, but most evidence points to azadirachtin as the most important active principle. Neem seeds typically contain 0.2–0.6 % azadirachtin by weight, so solvent partitions or other chemical processes are required to concentrate this active ingredient to the level of 10–50 % seen in the technical grade material used to produce commercial products.

Azadirachtin has two profound effects on insects. At the physiological level, azadirachtin blocks the synthesis and release of molting

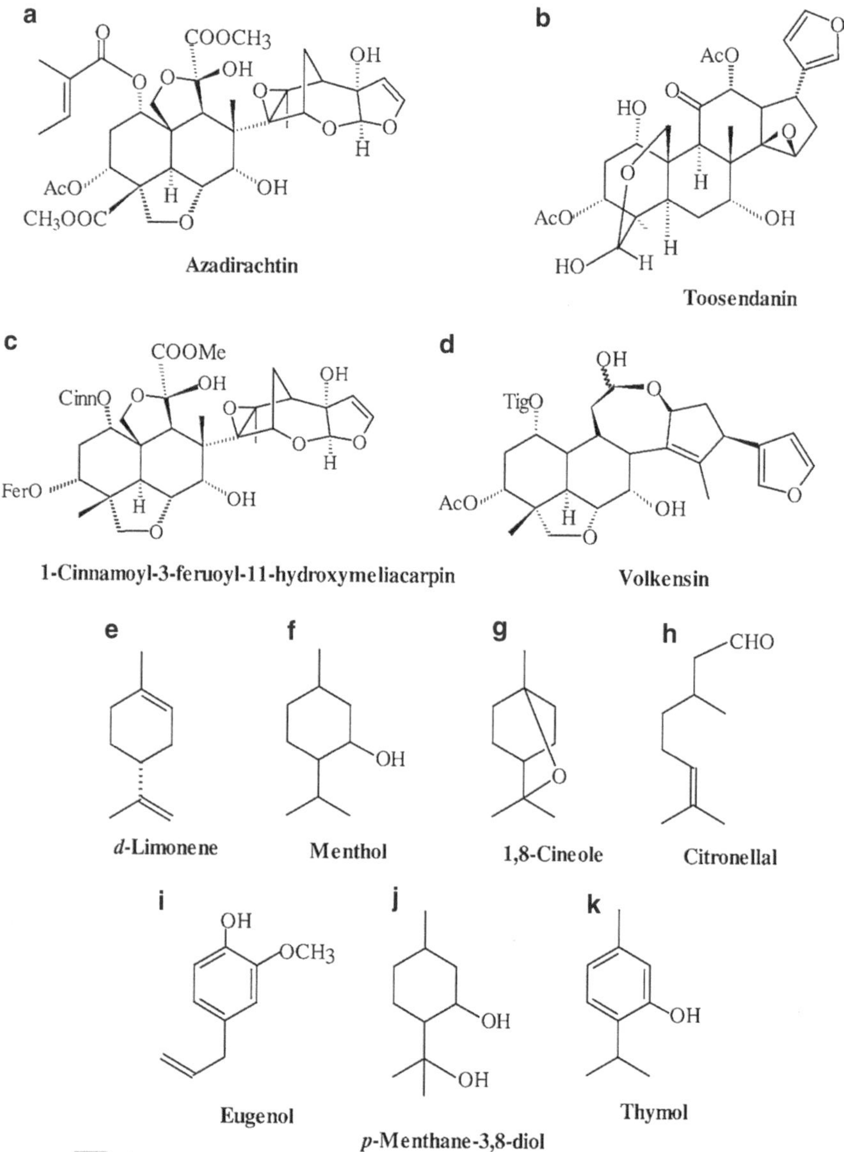

Fig. 19.2 Active constituents of some botanical insecticides from neem (*a*), *Melia* species (**b–d**), and selected plant essential oils (**e–k**): (**a**) Azadirachtin, (**b**) toosendanin, (**c**) 1-cinnamoyl-3-feruoyl-11-hydroxymeliacarpin, (**d**) volkensin, (**e**) *d*-limonene, (**f**) menthol, (**g**) 1,8-cineole, (**h**) citronellal, (**i**) eugenol, (**j**) *p*-menthane-3,8-diol, and (**k**) thymol

hormones (ecdysteroids) from the prothoracic gland, leading to incomplete ecdysis in immature insects. In adult female insects, a similar mechanism of action leads to sterility. In addition, azadirachtin is a potent antifeedant to many insects. The discovery of neem by western science is attributed to Heinrich Schmutterer, who observed that swarming desert locusts in Sudan defoliated almost all local floras except for some introduced neem trees (NRC 2000). Indeed, azadirachtin was first isolated based on its exceptional antifeedant activity in the desert locust, and this substance remains the most potent locust antifeedant discovered to date. Unlike pyrethrins, azadirachtin has defied total synthesis to this point. Promoted in the United States by Robert Larson (with assistance from the US Department of Agriculture), neem rapidly became the modern paradigm for the development of botanical insecticides.

Enthusiasm for neem was fostered by several international conferences in the 1980s and 1990s, and several volumes dedicated to neem and neem insecticides have been published (Schmutterer 2002). Unfortunately, neem's commercial success has fallen well short of the initial hype fueled by the explosive scientific literature surrounding it. In part, this is due to the relatively high cost of the refined product (Isman 2004) and the relatively slow action on pest insects. Nonetheless, several azadirachtin-based insecticides are sold in the United States and at least two of such products in the European Union. In California, azadirachtin-based insecticides constituted about one third of the botanicals used in agriculture in 2003 (600 kg). In practice, reliable efficacy is linked to the physiological action of azadirachtin as an insect growth regulator; the antifeedant effect, which is spectacular in the desert locust, is highly variable among pest species, and even those species initially deterred are often capable of rapid desensitization to azadirachtin.

The influence of azadirachtin on head protein of *Helicoverpa armigera* was reported by Neoliya et al. (2005, 2007). The protein content was higher during the active feeding stage and dropped at the later stage of the second larval instar, suggesting epidermal programming. In the control group, high protein content was observed in the fifth larval instar when larval transformation was nearing the completion. In the treated larvae, the trend was similar, but the protein content was very low when compared to the control groups.

What is clear is that azadirachtin is considered nontoxic to mammals (rat oral acute LD_{50} is >5,000 mg kg^{-1}), fish, and pollinators. The influence of azadirachtin on natural enemies is highly variable. Like pyrethrins, azadirachtin is rapidly degraded by sunlight. For example, on olives growing in Italy, azadirachtin has a half-life of approximately 20 h. On the other hand, azadirachtin has a systemic action in certain crop plants, greatly enhancing its efficacy and field persistence (Schmutterer 2002).

2.3 Plant Essential Oils

Steam distillation of aromatic plants yields essential oils, long used as fragrances and flavorings in the perfume and food industries, respectively, and more recently for aromatherapy and as herbal medicines. Plant essential oils are produced commercially from several botanical sources, many of which are members of the mint family (Lamiaceae). The oils are generally composed of complex mixtures of monoterpenes, biogenetically related phenols, and sesquiterpenes. Examples include 1,8-cineole, the major constituent of oils from rosemary (*Rosmarinus officinale*) and eucalyptus (*Eucalyptus globulus*), eugenol from clove oil (*Syzygium aromaticum*), thymol from garden thyme (*Thymus vulgaris*), and menthol from various species of mint (*Mentha* species) (Fig. 19.2). A number of source plants have been traditionally used for protection of stored commodities, especially in the Mediterranean region and in southern Asia, but interest in the oils was renewed with emerging demonstration of their fumigant and contact insecticidal activities to a wide range of pests in the 1990s (Isman 2000). The rapid action against some pests is indicative of a neurotoxic mode of action, and there is evidence for interference with the neuromodulator octopamine (Enan 2001; Kostyukovsky et al. 2002) by some oils and with GABA-gated chloride channels by others (Priestley et al. 2003).

Plant essential oils possess diversified insecticidal properties (Jayasekara et al. 2005). Under screening program of survey of bioactive agents for insect control, 31 essential oils from different botanicals (2 % in acetone) were studied for repellency and direct toxicity (insecticidal) against laboratory-bred houseflies, *Musca domestica* L (Singh and Singh 1991). The essential oils obtained from *Ocimum gratissimum* L., *Thymus serpyllum* L., *Illicium verum* Hooks f., *Myristica fragrans* Houtt., and *Curcuma amada* Roxb. showed 100 % repellent activity, and *Acorus calamus* L. and *Thymus serpyllum* L. showed about 40 % insecticidal activity. Results were compared with N, N-diethyl-meta-toluamide, dimethyl phthalate, malathion, pyrethrum extract, thymol, mineral oil, and piperonyl butoxide. Biochemical studies of these promising oils may emerge new leads in developing future pesticides.

Some of the purified terpenoid constituents of essential oils are moderately toxic to mammals, but, with few exceptions, the oils themselves or products based on oils are mostly nontoxic to mammals, birds, and fish (Isman 2000). However, as broad-spectrum insecticides, both pollinators and natural enemies are vulnerable to poisoning by products based on essential oils. Owing to their volatility, essential oils have limited persistence under field conditions; therefore, although natural enemies are susceptible via direct contact, predators and parasitoids reinvading a treated crop one or more days after treatment are unlikely to be poisoned by residue contact as often occurs with conventional insecticides. In the United States, commercial development of insecticides based on plant essential oils has been greatly facilitated by the exemption from registration for certain oils commonly used in processed foods and beverages. This opportunity has spurred the development of essential oil-based insecticides, fungicides, and herbicides for agricultural and industrial applications and for the consumer market, using rosemary oil, clove oil, and thyme oil as active ingredients. Interest in these products has been considerable, particularly for control of greenhouse pests and diseases and for control of domestic and veterinary pests, with several private companies (e.g., EcoSMART Technologies, Inc., United States) moving toward or into the marketplace. Another factor favoring the development of botanical insecticides based on plant essential oils is the relatively low cost of the active ingredients, a result of their extensive worldwide use as fragrances and flavorings. In contrast, pyrethrum and neem are used primarily for insecticide production.

3 Potential New Botanicals

3.1 Annonaceous Acetogenins

Botanical insecticides have been traditionally prepared from the seeds of tropical *Annona* species, members of the custard apple family (Annonaceae). These include the sweet sop (*A. squamosa*) and sour sop (*A. muricata*), important sources of fruit juices in Southeast Asia. Detailed investigations in the 1980s led to the isolation of a number of long-chain fatty acid derivatives, termed acetogenins, responsible for the insecticidal bioactivity. The major acetogenin obtained from the seeds of *A. squamosa* is annonin I, or squamocin, and a similar compound, asimicin (Fig. 19.1), was isolated from the bark of the American pawpaw tree, *Asimina triloba* (Johnson et al. 2000). McLaughlin and colleagues hold a US patent on insecticides based on acetogenins from *A. triloba*; Bayer AG (Germany) holds a similar patent based on *Annona acetogenins*. These compounds are slow-acting stomach poisons, particularly effective against chewing insects such as lepidopterans and the Colorado potato beetle (*Leptinotarsa decemlineata*).

Further investigations revealed that acetogenins have a mode of action identical to that of rotenone, i.e., they block energy production in mitochondria in both insects and mammals. In purity, certain acetogenins are toxic to mammals (LD_{50} is <20 mg kg^{-1}), an impediment to regulatory approval, even though standardized extracts from *Annona* seeds and *Asimina* bark are much less toxic. McLaughlin and associates (Johnson et al. 2000) have isolated hundreds of acetogenins from Annonaceae, and for many, their potential as anticancer agents exceeds their value as insecticides. In spite of the patents based on the insecticidal activities of these materials, no commercial development has proceeded with the exception of a head lice shampoo that contains a standardized pawpaw extract among its active ingredients. *Annona* seed extracts may prove more useful in tropical countries where the fruits are commonly consumed or used to produce fruit juice; in these cases, the seeds are the waste products. For example, Leatemia and Isman (2004) recently demonstrated that crude ethanolic extracts or even aqueous extracts of seeds from *A. squamosa* collected at several sites in eastern Indonesia are effective against the diamondback moth (*Plutella xylostella*).

3.2 Sucrose Esters

In the early 1990s, scientists at the US Department of Agriculture discovered that sugar esters naturally occurring in the foliage of wild tobacco (*Nicotiana gossei*) were insecticidal to certain soft-bodied insects and mites. Although patented, extraction of these substances on a commercial scale from plant biomass proved impractical, leading to the development of sucrose esters manufactured from sugar and fatty acids obtained from vegetable oils. AVA Chemical Ventures (United States) has patented and registered an insecticide/miticide based on C_8 and C_{10} fatty acid mono-, di-, and tri-esters of sucrose octanoate and sucrose dioctanoate. The product, first registered in 2002, contains 40 % active ingredient. Functionally, this product appears to differ little from the insecticidal soaps based on fatty acid salts developed in the 1980s, particularly potassium oleate. Both products are contact insecticides that kill small insects and mites through suffocation (by blocking the spiracles) or disruption of cuticular waxes and membranes in the integument, leading to desiccation. Although useful in home and garden products and in greenhouse production, the utility of these materials for agriculture remains to be seen.

3.3 *Melia* Extracts

The remarkable bioactivity of azadirachtin from the Indian neem tree (*Azadirachta indica*) led to the search for natural insecticides in the most closely related genus, *Melia*. Seeds from the chinaberry tree, *M. azedarach*, contain a number of triterpenoids, the meliacarpins (Fig. 19.2), that are similar but not identical to the azadirachtins, and these too have insect growth regulating bioactivities (Kraus 2002). But in spite of the abundance of chinaberry trees in Asia and other tropical and subtropical areas to which they were introduced, the development of commercial insecticides has not paralleled that of the neem insecticides. The main reason is the presence, in chinaberry seeds, of additional triterpenoids, the meliatoxins, that have demonstrated toxicity to mammals (Ascher et al. 2002). However, the chemistry of chinaberry varies considerably across its natural and introduced range, and seeds of *M. azedarach* growing in Argentina lack melia toxins but produce triterpenoids (most notably meliartenin) that are strong feeding deterrents to pest insects and could prove useful for pest management (Carpinella et al. 2003). Similar results have been obtained from South Africa using aqueous extracts of chinaberry leaves, presumably lacking melia toxins but efficacious against the diamondback moth (Charleston 2004).

In the early 1990s, a botanical insecticide produced in China was based on an extract of the bark of *Melia toosendan*, a tree considered by most taxonomists to be synonymous with *M. azedarach*. The extract contains a number of triterpenoids based on toosendanin (Fig. 19.2), a substance reported to be a stomach poison for chewing insects. Later studies suggest that this substance acts primarily as a feeding deterrent but can also serve as a synergist for conventional insecticides. Although relatively nontoxic to mammals, it is unclear whether this material remains in production or whether it is sufficiently efficacious as a stand-alone crop protectant. When *M. toosendan* came under scientific scrutiny, investigation of the East African *M. volkensii* demonstrated bioactivity in insects from seed extracts of this species. The active principles in *M. volkensii* include the triterpenoids alannin, also a major constituent of neem seed extracts and some novel triterpenoids such as volkensin (Fig. 19.2). Collectively, these function as feeding deterrents and stomach poisons with moderate efficacy against chewing insects and as a mosquito larvicide. Although a standardized seed extract has been made in quantities sufficient for research (Rembold and Mwangi 2002), commercial production appears unlikely owing to a lack of infrastructure for harvesting seeds in addition to regulatory impediments.

4 Insect Antifeedants and Repellents

4.1 Antifeedants

The possibility of using nontoxic deterrents and repellents as crop protectants is intuitively attractive (Bernays 1990). The concept of using insect antifeedants (feeding deterrents) gained strength in the 1970s and 1980s with the demonstration of the potent feeding deterrent effect of azadirachtin and neem seed extracts to a large number of pest species. Indeed, considerable literature, scientific and otherwise, touts neem as a successful demonstration of the antifeedant concept (Koul 2008). In reality, it is the physiological actions of azadirachtin that appear most reliably linked to field efficacy of neem insecticides; although purely behavioral effects cannot be ruled out, there is hardly any irrefutable evidence or documentation of field efficacy based on the antifeedant effects of neem alone. As an academic exercise, the discovery and demonstration of plant natural products as insect antifeedants has been unquestionably successful. In addition to the neem triterpenoids, extensive work has been performed on clerodane diterpenes from the Lamiaceae and sesquiterpene lactones from the Asteraceae (Gonzalez-Coloma et al. 2002; Klein Gebbinck et al. 2002).

On the other hand, not a single crop-protection product based unequivocally on feeding or oviposition deterrence has been commercialized. Two main problems are faced in the use of antifeedants in agriculture (Isman 2002). The first is interspecific variation in response—even closely related species can differ dramatically in behavioral responses to a substance limiting the range of pests affected by a particular antifeedant. Some substances that deter feeding by one pest can even serve as attractants or stimulants for other pests. The second is the behavioral plasticity in insects—pests can rapidly habituate to feeding deterrents, rendering them ineffective in a matter of hours. This has been recently demonstrated not only for pure substances like azadirachtin but also for complex mixtures (plant extracts). Whereas a highly mobile (flying) insect may leave a plant upon first encountering an antifeedant, a less mobile one (larva) may remain on the plant long enough for the deterrent response to wane. Such behavioral changes are important in light of the observation that some plant substances are initially feeding deterrents but lack toxicity if ingested. Azadirachtin is clearly an exception to this rule, as ingestion leads to deleterious physiological consequences, but many other compounds or extracts with demonstrated antifeedant effects lack toxicity when administered topically or via injection.

4.2 Repellents

For many chemists, an effective alternative to DEET (N, N-diethyl-*m*-toluamide) for personal protection against mosquitoes and biting flies is the holy grail (Fradin and Day 2002). In spite of five decades of research, no chemical has been found that provides the degree of protection against biting mosquitoes or persistence on human skin afforded by DEET (Peterson and Coats 2001). Concerns with the safety of DEET, especially to children, have resulted in the introduction of several plant oils as natural alternatives. Some personal repellents in the US marketplace contain oils of citronella, eucalyptus, or cedarwood as active ingredients; 2-phenethylpropionate, a constituent of peanut oil, and *p*-menthane-3,8-diol (obtained from a particular species of mint) (Fig. 19.2) are also used in consumer products. All of these materials can provide some protection, but the duration of their effect can be limited (often <1 h). In tropical areas where mosquito-borne disease is a threat (e.g., yellow fever, dengue, malaria), DEET probably remains the only reliable repellent. Oil of citronella or the constituent citronellal (Fig. 19.2) is also used in mosquito coils to repel mosquitoes from outdoor areas. Several veterinary products for flea and tick control on domestic pets contain *d*-limonene (from citrus peels; Fig. 19.2) as the active ingredient. Other uses for repellents under investigation include perimeter treatments of buildings to exclude termites and the use of essential oils to repel cockroaches from kitchens and flies from dairy barns (Maistrello et al. 2004). Another important use of plant essential oil constituents is in fumigation of beehives to manage economically important honeybee parasites, the Varroa mite (*Varroa jacobsoni*) and the tracheal mite (*Acarapis woodi*). In North America, menthol (from peppermint; Fig. 19.2) is widely used for this purpose, and in Europe, thymol (from garden thyme; Fig. 19.2) is most often used (Floris et al. 2004).

5 Botanical Formulations

Two main barriers for commercialization of botanical insecticides are as follows: sustainability of the botanical resource and standardization of chemically complex extracts (Weinzierl 2000). Besides these, availability and cost considerations are also important factors. Other drawbacks or limitations are the slow action of many botanicals—growers must gain confidence in insecticides that do not produce an immediate "knockdown" effect—and the lack of residual action for most botanicals.

5.1 Sustainability

To produce a botanical insecticide on a commercial scale, the source plant biomass must be obtainable on an agricultural scale and preferably not on a seasonal basis. Unless the plant in question is extremely abundant in nature, or already grown for another purpose (e.g., sweetsop, *Annona squamosa*, grown for its edible fruit and rosemary, *Rosmarinus officinale*, as a flavoring), it must be amenable to cultivation. Pyrethrum and neem meet this criterion; the latter has been extensively introduced into Africa, Australia, and Latin America, more so as a shade tree, windbreak, or source of firewood than for its yield of natural medicines or insecticides (Glynne-Jones 2001). Research aimed at producing azadirachtin from neem tissue culture provided proof of concept, but economic feasibility has yet to be attained.

In the not-too-distant future, it may be possible to produce botanical insecticides by "phytopharming," i.e., through genetic engineering of an existing field crop to produce high-value natural products originally isolated from a different botanical source. But as the progress in plant biotechnology continues at a rapid pace, it may prove just as easy to modify the plants we wish to protect from pests directly, such that they produce the natural product protectant constitutively, alleviating the need to obtain the desired botanical product through extraction, formulate it, and then apply it to the crop we wish to protect (Casanova et al. 2002). These sorts of technological advancements seem far more likely now than they did even a decade ago; however, the cost of these technologies will dictate that the traditional means of obtaining botanical insecticides and indeed their minor uses (on small acreage specialty crops) or uses in developing countries on lesser value crops will continue for many years to come. For example, neem seed oil had a long history of use in India for the production of soaps and low-grade industrial oil. When extraction companies began purchasing neem seeds in bulk to produce insecticides, the price of seeds increased tenfold (Schmutterer 2002). In contrast, certain plant essential oils have numerous uses as fragrances and flavorings, and the massive volumes required to satisfy these industries maintain low prices that make their use as insecticides attractive.

5.2 Standardization of Botanical Extracts

An often cited drawback to the adoption of botanical insecticides by growers is the variation in the performance of a particular product, even when prepared by the same process. Natural variation in the chemistry of a plant-based commodity should come as no surprise to anyone who enjoys coffee, tea, wine, or chocolate. Recent investigation of seed extracts from sweet sop (*A. squamosa*) collected in Indonesia demonstrated both geographical and annual variations in their insecticidal potency (Leatemia and Isman 2004). For a botanical insecticide to provide a reliable level of efficacy to the user, there must be some degree of chemical standardization, presumably based on the putative active ingredient(s). This has certainly been achieved with more refined products based on pyrethrum, neem, and rotenone, but crude preparations often contain low concentrations of active ingredients without adequate quantitation. To achieve standardization, the producer must have an analytical method and the equipment necessary for analysis and may need to mix or blend extracts from different sources, which requires storage facilities and is partially dependent on the inherent stability of the active principles in the source plant material or extracts thereof held in storage (Atkinson et al. 2004).

5.3 Botanical Formulations Developed at IPFT

5.3.1 Spreading Oil Formulations

The surface-spreading formulations are oil-based formulations which can be applied by dropping on water surface, and after application, the formulation spreads to the whole water surface within a few seconds. Basically these formulations are stable dispersion of water-insoluble liquid or solid in oil, and if it is a solid dispersion, the dispersed particles are wet, grinded into a fine particle suspension and stabilized with the help of wetting and dispersing agents. When these formulations are applied on the water surface, the active ingredient maintains a smooth networking film on the water surface. The films formed by these formulations are not continuous films but having holes which maintain the required oxygen level within the water for other aquatic organisms present. The layer of the active ingredient maintains on the water surface. The larvae of anopheles mosquitoes feed on the water surface, and the active ingredient produces mortality after ingestion.

Spreading oil formulations of neem oil may directly be dropped on the surface of water sources like lake, pond, etc., and the formulations will be nontoxic to nontarget organisms, fish and other aquatic species. The application does not require any spray, and good control of mosquito larvae may be obtained just by dropping it on the water surface. The formulations designed in the laboratory have the characteristics of maintaining the networking film throughout the water surface, and under natural conditions, the film remains in networking condition even after the winds and water waves.

Advantages

- Botanical based thus safe for nontarget organism
- Easy to pour and measure
- May give enhanced biological activity
- Nonflammable
- Low cost of formulation

Further we have the flexibility of maintaining high or low spreading pressure of the formulation during application by optimizing the ingredient ratios (Fig. 19.3).

Example IPFT developed neem oil based 10 % spreading oil formulation.

5.3.2 Floating Tablets

The floating tablets are slow release tablets which after application in water bodies floats on the surface of water due to low specific gravity and specific inert ingredients. It slowly releases the active ingredient into the water. The floating tablets offer a simple and practical approach to achieve increased surface-residence time for the dosage form and sustained active ingredient's release. It can reduce the frequency of dosing required for mosquito larval control and decrease variation in larvicidal concentration. Preparing the larvicides in a floating dosage form can control the extent of bioavailability for such poorly water-soluble active ingredients.

Floating tablet formulations of neem may directly be applied on surface of water sources like lake, pond, etc.; the formulations will be nontoxic to nontarget organisms, fishes and other aquatic species. The application does not require any spray, and good control of mosquito larvae may be obtained just by dropping it on the water surface. The release of active ingredient from the

Fig. 19.3 Neem oil based 10 % spreading oil formulation

formulations may be designed based upon the type and infestation of the target species.

Advantages

- Dust free
- Easy packaging
- Easy to handle and apply
- Long shelf life
- Adjustable release rates

These types of formulations are having advantages of having the action target specific with lower doses of the active ingredient compared to general formulations which mix throughout the water body and require high doses. By designing these types of floating tablets, we can make the tablets fast or slow release as per requirement. Another advantage of this type of product is that the total ingredients used for their preparation are bio-botanical-based products (Fig. 19.4).

Example IPFT developed floating tablet containing 10 % a.i. and patent has been filed.

5.3.3 Mosquito Coil

Mosquito coils are among the most popularly and widely used formulations in houses especially in rural areas because they are easy to use, effective, and inexpensive. The mosquito coils being produced by different companies are based on synthetic pyrethroids. They pose the risk of adverse effect on the health of human being due to the presence of synthetic pesticides. IPFT has developed typical mosquito coils based on neem and a very low-cost botanical product which enhances the bioefficacy of neem. Since the synthetic pesticide is not used, the formulation will not pose any risk to the health of human being. Composition is based on neem and other botanical based along with botanical synergist to improve knockdown and repellent properties.

Advantages

- Active components of plant origin
- Nontoxic and safe to human beings
- Easily and abundantly available raw materials
- Economical
- Good repellent and knockdown effect against mosquitoes

The coils developed in the laboratory were evaluated for their bioefficacy and were found to be more effective than the commercially available synthetic pyrethroid-based coil formulations. The enhancement of efficacy of neem-based coil formulations was achieved by using a botanical-based synergist, which make the neem-based coils more effective than synthetic pesticide-based commercially available formulation in the market (Fig. 19.5).

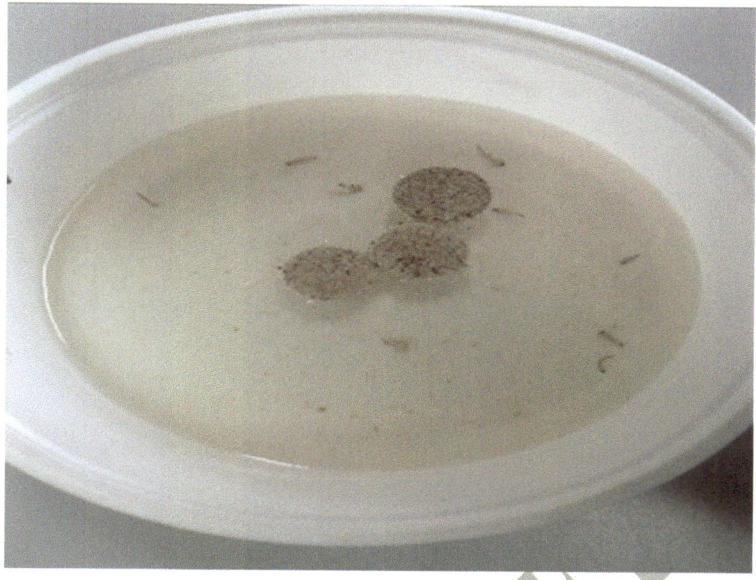

Fig. 19.4 Floating tablet

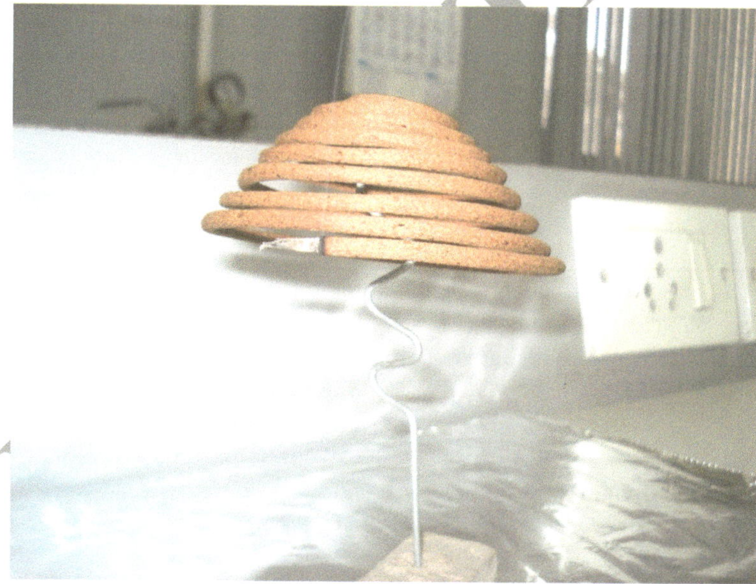

Fig. 19.5 Mosquito coil

Example IPFT developed mosquito coil containing 28.8 % neem kernel powder + 28.8 % a.i. botanical synergist, and patent has been filed.

5.3.4 Nanoemulsion

Botanical-based nanoemulsion may be a good alternative of synthetic pesticides for mosquito control spray application. Due to finer droplet size, total surface area is increased many folds so total coverage of target site also increases as results dose rate can be reduced to some extent.

Advantages

- Low dose rate
- No skin and eye irritation
- Less or no solvent
- Minimal toxicity
- No flammability

Fig. 19.6 Eucalyptus-oil-based nanoemulsion

A number of essential and neem-oil-based nanoemulsions have been prepared with insecticidal activity for mosquito and other insect pest control. These nanoemulsions may be a good alternative to any kinetically stable emulsion or suspension used for spray application (Fig. 19.6).

Example IPFT developed eucalyptus-oil-based nanoemulsion containing 30 % a.i, and bioefficacy study was conducted against store grain pest (*Tribolium* spp.).

5.3.5 Suspension Concentrate

Suspension concentrate formulations are micronized suspension of solid pesticides in a bulk medium-like water or oil. The formulations are being popularly used in agricultural as well as household applications due to advantages of good bioefficacy, easy dilution and dispersion, and dust freeness. The neem kernel powder along with the neem oil may be formulated as oil- or water-based suspension concentrate for indoor/outdoor spray application. By the application of specialized surfactants, the formulation will be dilutable with water for spray application. The process of wet grinding reduces the particle size to a very fine level so that the distribution on target surface is uniform and capable of producing good bioefficacy.

Fig. 19.7 Neem-oil-based SC formulation

Advantages

- Easy to pour and measure
- May give enhanced biological activity and rain-fastness
- Water- or oil-based dispersion
- Low cost of formulation

Further these formulations have better efficacy and surface coverage as compared to wettable powder (avg. particle size = 8–10 μm) because of its lower particle size (avg. particle size = 2–3 μm) (Fig. 19.7).

Example IPFT developed suspension concentrate formulation containing 30 % neem oil.

5.3.6 Insect Repellent Cream

Insect repellents are widely used in developed and developing countries for protection against mosquito bites. The advantage of skin repellents is their relative cheapness. Most of the insect repellent cream being produced by different manufacturers are chemical based and may cause skin problems to human being on prolonged use. Neem-based mosquito repellent cream may be effective in preventing mosquito bite. Due to no or minimum use of synthetic ingredients, it will be safer to the human being.

Advantages

- No skin and eye irritation
- Less or no solvent
- Minimal toxicity
- No flammability

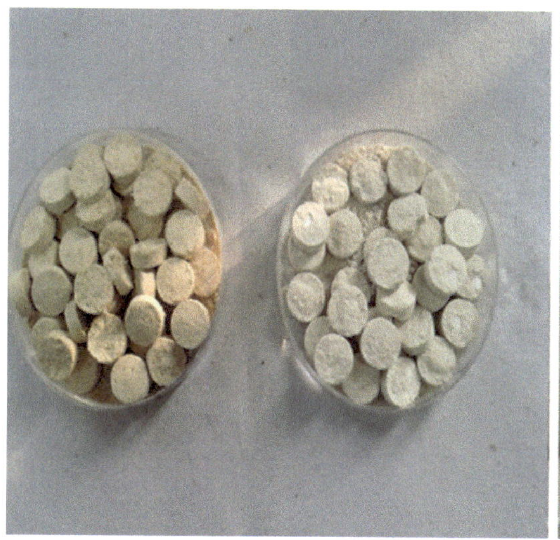

Fig. 19.8 Botanical-waste-based tablet formulation

These types of products are effective especially for soldiers deployed in forest areas or rural societies because of its easy transportation and effectiveness. The efficiency/efficacy of neem-based cream was enhanced by using other botanical-based ingredients which are safe to human being and biodegradable.

Example IPFT developed neem-oil-based insect repellent cream formulation.

5.3.7 Tablet Bait Formulation

Effervescent tablets have high stability and convenient application. The raw materials of the tablet are safe and nontoxic. Botanical-based tablets have the advantages of safety and can be used for cockroach control.

Advantages

- Dust free
- Easy packaging
- Easy to handle
- Long shelf life

The biggest advantage of these tablet formulations is their convenience in use, storage, and transportation. These types of tablet formulations can be directly applied to places where cockroaches are crawling like kitchen, storeroom, etc. (Fig. 19.8).

Example IPFT developed botanical-waste-based tablet formulation, and bioefficacy study was conducted against cockroach (*Periplaneta americana*).

5.4 Scope of Botanical Formulation

Over 100 neem-based products are marketed in India alone. Commercial neem-based products are marketed in a few African countries including Kenya, Benin, Nigeria, Senegal, and Ghana. Currently a number of commercial neem products are also registered and marketed in some developed countries such as the United States, Germany, Australia, Italy, Switzerland, Sweden, Denmark, Austria, Spain, and Israel. The other plant exploited commercially is *Acorus calamus*. A preparation containing 70 % β-asarone is marketed by Alrich of Germany. It must be emphasized that before any of the botanicals can be commercialized even for local production and consumption, it must be shown to be safe.

5.5 Problems Associated with Botanical Formulations

There is no doubt that the successful utilization of botanicals can help to control many of the world's destructive pests and diseases as well as reduce erosion, desertification, deforestation, and perhaps even control human population due to the antifertility action of some of them such as the neem. Although the possibilities of using botanical pesticides seem almost endless, so many details remain to be clarified. Many obstacles must be overcome, and many uncertainties clarified before their potential can be fully realized. These limitations seem surmountable; however, they present exciting challenges to the scientific and economic development communities. Solving the following obstacles and uncertainties may well bring a major new resource which will benefit much of the world. These obstacles include:

- Lack of experience and appreciation of the efficacy of botanicals for pest control. There are still doubts as to the effectiveness of plant-derived products (both "home-made" and commercial products) due to their slow action and lack of rapid knockdown effect.
- Genetic variability of plant species in different localities.
- Difficulty of registration and patenting of natural products and lack of standardization of botanical pesticide products.
- Economic uncertainties occasioned by seasonal supply of seeds, perennial nature of most botanical trees, and change in potency with location and time with respect to geographical limitations.
- Handling difficulties as there is no method for mechanizing the process of collecting, storing, or handling the seeds from some of the perennial trees.
- Instability of the active ingredients when exposed to direct sunlight.
- The usage of botanicals is still not held in high social esteem in many countries.
- Competition with synthetic pesticides through aggressive advertising by commercial pesticide dealers and commercial formulated botanicals are more expensive than synthetic insecticides and are not as widely available.
- Possible health hazards when seeds used in the preparation of the products are infected with the *Fungus aspergillus flavus* which produces aflatoxins, which is one of the most potent carcinogens known in the world.
- Rapid degradation, although desirable in some respects, creates the need for more precise timing or more frequent applications.
- Data on the effectiveness and long-term (chronic) mammalian toxicity are unavailable for some botanicals, and tolerances for some have not been established.

6 Current Trends in the Use of Botanicals

6.1 Pacific/Asia

India appears to embrace botanicals more than many other countries in the region, permitting all of the materials mentioned herein and allowing new products provisional registration, while toxicological and environmental data in support of full registration are acquired. New Zealand has registrations for pyrethrum, rotenone, and neem, whereas Australia has yet to approve neem in spite of almost two decades of research and development in that country. Likewise, neem has yet to be approved for use in the Philippines, where pyrethrum is the only approved botanical insecticide.

6.2 Europe

Although considered by many in the agrochemical industry to be especially restrictive with respect to pesticide registrations, the European Union permits the use of pyrethrum, neem, rotenone, and nicotine, along with "components of etheric oils of plant origin" (PAN 2004). Individual countries in Western Europe show considerable variation in the botanicals they permit. For example, Hungary permits pyrethrum and nicotine, although the latter is severely restricted. Denmark permits only pyrethrum and rotenone, Germany permits pyrethrum and neem, while the Netherlands permits pyrethrum alone. In spite of years of research on azadirachtin and neem in the United Kingdom, this botanical has never achieved registration there, leaving pyrethrum, rotenone, and nicotine as the only approved botanicals.

7 Trends and Changes in Registration

Given the ongoing negative perception of pesticides by the general public, government response to that perception, and increasing documentation of environmental contamination, it is hard to imagine pesticide regulators easing toxicological requirements for new pesticides, with the possible exception of certain plant oils and extracts widely used in human foods. Globalization of agricultural commodities will serve only to tighten restrictions on pesticide used in developing countries where fresh produce for export to wealthier countries is an important source of revenue. All produce imported into the European Union, United States, and Japan, for example, must comply with pesticide regulations in the respective importing country, meeting the same standards as their own domestic produce. As a result, pesticide regulations set in the wealthiest countries have global reach—they affect growers directly in developing countries who are forced to comply. In short, a lack of confidence in the safety of a specific botanical insecticide by the European Union could make that product unfavorable in a tropical country, even where it makes sense for poorer growers providing agricultural produce for their domestic markets, and perhaps where the botanical source material grows and could be inexpensively prepared for crop protection. If nothing else, this review should highlight the fact that few new botanical insecticides are likely to see commercialization on a meaningful scale in the near future, in spite of the continual discovery of plant natural products with bioactivity against insect pests.

7.1 Registration of Botanicals in India

In India, registration of plant biopesticide is approved by Central Insecticide Board (CIB) if the desired data requirement is fulfilled by the concerned company. For example, for the registration of neem-based plant biopesticide, the following data are required.

Data requirements for registration of neem-based botanical pesticides

A.	*Chemistry*	9(3b)/9(3)
1.	Name of the part of the plant(s) to be used for extraction of the active ingredients/ components	R
2.	Outline of the process of manufacture clearly identifying the chemicals as indicated in point (3) below	R
3.	(a) Neem extracts contain "azadirachtin" as one of the major active constituents. The concentration of azadirachtin in the formulated neem extract should contain not less than 1,500 ppm of azadirachtin a.i. in "kernel"-based formulations and 300 ppm in "neem-oil"-based formulations	R
	(b) When the insecticidal a.i. is other than azadirachtin, then the applicant has to indicate the name, quality, and quantity of that particular a.i. (s)	
4.	Chemical identity of the ingredient as stated at point (3) above	R
5.	Physicochemical properties	R
6.	Specifications of ingredient as indicated at point (3) above	R
7.	Method of analysis for azadirachtin or other insecticidal a.i. other than azadirachtin	R
8.	Analytical test report	R
10.	Shelf life claim or data	R
B.	*Bioefficacy*	
1.	Bio-effectiveness	R
2.	Phytotoxicity	R
3.	Compatibility with other chemicals	R
4.	Purpose of manufacture	R
5.	Direction concerning dosage	R
6.	Time of application	R
7.	Waiting period	R

8.	Application equipment	R
9.	Information regarding the registration status in other countries, if any	R
C.	*Toxicity: data on parameters 1–3 are required for 9 (3b) registrations*	
1.	Acute oral rat and mice	R
2.	Acute dermal	R
3.	Primary skin irritation, irritation to mucous membrane	R
4.	Neurobehavioral toxicity	R
5.	Reproductive toxicity	R
6.	Carcinogenicity	R
7.	Mutagenicity	R
8.	Effect on spray operators (health records) as per the protocol to be approved by the registration committee. (Not required as per decision of 318th meeting of RC held on 27-04-2011)	NR
D.	*Packaging and labeling*	
1.	Labels and leaflets as per IR 1971 existing norms	R
2.	Type of packaging (container content compatibility data)	R
3.	Manner of labeling (container content compatibility data)	R
4.	Specification for primary package	R
5.	Specification for secondary package	R
6.	Specification for transport package	R
7.	Manner of labeling	R
8.	Instructions for storage and use	R
9.	Information regarding disposal of used package	R
10.	Process of manufacturing or indicating material balance generation of wasted	R

R required, *NR* not required

8 Botanical Pesticides Registered Under the Insecticides Act 1968

A. Neem based

Sl. No.	Formulations	Crops	Target pests
1.	Azadirachtin 0.03 % EC	Cotton	Bollworm, aphids
		Rice	Leaf roller, stem borer, BPH
2.	Azadirachtin	Cotton	Whitefly, bollworm
	0.15 % EC	Rice	Thrips, stem borer, brown plant hopper, leaf folder
3.	Azadirachtin 0.3 % EC	Cotton	American bollworm
4.	Azadirachtin 1.0 % EC	Tomato	Fruit borer
		Brinjal	Fruit and shoot borer
		Tea	Thrips, red spider mites
5.	Azadirachtin	Tobacco	Tobacco caterpillar, aphids
	5.0 % EC	Tea	Caterpillar, pink mite
			Red spider mites, thrips
		Rice	Brown plant hopper, leaf folder, stem borer
		Cotton	Whitefly, leafhoppers, heliothis
		Cauliflower	Spodoptera, diamondback moth, aphids
		Bhindi	Leafhopper, whitefly, aphid, pod borer
		Tomato	Aphids, whitefly, fruit borer

B. Cymbopogon

Cymbopogon winterianus 20 % EC

C. Pyrethrum (pyrethrin)

(a) Pyrethrin 0.05 % w/w (household spray)
(b) Pyrethrum extract 2 % tech.
(c) Pyrethrum 0.2 % (household spray)

applicators and farm workers, for whom, arguably, the more demonstrable hazards occur.

9 Quality Regulation

Regulatory approval remains the most formidable barrier to the commercialization of new botanical insecticides. In many jurisdictions, no distinction is made between synthetic pesticides and biopesticides, including botanicals. Simply put the market for botanicals in industrialized countries—based mostly on uses in green house production and organic agriculture—is too small to generate sufficient profits to offset multimillion dollar regulatory costs. Unfortunately, this situation may prevent many "green" pesticides from reaching the marketplace in countries where the demand is greatest. I am not making the case that botanicals should be exempt from all regulatory scrutiny; as discussed above, nicotine is as toxic and hazardous as many synthetic insecticides, and strychnine, still used for rodent and insect control in some regions, is responsible for some human poisonings. Natural products can pose risks, and safety cannot be assumed (Trumble 2002). But most of the botanicals discussed in this review are characterized by low mammalian toxicity, reduced effects on nontarget organisms, and minimal environmental persistence.

As noted, several plant essential oils and their constituents are exempt from registration in the United States, attributed to their long use history as food and beverage flavorings or as culinary spices. This exemption has facilitated the rapid development and commercialization of insecticides based on these materials as active ingredients (Isman 2000). Although other jurisdictions have yet to follow the lead of the United States in this regard, there are proposals in some Asian countries to exempt some types of pesticides from registration for specific uses in public health, for example, in head lice preparations or for cockroach and fly suppression. It seems that regulatory agencies continue to focus their efforts on protecting the general public from miniscule traces of pesticides in the food supply rather than focusing on the safety of applicators and farm workers, for whom, arguably, the more demonstrable hazards occur.

10 Conclusion

In industrialized countries, it is hard to imagine botanicals playing a greater role than at present, except in organic food production. Organic production is estimated to be growing by 8–15 % per annum in Europe and in North America (NRC 2000), and it is in those marketplaces that botanicals face the fewest competitors. Even there, however, microbial insecticides and spinosad have proven efficacious and cost-effective (Bailey et al. 2010). Rather than considered as stand-alone products, botanicals might be better placed as products in crop protectant rotations, especially in light of the documented resistance of the diamondback moth to *Bacillus thuringiensis* and spinosad due to overuse (Zhao et al. 2002). In conventional agriculture, botanicals face tremendous competition from the newest generation of "reduced-risk" synthetic insecticides such as the neonicotinoids. Overall, it is hard not to conclude that the best role for botanicals in the wealthier countries is in public health (mosquito and cockroach abatement) and for consumer (home and garden) use.

The real benefits of botanical insecticides can be best realized in developing countries, where farmers may not be able to afford synthetic insecticides, and the traditional use of plants and plant derivatives for protection of stored products is long established (Fields et al. 2001). Even where synthetic insecticides are affordable to growers (e.g., through government subsidies), limited literacy and a lack of protective equipment result in thousands of accidental poisonings annually. Recently, attention has been paid to traditional plants used in West Africa for postharvest protection against insects (Belmain et al. 2001).

References

Antonious GF (2004) Residues and half lives of pyrethrins on field-grown pepper and tomato. J Environ Sci Health B39:491–503

Ascher KRS, Schmutterer H, Mazor M, Zebitz CPW, Naqvi SNH (2002) The Persian lilac or chinaberry tree: *Melia azedarach* L. In: Schmutterer H (ed) The neem tree. Neem Foundation, Mumbai, pp 770–820

Atkinson BL, Blackman AJ, Faber H (2004) The degradation of the natural pyrethrins in crop storage. J Agric Food Chem 52:280–287

Bailey KL, Boyetchko SM, Langle T (2010) Social and economic drivers shaping the future of biological control: a Canadian perspective on the factors affecting the development and use of microbial biopesticides. Biol Control 52:221–229

Bastiaans L, Paolini R, Baumann DT (2008) Focus on ecological weed management: what is hindering adoption? Weed Res 48:481–491

Belmain SR, Neal GE, Ray DE, Golob P (2001) Insecticidal and vertebrate toxicity associated with ethnobotanicals used as post-harvest protectants in Ghana. Food Chem Toxicol 39:287–291

Bernays EA (1990) Plant secondary compounds deterrent but not toxic to the grass specialist acridid *Locusta migratoria*: implications for the evolution of graminivory. Entomol Exp Appl 54:53–56

California Department of Pesticide Regulation (2005) Summary of pesticide use report data 2003, indexed by chemical. http://www.cdpr.ca.gov/

Carpinella MC, Defago MT, Valladares G, Palacios SM (2003) Antifeedant and insecticide properties of a limonoid from *Meliaazedarach* (Meliaceae) with potential use for pest management. J Agric Food Chem 51:369–374

Casanova H, Ortiz C, Peláez C, Vallejo A, Moreno ME, Acevedo M (2002) Insecticide formulations based on nicotine oleate stabilized by sodium caseinate. J Agric Food Chem 50:6389–6394

Casida JE, Quistad GB (1995) Pyrethrum flowers: production, chemistry, toxicology and uses. Oxford University Press, Oxford, p 356

Charleston DS (2004) Integrating biological control and botanical pesticides for management of *Plutella xylostella*. PhD thesis, Wageningen University, p 176

Enan E (2001) Insecticidal activity of essential oils: octopaminergic sites of action. Comp Biochem Physiol 130:325–337

Fields PG, Xie YS, Hou X (2001) Repellent effect of pea (*Pisum sativum*) fractions against stored-product pests. J Stored Prod Res 37:359–370

Floris I, Satta A, Cabras P, Garau VL, Angioni A (2004) Comparison between two thymol formulations in the control of *Varroa destructor*: effectiveness, persistence and residues. J Econ Entomol 97:187–191

Fradin MS, Day JF (2002) Comparative efficacy of insect repellents against mosquito bites. N Engl J Med 347:13–18

Gelernter WD (2005) Biological control products in a changing landscape. In: BCPC international congress proceedings. British Crop Protection Council, Alton, pp 293–300

Glynne-Jones A (2001) Pyrethrum. Pestic Outlook 12:195–198

Gonzalez-Coloma A, Valencia F, Martin N, Hoffmann JJ, Hutter L (2002) Sylphine sesquiterpenes as model insect antifeedants. J Chem Ecol 28:117–129

Hajek A (2004) Natural enemies: an introduction to biological control. Cambridge University Press, Cambridge, p 378

Heap I (2010) The international survey of herbicide resistant weeds. See www.weedscience.org/In.asp

Isman MB (2000) Plant essential oils for pest and disease management. Crop Prot 19:603–608

Isman MB (2002) Insect antifeedants. Pestic Outlook 13:152–157

Isman MB (2004) Factors limiting commercial success of neem insecticides in North America and Western Europe. In: Koul O, Wahab S (eds) *Neem*: today and in the new millennium. Kluwer, Dordrecht, pp 33–41

Isman MB (2005) Problems and opportunities for the commercialization of botanical insecticides. In: Regnault-Roger C, Philogène BJR, Vincent C (eds) Biopesticides of plant origin. Lavoisier, Paris, pp 283–291

Jayasekara TK, Stevenson PC, Hall DR, Belmain SR (2005) Effect of volatile constituents from *Securidaca long epedunculata* on insect pests of stored grain. J Chem Ecol 31:303–313

Johnson HA, Oberlies NH, Alali FQ, McLaughlin JE (2000) Thwarting resistance: annonaceous acetogenins as new pesticidal and antitumor agents. In: Cutler SJ, Cutler JG (eds) Biological active natural products: pharmaceuticals. CRC Press, Boca Raton, pp 173–183

Klein Gebbinck EA, Jansen BJM, De Groot A (2002) Insect antifeedant activity of clerodane diterpenes and related model compounds. Phytochemistry 61:737–770

Kostyukovsky M, Rafaeli A, Gileadi C, Demchenko N, Shaaya E (2002) Activation of octopaminergic receptors by essential oil constituents isolated from aromatic plants: possible mode of action against insect pests. Pest Manage Sci 58:1101–1106

Koul O (2008) Phyto chemicals and insect control: an anti feedant approach. Crit Rev Plant Sci 27(1):1–24

Kraus W (2002) Azadirachtin and other triterpenoids. In: Schmutterer H (ed) The neem tree. Neem Foundation, Mumbai, pp 39–111

Leatemia JA, Isman MB (2004) Efficacy of crude seed extracts of *Annona squamosa* against diamondback moth, *Plutella xylostella* L. Int J Pest Manage 50:129–133

Maistrello L, Henderson G, Laine RA (2004) Efficacy of vetiver oil and nootkatone as soil barriers against Formosan subterranean termite (Isoptera: Rhinotermitidae). J Econ Entomol 94:1532–1537

Marrone PG (2007) Barriers to adoption of biological control agents and biological pesticides. CAB Rev Perspect Agric Vet Sci Nutr Nat Resour 2(15):1–12, CABI Publishing, Wallingford

Millennium Ecosystem Assessment (2005) Ecosystems and human well-being: biodiversity synthesis. World Resources Institute, Washington, DC

Morse S, Ward A, McNamara N, Denholm I (2002) Exploring the factors that influence the uptake of botanical insecticides by farmers: a case study of tobacco-based products in Nigeria. Exp Agric 38:469–479

Moser R, Pertot I, Elad Y, Raffaelli R (2008) Farmers' attitudes towards the use of biocontrol agents in IPM strawberry production in three countries. Biol Control 47:125–132

National Research Council (2000) The future role of pesticides in US agriculture. National Academic Press, Washington, DC, p 301

Neoliya NK, Singh D, Sangawan RS (2005) Azadirachtin influences total head protein content of *Helicoverpa armigera* Hub larvae. Curr Sci 88(12):1889–1890

Neoliya N, Dwijendra S, Sangwan R (2007) Azadirachtin-based insecticides induce alteration in *Helicoverpa armigera* Hub. head polypeptide. Curr Sci 92(1):94–99

Pesticide Action Network (2004) Pesticide registration by country. http://www.pesticideinfo.org/Search Countries.jsp

Pesticides Safety Directorate (2008a) Assessment of the impact on crop protection in the UK of the "cut-off criteria" and substitution provisions in the proposed Regulation of the European Parliament and of the Council concerning the placing of plant protection products in the market. Pesticides Safety Directorate, New York

Pesticides Safety Directorate (2008b) Plant protection products regulation: agronomic implications of proposals in the EU. Pesticides Safety Directorate, New York

Peterson C, Coats J (2001) Insect repellents – past, present and future. Pestic Outlook 12:154–158

Pretty J (2008) Agricultural sustainability: concepts, principles and evidence. Philos Trans R Soc 363:447–465

Priestley CM, Williamson EM, Wafford KA, Sattelle DB (2003) Thymol, a constituent of thyme essential oil, is a positive allosteric modulator of human GABAA receptors and a homo-oligomeric GABA receptor from Drosophila melanogaster. Br J Pharmacol 140:1363–1372

Regnault-Roger C, Philogène BJR (2006) Biopesticides of plant origin. Annu Rev Entomol 51:45–66

Rembold H, Mwangi RW (2002) Meliavolkensii Gürke. In: Schmutterer H (ed) The neem tree. Neem Foundation, Mumbai, pp 827–832

Schmutterer H (ed) (2002) The neem tree. Neem Foundation, Mumbai, p 892

Singh D, Singh AK (1991) Repellent and insecticidal properties in essential oils against housefly, *Musca domestica* L. Insect Sci Appl (Int J Trop Insect Sci) 12(4):487–491

Speranza CI, Kiteme B, Wiesmann U (2008) Droughts and famines: the underlying factors and the causal links among agropastoral households in semiarid Makueni district. Kenya Glob Environ Change 18:220–233

Thacker JRM (2002) An introduction to arthropod pest control. Cambridge University Press, Cambridge, p 343

Trumble JT (2002) Caveat emptor: safety considerations for natural products used in arthropod control. Am Entomol 48:7–13

Van Emden HF, Service MW (2004) Pest and vector control. Cambridge University Press, Cambridge

Weinzierl RA (2000) Botanical insecticides, soaps, and oils. In: Rechcigl JE, Rechcigl NA (eds) Biological and biotechnological control of insect pests. CRC Press, Boca Raton, pp 101–121

Zhao JZ, Li YX, Collins HL, Gusukuma- Minuto L, Mau RFL (2002) Monitoring and characterization of diamondback moth (Lepidoptera: Plutellidae) resistance to spinosad. J Econ Entomol 95:430–436

Retraction Note to:

Progress in the Development of Plant Biopesticides for the Control of Arthropods of Veterinary Importance

Srikant Ghosh and Reghu Ravindran

S. Ghosh, Ph.D. (✉)
Entomology Laboratory, Parasitology Division,
Indian Veterinary Research Institute, Izatnagar
243122, Uttar Pradesh, India
e-mail: sghoshp@yahoo.co.in; sghoshtick@gmail.com

R. Ravindran, Ph.D.
Department of Parasitology, College of Veterinary
and Animal Sciences, Pookode, Lakkidi, Wayanad,
Kerala 673576, India

D. Singh (ed.), *Advances in Plant Biopesticides*,
DOI 10.1007/978-81-322-2006-0_11, © Springer India 2014

DOI 10.1007/978-81-322-2006-0_20

The authors are retracting their chapter entitled, "**Progress in the Development of Plant Biopesticides for the Control of Arthropods of Veterinary Importance**" published in contributed volume titled "**Advances in Plant Biopesticides**". Several expressions of this article were identical to those of previously published paper by Dr. Brooke Bissinger and Dr. Michael Roe in Pesticide Biochemistry and Physiology, 2010, 96th issue, pages 63–79, DOI: 10.1016/j.pestbp.2009.09.010. The authors would like to express their most sincere apology to Dr. Brooke Bissinger and Dr. Michael Roe and the journal Pesticide Biochemistry and Physiology and to the editors and readers of the contributed volume "**Advances in Plant Biopesticides**."

The online version of the original chapter can be found at
http://dx.doi.org/10.1007/978-81-322-2006-0_11

Retraction Note to:

Formulation, Registration, and Quality Regulation of Plant Biopesticides

D.K. Hazra, P.K. Patanjali, and S.K. Raza

D.K. Hazra, Ph.D. (✉)
AINP on Pesticide Residues, Department of Agricultural Chemicals, Bidhan Chandra Krishi Viswavidyalaya, Mohanpur, Nadia, West Bengal, India
e-mail: dipakipft@gmail.com

P.K. Patanjali, Ph.D. • S.K. Raza, Ph.D.
Institute of Pesticide Formulation Technology (IPFT), Gurgaon 122016, India

D. Singh (ed.), *Advances in Plant Biopesticides*,
DOI 10.1007/978-81-322-2006-0_19, © Springer India 2014

DOI 10.1007/978-81-322-2006-0_20

The authors are retracting their chapter entitled "**Formulation, Registration, and Quality Regulation of Plant Biopesticides**" published in contributed volume titled "**Advances in Plant Biopesticides**". Several expressions of this article were identical to those of previously published paper by Dr. Murray B. Isman in Annual Review of Entomology, 2006, 51st issue, pages 45–66, DOI: 10.1146/annurev.ento.51.110104.151146. The authors would like to express their most sincere apology to Dr. Murray B. Isman and the journal Annual Review of Entomology and to the editors and readers of the contributed volume "**Advances in Plant Biopesticides**."

The online version of the original chapter can be found at
http://dx.doi.org/10.1007/978-81-322-2006-0_19